T0317679

Agroclimatology:
Linking Agriculture to Climate

Jerry L. Hatfield, Mannava V.K. Sivakumar,
and John H. Prueger, editors

Agronomy Monographs 60

Copyright © 2019 by American Society of Agronomy
Crop Science Society of America
Soil Science Society of America

ALL RIGHTS RESERVED. No part of this publication may be reproduced or transmi ed in any form or by any means, electronic or mechanical, including photocopying, recording, or any information storage and retrieval system, without permission in writing from the publisher.

The views expressed in this publication represent those of the individual Editors and Authors. These views do not necessarily reflect endorsement by the Publisher(s). In addition, trade names are sometimes mentioned in this publication. No endorsement of these products by the Publisher(s) is intended, nor is any criticism implied of similar products not mentioned.

American Society of Agronomy
Crop Science Society of America
Soil Science Society of America
5585 Guilford Rd., Madison, WI 53711-5801 USA

agronomy.org • crops.org • soils.org

Agronomy Monograph Series
ISSN: 2156-3276 (online)
ISSN: 0065-4663 (print)
ISBN: 978-0-89118-358-7
ISBN: 978-0-89118-357-0
doi: 10.2134/agronmonogr60

Cover design: Patricia Scullion

Contents

Preface

Agroclimatology encompasses the study of the interactions between climatic parameters and biological systems (plants and animals) and is closely allied with agrometeorology which deals with the shorter time scale responses associated with meteorological conditions. Given the concern about the potential impacts of climate change on crop and animal production there is a renewed interest in agroclimatology and how an improved understanding of the linkages between the environment and biological systems could develop adaptive management strategies which would lead to increased food security.

There is a large range in agroclimatic indices designed to quantify how plants fit within their growing region. The most recognizable of these is the growing degree day (GDD) concept for plants that has been employed to determine the rate of phenological development of crops, For perennial crops, the use of the chilling hours represents how climate may impact productivity and distribution of different tree species. In animals, the use of the temperature humidity index (THI) has proven valuable in assessing the level of stress an animal may be exposed to at any time. The modifications of these indices have incorporated additional climatic parameters, e.g., solar radiation , precipitation, wind speed, and humidity because all of these factors directly affect plants, animals, insects, diseases, and humans. Understanding the changes in these climatic parameters builds the foundation for how we begin to understand and quantify agroclimatology. It is also important to understand how climatic variables are collected and analyzed so that anyone can understand how to effectively evaluate climatic data.

The importance of agroclimatology is linked to how different plant systems; field crops, vegetables, perennial crops, and pastures respond to the climate. This monograph includes chapters on these different crops and the effect of climate on the geographic distribution of these different plant systems and how climate affects their productivity. Development of simulation models makes it possible to link climatic data into short-term meteorological data to evaluate the potential impacts on the soil, plants, and pests over time. As we go forward in science, it will be imperative that we take a long-term view of our agricultural systems by using agroclimatic principles and linkages with our biological systems. If we are to achieve the goal of feeding the world while enhancing or environmental quality we need to understand the role of climate in these processes. This monograph is assembled to provide that understanding to foster discussion about the value of the agroclimatic time and space scales.

We express our deep gratitude to the authors who have contributed their expertise and energy to make this work a contribution to science. We value your contributions and insights into this topic. We thank the American Society of Agronomy, Crop Science society of America, and Soil Science Society of America for offering us this opportunity to develop this monograph and bring it publication. This would not be possible without the dedicated staff at headquarters who provide valuable efforts in making our efforts a reality.

Jerry L. Hatfield, co-editor. National Laboratory for Agriculture and the Environment

John H. Prueger, co-editor, National Laboratory for Agriculture and the Environment

M.V.K. Sivakumar, co-editor, Retired, World Meteorological Organization

Solar, Net, and Photosynthetic Radiation

J. Mark Blonquist Jr. and Bruce Bugbee*

Introduction

Shortwave and longwave radiation incident at Earth's surface are the source of available energy that controls key processes, including surface and atmospheric heating, evaporation, sublimation, transpiration, and photosynthesis. Shortwave radiation (approximately 280 to 4000 nm) is emitted by the sun, whereas longwave radiation (approximately 4000 to 100 000 nm) is emitted by molecules in the atmosphere and objects at the surface. There is no universally accepted definition of the cutoff wavelengths for shortwave or longwave radiation, but there is minimal shortwave radiation at wavelengths greater than 3000 nm.

Shortwave and longwave radiation at Earth's surface are spatially and temporally variable due to changes in position of the sun with respect to Earth's surface and changes in atmospheric conditions. Shortwave radiation accounts for a larger proportion of total radiation at the surface, but longwave radiation is a significant contributor. Ultimately, solar radiation is the source of available energy at Earth's surface and within Earth's atmosphere, driving weather and climate, as longwave radiation results from atmospheric and surface heating by shortwave radiation.

Radiation Theory

Radiation is energy transfer in the form of electromagnetic waves. All materials with temperature above absolute zero continuously emit electromagnetic radiation. The intensity and wavelengths of radiation emitted are dependent on temperature, according to Planck's Law, which describes the spectral (wavelength-dependent) distribution of electromagnetic radiation emitted by an object as a function of absolute temperature (T, in units of K) and emissivity (ε_λ):

$$B_\lambda = \frac{2\pi\varepsilon_\lambda hc^2}{\lambda^5\left[e^{\left(\frac{hc}{\lambda kT}\right)}-1\right]} \qquad [1]$$

Abbreviations: CF, calibration factor; DNI, direct normal irradiance; DHI, diffuse horizontal irradiance; ET, evapotranspiration; IR, infrared; IRR, infrared radiometer; LW_i, incoming longwave irradiance; LW_n, net longwave irradiance; LW_o, outgoing longwave irradiance; NIR, near infrared; PAR, photosynthetically active radiation; PPFD, photosynethic photon flux density; R_n, net radiation; SW_i, incoming shortwave irradiance; SW_o, outgoing shortwave irradiance; WRR, world radiometric reference; YPFD, yield photon flux density.

J.M. Blonquist, Jr., Apogee Instruments, Inc., Logan, UT; B. Bugbee, Dep. of Plants, Soils, and Climate, Utah State University, Logan, UT. *Corresponding author (bruce.bugbee@usu.edu)

doi:10.2134/agronmonogr60.2016.0001

© ASA, CSSA, and SSSA, 5585 Guilford Road, Madison, WI 53711, USA.
Agroclimatology: Linking Agriculture to Climate, Agronomy Monograph 60.
Jerry L. Hatfield, Mannava V.K. Sivakumar, John H. Prueger, editors.

where B_λ is radiation intensity (energy flux density) (W m^{-2} m^{-1}), h is Planck's constant (6.6261×10^{-34} J s), c is speed of light in vacuum (2.9979×10^{8} m s^{-1}), k is the Boltzmann constant (1.3807×10^{-23} J K^{-1}), and λ is wavelength (m). Emissivity (ε_λ, subscript λ denotes wavelength-dependent) is defined as the fraction of blackbody emission. A blackbody is an object with an emissivity of one, emitting the maximum amount of radiation for its temperature. Thus, emissivity is the ratio of energy emitted by an object to energy emitted by a blackbody at the same temperature. For most terrestrial surfaces, emissivities are near one.

In addition to being a stream of energy, radiation can also be described and quantified as a stream of elementary particles called photons, which are defined as a single quantum of radiation and can be thought of as discrete energy packets. Thus, radiation can be expressed in units of energy (typically Joules) or quantity (number of photons or often moles of photons). Both units of measurement are important in environmental applications, and the application will determine the appropriate units. For example, energy flux density drives evapotranspiration and photon flux density drives photosynthesis. The relationship between units of energy and units of quantity is:

$$B_\lambda = \frac{n_\lambda hc}{\lambda} \qquad [2]$$

where n_λ is number of photons, and h, c, and λ are as defined for Eq. [1]. When $n_\lambda = 1$, Eq. [2] yields energy content (J) of a single photon. To calculate energy content of a mole of photons (mol), Avogadro's number (6.0221×10^{23}) is input for n_λ.

Typically, radiation is measured in terms of flux density, a flux of energy or flux of photons over a unit area (often 1 m^2), where flux is flow of energy or photons per unit time (often 1 s). Typical units for energy flux are power (W = J s^{-1}), and power per unit area (W m^{-2} = J m^{-2} s^{-1}) for energy flux density. Typical units for photon flux are number of photons per unit time (μmol s^{-1}), and number of photons per unit time per unit area (μmol m^{-2} s^{-1}) for photon flux density. Sometimes the single word 'flux' is used to indicate units of time and area in the denominator, rather than just time. To avoid ambiguity, herein flux is defined as energy or quantity per unit time, and flux density is defined as energy or quantity per unit time per unit area.

Integration of Eq. [1] across an infinite wavelength range (λ ranging from 0 to ∞) yields the Stefan–Boltzmann Law, describing total energy flux density (E, in units of W m^{-2}) from an object as a function of absolute temperature:

$$E = \varepsilon \sigma T^4 \qquad [3]$$

where ε is broadband emissivity (effective emissivity for all wavelengths emitted), σ is the Stefan-Boltzmann constant (5.6704×10^{-8} W m^{-2} K^{-4}), and T is absolute temperature (K). Differentiation of Eq. [1] with respect to wavelength yields Wien's Displacement Law, relating the wavelength of peak emission (λ_{max}, in units of μm) for a blackbody radiator to absolute temperature:

$$\lambda_{max} = \frac{2898}{T} \qquad [4]$$

In energy units, flux density from all directions incident on a surface of unit area is called incident radiation flux density, or irradiance. Irradiance is often measured on a horizontal surface for the hemispherical field of view above the surface. Irradiance at Earth's surface measured with respect to a horizontal plane is often called global hemispherical irradiance or global irradiance. Total global irradiance at Earth's surface contains shortwave and longwave components.

Shortwave irradiance at Earth's surface is often defined as radiant energy in the 280 to 4000 nm wavelength range, but the cutoff at 4000 nm is somewhat arbitrary, with 4000 nm being an approximation of the point where shortwave and longwave radiation spectra overlap (Fig. 1). Wavelengths shorter than 280 nm are emitted by the sun and are considered shortwave radiation, but are absorbed by Earth's atmosphere (largely by ozone molecules) before reaching the surface. Shortwave radiation at Earth's surface is often subdivided into ultraviolet (UV, 280 to 400 nm), visible (400 to 700 nm), and near infrared (NIR, 700 to 4000 nm) wavelength ranges.

Global shortwave irradiance incident on Earth's surface (SW_i) is made up of direct and diffuse components. Direct irradiance is transmitted through the atmosphere without interacting with air molecules (no absorption or scattering) and is the major contributor to SW_i, approximately 90%, on clear days in the middle of the day. Diffuse irradiance interacts with air molecules and atmospheric constituents (e.g., clouds, aerosols, pollutants) and is scattered or reflected in the direction of Earth's surface. Diffuse irradiance contributes approximately 10% to SW_i on clear days in the middle of the day, but the contribution increases as the solar zenith angle increases. Diffuse irradiance also increases as cloud cover increases and is the only contributor to SW_i on overcast days. A fraction of diffuse irradiance is shortwave radiation reflected by Earth's surface and then scattered back toward the surface by the atmosphere. This component of diffuse irradiance increases as Earth's surface reflectivity increases.

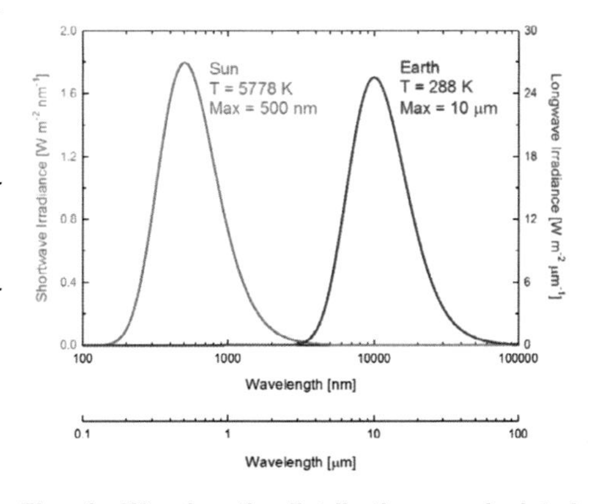

Fig. 1. Wavelength distributions, calculated with Eq. [1] (assuming $\varepsilon_\lambda = 1$ at all wavelengths for the sun and Earth), for shortwave irradiance incident at the top of Earth's atmosphere (sun was assumed to be a 5778 K blackbody and mean distance between Earth and sun was assumed) and longwave irradiance emitted from Earth (assumed to be a 288 K blackbody). The wavelengths of maximum emission (peaks in the distributions) were calculated with Eq. [4]. There is little overlap between the two distributions (intersection is between 3000 and 4000 nm), allowing for relatively distinct definitions of shortwave and longwave radiation. Note the change in units on the *y* axis scales for shortwave and longwave irradiance, where shortwave irradiance is per nm and longwave irradiance is per μm.

Longwave irradiance at Earth's surface is radiant energy in the thermal infrared (IR) wavelength range, typically defined as wavelengths greater than 4000 nm or 4 μm. Longwave radiation is energy emitted by objects with temperature that is not hot enough to result in shortwave radiation. Calculation of radiant energy emission for the sun and Earth's surface using Eq. [1], where the distance between Earth and sun was accounted for to yield top of atmosphere shortwave irradiance (extraterrestrial radiation), shows little overlap between the two distributions of wavelengths (Fig. 1). Global longwave irradiance incident on Earth's surface (LW_i) is only diffuse because the air molecules responsible for emitting longwave radiation are relatively evenly distributed in the atmosphere and emit radiation in all directions.

This chapter reviews radiation measurements made at Earth's surface and is divided into three sections, each reviewing a specific measurement common to agricultural applications: Global Shortwave Irradiance, Net Radiation, and Photosynthetically Active Radiation. Each section contains subsections that cover specific topics related to each measurement.

Global Shortwave Irradiance

Shortwave irradiance at the top of Earth's atmosphere on a plane perpendicular to the sun's rays at the mean distance between Earth and the sun is nearly constant, and is often called the solar constant. Traditional values of the solar constant range from 1365 to 1370 W m^{-2}, but data from a recent study indicate it is closer to 1361 W m^{-2} (Kopp and Lean, 2011). The solar constant is not a true constant, but varies with solar cycles (Steinhilber et al., 2009; Vieira et al., 2011). The distance between Earth and the sun varies with time of year, and is at a minimum in January and a maximum in July. Thus, top of atmosphere shortwave irradiance is seasonally variable.

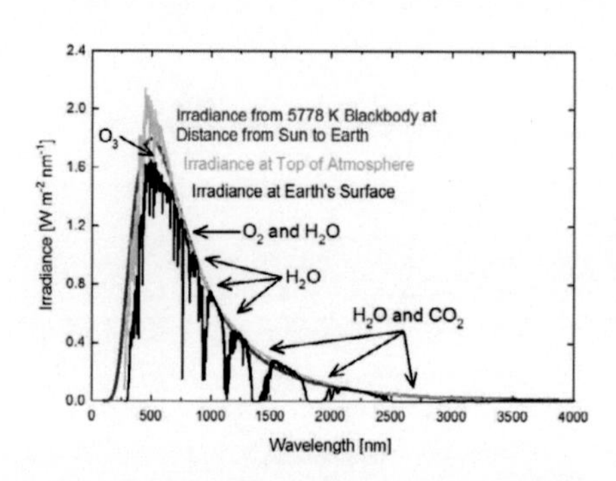

Fig. 2. Shortwave irradiance spectra from a 5778 K blackbody at the mean distance between the sun and Earth (calculated from Eq. [1]), at the top of Earth's atmosphere (American Society for Testing and Materials (ASTM) G173–03 Reference Solar Spectral Irradiance, derived from Simple Model of the Atmospheric Radiative Transfer of Sunshine (SMARTS) v. 2.9.2, available at: http://rredc.nrel.gov/solar/spectra/am1.5/ASTMG173/ASTMG173.html, verified 17 July 2016), and at Earth's surface on a clear day (measured on a clear day near solar noon in June in Logan, UT, with an Advanced Spectral Designs model FieldSpec Pro spectroradiometer). Absorption by atmospheric gases [ozone (O_3,) oxygen (O_2,) water vapor (H_2O), carbon dioxide (CO_2)] reduces radiation transmission and causes spectral differences in shortwave irradiance between top of atmosphere and Earth's surface.

Shortwave irradiance at the top of Earth's atmosphere approximates that calculated with Eq. [1] for a 5778 K blackbody at the mean distance between the sun and Earth, but there are some differences due to absorption and emission by gases in the outermost layer of the sun (Fig. 2). As shortwave radiation passes through Earth's atmosphere it is absorbed, reflected, and scattered by air molecules, clouds, aerosols, and particulate matter (e.g., dust, smoke, pollutants). Thus, shortwave irradiance at Earth's surface is less than top of atmosphere shortwave irradiance, particularly at certain wavelengths where absorption by atmospheric gases (ozone, oxygen, water vapor, carbon dioxide) is strong (Fig. 2). On clear days, 70 to 80% of top of atmosphere shortwave irradiance is transmitted to Earth's surface. The rest is absorbed and scattered by gases and particulates in the atmosphere. Shortwave irradiance at Earth's surface is highly variable in space and time, with the largest contributors to variability being time of year and time of day (Fig. 3), and atmospheric water content and degree of cloudiness. A typical value of global shortwave irradiance (SW_i) for a midlatitude location in summer on a clear day near solar noon is 1000 W m^{-2}. High, thin clouds can enhance SW_i by as much as 50% by reflection (Yordanov et al., 2012), whereas thick clouds reduce SW_i to approximately 10%.

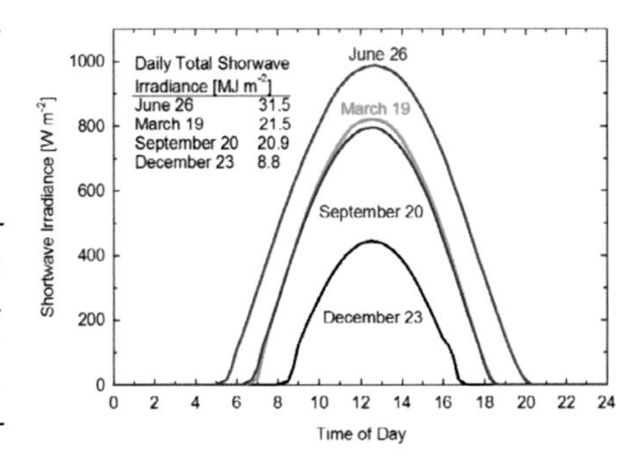

Fig. 3. Seasonal comparison of global shortwave irradiance in Logan, Utah, United States (41.77° N lat. 111.86° W long.), for days near solstices and equinoxes (clear sky days near solstices and equinoxes in 2014 were selected). Global shortwave irradiance was integrated over the course of the day to provide daily totals (energy flux density on a daily time scale) listed in the table.

Radiometers designed to measure SW_i are called pyranometers. There are two main types of pyranometer: blackbody thermopile and silicon-cell. Blackbody thermopile pyranometers use the thermoelectric effect (conversion of a temperature difference between two different metals or alloys to voltage) and silicon-cell pyranometers use the photoelectric effect (electron emission driven by photon absorption in a semiconductor) to generate electrical signals proportional to SW_i.

Blackbody Thermopile Pyranometers

Blackbody thermopile pyranometers use the combination of a glass dome (or two domes), blackbody absorber, and thermopile (multiple thermocouple junctions connected in series) transducer to produce a voltage signal proportional to incident shortwave radiation. Blackbody pyranometers with alternative transducers to thermopiles (e.g., platinum resistance thermometers) have also been built (Beaubien et al., 1998), but they are not commonly used. Voltage output by a thermopile is equal to the temperature difference between sensing (hot) and reference (cold) junctions multiplied by the number of thermocouple junctions in the thermopile and the Seebeck coefficient (thermocouple sensitivity, which is dependent on thermocouple type) of the thermocouples. The sensing junction of the thermopile is thermally bonded to the blackbody absorber and the reference junction of the thermopile is thermally bonded to the sensor housing, which is used as a heat sink. Radiation incident on the blackbody absorber (top surface of the detector plate) heats the detector plate and sensing junction above the internal reference junction, which is shaded from shortwave radiation. Voltage output by the thermopile is proportional to the temperature difference and heat flux between the blackbody surface and internal reference junction, and scales linearly with incident radiation. To minimize errors, heat flux to the absorber should only be from incident shortwave radiation and heat flux away from the absorber should only be via conduction to the thermopile. Heat gain by the blackbody surface not caused by incident shortwave radiation (e.g., absorption of longwave radiation) and heat loss by the blackbody absorber not caused by conduction to the thermopile (e.g., conduction to sensor housing adjacent to the absorber or convection caused by wind) cause changes in signal unrelated to shortwave irradiance.

Blackbody absorbers are sensitive to both shortwave and longwave radiation, so a filter must be used to block longwave radiation. Glass domes are typically used for this purpose, as transmittance of glass is near 100% and uniformly transmits wavelengths from about 290 to 3000 nm. Quartz domes are also sometimes used, and extend the transmittance range from 200 to 3600 nm. Domes protect the absorber surface from dust and moisture, which reduce absorptivity. Domes also provide an insulative barrier that reduces the influence of wind and thermal gradients on advective and convective heat loss from the absorber. Lower cost instruments have a single dome. Higher cost instruments have two domes, a smaller inner dome covered by a larger outer dome, which further reduce advective and convective heat losses from the absorber. The blackbody absorber should be thermally isolated from other sensor components to minimize conductive heat loss. The absorber is an efficient emitter of longwave radiation, and when exposed to a clear cold sky, the absorber and domes cool via longwave emission. This radiative heat loss, often referred to as a thermal offset (discussed below), varies with sky conditions. It is maximum under clear sky conditions and minimum when the sky is overcast.

Some thermopile pyranometers have black and white receiving surfaces (e.g., Eppley Laboratory model 8–48), where the temperature difference measured by the thermopile is between the black (absorbing) and white (reference) surfaces, rather than a black absorbing surface and internal reference point. The black surface has high shortwave absorptivity and heats up relative to the white surface, which has high shortwave reflectivity. The advantage of this design over a completely black surface is that the black and white surfaces are subject to the same conditions (exposure, wind, and thermal gradients), so a single dome can be used and heat exchange between black and white surfaces via advection and convection is minimal. Thermal offset is also minimized because black and white surfaces are both exposed to the sky. The disadvantage of this type of design is daily and seasonal changes in sun alignment with respect to black and white surfaces. Also, discoloration of the white reference surfaces alters shortwave reflectivity, which changes the calibration.

Silicon-cell Pyranometers

Silicon-cell pyranometers consist of a silicon photodiode mounted behind a diffuser. Silicon is a semiconductor that emits electrons (electrical current) in response to incident photons (radiation) of specific wavelength. The diffuser is designed to provide accurate angular response and is typically made of acrylic with a white colorant added to make it opaque. Some models output short circuit current from the photodiode directly, but in most models manufacturers add a shunt resistor to convert current to voltage. The electrical signal is linearly related to incident radiation within the range of silicon sensitivity. Silicon is generally sensitive to wavelengths from about 350 to 1100 nm. The transmittance of acrylic is fairly uniform across this range (Kerr et al., 1967).

The limited sensitivity of the detector (350 to 1100 nm) means that silicon-cell pyranometers subsample the solar spectrum and must be calibrated to estimate total shortwave irradiance. Changes in sky conditions can change proportions of SW_i within the silicon-cell sensitivity range. For example, 70 to 80% of SW_i is between 350 and 1100 nm on a clear day (Myers, 2011), and about 90% of SW_i is within this range on an overcast day due to absorption by water vapor at wavelengths greater than 1100 nm (Federer and Tanner, 1965; Kerr et al., 1967). This results in spectral errors for different sky conditions (discussed below). Silicon-cell pyranometers are most accurate in conditions similar to those during calibration (Habte et al., 2014).

While silicon-cell pyranometers have the disadvantage of spectral error, they have the advantage of lower cost, smaller size, and faster response. Fast response can be important for solar power applications (Sengupta et al., 2012), but is not necessarily an advantage for weather and climate measurements. Response time of thermopile pyranometers is determined by thermal mass and thermal conductivity of components, and is typically a few seconds, whereas the response of silicon-cell pyranometers is independent of these factors, and is less than 1 ms. A new blackbody thermopile pyranometer (EKO Instruments model MS-80) has a response time of 0.5 s, resulting largely from a small (low thermal mass) detector.

Pyranometer Classification

Three classes of pyranometers are defined based on performance characteristics and specifications. Two organizations, the World Meteorological Organization (WMO) and the International Organization for Standardization (ISO), have instituted similar classifications (Table 1). The names of the classes in the ISO classification can be confusing because similar names, Second Class and Secondary Standard, are used for classes. The lowest cost blackbody thermopile pyranometers are classified as Moderate Quality (WMO classification) and second class (ISO classification), and price increases as classification increases. Several models of blackbody thermopile pyranometers are available, for example:

Moderate Quality (WMO) or Second Class (ISO): EKO Instruments model MS-602, Hukseflux model LP02, Kipp & Zonen model CMP 3.

Good Quality (WMO) or First Class (ISO): EKO Instruments model MS-402, Hukseflux model SR11, Kipp & Zonen model CMP 6.

High Quality (WMO) or Secondary Standard (ISO): EKO Instruments model MS-802, Eppley Laboratory model SPP, Hukseflux model SR20, Kipp & Zonen model CMP 11.

Specifications for silicon-cell pyranometers (e.g., Apogee Instruments model SP-110, EKO Instruments model ML-01, Kipp & Zonen model SP Lite2, LI-COR model LI-200R, Skye Instruments model SKS 1110) compare favorably to specifications for Moderate and Good Quality classifications (WMO) and for Second Class and First Class classifications (ISO), but their limited spectral sensitivity means they do not meet the spectral selectivity specification necessary for WMO or ISO classification. Silicon-cell pyranometers are generally a third to half the cost of Moderate Quality and Second Class pyranometers. A pyranometer with a thermopile detector and acrylic diffuser is also available (Apogee Instruments model SP-510), with performance characteristics similar to blackbody thermopile pyranometers, but with a cost similar to silicon-cell pyranometers.

Table 1. Specifications for classification of pyranometers according to World Meteorological Organization (WMO) and International Organization for Standardization (ISO).

WMO classification	High quality	Good quality	Moderate quality
ISO classification	Secondary standard	First class	Second class
Specifications			
Response time, s (to 95% of final value)	< 15	< 30	< 60
Zero offset response, W m^{-2} (to 200 W m^{-2} net thermal radiation, ventilated)	7	15	30
Zero offset response, W m^{-2} (to 5 °C h^{-1} change in ambient temperature)	± 2	± 4	± 8
Resolution, W m^{-2} (smallest detectable change)	± 1	± 5	± 10
Stability, % (relative change in sensitivity per year)	± 0.8	± 1.5 (WMO) ± 1.6(ISO)	± 3.0 (WMO) ± 2.0 (ISO)
Nonlinearity, % (relative deviation from sensitivity at 500 W m^{-2} over range of 100 to 1000 W m^{-2})	± 0.5 (WMO) ± 0.2 (ISO)	± 1.0 (WMO) ± 0.5 (ISO)	± 3.0 (WMO) ± 2.0 (ISO)
Directional response, W m^{-2} (absolute deviation from 1000 W m^{-2} direct beam)	± 10	± 20	± 30
Tilt response, % (relative deviation due to tilt from horizontal to vertical at 1000 W m^{-2})	± 0.5	± 2.0	± 5.0
Temperature response, % (relative deviation over an interval of 50 °C)	± 2.0	± 4.0	± 8.0
Spectral selectivity, % (relative deviation of spectral sensitivity from mean spectral sensitivity)	± 2.0	± 5.0	± 10

Shortwave Irradiance Measurement Error Sources

Errors in SW_i measurements can be separated into two groups:

1. Sensor characteristics (calibration, directional, spectral, temperature response, and stability).
2. General use (installation, dirt/dust, and moisture).

Calibration Error

Calibration is accomplished by deriving a calibration factor (CF) that scales the signal output by a thermopile or photodiode detector to match a reference measurement of SW_i:

$$\mathrm{CF}=\frac{SW_i}{S} \qquad [5]$$

where S is signal output by the pyranometer (typically voltage, but amperage for some models) and CF is in units of W m^{-2} per unit of signal (V or A). Subsequent measurements are then made by rearranging Eq. [5] to solve for SW_i and multiplying the measured S by CF.

The reciprocal of CF is sensitivity and is often reported. Sensitivity provides an indication of the necessary resolution of the analog signal measurement. For example, a typical sensitivity for a blackbody thermopile pyranometer is 10 μV per W m^{-2}. This means the sensor outputs 10 μV for every 1 W m^{-2} of incident shortwave radiation. To yield SW_i measurement resolution of 1 W m^{-2}, a voltage resolution of 10 μV is required, and 1 μV resolution is required to achieve 0.1 W m^{-2}. This resolution is not available on all meters and dataloggers.

Analysis of the energy balance of a blackbody detector plate indicates multiple radiometer properties contribute to the voltage signal generated by the thermopile (Campbell et al., 1978; Campbell and Diak, 2005; Fairall et al., 1998; Ji and Tsay, 2000). For thermopile pyranometers these include shortwave transmittance of the dome, shortwave absorptance of the blackbody surface, thermal emittance of the blackbody surface, thermal conductivity of the detector plate, and temperature of the sensor body. For silicon-cell pyranometers, contributing factors are shortwave transmittance of the diffuser and shortwave absorptance of the silicon-cell. These factors are accounted for in calibration. Pyranometers should be calibrated in conditions similar to those in which they will be used. Alternatively, for laboratory calibrations, the reference pyranometer should be the same model as the pyranometer being calibrated.

Pyranometer calibrations should be traceable to the World Radiometric Reference (WRR) in Davos, Switzerland. The WRR is an absolute cavity radiometer that is self-calibrated by applying an electric current that duplicates the signal generated by incident solar radiation. Working reference absolute cavity radiometers are periodically calibrated against the WRR. Secondary standard pyranometers can then be calibrated against working reference cavity radiometers and can be used as transfer standards. Transfer standard calibrations are typically done outdoors over the course of a day or multiple days using component summation to measure reference SW_i. Component summation refers to independent measurements of the direct and diffuse components of SW_i and summation to yield SW_i. Secondary standard pyranometers with calibration traceable to the WRR should be used by pyranometer manufacturers for calibration.

Pyranometer calibration procedures differ among manufacturers, with outdoor and indoor calibration procedures in use. Outdoor calibration typically consists of simultaneous measurement of SW_i from component summation, or from a secondary standard pyranometer, and S from the pyranometer to be calibrated, over the course of a day. The advantage of this approach is characterization of CF with solar zenith angle. A component summation calibration method that accounts for thermal offset of thermopile pyranometers has been proposed (Reda et al., 2005). Often, CF at 45° zenith angle is used, but a zenith angle-dependent CF can be applied to account for changes in CF with zenith angle (Raïch et al., 2007). A short-term shade and unshade method has also been used, which allows for subtraction of the signal when the pyranometer is shaded from signal when the pyranometer is unshaded. This minimizes the influence of thermal offset (for thermopile pyranometers) on calibration because thermal offset should be the same for the shaded and unshaded condition (Philipona, 2002; Reda et al., 2005). This method requires accurate measurement of the direct component of SW_i because the difference between unshaded and shaded signal from the pyranometer is proportional to the direct component of SW_i. A shade and unshade method is often used for indoor calibration and requires SW_i measurement with a transfer standard pyranometer and S measurement with the pyranometer to be calibrated. Time is required for the pyranometers to equilibrate when shaded and unshaded, with 20 to 60 time constants being typical equilibration times. If reference SW_i is incorrect during calibration, error will be transferred to the pyranometers being calibrated. Calibration uncertainty is not reported by all manufacturers of blackbody thermopile pyranometers, but those with published specifications list between 1 and 2%. Calibration uncertainty for silicon-cell pyranometers is often reported as 5%. Temperature should be recorded during calibration, allowing for temperature correction once the pyranometer is deployed in the field.

Pyranometer manufacturers typically recommend recalibration at two-year intervals. Verification of calibration accuracy should be done at least annually, and can be done by comparison to a reference instrument in the same location. Another alternative method is to compare shortwave irradiance measurements to modeled global shortwave irradiance for clear sky conditions (SW_{ic}). Multiple clear sky global shortwave irradiance models are available (Atwater and Ball, 1978; Gueymard, 2008; Lefèvre et al., 2013; Meyers and Dale, 1983). A recent book by Myers (2013) discusses shortwave radiation modeling. A commonly used model in agricultural applications is contained within the net radiation submodel of the American Society of Civil Engineers (ASCE) standardized reference evapotranspiration equation (ASCE-EWRI, 2005). A user-friendly version of this model is available online at clearskycalculator.com (verified 17 July 2016). On clear days, SW_{ic} can be used as a reference for estimating pyranometer accuracy. Recalibration is recommended if measured SW_i consistently deviates more than 5% from SW_{ic} on clear sky days near solar noon. Blonquist et al. (2010) analyzed model accuracy for clear sky days and reported ± 3% as a reasonable estimate of accuracy for nonpolluted summer days near solar noon. This is consistent with uncertainty estimates for other shortwave irradiance models (Gueymard, 2012). In addition to SW_{ic} being a reference for recalibration requirements, the ratio SW_i/SW_{ic} provides an estimate of cloudiness.

Directional Error

Directional or angular response error results from imperfect cosine correction. Cosine correction means that the sensor is accurate at all incidence angles. Lambert's cosine law states that radiant intensity is directly proportional to the cosine of the angle between the incident radiation beam and a plane perpendicular to the receiving surface. A radiometer that accurately measures radiation according to Lambert's cosine law is said to be cosine corrected. Directional response is often called cosine response and directional error is often called cosine error.

Directional response of blackbody thermopile pyranometers is influenced by multiple radiometer properties, including spatial uniformity of the domes and blackbody absorber, and alignment of domes with respect to the absorber. Similarly, spatial uniformity and alignment of the diffuser with respect to the underlying silicon detector influences directional response in silicon-cell pyranometers. Blackbody and silicon-cell absorbers must be horizontal, the leveling device must be in the same plane as the absorber, and the sensor must be exactly level.

Directional response is often specified as deviation from true cosine response, where a radiation beam of known intensity is used to determine directional response in the laboratory. True cosine response is beam intensity at a zenith angle of zero multiplied by the cosine of the angle between the direct beam and sensor. A common directional response specification for pyranometers is deviation of less than 10 W m^{-2} from a direct beam of 1000 W m^{-2} up to an incidence angle of 80°. The cosine of 80° is 0.174, so irradiance from a 1000 W m^{-2} direct beam is 174 W m^{-2} at 80°. Thus, a pyranometer with this specification should measure within the range 164 to 184 W m^{-2} at a zenith angle of 80°. This specification can be interpreted in terms of relative error by dividing 10 W m^{-2} by 174 W m^{-2}. Thus, an absolute error of 10 W m^{-2} at an 80° incidence angle is a relative error of 5.7%. If the directional error specification is 20 W m^{-2} up to 80°, then relative error at 80° is double that for 10 W m^{-2} (11.4%). For a directional error specification of 5 W m^{-2}, relative error is half that at 80° (2.9%).

Another method of determining directional response is to compare SW_i measurements on a clear day against reference SW_i (often from component summation, the sum of independent measurements of direct and diffuse SW_i components as explained in the previous section). Reference SW_i must be assumed to represent true SW_i when using field measurements to determine directional response. Directional responses of two common second class blackbody thermopile pyranometers and three common silicon-cell pyranometers indicate errors less than 2% for solar zenith angles between 20° and 60°, and less than 5% for solar zenith angles less than 75° (Fig. 4). We have measured and compared directional response data for multiple replicates of the same pyranometer models at our outdoor calibration facility in Logan, Utah (UT), United States, and found similar results. Reference SW_i was mean SW_i calculated from four secondary standard pyranometers (Hukseflux model SR20, Kipp & Zonen models CMP 11, CM 11, and CM 21) calibrated at National Renewable Energy Laboratory (NREL). Habte et al. (2014) compared multiple thermopile and blackbody pyranometers to reference SW_i and found differences were typically less than 5% for zenith angles less than 60° under clear sky conditions. Differences increased for solar zenith angles greater than 60°.

Directional errors can be significant in applications where hourly (or higher frequency) data are required. Directional errors can also be significant at high latitudes in winter when the sun is low in the sky. A pyranometer with poor directional response may be calibrated to provide accurate measurements in the summer when zenith angles are low for much of the day at mid and high latitude locations, but may have much larger errors during winter months when zenith angles are always high. Over daily time scales, directional errors are reduced by calibrating pyranometers to daily total SW_i. Most of the daily total SW_i is received

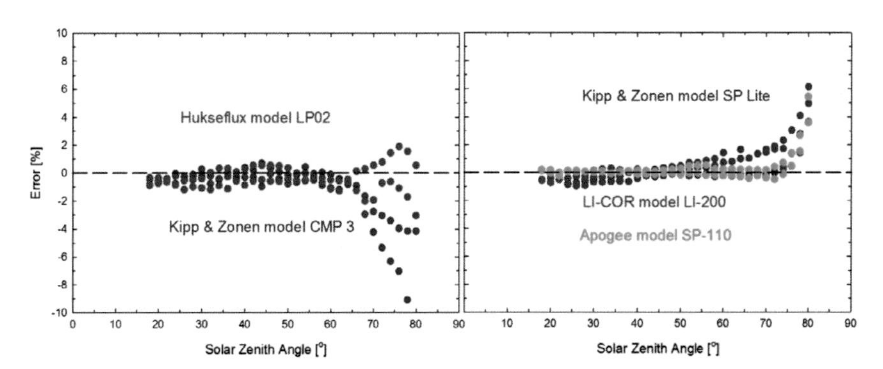

Fig. 4. Directional error (cosine error) for two common second class blackbody thermopile pyranometers (Hukseflux model LP02 and Kipp & Zonen model CMP 3) and three common silicon-cell pyranometers (Apogee Instruments model SP-110, Kipp & Zonen model SP Lite, and LI-COR model LI-200). Data were collected during a Broadband Outdoor Radiometer Calibration (BORCAL) at the National Renewable Energy Laboratory (NREL) in Golden, Colorado, United States.

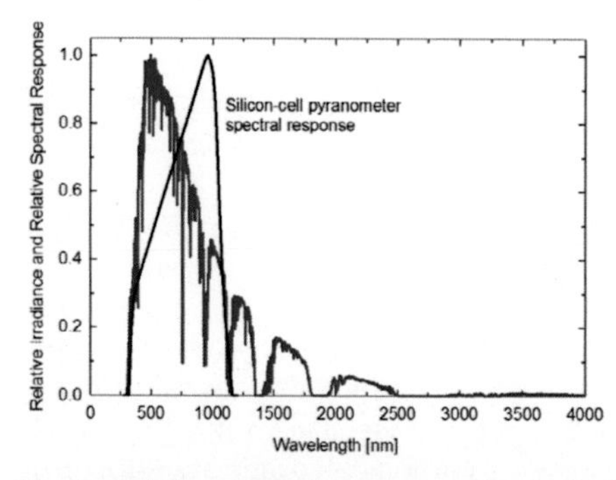

Fig. 5. Relative spectral response of a typical silicon-cell pyranometer (normalized to maximum response at 960 nm) compared to a relative global shortwave irradiance spectrum (normalized to maximum irradiance at 495 nm) at Earth's surface. Silicon-cell pyranometers subsample the shortwave spectrum (350–1100 nm) and thus have spectral errors when the spectrum changes.

in the middle of the day when solar zenith angles are low. Solar zenith angle dependent calibration factors have been used to minimize directional errors (King et al., 1997).

Spectral Error

Spectral error occurs when the detector is not uniformly sensitive to all wavelengths within the shortwave radiation spectrum, or when a dome is not uniformly transmissive to all wavelengths within the shortwave spectrum. Blackbody thermopile pyranometers have minimal spectral error because they are nearly uniformly sensitive to wavelengths from at least 290 to 3000 nm. Spectral sensitivity varies slightly among models due to different materials used to make the domes, as some materials are not uniformly transmissive near the lower and upper bounds of the shortwave spectrum (280 nm and 4000 nm, respectively), but this results in negligible errors. However, silicon-cell pyranometers can have large spectral errors because they subsample the shortwave spectrum (350 to 1100 nm), and are not equally sensitive within this range (Fig. 5). Subsampling and extrapolation are common in scientific measurements for prediction beyond the sample, but extrapolation must always be done with caution. Silicon-cell pyranometers are typically calibrated against blackbody thermopile pyranometers under clear sky conditions, thus they are accurate for clear sky conditions of similar humidity. However, changes in atmospheric air mass, humidity, clouds, dust, or pollution alter the shortwave irradiance spectrum and cause spectral errors.

Atmospheric air mass is the relative mass of the air column between the top of the atmosphere at the solar zenith and a point at Earth's surface. Atmospheric air mass changes as a function of solar zenith angle. At low solar zenith angles near solar noon (low atmospheric air mass), the sky is blue on clear days because Rayleigh scattering is more effective at scattering short wavelength (blue) radiation in the visible spectrum. At high solar zenith angles near the beginning and end of the day (high atmospheric air mass) solar radiation traverses a long atmospheric path before reaching Earth's surface. As a result, the sky looks red because Rayleigh scattering has selectively scattered the shorter wavelength (blue) radiation. Thus, from morning to midday to evening sky color changes from red to blue to red. This alters the signal output of silicon-cell pyranometers because they are more sensitive to red and near infrared radiation than to blue radiation (Fig. 5).

Directional or angular response of silicon-cell pyranometers includes both a directional and spectral component (directional response in the field is the combination of the true directional response as measured in the laboratory and spectral response of the sensor). Spectral error at high solar zenith angles (morning and evening) causes silicon-cell pyranometers to read high and directional error at high solar zenith angles causes most pyranometers (silicon-cell and blackbody thermopile) to read low. These two errors often cancel each other and yield small directional errors for silicon-cell pyranometers in the field (King and Myers, 1997; Klassen and Bugbee, 2005; Selcuk and Yellott, 1962), except for solar zenith angles greater than about 75° where spectral error dominates (Fig. 4).

Because it is challenging to separate spectral error and directional error, manufacturers of silicon-cell pyranometers optimize diffusers to achieve minimum error as a function of solar zenith angle. This amounts to designing sensors to intentionally read low at high angles of incidence in the laboratory to account for spectral error at high solar zenith angles in the field. Thus, spectral errors due to changes in air mass are largely accounted for in instrument design. Only at high atmospheric air mass (high solar zenith angles) do significant errors occur (Fig. 4, the sharp increase in silicon-cell pyranometer error at solar zenith angles greater than 75° is largely due to spectral error from the increasing proportion of red and near infrared wavelengths within the 350 to 1100 nm range).

Over the relatively narrow range of atmospheric vapor pressure variability, 0.2 to 2 kPa, in Logan, UT, silicon-cell pyranometer errors (relative to the mean of measurements from four secondary standard blackbody thermopile pyranometers) on clear days near solar noon were not statistically significant, and there was no correlation between error and atmospheric vapor pressure over this range. The spectral error may be larger over a wider range of vapor pressure, but it is likely smaller than the error under cloudy conditions (discussed below).

Spectral error can be calculated if sensor spectral sensitivity, spectrum of the radiation source the sensor was calibrated with, and spectrum of radiation source the sensor is measuring are available:

$$\text{Error} = \frac{\int S_\lambda I_{\lambda\text{Measurement}}\,d\lambda \int I_{\lambda\text{Calibration}}\,d\lambda}{\int S_\lambda I_{\lambda\text{Calibration}}\,d\lambda \int I_{\lambda\text{Measurement}}\,d\lambda} \qquad [6]$$

where S_λ is relative (normalized to maximum) pyranometer spectral sensitivity, $I_{\lambda\text{Measurement}}$ is relative (normalized to maximum) spectrum of radiation source being measured, $I_{\lambda\text{Calibration}}$ is relative (normalized to maximum) spectrum of radiation source the pyranometer was calibrated to, and λ is wavelength. As stated, silicon-cell pyranometers are usually calibrated against blackbody thermopile pyranometers under clear sky conditions, thus $I_{\lambda\text{Calibration}}$ is a clear sky spectrum. Silicon-cell pyranometers are therefore most accurate for clear sky conditions with humidity similar to that during calibration.

Spectral response varies slightly among models of silicon-cell pyranometers, due to diffuser materials and variability in silicon-cell options, but S_λ is similar among models because it is largely determined by the spectral properties of silicon. The silicon-cell pyranometer spectral response data from Fig. 5 were input into Eq. [6] to estimate spectral error for silicon-cell pyranometers under cloudy conditions. Using a clear sky calibration spectrum and overcast sky measurement spectrum (both measured during June 2013 in Logan, UT, with an ASD model FieldSpec Pro spectroradiometer) spectral error was predicted to be 9.6% high under overcast conditions. Kerr et al. (1967) studied this error and reported that SW_i on an overcast day was 11.3% of SW_i on a clear day and SW_i weighted according to the spectral response of silicon on the overcast day was 12.6%, which results in a spectral error of 11.5% (ratio of 12.6/11.3) on the overcast day if the

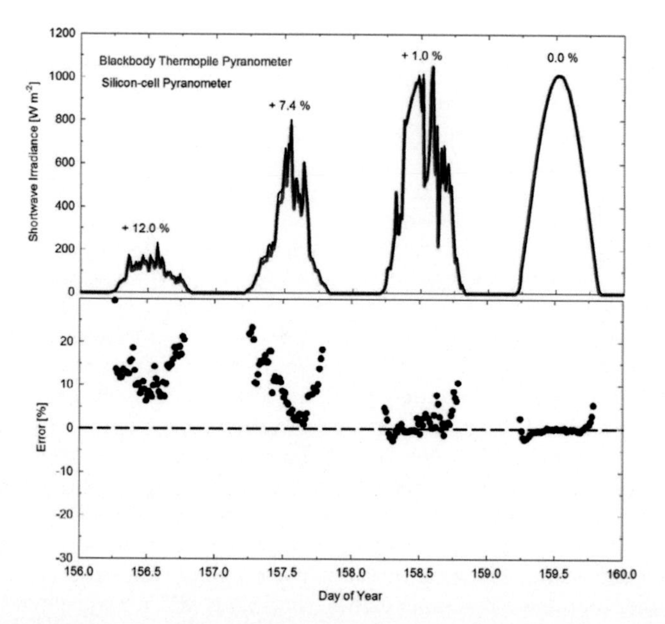

Fig. 6. Spectral error of a silicon-cell pyranometer caused by changes in cloudiness. The reference was a secondary standard blackbody thermopile pyranometer (Kipp & Zonen model CMP 11). This was compared with a silicon-cell pyranometer (Apogee Instruments model SP-110). Over a four-day period, conditions transitioned from overcast to clear sky. Errors for the silicon-cell pyranometer were calculated on daily total basis (daily total error shown in upper graph) and on a fifteen minute interval (lower graph). All silicon-cell pyranometers are subject to this spectral error.

sensor was calibrated on a clear day. This error is similar to the result of 9.6% calculated from Eq. [6] using the spectral data collected in Logan, UT. Shortwave irradiance measurements from silicon-cell pyranometers confirm these predicted errors for cloudy conditions. Measurements were made over a four-day period (5–8 June 2007, in Logan, UT) with a secondary standard thermopile pyranometer (Kipp & Zonen model CM 11) and a silicon-cell pyranometer (Apogee Instruments model SP-110). Spectral error for the thermopile pyranometer should be zero, so it was used to provide reference SW_i. Over the four-day period, conditions transitioned from complete cloud cover to clear sky, with spectral errors (calculated on a daily total basis) for the silicon-cell pyranometer declining from 12.0% under completely overcast conditions to 0.0% for clear sky conditions (Fig. 6). Error of a silicon-cell pyranometer (LI-COR model LI-200) relative to secondary standard thermopile pyranometers (mean of four replicates in the Apogee calibration facility) plotted versus cloudiness (ratio of measured SW_i to modeled clear sky SW_i) indicates errors typically less than 5% until the ratio of actual SW_i to clear sky SW_i (SW_i/SW_{ic}) declines below about 0.3, then errors increase (Fig. 7). Scatter in the error data results from different cloud types, as the magnitude of error will vary with cloud thickness and water content.

It is challenging to determine spectral error for silicon-cell pyranometers when sky conditions are dusty or polluted because solar radiation spectra or simultaneous data from a blackbody thermopile and silicon-cell pyranometers during dusty and polluted conditions are required. Throughout August 2015, Logan, UT, experienced a high level of atmospheric smoke from wildfires in the Pacific Northwest. Comparison of silicon-cell pyranometers to mean SW_i from four secondary standard blackbody thermopile pyranometers on a smoky day (visibility reduced to 6 km and PM 2.5 of 100 mg m^{-3}) and a clear day (visibility of 30 km and PM 2.5 of 5 mg m^{-3}) revealed differences of less than 2%. It is possible aerosols or dust will have a different effect.

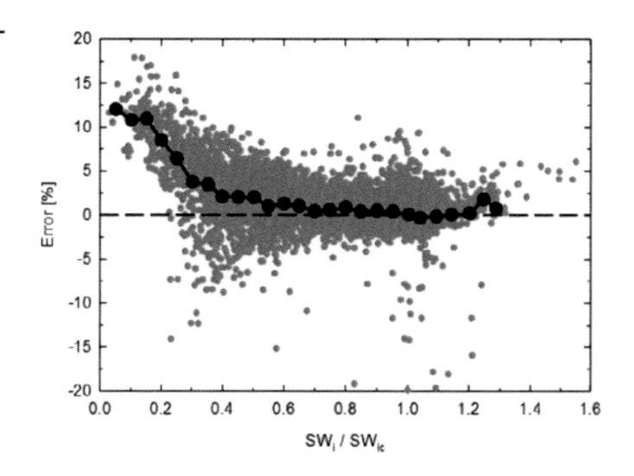

Fig. 7. Spectral error of a silicon-cell pyranometer as a function of cloudiness (SW_i/SW_{ic}, ratio of measured SW_i to modeled clear sky SW_i, often called the cloudiness index and serves as a surrogate variable for cloudiness). Error for the silicon-cell pyranometer (LI-COR model LI-200) was calculated relative to the mean of four secondary standard blackbody pyranometers. Black line is a bin average. All silicon-cell pyranometers will have this spectral error and follow a similar pattern with cloudiness. The magnitude of error varies with cloud thickness and water content, resulting in the scatter around the average error.

Temperature Error

Temperature error results from the temperature sensitivity of sensor components (thermopile or silicon-cell detector, resistor). Temperature sensitivity specifications are typically lower for blackbody thermopile pyranometers, particularly first class and secondary standard pyranometers, than for silicon-cell pyranometers. Most blackbody thermopile pyranometers are specified at less than 4% sensitivity from -10 to 40 °C (0.08% per °C, assuming temperature response is linear). Silicon-cell pyranometers are often specified at less than 0.15% per °C, about twice the sensitivity of many blackbody thermopile pyranometers. Secondary standard blackbody thermopiles are specified at 1.0% variability or lower for a temperature range of at least -10 to 40 °C. Some blackbody thermopile pyranometer manufacturers offer models with a temperature sensor and characterized response (e.g., EKO model MS-80, Hukseflux model SR20, or Kipp & Zonen model CMP 21), so users can correct for the temperature effect.

Blackbody thermopile pyranometers can also have thermal offset errors associated with net longwave radiation and rapid heating and/or cooling. If the dome and sensor body (housing) temperatures are different, this influences the energy balance of the detector plate and can cause measurement error. The longwave radiation balance at the blackbody surface is zero when the surface and glass dome are at the same temperature, but rapid heating or cooling and negative net longwave radiation (caused by exposure to the cold sky) result in signal gain or loss due to longwave energy transfer. Some models of blackbody thermopile pyranometers have ventilation units attached to them to minimize temperature differences between components.

There are two types of thermal offset errors: Zero Offset A and Zero Offset B. Both offsets are determined in the dark. Zero Offset A is response to thermal radiation of 200 W m^{-2} and Zero Offset B is the response to a temperature change of 5 °C per hour. Specifications for Zero Offset A range from 10 to 15 W m^{-2} for second class pyranometers to 3 to 6 W m^{-2} for secondary standard pyranometers. Specifications for Zero Offset B range from 4 to 6 W m^{-2} for second class pyranometers to

1 to 2 W m^{-2} for secondary standard pyranometers. Double domed sensors have lower offsets than single domed sensors because of the additional insulative layer of air.

Zero Offset A is dependent on the longwave radiation balance between the sky and dome. It is negative because the dome emits more longwave radiation than it is absorbing from the sky (dome is warmer than the sky), reducing its temperature relative to the blackbody absorber. Thus, the blackbody absorber emits more longwave radiation to the dome than it receives from the dome. Magnitude of this thermal offset is greatest during the day due to solar heating of the sensor (Ji and Tsay, 2010; Philipona, 2002), especially under clear sky and calm conditions, but it is also present at night (small negative signals are measured at night). Much lower offsets occur on cloudy days (Haeffelin et al., 2001). Subtraction of the nighttime signal has been used to partly correct for daytime heat loss due to radiative cooling (Dutton et al., 2001). Calibration of pyranometers should include correction for the thermal offset (Reda et al., 2005). Errors due to daytime radiative cooling can be corrected with measurements of net longwave irradiance (from a pyrgeometer) and ambient air temperature.

Thermal offsets are reduced by ventilation, which reduces temperature differences between components (dome and detector). Ventilation units that provide air flow over the outer dome are commonly available (e.g., EKO Instruments model MV-01, Eppley Laboratory model VEN, Hukseflux model VU01, Kipp & Zonen model CVF4), but they require 5 to 10 W of power. Internal ventilation, air flow between the inner and outer domes (e.g., Hukseflux model SR30), reduces thermal offset more than external ventilation and typically uses less power. Another method of reducing thermal offset is the use of a sapphire dome (e.g., Hukseflux model SR25), which has a much higher thermal conductivity than quartz and better matches the thermal conductivity of the sensor body, thus reducing the temperature difference between dome and detector. A design with a quartz diffuser over the blackbody detector and a single glass dome over the diffuser (e.g., EKO Instruments model MS-80) has also been used to reduce thermal offset by improved thermal coupling of the detector, diffuser, and dome, reducing the temperature difference.

Silicon-cell pyranometers operate via the photoelectric effect and are not subject to these thermal errors. Thus, they do not have specifications for Zero Offset A and Zero Offset B. They have been reported to have a negative temperature response (Klassen and Bugbee, 2005). This is based on the response of open circuit voltage of silicon-cells (solar panels operate in open circuit mode), which declines with increasing temperature (Osterwald, 1986). However, silicon-cell pyranometers operate in short circuit mode. This means that the temperature sensitivity of a silicon-cell is wavelength dependent, with a negative response below about 500 nm and a positive response above about 900 nm (Fig. 8). This makes temperature sensitivity dependent on the spectral intensity of the radiation source being measured. For sunlight, spectral intensity is greatest near 500 nm (Fig. 5), in the range where silicon-cell temperature response is slightly negative. Silicon-cell sensitivity peaks at about 960 nm (Fig. 5), where the temperature coefficient is positive. Sunlight includes radiation at all wavelengths across the silicon sensitivity range (350–1100 nm), making the temperature response of silicon-cell pyranometers for sunlight complex and dependent on the solar spectrum, silicon-cell sensitivity, and wavelength-dependent temperature coefficient.

To determine the temperature response silicon-cell pyranometers, four replicates of each of two models (Apogee Instruments model SP-110 and LI-COR model LI-200) were placed in a freezer and allowed to equilibrate. The sensors were then removed and immediately placed outside under clear sky conditions near solar noon. Sensor output was continuously monitored as they warmed over thirty minutes. Measurements were compared to shortwave radiation measurements from a reference blackbody thermopile pyranometer (Hukseflux model SR11) and plotted versus the silicon-cell pyranometer temperature (Fig. 9). Silicon-cell pyranometer temperature was continuously measured with thermistors mounted inside the four SP-110 pyranometers. The LI-200 pyranometers were assumed to be equal to the mean temperature from the SP-110 pyranometers. Results indicate a positive, linear temperature coefficient (Fig. 9) of 0.04 to 0.07% per °C under sunlight. These temperature coefficients are slightly lower than a previously reported value of 0.082% per °C for LI-COR model LI-200 pyranometers (King and Myers, 1997) and within the 0.04 to 0.10% per °C range reported by Kerr et al. (1967) for silicon-cells. The temperature response is challenging to measure accurately. One anomalous study found a much larger, nonlinear temperature response of LI-200 pyranometers, with a decline of about 5% from 25 °C to 10 °C (Raïch et al., 2007).

Blackbody thermopile and silicon-cell pyranometer temperature responses are usually considered negligible for most applications, as they are typically much smaller than other sources of error. For example, if the temperature coefficient was 0.05% per °C and daytime temperature variability was 20 °C (36 °F), then signal change due to temperature would be 1%. Temperature correction may be more applicable on a seasonal time scale, where temperature changes can be much larger than 20 °C. Temperature sensitivity

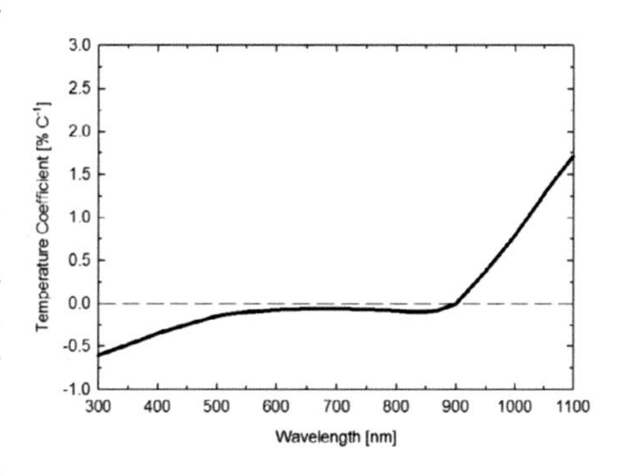

Fig. 8. Wavelength-dependent temperature coefficient of a typical silicon-cell in short-circuit mode.

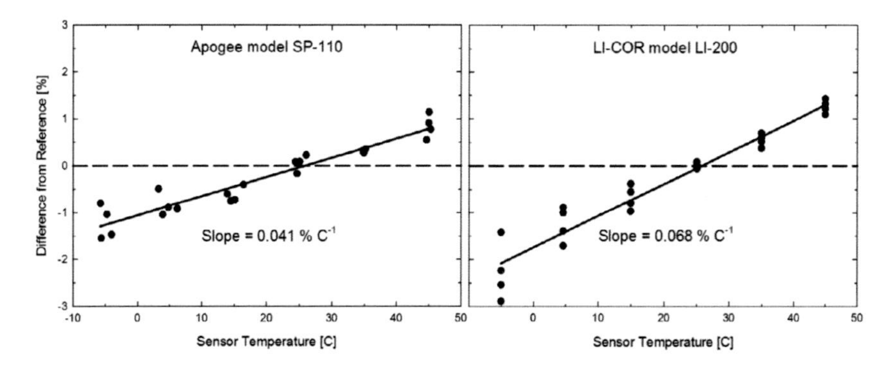

Fig. 9. The temperature response of two silicon-cell pyranometers (Apogee Instruments model SP-110 and LI-COR model LI-200). The slope is the mean temperature response of four replicate sensors.

of silicon-cell pyranometers is linear (Fig. 9), so the signal can be corrected for temperature changes if a measurement or estimate of detector temperature is available. All silicon-cell pyranometers are likely to have positive temperature coefficients with a similar slope when measuring sunlight because the measured data match the expected temperature sensitivity based on the wavelength-dependent temperature coefficient (Fig. 8). These results will be different if the radiation source is not the sun. As an example, the temperature response of ten replicate silicon-cell pyranometers (Apogee Instruments model SP-110) was determined in a temperature-controlled chamber under a cool white light emitting diode (LED) lamp with a high fraction of blue radiation below 500 nm and negligible output at wavelengths greater than 800 nm. As expected, based on the wavelength-dependent temperature sensitivity of silicon-cells (Fig. 8), the temperature coefficient was negative (-0.043% per °C).

Stability

Long-term stability of pyranometers is dependent on optical stability (glass dome and blackbody absorber surface for blackbody thermopile pyranometers, and acrylic diffuser for silicon-cell pyranometers) and electrical stability (thermopile or silicon-cell detectors, resistors, and solder joints). Some blackbody thermopile pyranometers are sensitive to the stability of the blackbody surface, which is subject to fading and discoloration with exposure to shortwave radiation. In a recent publication, Wood et al. (2015) measured the drift in two replicate Eppley model PSP pyranometers. Discoloration (greening) of the blackbody surface on PSP pyranometers caused downward drift in signal, leading to low measurements of SW_i. An initial stable period was observed, which varied among replicates. Similar rates of downward drift, approximately -1.5% per yr, were measured once drift started to occur. Downward drift of Eppley PSP pyranometers was also reported in an earlier study, but at lower rates of -0.4 to -1.0% per yr (Riihimaki and Vignola, 2008). Downward drift of Eppley PSP pyranometers was found to be a linear function of exposure to shortwave radiation, thus cumulative exposure can be used to predict signal decline as a result of fading or discoloration of the blackbody detector.

In addition to documenting drift in Eppley PSP pyranometers, Wood et al. (2015) proposed a method that uses time series of three different ratios (SW_i to extraterrestrial shortwave irradiance, PAR to SW_i and PAR to extraterrestrial shortwave irradiance) to detect drift in pyranometers. Using the proposed method and time series of radiation ratios, Wood et al. (2015) measured downward drift in multiple Eppley PSP pyranometers and upward drift in two Kipp and Zonen CM 3 pyranometers at field research sites in a measurement network. Drift of PSP pyranometers was consistent with fading of the blackbody detector. The exact cause of upward drift in CM 3 pyranometers was not determined, but was possibly attributed to increased electrical resistance. As discussed above, comparison with modeled SW_i for clear sky conditions can provide a reasonable estimate of drift and need for recalibration.

We have four secondary standard pyranometers (Hukseflux model SR20, Kipp & Zonen models CMP 11, CM 11, and CM 21) in our calibration facility at Apogee Instruments in Logan, UT. Each pyranometer is sent to NREL every two to three years for recalibration. The oldest one is the CM 21 (purchased in 2004), and this instrument had a large decline of 1.8% between June 2008 and

May 2011. Two calibrations since May 2011 (June 2013 and June 2015) indicate that the pyranometer is stable. After June 2008, NREL started using an updated method to account for pyranometer thermal offset (Reda et al., 2005). This particular CM 21 has a large thermal offset compared with the other three secondary standard pyranometers. It is likely that use of the new calibration method caused apparent drift, rather than actual change in calibration, as the new calibration method better accounted for the thermal offset. The other three secondary standard pyranometers have been stable at less than 1% change between calibrations. Visual analysis indicates no discoloration of the blackbody detectors on any of these four pyranometers. Multiple replicates of three models of silicon-cell pyranometers (Kipp & Zonen model SP-Lite, LI-COR model LI-200, and Apogee Instruments model SP-110) were compared to the mean of these four secondary standard pyranometers over a two-year period (August 2013–August 2015). Data for clear sky conditions indicate none of the silicon-cell pyranometers has drifted by more than 2% per yr. Only one of the sensors drifted by more than 1% per yr, an SP-Lite drifted down by -1.2% per yr. Tanner (2001) found that the change in calibration for 520 LI-200 pyranometers was less than 2% per yr for 86% of sensors, and only 6% of pyranometers drifted by more than 3% per yr. Geuder et al. (2014) reported change in calibration for 30 LI-200 pyranometers, with only one drifting by more than 2% per year.

Drift problems can be temporarily corrected by recalibration, but regular recalibration is required to account for continued drift. It should be noted that recalibration only partially fixes the problem in the case of physical changes in sensor optics (e.g., discoloration of blackbody surface). Wood et al. (2015) reported that discoloration of blackbody detector surfaces resulted in changes in spectral response, meaning pyranometers with discolored blackbody surfaces were no longer equally sensitive to all wavelengths within the shortwave spectrum. This condition results in spectral errors when measurements are made in conditions significantly different than conditions during calibration. However, to our knowledge, the spectral absorptivity of a discolored blackbody surface has not been published, so the magnitude of spectral error can't be quantified using Eq. [6].

General Use

The most common field errors are improper mounting, inaccurate leveling, and occlusion of the dome or diffuser. Pyranometers should be mounted in an open area away from buildings, trees, and other structures that may obstruct the field of view. On a weather station, pyranometers should be mounted on the south side of the tower in the northern hemisphere and north side of the tower in the southern hemisphere. Other sensors should not obstruct the field of view.

Inaccurate leveling can result from improper installation, drift in mounting hardware following installation, or a level bubble that is not in the exact same plane as the detector. Leveling errors can potentially be detected by comparing measured SW_i to modeled clear sky SW_i (Menyhart et al., 2015).

Occlusion is typically caused by residual precipitation, condensation, or dust and/or debris deposition caused by wind or birds. Dust occlusion can be particularly bad when the dome or diffuser is wet, following dew deposition or precipitation. Glass domes on blackbody thermopile pyranometers and diffusers on some models of silicon-cell pyranometers are dome-shaped, and can be self-cleaning. Periodic cleaning is recommended, and frequency of cleaning is dependent on local conditions. Some pyranometer models include a heater and/or ventilation unit to minimize errors due to dew, frost, rain, and snow. Power requirements for heated and/or ventilated pyranometers vary over a 100-fold range, from 0.2 to 20 W. Power consumption increases with size of the pyranometer, number of resistance heaters, and size of the fan. For units with a heater and ventilator, approximately half the power goes to heating and the other half to ventilation. Ventilator filters should be cleaned or replaced at least annually to maintain proper air flow. The necessary interval varies widely depending on moisture and dust in the air.

Measurement Uncertainty

Collectively, the sources of error described above contribute to significant uncertainty in SW_i measurements. A recent report indicated measurement uncertainty of about 4% and 8% for SW_i measurements with blackbody thermopile and silicon-cell pyranometers, respectively (Reda, 2011). The uncertainty calculation method proposed by Reda (2011) was based on the International Guidelines of Uncertainty in Measurement (JCGM/WG 1, 2008). The main contributor to the 4% greater uncertainty estimate for silicon-cell pyranometers is their limited spectral response. Aside from this, calibration uncertainty was the largest contributor to total measurement uncertainty for both types of pyranometers, and was estimated at nearly one third of total uncertainty. When solar zenith angle–dependent calibration factors were used, rather than constant calibrations factors, total measurement uncertainty was reduced by about half. The uncertainty estimates of Reda (2011) are similar to those reported by Klassen and Bugbee (2005), 2 to 3% for thermopile pyranometers and 3 to 6% for silicon-cell pyranometers. These estimates provide an approximation of uncertainty in SW_i measurements with pyranometers.

Direct and Diffuse Shortwave Irradiance

This section, Shortwave Irradiance, has discussed global horizontal shortwave irradiance (SW_i) measurements with pyranometers. As discussed in the Theory section, SW_i is the sum of direct and diffuse components. The direct component is determined by measuring direct normal irradiance (DNI, irradiance on a plane perpendicular to sun) with a pyrheliometer and multiplying by the cosine of the solar zenith angle. The diffuse component is called diffuse horizontal irradiance (DHI, radiation emanating from entire hemisphere of sky when the solar disk is blocked from the field of view) and is measured with a pyranometer shaded by a solar tracking disk. Summation of DNI and DHI is referred to as the component summation method for determination of SW_i and has been reported as more accurate than SW_i measurement with pyranometers (Michalsky et al., 1999). However, measurements of DNI and DHI with pyrheliometers and shaded pyranometers, respectively, are complex, expensive, and not common on agricultural weather stations or in agricultural measurement networks. Vignola et al. (2012) discussed DNI measurements with pyrheliometers and DHI measurements with shaded pyranometers in detail.

Net Radiation

Net radiation (R_n) at Earth's surface is defined as the difference between incoming (downwelling) and outgoing (upwelling) irradiance, and consists of shortwave and longwave components:

$$R_n = (SW_i - SW_o) + (LW_i - LW_o) \quad [7]$$

where SW_i is incoming (downwelling) shortwave irradiance (global shortwave), SW_o is outgoing (upwelling) shortwave irradiance (reflected shortwave), LW_i is incoming (downwelling) longwave irradiance (global longwave), LW_o is outgoing (upwelling) longwave irradiance (emitted longwave), $SW_i - SW_o$ is net shortwave irradiance (SW_n), $LW_i - LW_o$ is net longwave irradiance (LW_n), and all terms are expressed as energy flux densities (typically in units of W m^{-2}). While net radiation is the name universally used to describe net irradiance at Earth's surface, it should be recognized that all terms in Eq. [7] are irradiance and R_n could be referred to as net irradiance. Typical values of the four components of R_n for clear sky conditions near solar noon in mid-summer and mid-winter provide approximations for magnitudes of expected irradiances (Fig. 10).

Net radiation is a major component of the surface energy balance:

$$R_n - G = H + LE + A_n + S \quad [8]$$

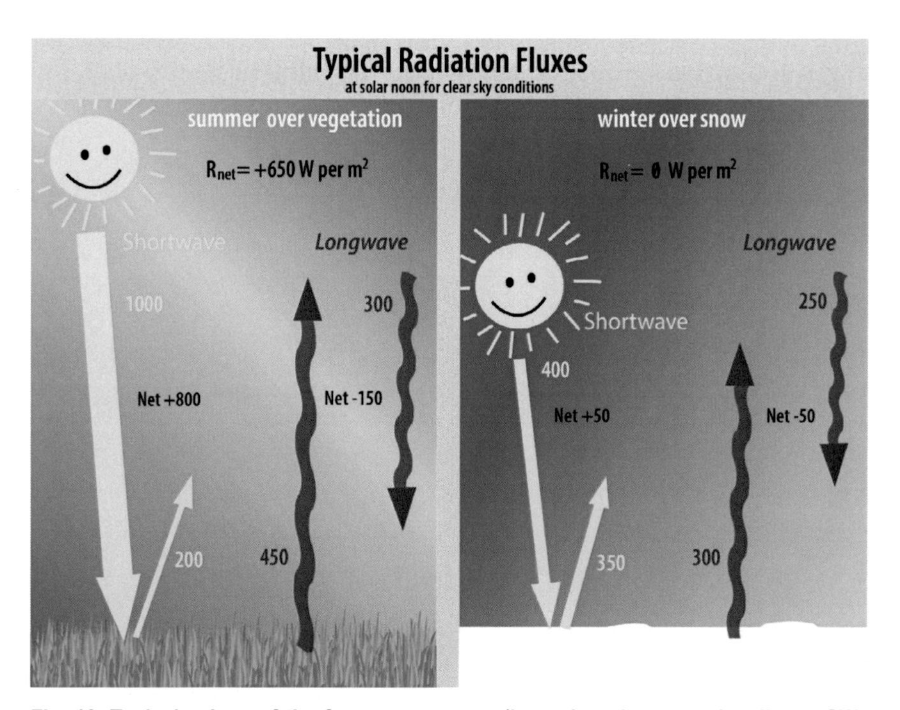

Fig. 10. Typical values of the four components (incoming shortwave irradiance SW_i, outgoing shortwave irradiance SW_o, incoming longwave irradiance LW_i, and outgoing longwave irradiance LW_o) of net radiation (R_n) for clear sky conditions near solar noon in mid-summer and mid-winter.

Table 2. Albedo (α, broadband shortwave reflectivity, SW_o/SW_i) of some representative terrestrial surfaces (Campbell and Norman, 1998).

Surface	Albedo (α)
Grass field	0.24–0.26
Wheat field	0.16–0.26
Corn field	0.18–0.22
Deciduous forest	0.10–0.20
Coniferous forest	0.05–0.15
Tundra	0.15–0.20
Steppe	0.20
Fresh snow	0.75–0.95
Old Snow	0.40–0.70
Wet, dark soil	0.08
Dry, dark soil	0.13
Wet, light soil	0.10
Dry, light soil	0.18
Dry, white sand	0.35
Urban area (average)	0.15

where G is ground heat flux, H is sensible heat flux, LE is latent heat flux, A_n is net photosynthesis, S is heat storage (typically negligible in crop canopies, but must be accounted for in forest canopies), and like R_n, all terms are flux densities (typically W m^{-2}). The difference between R_n and G ($R_n - G$) is defined as available energy and is the energy at the surface available to drive surface processes (e.g., heating, evapotranspiration).

In many cases, R_n is the largest term in Eq. [8], thus accurate R_n measurements or estimates are essential to surface energy balance studies. Surface energy imbalance (meaning, measurement of $R_n - G$ is not balanced by the sum of measurements of terms on the right-hand side of Eq. [8]) has received considerable attention (e.g., Foken, 2008; Leuning et al., 2012; Wilson, 2002). Some studies have found R_n measurements have lower uncertainty than turbulent flux measurements (H and LE) in surface energy balance experiments (Twine et al., 2000), while other studies have suggested energy balance studies are limited by R_n accuracy and there is a need for improved net radiometer designs and calibration procedures (Kustas et al., 1998).

Shortwave Irradiance

Measurement of SW_i was discussed in the previous chapter section, Global Shortwave Irradiance. Measurement of SW_o requires inversion of a pyranometer to measure shortwave irradiance reflected from the ground surface. The same considerations for measurement of SW_i apply to measurement of SW_o. Some additional specifics for SW_o measurement should also be considered. Downward-looking pyranometers should be equipped with a shade ring and mounted so as to eliminate any shortwave radiation not reflected by the surface from reaching the detector (the downward-looking pyranometer shouldn't 'see' above the horizon). Nearly all terrestrial surfaces are diffuse reflectors, meaning reflected radiation is reflected in all directions, and are relatively isotropic (not directionally dependent), even if radiation from the source illuminating the surface (sun) is directionally dependent (a clear day). As a result, directional response of pyranometers used to measure SW_o is less critical than directional response of pyranometers used to measure SW_i. As long as the distribution of angles of reflected shortwave radiation is relatively constant for diverse surface conditions (plant canopy, soil, or snow), sensors with fairly poor directional response can still yield accurate SW_o measurements if they are calibrated over a diffuse reflecting surface.

The ratio SW_o/SW_i quantifies broadband surface shortwave reflectivity and is called albedo. Albedo of terrestrial surfaces varies over a wide range (Table 2). Albedo also changes with sky conditions. On a clear day albedo is highest when the sun is near the horizon and lowest near solar noon. On an overcast day albedo is nearly constant, assuming the underlying surface doesn't change over the course of the day. The combination of an upward-looking pyranometer and downward-looking pyranometer (typically of the same model), a net shortwave radiometer, is often called an albedometer. A single pyranometer mounted on a rotating device can also be used as a net shortwave radiometer. When measuring SW_n or albedo with this type of device, the pyranometer must be returned to a level position each time the mechanism is rotated to measure SW_i. Also, sky conditions (degree of cloudiness) must be constant for short-term measurements because SW_i and SW_o will not be measured at the same time.

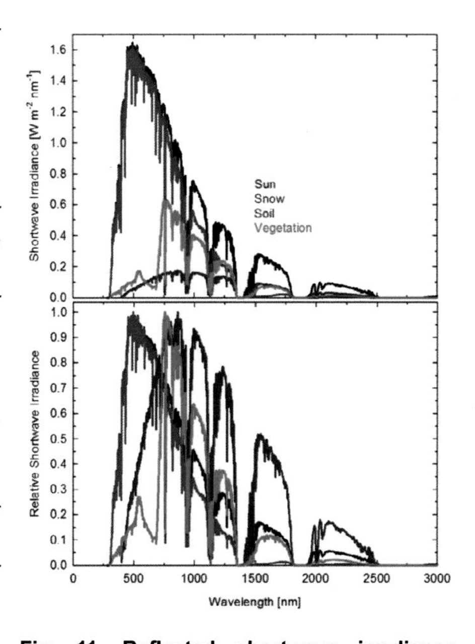

Fig. 11. Reflected shortwave irradiance spectra for snow, soil, and vegetation compared to an incoming shortwave irradiance spectrum (absolute values of irradiance are shown in top graph, relative values of irradiance are shown in bottom graph and were calculated by normalizing to the maximum value). Reflectance data were taken from the Advanced Spaceborne Thermal Emission and Reflection (ASTER) spectral library (Baldridge et al., 2009). If a silicon-cell pyranometer is used to measure reflected shortwave irradiance (SW_o) these differences in spectral reflectance cause errors in the measurement (Table 3).

Silicon-cell pyranometers should not be used for SW_o measurement. The spectrum of reflected radiation from terrestrial surfaces is different than the spectrum of incoming radiation (Fig. 11), and the limited and uneven spectral sensitivity of silicon-cell pyranometers (Fig. 5) will be subject to large errors (Table 3). Ross and Sulev (2000) reported errors of 20 to 40% for measurements of SW_o and measurements of shortwave irradiance transmitted below plant canopies with silicon-cell pyranometers. While it is possible to calibrate a silicon-cell pyranometer for specific surface conditions, over the course of a year an agricultural field may change from bare soil, to partial canopy cover, to full green canopy cover, to stubble (following harvest)

Table 3. Spectral errors in measurement of reflected shortwave irradiance (SW_o) with silicon-cell pyranometers calibrated to clear sky conditions. Equation [6] was used to calculate errors.

Surface	Error
	%
Grass canopy	14.6
Deciduous canopy	16.0
Conifer canopy	19.2
Agricultural soil	-12.1
Forest soil	-4.1
Desert soil	3.0
Water	6.6
Ice	0.3
Snow	13.7

or senesced canopy, to bare soil, to snow. Each of these surfaces has a characteristic reflected shortwave irradiance spectrum. A blackbody thermopile pyranometer, with equal sensitivity to all shortwave wavelengths, eliminates spectral error in SW_o measurements under all conditions.

Longwave Irradiance

Longwave irradiance is measured with pyrgeometers (e.g., EKO Instruments model MS-202, Eppley Laboratory model PIR, Hukseflux model IR20, or Kipp & Zonen model CGR 4), which are broadband radiometers similar to blackbody thermopile pyranometers, but fitted with a silicon dome or flat window (instead of a quartz or glass dome) and an interference filter to block shortwave radiation at wavelengths less than about 4.5 μm. Silicon transmits infrared wavelengths and has cutoff between 40 and 50 μm. Thus, the combination of shortwave blocking filter and silicon window and/or dome span most of the range of infrared wavelengths emitted by Earth's surface and Earth's atmosphere (Fig. 12). Longwave radiation spectra for terrestrial surfaces (soil, plant canopy, or snow) are similar to blackbody spectra calculated with Eq. [1], as emissivities for terrestrial surfaces are often near one. However, under clear sky conditions atmospheric longwave radiation spectra are different because air molecules are selective absorbers and emitters, meaning sky absorptivity and emissivity are variable and much less than one at certain wavelengths. Thus, when the sky is clear it appears colder than the surface. Longwave radiation spectra for clear and cloudy sky conditions are also

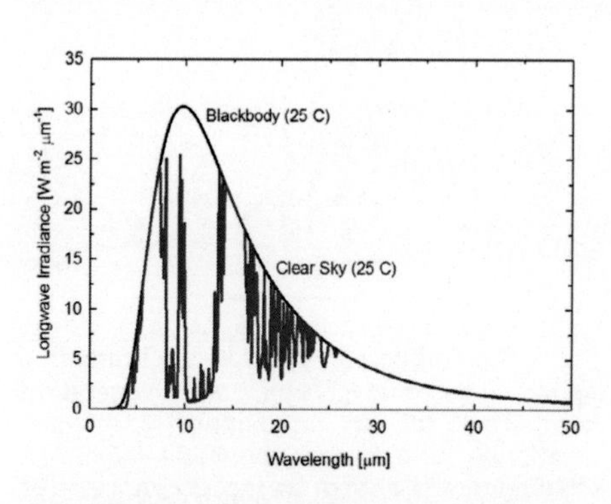

Fig. 12. Clear sky longwave irradiance for a mid-latitude atmosphere at 25 °C compared to blackbody irradiance at 25 °C. Many terrestrial surfaces (soil, plant canopy, snow) are nearly blackbody radiators (emissivities are near one), but a clear sky is unique because of selective absorption and emission by atmospheric gases. An overcast sky approximates a blackbody emitter and irradiance is similar to that shown for the blackbody.

different, as the water vapor in clouds makes them nearly blackbody emitters like terrestrial surfaces. As a result, longwave irradiance spectra for overcast conditions can be approximated with Eq. [1] using the effective temperature of the bottom of the clouds. A typical value of incoming global longwave irradiance (LW_i) for a mid-latitude location in summer on a clear day near solar noon is 300 W m^{-2}. A typical value for a cloudy summer day is 400 W m^{-2}. Typical values of outgoing longwave irradiance (LW_o) can be calculated from surface temperature with Eq. [3].

The instrument properties contributing to the energy balance of the absorbing surface of the detector plate in a pyrgeometer are longwave reflectance and transmittance of the dome or window (hereafter referred to as a dome, even though many instruments have flat windows), thermal conductance of the detector plate, and sensor body temperature. These factors are accounted for during calibration, but conditions dissimilar to conditions during calibration can cause measurement errors if there are imperfections in instruments properties (e.g., longwave transmittance of the dome is less than 100%). Pyrgeometer calibration is accomplished by deriving a calibration factor (CF) that scales the signal output by the thermopile to match a measurement of the longwave radiation balance at the detector surface ($LW_{in} - LW_{out}$):

$$\mathrm{CF} = \frac{\mathrm{LW_{in}} - \mathrm{LW_{out}}}{S} \qquad [9]$$

where LW_{in} is incoming longwave irradiance (W m^{-2}) absorbed by the blackbody surface, LW_{out} is outgoing longwave irradiance (W m^{-2}) emitted by the blackbody surface, S is voltage signal output by the thermopile, and CF is in units of W m^{-2} per μV. The reciprocal of CF is pyrgeometer sensitivity. Typical sensitivity for a pyrgeometer is 10 μV per W m^{-2}, meaning signal output is 10 μV for every 1 W m^{-2} difference between LW_{in} and LW_{out}. Resolution of the analog signal measurement must be 10 μV to provide LW_{in} measurement resolution of 1 W m^{-2} and 1 μV to provide 0.1 W m^{-2}.

In Eq. [9] $LW_{out} = \sigma T_D^4$, where T_D is detector temperature in units of K, and is measured with an internal temperature sensor (thermistor or platinum resistance thermometer). The internal temperature sensor should be thermally coupled to the detector plate to provide an accurate estimate of detector temperature. The detector plate and internal temperature sensor should be thermally isolated from the sensor housing adjacent to the plate to minimize the influence of heat transport by conduction. Once a pyrgeometer is calibrated, subsequent measurements of LW_i or LW_o are made by rearranging Eq. [9] to solve for LW_{in} and inputting CF, measured detector temperature, and measured S. Signal from a pyrgeometer directed toward the sky to measure LW_i is typically negative because LW_{in} is typically less than LW_{out}. Signal is often near zero for a pyrgeometer directed toward the ground to measure LW_o because LW_{in} and LW_{out} are similar in magnitude.

Similar to blackbody thermopile pyranometers, error is introduced in measurements of longwave irradiance if the pyrgeometer dome and reference junction temperature are different. This error can be accounted for if a measurement of dome temperature (T_{Dome}) is available:

$$\mathrm{LW_{in}} = \mathrm{CF} \cdot S + \sigma T_D^4 - k\sigma\left(T_{\mathrm{Dome}}^4 - T_{\mathrm{Detector}}^4\right) \qquad [10]$$

where k is factor related to the thermal coupling between dome and detector and must be determined during calibration. Some pyrgeometer models (e.g., Eppley Laboratory model PIR) have a second temperature sensor to provide a measurement of T_{Dome}. Good thermal coupling between dome and detector minimizes k and makes the term on the right-hand side of Eq. [10] negligible. It is also possible to shade pyrgeometer domes to minimize the temperature difference between dome and detector. Shading also reduces transmission of shortwave wavelengths if the shortwave blocking filter is imperfect.

Theoretically, pyrgeometers should equally weight all wavelengths of radiation between approximately 4.5 and 50 μm. In practice, pyrgeometers are not uniformly sensitive to wavelengths within this range because of absorption bands of silicon. This must be accounted for in the calibration procedure, meaning the longwave transmittance of the silicon dome must be accounted for when deriving CF (Eq. [9]). There are two methods for calibrating pyrgeometers: calibration against a reference pyrgeometer under outdoor conditions, or calibration against a blackbody radiation source in the laboratory. Outdoor calibrations should be conducted at night to avoid the influence of solar heating of the silicon dome (Gröbner and Los, 2007). Data from clear and cloudy nights should be included to provide a range of conditions. For laboratory calibration, temperature of a blackbody can be controlled to produce longwave irradiances characteristic of outdoor conditions, and temperature measurement of the blackbody surface can be used to determine emitted longwave irradiance. Pyrgeometer body temperature and blackbody source temperature should be varied to span the range of the conditions expected at the site where the pyrgeometer will be used (Philipona et al., 1995). For pyrgeometers intended to measure LW_i, Philipona et al. (1998) suggested that blackbody calibration source temperature should be 10 to 25 °C below pyrgeometer body temperature. Gröbner and Los (2007) proposed a laboratory calibration method that weights radiation from a blackbody source according to the spectral response of the pyrgeometer to be calibrated. They reported that differences between this method and direct comparison to the World Infrared Standard Group (WISG) under outdoor conditions were less than 1%. The drawback of this method is that the spectral response of the pyrgeometer to be calibrated must be known. Another approach for laboratory calibration is use of a reference pyrgeometer to measure longwave irradiance from the blackbody source and serve as a transfer standard. The reference pyrgeometer should be identical to the pyrgeometer to be calibrated, and it should be calibrated against reference pyrgeometers outdoors. This method assumes that variability in silicon domes is negligible among replicate pyrgeometers of the same model.

Multiple studies have pointed out the lack of a world standard for broadband longwave irradiance measurement that is analogous to the World Radiation Reference for broadband shortwave irradiance (Blonquist et al., 2009a; Brotzge and Duchon, 2000; Halldin and Lindroth, 1992; Ohmura et al., 1998). Progress toward a longwave irradiance standard has been made through the establishment of an interim World Infrared Standard Group (WISG), consisting of four pyrgeometers calibrated against an absolute sky-scanning radiometer (Marty et al., 2003; Philipona et al., 2001). Outdoor calibration to reference pyrgeometers traceable to the WISG has been reported as the current best practice for pyrgeometer calibration (Vignola et al., 2012).

Longwave irradiance is diffuse, whether the radiation source is the sky or Earth's surface. Land surface temperature differences are relatively small for a uniform surface (e.g., full cover crop canopy). Even for nonuniform land surfaces, elements at the surface are often within a few degrees of each other. For clear skies, however, there are temperature and aerosol gradients across the hemisphere of sky (Unsworth and Monteith, 1975). For example, sky temperature for clear conditions is coldest in a direction perpendicular to the surface and gets warmer toward the horizons. Also, partly cloudy skies can produce temperature differences across the hemisphere of sky because clouds are much closer to blackbodies than clear sky. Despite nonuniform sky conditions, angular distribution of longwave radiation has been reported to be similar for clear and overcast skies (Unsworth and Monteith, 1975). Thus, outdoor calibration of pyrgeometers accounts for imperfect directional response. It has also been reported that the distribution of angles of longwave radiation emitted by the atmosphere and a blackbody cavity are similar (Gröbner and Los, 2007), accounting for imperfect directional response of a pyrgeometer if it is calibrated with a blackbody source in the laboratory. Thus, directional response of pyrgeometers is less critical than directional response for upward-looking pyranometers.

Errors from a pyrgeometer with imperfect directional response are typically small compared to the potentially large sources of error due to solar heating of the dome, imperfect spectral sensitivity, and imperfect shortwave radiation blocking. A pyrgeometer with imperfect shortwave radiation blocking due to a filter that is partially transparent to some solar wavelengths will yield high LW_i or LW_o measurements because some fraction of shortwave radiation is transmitted to the blackbody absorber and perceived as longwave radiation.

Solar heating of the dome causes it to be warmer than the underlying blackbody absorber, resulting in increased radiation toward the blackbody surface and errors in LW_i measurement. This has been well documented for Eppley model PIR pyrgeometers with an older KRS-5 dome (Albrecht and Cox, 1977; Enz et al., 1975) and newer versions with a silicon dome (Alados-Arboledas et al., 1988; Perez and Alados-Arboledas, 1999; Udo, 2000). This can result in large LW_i measurement errors on clear and intermittently cloudy days, especially if there is little wind (natural ventilation). Solar heating of the dome can be reduced by shading the dome (Alados-Arboledas et al., 1988; Enz et al., 1975) and/or ventilating the dome (Enz et al., 1975; Perez and Alados-Arboledas, 1999). Corrections for the solar heating effect have also been developed (Alados-Arboledas et al., 1988; Oliveira et al., 2006). Solar heating of the dome should be minimized in pyrgeometer construction by maximizing thermal coupling between the dome and detector plate. A recent study found LW_i measurements from an unshaded Kipp & Zonen model CGR 4 pyrgeometer matched LW_i measurements from a shaded Eppley PIR within the range of measurement uncertainty, indicating that the design of the CGR 4 reduces the solar heating effect (Meloni et al., 2012). Some pyrgeometer manufacturers offer shading devices (e.g., Kipp & Zonen model CM 121B/C, EKO Instruments model RSR-01).

Nonuniform spectral response has been shown to cause errors of about 2% for the range of integrated water vapor content of the atmosphere (Gröbner and Los, 2007; Miskolczi and Guzzi, 1993). Spectral errors for pyrgeometers can be calculated with Eq. [6] if the spectral transmittance of the pyrgeometer dome is available, along with spectra for the radiation source used during calibration and radiation source being measured. Data from Fig. 12 and a spectral response for a silicon dome were input into Eq. [6] to provide an estimate of pyrgeometer spectral error for clear and overcast sky conditions. If error for measurement of a blackbody (overcast sky) was assumed to be zero, spectral error for clear sky results in low measurements by approximately 2%. This is a theoretical value. Differences between LW_i measurements from field calibrated pyrgeometers and an absolute sky scanning radiometer have been reported as less than 2 W m^{-2} for nighttime conditions (Marty et al., 2003; Philipona et al., 2001), indicating small spectral errors for spectral differences between clear and cloudy skies.

Net Radiometer Designs

Instruments designed to measure R_n are called net radiometers. There are four basic designs of net radiometer. The most complex is a four-component net radiometer (e.g., Apogee Instruments model SN-500, EKO Instruments model MR-60, Hukseflux model NR01, Kipp & Zonen model CNR 4), which consists of four individual radiometers (upward-looking pyranometer, downward-looking pyranometer, upward-looking pyrgeometer, and downward-looking pyrgeometer) to independently measure the four components of R_n (SW_i, SW_o, LW_i, and LW_o). The radiometers are usually mounted in a single housing, but a four-component net radiometer can be assembled by deploying individual radiometers to measure SW_i, SW_o, LW_i, and LW_o at the same location. The information in preceding sections regarding shortwave and longwave irradiance measurements applies directly to pyranometers and pyrgeometers used in four-component net radiometers.

A net radiometer model similar to a four component instrument consists of four independent blackbody absorbers, but only two thermopiles, one for SW_n mounted between the upward- and downward-looking shortwave absorbers and one for LW_n mounted between the upward- and downward-looking longwave absorbers (e.g., Kipp & Zonen model CNR 2, which was discontinued in 2011). A challenge with this type of net radiometer design is matching the upward- and downward-looking detectors (combination of filter, absorber, and detector plate) so sensitivity is equal. If sensitivities of upward- and downward-looking detectors are not equal, then changing proportions of SW_i and SW_o, and LW_i and LW_o, will cause errors in measurements of SW_n and LW_n, respectively.

An instrument that measures all wavelengths, shortwave and longwave, incident on the absorber is called an all-wave radiometer or pyrradiometer (e.g., Philipp Schenk model 240–8111). Pyrradiometers consist of a blackbody surface and thermopile covered by a dome that transmits shortwave and longwave radiation (typically polyethylene). Thus, another type of net radiometer is the combination of an upward-looking pyrradiometer and downward-looking pyrradiometer, which independently measure incoming all-wave irradiance and outgoing all-wave irradiance. A challenge with this type of net radiometer is unequal dome transmittance for shortwave and longwave radiation. If dome transmittance is not equal for shortwave and longwave radiation, changing proportions of shortwave and longwave cause measurement errors.

The simplest type of net radiometer is a net all-wave radiometer, sometimes called a net pyrradiometer, which consists of a single thermopile fitted between two blackbody absorbers, one upward-looking and one downward-looking (e.g., EKO Instruments model MF-11, Hukseflux model NR02, Kipp & Zonen model NR Lite2, Radiation and Energy Balance Systems (REBS) model Q*7.1). The thermopile produces a voltage proportional to the difference between incoming and outgoing all-wave irradiance (temperature difference between upward- and downward-looking detectors). The challenges of matching detectors and unequal transmittance of shortwave and longwave radiation apply to net all-wave radiometers.

The energy balance of detector plates in pyrradiometers and net pyrradiometers is dependent on the shortwave and longwave transmittance of the domes and shortwave and longwave absorptance of the detector surface. Theoretically, transmittance and absorptance should be 100% for all wavelengths. Differences in transmittance and absorptance, particularly differences in shortwave and longwave transmittance and absorptance, will cause errors with changes in irradiance spectra (e.g., clear versus cloudy sky, day versus night). Measurements of shortwave and longwave transmittance of multiple different polyethylene domes revealed lower longwave transmittance relative to shortwave transmittance in every case (Campbell and Diak, 2005). On average, transmittance of longwave was lower than shortwave by 14% (range was 4 to 25%), with greater differences for thicker domes. One solution is to use thinner domes, but they are not as rugged and may require more maintenance. Matching upward- and downward-looking detectors for a net pyrradiometer is also challenging. If sensitivity of the two detectors is not matched, changing proportions of incoming and outgoing radiation will cause errors in measurement of R_n.

Net Radiometer Comparisons

In the most recent net radiometer study we are aware of, Blonquist et al. (2009a) compared five different net radiometer models (Kipp & Zonen model CNR 1, Hukseflux model NR01, Kipp & Zonen model CNR 2, Kipp & Zonen model NR Lite, and Radiation and Energy Balance Systems model Q*7.1) in the field. There is not a standard for R_n measurement, so the means of measurements from the four-component net radiometers (Kipp & Zonen model CNR 1 and Hukseflux model NR01) were used to calculate reference R_n and was used for comparison of all individual radiometers. Measurements of SW_i and LW_o from the four-component radiometers closely matched independent reference measurements (SW_i from a Kipp & Zonen model CM 11 pyranometer and LW_o from an Apogee Instruments model SI-111 infrared radiometer). Measurements of SW_i were typically within 1% of the reference (except at low solar zenith angles) and measurements of LW_o were typically within 2% of the reference (see Fig. 14 in Surface Temperature section). Duchon and Wilk (1994) also compared downward-looking pyrgeometers to an infrared radiometer and found close agreement. In addition to comparing radiometers to reference SW_i and LW_o, Blonquist et al. (2009a) flipped all radiometers during the day (near solar noon on a clear day) and found the SW_o and LW_i radiometers matched SW_i and LW_o radiometers within 1% in all cases but one (which was approximately 2%). Based on these results, it was judged that R_n calculated from mean component measurements from the four-component net radiometers was a reasonable R_n measurement for use as reference R_n.

Blonquist et al. (2009a) reported that four-component net radiometers were the most accurate under all conditions (day, night, clear, cloudy). A difference of approximately 5% in LW_i measurements from the Kipp & Zonen CNR 1 and Hukseflux NR01 radiometers was measured, likely due to differences in calibration procedures used by manufacturers. In a similar net radiometer study, Brotzge and Duchon (2000) also found differences between longwave measurements. As detailed earlier, different techniques are available to calibrate pyrgeometers and yield different results depending on how radiometer spectral response is accounted for (Gröbner and Los, 2007). In a round-robin pyrgeometer calibration experiment, some laboratories derived the same calibration factors (within limits of uncertainty) and other laboratories did not (Philipona et al., 1998). Differences in SW_i, SW_o, and LW_o from the two four-component net radiometer models were typically 1 to 2%. Michel et al. (2008) found larger error in shortwave measurements than longwave measurements when comparing four-component net radiometers.

Blonquist et al. (2009a) found the two net all-wave radiometers (Kipp & Zonen model NR-Lite and REBS model Q*7.1) were the least accurate, and the Kipp & Zonen model CNR 2 (discontinued in 2011) was intermediate in accuracy. Net all-wave radiometers tended to measure R_n approximately 2 to 4% low during the day and approximately 15 to 30% low in magnitude at night. Others have reported similar low readings relative to four-component R_n measurements (Brotzge and Duchon, 2000; Cobos and Baker, 2003). Both models were less sensitive to longwave than shortwave, by approximately 20% and 30% for the NR-lite and Q*7.1, respectively. Others have also reported lower longwave sensitivity for net all-wave radiometers (Brotzge and Duchon, 2000; Cobos and Baker, 2003; Duchon and Wilk, 1994; Field et al., 1992; Halldin and Lindroth, 1992). This means the radiometers will be most accurate under conditions similar to conditions during calibration and error will increase as proportions of shortwave and longwave radiation deviate from those during calibration. This explains why both radiometers measured low in magnitude at night, as there is no shortwave radiation at night. Others have measured differences between clear and cloudy skies (Cobos and Baker, 2003; Field et al., 1992). Separate calibrations for day and night have been proposed (Brotzge and Duchon, 2000; Fritschen and Fritschen, 2007), and others have recommended field calibration of net all-wave radiometers under conditions similar to those the radiometers will be used in (Halldin and Lindroth, 1992; Kustas et al., 1998).

Sensitivities of upward- and downward-looking detectors were less than 1% different for all NR Lite radiometers, but upward-looking detectors were approximately 3% more sensitive than downward-looking detectors on all Q*7.1 radiometers (Blonquist et al., 2009a). This may have been due to increased exposure to shortwave for upward-looking detectors, as matching of detectors wasn't measured until after radiometers had been deployed in the field for a few weeks. Similar to differential sensitivity to shortwave and longwave radiation, differential sensitivity between upward- and downward-looking detectors results in error when proportions of incoming and outgoing radiation are significantly different than those during calibration (e.g., plant canopy surface compared to snow covered surface). Results from the CNR 2 highlight the importance of matching upward- and downward-looking detectors. Two of the three CNR 2 SW_n radiometers had mismatched detectors, on the order of 6 to 8%. This did not lead to large errors when the radiometers were deployed over a vegetated surface, indicating calibration conditions were similar to a vegetated surface. However, when going from a vegetated surface (albedo near 0.20) to a snow surface (albedo near 0.80), detector mismatching of 5 to 10% leads to SW_n errors in the range of 20 to 40%. Errors are highly dependent on albedo and get worse as albedo increases. Mismatching of the longwave detectors on the CNR 2 was much lower, approximately 2%. This would yield much smaller LW_n errors, as relative proportions of LW_i and

LW_o typically don't change as much as relative proportions of SW_i and SW_o when going from one surface (or sky) condition to another.

The NR Lite net radiometers were extremely sensitive to precipitation and dew and/or frost on the detectors because the detectors are unshielded (Blonquist et al., 2009a). Thus, energy balance of the detector is altered by the presence of water. Data collected when detector surfaces were wet were not accurate. Others have reported this as well (Brotzge and Duchon, 2000; Cobos and Baker, 2003). Wind also influences the energy balance of the detector surfaces and causes errors if not accounted for (Brotzge and Duchon, 2000; Cobos and Baker, 2003; Fritschen and Fritschen, 2007). The NR Lite and Q*7.1 net radiometers have manufacturer-supplied wind speed correction equations that should be applied.

Since the net radiometer comparison of Blonquist et al. (2009a), Kipp & Zonen upgraded the CNR 1 to the CNR 4 and the NR Lite to the NR Lite2. Improvements in the CNR 4 are a curved dome and better thermal coupling on the upward-looking pyrgeometer, reference temperature sensor placed closer to pyrgeometer detectors, solar shield, and lighter weight. A heating and ventilation unit is now available as a built-in option. Halldin and Lindroth (1992) recommended heating and ventilation under all conditions. Malek (2008) found unheated and unventilated net radiometers measured higher than a heated and ventilated unit when dew/frost was on the domes (enhancing LW_i), but lower when snow covered the domes (reducing SW_i). Michel et al. (2008) found that a heated and ventilated CNR 1 net radiometer did not match a four-component Eppley system in the field, but could be field calibrated to match. Improvements in the NR Lite2 are refinements to specifications. In addition, other net radiometers, beyond those compared by Blonquist et al. (2009a), are available and some new net radiometers have been released since the field data were collected in 2007.

Rather than serve as a comprehensive review of all available instruments, information contained in this section should serve as a guide to users regarding strengths and weaknesses of instrument types, and challenges of making R_n measurements with any instrument. To summarize, four-component net radiometers provide the most information, but the disadvantage is high cost and requirement of multiple datalogger channels (at least five differential channels) to make the measurements. Paired all-wave radiometers and net all-wave radiometers are lower cost than four-component instruments for measuring R_n, but they are subject to sources of error that typically make them less accurate than four-component instruments. In addition, detectors (combination of domes, absorbing surfaces, and detector plates) may have different sensitivities to shortwave and longwave radiation. Mismatching of upward-looking and downward-looking detectors can also cause errors in net all-wave radiometers. Blonquist et al. (2009a) concluded that R_n measurements from four-component net radiometers are the most accurate, assuming accurate calibration of individual radiometers, and are preferred over other R_n measurement methods (as long as cost is not limiting). Similar conclusions have been reported in other studies comparing R_n from four-way net radiometers and net all-wave radiometers (Kohsiek et al., 2007).

Modeling Net Radiation

Due to the cost of net radiometers, and in some cases number of datalogger channels required to make measurements with net radiometers, R_n is often modeled. Perhaps the most common application of R_n modeling is in evapotranspiration

(ET) prediction from automated weather stations, as R_n is one of the primary drivers of ET, but it is uncommon to find an ET weather station with a net radiometer. Most R_n models use a measurement of SW_i from a pyranometer, an assumed (typically constant) value of albedo to calculate SW_o from SW_i, and air temperature, humidity, and cloudiness measurements/estimates to estimate LW_n. An example of this approach is the R_n sub-model in the ASCE Standardized Reference Evapotranspiration Equation (ASCE-EWRI, 2005). Comparison of R_n modeled with this approach to measured R_n indicated R_n measurements from multiple net radiometer models were more accurate than R_n estimates from this particular R_n prediction model (Blonquist et al., 2010; Blonquist et al., 2009a).

A better approach to model R_n for vegetated surfaces may be to measure SW_i, estimate albedo from a simple plant canopy radiative transfer model (Campbell and Norman, 1998), measure surface temperature with an IR radiometer from which LW_o can be calculated (see next section), and estimate LW_i with a prediction model. While the albedo estimate requires some lengthy equations, it yields a temporally variable albedo for clear sky conditions, consistent with measured values. Measurement of LW_o (from surface temperature measurement) eliminates the need to model LW_n ($LW_i - LW_o$), which was found to be the largest source of error in a common R_n model (Blonquist et al., 2010), in favor of estimating only the LW_i component of LW_n from a model. Multiple LW_i models have been evaluated and assessed for accuracy relative to LW_i measurements (Flerchinger et al., 2009).

Surface Temperature

Radiometers used to remotely measure surface temperature are often called infrared thermometers (IRTs) (e.g., Apogee Instruments model SI-111, Everest Interscience model ENVIRO-THERM), but infrared radiometer (IRR) is a better descriptor because thermal infrared radiation emitted from the surface of interest is being detected and then converted to surface temperature (Huband, 1985a). Infrared radiometers function exactly the same way as pyrgeometers, where the output signal is dependent on the longwave radiation balance of the detector (difference between absorbed and emitted longwave radiation). The difference between an IRR and pyrgeometer is the filter and field of view. Pyrgeometers have a wide field of view (150°-180°), whereas IRRs typically have a narrow field of view (60° or less). Pyrgeometers have a filter sensitive to broad range (4.5 to 50 μm), whereas IRRs have a filter that is only sensitive to a narrow range of wavelengths within the thermal infrared range. IRRs should have filters that approximate the so-called atmospheric window, the wavelength range where the atmosphere is relatively transparent to thermal infrared radiation, typically defined as 8 to 14 μm (Fig. 13). Between 8 and 14 μm air molecules emit and absorb little thermal infrared radiation. As a result, IRRs 'see through' the atmosphere and sense radiation emitted by the surface.

As with pyrgeometers, the signal output from an IRR is proportional to the incoming longwave radiation absorbed by the detector and the longwave radiation emitted by the detector. With IRRs, temperature is the desired quantity, rather than radiation, thus Eq. [9] is written in terms of temperature using the relationship between radiation emission and temperature (Stefan–Boltzmann Law) (Fuchs and Tanner, 1966; Kalma et al., 1988):

$$\mathrm{CF} = \frac{\sigma T_T^4 - \sigma T_D^4}{S} \qquad [11]$$

where T_T is target temperature and σT_T^4 is radiation emitted by the target and absorbed by the detector, T_D is detector temperature and σT_D^4 is radiation emitted by the detector, and CF is a calibration factor determined during radiometer calibration. Calibration of IRRs is typically done by the manufacturer, where a blackbody or near blackbody radiation source of known temperature is used to provide a range of T_T values for different values of T_D (Fuchs and Tanner, 1966; Amiro et al., 1983; Huband, 1985b; Kalma et al., 1988). Following calibration, subsequent surface temperature measurements are made by rearranging Eq. [11] to solve for T_T and inputting measurements of S and T_D.

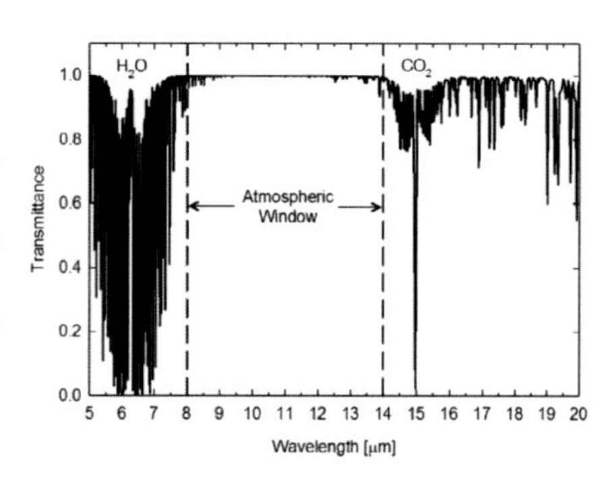

Fig. 13. Atmospheric transmittance from 5 to 20 μm. The atmospheric window is typically defined as the range from 8 to 14 μm. These data were modeled with MODTRAN assuming a typical mid-latitude summer atmosphere at a distance of two meters from the surface (two meters represents a typical IRR mounting height above the surface). At wavelengths less than 8 μm, water vapor (H_2O) absorption and emission can cause interference, and at wavelengths greater than 14 μm carbon dioxide (CO_2) absorption and emission can cause interference.

Unless the surface being measured is a blackbody (ε = 1), target temperature (T_T in Eq. [11]) measured by an IRR is an apparent surface temperature, often called brightness temperature, which can be thought of as temperature of a blackbody emitting the same radiance as that observed by the radiometer (Norman and Becker, 1995). Actual surface temperature of the elements within the field of view of the radiometer, called radiometric temperature, is determined by making a correction for the influence of surface emissivity. Thermal infrared radiation incident on the radiometer detector (σT_B^4) includes radiation emitted by the surface (σT_R^4) and reflected radiation from the background (σT_{Sky}^4):

$$\sigma T_B^4 = \varepsilon \sigma T_R^4 - (1-\varepsilon)\sigma T_{Sky}^4 \quad [12]$$

where T_B is brightness temperature (K), T_R is surface radiometric temperature (K), and T_{Sky} is sky temperature (K) in the 8 to 14 μm wavelength range. Radiation emitted by the surface is usually a large fraction, equal to surface ε, of total radiation incident on the detector because most terrestrial surfaces are nearly blackbodies (Table 4).

Radiation reflected from the background is a small fraction of total radiation, equal to 1 - ε. In outdoor settings, background radiation is sky radiation in the 8 to 14 μm range. Sky temperature in the 8 to 14 μm range is far colder than surface temperature on clear days, thus the small reflected fraction from the background can have a large impact on radiometric temperature and must be accounted for. It can be measured by directing an IRR toward the sky, or it can be predicted with a model (Idso, 1981; Kimball et al., 1982). We have successfully used a simple model derived for the high-elevation, semiarid climate in Logan, UT:

$$T_{Sky} = T_{Air} + 50 f_{Clouds} - 60 \quad [13]$$

Table 4. Emissivities (ε, broadband longwave emittance) for some terrestrial and man-made surfaces (Campbell and Norman, 1998).

Surface	Emissivity (ε)
Plant leaves	0.94–0.99
Soil	0.93–0.96
Water	0.96–0.98
Concrete	0.88–0.93
Aluminum foil	0.06

where T_{Air} is air temperature (K) and f_{Clouds} is fraction of cloud cover. The coefficients 50 and 60 in Eq. [13] may not be representative of all climates, but can be determined for a given location by simultaneous measurement or estimation of T_{Sky}, T_{Air}, and f_{Clouds}.

By rearranging Eq. [12], the influence of surface ε and reflected radiation from the background can be corrected, if ε and background (sky) temperature are measured or estimated:

$$T_R = \sqrt[4]{\frac{T_B^{\ 4} - (1-\varepsilon) T_{Sky}^{\ \ 4}}{\varepsilon}} \quad [14]$$

where T_T determined from Eq. [11] is input as T_B. Equations [12] and [14] assume constant ε at all wavelengths, an infinite wavelength range for radiation emission, and equal sensitivity to all wavelengths for the IRR. These assumptions are not strictly valid, as wavelength ranges measured by IRRs are typically close to the atmospheric window of 8 to 14 μm, IRR sensitivity is not equal across all wavelengths, and ε varies with wavelength for some surfaces. However, Eq. [14] is a reasonable approximation because background radiation is a small fraction (1- ε) of total radiation for land surfaces and ε varies little with wavelength with the 8 to 14 μm range for most land surfaces (soils with significant amounts of quartz are an exception). IRRs should have highly uniform sensitivity to wavelengths within the 8 to 14 μm range.

To demonstrate the importance of accounting for reflected background radiation, some typical values can be input into Eq. [14]. For a full cover plant canopy, ε is higher than that for individual leaves, and is often 0.97 to 0.99 (Fuchs and Tanner, 1966; Campbell and Norman, 1998). On a clear day, T_B = 25 °C and T_{Sky} = -40 °C are reasonable values. Assuming ε = 0.98, then T_R = 25.95 °C, nearly a whole degree warmer than T_B (apparent surface temperature). On a cloudy day, T_B = 25 °C and T_{Sky} = 15 °C are reasonable values. Again assuming ε = 0.98, then T_R = 25.19 °C, much closer to T_B. The difference between T_R and T_B will be larger if emissivity is smaller and if the difference between T_B and T_{Sky} is larger. If T_B and T_{Sky} are equal, T_R is also equal and correction for reflected background radiation is not necessary, but this only occurs on very cloudy or overcast days.

Sometimes an emissivity correction is applied where signal returned by an IRR is divided by surface emissivity, and some IRRs have an emissivity dial that makes this adjustment. Campbell and Diak (2005) suggested this type of correction is a holdover from days when IRRs were used to determine temperature of molten metal in furnaces. However, this correction does not account for reflected radiation and should not be used to calculate radiometric temperature in environmental applications, thus the emissivity dial should be set to 1.00. Under clear sky conditions, this correction reduces error relative to no correction for emissivity, but under cloudy conditions it may cause larger error than no emissivity correction. These effects are shown graphically in Appendix B of Blonquist et al. (2009b).

Infrared radiometers directed toward Earth's surface can be used as pyrgeometers because terrestrial surfaces are approximately blackbody emitters. Thus, T_T from Eq. [11] can be input into the Stefan–Boltzmann equation, along with ε equal to one, to calculate LW_o. If the surface is not uniform, LW_o from the IRR will be representative of the surface elements within the field of view of the IRR (IRRs typically have much narrower fields of view than pyrgeometers). Data comparing LW_o from an IRR (Apogee Instruments model SI-111) and pyrgeometer (Kipp & Zonen model CGR 3) for a clear and cloudy day over turfgrass in Logan, UT, indicate differences less than 3% (Fig. 14). IRRs with a filter that approximates the atmospheric window cannot be directed toward the sky and used as pyrgeometers. Sky temperature measured within this wavelength range will be much colder than actual sky temperature, thus LW_i calculated from this sky temperature will be much lower than actual LW_i.

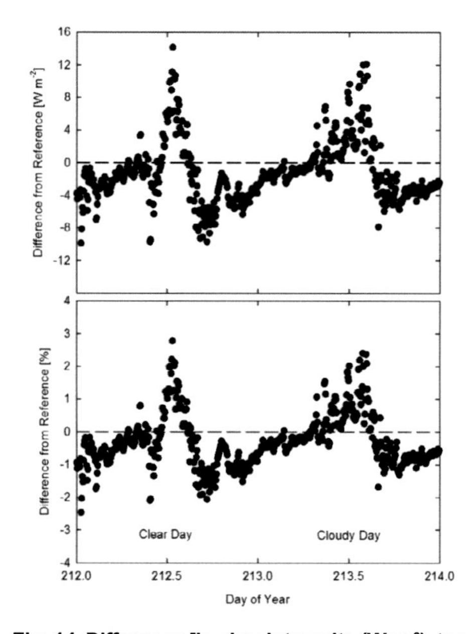

Fig. 14. Difference [in absolute units (W m^{-2}), top graph, and relative units (%), bottom graph) in outgoing longwave radiation calculated from surface temperature measured with an infrared radiometer (Apogee Instruments model SI-111) relative to a pyrgeometer (Kipp & Zonen model CGR 3). Day of year 212 was a clear day and day of year 213 was a cloudy day.

Photosynthetically Active Radiation

Photosynthetically active radiation (PAR) is the subset of shortwave radiation that drives photosynthesis. PAR should be quantified in units that express number of photons (photon flux density) rather than energy content of radiation (energy flux density) because one photon excites one electron (Stark–Einstein Law) in chlorophyll molecules in plant leaves, independent of the energy of the photon (energy content of photons is dependent on wavelength, with photons of shorter wavelength having higher energy, see Eq. [2]). As a result, equal energy flux densities of radiation of different wavelengths will not yield equal photon flux densities.

Instantaneous PAR is typically reported as photon flux density in units of micromoles of photons per square meter of area per second (μmol m^{-2} s^{-1}) between 400 and 700 nm. Daily total PAR is typically reported in units of moles of photons per square meter per day (mol m^{-2} d^{-1}) between 400 and 700 nm, and is often called daily light integral (DLI).

Photosynthesis does not respond equally to all photons due to the combination of spectral absorptivity of plant leaves (absorptivity is higher for blue and red

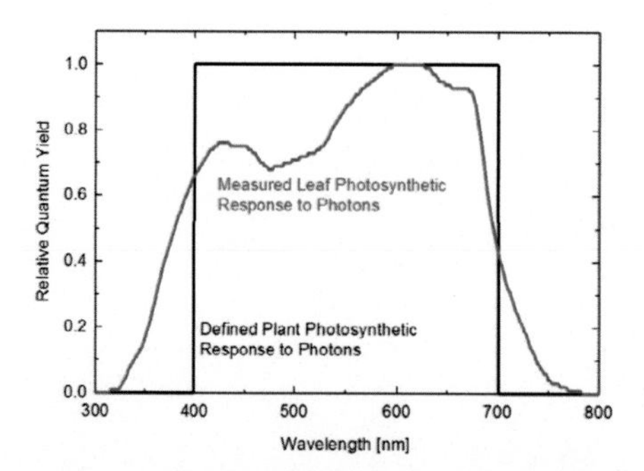

Fig. 15. Measured single leaf relative quantum yield and defined plant relative quantum yield. Quantum yield is moles of carbon fixed per mole of photosynthetic photons absorbed. Relative quantum yields were calculated by normalizing to maximum values giving the relative photosynthetic response photons, which can be thought of as photosynthetic efficiency. Single leaf measurements were made at low intensity in monochromatic radiation and the plot shown is the mean of 22 species of plants grown in the field [from McCree (1972a); Inada (1976) derived a similar curve with data from 33 species, but with a narrower blue peak and broader red peak]. The defined relative plant photosynthetic response to photons gives equal weight to all photons between 400 and 700 nm, a somewhat arbitrary, but simpler and almost universally used, definition.

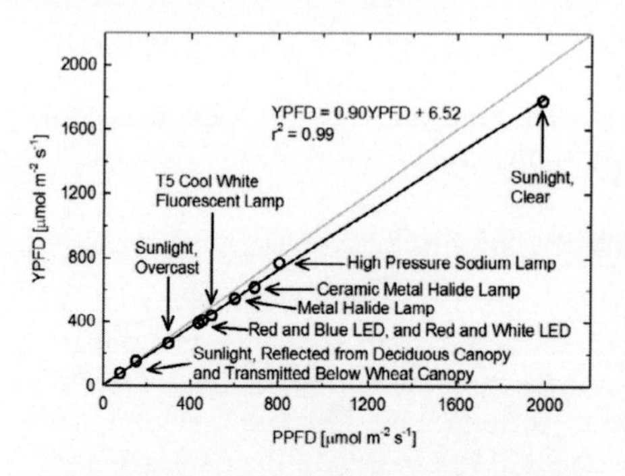

Fig. 16. Correlation between photosynthetic photon flux density (PPFD) and yield photon flux density (YPFD) for multiple radiation sources. Yield photon flux density is about 90% of PPFD. Spectral photon flux density measurements were made with a spectroradiometer (Apogee Instruments model PS-200) and appropriate weighting factors were used to calculate PPFD (defined plant photosynthetic response to photons in Fig. 15) and YPFD (measured leaf photosynthetic response to photons in Fig. 15).

photons than green photons) and absorption of PAR by nonphotosynthetic pigments. In monochromatic radiation, photons from approximately 600 to 630 nm are the most efficient (Fig. 15) (McCree, 1972a; Inada, 1976). One potential definition of PAR is weighting photon flux density (μmol m^{-2} s^{-1}) at each wavelength between 300 and 800 nm by relative quantum yield (quantum yield is moles of carbon fixed per mole of photosynthetic photons absorbed; relative quantum yield is calculated by normalizing by the maximum value) measured in monochromatic radiation and summing the result. This is called yield photon flux density (YPFD) (μmol m^{-2} s^{-1}) (Sager et al., 1988). This definition, however, is not generally appropriate for broad spectrum radiation sources over longer time intervals in whole plants. Measurements used to generate the relative quantum yield data were made on single leaves under low intensity monochromatic radiation levels and at short time scales (McCree, 1972a; Inada, 1976). Whole plants and plant canopies have multiple leaf layers and photosynthetic pigments may adapt to their radiation environment. Also, relative quantum yield shown in Fig. 15 is the mean from 22 species grown in the field (McCree, 1972a); there was some variability between species (McCree, 1972a; Inada, 1976). In addition, mean relative quantum yield for the same species grown in growth chambers was similar, but there were significant differences at shorter wavelengths less than 450 nm.

McCree (1972b) found that equally weighting all photons between 400 and 700 nm and summing the result, defined as photosynthetic photon flux density (PPFD) (μmol m^{-2} s^{-1}), was well correlated to photosynthesis, and was similar to correlation between YPFD and photosynthesis. As a matter of practicality, PPFD is a simpler definition of PAR, and it is easier to construct a sensor with spectral response that matches PPFD weighting factors. At the same time as McCree's work, others had proposed PPFD as an accurate measure of PAR and built sensors that approximated the PPFD weighting factors (Biggs et al., 1971; Federer and Tanner, 1966). Correlation between PPFD and YPFD measurements for several radiation sources is very high (Fig. 16). As an approximation, YPFD = 0.90*PPFD. Based on all these factors, PAR is almost universally defined as PPFD rather than YPFD. The only radiation sources shown in Fig. 16 that don't fall on the regression line are the high pressure sodium (HPS) lamp, reflection from a plant canopy, and transmission below a plant canopy. A large fraction of radiation from HPS lamps is in the red range of wavelengths where the YPFD weighting factors are at or near one. The factor for converting PPFD to YPFD for HPS lamps is 0.95, rather than 0.90. The factor for converting PPFD to YPFD for reflected and transmitted photons is 1.00.

Quantum Sensors

The simplest and most common way to measure PPFD is with a quantum sensor (e.g., Apogee Instruments model SQ-500, EKO Instruments model ML-020P, Kipp & Zonen model PQS 1, LI-COR model LI-190R, Skye Instruments model SKP 215), so called because a photon is a single quantum of radiation. Quantum sensors are also called PAR sensors, which is a more intuitive name. Standard quantum sensors consist of a combination of a photodetector, interference filter(s), and diffuser, all mounted in a weatherproof housing. The photodetector is often a silicon-cell photodiode, but other photodetectors have been used (Mims, 2003; Pontailler, 1990). The major difference between a quantum sensor and silicon-cell pyranometer is the interference filter. The purpose of the filter is to provide equal response to photons between 400 and 700 nm and block photons outside this range [filter(s) matches sensor response to defined plant photosynthetic response to photons shown in Fig. 15].

Quantum sensors designed for measuring PPFD underneath plant canopies are called line quantum sensors. Line quantum sensors provide an average PPFD measurement along the length of a bar (line) where multiple detectors or a single linear diffuser and light transmission mechanism are mounted. An average PPFD measurement along a line is highly beneficial underneath plant canopies because they are heterogeneous, making light transmission and under-canopy PPFD nonuniform.

The combination of diffuser transmittance, interference filter(s) transmittance, and photodetector sensitivity yields spectral response of a quantum sensor. A perfect photodetector, filter, and/or diffuser combination would exactly reproduce the defined plant photosynthetic response to photons (Fig. 15), but this is challenging in practice. Mismatch between the defined plant photosynthetic response and sensor spectral response results in spectral error when the sensor is used to measure radiation from sources with a different spectrum than the radiation source used to calibrate the sensor (Federer and Tanner, 1966; Ross and Sulev, 2000). This concept is exactly the same as spectral error resulting from the imperfect spectral response of a silicon-cell pyranometer.

To demonstrate spectral errors, spectral responses of six models of quantum sensors (Apogee Instruments models SQ-110 and SQ-500, Kipp & Zonen model PQS 1, LI-COR models LI-190 and LI-190R, Skye Instruments model SKP 215;

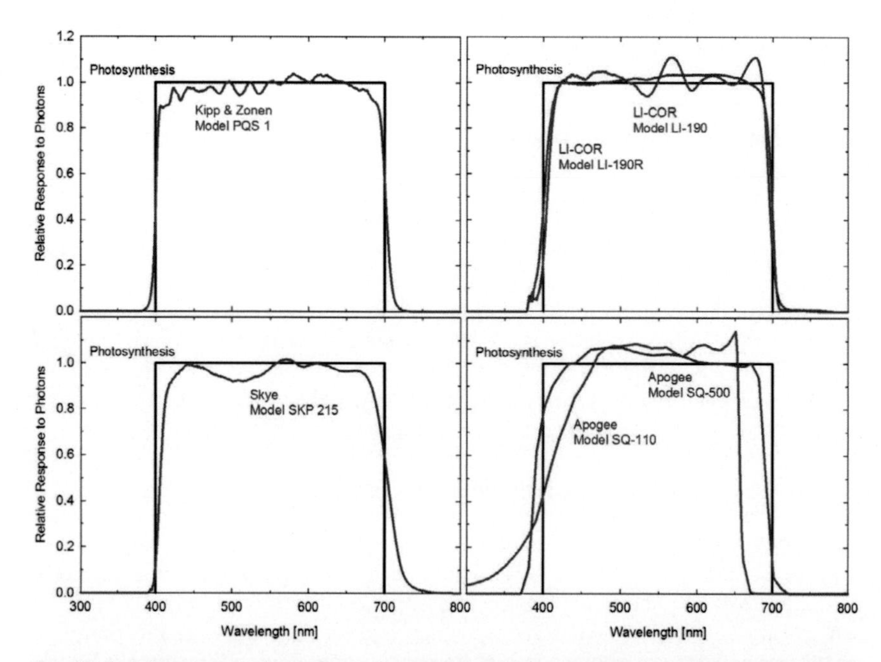

Fig. 17. Relative spectral response (quantum sensor response is signal output by the sensor per μmol of photons incident on the sensor, and relative responses were derived by normalizing by a mean value calculated from data in the 400 and 700 nm range) of six models of quantum sensors (Apogee Instruments models SQ-110 and SQ-500, Kipp & Zonen model PQS 1, LI-COR models LI-190 and LI-190R, Skye Instruments model SKP 215,) compared to the defined plant photosynthetic response to photons (labeled photosynthesis). Sensor spectral response data were used to calculate spectral errors with Eq. [15] (Table 5). Sensor spectral response data were obtained from manufacturers.

spectral response data were provided by sensor manufacturers) were compared to the defined plant photosynthetic response to photons (Fig. 17). The quantum sensor models are designed to provide equal weighting to photons between 400 and 700 nm and block photons outside this range, but they deviate by as much as ± 10% at certain wavelengths and don't have exact cutoffs at 400 and 700 nm. As with silicon-cell pyranometers, spectral error can be quantified for any quantum sensor used to measure any radiation source as long as sensor spectral response (S_λ), calibration source spectral output ($I_{\lambda\text{Calibration}}$), and measured radiation source spectral output ($I_{\lambda\text{Measurement}}$) are known (Federer and Tanner, 1966; Ross and Sulev, 2000):

$$\text{Error} = \frac{\int S_\lambda I_{\lambda\text{Measurement}} d\lambda \int_{400}^{700} I_{\lambda\text{Calibration}} d\lambda}{\int S_\lambda I_{\lambda\text{Calibration}} d\lambda \int_{400}^{700} I_{\lambda\text{Measurement}} d\lambda} \qquad [15]$$

where the integral from 400 to 700 is for the defined plant photosynthetic response to photons. Spectral errors for different radiation sources were calculated with Eq. [15] for the quantum sensor spectral responses shown in Fig. 17 (Table 5). Data indicate errors typically less than 3% for sunlight in multiple conditions (clear, cloudy, reflected from plant canopies, transmitted below plant canopies) and common broad spectrum electric lamps (cool white fluorescent, metal halide, high pressure sodium), but errors tend to be larger for single color (narrowband) light emitting diodes (LEDs), and mixtures of LEDs. Barnes et al. (1993) reported similar results for direct comparison of quantum sensor measurements of PPFD to PPFD measurements from a spectroradiometer. For the best spectral accuracy, Barnes et al. (1993) suggested calibrating quantum sensors against a spectroradiometer for specific radiation sources.

In addition to spectral errors, quantum sensors have the same errors discussed for pyranometers (calibration error, directional error, temperature sensitivity, long-term stability, general use error). To evaluate these errors, approximately two and half years of continuous data were collected for replicates of three models of quantum sensors (Apogee Instruments model SQ-110, Kipp & Zonen model PQS 1, and LI-COR model LI-190) deployed outside in Logan, UT. There is not a reference standard for PAR measurements, so the mean of two replicates of each quantum sensor was used as a reference for comparison. Relative to this reference PPFD, angular differences were typically less than 3% for solar zenith angles between 20° and 60°, and less than 6% at a solar zenith angle of 75°. Drift rates for all sensors deployed on the rooftop, relative to the reference PPFD, were less than 1% per year (except for one LI-190, which appeared to be influenced by moisture intrusion).

Table 5. Relative spectral errors (%) for six models of quantum sensors (Apogee Instruments models SQ-110 and SQ-500, Kipp & Zonen model PQS 1, LI-COR models LI-190 and LI-190R, Skye Instruments model SKP 215) for multiple radiation sources.

Radiation Source	Kipp & Zonen Model PQS 1	LI-COR Model LI-190	LI-COR Model LI-190R	Skye Model SKP 215	Apogee Model SQ-110‡	Apogee Model SQ-500
		Sunlight				
Clear sky†	0.0	0.0	0.0	0.0	0.0	0.0
Overcast sky	-0.2	0.0	-0.2	-0.4	1.4	0.5
Reflected from grass canopy	1.8	0.3	1.5	6.6	5.7	0.0
Transmitted below wheat canopy	0.6	0.3	0.7	4.1	6.4	1.1
		Common electric lamps				
Cool white fluorescent, T5	1.1	0.1	1.6	0.2	0.0	2.2
Metal halide	-0.1	0.3	0.3	-2.1	-3.7	3.1
Ceramic metal halide	0.5	0.6	1.5	-0.3	-6.0	1.9
High pressure sodium	2.3	0.9	2.9	1.5	0.8	2.2
		Light emitting diodes (LEDs)				
Blue (448 nm peak)	-2.2	2.0	-0.4	0.8	-12.7	3.0
Green (524 nm peak)	-1.0	-1.6	2.0	-3.1	8.0	5.2
Red (635 nm peak)	2.7	0.9	3.4	0.7	4.8	0.2
Red (668 nm peak)	-1.1	4.5	0.5	-1.4	-79.1	-1.9
Red and blue mix (84% Red,16% Blue)	-1.1	3.8	0.5	-1.1	-65.3	-1.2
Red and white mix (79% Red, 21% White)	-1.0	3.6	0.7	-1.4	-60.3	-0.8
Cool white fluorescent	0.2	1.2	1.8	0.3	-4.6	2.2

† Sunlight under clear sky conditions was used as the reference ($I_{ICalibration}$ from Eq. [15]).

‡ Assumes separate calibrations for sunlight and electric light (T5 cool white fluorescent was used as electric light calibration reference).

Temperature response from the rooftop data was difficult to measure due to simultaneous changes in solar zenith angle, sky conditions (degree of cloudiness), and temperature. The quantum sensor models have temperature sensitivity specifications ranging from -0.15 to 0.15% per °C. The data suggested that all of the sensors met their temperature sensitivity specifications. Subsequently, temperature sensitivity for three quantum sensor models (Apogee Instruments models SQ-110 and SQ-500, LI-COR model LI-190) was measured by placing them underneath a radiation source (cool white fluorescent LED bulb) inside a temperature controlled chamber and varying the temperature across a wide range (-20 to 50 °C). Data were collected at multiple temperature steps across the range, and at each temperature sensors were allowed to equilibrate for thirty minutes. Reference radiation intensity was measured with a spectroradiometer (Apogee Instruments model PS-200) mounted outside the temperature chamber to keep it at room temperature (only the sensing head and fiber optic cable were placed inside the temperature chamber). Two of the sensor models (LI-190 and SQ-500) had negative slopes (signal decreased as temperature increased) and were less than -0.1% per °C, within specifications provided by manufacturers. This is the expected result based on inference from the wavelength-dependent temperature sensitivity of silicon cells (Fig. 8), where the temperature coefficient is negative for all wavelengths measured by the quantum sensors. Most quantum sensors have relatively uniform sensitivity from 400 to 700 nm. The temperature coefficient of silicon is relatively uniform and near zero from 500 to 700 nm, but is negative at all wavelengths less than 500 nm. This means quantum sensors built with silicon-cell photodiodes should have negative and small temperature sensitivity (less than -0.1% per °C), unless the radiation source being measured outputs a significant amount of blue radiation (wavelengths less than 500 nm). The other model (SQ-110) is not constructed with a silicon-cell photodiode, but uses a gallium arsenide phosphide photodetector, and had a small positive slope (less than 0.1% per °C).

Spectroradiometers

While quantum sensors are typically used to measure PPFD, spectroradiometers have the potential to be the most accurate PPFD sensors because they separate radiation into individual wavelengths and independently measure intensity at multiple wavelengths within the range of interest, in this case the defined PAR range. The major drawback of spectroradiometers is cost, which is approximately an order of magnitude higher than quantum sensors. A few hundred dollars is common for quantum sensors, whereas a few thousand dollars is common for spectroradiometers. However, for radiation sources with unique spectra, such as LEDs with narrowband output, spectroradiometers are often used to measure PPFD because of better spectral accuracy. The main components of spectroradiometers are a prism or diffraction grating to separate radiation into individual wavelengths, a detector or detector array to measure radiation intensity of the different wavelengths, and circuitry to digitize the signal. Spectral errors of quantum sensors (Table 5) are eliminated when measuring PPFD with a spectroradiometer. All other sources of error (imperfect directional response, temperature sensitivity, etc.) must still be considered and will be dependent on the specific spectroradiometer model used to measure PPFD. Spectroradiometers can also be used to measure YPFD by multiplying photon flux density measurements by measured leaf photosynthetic response weighting factors (Fig. 15) and summing the result. As previously indicated, YPFD measurements may not be appropriate for broadband radiation sources and are not commonly made or reported.

Modeling Photosynthetically Active Radiation

Quantum sensors are widely available and relatively inexpensive, thus measurement of PPFD is common, particularly in laboratories, greenhouses, growth chambers, and at flux tower sites. However, most weather stations do not include quantum sensors. As a result, there is considerable interest in modeling PPFD (Aguiar et al., 2011; Alados et al., 1996; Hu et al., 2007a; Xia et al., 2008), and it is often estimated from global shortwave irradiance (SW_i) measurements because pyranometers are common on weather stations. The modeling approach calculates PPFD from measured SW_i by multiplying by a factor that accounts for the fraction of PAR in SW_i (PAR/SW_i) and the average energy content of photons in the photosynthetically active range ($E_{Content}$):

$$PPFD_{Model} = SW_i \frac{PAR/SW_i}{E_{Content}} \qquad [16]$$

where PAR is in energy units of W m^{-2}, making the ratio PAR/SW_i unitless, and $E_{Content}$ is in units of J µmol^{-1}. Both PAR/SW_i and $E_{Content}$ are dependent on the solar spectrum, which varies with solar zenith angle and atmospheric conditions (e.g., degree of cloudiness, water vapor content). Solar spectral measurements made with a spectroradiometer (Advanced Spectral Designs model FieldSpec Pro) on a clear summer day in Logan, UT, yielded mean $E_{Content}$ of 0.221 J µmol^{-1} with approximately 1% variability for solar zenith angles less than 70°. Ross and Sulev (2000) reported a similar value, 0.219 J µmol^{-1}, for clear sky conditions. Akitsu et al. (2015) reported a mean value of 0.219 J µmol^{-1} with 3% variability. Spectral measurements on an overcast summer day in Logan, UT, yielded a value of 0.225 J µmol^{-1}.

Variability in PAR/SW_i for clear sky conditions was measured by rearranging Eq. [16] to solve for PAR/SW_i, inputting measured PPFD and SW_i from reference quantum sensors (mean of two Apogee model SQ-110, two Kipp & Zonen model PQS 1, and two LI-COR model LI-190 quantum sensors) and pyranometers (mean of four secondary standard blackbody thermopile pyranometers), and setting $E_{content}$ equal to 0.221 J µmol^{-1} (mean value for clear conditions). Results indicate a decline from about 0.45 at a solar zenith angle of 20° to 0.415 at a solar zenith angle of 80°, but scatter around the trendline is high (± 0.02) (Fig. 18), likely due to variable humidity in the atmosphere. A decline in SW_i/PAR with solar zenith angle is the expected trend based on spectral shifts of the solar spectrum. A greater proportion of shorter solar wavelengths (within PAR range) are filtered relative to longer solar wavelengths (outside PAR range) when the atmospheric path is long at high solar zenith angles.

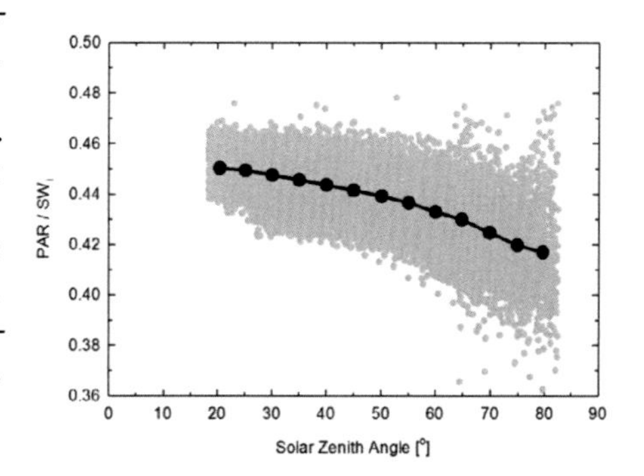

Fig. 18. Solar zenith angle-dependent variability of the ratio of photosynthetically active radiation (PAR) to global shortwave irradiance (SW_i) on a horizontal surface for clear sky conditions in Logan, Utah, United States. Black line is a bin average. The ratio PAR/SW_i decreases as solar zenith angle increases due to increasing atmospheric air mass and greater filtering of wavelengths within the PAR range relative to those outside the PAR range.

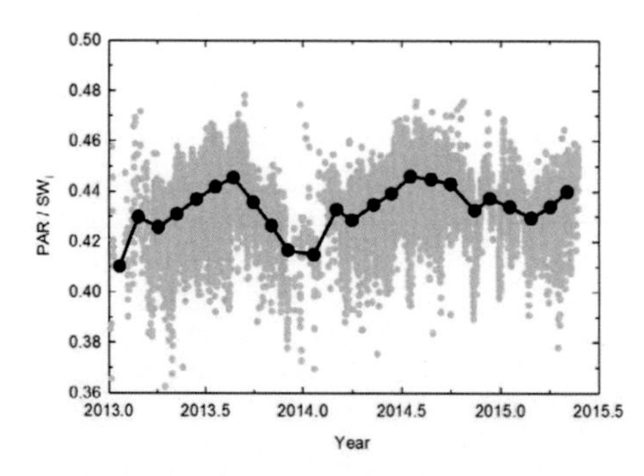

Fig. 19. Seasonal variability of the ratio of photosynthetically active radiation (PAR) to global shortwave irradiance (SW_i) on a horizontal surface for clear sky conditions in Logan, Utah, United States. Black line is a bin average. The ratio PAR/SW_i decreased in winter months and increased in summer months due to seasonal changes in atmospheric air mass and water vapor content.

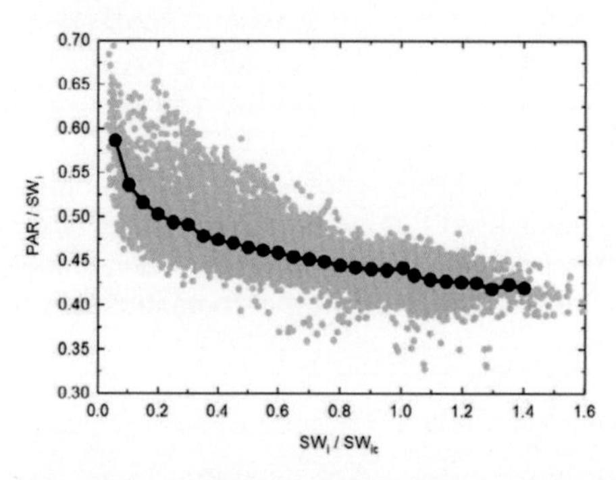

Fig. 20. Variability in the ratio of photosynthetically active radiation (PAR) to global shortwave irradiance (SW_i) with cloudiness (the ratio of SW_i to clear sky global shortwave irradiance, SW_{ic}, is often called the cloudiness index and serves as a surrogate variable for cloudiness). Black line is a bin average. The ratio PAR/SW_i increased as SW_i/SW_{ic} decreased because clouds are more effective at filtering near infrared radiation than radiation within the PAR range.

Seasonal variability of PAR/SW_i for clear sky conditions in Logan, UT, indicates a range of about 0.40 to 0.47, with low values occurring in winter and high values occurring in summer (Fig. 19), consistent with measurements from other locations (Alados et al., 1996; Hu et al., 2007b; Jacovides et al., 2007). Again, this is the expected trend based on spectral shifts of the solar spectrum. There is a greater proportion of red and near infrared radiation relative to blue radiation when solar zenith angle is relatively high during winter. The opposite occurs in summer. Water vapor content of the atmosphere likely influences seasonal trends as well because there is typically less water vapor in the atmosphere in winter and more in summer. A recent study also found a range of 0.40 to 0.47, with PAR/SW_i increasing as atmospheric water vapor content increased (Akitsu et al., 2015), consistent with other studies (Bat-Oyun et al., 2012; Hu et al., 2007b; Weiss and Norman, 1985).

For all sky conditions, PAR/SW_i increased as the sky transitioned from clear to cloudy (Fig. 20), consistent with the results from other studies (Escobedo et al., 2009; Jacovides et al., 2007). The mean value for clear sky conditions in Logan, UT, was 0.441, similar to that reported by others (Bat-Oyun et al., 2012; Jacovides et al., 2003; Meek et al., 1984; Weiss and Norman, 1985). Below a cloudiness index (SW_i/SW_{ic}) of about 0.3, PAR/SW_i is greater than 0.50 (Fig. 20). Estimates of PAR/SW_i can be derived from air mass and precipitable water (González and Calbó, 2002). Dust and aerosols can also have a large impact on PAR/SW_i (Bat-Oyun et al., 2012; Jacovides et al., 2003).

Modeled PPFD, assuming a constant PAR/SW_i = 0.45 (mean for all sky conditions) and constant E_{Content} = 0.223 (mean of clear sky and overcast values), for two and a half years in Logan, UT, indicate deviation from measured PPFD (mean of value calculated from six reference quantum sensors) was typically less than 5%, unless the sun was low in the sky (solar zenith angle greater than 60°) or sky was very cloudy ($SW_i/SW_{ic} < 0.3$). However, measurements from a quantum sensor over the same two and a half year period were much less variable and provided more accurate PPFD (Fig. 21). The simple model presented here (Eq. [16]) could be improved by accounting for solar zenith angle-dependent variability in PAR/SW_i for clear sky conditions (Fig. 18) and variability in PAR/SW_i with cloudiness (Fig. 20). However, scatter around trend lines in Fig. 18 and 20 indicates that even with these improvements, modeled PPFD likely won't be as accurate as measured PPFD.

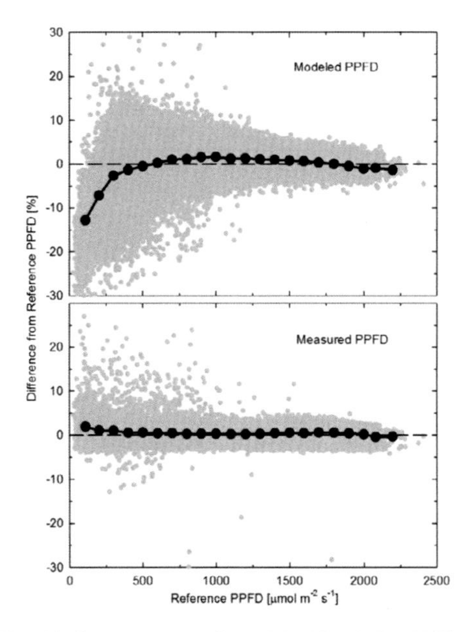

Fig. 21. Comparison of modeled (with Eq. [16]) and measured (with Apogee model SQ-110 quantum sensor) PPFD. Reference PPFD was mean calculated from six quantum sensors (two replicates each of Apogee Instruments model SQ-110, Kipp & Zonen model PQS 1, LI-COR model LI-190). Black lines are bin averages. Data were collected in Logan, UT, over a two and a half year time frame.

Summary

Radiation measurements are a key component on weather stations in agricultural weather and climate networks, and on flux towers. The radiation measurements made are dependent on the objectives of the measurement network and/or research site. At a minimum, global shortwave irradiance should be measured with a pyranometer. Silicon-cell pyranometers are often used for routine measurements, but thermopile pyranometers are more accurate for all sky conditions. For surface energy balance quantification, net radiation is required, with net radiation measurement from a four-component net radiometer typically being the most accurate method to obtain net radiation. Photosynthetically active radiation (PAR) is required in crop modeling and photosynthesis research. While PAR can be estimated from global shortwave irradiance measurements from a pyranometer, quantum sensors are more accurate and their use is becoming more common.

In addition to the information presented and reviewed in this chapter, the subjects of shortwave, net, and photosynthetically active radiation measurement have been treated extensively in the literature and this chapter is by no means exhaustive. For an excellent reference, particularly for shortwave and longwave irradiance measurements with pyranometers and pyrgeometers, respectively, see the recent book Solar and Infrared Radiation Measurements by Vignola et al. (2012).

Acknowledgments

Thanks to Victor Cassella at Kipp & Zonen, Dave Johnson at LI-COR, and Adam Taylor at Skye Instruments for providing quantum sensor spectral data shown in Figure 17.

References

Aguiar, L.J.G., G.R. Fischer, R.J. Ladle, A.C.M. Malhado, F.B. Justino, R.G. Aguiar, and J.M.N. da Costa. 2011. Modeling the photosynthetically active radiation in south west Amazonia under all sky conditions. Theor. Appl. Climatol. doi:10.1007/s00704-011-0556-z

Akitsu, T., A. Kume, Y. Hirose, O. Ijima, and K.N. Nasahara. 2015. On the stability of radiometric ratios of photosynthetically active radiation to global solar radiation in Tsukuba, Japan. Agric. For. Meteorol. 209-210:59–68. doi:10.1016/j.agrformet.2015.04.026

Alados, I., I. Foyo-Moreno, and L. Alados-Arboledas. 1996. Photosynthetically active radiation: Measurements and modeling. Agric. For. Meteorol. 78:121–131. doi:10.1016/0168-1923(95)02245-7

Alados-Arboledas, L., J. Vida, and J.I. Jiménez. 1988. Effects of solar radiation on the performance of pyrgeometers with silicon domes. J. Atmos. Ocean. Technol. 5:666–670. doi:10.1175/1520-0426(1988)005<0666:EOSROT>2.0.CO;2

Albrecht, B., and S.K. Cox. 1977. Procedures for improving pyrgeometer performance. J. Appl. Meteorol. 16:188–197. doi:10.1175/1520-0450(1977)016<0190:PFIPP>2.0.CO;2

Amiro, B.D., G.W. Thurtell, and T.J. Gillespie. 1983. A small infrared thermometer for measuring leaf temperature in leaf chambers. J. Exp. Bot. 34:1569–1576. doi:10.1093/jxb/34.11.1569

ASCE-EWRI. 2005. The ASCE standardized reference evapotranspiration equation. Report of the Task Committee on Standardization of Reference Evapotranspiration. Environmental and Water Resources Institute of the American Society of Civil Engineers, Reston, VA.

Atwater, M.A., and J.T. Ball, 1978. A numerical solar radiation model based on standard meteorological observations. Sol. Energy 21:163-170. doi:10.1016/0038-092X(78)90018-X

Baldridge, A.M., S.J. Hook, C.I. Grove, and G. Rivera. 2009. The ASTER spectral library version 2.0. Remote Sens. Environ. 113:711–715. doi:10.1016/j.rse.2008.11.007

Barnes, C., T. Tibbits, J. Sager, G. Deitzer, D. Bubenheim, G. Koerner, and B. Bugbee. 1993. Accuracy of quantum sensors measuring yield photon flux and photosynthetic photon flux. HortScience 28:1197–1200.

Bat-Oyun, T., M. Shinoda, and M. Tsubo. 2012. Effects of cloud, atmospheric water vapor, and dust on photosynthetically active radiation and total solar radiation in a Mongolian grassland. J. Arid Land 4:349–356. doi:10.3724/SP.J.1227.2012.00349

Beaubien, D.J., A. Bisberg, and A.F. Beaubien. 1998. Investigations in pyranometer design. J. Atmos. Ocean. Technol. 15:677–686. doi:10.1175/1520-0426(1998)015<0677:IIPD>2.0.CO;2

Biggs, W., A.R. Edison, J.D. Eastin, K.W. Brown, J.W. Maranville, and M.D. Clegg. 1971. Photosynthesis light sensor and meter. Ecology 52:125–131. doi:10.2307/1934743

Blonquist, J.M., Jr., B.D. Tanner, and B. Bugbee. 2009a. Evaluation of measurement accuracy and comparison of two new and three traditional net radiometers. Agric. For. Meteorol. 149:1709–1721. doi:10.1016/j.agrformet.2009.05.015

Blonquist, J.M., Jr., J.M. Norman, and B. Bugbee. 2009b. Automated measurement of canopy stomatal conductance based on infrared temperature. Agric. For. Meteorol. 149:1931–1945. doi:10.1016/j.agrformet.2009.06.021

Blonquist, J.M., Jr., R.G. Allen, and B. Bugbee. 2010. An evaluation of the net radiation sub-model in the ASCE standardized reference evapotranspiration equation: Implications for evapotranspiration prediction. Agric. Water Manage. 97:1026–1038. doi:10.1016/j.agwat.2010.02.008

Brotzge, J.A., and C.E. Duchon. 2000. A field comparison among a domeless net radiometer, two four-component net radiometers, and a domed net radiometer. J. Atmos. Ocean. Technol. 17:1569–1582. doi:10.1175/1520-0426(2000)017<1569:AFCAAD>2.0.CO;2

Campbell, G.S., and G.R. Diak. 2005. Net and thermal radiation estimation and measurement. In: J.L. Hatfield and J.M. Baker, editors, Micrometeorology in Agricultural Systems. ASA, Madison, WI. p. 59–92.

Campbell, G.S., and J.M. Norman. 1998. An introduction to environmental biophysics. Springer-Verlag, New York. doi:10.1007/978-1-4612-1626-1

Campbell, G.S., J.N. Mugaas, and J.R. King. 1978. Measurement of long-wave radiant flux in organismal energy budgets: A comparison of three methods. Ecology 59:1277–1281. doi:10.2307/1938242

Cobos, D.R., and J.M. Baker. 2003. Evaluation and modification of a domeless net radiometer. Agron. J. 95:177–183. doi:10.2134/agronj2003.0177

Duchon, C.E., and G.E. Wilk. 1994. Field comparisons of direct and component measurements of net radiation under clear skies. J. Appl. Meteorol. 33:245–251. doi:10.1175/1520-0450(1994)033<0245:FCODAC>2.0.CO;2

Dutton, E.G., J.J. Michalsky, T. Stoffel, B.W. Forgan, J. Hickey, D.W. Nelson, T.L. Alberta, and I. Reda. 2001. Measurement of broadband diffuse solar irradiance using current commercial instrumentation with a correction for thermal offset errors. J. Atmos. Ocean. Technol. 18:297–314. doi:10.1175/1520-0426(2001)018<0297:MOBDSI>2.0.CO;2

Enz, J.W., J.C. Klink, and D.G. Baker. 1975. Solar radiation effects on pyrgeometer performance. J. Appl. Meteorol. 14:1297–1302. doi:10.1175/1520-0450(1975)014<1297:SREOPP>2.0.CO;2

Escobedo, J.F., E.N. Gomes, A.P. Oliveira, and J. Soares. 2009. Modeling hourly and daily fractions of UV, PAR, and NIR to global solar radiation under various sky conditions at Botucatu, Brazil. Appl. Energy 86:299–309. doi:10.1016/j.apenergy.2008.04.013

Fairall, C.W., P.O.G. Persson, E.F. Bradley, R.E. Payne, and S.P. Anderson. 1998. A new look at calibration and use of Eppley precision infrared radiometers. Part I: Theory and application. J. Atmos. Ocean. Technol. 15:1229–1242. doi:10.1175/1520-0426(1998)015<1229:ANLACA>2.0.CO;2

Federer, C.A., and C.B. Tanner. 1965. A simple integrating pyranometer for measuring daily solar radiation. J. Geophys. Res. 70:2301–2306. doi:10.1029/JZ070i010p02301

Federer, C.A., and C.B. Tanner. 1966. Sensors for measuring light available for photosynthesis. Ecology 47:654–657. doi:10.2307/1933948

Field, R.T., L.J. Fritschen, E.T. Kanemasu, E.A. Smith, J.B. Stewart, S.B. Verma, and W.P. Kustas. 1992. Calibration, comparison, and correction of net radiation instruments used during FIFE. J. Geophys. Res. 97:18,681–18,695. doi:10.1029/91JD03171

Flerchinger, G.N., W. Xaio, D. Marks, T.J. Sauer, and Q. Yu. 2009. Comparison of algorithms for incoming atmospheric longwave radiation. Water Resour. Res. 45:W03423. doi:10.1029/2008WR007394

Foken, T. 2008. The energy balance closure problem– An overview. Ecol. Appl. 18:1351–1367. doi:10.1890/06-0922.1

Fritschen, L.J., and C.L. Fritschen. 2007. Calibration of shielded net radiometers. Agron. J. 99:297–303. doi:10.2134/agronj2005.0077s

Fuchs, M., and C.B. Tanner. 1966. Infrared thermometry of vegetation. Agron. J. 58:597–601. doi:10.2134/agronj1966.00021962005800060014x

Geuder, N., R. Affolter, B. Kraas, and S. Wilbert. 2014. Long-term behavior, accuracy and drift of LI-200 pyranometers as radiation sensors in Rotating Shadowband Irradiometers (RSI). Energy Procedia 49:2330–2339. doi:10.1016/j.egypro.2014.03.247

González, J.A., and J. Calbó. 2002. Modelled and measured ratio of PAR to global radiation under cloudless skies. Agric. For. Meteorol. 110:319–325. doi:10.1016/S0168-1923(01)00291-X

Gröbner, J., and A. Los. 2007. Laboratory calibration of pyrgeometers with known spectral responsivities. Appl. Opt. 46:7419–7425. doi:10.1364/AO.46.007419

Gueymard, C.A. 2008. REST2: High-performance solar radiation model for cloudless-sky irradiance, illuminance, and photosynthetically active radiation– Validation with a benchmark dataset. Sol. Energy 82:272–285. doi:10.1016/j.solener.2007.04.008

Gueymard, C.A. 2012. Clear-sky irradiance predictions for solar resource mapping and large-scale applications: Improved validation methodology and performance analysis of 18 broadband radiative models. Sol. Energy 86:2145–2169. doi:10.1016/j.solener.2011.11.011

Habte, A., S. Wilcox, and T. Stoffel. 2014. Evaluation of radiometers deployed at the National Renewable Energy Laboratory's Solar Radiation Research Laboratory. Technical Report NREL/TP-5D00-60896. National Renewable Energy Laboratory, Golden, CO.

Haeffelin, M., S. Kato, A.M. Smith, C.K. Rutledge, T.P. Charlock, and J.R. Mahan. 2001. Determination of the thermal offset of the Eppley precision spectral pyranometer. Appl. Opt. 40:472–484. doi:10.1364/AO.40.000472

Halldin, S., and A. Lindroth. 1992. Errors in net radiometry: Comparison and evaluation of six radiometer designs. J. Atmos. Ocean. Technol. 9:762–783. doi:10.1175/1520-0426(1992)009<0762:EINRCA>2.0.CO;2

Hu, B., Y. Wang, and G. Liu. 2007a. Measurements and estimations of photosynthetically active radiation in Beijing. Atmos. Res. 85:361–371. doi:10.1016/j.atmosres.2007.02.005

Hu, B., Y. Wang, and G. Liu. 2007b. Spatiotemporal characteristics of photosynthetically active radiation in China. J. Geophys. Res. 112:1–12. doi:10.1029/2006JD007965

Huband, N.D.S. 1985a. An infra-red radiometer for measuring surface temperature in the field. Part I. Design and Construction. Agric. For. Meteorol. 34:215–226. doi:10.1016/0168-1923(85)90021-8

Huband, N.D.S. 1985b. An infra-red radiometer for measuring surface temperature in the field Part II. Calibration and performance. Agric. For. Meteorol. 34:227–233. doi:10.1016/0168-1923(85)90022-X

Idso, S.B. 1981. A set of equations for full spectrum and 8- to 14-μm and 10.5- to 12.5-μm thermal radiation from cloudless skies. Water Resour. Res. 17:295–304. doi:10.1029/WR017i002p00295

Inada, K. 1976. Action spectra for photosynthesis in higher plants. Plant Cell Physiol. 17:355–365.

Jacovides, C.P., F.S. Tymvios, D.N. Asimakopoulos, K.M. Theofilou, and S. Pashiardes. 2003. Global photosynthetically active radiation and its relationship with global solar radiation in the Eastern Mediterranean basin. Theor. Appl. Climatol. 74:227–233. doi:10.1007/s00704-002-0685-5

Jacovides, C.P., F.S. Tymvios, V.D. Assimakopoulos, and N.A. Kaltsounides. 2007. The dependence of global and diffuse PAR radiation components on sky conditions at Athens, Greece. Agric. For. Meteorol. 143:277–287. doi:10.1016/j.agrformet.2007.01.004

JCGM/WG 1, 2008. Guide to the expression of uncertainty in measurement. Working Group 1 of the Joint Committee for Guides in Metrology. http://www.bipm.org/utils/common/documents/jcgm/JCGM_100_2008_E.pdf (verified 17 July 2016).

Ji, Q., and S.-C. Tsay. 2000. On the dome effect of Eppley pyrgeometers and pyranometers. Geophys. Res. Lett. 27:971–974. doi:10.1029/1999GL011093

Ji, Q., and S.-C. Tsay. 2010. A novel nonintrusive method to resolve the thermal dome effect of pyranometers: Instrumentation and observational basis. J. Geophys. Res. doi:10.1029/2009JD013483

Kalma, J.D., H. Alksnis, and G.P. Laughlin. 1988. Calibration of small infra-red surface temperature transducers. Agric. For. Meteorol. 43:83–98. doi:10.1016/0168-1923(88)90008-1

Kerr, J.P., G.W. Thurtell, and C.B. Tanner. 1967. An integrating pyranometer for climatological observer stations and mesoscale networks. J. Appl. Meteorol. 6:688–694. doi:10.1175/1520-0450(1967)006<0688:AIPFCO>2.0.CO;2

Kimball, B.A., S.B. Idso, and J.K. Aase. 1982. A model of thermal radiation from partly cloudy and overcast skies. Water Resour. Res. 18:931–936. doi:10.1029/WR018i004p00931

King, D.L., and D.R. Myers. 1997. Silicon-photodiode pyranometers: Operational characteristics, historical experiences, and new calibration procedures. In: Proceedings from the 26th IEEE Photovoltaic Specialists Conference, 29 Sept.–3 Oct. 1997, Anaheim, CA. IEEE, New York. doi:10.1109/PVSC.1997.654323

King, D.L., J.A. Kratochvil, and W.E. Boyson. 1997. Measuring solar and spectral angle-of-incidence effects on photovoltaic modules and solar irradiance sensors. In: Proceedings from the 26th IEEE Photovoltaic Specialists Conference, Sept. 29-Oct. 3, 1997, Anaheim, CA. IEEE, New York, New York. doi:10.1109/PVSC.1997.654283

Klassen, S., and B. Bugbee. 2005. Shortwave Radiation. In: J.L. Hatfield and J.M. Baker, editors, Micrometeorology in agricultural systems. ASA, Madison, WI. p. 43–57.

Kopp, G., and J.L. Lean. 2011. A new, lower value of total solar irradiance: Evidence and climate significance. Geophys. Res. Lett. 38:1–7. doi:10.1029/2010GL045777

Kohsiek, W., C. Liebethal, T. Foken, R. Vogt, S.P. Oncley, C. Bernhofer, and H.A.R. Debruin. 2007. The energy balance experiment EBEX-2000. Part III: Behaviour and quality of the radiation measurements. Boundary-Layer Meteorol. 123:55–75. doi:10.1007/s10546-006-9135-8

Kustas, W.P., J.H. Prueger, L.E. Hipps, J.L. Hatfield, and D. Meek. 1998. Inconsistencies in net radiation estimates from use of several models of instruments in a desert environment. Agric. For. Meteorol. 90:257–263. doi:10.1016/S0168-1923(98)00062-8

Lefèvre, M., A. Oumbe, P. Blanc, B. Espinar, B. Gschwind, Z. Qu, L. Wald, M. Schroedter-Homscheidt, C. Hoyer-Klick, A. Arola, A. Benedetti, J.W. Kaiser, and J.-J. Morcrette. 2013. McClear: A new model estimating downwelling solar radiation at ground level in clear-sky conditions. Atmos. Meas. Tech. 6:2403–2418. doi:10.5194/amt-6-2403-2013

Leuning, R., E. van Gorsel, W.J. Massman, and P.R. Isaac. 2012. Reflections on the surface energy imbalance problem. Agric. For. Meteorol. 156:65–74. doi:10.1016/j.agrformet.2011.12.002

Malek, E. 2008. The daily and annual effects of dew, frost, and snow on a non-ventilated net radiometer. Atmos. Res. 89:243–251. doi:10.1016/j.atmosres.2008.02.006

Marty, C., R. Philipona, J. Delamere, E.G. Dutton, J. Michalsky, K. Stamnes, R. Storvold, T. Stoffel, S.A. Clough, and E.J. Mlawer. 2003. Downward longwave irradiance uncertainty under arctic atmospheres: Measurements and modeling. J. Geophys. Res. 108:4358–4369. doi:10.1029/2002JD002937

McCree, K.J. 1972a. The action spectrum, absorptance and quantum yield of photosynthesis in crop plants. Agric. Meteorol. 9:191–216. doi:10.1016/0002-1571(71)90022-7

McCree, K.J. 1972b. Test of current definitions of photosynthetically active radiation against leaf photosynthesis data. Agric. Meteorol. 10:443–453. doi:10.1016/0002-1571(72)90045-3

Meek, D.W., J.L. Hatfield, T.A. Howell, S.B. Idso, and R.J. Reginato. 1984. A generalized relationship between photosynthetically active radiation and solar radiation. Agron. J. 76:939–945. doi:10.2134/agronj1984.00021962007600060018x

Meloni, D., C. Di Biagio, A. Di Sarra, F. Monteleone, G. Pace, and H.M. Sferlazzo. 2012. Accounting for the solar radiation influence on downward longware irradiance measurements by pyrgeometers. J. Atmos. Ocean. Technol. 29:1629–1643. doi:10.1175/JTECH-D-11-00216.1

Menyhart, L., A. Anda, and Z. Nagy. 2015. A new method for checking the leveling of pyranometers. Sol. Energy 120:25–34. doi:10.1016/j.solener.2015.06.033

Meyers, T.P., and R.F. Dale. 1983. Predicting daily insolation with hourly cloud height and coverage. J. Clim. Appl. Meteorol. 22:537–545. doi:10.1175/1520-0450(1983)022<0537:PDIWHC>2.0.CO;2

Michalsky, J., E. Dutton, M. Rubes, D. Nelson, T. Stoffel, M. Wesley, M. Splitt, and J. DeLuisi. 1999. Optimal measurement of surface shortwave irradiance using current instrumentation. J. Atmos. Ocean. Technol. 16:55–69. doi:10.1175/1520-0426(1999)016<0055:OMOSSI>2.0.CO;2

Michel, D., R. Philipona, C. Ruckstuhl, R. Vogt, and L. Vuilleumier. 2008. Performance and uncertainty of CNR 1 net radiometers during a one-year field comparison. J. Atmos. Ocean. Technol. 25:442–451. doi:10.1175/2007JTECHA973.1

Mims, F.M., III. 2003. A 5-year study of a new kind of photosynthetically active radiation sensor. Photochem. Photobiol. 77:30–33. doi:10.1562/0031-8655(2003)077<0030:AYSOAN>2.0.CO;2

Miskolczi, F., and R. Guzzi. 1993. Effect of nonuniform spectral dome transmittance on the accuracy of infrared radiation measurements using shielded pyrradiometers and pyrgeometers. Appl. Opt. 32:3257–3265. doi:10.1364/AO.32.003257

Myers, D.R. 2011. Quantitative analysis of spectral impacts on silicon photodiode radiometers. NREL/CP-5500-50936. National Renewable Energy Laboratory Conference Paper, Golden, CO.

Myers, D.R. 2013. Solar radiation practical modeling for renewable energy applications. CRC Press, Taylor & Francis Group, Boca Raton, FL.

Norman, J.M., and F. Becker. 1995. Terminology in thermal infrared remote sensing of natural surfaces. Agric. For. Meteorol. 77:153–166. doi:10.1016/0168-1923(95)02259-Z

Ohmura, A., E.G. Dutton, B. Forgan, C. Fröhlich, H. Gilgen, H. Hegner, A. Heimo, G. König-Langlo, B. McArthur, G. Müller, R. Philipona, R. Pinker, C.H. Whitlock, K. Dehne, and M. Wild. 1998. Baseline surface radiation network ()BSRN/WCRP): New precision radiometry for climate research. Bull. Am. Meteorol. Soc. 79:2115–2136. doi:10.1175/1520-0477(1998)079<2115:BSRNBW>2.0.CO;2

Oliveira, A.P., J. Soares, M.Z. Božnar, P. Mlakar, and J. Escobedo. 2006. An application of neural network technique to correct the dome temperature effects on pyrgeometer measurements. J. Atmos. Ocean. Technol. 23:80–89. doi:10.1175/JTECH1829.1

Osterwald, C.R. 1986. Translation of device performance measurements to reference conditions. Sol. Cells 18:269–279. doi:10.1016/0379-6787(86)90126-2

Pérez, M., and L. Alados-Arboledas. 1999. Effects of natural ventilation and solar radiation on the performance of pyrgeometers. J. Atmos. Ocean. Technol. 16:174–180. doi:10.1175/1520-0426(1999)016<0174:EONVAS>2.0.CO;2

Philipona, R. 2002. Underestimation of solar global and diffuse radiation measured at Earth's surface. J. Geophys. Res. doi:10.1029/2002JD002396

Philipona, R., C. Fröhlich, and C. Betz. 1995. Characterization of pyrgeometers and the accuracy of atmospheric longwave radiation measurements. Appl. Opt. 34:1598–1605. doi:10.1364/AO.34.001598

Philipona, R., C. Fröhlich, K. Dehne, J. DeLuisi, J. Augustine, E. Dutton, D. Nelson, B. Forgan, P. Novotny, J. Hickey, S.P. Love, S. Bender, B. McArthur, A. Ohmura, J.H. Seymour, J.S. Foot, M. Shiobara, F.P.J. Valero, and A.W. Strawa. 1998. The baseline surface radiation network pyrgeometer round-robin calibration experiment. J. Atmos. Ocean. Technol. 15:687–696. doi:10.1175/1520-0426(1998)015<0687:TBSRNP>2.0.CO;2

Philipona, R., E.G. Dutton, T. Stoffel, J. Michalsky, I. Reda, A. Stifter, P. Wendling, N. Wood, S.A. Clough, E.J. Mlawer, G. Anderson, H.E. Revercomb, and T.R. Shippert. 2001. Atmospheric longwave irradiance uncertainty: Pyrgeometers compared to an absolute sky-scanning radiometer, atmospheric emitted radiance interferometer, and radiation transfer model calculations. J. Geophys. Res. 106:28129–28141.

Pontailler, J.-Y. 1990. A cheap quantum sensor using a gallium arsenide photodiode. Funct. Ecol. 4:591–596. doi:10.2307/2389327

Raïch, A., J.A. González, and J. Calbó. 2007. Effects of solar height, cloudiness and temperature on silicon pyranometer measurements. Tethys 4:11–18. doi:10.3369/tethys.2007.4.02

Reda, I., J. Hickey, C. Long, D. Myers, T. Stoffel, S. Wilcox, J.J. Michalsky, E.G. Dutton, and D. Nelson. 2005. Using a blackbody to calculate net longwave responsivity of shortwave solar pyranometers to correct for their thermal offset error during outdoor calibration using the component sum method. J. Atmos. Ocean. Technol. 22:1531–1540. doi:10.1175/JTECH1782.1

Reda, I. 2011. Method to calculate uncertainties in measuring shortwave solar irradiance using thermopile and semiconductor solar radiometers. Technical Report NREL/TP-3B10-52194. National Renewable Energy Laboratory, Golden, CO. doi:10.2172/1021250

Riihimaki, L., and F. Vignola. 2008. Establishing a consistent calibration record for Eppley PSPs. In: Proceedings of the 37th American Solar Energy Society Conference, San Diego, CA. 3–8 May 2008. American Solar Energy Society, Boulder, CO.

Ross, J., and M. Sulev. 2000. Sources of errors in measurements of PAR. Agric. For. Meteorol. 100:103–125. doi:10.1016/S0168-1923(99)00144-6

Sager, J.C., W.O. Smith, J.L. Edwards, and K.L. Cyr. 1988. Photosynthetic efficiency and phytochrome photoequilibria determination using spectral data. Trans. ASAE 31:1882–1889. doi:10.13031/2013.30952

Selcuk, K., and J.I. Yellott. 1962. Measurement of direct, diffuse and total solar radiation with silicon photovoltaic cells. Sol. Energy 6:155–163. doi:10.1016/0038-092X(62)90127-5

Sengupta, M., P. Gotseff, D. Myers, and T. Stoffel. 2012. Performance testing using silicon devices– Analysis and accuracy. Conference Paper NREL/CP-5500-54251. National Renewable Energy Laboratory, Golden, CO. doi:10.1109/PVSC.2012.6318278

Steinhilber, F., J. Beer, and C. Fröhlich. 2009. Total solar irradiance during the Holocene. Geophys. Res. Lett. 36:L19704. doi:10.1029/2009GL040142

Tanner, B.D. 2001. Evolution of automated weather station technology through the 1980s and 1990s. In: K.G. Hubbard and M.V.K. Sivakumar, editors, Automated weather stations for applications in agriculture and water resources management: Current uses and future perspectives. WMO/TD number 1074. World Meteorological Organization, Geneva, Switzerland. p. 3–20.

Twine, T.E., W.P. Kustas, J.M. Norman, D.R. Cook, P.R. Houser, T.P. Meyers, J.H. Prueger, P.J. Starks, and M.L. Wesely. 2000. Correcting eddy-covariance flux underestimates over a grassland. Agric. For. Meteorol. 103:279–300. doi:10.1016/S0168-1923(00)00123-4

Udo, S.O. 2000. Quantification of solar heating of the dome of a pyrgeometer for a tropical location: Ilorin, Nigeria. J. Atmos. Ocean. Technol. 17:995–1000. doi:10.1175/1520-0426(2000)017<0995:QOSHOT>2.0.CO;2

Unsworth, M.H., and J.L. Monteith. 1975. Long-wave radiation at the ground: I. Angular distribution of incoming radiation. Q. J. R. Meteorol. Soc. 101:1029–1030. doi:10.1002/qj.49710143029

Vieira, L.E.A., S.K. Solanki, N.A. Krivova, and I. Usoskin. 2011. Evolution of the solar irradiance during the Holocene. Astron. Astrophys. 531:A6–A26. doi:10.1051/0004-6361/201015843

Vignola, F., J. Michalsky, and T. Stoffel. 2012. Solar and infrared radiation measurements. CRC Press, Taylor & Francis, Boca Raton, FL.

Weiss, A., and J.M. Norman. 1985. Partitioning solar radiation into direct and diffuse, visible and near-infrared components. Agric. For. Meteorol. 34:205–213. doi:10.1016/0168-1923(85)90020-6

Wilson, K.B. 2002. Energy balance closure at FLUXNET sites. Agric. For. Meteorol. 113:223–243. doi:10.1016/S0168-1923(02)00109-0

Wood, J.D., T.J. Griffis, and J.M. Baker. 2015. Detecting drift bias and exposure errors in solar and photosynthetically active radiation data. Agric. For. Meteorol. 206:33–44. doi:10.1016/j.agrformet.2015.02.015

Xia, X., Z. Li, P. Wang, M. Cribb, H. Chen, and Y. Zhao. 2008. Analysis of photosynthetic photon flux density and its parameterization in northern China. Agric. For. Meteorol. 148:1101–1108. doi:10.1016/j.agrformet.2008.02.008

Yordanov, G.H., O.-M. Midtgård, T.O. Saetre, H.K. Nielsen, and L.E. Norum. 2012. Overirradiance (cloud enhancement) events at high latitudes. In: Photovoltaic Specialists Conference, Vol. 2, Austin, TX. 3–8 June 2012 IEEE, New York. doi: 10.1109/PVSC-Vol 2.2013.6656797

Air Temperature

J. Mark Blonquist Jr. and Bruce Bugbee*

Introduction

The properties of materials and nearly all biological, chemical, and physical processes are temperature dependent. As a result, air temperature is probably the most widely measured environmental variable, and multiple sensors are available to measure it. The first thermometers were developed in the 1600s, but accurate measurement of air temperature remains a challenging task today. Instrument and method development, and error quantification, are active areas of research (Clark et al., 2013; Holden et al., 2013; Lopardo et al., 2014; Thomas and Smoot, 2013; Young et al., 2014).

Temperature indicates the relative degree of 'hotness' or 'coldness' of an object, material, or fluid. More specifically, temperature is a measure of the thermal energy of a substance. Thermal energy is associated with internal kinetic energy, or energy of motion of the atoms and molecules making up the substance. Higher temperatures correspond to higher kinetic energy (faster motion of atoms and molecules), whereas colder temperatures correspond to lower kinetic energy (slower motion of atoms and molecules).

Unlike temperature of solids and liquids, air temperature is challenging to measure because air has extremely low thermal mass and thermal conductivity. This means it is difficult to get a sensor with larger thermal mass into thermal equilibrium with air. Here we review the current state of sensors and techniques for automated measurement of air temperature.

Types of Sensors

Sensors for automated air temperature measurement include thermocouples, thermistors, platinum resistance thermometers (PRTs), integrated circuit (IC) sensors, and sonic thermometers, with thermocouples, thermistors, and PRTs being the most commonly used (Fig. 1). Each has associated advantages and disadvantages, and each will be treated separately.

Thermocouples

A thermocouple consists of two different metals or alloys connected at the ends to form a simple electrical circuit (current loop). A temperature difference (thermal energy difference) between the two ends of the circuit produces a voltage (called an electromotive force) that is proportional to the temperature difference. This is called the Seebeck effect. The magnitude of the voltage produced depends

---

Abbreviations: IC, Integrated Circuit; PRT, Platinum Resistance Thermometer;

J.M. Blonquist Jr., Apogee Instruments, Inc., 721 W. 1800 N, Logan, UT 84321; B. Bugbee, Dep. of Plants, Soils, and Climate, Utah State University, Logan, UT 84322. *Corresponding author (bruce.bugbee@usu.edu)

doi:10.2134/agronmonogr60.2016.0012

© ASA, CSSA, and SSSA, 5585 Guilford Road, Madison, WI 53711, USA.
Agroclimatology: Linking Agriculture to Climate, Agronomy Monograph 60.
Jerry L. Hatfield, Mannava V.K. Sivakumar, John H. Prueger, editors.

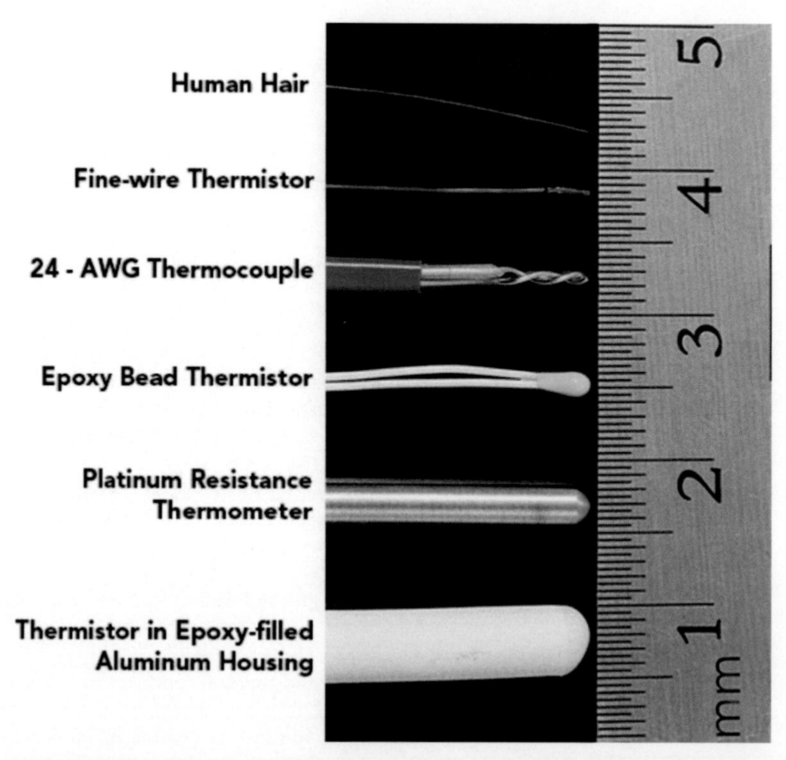

Fig. 1. Size comparison of five temperature sensors. From top to bottom: human hair (for scale), fine wire ceramic thermistor with thin epoxy coating, 24-AWG (0.51 mm; AWG stands for American wire gauge) type-E thermocouple (multiple wire diameters are available), ceramic thermistor with epoxy bead coating, PRT in an 1/8 inch (3.18 mm) diameter stainless steel sheath, and thermistor in an epoxy-filled aluminum housing.

on the temperature difference between the sample and reference ends, or junctions, of the circuit and the two metals used to create the circuit. The sample junction is the connection between the metals at the point where temperature is being measured, and the reference junction is the connection between the metals at the point where voltage is measured. Sometimes the sample and reference junctions are referred to as the 'hot' and 'cold' junctions, respectively, but this can be misleading because the sample junction can be hotter or colder than the reference junction. Thermocouple sample junctions are sometimes coated with epoxy for electrical isolation and waterproofing. White paint has also been used to maximize shortwave radiation reflectivity (Christian and Tracy, 1985). For air temperature measurements the epoxy coating is not essential, but it prevents corrosion and improves durability.

The voltage measured at the reference junction is nonlinearly related to the temperature difference between the junctions, and a polynomial with thermocouple-specific coefficients is typically used to convert the voltage to the temperature difference. There are multiple types of thermocouples, each with a unique combination of metals or alloys used to create the circuit. The magnitude of the voltage per degree difference (sensitivity) varies with the type of thermocouple (Table 1). Type-E thermocouples are generally considered the best option for environmental measurements because of high sensitivity and low thermal conductivity of the wire (discussed below).

Thermocouples measure the temperature difference between the sample and reference junctions, not the actual temperature at the sample junction. The temperature at the reference junction must be known to calculate the temperature at the sample junction. One of the challenges with thermocouples is accurate measurement of the temperature at the reference junction, in addition to accurately measuring the voltage produced by the temperature difference. Some meters (e.g., dataloggers) have an internal temperature sensor (thermistor or PRT) near the reference junction (directly under the wiring panel where the lead wires of the thermocouple are connected to the meter) to measure reference temperature. The voltage change with temperature (sensitivity) of thermocouples is extremely small (Table 1), so the voltage measurement resolution and accuracy of the meter must be on the order of one micro-volt (μV). Due to the small voltage output from thermocouples, electrical interference can cause errors when thermocouples are used in electrically noisy environments (e.g., near electric lights or radio equipment).

Table 1. The four most common thermocouples used for environmental temperature measurement.

Thermocouple	Metals or Alloys	Approximate Voltage Difference Between Junctions (Sensitivity)
Type-T	Copper (+)/Constantan (-)	40 mV $^{\circ}C^{-1}$
Type-E	Chromel (+)/Constantan (-)	60 mV $^{\circ}C^{-1}$
Type-K	Chromel (+)/Alumel (-)	40 mV $^{\circ}C^{-1}$
Type-J	Iron (+)/Constantan (-)	51 mV $^{\circ}C^{-1}$

Thermistors

A thermistor is an electrical resistor, often ceramic or metal oxide, where resistance changes with temperature. Thermistors with a positive temperature coefficient increase in resistance with increasing temperature, whereas thermistors with a negative temperature coefficient decrease in resistance with increasing temperature. Thermistors with a negative temperature coefficient are more common. The relationship between temperature and resistance is nonlinear, and is described by a standard fitting equation, the most common being the Steinhart–Hart equation (Steinhart and Hart, 1968). The β-parameter equation is also used. Thermistors are typically sealed in a water-resistant or waterproof material, with epoxy and glass being common.

Thermistor resistance (R_T, in Ω) is determined with a half-bridge measurement. The circuit is a voltage divider, with the thermistor and a bridge resistor of known and fixed resistance (R_B, in Ω) connected in series. An excitation voltage (V_{EX}, in V) is applied across the resistors (bridge resistor and thermistor in series) and voltage is measured across the bridge resistor (V_B, in V). Thermistor resistance is then calculated from the voltages and R_B:

$$R_T = R_B\left(\frac{V_{EX}}{V_B} - 1\right) \quad [1]$$

where V_B is always less than V_{EX} as a result of the voltage divider circuit. Thermistor temperature (T_K, in Kelvin) can be calculated from R_T with the Steinhart–Hart equation:

$$T_K = \frac{1}{A + B\ln(R_T) + C[\ln(R_T)]^3} \quad [2]$$

where A, B, and C are thermistor-specific coefficients, or the β-parameter equation (sometimes called B-parameter):

$$T_K = \frac{\beta}{\ln\left(\frac{R_T}{R_0 e^{-\beta/T_0}}\right)} \quad [3]$$

where R_0 is resistance at temperature T_0 (298.15 K or 25 °C) and β is a thermistor-specific coefficient.

Excitation voltage is required to make the half-bridge resistance measurement, thus a small amount of power is required to measure temperature with thermistors. Current flow (I, in A) through the circuit is calculated from Ohm's Law:

$$I = \frac{V_{EX}}{R_B + R_T} \quad [4]$$

where $R_B + R_T$ is the total resistance of the circuit. Power consumed to make the temperature measurement (P, in W) is calculated by multiplying I by V_{EX}:

$$P = IV_{EX} \quad [5]$$

where the magnitude of I and P are dependent on V_{EX}, R_B, and R_T. Based on optimized combinations of R_B and V_{EX} for a typical thermistor (R_T versus temperature varies from thermistor to thermistor), approximate values for I and P are 0.1 mA and 0.2 mW, respectively.

Electrical current flowing through a thermistor produces heat, raising the temperature of the thermistor above air temperature. Self-heating errors for a thermistor are related to power dissipation as a result of current flow. Self-heating errors are typically minimized by intermittently powering the thermistor. They can also be reduced by decreasing current flow, which is reduced by decreasing V_{EX}. The magnitude of self-heating is calculated by dividing power input to the thermistor by the thermistor heat dissipation constant (W °C^{-1}). Thermistor heat dissipation constants are often measured in either still air or stirred oil. Heat dissipation is much higher for stirred oil because of higher thermal conductivity of the fluid and fluid motion. Thus, self-heating error in stirred oil is much smaller than in still air. Still air is a worst case condition for heat dissipation and self-heating error. A reasonable value for a thermistor heat dissipation constant in still air is 1 mW °C^{-1}. A reasonable value for power input to a thermistor is 0.06 mW (calculated from Eq. [5], where V_{EX} is voltage across the thermistor instead of excitation voltage). With these estimates, self-heating in still air, even with continuous excitation, is only 0.06 °C. Heat dissipation is far more efficient when air is moving, so self-heating errors should be negligible for most environmental applications.

Power consumption and self-heating can be reduced by decreasing V_{EX}. However, decreasing V_{EX} also reduces the voltage across the bridge (V_B in Eq. [1], which must be measured to determine temperature). As a result, the combination of thermistor, bridge resistor, and excitation voltage must be optimized to minimize power consumption and self-heating, and maximize voltage output and temperature measurement resolution. Typically, companies selling thermistors for air temperature measurement add the bridge resistor to the electrical circuit and suggest an optimum excitation voltage for the specific thermistor.

The bridge resistor and length of cable between the thermistor and meter (e.g., datalogger) influence the accuracy of the temperature measurement. Resistance of the bridge resistor (R_B in Eq. [1]) must be accurately determined and stable with temperature and time because it is required to calculate resistance of the thermistor. Long lengths of cable (tens to hundreds of meters) have non-negligible resistance and will contribute to total resistance when connected in series with the thermistor and bridge resistor. As an example, resistance for a common thermistor for air temperature measurement (10 kΩ at 25 °C) changes by about 30 Ω per 0.1 °C at 30 °C and 20 Ω per 0.1 °C at 40 °C. If resistance of a long cable added 20 to 30 Ω, errors would approach 0.1 °C for air temperatures in the 30 to 40 °C range. For reference, the resistance of 24-AWG (0.51-mm diam.) copper wire is about 0.053 Ω per meter. It would take about 189 m of this lead wire to produce a resistance of 20 Ω (20 Ω/0.053 Ω per meter = 378 m, but there is a wire connected to each side of the thermistor and each wire contributes to the resistance, thus 378 m is divided by two to yield 189 m). Wire resistance changes with conductor material (copper is common, but other metals and alloys are used) and diameter (smaller diameter wires have higher resistance). Thermistor resistance change with temperature is typically higher at lower temperatures, thus error from long lead wire at lower temperatures is smaller. Thermistors used with long cables can be measured in four-wire half-bridge configurations to eliminate the effects of cable resistance (this is detailed in the next section for PRTs). Reference thermistors used in laboratories for calibration of air temperature sensors are often measured in a four-wire half-bridge configuration.

Platinum Resistance Thermometers

Like thermistors, platinum resistance thermometers (PRTs) operate by resistance change with temperature, but are made with a platinum sensing element, which is often a coiled wire or sometimes a thin film on a ceramic or plastic substrate. Thin film elements are becoming more common because they require less platinum and are lower cost. Other metals have been used in resistance temperature detectors (e.g., copper, nickel), but platinum is preferred because of high stability and wide temperature range. Platinum resistance thermometers are characterized by resistance at 0 °C, with the two most common being 100 Ω and 1000 Ω, referred to as a PT100 and PT1000, respectively.

Unlike most thermistors, the resistance of PRTs increases with increasing temperature (positive temperature coefficient), and PRTs are far more linear than thermistors over a wide temperature range. However, PRTs produce extremely small changes in resistance with temperature, which means a high resolution meter is required to make the measurement. Resistance change with temperature (sensitivity) of the PRT is quantified by the temperature coefficient of resistance (α, in $\Omega\ \Omega^{-1}\ ^{\circ}C^{-1}$):

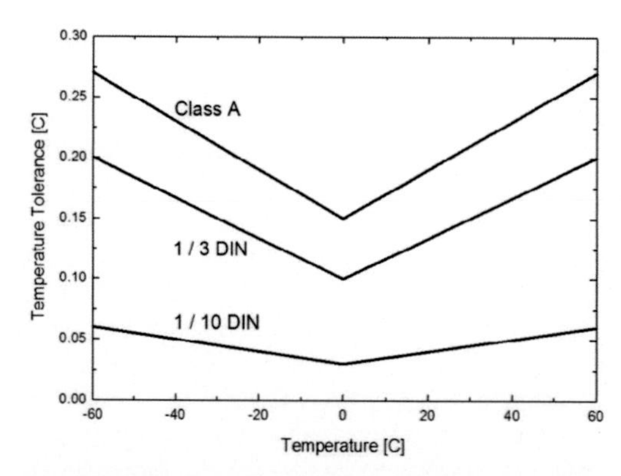

Fig. 2. Temperature tolerance for three accuracy classifications of PRTs. Temperature tolerance specifications (y axis) are defined by International Electrotechnical Commission (2017). Some lower accuracy classes (Class B and Class C) are not shown. There is also a 1/5 DIN class, which falls between 1/3 and 1/10 DIN (DIN stands for Deutsches Institut für Normung, translating to German Institute for Standardization).

$$\alpha = \frac{R_{100} - R_0}{100 R_0} \quad [6]$$

where R_{100} is resistance at 100 °C, R_0 is resistance at 0 °C, and 100 in the denominator is 100 °C. The coefficient α specifies the average resistance change of the PRT from 0 to 100 °C. Common α values are 0.00385 and 0.00392 $\Omega\ \Omega^{-1}\ °C^{-1}$, which indicate an average change in resistance of 0.385 Ω or 0.392 Ω per °C between 0 and 100 °C. These values of α also indicate the resistance at 100 °C is 138.5 Ω or 139.2 Ω, assuming a resistance of 100 Ω at 0 °C (PT100).

The platinum element of a PRT is usually contained in a stainless steel sheath, making PRTs rugged and weatherproof. Multiple stainless steel sheath sizes are available (1/8-inch diam. stainless steel sheath PRT shown in Fig. 1 is small compared to most sheaths), with larger diameter and/or longer sheaths yielding slower response time of the PRT. Multiple accuracy classes of PRTs are available (Fig. 2), with cost increasing as accuracy increases.

There are three PRT wire configurations: two-wire, three-wire, and four-wire, and multiple ways (electrical circuits) to measure each configuration. In the two-wire current excitation circuit a constant excitation current (I_{EX}, in A) is input to the PRT across the same wires where voltage across the PRT (V_{PRT}, in V) is measured using a differential voltage measurement. Resistance of the PRT (R_{PRT}, Ω) is calculated from V_{PRT}/I_{EX} (Ohm's Law). Measurement of V_{PRT} and input of I_{EX} must be accurate, as errors in the determination of R_{PRT} result in errors in temperature. Errors in V_{PRT} and I_{EX} can cause large temperature errors because the change in resistance with temperature (sensitivity) of PRTs is small. The half-bridge measurement described for thermistors (Eq. [1]) can be used to measure PRTs in the two-wire half-bridge circuit if a bridge resistor is included in series with the PRT. Two-wire is the simplest PRT configuration, but resistance in the lead wires connecting the PRT to the meter causes measured voltage to be higher than the voltage across the PRT, resulting in temperature errors. Sensitivity for many common PRTs is 0.385 Ω per °C ($\alpha = 0.00385\ \Omega\ \Omega^{-1}\ °C^{-1}$), resulting in errors of 1 °C for every 0.385 Ω of resistance added by lead wire. Resistance of 24-AWG (0.51-mm diam.) copper lead wire is about 0.053 Ω per meter, resulting in errors of about 1 °C for about 3.6 m of lead wire (0.385 Ω/0.053 Ω per meter = 7.2 m, but there is a wire connected to each side of the PRT and each wire contributes to the resistance, thus 7.2 m is divided by two to yield 3.6 m). Additional errors are caused by temperature-induced changes in the resistance of the lead wires, and thermal voltages generated by dissimilar lead wire and PRT metals. These effects cause resistance changes that are not associated with the PRT. For these reasons the two-wire configuration is not common in environmental applications.

A three-wire current excitation circuit can be used to measure PRTs. Similar to the two-wire current excitation circuit, I_{EX} is applied across the PRT and V_{PRT} is measured across the PRT, but no current flows on one of the wires used for the voltage measurement, thus error from wire resistance is cut in half compared to the two-wire current excitation circuit. A three-wire half-bridge circuit can also be used to measure PRTs. In this circuit the PRT and a bridge resistor of known resistance are connected in series. An excitation voltage is applied across the resistors (bridge resistor and PRT in series) and voltage is measured across the PRT on two different wires, each connected to the same end of the PRT. Each voltage measurement uses a single-ended measurement. This accounts for lead wire resistance by assuming the resistance of the two wires over which the voltage measurements are made is the same. The impact of lead wire resistance is eliminated if lead wire resistances are indeed equal, but the challenge is matching of the wires. Error in the temperature measurement results if wires are not the exact same length or small resistance differences between lead wires are present. Measurements with PRTs with three-wire configuration are more common than those with PRTs with two-wire configuration, but air temperature measurements with PRTs are typically made with PRTs with four-wire configuration.

There are three ways to measure PRTs with four-wire configuration. In the four-wire current excitation circuit I_{EX} is applied across the PRT with two of the wires, V_{PRT} is measured across the other two wires, and R_{PRT} is calculated from V_{PRT}/I_{EX} (Ohm's Law). This requires a one differential voltage measurement. Current doesn't flow in the wires where voltage is measured, so resistance of the lead wires does not influence the measurement. In the four-wire half-bridge circuit the PRT and a bridge resistor of known and fixed resistance (R_B, in Ω) are connected in series. An excitation voltage (V_{EX}, in V) is applied across the resistors (bridge resistor and PRT in series) and voltages are measured across the bridge resistor (V_B, in V) and PRT (V_{PRT}, in V). Resistance from the PRT (R_{PRT}, in Ω) is then calculated from the voltage measurements and R_B:

$$R_{PRT} = R_B \frac{V_{PRT}}{V_B} \qquad [7]$$

where the ratio of voltages (V_{PRT}/V_B) is equal to the ratio of resistances (R_{PRT}/R_B). The four-wire half-bridge circuit accounts for resistance of the lead wires, like the four-wire current excitation circuit. These two circuits yield the highest accuracy temperature measurements with PRTs, but the four-wire half-bridge circuit requires two differential voltage measurements, whereas the four-wire current excitation circuit only requires one. A four-wire full-bridge configuration is also an option. It only requires one differential voltage measurement, but isn't as accurate as the four-wire current excitation and half-bridge circuits unless the two bridge resistors required for the full-bridge circuit are perfectly matched.

As with thermistors, resistance of bridge resistors must be accurately determined and stable with temperature and time. Unlike most thermistors, resistance of PRTs is relatively small (often 100 Ω or 1000 Ω at 0 °C) and resistance change with temperature is very small (as described above, α for a PRT indicates the average resistance change per degree °C between 0 and 100 °C; values of 0.385 and 0.392 Ω per °C are common). Thus, errors in bridge resistance, changes in bridge resistance, or resistance added by lead wires, on the order of small fractions of an Ohm, can have large impacts on temperature measurements. Also like thermistors, PRTs are subject to self-heating errors because electrical current is flowing through the PRT. Self-heating is minimized by minimizing the excitation current or voltage.

Temperature is typically related to R_{PRT} with the Callendar–Van Dusen Equation (Callendar, 1887; Van Dusen, 1925). The most common solutions to the Callendar–Van Dusen Equation separate the temperature scale into two parts, with 0 °C being the dividing line. When R_{PRT}/R_0 is less than one (temperature is below 0 °C), where R_0 is the resistance of the PRT at 0 °C, then PRT temperature (T_{PRT}, in °C) equals:

$$T_{PRT} = gK^4 + hK^3 + iK^2 + jK \quad [8]$$

where $K = (R_{PRT}/R_0) - 1$ and g, h, i, and j are PRT–specific coefficients (coefficients vary with α). When R_{PRT}/R_0 is greater than or equal to 1 (temperature is above 0 °C), then T_{PRT} equals:

$$T_{PRT} = \frac{\sqrt{d(R_{PRT}/R_0) + e} - a}{f} \quad [9]$$

where a, d, e, and f are PRT–specific coefficients. Different equations for conversion of R_{PRT} to temperature have also been used. Many datalogger manufacturers have preprogrammed instructions for PRTs, which use the equations above or similar to calculate temperature from R_{PRT}. The only required input is α for the PRT (to determine coefficients).

As with thermistors, PRTs require power because excitation voltage is input to determine resistance. Power requirements for PRTs are small, with 0.4 mW being a typical value. Self-heating of PRTs is present, and dependent on the voltage across the PRT. This is typically much smaller than that for a thermistor, thus self-heating of PRTs is much less than the PRT temperature tolerance (Fig. 2) and can be considered negligible. Companies selling PRTs for use as air temperature sensors typically add the necessary resistors for the different configurations and suggest an optimum excitation voltage to minimize self-heating and maximize temperature measurement resolution.

Advantages and Disadvantages of Thermocouples, Thermistors, and Platinum Resistance Thermometers

Changes in signal (voltage, resistance) from a temperature sensor caused by changes in air temperature must be measureable, repeatable, and stable. Thermocouples, thermistors, and PRTs all require a meter (e.g., datalogger) to measure the electrical signal and convert it to temperature. This means several of the advantages and disadvantages depend on the data acquisition device. Dataloggers are desirable for environmental monitoring, where automated data collection of high frequency and/or long-term data sets are often required. The advantages and disadvantages of thermocouples, thermistors, and PRTs are summarized (Table 2). The advantages and disadvantages of PRTs are similar to or the same as those for

Table 2. Advantages and disadvantages of thermocouples, thermistors, and platinum resistance thermometers (PRTs) when used for air temperature sensors.

Sensor Type	Advantages	Disadvantages
Thermocouple	Does not require excitation voltage	Requires accurate reference temperature measurement
	No self-heating	Small signal change per °C change
	Multiple sensors can be made from one roll of wire	Requires differential channel on datalogger
	No compensation for long lead wires is required	Expensive wire; variability in wire from batch to batch
Thermistor	Reference temperature not required	Requires excitation voltage
	Large signal change per °C change	Self-heating error from continuously applied excitation voltage
	Only requires single-ended channel on datalogger	Can drift over time if not enclosed in water proof material
	Inexpensive wire	Long lead wires increase resistance
PRT	Reference temperature not required	Requires excitation voltage
	Inexpensive wire	Self-heating error from continuously applied excitation voltage
		Small signal change per °C change
		Requires at least one differential channel on datalogger
		Long lead wires must be compensated for with a four-wire configuration

thermistors, except that the resistance change with temperature (sensitivity) of PRTs is much smaller and more difficult to measure, PRTs typically require at least one differential channel on a datalogger, and PRTs typically have slower response time because the housing is often much larger than that for thermistors (Fig. 1).

Some of the advantages and disadvantages depend on the circumstances of the specific measurement and application. For example, the power requirement of thermistors and PRTs is extremely small. As stated above, typical current draw for a thermistor is approximately 0.1 mA, but many dataloggers can source several mA. A common datalogger used in environmental monitoring (Campbell Scientific model CR1000) can source 25 mA. Based on this specification, this specific datalogger could accommodate 250 thermistors (or PRTs) if there were enough measurement channels available.

The reference temperature required for thermocouples is available on thermocouple-specific meters and on many dataloggers. Accurate reference temperature measurements are then dependent on the accuracy of the reference sensor. This is a thermistor or PRT. Periodic recalibration of the meter or datalogger is recommended to ensure the internal temperature sensor is accurate. Also, the datalogger wiring panel (where the thermocouples are connected) must be maintained isothermal. This is best accomplished by installing the datalogger in an insulated, weatherproof box that shields it from solar radiation.

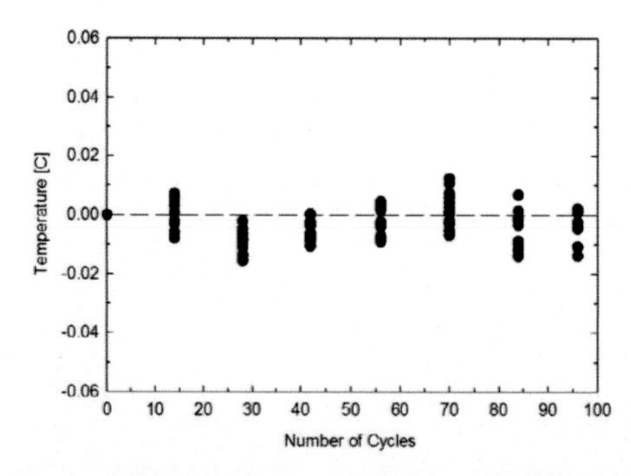

Fig. 3. Stability of twelve replicate thermistors (Measurement Specialties model 10K3A1IA) in a thermal cycling chamber. Cycles were -20 to 60 °C, with a vapor pressure range of 0.5 to 12 kPa (condensing humidity). Thermistor stability was periodically checked by removing the thermistors and measuring temperature in a slush bath (assumed to be 0 °C). The small down and up trend of the data are due to slight differences in the slush bath from one verification to the next.

The small output signal of thermocouples and PRTs relative to the large output signal of thermistors is only a disadvantage when a low-resolution datalogger is used. Some dataloggers have adequate resolution to make accurate thermocouple and PRT measurements, but the requirement of a differential channel is always a disadvantage. This means twice as many thermistors can be connected to the same number of datalogger channels (or four times the number of thermistors if PRTs are measured in the four-wire half-bridge configuration).

Nonlinearity is typically listed as a disadvantage of thermistors (e.g., Hubbard and Hollinger, 2005), but as long as the nonlinearity is repeatable, many dataloggers can be programmed to convert resistance to temperature using the Steinhart–Hart or β-parameter equation. These equations are not difficult to implement in software (e.g., spreadsheet) if the datalogger is not programmable. Also, thermistor and/or bridge resistor combinations that yield a more linear relationship between resistance and temperature are available and have been used in meteorological measurement networks (less commonly used now because of increased measurement resolution and processing capability of data acquisition systems).

Platinum resistance thermometers have the reputation of being very stable over time, but the platinum sensing element is relatively fragile when compared to a thermistor. Shift or drift in the resistance to temperature relationship of a PRT can be caused by thermal cycling or mechanical shock displacing the sensing element.

Historically, thermistors have been considered less stable than PRTs, but encasing thermistors in an epoxy or glass housing to keep moisture away from the sensing element gives them stability similar to PRTs. When exposed, oxidation of the sensing element can occur, leading to drift in the resistance to temperature relationship of the thermistor. Thermistors sealed in weatherproof housings can be very stable when measurements are made at environmental temperatures. Twelve replicate 2-mm diam. thermistors in epoxy housings (Measurement Specialties model 10K3A1IA) were thermal cycled 96 times from -20 to 60 °C, with a corresponding vapor pressure range of 0.5 to 12 kPa, in a test chamber. The thermistors were stable within 0.015 °C (Fig. 3), which is within the uncertainty range of the experiment. A second group of twelve replicates of the same model of thermistor were thermal cycled (same conditions as the initial group) 584 times. Temperature measurements were only made at the beginning and end of thermal cycling. None of the thermistors drifted by more than 0.015 °C. Nine replicates of the same model of thermistors were also deployed in radiation shields outdoors in Logan, Utah (UT), United States, for over two years and no detectable drift was measured.

Response time for air temperature sensors is important for applications where high frequency data are required, and can be critical for accurate measurements of minimum and maximum temperatures. Air temperature measurements from electronic sensors are typically made at relatively high frequencies (every few seconds), but they are often averaged over longer time intervals (every few minutes). Response time does not always need to be rapid for temperature measurements averaged over a longer time interval. Response time is a function of thermal mass of the sensor. More thermal mass means longer response times because equilibration with air is slower. Response time of thermocouples can be reduced by decreasing the diameter of the wire, but the tradeoff is a more fragile sample junction. Response times of thermistors and PRTs can be reduced by mounting the resistive element in a smaller casing, as long as it is rugged and weatherproof. Response time is also influenced by exposure and type of shield sensors are housed in (discussed below). Response times are faster in active (fan-ventilated) shields.

Thermocouples, thermistors, and PRTs all have errors caused by heat conduction down the cable to the sensing element. This error has been documented for type-T thermocouples used to measure leaf temperatures (Tarnopolsky and Seginer, 1999), but the concept is the same for air temperature measurements. If the lead cable is exposed to a heat source, such as solar radiation, it warms and heat is conducted to the sensing element. Heat conduction can be reduced by building and/or using sensors with wire material that has low thermal conductivity. For example, the thermal conductivity of copper is 386 W m^{-1} K^{-1}, whereas thermal conductivity for chromel and constantan (two common metals used in thermocouples) is about twenty times less at 19 and 21 W m^{-1} K^{-1}, respectively. For this reason, copper wire should be avoided in air temperature sensors. Heat conduction can also be reduced by shielding cables from direct radiation (e.g., cables are often mounted on the bottom of cross-arms on weather stations to reduce absorption of solar radiation).

Integrated Circuit Sensors

An integrated circuit (IC) temperature sensor is a semiconductor component integrated into a circuit board. Signal (current or voltage) from the semiconductor is temperature dependent, and can be related to absolute temperature. Many circuit boards have IC sensors onboard, but their use as air temperature sensors has not been widespread. Accuracy specifications for IC temperature sensors are typically lower than other sensors (some new IC sensors have accuracy specifications similar to traditional temperature sensors). Also, circuit boards generate heat, which can influence the temperature of components on the board, including IC temperature sensors. The advantage of IC temperature sensors is low cost.

Sonic Thermometers

Sonic anemometers measure the travel time of acoustic signals over a fixed distance. Travel time of an acoustic signal is linearly dependent on the wind velocity component along the distance traveled. Sonic anemometers have been used for decades to measure wind speed, especially in micrometeorological studies (Kaimal and Businger, 1963; Mitsuta, 1966; Schotland, 1955; Suomi, 1957).

The speed of sound in air (c) is dependent on air temperature (T, in units of K):

$$c^2 = \gamma R_{\text{Specific}} T \left(1 + 0.32\chi_w\right) \qquad [10]$$

where γ is the heat capacity ratio for air (ratio of heat capacity at constant pressure to heat capacity at constant volume, 1.4 for dry air at 20 °C), R_{Specific} is the specific gas constant for air (universal gas constant divided by the molar mass of air, 287 J kg^{-1} K^{-1}), and χ_w is the water vapor mole fraction (ratio of moles of water vapor in air to moles of air, in units of mol mol^{-1}). Water vapor increases the speed of sound in air and the term $(1 + 0.32\chi_w)$ accounts for this effect. Rearrangement of Eq. [10] to solve for T provides a measurement of air temperature from sonic measurement of c and measurement of humidity (χ_w). Thus, a sonic anemometer can serve as a sonic thermometer if a humidity measurement is available. A sonic thermometer is directly connected to first principles because the speed of sound in a gas (air in this case) is directly related to the thermodynamic temperature of the gas. Sonic thermometry has been proposed as a means of measuring air temperature (Barrett and Suomi, 1949; Pardue and Hedrich, 1956), often for micrometeorological studies (Kaimal and Gaynor, 1991; Schotanus et al., 1983).

The advantage of sonic thermometry is the absence of a physical sensor (thermal mass) that must equilibrate with air. This provides rapid response time and eliminates radiant heating of the sensor. This means temperature derived from sonic measurements can be used as a reference to determine the influence of radiant heating on physical sensors (discussed below). The disadvantages of sonic thermometry are the high cost and high power requirements of sonic anemometers, and requirement of a humidity measurement and correction. The humidity term $(1 + 0.32\chi_w)$ is in the denominator when Eq. [10] is rearranged to solve for temperature, so the temperature measurement is high if χ_w is low and low if χ_w is high. At a χ_w value of 15 mmol mol^{-1} (characteristic of the semiarid climate of Logan, UT, in summer months), an error of 10% in χ_w causes an error of 0.14 °C in air temperature. At a χ_w value of 5 mmol mol^{-1} (characteristic of the semiarid climate of Logan, UT, in winter months), an error of 10% in χ_w causes an error of 0.04 °C. Error scales with χ_w, thus air temperature errors resulting from inaccurate χ_w measurement will be higher for summer months and lower for winter months, and higher for humid places and lower for arid places. Absolute accuracy of measurement of the speed of sound in air (c) also affects the temperature measurement. Inaccuracy of the sonic anemometer can result from small changes in sonic sensor spacing with temperature, sensor drift with temperature, and distortion from strong cross wind.

Housing Air Temperature Sensors

The challenge of accurate air temperature measurement is far greater than having an accurate sensor, as sensors must be in thermal equilibrium with air. Housings for sensors should minimize heat gains and losses due to conduction and radiation, enhance coupling to air via convective currents, and protect sensors from snow and ice accumulation. Radiation-induced heating increases as wind speed (convection) decreases and as radiation load increases (Fig. 4; Bugbee et al., 1996). The housing for an air temperature sensor must shield it from shortwave (solar) radiant heating and longwave radiant cooling. A temperature sensor should also be thermally isolated from the housing to minimize heat transport to and from the sensor by conduction. The housing should provide ventilation so the temperature sensor is in thermal equilibrium with the air. In addition, the housing should keep precipitation off the sensor, as precipitation causes evaporative cooling. Conversely, condensation on sensors causes warming, and when condensed water subsequently evaporates it cools the sensor.

Radiation shields for air temperature sensors should be in a location with representative air temperature (tops of buildings and areas where they will be influenced by reflection of solar radiation should be avoided). Conditions in microenvironments have the potential to be very different from surrounding conditions. Typical mounting heights for air temperature sensors are 1.2 to 2.0 m above the ground. Radiation shields should be mounted over vegetation.

Temperature sensors on automated weather stations are typically shielded from solar radiation by either a passive (naturally-ventilated) or an active (fan-ventilated) housing (Fig. 5). Passive radiation shields are louvered enclosures that rely on natural ventilation from wind to dissipate absorbed solar energy and equilibrate the sensor to air. Active radiation shields dissipate absorbed solar energy and maintain equilibrium with air through fan ventilation.

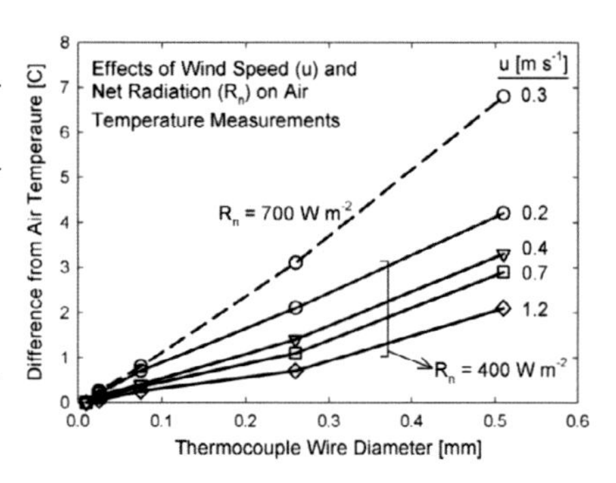

Fig. 4. Effects of wind speed (*u*) (m s^{-1}) and net radiation (R_n) (W m^{-2}) on air temperature measurements from unshielded type-E thermocouples as a function of thermocouple wire diameter (mm). The wire diameters from smallest to largest are 56-AWG (0.01 mm), 50-AWG (0.03 mm), 40-AWG (0.08 mm), 30-AWG (0.25 mm), and 24-AWG (0.51 mm). Difference from air temperature (*y* axis) is error of each thermocouple relative to actual air temperature. Data points represent the mean of three thermocouples (standard deviation was smaller than symbol size). Dashed line is the difference for greater R_n (700 W m^{-2}), indicating the difference from air temperature is proportional to radiation load. At a given *u*, temperature difference increases approximately linearly with increasing wire diameter [data from Bugbee et al. (1996)].

Passive Shields

Passive shields are simple, low-cost, and do not require power, but they warm above air temperature in low wind or high solar radiation (Fig. 6). Warming is increased when there is snow on the ground due to higher albedo and increased reflected solar radiation (Nakamura and Mahrt, 2005), resulting in more incident radiation from below. Errors as high as 10 °C have been reported in passive shields over snow (Genthon et al., 2011; Huwald et al., 2009). Warming is also increased when the sun is low in

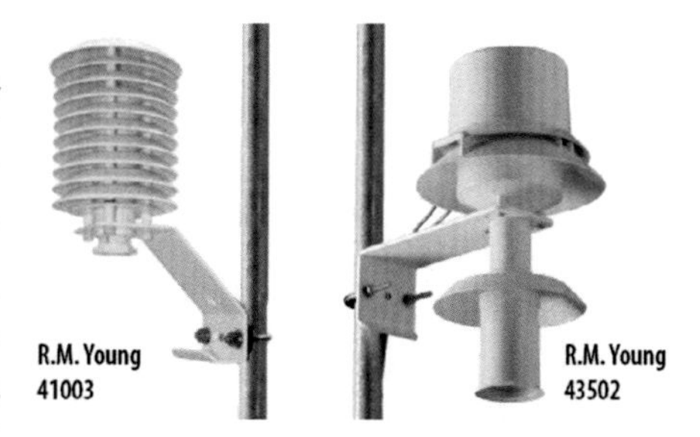

Fig. 5. Passive (left) and active (right) radiation shields. Both models (R. M. Young 41003 and R. M. Young 43502) are shown only as examples. Multiple models of passive and active shields are available from several manufacturers.

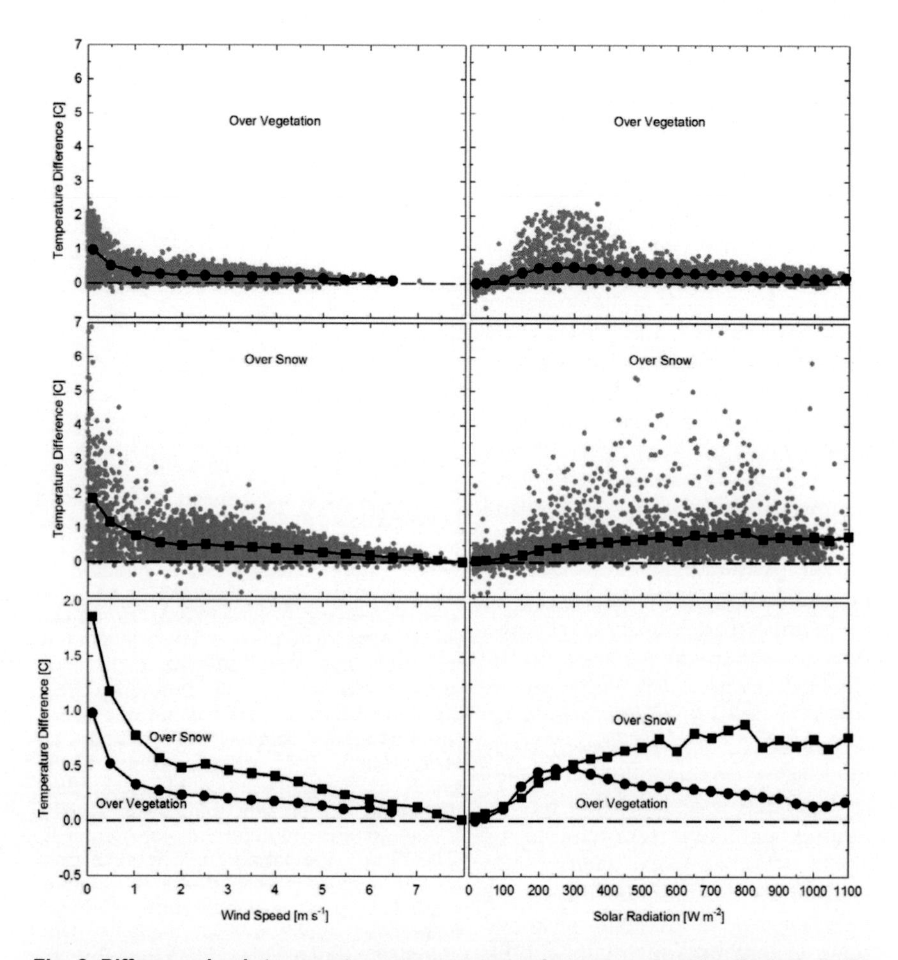

Fig. 6. Difference in air temperature measurements between a passive (6-plate, R. M. Young model 41303) and active (Apogee Instruments model TS-100) shield housing identical air temperature sensors (small thermistor, Apogee Instruments model ST-110). Top graph is daytime data collected over vegetation (no snow on the ground) and middle graph is daytime data collected over snow. Black lines are bin averages, and bottom graph shows a comparison of bin averages for conditions of no snow and snow (note y axis scale range is reduced to clearly show differences). Graphs on left hand side are differences with wind speed and graphs on right hand side are differences with solar radiation.

the sky (high solar zenith angles) because more radiation reaches the air temperature sensor through the open sides of passive shields. This is why temperature differences of a passive shield relative to an active shield over vegetation are often greatest from about 200 to 300 W m^{-2} and decline as solar radiation increases (Fig. 6).

Several models of passive radiation shields are available and not all models perform the same. Comparison of three models (R. M. Young model 41003, MetSpec model RAD 16 Mk 1, MetSpec model RAD 16 Mk 2) indicated mean differences of a few tenths of a degree (°C) at low wind speed (Fig. 7). Trends are similar for all passive shields, with the largest air temperature errors in conditions

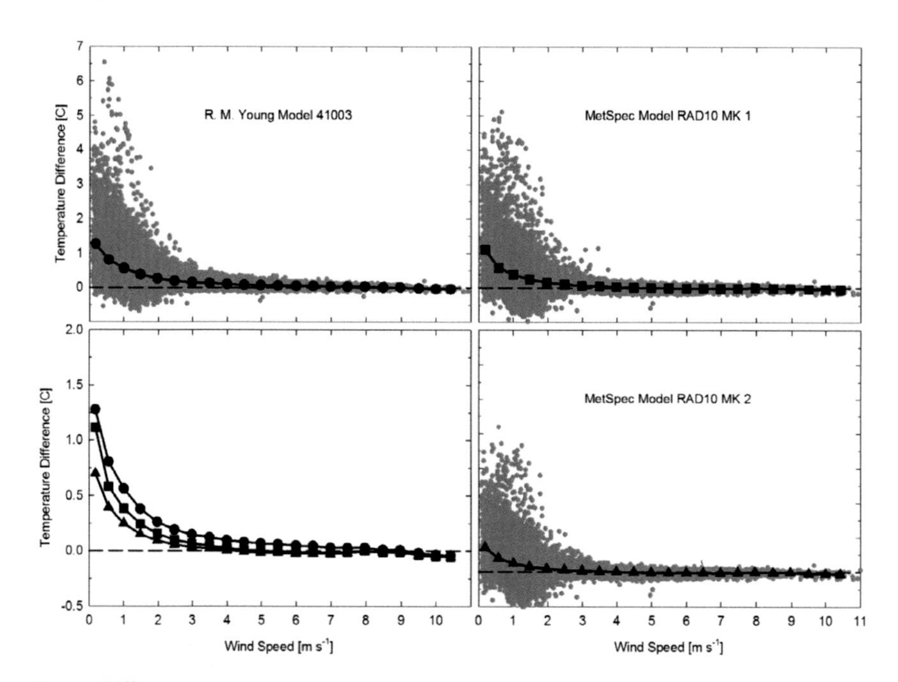

Fig. 7. Difference in air temperature measurements between three models of passive shields (R. M. Young model 41003, MetSpec model RAD10 Mk 1, MetSpec model RAD10 Mk2) and an active shield (Apogee Instruments model TS-100). Air temperature sensors were identical among housings (small thermistor, Campbell Scientific model 109SS). Data were collected over two years and include daytime measurements over vegetation and snow. Black lines are bin averages, and are compared in the graph in the lower left-hand corner (note *y* axis scale range is reduced to clearly show differences).

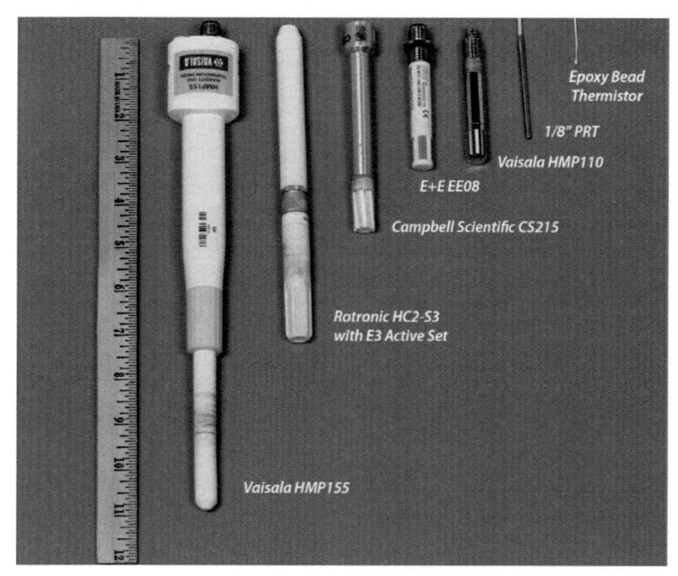

Fig. 8. Size comparison of some common temperature and relative humidity probes and two stand-alone temperature sensors (PRT and thermistor). Ruler scale is inches.

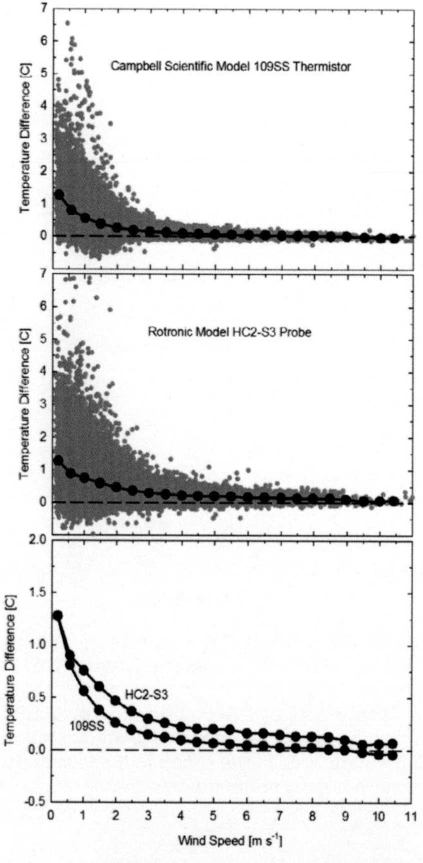

Fig. 9. Difference in air temperature measurements between two replicates of the same passive shield model (R. M. Young model 41003) and an active shield (Apogee Instruments model TS-100). One of the passive shields housed a small thermistor (Campbell Scientific model 109SS) and the other housed a combined temperature and/or relative humidity probe (Rotronic model HC2-S3). Data were collected over a two year period and include daytime measurements over vegetation and snow. Black lines are bin averages, and are compared in the bottom graph (note *y*-axis scale range is reduced to clearly show differences).

of high solar zenith angle (morning and evening) and low wind speed. Lopardo et al. (2014) found that errors were larger with older passive radiation shields (by as much as 1.6 °C for a five-year-old shield), presumably because of decreased shortwave reflectivity caused by aging. Temperature sensors in passive shields are better equilibrated to air under high wind speeds. A wind speed of 4 m s^{-1} represents an approximate threshold for low albedo (surface reflectively) conditions over a vegetative surface. At wind speeds greater than about 4 m s^{-1}, air temperature measurements in three passive shield models matched measurements from active shields within about 0.1 °C (Fig. 7). Data from Tanner et al. (1996) indicate a wind speed of about 4 m s^{-1} is the point where passive shields match active shields. Comparison of passive shields from one weather network to active shields from another weather network revealed passive shields measured warmer daily maximum temperatures (mean was 0.48 °C) and colder daily minimum temperatures (mean was -0.36 °C) (Leeper et al., 2015).

The magnitude of temperature errors caused by radiant heating of sensors in passive shields is highly dependent on the surface area of the sensor. Many weather stations have combined relative humidity and temperature sensors, which are much larger (more surface area) than stand-alone air temperature sensors (Fig. 8). Air temperature errors generally increase with increasing surface area of the sensor (Fig. 9). Tanner (2001) reported similar results, where a common temperature and/or RH probe (Vaisala model HMP35C) was about 0.5 °C warmer than a medium sized thermistor (Campbell Scientific model 107). Fine-wire thermocouples have been used in passive shields because surface area is minimized (Kurzeja, 2010). Thermal mass of temperature sensors has a major impact on sensor response time. Sensors with small thermal mass equilibrate and respond to changes quicker and are necessary for applications requiring high frequency air temperature measurements.

Equations to correct air temperature measurements in passive shields have been proposed, but often require measurement of wind speed and solar radiation, and are applicable to specific shield designs (Mauder et al., 2008). Corrections that don't require additional meteorological measurements have also been proposed, such as air temperature adjustment based on the difference between air temperature and interior plate temperature differences (Kurzeja, 2010). Others have suggested modifying traditional multi-plate passive shields to include a small fan that can be operated under specific conditions, but utilize natural aspiration when wind speeds are above an established threshold (Richardson et al., 1999).

Active Shields

Warming of air temperature sensors above actual air temperature is minimized with active shields, but power is required for the fan. The power requirement for active shields ranges from one to six watts (80–500 mA at 12 V DC). For solar-powered weather stations this can be a major fraction of power usage for the entire station and has typically required a large solar panel and large battery. Power requirement and cost are disadvantages of active shields (Table 3), and they have led to the use of less accurate passive shields on many solar-powered stations. Also, the fan motor can heat air as it passes by the fan. Active shields should be constructed to avoid recirculation of heated air back into the shield.

There is no reference standard for the elimination of radiation-induced temperature increase of a sensor for air temperature measurement, but well-designed active shields minimize this effect. Radiation-induced temperature increase for active shields was analyzed in long-term experiments over snow and grass surfaces by comparing temperature measurements from three models of active radiation shields (Apogee Instruments model TS-100, Met One model 076B, and R. M. Young model 43502). The study included two replicates of each shield model, with matching 2-mm diam. calibrated thermistors in all shields (Apogee Instruments model ST-110). Continuous measurements for two months in summer and two months in winter in Logan, UT, indicated that mean differences among shields were less than 0.1 °C in summer when measurements were made over vegetation, but were as high as 0.4 °C for the R. M. Young shield in winter when measurements were made over snow (Fig. 10). Differences increased with increasing solar radiation, particularly during winter months when there was snow (high reflectivity) on the ground. The Met One model 076B shield was used as a reference because temperatures from this shield tended to be slightly cooler than temperatures from the other two shield models over vegetation.

Table 3. Advantages and disadvantages of passive (naturally-ventilated) and active (fan-ventilated) radiation shields.

Shield type	Advantages	Disadvantages
Passive (naturally-ventilated)	Do not require power	Less accurate (overheat in low wind or high solar radiation)
	Lower cost	
Active (fan-ventilated)	More accurate	Require power
		Higher cost

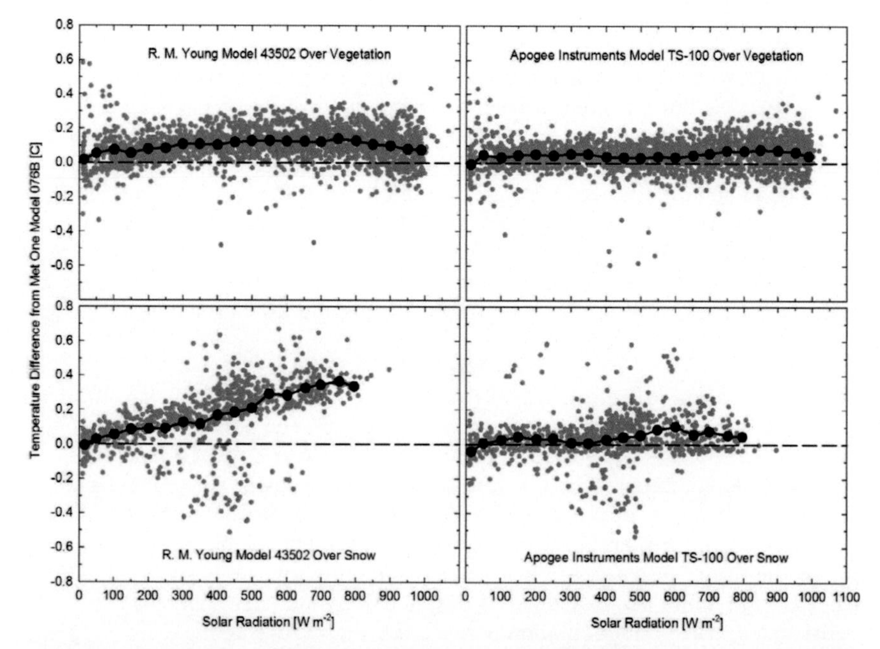

Fig. 10. Air temperature differences of two active shields (R. M. Young model 43502, left side, and Apogee Instruments model TS-100, right side) from a reference active shield (Met One model 076B, used as the reference because it tended to measure slightly cooler than the other two models over vegetation). Top graphs are data over vegetation (66 d from summer 2012) and bottom graphs are data over snow (64 d from winter 2013). A small thermistor (Apogee Instruments model ST-110) was used to measure temperature in all shields. All data are from daytime under clear sky conditions and low wind speed (less than 2 m s^{-1}). Black lines are bin averages. All sensors were within 0.05 °C at night, indicating minimal differences among sensors.

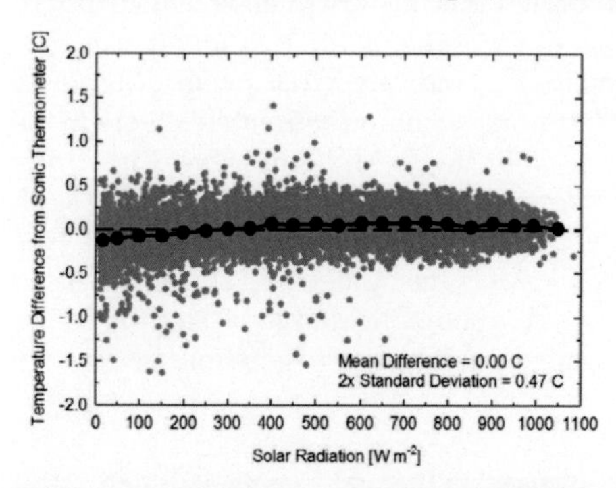

Fig. 11. Differences of air temperature measured in an active shield (Apogee Instruments model TS-100 with an Apogee Instruments model ST-110 thermistor) to air temperature from a sonic thermometer (Campbell Scientific model IRGASON). Data are from daytime only and were collected over a one-year period in Logan, Utah (United States). Black line is a bin average.

At higher wind speeds (greater than 3 m s^{-1}) during daytime, sensors in the Apogee model TS-100 read slightly cooler on average (-0.05 to -0.1 °C) than sensors in the Met One model 076B and R. M. Young model 43502 shields, possibly due to the ninety degree inlet orifice of the 076B and 43502 shields, which may have reduced the inlet air velocity, and thus ventilation of the thermistor, during high cross winds. Tanner et al. (1996) found passive shields read 0.0 to 0.3 °C cooler than an active shield with a ninety degree inlet at wind speeds greater than about 4 m s^{-1}, indicating poor ventilation of the temperature sensor inside the active shield during higher wind speed.

As mentioned, there is not a reference standard for radiation-induced temperature increase of air temperature sensors, but measurements from sensors in radiation shields can be compared with measurements from sonic thermometry, which does not require equilibration of a physical sensor with the surrounding air. This comparison was done using daytime data from a small thermistor (Apogee Instruments model ST-110) in an active shield (Apogee Instruments model TS-100) and a sonic thermometer (Campbell Scientific model IRGASON, which combines a sonic anemometer and infrared gas analyzer to measure water vapor) over a one-year period in Logan, UT (Fig. 11). Measurements from the active shield were slightly cooler than the sonic thermometer when solar radiation was less than about 350 W m^{-2} and slightly warmer when solar radiation was greater than 350 W m^{-2}. Average differences were 0.08 °C or less across the range of solar radiation from 50 to 1050 W m^{-2}. Results were the same when data were analyzed for low wind speeds (less than 2 m s^{-1}). There was a small seasonal dependence of the temperature difference, with the active shield slightly cooler (0.09 °C on average) than the sonic thermometer in summer and fall and slightly warmer (0.09 °C on average) than the sonic thermometer in winter and spring. The seasonal dependence appears unrelated to solar radiation intensity, solar zenith angle, and snow on the ground. It is possible the seasonal dependence was related to the humidity measurement that goes into the humidity correction, as the seasonal trend is similar to the trend in water vapor mole fraction (χ_w). The sonic thermometer used for the comparison does not have a specified accuracy, but the close match to the thermistor (accuracy specification of ± 0.1 °C) in the active shield suggests sonic thermometry has potential as a reference for temperature measurements with physical sensors.

Errors Caused by Rime

Rime can cause large errors in air temperature measurements. Soft rime is relatively common in Logan, UT, on cold, clear winter days when strong air temperature inversions occur. Soft rime is made of tiny ice particles with pockets of air between them, giving it a white color and feathery, needle-like structure. There is relatively poor cohesion between adjacent ice particles, due to rapid freezing of individual super-cooled water droplets when soft rime is formed. This makes soft rime fragile, and easy to remove from surfaces. Soft rime is sometimes called 'snow feathers' because of the feathery appearance of the white ice

Fig. 12. Photo of three models of active shields (Met One 076B, R. M. Young 43502, Apogee Instruments TS-100) with soft rime deposits during a winter temperature inversion in Logan, Utah (United States).

needles and granules that it is composed of. On days when soft rime occurs, the fan on active shields draws it into the shields, sometimes filling the shield (Fig. 12).

Measurements made on days when active shields were full of soft rime indicate Met One model 076B and R. M. Young model 43502 shields tended to read colder than actual air temperature and Apogee model TS-100 shields tended to read warmer than actual air temperature. Actual air temperature was determined by clearing a Met One 076B shield of rime and using the subsequent temperature measurements as the reference. The magnitude of errors appeared to be dependent on the amount of rime inside the shields, with errors of 0.5 to 1.5 °C being typical.

Soft rime also forms on passive shields, but it was difficult to determine temperature errors because soft rime in Logan, UT, occurs on days when passive shield errors are already at a maximum (low wind, clear sky, snow on the ground). However, data indicate passive shield errors were similar to those reported above for conditions of low wind over snow (Fig. 6), independent of whether rime was present or not. Other types of rime and frost would likely have similar impact as soft rime on both active and passive shields. Blowing snow can fill the spaces between the plates on passive shields and significantly reduce ventilation, causing errors similar to those reported for soft rime and detailed above (Fig. 6). Active shields work better in blowing snow as long as snow does not clog the inlet.

Summary

Accurate air temperature measurement remains challenging, despite decades of research and development to improve instruments and methods. Thermocouples, thermistors, and platinum resistance thermometers (PRTs) have all been used for air temperature measurement in environmental applications, and each have associated advantages and disadvantages. Platinum resistance thermometers have the reputation as the preferred sensor for air temperature measurement due to high accuracy and stability, but modern glass or epoxy coated thermistors are similarly stable, especially at temperatures below 60 °C. Thermistors have high signal-to-noise ratio, are easy to use and low cost, and have accuracy similar to PRTs. Thermocouples have often been used for research purposes, but are becoming less common for air temperature measurement because of the requirement of accurate measurement of reference temperature (datalogger panel temperature).

Methods for shielding and ventilation of the air temperature sensor can be more important than sensor type. Passive, natural ventilation reduces accuracy in conditions of high solar load or low wind speed. Active, fan ventilation improves accuracy compared to passive shields, but increases the cost and power requirement. Sonic thermometry has the potential to be the most accurate because it does not require equilibration of a physical sensor with air, but requires accurate measurements of speed of sound in air and humidity, and the technology is expensive.

Air temperature measurements are an essential component of weather monitoring and climate research worldwide, and will continue to be challenging given the tradeoffs between accuracy, power consumption, and costs of the instrument options.

Acknowledgments

Thanks to Jobie Carlisle at the Utah Climate Center for providing the data shown in Figure 6, Bart Nef at Campbell Scientific for providing the data shown in Figures 7 and 9, Ivan Bogoev at Campbell Scientific for providing the data shown in Figure 11 and reviewing a draft of the manuscript, and Alan Hinckley at Campbell Scientific for reviewing a draft of the manuscript. Special thanks to Andrew Sandford at Campbell Scientific Limited for reviewing the manuscript and providing multiple suggestions that significantly improved it.

References

Barrett, E.W., and V.E. Suomi. 1949. Preliminary report on temperature measurement by sonic means. J. Meteorol. 6:273–276. doi:10.1175/1520-0469(1949)006<0273:PROTMB>2.0.CO;2

Bugbee, B., O. Monje, and B. Tanner. 1996. Quantifying energy and mass transfer in crop canopies: Sensors for measurement of temperature and air velocity. Adv. Space Res. 18:149–156. doi:10.1016/0273-1177(95)00871-B

Callendar, H.L. 1887. On the practical measurement of temperature: Experiments made at the Cavendish Laboratory, Cambridge. Philos. Trans. R. Soc. Lond. A 178:161–230. doi:10.1098/rsta.1887.0006

Christian, K.A., and C.R. Tracy. 1985. Measuring air temperature in field studies. J. Therm. Biol. 10:55–56. doi:10.1016/0306-4565(85)90012-9

Clark, M.R., D.S. Lee, and T.P. Legg. 2013. A comparison of screen temperature as measured by two Met Office observing systems. Int. J. Climatol. 34:2269–2277. doi:10.1002/joc.3836

Genthon, C., D. Six, V. Favier, M. Lazzara, and L. Keller. 2011. Atmospheric temperature measurement biases on the Antarctic plateau. J. Atmos. Ocean. Technol. 28:1598–1605. doi:10.1175/JTECH-D-11-00095.1

Holden, Z.A., A.E. Klene, R.F. Keefe, and G.G. Moisen. 2013. Design and evaluation of an inexpensive radiation shield for monitoring surface air temperatures. Agric. For. Meteorol. 180:281–286. doi:10.1016/j.agrformet.2013.06.011

Hubbard, K.G., and S.E. Hollinger. 2005. Standard meteorological measurements. In: J.L. Hatfield and J.M. Baker, editors, Micrometeorology in agricultural systems. American Society of Agronomy, Madison, Wisconsin. p. 1–30.

Huwald, H., C.W. Higgins, M.-O. Boldi, E. Bou-Zeid, M. Lehning, and M.B. Parlange. 2009. Albedo effect on radiative errors in air temperature measurements. Water Resour. Res. doi:10.1029/2008WR007600

International Electrotechnical Commission. 2017. IEC 60751:2008. Industrial platinum resistance thermometers and platinum temperature sensors. International Electrotechnical Commission. https://webstore.iec.ch/publication/3400 (verified 10 Oct. 2017). [*2017 is year accessed*].

Kaimal, J.C., and J.A. Businger. 1963. A continuous wave sonic anemometer-thermometer. J. Appl. Meteorol. 2:156–164. doi:10.1175/1520-0450(1963)002<0156:ACWSAT>2.0.CO;2

Kaimal, J.C., and J.E. Gaynor. 1991. Another look at sonic thermometry. Boundary-Layer Meteorol. 56:401–410. doi:10.1007/BF00119215

Kurzeja, R. 2010. Accurate temperature measurements in a naturally-aspirated radiation shield. Boundary-Layer Meteorol. 134:181–193. doi:10.1007/s10546-009-9430-2

Leeper, R.D., J. Rennie, and M.A. Palecki. 2015. Observational perspectives from U.S. Climate Reference Network (USCRN) and Cooperative Observer Program (COOP) Network: Temperature and precipitation comparison. J. Atmos. Ocean. Technol. 32:703–721. doi:10.1175/JTECH-D-14-00172.1

Lopardo, G., F. Bertiglia, S. Curci, G. Roggero, and A. Merlone. 2014. Comparative analysis of the influence of solar radiation screen ageing on temperature measurements by means of weather stations. Int. J. Climatol. 34:1297–1310. doi:10.1002/joc.3765

Mauder, M., R.L. Desjardins, Z. Gao, and R. van Haarlem. 2008. Errors of naturally ventilated air temperature measurements in a spatial observation network. J. Atmos. Ocean. Technol. 25:2145–2151. doi:10.1175/2008JTECHA1046.1

Mitsuta, Y. 1966. Sonic anemometer-thermometer for general use. J. Meteorol. Soc. Jpn. 44:12–24. doi:10.2151/jmsj1965.44.1_12

Nakamura, R., and L. Mahrt. 2005. Air temperature measurement errors in naturally ventilated radiation shields. J. Atmos. Ocean. Technol. 22:1046–1058. doi:10.1175/JTECH1762.1

Pardue, D.R., and A.L. Hedrich. 1956. Absolute method for sound intensity measurement. Rev. Sci. Instrum. 27:631. doi:10.1063/1.1715654

Richardson, S.J., F.V. Brock, S.R. Semmer, and C. Jirak. 1999. Minimizing errors associated with multiplate radiation shields. J. Atmos. Ocean. Technol. 16:1862–1872. doi:10.1175/1520-0426(1999)016<1862:MEAWMR>2.0.CO;2

Schotanus, P., F.T.M. Nieuwstadt, and H.A.R. De Bruin. 1983. Temperature measurement with a sonic anemometer and its application to heat and moisture fluxes. Boundary-Layer Meteorol. 26:81–93. doi:10.1007/BF00164332

Schotland, R.M. 1955. The measurement of wind velocity by sonic means. J. Meteorol. 12:386–390. doi:10.1175/1520-0469(1955)012<0386:TMOWVB>2.0.CO;2

Steinhart, J.S., and S.R. Hart. 1968. Calibration curves for thermistors. Deep-Sea Res. and Oceanogr. Abstr. 15:497–503. doi:10.1016/0011-7471(68)90057-0

Suomi, V.E. 1957. Energy budget studies and development of the sonic anemometer for spectrum analysis. AFCRC Technical Report 56-274, University of Wisconsin, Department of Meteorology, Madison, WI.

Tanner, B.D. 2001. Evolution of automated weather station technology through the 1980s and 1990s. In: K.G. Hubbard and M.V.K. Sivakumar, editors, Automated weather stations for applications in agriculture and water resources management: Current uses and future perspectives. WMO/TD number 1074. World Meteorological Organization, Geneva, Switzerland. p. 3–20.

Tanner, B.D., E. Swiatek, and C. Maughan. 1996. Field comparisons of naturally ventilated and aspirated radiation shields for weather station air temperature measurements. Conference on Agricultural and Forest Meteorology with Symposium on Fire and Forest Meteorology 22:297–300.

Tarnopolsky, M., and I. Seginer. 1999. Leaf temperature error from heat conduction along thermocouple wires. Agric. For. Meteorol. 93:185–194. doi:10.1016/S0168-1923(98)00123-3

Thomas, C.K., and A.R. Smoot. 2013. An effective, economic, aspirated radiation shield for air temperature observations and its spatial gradients. J. Atmos. Ocean. Technol. 30:526–527. doi:10.1175/JTECH-D-12-00044.1

Van Dusen, M.S. 1925. Platinum-resistance thermometry at low temperatures. J. Am. Chem. Soc. 47:326–332. doi:10.1021/ja01679a007

Young, D.T., L. Chapman, C.L. Muller, and X.-M. Cai. 2014. A low-cost wireless temperature sensor: Evaluation for use in environmental monitoring applications. J. Atmos. Ocean. Technol. 31:938–944. doi:10.1175/JTECH-D-13-00217.1

Soil Temperature and Heat Flux

Thomas J. Sauer* and Xiaoyang Peng

Agroclimatology is a specialization within the field of climatology that focuses on terrestrial, near-surface processes as they relate to crop, livestock, and forest production. Climate near the ground is influenced by the coupling between surface conditions as affected by elevation, latitude, slope aspect, and vegetative cover with large-scale meteorological forces including air and ocean currents. Crop growth and production are strongly affected by these climatic factors while energy exchange within and above the crop canopy in turn influences the local microclimate. Soil properties play an important role in affecting climate near the surface as soil water content, color, texture, and density affect the partitioning of incoming energy between warming the soil, evaporation of water from the soil and plants, and warming of the air above the ground surface.

Temperature is a key factor affecting the rate of all soil biological, chemical, and physical processes essential for plant growth and the provisioning of ecosystem services (van Bavel, 1972; Jury et al., 1991; McInnes, 2002). The rate at which heat is transferred in a soil determines how fast soil temperature changes during a day or with the seasons. Soil heat flux, the rate of thermal energy transfer through a unit area of soil over time, is an important parameter because it effectively couples energy transfer processes at the surface, the surface energy balance, with energy transfer processes in the soil, the soil thermal regime. The soil surface is therefore the pivotal interface between surface and subsurface energy transfer processes (Sauer and Norman, 1995; Sauer et al., 1995). Global climate change effects on temperature and precipitation patterns have important impacts on surface soil conditions that in turn have significant implications for crop growth and food security.

Temperature ranges for optimal growth of many economically important crops are now well known (Sánchez et al., 2014; Yamori et al., 2014). Growers and plant breeders are aware of these temperature ranges as evidenced by the recent planting of commodity crops in non-traditional areas of production, for example, increased corn (*Zea mays* L.) and soybean (*Glycine max* L.) production in

---

Abbreviations: LAI, leaf area index; RTD, resistance temperature detectors; VI, vegetation index.

T.J. Sauer, U.S. Department of Agriculture– Agricultural Research Service, National Laboratory for Agriculture and the Environment, 1015 North University Boulevard, Ames, IA 50011-3611; X. Peng, China Agricultural University No 2, Yuan Ming Yuan Xi Lu, Beijing, China 100193 *Corresponding Author (tom.sauer@ars.usda.gov)

doi:10.2134/agronmonogr60.2016.0024

Mention of trade names or commercial products in this publication is solely for the purpose of providing specific information and does not imply recommendation or endorsement by the U.S. Department of Agriculture (USDA). USDA is an equal opportunity provider and employer.

© ASA, CSSA, and SSSA, 5585 Guilford Road, Madison, WI 53711, USA.
Agroclimatology: Linking Agriculture to Climate, Agronomy Monograph 60.
Jerry L. Hatfield, Mannava V.K. Sivakumar, John H. Prueger, editors.

the northern Great Plains of the United States. Changes in crop distribution are analogous to observations of plant species migrations to higher elevations and latitudes in natural ecosystems (Linderholm, 2006; Harsch et al., 2009; Woodall et al., 2009). On shorter time scales, changes in seasonal and even daily temperature and moisture patterns have been found to have important implications for crop production, especially during the reproductive phase of crop growth (Porter and Gawith, 1999; Zinn et al., 2010; Bita and Gerats, 2013; Hatfield and Prueger, 2015). Some of these effects are primarily in response to changes in mean temperature and annual precipitation while extremes in precipitation and temperature present additional challenges for agricultural systems (Olesen and Bindi, 2002; Luo, 2011). Schlenker and Roberts (2009) predicted that U.S. corn and soybean production would increase with temperature up to 29 °C for corn and 30 °C for soybean but would then decrease by 30 to 80% depending on future warming scenarios.

Changes in mean and extreme temperatures are often, directly or indirectly, accompanied by altered soil water conditions (insufficient or excess) or biotic stresses (weed competition, insects, and pathogens). There is a complex interaction of abiotic (temperature and moisture stress) with simultaneous biotic stress (weeds, insects, and disease) that can create significant challenges for the development of viable control and mitigation options through crop management strategies or plant breeding (Lipiec et al., 2013; Suzuki et al., 2014; Ramegowda and Senthil-Kumar, 2015). Understanding the soil thermal regime, its interaction with plant available water, and practices available to manage soil temperature and water content for optimum crop production are critical to any climate change adaptation strategy (Kang et al., 2009). At the global scale, the key role of soil in agricultural systems within the context of the water, food, and energy nexus continues to be an active area of discussion (Hatfield et al., 2017).

In this chapter we introduce the soil thermal regime and surface energy balance concepts. Included are brief discussions of relevant theory and measurement techniques for soil temperature and thermal properties. Examples of soil temperature and heat flux data are presented and the effects of microclimate and soil properties on these parameters are discussed. A section is devoted to the unique characteristics of paddy rice systems. The chapter concludes with a brief summary of topics covered.

Soil Temperature and Heat Flow and the Surface Energy Balance

Solar radiation is the driving force behind the thermal regime of terrestrial surfaces. The diurnal or diel variation in shortwave radiation creates a sinusoidal soil temperature pattern described by

$$T(z,t) = T_{av} + A_0 \exp(-z/d)\,[\sin(wt - z/d)] \qquad [1]$$

where $T(z,t)$ is soil temperature at depth z and time t (°C), T_{av} is the average soil temperature (°C), A_0 is the amplitude of the surface fluctuations (°C), d is the damping depth (m), and ω is the angular frequency in radians equal to $2\pi/P$ where P is the period. Eq. [1] describes soil temperature as a sine wave with an amplitude that decreases with depth and a phase lag that increases with depth. Figure 1 shows soil temperature at four depths over two intervals in a reconstructed prairie and a corn field in central Iowa (Sauer, unpublished data, 2007). In mid-May (Fig. 1a and 1c) the temperature amplitude at the 6 cm depth is approximately 7 °C for both prairie and corn sites due to the exposed soil surfaces between the small plants in spring. At 50 cm, the soil temperature is near 13 °C and slowly increasing. By mid-August (Fig. 1b and 1d) the temperature at 50 cm has increased to ~ 20 °C. With fully developed canopies, less solar radiation is reaching the soil surface and the amplitude of the 6 cm temperature wave is reduced to < 5 °C. On all days, the daily peak soil temperature at 20 cm lags the peak at 6 cm by ~ 3 h. and has less than half of the amplitude of the temperature wave at 6 cm. While the idealized pattern of soil temperature propagation represented by Eq. [1] is conceptually useful, in most instances variation in solar radiation (i.e., clouds and canopy shading) and soil properties with depth (especially moisture content) result in a considerably more complex soil thermal regimes.

Conduction is the principal mode of energy transfer in soils, although energy may also be transferred by radiation or convection in shallow layers. Heat conduction in soil is analogous to heat conduction in a solid as described by Fourier's Law:

$$G=-\lambda dT/dz \qquad [2]$$

where G is the heat flux (W m^{-2}), λ is the thermal conductivity of the material (W m^{-1} K^{-1}) and dT/dz is the temperature gradient within the solid (K m^{-1}). Eq. [2] applies to a homogeneous solid under steady state conditions with constant thermal conductivity. Applying Eq. [2] to a porous, three-phase medium like soil

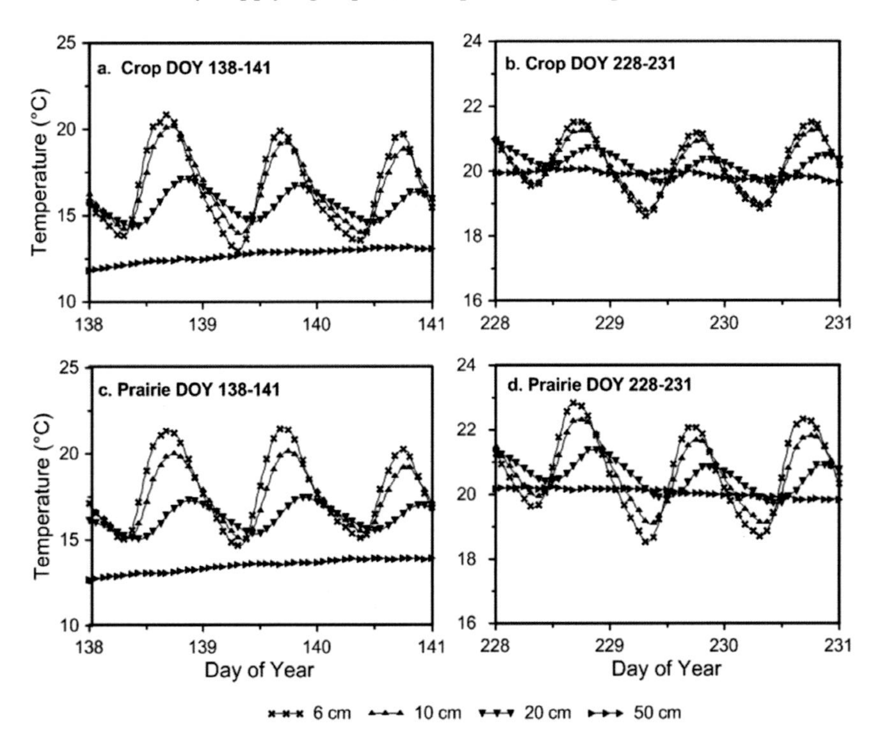

Fig. 1. Soil temperature measured with thermocouples at 6, 10, 20, and 50 cm depths beneath a corn canopy in May (a) and August (b) 2007 and reconstructed prairie during the same intervals (c and d) in central Iowa.

involves some challenges. Mineral type, particle size, amount of organic matter, water content, and bulk density all have a significant effect on soil thermal conductivity (Abu-Hamdeh and Reeder, 2000; Bristow, 2002; Smits et al., 2010). Although mineralogy, particle size, and organic matter content are relatively static properties, they often vary spatially horizontally and with depth. Water content and bulk density are much more dynamic properties that, although also spatially variable, vary temporally in response to wetting and drying cycles.

Soil heat flux is often considered within the context of a surface energy balance

$$R_n - G = LE + H \quad [3]$$

where R_n is the net all-wave radiation, LE is the latent heat flux, and H is the sensible heat flux, respectively (all in W m^{-2}). Each term except for R_n is defined as positive when energy transfer is away from the soil surface. Much of the solar energy absorbed by the soil during the day is emitted to the atmosphere at night through terrestrial longwave radiation. G is often the smallest component of the daily surface energy balance and has, in some cases, been ignored. However, there are often significant transfers of energy into and out of the soil during the course of a day and failure to include G in energy balance determinations, especially over short (e.g., hourly) intervals, leads to significant errors. Over an annual cycle, G exhibits a sinusoidal pattern similar to soil temperature coinciding with warming in the spring and summer and cooling in the fall and winter.

Soil heat flux is also a necessary input for many evaporation measurement and prediction techniques as the available energy ($R_n - G$) is partitioned between the turbulent fluxes ($LE + H$). For instance, evaporation measured with the Bowen ratio energy balance approach (Bowen, 1926), requires accurate $R_n - G$ to prevent large errors (Malek, 1993; de Silans et al., 1997). The commonly used Penman–Monteith (Penman, 1948; Monteith, 1965; Allen et al., 1998) and Priestley–Taylor (Priestley and Taylor, 1972) approaches for predicting evapotranspiration both require available energy as an input.

Equation [3] is often applied at the field scale, but the study of global climate change has led to some efforts to estimate the annual mean energy budget for the entire planet. Figure 2 shows one estimate of the annual mean energy budget of the earth for the period from March 2000 to May 2004 (Trenberth et al., 2009). The term "global warming" for terrestrial surfaces is essentially an acknowledgment of increasing G due to greenhouse gas-induced changes in shortwave and longwave transfer in the upper atmosphere (Trenberth and Stepaniak, 2004). Hansen et al. (2005) estimated that the global energy imbalance meant that the Earth was currently absorbing 0.85 W m^{-2} more energy from the sun that was being emitted to space. Huang (2006) used meteorological records to estimate annual heat budgets for the continental land masses except Antarctica from 1851 to 2004 and also concluded that there has been recent intensified heating.

Another method of climate reconstruction utilizes analysis of bedrock temperature profiles measured at high resolution (0.01 °C) at ~10 m intervals inside boreholes hundreds of meters into the earth (Pollack and Huang, 2000). Using this approach, Beltrami et al. (2000) estimated the average soil heat flux into the ground in eastern Canada over the last 1000 yr was approximately 2.8 mW m^{-2}. However, a significant change has occurred recently with the average G over the last 100 yr of 74.0 mW m^{-2}. Global analyses of data from hundreds of boreholes found a ~ 1 °C increase in temperature over the last 500 yr and 0.5 °C in the 20th century alone (Pollack et al., 1998; Huang et al., 2000). These interpretations of soil temperature and G data demonstrate the slow yet measurable changes in terrestrial heat storage that have occurred in the last several centuries.

Measurement of Soil Temperature and Thermal Properties

Soil temperature is most often measured using thermocouples with some measurements made with thermistors or resistance temperature detectors (RTDs). Thermocouples are made from wires composed of two dissimilar metals, often one or more of which is an alloy, that when joined in a closed circuit produce an electromotive force (emf) that is unique to the metal pair (couple) and proportional to the temperature difference between the two wire junctions. This Seebeck voltage (mV or μV °C^{-1}) can be accurately monitored with modern dataloggers and converted to temperature using an independent reference temperature provided by the datalogger. Accuracy of thermocouple temperature measurements is ~ 0.75% of the measured temperature. Thermistors and RTDs utilize changes in electrical resistance with temperature for resistors or metals like platinum, copper, and nickel (McInnes, 2002). Bridge circuits are used to measure the change in resistance of the sensor elements.

Thermocouples, thermistors, and RTDs all need to be enclosed in a waterproof material before being inserted into the soil. Typically, a trench is excavated to the desired depth of measurement, the sensors carefully inserted into the side wall, and the trench backfilled without damage to the sensors and the wires that connect them to a datalogger. Although not as sensitive as thermistors or RTDs, thermocouples have sufficient accuracy for most applications, are affordable, and are durable with an acceptable level of failure under field conditions. Thermistors or RTDs are more expensive and stable, but less durable; they are preferred when greater accuracy is required (0.5% of measured temperature). Such applications

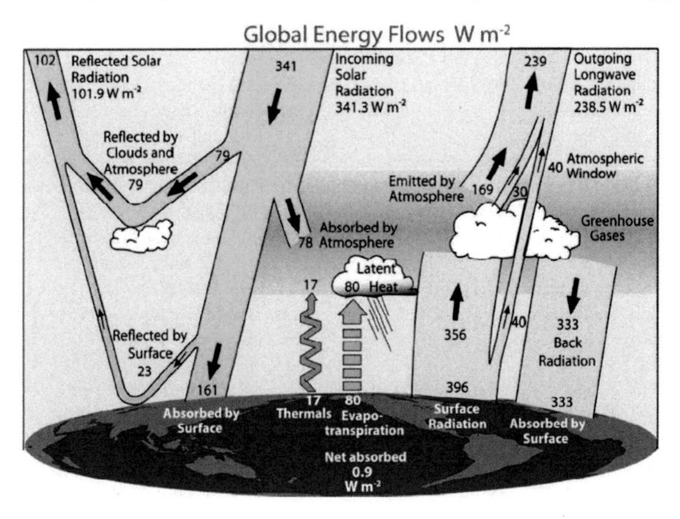

Fig. 2. Global annual mean energy budget from March 2000 to May 2004 from Trenberth et al. (2009). American Meteorological Society. Used with permission.

include temperature change over short time intervals and/or small depth increments, especially near the soil surface (Kool et al., 2016).

Soil thermal conductivity (l_s) is a physical property relating to how quickly the soil can conduct thermal energy (units of W m^{-1} K^{-1}) while the heat capacity of soil (C_s) is defined as the amount of energy required to raise a unit volume of soil one degree Kelvin (units of MJ m^{-3} K^{-1}). Values of l_s vary significantly with soil water content owing to the large difference in λ between water and air (0.57 vs. 0.025 W m^{-1} K^{-1} at 283K, respectively). Due to changes in water content, l_s values in the field can change by a factor of 2 to 5. Such large changes can occur over short intervals at shallow depths where the soil water content changes rapidly due to wetting by infiltration of precipitation or irrigation water or drying by evaporation and plant water uptake (Abu-Hamdeh and Reeder, 2000; Abu-Hamdeh, 2003). C_s is also strongly influenced by soil water content as C values for the solid, water, and air fractions of soil are 2.1 to 2.5, 4.2, and 0.0012 MJ m^{-3} K^{-1}, respectively. de Vries (1963) provided a widely-employed technique for estimating C_s based on the volume fractions of the mineral constituents and water and their respective heat capacities

$$C_s = 2.35\,(1-f) + 4.18\,\theta \qquad [4]$$

where f is the porosity and θ is the volumetric water content (m^3 m^{-3}).

Measurement of soil thermal properties in the laboratory and field are now commonly accomplished using the dual-probe heat-pulse technique (Bristow et al., 1994; Ochsner et al., 2001; Bristow et al., 2001; Smits et al., 2010). This technique is based on a solution of de Vries (1952) for the change in temperature as a function of time at a radial distance from the heat pulse source

$$T(r,t)=\frac{q}{4\pi\alpha C_s}\left[\mathrm{Ei}\left(\frac{-r^2}{4\alpha(t-t_0)}\right)-\mathrm{Ei}\left(\frac{-r^2}{4\alpha t}\right)\right] \quad t > t_0 \qquad [5]$$

where r is the radial distance (m), t is time (s), t_0 is the duration of the heat pulse (s), q is the quantity of energy applied (W m^{-1}), α is the soil thermal diffusivity equal to l_s/C_s (m^2 s^{-1}), and $Ei(-x)$ is the exponential integral (Ochsner et al., 2001). Welch et al. (1996) presented a nonlinear regression technique to fit an analytical solution of Eq. [5] to the temperature increase versus time data from the sensor that produces estimates of C_s, α, and l_s. This general approach is now widely used with variations in sensor design and data analysis techniques.

Measurement of Soil Heat Flux

The most common G measurement technique is the flux plate method (Sauer, 2002; Sauer and Horton, 2005). Soil heat flux plates, which are sometimes referred to as heat flow meters or heat flow transducers, are small disc-shaped sensors that are installed horizontally in the soil at a depth of 1 to 10 cm. This method is popular because the plates are relatively inexpensive and easy to use. Most flux plates contain multiple thermocouples wired in series (a thermopile) that produces an emf proportional to the temperature difference across the plate. The voltage signal from a calibrated plate of known and fixed λ can then be used to measure heat flow through the plate and infer G for the surrounding soil.

Although flux plates are simple to use, there are well known shortcomings of this approach. A solid plate of constant λ placed near the soil surface will distort heat flow when the plate λ is significantly different from l_s. Philip (1961) recognized this problem and proposed design criteria to limit errors by plate measurements of *G*. His equation for the ratio of heat flow through a plate to heat flow in the soil is:

$$G_m/G=\varepsilon/[1+(\varepsilon-1)H] \quad [6]$$

where G_m is the heat flux through the plate, ε is the ratio of the plate (l_m) to the soil thermal conductivity (l_m/l_s), and *H* is an empirical factor:

$$H=1-(\beta b) \quad [7]$$

where β is a dimensionless geometric constant that depends on plate shape and *b* is the plate thickness divided by side length (square plate) or plate thickness divided by plate diameter (circular plate). Figure 3 illustrates the relative performance of commercial heat flux plates having different dimensions and l_m on *G* measured at a depth of 6 cm beneath a maize canopy near Ames, IA. For commercially available flux plates, Eq. [6] predicts that as the difference between l_m and l_s increases, measurement errors up to ~ 50% in *G* can occur. Thermal conductivity for mineral soils ranges from ~ 0.4 to 1.2 W m^{-1} K^{-1} under typical field conditions. Commercially available plates have l_m from 0.2 to 1.3 W m^{-1} K^{-1}. Using

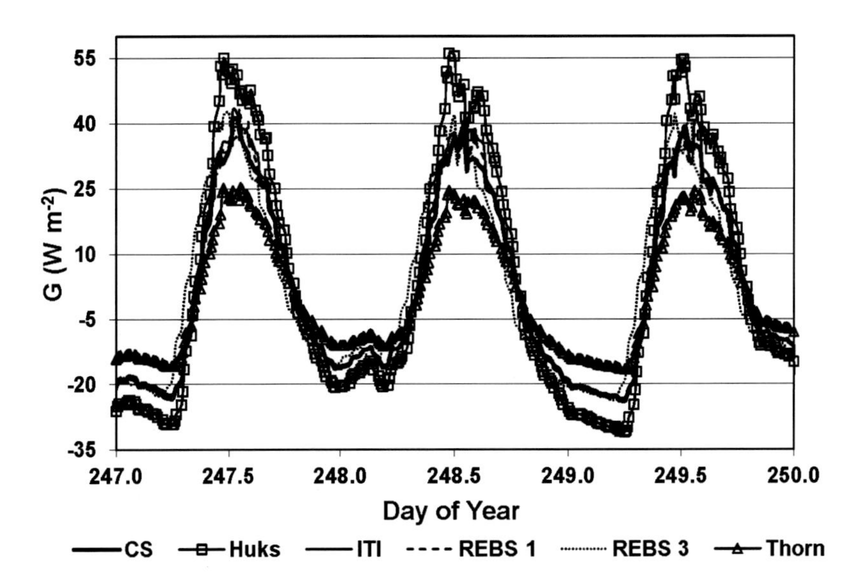

Fig. 3. Soil heat flux (G) measured with flux plates of different designs at 6 cm depth beneath a maize canopy near Ames, IA. CS = Carter-Scott Manufacturing† model CN3, Huks = Hukseflux Thermal Sensors model HFP01SC, ITI = International Thermal Instrument model GHT-1C, REBS-1 = Radiation and Energy Balance Systems model HFT-1, REBS-3 = Radiation and Energy Balance Systems model HFT-3, and Thorn = C. W. Thornthwaite Associates model 610.

a flux plate with l_m similar to the expected l_s will reduce errors associated with heat flow divergence, however, especially if there are large changes in soil water content then relatively large errors in plate estimates of G are inevitable.

Sauer et al. (2003) completed a series of laboratory experiments with six different soil heat flux plates and found that all plates routinely underestimated G in wet and dry sand. Underestimates ranged from 2.4 to 38.5% in saturated sand and from 13.1 to 73.2% in dry sand. Application of the Philip correction generally improved the G estimates. Field tests of several of the same plate designs again produced underestimates of G from 18 to 66% (Ochsner et al., 2006). Poor plate performance was attributed to low l_m compared to l_s, thermal contact resistance, and latent heat transfer effects.

Fuchs and Hadas (1973) identified thermal contact resistance as a potential source of error with flux plates. Soil texture, structure, water content, and plate installation procedure can all affect thermal contact between the plate caused by air gaps between the sensor and the soil particles. Sauer et al. (2007) addressed thermal contact resistance effects on plate performance by creating air gaps of known dimensions on the plate surface and by coating the plates with a thermal heat sink compound to reduce contact resistance. The authors concluded that errors associated with thermal contact resistance were < 10% in moist, medium-textured soils but were greater in dry sands.

Soil heat flux is also measured by gradient and calorimetric techniques that include measurement of soil temperature and thermal conductivity or heat capacity. In addition to the discussion above, detailed descriptions of techniques for measurement and estimation of soil thermal properties can be found in McInnes (2002), Bristow (2002), and Kluitenberg (2002). Farouki (1986) also provides a comprehensive review of soil thermal properties including measurement techniques and data for a large number of soils.

Of the other methods that have been used for G measurement, only the gradient method has been widely used (Jackson and Taylor, 1965; Fuchs, 1986; Sauer, 2002; Ochsner et al., 2006; Heitman et al., 2010). This method is an application of Eq. [2] with soil temperature measured at two or more depths that is combined with l_s measured in the soil between the depths of temperature measurements. The gradient method is conceptually simple, and with the significant improvement in in situ measurement of l_s with the dual-probe heat-pulse technique, is now a viable alternative to the plate method.

Heat flux plate or gradient measurements of G at the depth of placement often need to be supplemented with data on the amount of heat stored in the soil above the measurement depth to obtain G at the soil surface (G_0) needed for the surface energy balance (Eq. [3]). The calorimetric method is often used to determine the change in heat storage in the soil above the plate during the measurement interval (Fuchs and Tanner, 1968; Sauer, 2002; Sauer and Horton, 2005). Errors as large as 80 W m^{-2} in daytime, half-hour average G_0 in a sugarcane field in Australia occurred when the change in heat storage was ignored (Mayocchi and Bristow, 1995). Ochsner et al. (2007) provides a thorough analysis of methods and recommendations for measurement of heat storage in soil near the surface to obtain accurate G_0 values.

Estimation and Prediction of G

The importance of G to surface energy balance and evaporation investigations has led to the development of multiple prediction techniques for use when measured values are unavailable. These techniques rely on surrogate micrometeorological data, remote sensing parameters, or soil thermal properties. One of the simplest approaches involves the ratio of G to R_n. This ratio is an indicator of the relative proportion of available energy that is used to warm the soil. During spring, G/R_n for the crop surface changes as the soil transitions from wet and cold with no plant canopy to warm and moist with increasing shade provided by the growing crop. Soil heat flux typically represents 1 to 10% of R_n for growing crops (Sauer and Horton, 2005). G/R_n can exceed 50% in the fall or spring when R_n is low and the soil is cooling/warming or in arid climates when there is little vegetative cover.

Soil heat flux during a tillage trial in Minnesota was measured with three soil heat flux plates (HFT-1, Radiation and Energy Balance Systems, Seattle, WA) placed in the crop row at 0.05 m depth (Sauer and Eash, unpublished data, 2001). Net radiation was measured with a net radiometer (Q*6, Radiation and Energy Balance Systems, Seattle, WA) installed 1.5 m above the soil surface. For the 2000 growing season, significant differences in G_0/R_n were measured between each tillage system with mean values for the entire measurement interval of 0.23, 0.20, and 0.17 for no till, chisel, and strip till, respectively (Fig. 4). The 2000 G_0/R_n values, however, showed no trend with time after planting of the corn. Note that values in Fig. 4 are only for daylight hours ($R_n > +100$ W m^{-2}) on days without rain. In contrast to the 2000 data, in 2001 there were no significant differences between tillage systems, but a clear decrease in G_0/R_n with time was observed. Mean values of G_0/R_n for the strip till and chisel tillage systems decreased from 0.21 to 0.10 from 2 to 56 d after planting.

The significantly higher G_0/R_n for no till in 2000 was due to lower R_n observed for the no till system (daily average 9.47 MJ m^{-2} compared to 9.79 and 9.90 for strip till and chisel, respectively). The no till soil was the coolest of the three tillage systems, which would reduce the terrestrial longwave radiation and therefore reduce R_n. This system also had the greatest residue cover (78%), which would increase the amount of reflected shortwave radiation and also reduce R_n. These relatively small differences along with slightly higher daily average G_0 (0.68 MJ m^{-2} vs. 0.45 and 0.66 for strip till and chisel, respectively) combined to produce the significantly greater G_0/R_n in 2000. Although the average daily values of G_0/R_n for all tillage systems in 2000 are comparable to those observed in 2001, it is not clear why the ratio failed to decrease with canopy growth as was observed in 2001.

Sauer and Horton (2005) provide a summary table of example values of G/R_n for a variety of agricultural surfaces. G/R_n is relatively high (> ~0.15) for bare soils and sparse canopies and lower for fully-developed crop canopies and forests, which have greater attenuation of R_n within the canopy. The ratio has been found to be sensitive to changing soil water content and canopy density (Idso et al., 1975; Clothier et al., 1986; Oliver et al., 1987), thus using G/R_n ratios to estimate G are primarily useful only as a first approximation.

Several more sophisticated approaches to estimating G include Daughtry et al. (1990) and Kustas and Daughtry (1990) who extended the work of Clothier et al. (1986) to utilize multispectral reflectance data to determine an index based on the difference between the near infrared and red reflectance divided by their sum. Use

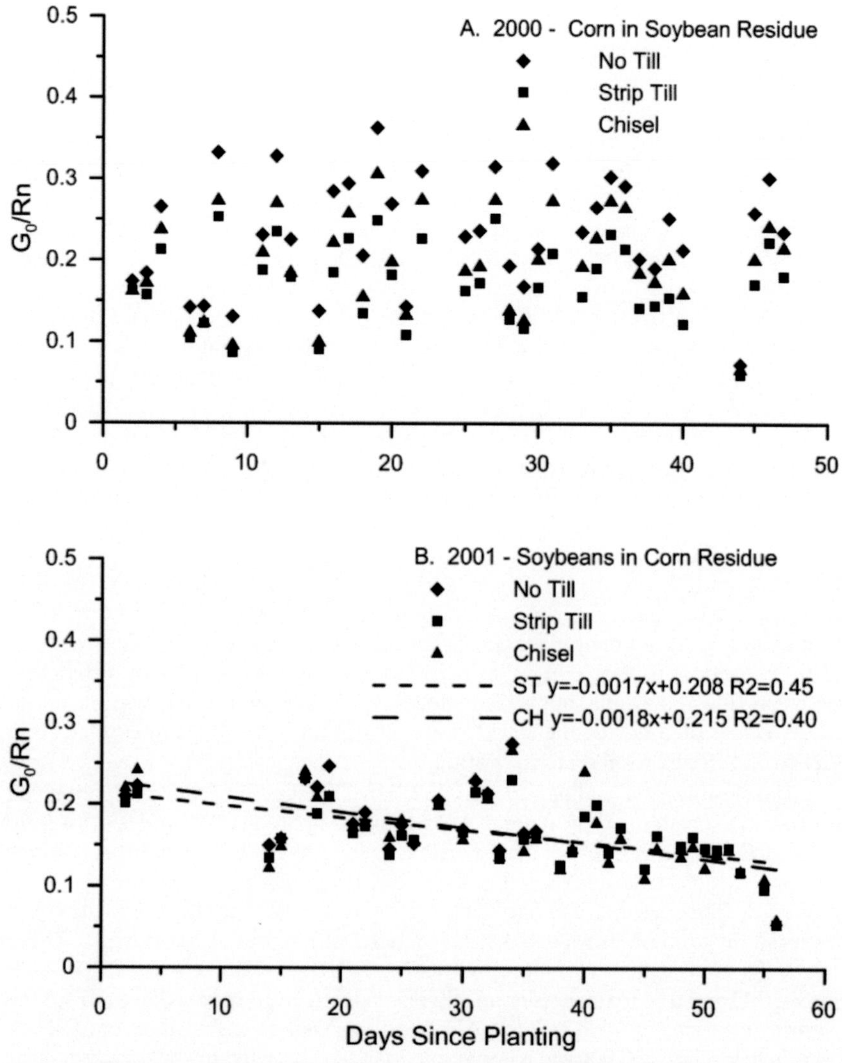

Fig. 4. Daily average G_0/R_n for three tillage systems following planting of corn into soybean residue (2000) and soybean into corn residue (2001) (Sauer and Eash, unpublished data).

of the normalized difference vegetation index (NDVI) resulted in improved estimates of G from R_n for fields with bare soil and crop canopies at different growth stages. Kustas et al. (1993) further explored nonlinear relationships between G/R_n and vegetation indices (VIs) and concluded that a power function was more appropriate than the previously derived linear relationships between G/R_n ratios

$$G/R_n = aVI^b \quad [8]$$

where a and b are fitted constants.

Horton and Wierenga (1983) developed a harmonic analysis of soil temperature at one depth to estimate *G*, if α is known, or at two depths if α is unknown. Sharratt et al. (1992) estimated *G* from hourly soil temperature data at three depths using a finite-difference solution to the transient heat flow equation. Wang and Bras (1999) established an explicit relationship between surface temperature and soil heat flux by applying fractional calculus. The *G* estimated with their technique compared well with field data from Kansas, United States and from Brazil although they assumed that the soil thermal properties were uniform and independent of soil water content and temperature.

Soil Temperature and Heat Flux Inarable Cropping Systems

Among agricultural management practices, irrigation, drainage, tillage, and residue management are among those that strongly affect the soil thermal regime (Al-Darby and Lowery, 1987; Bristow, 1988; Unger, 1988). Tillage loosens the surface soil layers, thereby lowering their bulk density and resulting in lower l_s. Reduced *G* has been observed in tilled soil compared to no-tilled and non-compacted soil (Sauer et al., 1996; Azooz et al., 1997; Richard and Cellier, 1998). Dry crop residue has a low λ and, whether on the soil surface or incorporated into the soil by tillage, can inhibit heat transfer when present in sufficient quantity. Crop residue on the soil surface also has shortwave reflectivity that is higher than most soils and provides a barrier to water vapor transfer (Gausman et al., 1975; Horton et al., 1996; Sauer et al., 1997). Soils with significant residue cover tend to have higher water contents, lower temperatures, and reduced *G*. These effects on the soil thermal regime have pronounced implications for the surface energy balance and evaporation.

Figure 5 presents soil temperature data from the tillage trial in Minnesota for a period after planting of corn or soybean crops. Soil temperature was measured with 20 0.005-m diam. Type T (copper-constantan) thermocouples placed directly in the crop row at 0.05 m depth. The thermocouples were placed in a 5 × 4 grid pattern by installing four thermocouples 0.76 m apart in five rows near the center of each research plot. Average in-row soil temperature was very uniform for the three tillage systems in 2000 when corn was grown following soybean. Daily mean soil temperatures differed by < 1 °C on every day during the measurement interval and showed no tendency toward any separation with time or soil drying. In 2001, with soybean following corn, soil temperature for the chisel system was consistently ~ 1 °C warmer than the no-till and strip-till systems. This temperature difference tended to increase with time and soil drying with a maximum difference of 2.45 °C between chisel and no till on Day 190.

Over twice as much precipitation fell during the 2000 measurement interval (199 mm) as compared with 2001 (94 mm), which resulted in more moist soil conditions and suppressed soil warming. In 2001, most of the precipitation fell within 30 d of planting, after which soil temperatures increased and soil water content steadily decreased. Daily average, maximum, or minimum soil temperature by tillage system were not significantly different in either year. Daily temperature range (max - min) for the chisel system was significantly higher than strip till in 2001 with no till intermediate (mean values for the entire measurement interval were 6.39, 5.29, and 5.96 °C, respectively).

The very small differences between tillage systems for soil temperature in 2000 are likely the result of similar in-row conditions when corn was planted into soybean residue. With the low amount of residue on the surface even for the no

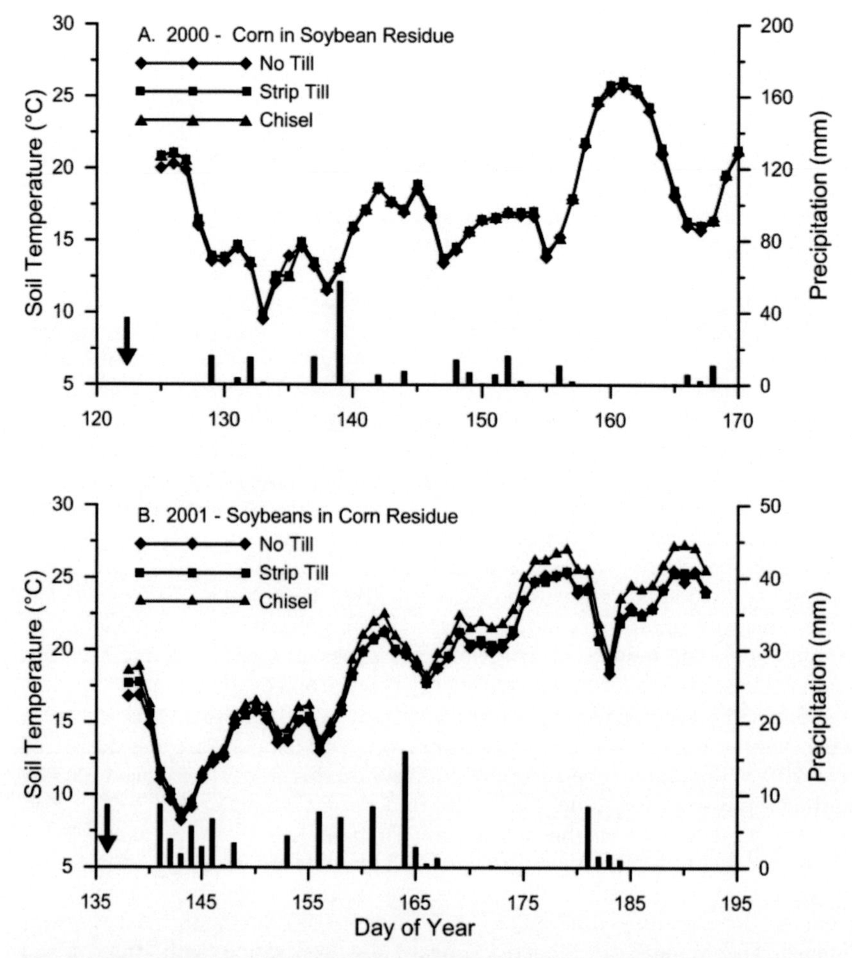

Fig. 5. Daily average soil temperature at 0.05 m and precipitation for three tillage systems following planting of corn into soybean residue (2000) and soybean into corn residue (2001) (Sauer and Eash, unpublished data).

till and strip till systems and similar clearing of residue from the row area during planting, soil temperature in the seed zone is remarkably similar for the three tillage systems. In 2001, the chisel system produced higher soil temperatures which, although not statistically significant, were clearly evident and consistent. These findings can be interpreted to indicate that the greater amount of corn residue on the soil surface in the interrow area in the no till and strip till systems was not suppressing intrarow seed zone soil temperature.

For agricultural systems, the properties of the land surface and crop canopy affect the partitioning of incident radiation and the magnitude of energy that is used to warm the soil, warm the aboveground air, or evaporate water. The intricate relationships between soil and air temperature, soil water content, canopy characteristics, residue cover, and wind speed can have significant impact on the direction and magnitude of G. Daytime peak hourly values of G_0 for a bare, dry soil in midsummer, could be in excess of 300 W m^{-2} (Fuchs and Hadas, 1972; Rao et al., 1977; Enz et al., 1988). Hourly G_0 for a moist soil beneath a plant canopy, residue layer, or snow cover can be less than 20 W m^{-2}.

Soil heat flux data and interpretations are available for wide variety of surfaces including grasslands (Rosset et al., 1997; Bremer and Ham, 1999; Twine et al., 2000; Jacobs et al., 2011), vineyards and orchards (Fritton et al., 1976; Fritton and Martsolf, 1980; Glenn and Welker, 1987; Heilman et al., 1994), forests (McCaughey and Saxton, 1988;Tamai et al., 1998; Wilson et al., 2000; Ogée et al., 2001), small grains (Lourence and Pruitt, 1971; Choudhury et al., 1987; Kimball et al., 1999), and row crops with varying canopy cover (Brown and Covey, 1966; Ham et al., 1991; Ham and Kluitenberg, 1993; Sauer et al., 1998; Hernandez-Ramirez et al., 2009). Additional analyses are available for non-crop surfaces including G beneath reclining sheep (Gatenby, 1977), sparse or desert canopies (Tuzet et al., 1997; Verhoef et al., 1999; Kustas et al., 2000), or on sloping terrain (Oliver, 1992).

Soil Temperature and Heat Flux in Paddy Rice Systems

Many surface energy balance studies have focused on cereal crops, such as wheat (*Triticum* spp.) and maize, which dominate both the United States and European agricultural sectors. Surface energy balance studies of rice (*Oryza sativa* L.), another of the world's major crops, are much fewer than for arable cropping systems. Rice feeds over half of the world population and about 90% of the world's rice is grown and consumed in Asia (USDA-FAS, 2002). European rice production is relatively modest (about 3 million tons per year) and Italy is the leading producer with over half of the total production. Traditional paddy rice occupies 54% of world rice area (IRRI, 2009) and consumed about 45% of the total freshwater for irrigation agriculture in Asia. The demands of increasing rice production (to meet the food demand of the exploding population) and of improving rice water productivity (to meet the growing water scarcity in agriculture) has led to increased interest in surface energy balance studies for rice agricultural systems.

Paddy rice production, with periods of flooding with shallow water ponding on the soil surface and saturated soils, provides more complicated surface energy balance conditions. Paddy rice fields sometimes have the additional feature of the energy storage in the paddy water layer (ΔS_w) above the soil surface. In paddy rice systems, G is observed to be the smallest surface energy balance component on a daily basis. Harazono et al. (1998) found that daytime average G_0 accounted for approximately 8% of the R_n under drained conditions over a rice canopy in central Japan. Tsai et al. (2007) conducted an energy balance study over a flooded rice paddy in the Taichung basin in central Taiwan and found the daily average G at 8 cm was less than 1% of the daily R_n. Maruyama and Kuwagata (2010) observed G at the water surface to be approximately 10% of the daily R_n over flooded rice fields in representative rice-growing areas in Japan. Hossen et al. (2012) observed the diurnal and seasonal variability of G at the soil or water surface (when flooded) in an irrigated and rain-fed double-cropping paddy field in Bangladesh. The diurnal course of half-hourly G (ranging from about −50 to 100 W m^{-2}) suggested G was a minor energy-balance component. This diurnal pattern was observed in both flooded cropping periods and summer fallow periods (with a flooded field and ratoon crop and weeds). The energy partitioning was different in the winter fallow period, when R_n was equally partitioned to G and other components under bare and dry field conditions.

Short-term (i.e., hourly or half-hourly) G could be large and represent a large proportion of the R_n over rice paddy fields. Masseroni et al. (2014) reports half-hour G measurements at the soil or water surface in both continuous flooded and intermittent-irrigated rice fields from 5 July to 10 July 2013. In the flooded field, daily peak G values ranged from 200 to 300 W m^{-2} (representing about 35 to 45% of the corresponding peak R_n). The intermittently irrigated field had smaller daily peak G values (from about 170 to 200 W m^{-2}), which accounted for about 25 to 30% of the R_n. During the nighttime, G was of smaller magnitude but was the dominant energy contributor to the surface.

Harazono et al. (1998) observed that for a rice field with a standing water layer of about 0.05 m, ΔS_w was as large as G_0 and the ratio of daytime average G_0 to R_n increased to about 12% after adding the ΔS_w. Tsai et al. (2007) found that the average daytime G at 8 cm represented 26% of the average R_n and the average ΔS_w was 5% of R_n. Hossen et al. (2012) reported that the energy balance closure in the flooded period over a double-cropping paddy field was improved by 5% by considering the ΔS_w. Masseroni et al. (2014) observed the ΔS_w in a flooded rice field ranged from about -100 to 150 W m^{-2} in the early rice growing period and there was a phase shift between the daily peak G (measured with heat flux plates 8 cm under the soil surface) and R_n.

Irrigation pattern and growth stage of the rice affect the surface energy partitioning and microclimate (e.g., soil and standing water temperatures) over a rice paddy field. Alberto et al. (2009) compared the seasonal variations of soil temperature (at 5-cm below the soil surface) between a flooded and a non-flooded rice field in the Philippines. With a standing water layer 3 to 5 cm deep, daytime soil temperature in the flooded field was significantly lower than that of the non-flooded field and the reverse relationship was observed at night. Similar results were obtained by Miyata et al. (2000) in an intermittently flooded paddy field. This is due to the greater C_s in a flooded field, which makes it warm and cool more slowly than a non-flooded field. When the flooded field was drained, the daytime soil temperature increased faster and was greater than that of the non-flooded field. This is because the soil in the flooded field is darker and contains reduced compounds with low C_s (compared with its oxidized species in non-flooded field (Dean, 1999). The flooded field absorbed more solar radiation and had relatively higher and faster increases in soil temperature during the daytime.

Maruyama and Kuwagata (2010) observed that the variation of daily average G at the soil or water surface was reduced with rice growth from -25 to 25 W m^{-2} to about -10 to 10 W m^{-2}. Similarly, the difference between the standing water layer temperature (T_w) and air temperature (T_a) reduced with rice growth gradually from 2 to 3 °C ($T_w > T_a$ just after transplanting) to a negligible amount before maturity, which was mainly attributed to the decreased solar radiation transmissivity of the rice canopy with increased leaf area index (LAI). Jia et al. (2010) studied the surface energy balance of a rice paddy in northeast China and observed the ratio of daily mean G at 4.5 cm to R_n to decrease with the rice growth (from 0.14 before transplanting to -0.08 immediately after harvest). This was attributed to the increased LAI and canopy conductance and consequently increased incident energy partitioning into LE rather than G. Masseroni et al. (2014) found that in the early rice growing period (LAI < 2), half-hourly G at the soil surface in a non-flooded field ranged from -100 to 200 W m^{-2}, which is relatively smaller than G at the water surface in a flooded field (range -150 to 250 W m^{-2}). In the late rice growing period (LAI > 4) with extended vegetative cover, variation of G in both rice fields was reduced to a similar range of about -100 to 120 W m^{-2}.

To summarize previous studies of surface energy balance in rice fields:

1) On a daily basis, G is observed to be the smallest surface energy balance component, typically representing from 1 to 10% of R_n. Energy input by R_n was mostly consumed by LE with a LE/R_n ratio of about 0.6 to 0.8. Short-term (i.e., half-hourly) daytime G could be in excess of 250–300 W m^{-2} and represent nearly half of the R_n.
2) For flooded rice fields, ΔS_w was observed to be non-negligible. Considering ΔS_w can help improve energy balance closure by better accounting of energy storage and alignment of peak G_0 and R_n. This is a unique characteristic of surface energy balance studies for paddy rice systems.
3) Irrigation pattern and growth stage of rice have a large influence on surface energy partitioning and microclimate in rice paddy fields. Flooded rice fields typically have higher daytime G at the water surface and lower soil temperature than non-flooded fields. With rice growth, more incident energy is partitioned to LE instead of G or H, and thus, decreased variability in G/R_n and G is observed.

A critical component in previous studies for rice agricultural systems is the lack of information on heat storage changes between the soil surface and the G measurement depth (ΔS_s) when calculating the G_0 from G. Failure to include this storage term may result in inaccurate G_0, and thus, poor surface energy balance closure. Ochsner et al. (2007) showed that when the reference depth is sufficiently deep to permit accurate heat flux measurements, heat storage is too large to neglect. As for the flooded rice field with a standing water layer above the soil surface, high water content in the near-surface soil layer can lead to large soil C_s and thus large ΔS_s. Tsai et al. (2007) showed a daytime average ΔS_s of 40 W m^{-2} (representing 12% of the R_n) over a flooded rice field. However, many surface energy balance studies performed in flooded rice fields ignored ΔS_s when calculating G (Harazono et al., 1998; Maruyama and Kuwagata, 2010). Hossen et al. (2012) did not consider the heat storage between the soil surface and soil heat flux plate placement depth (5 cm), and they suggested this omission contributed to the low surface energy balance closure (about 75%) over a flooded rice field.

Summary

Climate near the ground is influenced by the coupling between surface conditions and large-scale meteorological forces. Crop growth and production are strongly affected by the regional climate while energy exchange within and above the crop canopy influences the local microclimate. The soil surface is the pivotal interface between surface and subsurface energy transfer processes. Soil management practices can affect the climate near the surface as soil properties affect the partitioning of incoming energy between warming the soil, evapotranspiration, and warming of the air above the ground surface. This energy partitioning and the soil thermal regime affect soil temperature, which is a key factor affecting the rate of all biological, chemical, and physical processes essential for plant growth and the provisioning of ecosystem services.

Global climate change effects on temperature and precipitation patterns have important impacts on surface soil conditions that in turn have significant implications for crop growth and food security. At the global scale, the key role of soil in agricultural systems within the context of the water, food, and energy nexus continues to be an active area of discussion. Greater understanding of soil temperature and heat flux in different cropping systems enables development of crop and soil management practices to adapt to a changing climate. Continued improvement in measurement techniques and models of soil heat flux is needed to provide the knowledge and data necessary to optimize soil management for future farming systems.

References

Abu-Hamdeh, N.H. 2003. Thermal properties of soils as affected by density and water content. Biosystems Eng. 86:97–102. doi:10.1016/S1537-5110(03)00112-0

Abu-Hamdeh, N.H., and R.C. Reeder. 2000. Soil thermal conductivity: Effects of density, moisture, salt concentration, and organic matter. Soil Sci. Soc. Am. J. 64:1285–1290. doi:10.2136/sssaj2000.6441285x

Alberto, M.C.R., R. Wassmann, T. Hirano, A. Miyata, A. Kumar, A. Padre, and M. Amante. 2009. CO2/heat fluxes in rice fields: Comparative assessment of flooded and non-flooded fields in the Philippines. Agric. For. Meteorol. 149:1737–1750. doi:10.1016/j.agrformet.2009.06.003

Al-Darby, A.M., and B. Lowery. 1987. Seed zone soil temperature and early corn growth with three conservation tillage systems. Soil Sci. Soc. Am. J. 51:768–774. doi:10.2136/sssaj1987.03615995005100030035x

Allen, R.G., L.S. Pereira, D. Raes, and M. Smith. 1998. Crop evapotranspiration– guidelines for computing crop water requirements. Irrigation and Drainage Paper 56. Food and Agriculture Organization of the United Nations, Rome.

Azooz, R.H., B. Lowery, T.C. Daniel, and M.A. Arshad. 1997. Impact of tillage and residue management on soil heat flux. Agric. For. Meteorol. 84:207–222. doi:10.1016/S0168-1923(96)02364-7

Beltrami, H., J. Wang, and R.L. Bras. 2000. Energy balance at the earth's surface: Heat flux history in eastern Canada. Geophys. Res. Lett. 27:3385–3388. doi:10.1029/2000GL008483

Bita, C.E., and T. Gerats. 2013. Plant tolerance to high temperature in a changing environment: Scientific fundamentals and production of heat stress-tolerant crops. Front. Plant Sci. doi:10.3389/fpls.2013.00273

Bowen, I.S. 1926. The ratio of heat losses by conduction and by evaporation from any water surface. Phys. Rev. 27:779–787. doi:10.1103/PhysRev.27.779

Bremer, D.J., and J.M. Ham. 1999. Effect of spring burning on the surface energy balance in a tallgrass prairie. Agric. For. Meteorol. 97:43–54. doi:10.1016/S0168-1923(99)00034-9

Bristow, K.L. 1988. The role of mulch and its architecture in modifying soil temperature. Aust. J. Soil Res. 26:269–280. doi:10.1071/SR9880269

Bristow, K.L. 2002. Thermal conductivity. In: G.C. Topp and J.H. Dane, editors, Methods of Soil Analysis. Part 4. Physical Methods. SSSA Book Series, no. 5. SSSA, Madison, WI. p. 1209–1226.

Bristow, K.L., G.J. Kluitenberg, C.J. Goding, and T.S. Fitzgerald. 2001. A small multi-needle probe for measuring soil thermal properties, water content and electrical conductivity. Comput. Electron. Agric. 31:265–280. doi:10.1016/S0168-1699(00)00186-1

Bristow, K.L., G.J. Kluitenberg, and R. Horton. 1994. Measurement of soil thermal properties with a dual-probe heat-pulse technique. Soil Sci. Soc. Am. J. 58:1288–1294. doi:10.2136/sssaj1994.03615995005800050002x

Brown, K.W., and W. Covey. 1966. The energy-budget evaluation of the micro-meteorological transfer processes within a cornfield. Agric. Meteorol. 3:73–96. doi:10.1016/0002-1571(66)90006-9

Choudhury, B.J., S.B. Idso, and J.R. Reginato. 1987. Analysis of an empirical model for soil heat flux under a growing wheat crop for estimating evaporation by an infrared-temperature based energy balance equation. Agric. For. Meteorol. 39:283–297. doi:10.1016/0168-1923(87)90021-9

Clothier, B. E., K. L. Clawson, P. J. Pinter, Jr., M. S. Moran, R. J. Reginato, and R.D. Jackson. 1986. Estimation of soil heat flux from net radiation during the growth of alfalfa. Agric. For. Meteorol. 37:319-329.

Daughtry, C.S.T., W.P. Kustas, M.S. Moran, P.J. Pinter, Jr., R.D. Jackson, P.W. Brown, W.D. Nichols, and L.W. Gay. 1990. Spectral estimates of net radiation and soil heat flux. Remote Sens. Environ. 32:111–124. doi:10.1016/0034-4257(90)90012-B

Dean, J.A. 1999. Thermodynamic properties: Enthalpies and Gibbs (free) energies of formation, entropies and heat capacities of elements and compounds. In: J. Spaight, editor, Lange's handbook of chemistry, 15th ed. McGraw-Hill, Inc., Burr Ridge, IL.

de Silans, A.P., B.A. Monteny, and J.P. Lhomme. 1997. The correction of soil heat flux measurements to derive an accurate surface energy balance by the Bowen ratio method. J. Hydrol. 188-189:453–465. doi:10.1016/S0022-1694(96)03187-3

de Vries, D.A. 1952. A nonstationary method for determining thermal conductivity of soil in situ. Soil Sci. 73:83–90. doi:10.1097/00010694-195202000-00001

de Vries, D.A. 1963. Thermal properties of soils. In: W.R. van Wijk, editor, Physics of plant environment. North-Holland Publishing Co., Amsterdam. p. 210–235.

Enz, J.W., L.J. Brun, and J.K. Larsen. 1988. Evaporation and energy balance for bare and stubble covered soil. Agric. For. Meteorol. 43:59–70. doi:10.1016/0168-1923(88)90006-8

Farouki, O.T. 1986. Thermal properties of soils. Series on rock and soil mechanics Vol. 11. Trans Tech Publications, Clausthal-Zellerfield, Germany. p. 136.

Fritton, D.D., and J.D. Martsolf. 1980. Soil heat flow under and orchard heater. Soil Sci. Soc. Am. J. 44:13–16. doi:10.2136/sssaj1980.03615995004400010003x

Fritton, D.D., J.D. Martsolf, and W.J. Busscher. 1976. Spatial distribution of soil heat flux under a sour cherry tree. Soil Sci. Soc. Am. J. 40:644–647. doi:10.2136/sssaj1976.03615995004000050015x

Fuchs, M. 1986. Heat flux. p. 957-968 In: A. Klute, editor, Methods of soil analysis: Part 1, Physical and mineralogical methods. 2nd Ed. Soil Science Society of America, Book Series No. 5, Madison, WI.

Fuchs, M., and A. Hadas. 1972. The heat flux density in a non-homogeneous bare loessial soil. Boundary-Layer Meteorol. 3:191–200. doi:10.1007/BF02033918

Fuchs, M., and A. Hadas. 1973. Analysis and performance of an improved soil heat flux transducer. Soil Sci. Soc. Am. Proc. 37:173–175. doi:10.2136/sssaj1973.03615995003700020009x

Fuchs, M., and C.B. Tanner. 1968. Calibration and field test of soil heat flux plates. Soil Sci. Soc. Am. Proc. 32:326–328. doi:10.2136/sssaj1968.03615995003200030021x

Gatenby, R.M. 1977. Conduction of heat from sheep to ground. Agric. Meteorol. 18:387–400. doi:10.1016/0002-1571(77)90034-6

Gausman, H.W., A.H. Gerbermann, C.L. Wiegand, R.W. Leamer, R.R. Rodriguez, and J.R. Noriega. 1975. Reflectance differences between crop residues and bare soils. Soil Sci. Soc. Am. Proc. 39:752–755. doi:10.2136/sssaj1975.03615995003900040043x

Glenn, D.M., and W.V. Welker. 1987. Soil management effects on soil temperature and heat flux in a young peach orchard. Soil Sci. 143:372–380. doi:10.1097/00010694-198705000-00007

Ham, J.M., J.L. Heilman, and R.J. Lascano. 1991. Soil and canopy energy balances of a row crop at partial cover. Agron. J. 83:744–753. doi:10.2134/agronj1991.00021962008300040019x

Ham, J.M., and G.J. Kluitenberg. 1993. Positional variation in the soil energy balance beneath a row-crop canopy. Agric. For. Meteorol. 63:73–92. doi:10.1016/0168-1923(93)90023-B

Hansen, J., L. Nazarenko, R. Ruedy, M. Sato, J. Willis, A. Del Genio, D. Koch, A. Lacis, K. Lo, S. Menon, T. Novakov, J. Perlwitz, G. Russell, G. A. Schmidt, and N. Tausnev. 2005. Earth's energy imbalance: Confirmation and implications. Science 308:1431–1435. doi:10.1126/science.1110252

Harazono, Y., J. Kim, A. Miyata, T. Choi, J.I. Yun, and J.W. Kim. 1998. Measurement of energy budget components during the International Rice Experiment (IREX) in Japan. Hydrol. Processes 12:2081–2092. doi:10.1002/(SICI)1099-1085(19981030)12:13/14<2081::AID-HYP721>3.0.CO;2-M

Harsch, M.A., P.E. Hulme, M.S. McGlone, and R.P. Duncan. 2009. Are treelines advancing? A global meta-analysis of treeline response to climate warming. Ecol. Lett. 12:1040–1049. doi:10.1111/j.1461-0248.2009.01355.x

Hatfield, J.L., and J.H. Prueger. 2015. Temperature extremes: Effect on plant growth and development. Weather and Climate Extremes 10:4–10. doi:10.1016/j.wace.2015.08.001

Hatfield, J.L., T.J. Sauer, and R.M. Cruse. 2017. Soil: The forgotten piece of the water, food, energy nexus. Adv. Agron. 143:1–46.

Heilman, J.L., K.J. McInnes, M.J. Savage, R.W. Gesch, and R.J. Lascano. 1994. Soil and canopy energy balances in a west Texas vineyard. Agric. For. Meteorol. 71:99–114. doi:10.1016/0168-1923(94)90102-3

Heitman, J.L., R. Horton, T.J. Sauer, T.S. Ren, and X. Xiao. 2010. Latent heat in soil heat flux measurements. Agric. For. Meteorol. 150:1147–1153. doi:10.1016/j.agrformet.2010.04.017

Hernandez-Ramirez, G., J.L. Hatfield, J.H. Prueger, and T.J. Sauer. 2009. Energy balance and turbulent flux partitioning in a corn-soybean rotation in the Midwestern US. Theor. Appl. Climatol. 10.1007/s00704-009-0169-y

Horton, R., K.L. Bristow, G.J. Kluitenberg, and T.J. Sauer. 1996. Crop residue effects on surface radiation and energy balance– review. Theor. Appl. Climatol. 54:27–37. doi:10.1007/BF00863556

Horton, R., and P.J. Wierenga. 1983. Estimating the soil heat flux from observations of soil temperature near the surface. Soil Sci. Soc. Am. J. 47:14–20. doi:10.2136/sssaj1983.03615995004700010003x

Hossen, M., M. Mano, A. Miyata, M. Baten, and T. Hiyama. 2012. Surface energy partitioning and evapotranspiration over a double-cropping paddy field in Bangladesh. Hydrol. Processes 26:1311–1320. doi:10.1002/hyp.8232

Huang, S. 2006. 1851-2004 annual heat budget of the continental landmasses. Geophys. Res. Lett. 10.1029/2005GL025300

Huang, S., H.N. Pollack, and P. Shen. 2000. Temperature trends over the past five centuries reconstructed from borehole temperatures. Nature 403:756–758.

Idso, S.B., J.K. Aase, and R.D. Jackson. 1975. Net radiation-soil heat flux relations as influenced by soil water content variations. Boundary-Layer Meteorol. 9:113–122. doi:10.1007/BF00232257

IRRI. 2009. Rice ecosystems. International Rice Research Institute, Los Baños, The Philippines. http://www.ipni.net/ppiweb/filelib.nsf/0/6191D544DF714DEF48257074002E78E6/$file/Rice%20HB%20p2-5.pdf (verified 15 June 2018).

Jackson, R.D., and S.A. Taylor. 1965. Heat transfer. In: C.A. Black and D.D. Evans, editors, Methods of soil analysis: Part 1. Agronomy Monograph No. 9. American Society of Agronomy, Madison, WI. p. 349–356.

Jacobs, A.F.G., B.G. Heusinkveld, and A.A.M. Holtslag. 2011. Long-term record and analysis of soil temperatures and soil heat fluxes in a grassland area, The Netherlands. Agric. For. Meteorol. 151:774–780. doi:10.1016/j.agrformet.2011.01.002

Jia, Z.J., W. Zhang, and Y. Huang. 2010. Analysis of energy flux in rice paddy in the Sanjiang Plain. Chin. J. Eco Agric. 18:820–826. doi:10.3724/SP.J.1011.2010.00820

Jury, W.A., W.R. Gardner, and W.H. Gardner. 1991. Soil physics. 5th Ed. John Wiley & Sons, New York. p. 328.

Kang, Y., S. Khan, and X. Ma. 2009. Climate change impacts on crop yield, crop water productivity and food security– A review. Prog. Nat. Sci. 19:1665–1674. doi:10.1016/j.pnsc.2009.08.001

Kimball, B.A., R.L. LaMorte, P.J. Pinter, Jr., G.W. Wall, D.J. Hunsaker, F.J. Adamsen, S.W. Leavitt, T.L. Thompson, A.D. Matthias, and T.J. Brooks. 1999. Free-air CO2 enrichment and soil nitrogen effects on energy balance and evapotranspiration of wheat. Water Resour. Res. 35:1179–1190. doi:10.1029/1998WR900115

Kluitenberg, G.J. 2002. Heat capacity and specific heat. In: J.H. Dane and G.C. Topp, editors, Methods of soil analysis. Part 4. Physical methods. SSSA Book Series, No. 5. SSSA, Madison, WI. p. 1201–1208.

Kool, D., J.L. Heitman, N. Lazarovitch, N. Agam, T.J. Sauer, and A. Ben-Gal. 2016. In: situ thermistor calibrations for improved measurement of soil temperature gradients. Soil Sci. Soc. Am. J. 80:1514–1519. doi:10.2136/sssaj2016.05.0134

Kustas, W.P., and C.S.T. Daughtry. 1990. Estimation of the soil heat flux/net radiation ratio from spectral data. Agric. For. Meteorol. 49:205–223. doi:10.1016/0168-1923(90)90033-3

Kustas, W.P., C.S.T. Daughtry, and P.J. Van Oevelen. 1993. Analytical treatment of the relationships between soil heat flux/net radiation ratio and vegetation indices. Remote Sens. Environ. 46:319–330. doi:10.1016/0034-4257(93)90052-Y

Kustas, W.P., J.H. Prueger, J.L. Hatfield, K. Ramalingam, and L.E. Hipps. 2000. Variability in soil heat flux from a mesquite dune site. Agric. For. Meteorol. 103:249–264. doi:10.1016/S0168-1923(00)00131-3

Linderholm, H.W. 2006. Growing season changes in the last century. Agric. For. Meteorol. 137:1–14. doi:10.1016/j.agrformet.2006.03.006

Lipiec, J., C. Doussan, A. Nosalewicz, and K. Kondracka. 2013. Effect of drought and heat stresses on plant growth and yield: A review. Int. Agrophys. 27:463–477. doi:10.2478/intag-2013-0017

Lourence, F.J., and W.O. Pruitt. 1971. Energy balance and water use of rice grown in the Central Valley of California. Agron. J. 63:827–832. doi:10.2134/agronj1971.00021962006300060003x

Luo, Q. 2011. Temperature thresholds and crop production: A review. Clim. Change 109:583–598. doi:10.1007/s10584-011-0028-6

Malek, E. 1993. Rapid changes of the surface soil heat flux and its effects on the estimation of evapotranspiration. J. Hydrol. 142:89–97. doi:10.1016/0022-1694(93)90006-U

Maruyama, A., and T. Kuwagata. 2010. Coupling land surface and crop growth models to estimate the effects of changes in the growing season on energy balance and water use of rice paddies. Agric. For. Meteorol. 150:919–930. doi:10.1016/j.agrformet.2010.02.011

Masseroni, D., A. Facchi, M. Romani, E.A. Chiaradia, O. Gharsallah, and C. Gandolfi. 2014. Surface energy flux measurements in a flooded and an aerobic rice field using a single eddy-covariance system. Paddy Water Environ. 13(4):1–20.

Mayocchi, C.L., and K.L. Bristow. 1995. Soil surface heat flux: Some general questions and comments on measurements. Agric. For. Meteorol. 75:43–50. doi:10.1016/0168-1923(94)02198-S

McCaughey, J.H., and W.L. Saxton. 1988. Energy balance storage terms in a mixed forest. Agric. For. Meteorol. 44:1–18. doi:10.1016/0168-1923(88)90029-9

McInnes, K.J. 2002. Temperature. In: J.H. Dane and G.C. Topp, editors, Methods of soil analysis. Part 4. Physical methods. SSSA Book Series, no. 5. SSSA, Madison, WI. p. 1183–1199.

Miyata, A., R. Leuning, O.T. Denmead, J. Kim, and Y. Harazono. 2000. Carbon dioxide and methane fluxes from an intermittently flooded paddy field. Agric. For. Meteorol. 102:287–303. doi:10.1016/S0168-1923(00)00092-7

Monteith, J.L. 1965. Evaporation and environment. Symp. Soc. Exp. Biol. 19:205–234.

Ochsner, T.E., R. Horton, and T. Ren. 2001. A new perspective on soil thermal properties. Soil Sci. Soc. Am. J. 65:1641–1647. doi:10.2136/sssaj2001.1641

Ochsner, T.E., T.J. Sauer, and R. Horton. 2006. Field tests of the soil heat flux plate method and some alternatives. Agron. J. 98:1005–1014. doi:10.2134/agronj2005.0249

Ochsner, T.E., T.J. Sauer, and R. Horton. 2007. Soil heat storage measurements in energy balance studies. Agron. J. 99:311–319. doi:10.2134/agronj2005.0103S

Ogée, J., E. Lamaud, Y. Brunet, P. Berbigier, and J.M. Bonnefond. 2001. A long-term study of soil heat flux under a forest canopy. Agric. For. Meteorol. 106:173–186. doi:10.1016/S0168-1923(00)00214-8

Olesen, J.E., and M. Bindi. 2002. Consequences of climate change for European agricultural productivity, land use and policy. Eur. J. Agron. 16:239–262. doi:10.1016/S1161-0301(02)00004-7

Oliver, H. R. 1992. Studies of surface energy balance of sloping terrain. Int. J. Climatol. 12:55-68.

Oliver, S.A., H.R. Oliver, J.S. Wallace, and A.M. Roberts. 1987. Soil heat flux and temperature variation with vegetation, soil type and climate. Agric. For. Meteorol. 39:257–269. doi:10.1016/0168-1923(87)90042-6

Penman, H.L. 1948. Natural evaporation from open water, bare soil, and grass. Proc. R. Soc. Lond. A 193:120–145. doi:10.1098/rspa.1948.0037

Philip, J.R. 1961. The theory of heat flux meters. J. Geophys. Res. 66:571–579. doi:10.1029/JZ066i002p00571

Pollack, H.N., and S. Huang. 2000. Climate reconstruction from subsurface temperatures. Annu. Rev. Earth Planet. Sci. 28:339–365. doi:10.1146/annurev.earth.28.1.339

Pollack, H.N., S. Huang, and P. Shen. 1998. Climate change record in subsurface temperatures: A global perspective. Science 282:279–281. doi:10.1126/science.282.5387.279

Porter, J.R., and M. Gawith. 1999. Temperatures and the growth and development of wheat: A review. Eur. J. Agron. 10:23–36. doi:10.1016/S1161-0301(98)00047-1

Priestley, C.H.B., and R.J. Taylor. 1972. On the assessment of surface heat flux and evaporation using large-scale parameters. Mon. Weather Rev. 100:81–92. doi:10.1175/1520-0493(1972)100<0081:OTAOSH>2.3.CO;2

Ramegowda, V., and M. Senthil-Kumar. 2015. The interactive effects of simultaneous biotic and abiotic stresses on plants: Mechanistic understanding from drought and pathogen combination. J. Plant Physiol. 176:47–54. doi:10.1016/j.jplph.2014.11.008

Rao, G.R., R.V. Ramamohan, and B.V.R. Rao. 1977. Soil heat flux studies in the bare red laterite soil during winter and summer months at Bangalore. Ann. Arid Zone 16:5–12.

Richard, G., and P. Cellier. 1998. Effect of tillage on bare soil energy balance and thermal regime: An experimental study. Agronomie 18:163–181. doi:10.1051/agro:19980301

Rosset, M., M. Riedo, A. Grub, M. Geissmann, and J. Fuhrer. 1997. Seasonal variation in radiation and energy balances of permanent pastures at different altitudes. Agric. For. Meteorol. 86:245–258. doi:10.1016/S0168-1923(96)02423-9

Sánchez, B., A. Rasmussen, and J.R. Porter. 2014. Temperatures and the growth and development of maize and rice: A review. Glob. Change Biol. 20:408–417. doi:10.1111/gcb.12389

Sauer, T.J. 2002. Heat flux density. In: J.H. Dane and G.C. Topp, editors, Methods of soil analysis. Part 4. physical methods. SSSA Book Series, no. 5. SSSA, Madison, WI. p. 1233–1248.

Sauer, T.J., J.L. Hatfield, and J.H. Prueger. 1996. Corn residue age and placement effects on evaporation and soil thermal regime. Soil Sci. Soc. Am. J. 60:1558–1564. doi:10.2136/sssaj1996.03615995006000050039x

Sauer, T.J., J.L. Hatfield, and J.H. Prueger. 1997. Over-winter changes in radiant energy exchange of a corn residue-covered surface. Agric. For. Meteorol. 85:279–287. doi:10.1016/S0168-1923(96)02386-6

Sauer, T.J., J.L. Hatfield, J.H. Prueger, and J.M. Norman. 1998. Surface energy balance of a corn residue-covered field. Agric. For. Meteorol. 89:155–168. doi:10.1016/S0168-1923(97)00090-7

Sauer, T.J., and R. Horton. 2005. Soil heat flux. In: J.L. Hatfield and J.M. Baker, editors, Micrometeorological measurements in agricultural systems. Agronomy Monograph No. 47. ASA, Madison, WI. p. 131–154.

Sauer, T.J., D.W. Meek, T.E. Ochsner, A.R. Harris, and R. Horton. 2003. Errors in heat flux measurement by flux plates of contrasting design and thermal conductivity. Vadose Zone J. 2:580–588. doi:10.2136/vzj2003.5800

Sauer, T.J., and J.M. Norman. 1995. Simulated canopy microclimate using estimated below-canopy soil surface transfer coefficients. Agric. For. Meteorol. 75:135–160. doi:10.1016/0168-1923(94)02208-2

Sauer, T.J., J.M. Norman, C.B. Tanner, and T.B. Wilson. 1995. Measurement of heat and vapor transfer coefficients at the soil surface beneath a maize canopy using source plates. Agric. For. Meteorol. 75:161–189. doi:10.1016/0168-1923(94)02209-3

Sauer, T.J., T.E. Ochsner, and R. Horton. 2007. Soil heat flux plates: Heat flow distortion and thermal contact resistance. Agron. J. 99:304–310. doi:10.2134/agronj2005.0038s

Schlenker, W., and M.J. Roberts. 2009. Nonlinear temperature effects indicate severe damages to U.S. crop yields under climate change. Proc. Natl. Acad. Sci. USA 106:15594–15598. doi:10.1073/pnas.0906865106

Sharratt, B.S., G.S. Campbell, and D.N. Glenn. 1992. Soil heat flux estimation based on the finite difference form of the transient heat flow equation. Agric. For. Meteorol. 61:95–111. doi:10.1016/0168-1923(92)90027-2

Smits, K.M., T. Sakaki, A. Limsuwat, and T.H. Illangasekare. 2010. Thermal conductivity of sands under varying moisture and porosity in drainage-wetting cycles. Vadose Zone Research 9:1–9.

Suzuki, N., R.M. Rivero, V. Shulaev, E. Blumwald, and R. Mittler. 2014. Abiotic and biotic stress combinations. New Phytol. 203:32–43. doi:10.1111/nph.12797

Tamai, K., T. Abe, M. Araki, and H. Ito. 1998. Radiation budget, soil heat flux and latent heat flux at the forest floor in warm, temperate mixed forest. Hydrol. Processes 12:2105–2114. doi:10.1002/(SICI)1099-1085(19981030)12:13/14<2105::AID-HYP723>3.0.CO;2-9

Trenberth, K.E., J.T. Fasullo, and J. Kiehl. 2009. Earth's global energy budget. Bull. Am. Meteorol. Soc. 90:311–324. doi:10.1175/2008BAMS2634.1

Trenberth, K.E., and D.P. Stepaniak. 2004. The flow of energy through the earth's climate system. Q. J. R. Meteorol. Soc. 130:2677–2701. doi:10.1256/qj.04.83

Tsai, J.L., B.J. Tsuang, P.S. Lu, M.H. Yao, and Y. Shen. 2007. Surface energy components and land characteristics of a rice paddy. J. Appl. Meteorol. Climatol. 46:1879–1900. doi:10.1175/2007JAMC1568.1

Tuzet, A., J.-F. Castell, A. Perrier, and O. Zurfluh. 1997. Flux heterogeneity and evapotranspiration partitioning in a sparse crop: The fallow savanna. J. Hydrol. 188-189:482–493. doi:10.1016/S0022-1694(96)03189-7

Twine, T.E., W.P. Kustas, J.M. Norman, D.R. Cook, P.R. Houser, T.P. Meyers, J.H. Prueger, P.J. Starks, and M.L. Wesely. 2000. Correcting eddy-covariance flux underestimates over a grassland. Agric. For. Meteorol. 103:279–300. doi:10.1016/S0168-1923(00)00123-4

Unger, P.W. 1988. Residue management effects on soil temperature. Soil Sci. Soc. Am. J. 52:1777–1782. doi:10.2136/sssaj1988.03615995005200060047x

USDA Foreign Agricultural Service, 2002. Rice Consumption and Production. USDA Foreign Agricultural Service, Washington, D.C.

van Bavel, C.H.M. 1972. Soil temperature and crop growth. In: D. Hillel, editor, Optimizing the soil physical environment toward greater crop yields. Academic Press, New York. p. 23–33.

Verhoef, A., S.J. Allen, and C.R. Lloyd. 1999. Seasonal variation of surface energy balance over two Sahelian surfaces. Int. J. Climatol. 19:1267–1277.

Wang, J., and R.L. Bras. 1999. Ground heat flux estimated from surface soil temperature. J. Hydrol. 216:214–226. doi:10.1016/S0022-1694(99)00008-6

Welch, S.M., G.J. Kluitenberg, and K.L. Bristow. 1996. Rapid numerical estimation of soil thermal properties for a broad class of heat-pulse emitter geometries. Meas. Sci. Technol. 7:932–938. doi:10.1088/0957-0233/7/6/012

Wilson, K.B., P.J. Hanson, and D.D. Baldocchi. 2000. Factors controlling evaporation and energy partitioning beneath a deciduous forest over an annual cycle. Agric. For. Meteorol. 102:83–103. doi:10.1016/S0168-1923(00)00124-6

Woodall, C.W., C.M. Oswalt, J.A. Westfall, C.H. Perry, M.D. Nelson, and A.O. Finley. 2009. An indicator of tree migration in forest of the eastern United States. For. Ecol. Manage. 257:1434–1444. doi:10.1016/j.foreco.2008.12.013

Yamori, W., K. Hikosaka, and D.A. Way. 2014. Temperature response of photosynthesis in C3, C4, and CAM plants: Temperature acclimation and temperature adaptation. Photosynth. Res. 119:101–117. doi:10.1007/s11120-013-9874-6

Zinn, K.E., M. Tunc-Ozdemir, and J.F. Harper. 2010. Temperature stress and plant sexual reproduction: Uncovering the weakest links. J. Exp. Bot. doi:10.1093/jxb/erq053

Atmospheric Humidity

John M. Baker* and Timothy J. Griffis

Abstract

Water vapor is a crucial atmospheric constituent. The cycling of water between atmosphere and biosphere transfers energy throughout the earth system via precipitation and evaporation, and the distribution of precipitation determines the distribution of biological productivity. The density difference between moist and dry air helps to drive the convective motion so important to atmospheric circulation. In addition, water vapor's numerous thermal infrared absorption bands make it the predominant source of down-welling long-wave radiation. Atmospheric humidity is thus a key determinant of terrestrial temperature, and potential changes in humidity represent a crucial and poorly understood feedback in a changing climate. There are a number of possible ways to measure atmospheric humidity, with the method of choice dependent on the application. In general, optical methods are favored for flux measurements, such as eddy covariance, where fast response is critical, while polymer-based capacitive sensors are preferable for routine relative humidity (RH) measurements in weather stations because of their low cost and long-term stability. Recently developed tunable diode lasers that can separately measure the isotopologues of water provide a particularly powerful diagnostic tool for a variety of applications, since certain processes impart isotopic discrimination to water vapor in predictable ways.

Water is a primary component of the Earth's atmosphere, with a molar concentration exceeded only by N_2 and O_2. All biological activity depends on water, and primary production by plants is particularly sensitive to the availability of water to drive growth and replenish what is lost by evaporation. The evaporation of water requires an unusually large amount of energy (~2.5 kJ kg^{-1}) and an equivalent amount is liberated on condensation. Consequently, atmospheric transport of water can transfer prodigious quantities of energy from sites of evaporation to sites of precipitation over relatively short time scales; the mean atmospheric lifetime of water molecules is on the order of 10 d. This concomitant atmospheric circulation of water and energy, in concert with ocean currents,

---

Abbreviations: IRGA, infrared gas analyzer; RH, relative humidity; UV, ultraviolet.

John M. Baker, USDA-ARS Soil & Water Management Unit, Dep. of Soil, Water & Climate, Univ. of Minnesota, 1991 Upper Buford Circle, St. Paul MN 55108. Timothy J. Griffis, Dep. of Soil, Water & Climate, Univ. of Minnesota, 1991 Upper Buford Circle, St. Paul, MN 55108 (tgriffis@umn.edu). *Corresponding author (John.baker@ars.usda.gov).

doi:10.2134/agronmonogr60.2015.0031

© ASA, CSSA, and SSSA, 5585 Guilford Road, Madison, WI 53711, USA.
Agroclimatology: Linking Agriculture to Climate, Agronomy Monograph 60.
Jerry L. Hatfield, Mannava V.K. Sivakumar, John H. Prueger, editors.

helps to dissipate the large latitudinal disparity in incident solar radiation. This topic is addressed in the Role of Humidity in the Atmosphere section.

The radiative properties of water vapor are also a critical factor in maintaining the habitability of Earth. Water vapor is nearly transparent to solar radiation but absorbs strongly at a number of wavelengths within the thermal infrared (8–14 μm) portion of the electromagnetic spectrum and reradiates some of the absorbed energy back toward the surface. It is in fact the dominant source of down-welling thermal infrared radiation, accounting for ~50% of long-wave radiative forcing (Schmidt et al., 2010), in the absence of which the equilibrium temperature of the Earth would be much lower and too cold to support life as we know it. There is widespread concern, and clear evidence, that the increasing concentrations of other greenhouse gases (CO_2, N_2O, and CH_4) are causing global temperatures to increase. The accuracy of models in predicting the extent of this warming depends on proper forecasting of changes in atmospheric water vapor, since it is the primary greenhouse gas and one of the strongest or most sensitive positive feedback factors identified in climate models. The Trends in Humidity Associated with Climatic Change section contains a discussion of measured and expected trends in both mean global atmospheric humidity and changes in spatial distribution. The latter are crucial because they determine precipitation distribution, which is anticipated to be the most critical local manifestation of climatic change for many agricultural areas.

Water vapor transport processes are driven by gradients, so atmospheric humidity measurements are critical inputs to models at a variety of scales. They are also necessary for assessing comfort levels for both humans and animals and for predicting the occurrence of various plant diseases. However, atmospheric humidity can be expressed and measured in several different ways, so the choice of an appropriate method or instrument is nontrivial. The Measurement of Humidity section describes the most common types of instruments and how they are typically applied. Of course, it is often necessary to rely on data collected by others, such as National Weather Service records, where the measured variable may not be the humidity parameter of interest. Humidity terms and equations for deriving one humidity variable from another are provided in Table 1 and the following section.

The dependence of saturation vapor pressure (e_s) on temperature can be derived from thermodynamic considerations (Brutsaert, 1984) and can be closely approximated with one of several equations including those developed by Tetens (1930), Goff and Gratch (1946), Murray (1967), and Lowe (1977). Buck (1981) reported several versions varying in accuracy and computational speed including a formulation that accounts for the slight enhancement of vapor pressure in moist air compared with that of pure water vapor (Ham, 2005):

$$e_s(T_C) = \left[1.0007 + \left(3.46\times10^{-5}P\right)\right]0.61121\exp\left(\frac{17.502T}{T+240.97}\right) \quad [1]$$

where T is air temperature (°C) and P is air pressure (kPa). Equation [1] describes the saturation vapor pressure in air over pure water. Below 0°C, the saturation vapor pressure over ice is increasingly depressed relative to that over supercooled water, so the coefficients are different:

Table 1. Definition of terms.

Term	Definition
Absolute humidity or vapor density (χ)	Mass of water vapor per unit volume of air, kg m^{-3}
Dew point (T_D)	The temperature at which air in equilibrium with a water surface is saturated with respect to water vapor, K
Frost point (T_f)	The temperature at which air in equilibrium with an ice surface is saturated with respect to water vapor, K
Isotopologue	Molecules that have a unique isotope composition or number of isotope substitutions
Latent heat of vaporization (λ)	The enthalpy change accompanying the transformation of water from liquid to vapor, kJ kg^{-1}
Mixing ratio (m_w)	Mass of water vapor per unit mass of dry air, kg kg^{-1}
Mole fraction (X_w)	mol_{H2O}/mol_{air}, unitless
Relative humidity (RH)	$(e/e_s) \times 100$, %
Saturation vapor pressure ($e_s[T]$)	Equilibrium vapor pressure of water at a specified temperature, kPa
Specific humidity (q)	Mass of water vapor per unit mass of moist air, kg kg^{-1}
Vapor pressure (e)	Partial pressure of water vapor in a mixture (e.g. air), kPa
Wet bulb temperature (T_w)	The minimum temperature obtainable by evaporative cooling of a fully wetted, well-ventilated surface.

$$e_s(T)=\left[1.0003+\left(4.18\times10^{-5}P\right)\right]0.61115\exp\left(\frac{22.452T}{T+272.55}\right) \quad [2]$$

Vapor pressure and vapor density in air are related by an equation of state:

$$\rho_V=\frac{0.622e}{R_d T_K} \quad [3]$$

where R_d is the gas constant for dry air (287.04 J mol^{-1} K^{-1}), 0.622 is the ratio of the molecular masses of water and dry air, and T is temperature (K). Vapor pressure can be calculated from the air temperature and wet bulb temperature:

$$e=e_s(T_W)-\gamma(T_a-T_W) \quad [4]$$

where γ, despite a dependence on pressure and a slight variation with temperature, is known as the psychrometric constant (kPa $°C^{-1}$). It is given by the following, in which C_p is the specific heat at constant pressure for dry air (1.005 kJ kg^{-1} K^{-1}):

$$\gamma=\frac{1.61PC_P}{\lambda} \quad [5]$$

The latent heat of vaporization (kJ kg^{-1}), is closely approximated by Eq. [6], where T is in degrees K:

$$\lambda=3146.22-2.3684T \quad [6]$$

The Role of Humidity in the Atmosphere

Water plays a fundamental role in atmospheric circulation, transferring energy poleward from the tropics and also from water bodies to land surfaces. Brutsaert (1984) summarized various estimates of global energy and water balances for land and oceans. They indicated that mean global net radiation over land is only ~60% of that over oceans, and that the dissipation of that energy is much more weighted in favor of latent vs. sensible heating over the oceans, so that mean annual evaporation per unit surface area is three times greater from the oceans than it is from land. Mean annual precipitation is also greater over the oceans, but that ratio is much smaller—on the order of 1.5:1—so that precipitation exceeds evaporation over land, while the opposite is true for the oceans, with the difference between precipitation and evaporation in each case equaling the annual return flow (runoff) from the continents to the oceans.

The amount of water vapor in the atmosphere plays an important role with respect to buoyancy, atmospheric stability, and convective precipitation. As water vapor increases in the lower atmosphere, there is greater potential for unstable atmospheric conditions (greater buoyancy of a displaced air parcel) because the moist adiabatic lapse rate (~4 K 1000 m^{-1}) is much less than the unsaturated rate (~9.8 K 1000 m^{-1}). There is growing evidence that the frequency and magnitude of convective precipitation events are increasing as a consequence of recent surface warming and evaporation (Trenberth et al., 2007; Trenberth, 2011; Min et al., 2011). This suggests that the amount of precipitation recycling over land (i.e., the fraction of terrestrial evapotranspiration water observed in precipitation water) will increase in a warmer and wetter world (Vallet-Coulomb et al., 2008). However, some of these processes and feedbacks are still poorly quantified and modeled and represent some of the important grand challenges in water cycle science (Trenberth and Asrar, 2014).

Trends in Humidity Associated with Climatic Change

The primary metric for global climate change is mean global temperature, but atmospheric temperature and humidity are linked. In accordance with the Clausius–Clapeyron equation, the moisture-holding capacity of the atmosphere increases by ~7% for each degree increase in mean global temperature. Most climate models assume (and data thus far confirm) that in a mean global sense, RH is conserved as temperature increases, so that absolute humidity changes by ~7% with each degree as well (Trenberth, 2011). There is independent evidence that water vapor is increasing over the oceans (Santer et al., 2007) and over the continents (Dai, 2006) in response to surface warming. Global precipitation is much less sensitive to temperature, since it must balance with global evapotranspiration, which primarily depends on net radiation. Increases in global precipitation are anticipated to be on the order of 1 to 2% $°C^{-1}$ but more concentrated temporally and spatially, that is, wet areas are predicted to get wetter and dry areas drier with more frequent occurrence of both drought and extreme precipitation events (Held and Soden, 2006; Trenberth, 2011).

Simulations reported by Harding and Snyder (2014) support this notion, at least for the central United States. They used dynamical downscaling of two global climate models to examine trends in warm-season precipitation and concluded that rainfall would be less frequent but more intense with the greatest increase in high-rainfall events in April associated with intensification of the low-level jet, a mesoscale southerly wind pattern responsible for transferring large quantities of moisture from the Gulf of Mexico into the continental United States. Subsequent weakening of the jet in late summer is projected to be a factor in increasing the frequency of drought. Other anthropogenic factors may affect humidity transport as well. There is some evidence that widespread irrigation in the US Great Plains has affected atmospheric circulation patterns and downwind precipitation patterns. Irrigation increases the amounts of precipitable water and convective available potential energy downwind, thus favoring increased precipitation (Sacks et al., 2009; Harding and Snyder, 2012). However, this is partially offset by the cooling effect of evapotranspiration.

Measurement of Humidity

Neither of the two fundamental measures of humidity—vapor pressure and mixing ratio—is directly and routinely measured with commercially available sensors or instruments. Rather, one of the other humidity parameters is measured, together with temperature or atmospheric pressure if necessary, and then the relevant equations given here can be used to determine the humidity parameters of interest. Humidity measurement has been a critical concern in several disciplines for many years. Entire books have been written on the subject of humidity measurement (e.g.,– Instrument Society of America, 1986; Wiederhold, 1997). It is beyond the scope of this chapter to provide exhaustive details on all possible techniques; instead, general information is provided about the most commonly used methods in micrometeorological applications.

Psychrometry

Psychrometers rely on the simultaneous measurement of wet bulb and air (dry bulb) temperatures under shaded, well-aspirated conditions from which vapor pressure can be calculated with Eq. [4] and [5], provided barometric pressure is also measured or known. The wet bulb, as its name implies, must be kept wet, which can be a challenge for long-term, unattended measurement. Aspirated psychrometers were widely used in micrometeorological research for many years (e.g., Brown and Covey, 1966; Lourence and Pruitt, 1969; Oke and East, 1971), particularly for measuring latent heat flux by both aerodynamic (profile) methods and by Bowen ratio, but are much less common today. However, they still provide an inexpensive option for micrometeorological research, since they can be fabricated by the end user with off-the-shelf parts. Psychrometers that must be operated for extended periods and logged automatically typically use thermistors or thermocouples for the temperature measurement, and the wet bulb wick can be passively supplied with water from a reservoir or actively wetted with a small pump. Portable psychrometers can be used for point-in-time humidity measurements, for example, for checking the accuracy of a humidity sensor when visiting an automated weather station or for characterizing ambient conditions on a playing field. They typically contain a squeeze bottle of deionized water that is used to wet a cloth wick that covers the bulb of one of the thermometers. In manual units, known as sling psychrometers, aspiration is provided by rapidly rotating the unit as if preparing to lasso a steer. In battery-powered units, a small fan provides the ventilation.

Dew Point

Dew cells contain lithium chloride (LiCl), a hygroscopic salt. For any ambient dew point there is a unique temperature, the LiCl dew point, at which saturated LiCl will be in equilibrium with the surrounding air (Tanner and Suomi, 1956). The relationship between dew point and the LiCl dew point is well known, so one can be readily determined from knowledge of the other. In a LiCl dew cell, two wires are wrapped around the LiCl cell and connected to a power source. As long as the vapor pressure of the surrounding air exceeds the water vapor pressure of a saturated LiCl solution, the salt will deliquesce. As it approaches equilibrium, its electrical resistance abruptly decreases, allowing current to flow, which in turn causes Joule heating that warms the solution, driving off water until an equilibrium temperature is reached. The absolute accuracy of LiCl dew cells is generally inferior to that of a chilled-mirror hygrometer because it is essentially a heat balance instrument; at the equilibrium point, the heat produced by Joule resistance is balanced by conductive, convective, and radiative losses from the cell, which can vary with ambient conditions. When used outdoors, they must be placed in a radiation shield and aspirated, but if the ventilation is too great, performance is unstable (Acheson, 1963). While LiCl dew cells were once widely used, they have been largely superseded by optical dew point sensors and are now rarely found.

Optical dew point sensors, commonly called chilled-mirror hygrometers or dew point hygrometers, contain a mirror, a light source, and a detector oriented such that the detector "sees" the light after it reflects off the mirror. The mirror is thermally bonded to a Peltier device that can either heat or cool it. Control circuitry and software is used to maintain the temperature of the mirror at exactly the temperature where dew forms, which alters the surface reflectance and thus the signal output from the optical detector. The primary advantage of optical dew point sensors is the fundamental nature of the measurement. There are a few notable disadvantages. One of them is cost; they tend to be more expensive than many others types of humidity sensors. Another issue is contamination of the mirror, which can cause systematic errors. To maintain adequate dynamic response, dew point sensors are usually plumbed to a moving stream of air or enclosed in a ventilated housing, so they are susceptible to contamination with dust, salts, and other foreign objects that can interfere with operation or cause systematic errors. It is good practice to filter incoming air to minimize such problems, but inevitably, the mirror must be cleaned periodically for optimal performance. Instability can be a problem at temperatures <0°C, where the control algorithm can potentially settle on either the dew point or the frost point.

Thermal dew point sensors also use the Peltier effect and control circuits to maintain a sensing element at the dew point temperature, but in this case, control is based on the thermal properties of the sensing surface. When dew forms, it alters the heat capacity and thermal conductivity of the surface membrane, which can be detected with temperature sensors embedded in the membrane (Kunze et al., 2012).

Absolute Humidity

Absolute humidity, or vapor density, is measured optically, taking advantage of the fact that water vapor absorbs radiation at many wavelengths. Some techniques are broadband in nature, while others use narrow bandwidths corresponding to specific absorption lines. In general, optical methods have rapid response times that favor their use in gas exchange research.

Nondispersive infrared gas analyzers, often referred to as IRGAs, are broadband instruments, although they typically contain two band-pass filters that allow them to measure both absolute humidity and carbon dioxide concentration. These can be incorporated in a variety of configurations depending on the objective. Benchtop units can be plumbed with pumps and tubing to measure absolute humidity in multiple growth chambers. Field-portable units can be used to switch between sample lines from two or more heights for Bowen ratio or aerodynamic measurements of latent heat flux, and the dynamic response of such instruments is sufficient that they can be used for eddy covariance measurements provided the sample flow rate is high enough to ensure fully turbulent flow. The latter issue is avoided with open-path IRGAs (Auble and Meyers, 1992) that have become widely used in micrometeorological research (Baldocchi et al., 1996). These approaches are described in detail elsewhere. As intimated above, the rapid dynamic response of the IRGA is its greatest advantage. They must be calibrated on a regular basis for both zero and span, for which it is necessary to provide dry air and air with a known water vapor concentration. The former can be easily produced by scrubbing air with a desiccant such as magnesium perchlorate; provision of a span gas is more challenging or at least more expensive. Dew point generators are commercially available, but it is also possible to construct a source of known vapor pressure (e.g., Tanner and Suomi, 1956; Cortes et al., 1991) or known mixing ratio (Baker and Griffis, 2010) if cost is a consideration.

Ultraviolet absorption hygrometers also measure absolute humidity by detecting absorption of radiation but in a different region of the electromagnetic spectrum. Tillman (1965) presented ultraviolet (UV) and infrared absorption spectra for water vapor and oxygen; described appropriate sources, windows, and detectors; and explored the various problems and potential solutions for their use in humidity measurement. At the same time, Randall et al. (1965) described a specific application of UV absorption, the Lyman-α hygrometer, an open-path instrument based on water vapor absorption of Lyman-α radiation, emitted at 121.6 nm. Numerous subsequent improvements and modifications have been made by others including Buck (1976) and Foken et al. (1998). On the source side, a hydrogen lamp emits the radiation through a lithium fluoride window that is transparent at the wavelength of interest. A similar window on the detector passes the radiation to a nitric oxide detector, in which photoionization causes a current to flow between electrodes. In accordance with Beers Law, the signal is attenuated by water vapor, allowing determination of concentration. A correction can be made for oxygen, which absorbs weakly at 121.6 nm. Lyman-α hygrometers are widely used in radiosondes and aircraft-based measurements primarily because they have a wide dynamic range including good sensitivity at the low vapor densities encountered at high altitudes, and they can be made quite small. Because of their fast response, Lyman-α hygrometers have also been widely used in micrometeorological research (e.g., Redford et al., 1980). A major challenge with Lyman-α hygrometers is that their operating characteristics change with time, necessitating frequent calibration. These changes include loss of hydrogen from the source lamp, deterioration of the windows as a result of reaction with hydrogen, and contamination of the electrodes in the detector. Foken et al. (1998) developed a novel unit with adjustable path length that facilitated in situ, automated calibration, but the usage of Lyman-α hygrometers in eddy covariance has become much less common with the advent of robust, commercially available open-path IRGAs.

Krypton hygrometers are open-path instruments that also rely on UV absorption by water vapor, in this case at two lines, 116.49 and 123.58 nm (Campbell and Tanner, 1985). They avoid some of the limitations of Lyman-α hygrometers, since they use a krypton glow tube as a radiation source rather than hydrogen. However, oxygen absorbs more strongly at these wavelengths than at the Lyman-α line, so it must be accounted for. The krypton hygrometer uses the same nitric oxide detector as the Lyman-α hygrometer, and the operating principles are quite similar. It is less sensitive than the Lyman-α, but its calibration is more stable and its radiation source has a longer lifetime. There is at least one commercial source of krypton hygrometers, and the primary application is eddy covariance measurements of latent heat flux.

Isotope Composition of Water Vapor

The isotope composition of water vapor can provide important information regarding its source contributions and recent hydrometeorological history. Its measurement is essential in isotope applications that attempt to partition evapotranspiration into its components of soil water evaporation and plant transpiration.

Over the past decade, considerable progress has been achieved in measuring the isotope composition of water vapor using optical (laser-based) techniques (Griffis, 2013; Wen et al., 2013; Lee et al., 2005). These methods exploit the fact that different isotopologues (i.e., molecules with unique isotope substitutions) have different molecular absorption characteristics as determined by their rotational–vibrational frequencies. Laser-based techniques, such as tunable diode laser spectroscopy and Fourier-transform infrared spectroscopy, have the sensitivity and selectivity to accurately measure the concentrations of the most relevant isotopologues (^{18}O-H_2O, ^{1}H-H_2O, and ^{2}H-H_2O) under ambient conditions. Each of the water isotopologues can be measured at wavenumbers in the vicinity of 1500, 2800, and 3600 cm^{-1}. The absorption region that is selected will depend on the radiation source type and detector configuration.

The measured concentration of each isotoplogue is usually expressed as a ratio in familiar delta notation (δ):

$$\delta = \frac{R_s - R_{std}}{R_{std}} \qquad [7]$$

where R_s is the sample molar ratio of the heavy to light isotope and R_{std} is the standard molar ratio (i.e., expressed relative to the Vienna Standard Mean Ocean or Vienna Pee Dee Belemnite scales). The δ value is often expressed as parts per thousand (per mil) by multiplying the ratio by 10^3.

One of the important challenges in measuring the isotope composition of water vapor is the need for routine and accurate calibration under field conditions. Unlike trace gases such as carbon dioxide or nitrous oxide, the researcher cannot purchase prepared water vapor standards. This challenge has been met with the development of innovative dripper systems (Lee et al., 2005; Griffis et al., 2010), Rayleigh distillation devices (Baker and Griffis, 2010), and nebulizers (Rambo et al., 2011) that can be used to correct the individual concentrations of isotoplogues via postprocessing. The essence of these calibration techniques is to produce, in real time, a water vapor sample that has a known isotope ratio and with a total water vapor concentration that tracks the ambient condition. In the case of a dripper system, this is accomplished by vaporizing a small water droplet of known isotope ratio that is precisely mixed with ultra-dry air.

Relative Humidity

Relative humidity is probably the most commonly measured humidity parameter, which may seem odd, since it is at first glance a derived value—the ratio e/e_0—rather than a fundamental property. The explanation for this apparent paradox lies in the fact that the amount of water contained by a porous medium or sorbed to a surface is a function of the Gibbs' free energy of water, also known as water potential (ψ), which is in turn related to RH by the following equation:

$$\psi = \frac{RT_k}{M} \ln\left(\frac{e}{e_0}\right) \quad [8]$$

where R is the universal gas constant and M is the partial molar mass of water. This can be taken advantage of in a variety of ways by incorporating a porous or adsorptive material in a circuit and measuring an electrical property that changes in response to changes in adsorbed water, typically either resistance or capacitance, along with temperature. There are a wide variety of such sensors now available, ranging from large, robust units designed for long-term use in weather stations to tiny integrated circuit sensors that can be embedded in confined areas such as the inner cavity of a porometer. Rittersma (2002) described a number of RH sensors with extensive explanations of their theory and operation, and a more recent review was conducted by Farahani et al. (2014). One area in which the performance of different RH sensors varies substantially is in their performance in cold environments (Makkonen and Laakso, 2005). All RH sensors must be calibrated, and their calibrations should be checked periodically. Calibration can be performed by suspending a sensor above two or more saturated salt solutions, although there are also commercially available humidity generators that are specifically designed for calibrating RH sensors. The RH of several common saturated salts at 25°C are given in Table 2.

Humidity and Animal Comfort

It has long been recognized that humidity can affect the comfort level of animals, much as it does humans. Humidity effects are generally studied in tandem with temperature in the form of a heat index. Much of this research has focused on dairy cows, where day to day changes in performance are routinely monitored.

Table 2. Relative humidity (RH) of air above saturated salt solutions at 25°C.

Salt	RH at 25°C
	%
LiCl	11.1
MgCl	32.8
$MgNO_3$	52.9
NaCl	75.3
KCl	84.3
KNO_3	92.5
KSO_4	97.3

Hot, humid conditions noticeably affect both dry matter intake and milk production (West et al., 2003). Various formulas have been used to quantify heat and humidity stress; one of the most popular was developed by Steadman (1979). It uses air temperature and wet bulb in an equation that estimates an equivalent, or "feels like," temperature that he referred to as a heat index. Subsequent variations have used air temperature and RH or air temperature and dewpoint. Bohmanova et al. (2007) evaluated several formulations against milk production statistics in an arid environment (Phoenix, AZ) and a humid environment (Baton Rouge, LA) and found that indices with heavier weights on humidity performed better at the humid location and indices with less weighting for humidity performed better at the arid site. The conclusion from this and similar studies is that humidity does not impact cows directly, but rather it affects their ability to maintain an optimal temperature by limiting their ability to shed heat through sweating. Since species differ in their ability to dissipate heat by sweating, their sensitivity to humidity varies as well. Swine for instance, do not possess sweat glands; consequently, they are much more sensitive to high air temperatures in arid regions than cows and less sensitive in humid regions. Presumably, other animals that shed heat by sweating (horses, goats, sheep) are more similar to cows in their humidity sensitivities, while those that, like pigs, depend solely on panting for latent heat loss (e.g., poultry) respond to humidity and temperature more in common with swine. Rodrigues et al. (2011) noted the empiricism in heat index formulations and suggested a more fundamental relationship between animal comfort and enthalpy of the air, the latter a readily calculated function of temperature, humidity, and barometric pressure.

Humidity Effects on Plant Disease

High RH, and particularly dew formation, favors the development of a variety of plant diseases (Agrios, 2005). It is the initiation of disease (e.g., germination of spores) that is particularly sensitive to humidity; once infection occurs, most pathogens become independent of ambient humidity. As a result, plant pathologists frequently use growth chambers in which dew formation can be induced to generate pathogen populations to study accompanying diseases (e.g., Clifford and Harris, 1981). Among the plant diseases in which humidity is a factor are leaf rust in wheat (*Triticum aestivum* L.) (Kumar, 2014), brown rust in sugarcane (*Saccharum officinarum* L.) (Barrera et al., 2013), pecan [*Carya illinoinensis* (Wangenh.) K. Koch] scab (Payne and Smith, 2012), and downy mildew in grapevine (*Vitis vinifera* L.) (Dalla Marta et al., 2005). A number of models have been developed to predict disease occurrence and severity (e.g., Huber and Gillespie, 1992; Bregaglio et al., 2011; Cooley et al., 2011; Leca et al., 2011) using humidity as an input, typically either duration of high RH values or duration of dew on leaves. The latter can be measured with commercially available leaf wetness sensors. The role of humidity in the incidence of disease has also been incorporated into models designed to predict the impacts of climate change on crop growth and food production (Elad and Pertot, 2013; Caubel et al., 2012).

References

Acheson, D.T. 1963. Some limitations and errors inherent in the use of the dew cell for measurement of atmospheric dew points. Mon. Weather Rev. 91:227–234. doi:10.1175/1520-0493(1963)091<0227:SLAEII>2.3.CO;2

Agrios, G.N. 2005. Plant pathology, 5th ed. Elsevier Academic Press. Burlington MA.

Auble, D., and T. Meyers. 1992. An open path, fast response infrared-absorption gas analyzer for H_2O and CO_2. Boundary-Layer Meteorol. 59:243–256. doi:10.1007/BF00119815

Baker, J.M., and T.J. Griffis. 2010. A simple, accurate, field-portable mixing ratio generator and Rrayleigh distillation device. Agric. For. Meteorol. 150:1607–1611. doi:10.1016/j.agrformet.2010.08.008

Baldocchi, D., R. Valentini, S. Running, W. Oechel, and R. Dahlman. 1996. Strategies for measuring and modelling carbon dioxide and water vapour fluxes over terrestrial ecosystems. Glob. Change Biol. 2:159–168. doi:10.1111/j.1365-2486.1996.tb00069.x

Barrera, W., J. Hoy, and B. Li. 2013. Effects of temperature and moisture variables on brown rust epidemics in sugarcane. J. Phytopathol. 161:98–106. doi:10.1111/jph.12035

Bohmanova, J., I. Misztal, and J.B. Cole. 2007. Temperature-humidity indices as indicators of milk production losses due to heat stress. J. Dairy Sci. 90:1947–1956. doi:10.3168/jds.2006-513

Bregaglio, S., M. Donatelli, R. Confalonieri, M. Acutis, and S. Orlandini. 2011. Multi metric evaluation of leaf wetness models for large-area application of plant disease models. Agric. For. Meteorol. 151:1163–1172. doi:10.1016/j.agrformet.2011.04.003

Brown, K.W., and W. Covey. 1966. The energy-budget evaluation of the micrometeorological transfer processes within a cornfield. Agric. Meteorol. 3:73–96. doi:10.1016/0002-1571(66)90006-9

Brutsaert, W. 1984. Evaporation into the atmosphere. Kluwer Acad. Publ. Dordrecht, the Netherlands.

Buck, A. 1976. The variable path Lyman-alpha hygrometer and its operating characteristics. Bull. Am. Meteorol. Soc. 57:1113–1118. doi:10.1175/1520-0477(1976)057<1113:TVPL AH>2.0.CO;2

Buck, A.L. 1981. New equations for computing vapor pressure and enhancement factor. J. Appl. Meteorol. 20:1527–1532. doi:10.1175/1520-0450(1981)020<1527:NEFCVP>2.0.CO;2

Campbell, G.S., and B.D. Tanner. 1985. A krypton hygrometer for measurement of atmospheric water vapor concentration. In: A. Wexler, editor, Humidity and moisture: Measurement and control in science and industry. Reinhold Publishing, London. p. 609–612.

Caubel, J., M. Launay, C. Lannou, and N. Brisson. 2012. Generic response functions to simulate climate-based processes in models for the development of airborne fungal crop pathogens. Ecol. Modell. 242:92–104. doi:10.1016/j.ecolmodel.2012.05.012

Clifford, B., and R. Harris. 1981. Controlled environment studies of the epidemic potential of *Puccinia recondita* f.sp *tritici* on wheat in Britain. Trans. Br. Mycol. Soc. 77:351–358. doi:10.1016/S0007-1536(81)80037-X

Cooley, D.R., D.A. Rosenberger, M.L. Gleason, G. Koehler, K. Cox, J.M. Clements, T.B. Sutton, A. Madeiras, and J.R. Hartman. 2011. Variability among forecast models for the apple sooty blotch/flyspeck disease complex. Plant Dis. 95:1179–1186. doi:10.1094/PDIS-03-11-0248

Cortes, P., C. Reece, and G. Campbell. 1991. A simple and accurate apparatus for the generation of a calibrated water-vapor pressure. Agric. For. Meteorol. 57:27–33. doi:10.1016/0168-1923(91)90076-3

Dalla Marta, A., R.D. Magarey, and S. Oriandini. 2005. Modelling leaf wetness duration and downy mildew simulation on grapevine in Italy. Agric. For. Meteorol. 132:84–95. doi:10.1016/j.agrformet.2005.07.003

Dai, A. 2006. Recent climatology, variability, and trends in global surface humidity. J. Clim. 19:3589–3606. doi:10.1175/JCLI3816.1

Elad, Y., and I. Pertot. 2013. Climate change impact on plant pathogens and plant diseases. In: M. Kang and S.S. Banga, editors, Combating climate change: An agricultural perspective. CRC Press, Boca Raton, FL. p. 183–211.

Farahani, H., R. Wagiran, and M.N. Hamidon. 2014. Humidity sensors principle, mechanism, and fabrication technologies: A comprehensive review. Sensors (Basel Switzerland) 14:7881–7939. doi:10.3390/s140507881

Foken, T., A.L. Buck, R.A. Nye, and R.D. Horn. 1998. A Lyman-alpha hygrometer with variable path length. J. Atmos. Ocean. Technol. 15:211–214. doi:10.1175/1520-0426(1998)015<0211:ALAHWV>2.0.CO;2

Goff, S.A., and S. Gratch. 1946. Low-pressure properties of water from −160 to 212 °F. J. Am. Soc. Heat. Vent. Eng. 52:95–121.

Griffis, T.J., S.D. Sargent, X. Lee, J.M. Baker, J. Greene, M. Erickson, X. Zhang, K. Billmark, N. Schultz, W. Xiao, and N. Hu. 2010. Determining the oxygen isotope composition of evapotranspiration using eddy covariance. Boundary-Layer Meteorol. 137(2):307–326. doi:10.1007/s10546-010-9529-5

Griffis, T. 2013. Tracing the flow of carbon dioxide and water vapor between the biosphere and atmosphere: A review of optical isotope techniques and their application. Agric. For. Meteorol. 174-175:85–109. doi:10.1016/j.agrformet.2013.02.009

Ham, J.M. 2005. Useful equations in micrometeorology. In: J.L. Hatfield and J.M. Baker, editors, Micrometeorology in agricultural systems. ASA, Madison, WI. p. 533–560.

Harding, K.J., and P.K. Snyder. 2012. Modeling the atmospheric response to irrigation in the Great Plains. Part I: General impacts on precipitation and the energy budget. J. Hydrometeor. 13:1667–1686. doi:10.1175/JHM-D-11-098.1

Harding, K.J., and P.K. Snyder. 2014. Examining future changes in the character of central U.S. warm-season precipitation using dynamical downscaling. J. Geophys. Res.: Atmos. 119:13116–13136.

Held, I.M., and B.J. Soden. 2006. Robust responses of the hydrological cycle to global warming. J. Clim. 19:5686–5699. doi:10.1175/JCLI3990.1

Huber, L., and T. Gillespie. 1992. Modeling leaf wetness in relation to plant-disease epidemiology. Annu. Rev. Phytopathol. 30:553–577. doi:10.1146/annurev.py.30.090192.003005

Instrument Society of America. 1986. Moisture and humidity: Proc. of the 1985 Int. Symp. on Moisture and Humidity. ISA, Research Triangle Park, NC.

Kumar, P.V. 2014. Development of weather-based prediction models for leaf rust in wheat in the Indo-Gangetic plains of India. Eur. J. Plant Pathol. 140:429–440. doi:10.1007/s10658-014-0478-6

Kunze, M., J. Merz, W. Hummel, H. Glosch, S. Messner, and R. Zengerle. 2012. A micro dew point sensor with a thermal detection principle. Meas. Sci. Technol. 23:014004. doi:10.1088/0957-0233/23/1/014004

Leca, A., L. Parisi, A. Lacointe, and M. Saudreau. 2011. Comparison of Penman–Monteith and non-linear energy balance approaches for estimating leaf wetness duration and apple scab infection. Agric. For. Meteorol. 151:1158–1162. doi:10.1016/j.agrformet.2011.04.010

Lee, X., S. Sargent, R. Smith, and B. Tanner. 2005. In-situ measurement of the water vapor $^{18}O/^{16}O$ isotope ratio for atmospheric and ecological applications. J. Atmos. Ocean. Technol. 22:555–565. doi:10.1175/JTECH1719.1

Lourence, F.J., and W.O. Pruitt. 1969. A psychrometer system for micrometeorology profile determination. J. Appl. Meteorol. 8:492–498. doi:10.1175/1520-0450(1969)008<0492:APS FMP>2.0.CO;2

Lowe, P.R. 1977. An approximating polynomial for the computation of saturation vapor pressure. J. Appl. Meteorol. 16:100–103. doi:10.1175/1520-0450(1977)016<0100:AAPFTC >2.0.CO;2

Makkonen, L., and T. Laakso. 2005. Humidity measurements in cold and humid environments. Boundary-Layer Meteorol. 116:131–147. doi:10.1007/s10546-004-7955-y

Min, S., X. Zhang, F. Zwiers, and G. Hegerl. 2011. Human contribution to more-intense precipitation extremes. Nature 470:378–381. doi:10.1038/nature09763

Murray, F.W. 1967. On the computation of saturation vapor pressure. J. Appl. Meteorol. 6:203–204. doi:10.1175/1520-0450(1967)006<0203:OTCOSV>2.0.CO;2

Oke, T.R., and C. East. 1971. The urban boundary layer in Montreal. Boundary-Layer Meteorol. 1:411–437. doi:10.1007/BF00184781

Payne, A.F., and D.L. Smith. 2012. Development and evaluation of two pecan scab prediction models. Plant Dis. 96:117–123. doi:10.1094/PDIS-03-11-0202

Rambo, J., C.T. Lai, J. Farlin, M. Schroeder, and K. Bible. 2011. On-site calibration for high precision measurements of water vapor isotope ratios using off axis cavity-enhanced absorption spectroscopy. J. Atmos. Ocean. Technol. 28:1448–1457. doi:10.1175/JTECH-D-11-00053.1

Randall, D.L., T.E. Hanley, and O.K. Larison. 1965. The NRL Lyman-alpha humidiometer. In: A. Wexler, editor, Humidity and moisture: Measurement and control in science and industry. Reinhold Publishing, London. p. 444–454.

Redford, T., S. Verma, and N. Rosenberg. 1980. Humidity fluctuations over a vegetated surface measured with a Lyman-alpha hygrometer and a fine-wire thermocouple psychrometer. J. Appl. Meteorol. 19:860–867. doi:10.1175/1520-0450(1980)019<0860:HFOAVS>2.0.CO;2

Rittersma, Z. 2002. Recent achievements in miniaturised humidity sensors: A review of transduction techniques. Sens. Actuators, A 96:196–210.

Rodrigues, V.C., I.J. Oliveira da Silva, F.M. Correa Vieira, and S.T. Nascimento. 2011. A correct enthalpy relationship as thermal comfort index for livestock. Int. J. Biometeorol. 55:455–459. doi:10.1007/s00484-010-0344-y

Sacks, W.J., B.I. Cook, N. Buenning, S. Levis, and J.H. Helkowski. 2009. Effects of global irrigation on the near-surface climate. Clim. Dyn. 33:159–175. doi:10.1007/s00382-008-0445-z

Santer, B.D., C. Mears, F J. Wentz, K.E. Taylor, P.J. Gleckler, T.M.L. Wigley, T.P. Barnett, J.S. Boyle, W. Brueggemann, N.P. Gillett, S.A. Klein, G.A. Meehl, T. Nozawa, D.W. Pierce, P.A. Stott, W.M. Washington, and M.F. Wehner. 2007. Identification of human-induced changes in atmospheric moisture content, Proc. Natl. Acad. Sci. USA, 104:15248–15253, doi:10.1073/pnas.0702872104

Schmidt, G., R. Ruedy, R. Miller, and A. Lacis. 2010. Attribution of the present-day total greenhouse effect. J. Geophys. Res.: Atmos. 115:D20106.

Steadman, R. 1979. Assessment of sultriness. 1. Temperature–humidity index based on human physiology and clothing science. J. Appl. Meteorol. 18:861–873. doi:10.1175/1520-0450(1979)018<0861:TAOSPI>2.0.CO;2

Tanner, C.B., and V.E. Suomi. 1956. Lithium chloride Dewcel properties and use for dew-point and vapor-pressure gradient measurements. Eos Trans. AGU 37:413–420. doi:10.1029/TR037i004p00413

Tetens, O. 1930. Uber einige meteorologische Begriffe. Z. Geophys. 6:297–309.

Tillman, J.E. 1965. Water vapor density measurements utilizing the absorption of vacuum ultraviolet and infrared radiation. In: A. Wexler, editor, Humidity and moisture: Measurement and control in science and industry. Reinhold Publishing, London. p. 428–443.

Trenberth, K. 2011. Changes in precipitation with climate change. Climate Res. 47:123–138.

Trenberth, K., and G. Asrar. 2014. Challenges and opportunities in water cycle research: WCRP contributions. Surv. Geophys. 35:515–532. doi:10.1007/s10712-012-9214-y

Trenberth, K., P. Jones, P. Ambenje, R. Bojariu, D. Easterling, A.K. Tank, D. Parker, F. Rahimzadeh, J. Renwick, M. Rusticucci, B. Sonden, and P. Zhai. 2007. Observations: Surface and atmospheric climate change. In: S. Solomon, D. Qin, M. Manning, Z. Chen, M. Marquis, K.B. Averyt, M. Tignor, and H.L. Miller, editors, Climate change 2007: The physical science basis. Contribution of Working Group 1 to the Fourth Assessment Report of the Intergovernmental Panel on Climate Change, Cambridge Univ. Press, Cambridge. p. 235–336.

Vallet-Coulomb, C., F. Gasse, and C. Sonzogni. 2008. Seasonal evolution of the isotopic composition of atmospheric water vapour above a tropical lake: Deuterium excess and implication for water recycling. Geochim. Cosmochim. Acta 72:4661–4674. doi:10.1016/j.gca.2008.06.025

Wen, X.F., Y. Meng, X.Y. Zhang, X.M. Sun, and X. Lee. 2013. Evaluating calibration strategies for isotope ratio infrared spectroscopy for atmospheric $^{13}CO_2/^{12}CO_2$ measurement. Atmos. Meas. Tech. 6:795–823. doi:10.5194/amtd-6-795-2013

West, J.W., B.G. Mullinix, and J.K. Bernard. 2003. Effects of hot, humid weather on milk temperature, dry matter intake, and milk yield of lactating dairy cows. J. Dairy Sci. 86:232–242. doi:10.3168/jds.S0022-0302(03)73602-9

Wiederhold, P. 1997. Water vapor measurement. CRC Press, New York.

A Brief Overview of Approaches for Measuring Evapotranspiration

J.G. Alfieri,* W.P. Kustas, and M.C. Anderson

Abstract

Information regarding evapotranspiration, the combined pathways of evaporative water loss from both the soil and vegetation, is critical to a broad range of applications from managing irrigation to maximizing both crop yield and quality and ensuring the sustainability of water resources. Evapotranspiration is controlled by multiple interconnected processes. These include the amount of available water in the soil, the atmosphere's ability to take up water vapor, and both the type and distribution of vegetation. A number of techniques have been developed to measure evapotranspiration. Each of the measurement techniques is built around its own unique set of theoretical and practical considerations and each has differing strengths and weaknesses. The overarching objective of this chapter is to provide an overview of the different methods for monitoring evapotranspiration using within-field and remote sensing-based approaches. Using lysimetry and the eddy covariance method as representative techniques, two broad approaches for monitoring evapotranspiration in situ, are discussed. Lysimetry is an example of the mass balance approach that leverages the principle of the conservation of mass to estimate evapotranspiration by measuring changes in the amount of water stored in the soil. Eddy covariance, which uses the relationship between turbulent air flow and humidity to determine evapotranspiration, is discussed as an example of the micrometeorological approach. Approaches that do not use field measurements are also available. For example, the ALEXI/DisALEXI modeling system is discussed as one remote sensing-based approach that can be used to obtain spatially-distributed estimates of evapotranspiration using remotely-sensed data from airborne or satellite platforms.

Information regarding evapotranspiration, the combined pathways of evaporative water loss from the land surface to the atmosphere, is critical to a broad range of applications ranging from predicting the weather and assessing fire risk in forested ecosystems to describing the epidemiology of bacterial and fungal diseases. In the context of agriculture, information about evapotranspiration is key to optimizing irrigation practices to maximize both crop yield and quality while ensuring sustainability. Moreover, as discussed by the Food and Agriculture Organization of the United Nations, it is also important for food security, rural development, and water and natural resource management (UN FAO, 2002).

The competing demands of urban, industrial, and agricultural communities already exceed the available supply of fresh water in many parts of the world.

J.G. Alfieri, W.P. Kustas, and M.C. Anderson, USDA-ARS Hydrology and Remote Sensing Laboratory, Beltsville, MD 20705-2350. *Corresponding author (joe.alfieri@ars.usda.gov)

doi:10.2134/agronmonogr60.2016.0034

© ASA, CSSA, and SSSA, 5585 Guilford Road, Madison, WI 53711, USA.
Agroclimatology: Linking Agriculture to Climate, Agronomy Monograph 60.
Jerry L. Hatfield, Mannava V.K. Sivakumar, John H. Prueger, editors.

For example, water scarcity is already a significant issue in such diverse regions as northern Africa, southern Europe, Australia, India, China, and the western United States (Wallace, 2000; Qadir et al., 2003; Hanasaki et al., 2008; Perry et al., 2009; Scanlon et al., 2010; Oweis et al., 2011; Anderson et al., 2012; Tarjuelo et al., 2015; Iglesias and Garrote, 2015). In the coming decades, meeting the competing needs for fresh water is expected to become more difficult as the world population increases and with it the demand for both fresh water and agricultural products. Based on projections from the United Nations, the world population is anticipated to be near 9 billion people by 2050 (UN ESA, 2011). As a result, agricultural productivity will need to increase by 60% or more with commensurate increases in the demand for water for agricultural use (UN FAO, 2009; Alexandratos and Bruinsma, 2012).

Agricultural production is by far the largest consumer of water resources. While there is substantial regional variability ranging from approximately 25% in Europe to nearly 85% in Asia and Africa, irrigation accounts for 70% of global water use; in 2010, this amounted to 2700 km^3 of water (Frenken and Gillet, 2012; see also the UN FAO AQUASTAT database available online at http://www.fao.org/nr/aquastat). Because it is the largest user of water, agricultural irrigation in particular has become a critical focus for water conservation efforts (Howell, 2001; Schultz and De Wrachien, 2002; Bastiaanssen et al., 2007; Turral et al., 2010). At the same time, irrigation is critical to meeting the demands for agricultural products including food, fiber, and fuel. While less than 20% of the world's croplands (approximately 400 million ha) are irrigated, they account for more than 40% of agricultural production (UN FAO, 2003; Thenkabail et al., 2010). Evapotranspiration data are essential to developing the policy and practices needed for balancing these goals.

Evapotranspiration data are also important for ensuring crop quality and yield. As an example, maintaining correct plant–water relations is key to ensuring crop quality in viticulture. To achieve the proper sugar content, pigment formation, and acidity, a modest water deficit needs to be maintained when growing wine grapes (Bravdo et al., 1985; Esteban et al., 2001; Basile et al., 2014). Excessive water availability not only reduces grape quality, it also leads to excessive leaf growth at the expense of the maturation of the woody vines. Excessive water stress, on the other hand, will hinder vine growth, sugar production, and yield (Escalona et al., 1999; Santos et al., 2003). Accurate information regarding the vine water status is needed to maintain the optimal level of water stress; in turn, evapotranspiration data are essential for characterizing the vine water use and status.

Similarly, the quality and yield of wheat (*Triticum aestivum* L.), like many other cereal crops, is highly sensitive to water availability. In this case, water stress not only decreases yield, it also impacts the relative protein and starch content of the grain (Guttieri et al., 2001; Erekul and Kohn, 2006). Water stress increases the production of proteins including gluten while reducing the production and storage of starches within the grain (Panozzo and Eagles, 2000; Ozturk and Aydin, 2004). Again, maintaining the correct crop–water relationship is critical to ensuring that the wheat has the proper proportion of starch and protein.

In addition to introducing the physical processes and environmental factors controlling evapotranspiration, this chapter provides an overview of different methods for monitoring evapotranspiration using within-field and remote sensing-based approaches. Although there are many techniques for collecting within-field measurements of evapotranspiration [see, for example, the text by

Foken (2008a) for detailed discussions of the various methods] lysimetry and the eddy covariance method are discussed here as representative techniques for two broad approaches for monitoring evapotranspiration in situ. Lysimetry is discussed as an example of the mass balance approach while eddy covariance is used as an example of the micrometeorological approach. Similarly, the ALEXI/DisALEXI modeling system is discussed as an example of remote sensing-based approaches that use imagery from airborne or satellite platforms to obtain spatially-distributed estimates of evapotranspiration.

Controls on Evapotranspiration

Evapotranspiration represents the total water loss from the surface due to the combined processes of direct evaporation and transpiration. Since it is the dominant pathway, direct evaporation typically refers to the evaporative loss of the water contained in the soil; however, evaporation also includes the intermittent evaporative loss of free water such as dew or rain intercepted by the plant canopy. Transpiration is the evaporative loss of water from plant through leaf pores (stomata). Not unexpectedly, since it links numerous biogeochemical and biogeophysical processes (for example, the water, carbon, energy, and nutrient cycles) evapotranspiration is regulated by a complex network of interconnected processes and feedbacks. Nonetheless, in broad terms, evapotranspiration is controlled by three factors: i) the availability of water; ii) the amount of available energy; and iii) the ability of the atmosphere to take up the additional water (Alfieri et al., 2007). While these factors are common to both evaporation and transpiration, the environmental conditions and processes controlling how easily water can be transported to the atmosphere are unique to each transport pathway.

Factors Common to Both Evaporation and Transpiration

Available Water

Typically, the water stored in the soil, which is commonly referred to as either the soil water or soil moisture content, determines the amount of water that is available for evapotranspiration. In turn, the soil moisture content is defined by the amount of water gained by the system through precipitation and irrigation and the amount of water loss due to surface runoff, drainage, and evapotranspiration itself. Following the principle of conservation of mass, this can be expressed as the well-known water budget relationship:

$$\Delta S=(P+I)-(R+D+ET) \qquad [2.1]$$

where ΔS is the change in the amount of water stored in the soil, P and I are the water inputs representing precipitation and irrigation, respectively, and R, D, and ET are the water outputs representing runoff, drainage, and evapotranspiration, respectively. These quantities can be expressed in terms of either equivalent depth of water, for example mm d^{-1}, or energy, for example W m^{-2}, depending on the application and timeframe of interest.

Similarly, both runoff and drainage are controlled by the ability of water to infiltrate and flow through the soil. In turn the movement of water in the soil is largely controlled by the soil physical and chemical properties. For example, the infiltration and redistribution tends to occur more slowing in fine-textured soils, that is, soils composed primarily of clay, due to their smaller pore size compared to their coarser-textured counterparts. At the same time, because fine-textured soils also tend to have greater surface area relative to pore space, they are also better able to retain water (Dingman, 2002). The infiltration and redistribution of water in the soil is also influenced by soil compaction, organic content, and management practices. For instance, as pointed out by Radford et al. (2001), Hamza and Anderson (2005), and Pietola et al. (2005), among others, compaction can significantly reduce the soil pore space and thereby limit the ability of the water to penetrate into and move through the soil. In contrast, the presence of organic matter tends to loosen soil and promote the formation of soil aggregates that enhance infiltration (Franzluebbers, 2002; Saxton and Rawls, 2006).

Available Energy

Likewise, the amount of energy available to drive evapotranspiration depends on numerous factors. Although there are a number of minor terms, such as the energy consumed by photosynthesis, the available energy is typical defined as the net radiation, i.e., the amount of radiative energy absorbed at the surface, less the soil heat flux, i.e., the amount of energy conducted into the soil (Meyers and Hollinger, 2004). The first of these quantities, net radiation, is often expressed as terms of the radiation budget:

$$R_n = K_{\downarrow} - K_{\uparrow} + L_{\downarrow} - L_{\uparrow} \quad [2.2]$$

where R_n is the net radiation, $K_{\downarrow}$ is the incoming solar (shortwave) radiation incident to the surface, $K_{\uparrow}$ is the reflected solar (shortwave) radiation, $L_{\downarrow}$ is the incoming long-wave radiation, and $L_{\uparrow}$ is the outgoing or terrestrial long-wave radiation. Although they are most often expressed in terms of energy, for example, W m^{-2}, the components of the radiation budget, like the components of the water budget, can be express either in terms of equivalent depth or energy.

Based on recent analyses (Kopp and Lean, 2011), the incident solar radiation at the top of the atmosphere is approximately 1360 W m^{-2}; this quantity is commonly referred to as the solar constant. However, the amount of energy that reaches the surface is typically reduced by 15% to 25% as it passes through the atmosphere. Under clear sky conditions, the degree of attenuation varies over time as a function of solar position, path length through the atmosphere, and atmospheric transmittance. The atmospheric transmittance, in turn, varies with water vapor, ozone, and aerosol content. The first two of these quantities reduce the incoming solar radiation through absorption, primarily in the near infrared and UV portion of the spectrum. Aerosols reduce the incoming solar radiation through both absorption and scattering (Monteith and Unsworth, 2008). While aerosols are the main reason for the decrease in incident solar radiation, the portion of the spectrum affected varies depending on the size of the particulates (Kim and Ramanathan, 2008; Eltbaakh et al., 2012).

The amount of solar radiation reflected from the surface varies over time as a function of solar angle and surface conditions such as surface optical properties. For example, the surface reflectance or albedo of the surface increases from a minimum at solar noon as the solar angle becomes more oblique; this is evident through the work of Song (1998), for instance, who found that the albedo of maize (*Zea mays* L.) varied by 36% over the course of the day. The optical properties of the surface change largely in response to changes in plant water content and surface roughness. For vegetated surfaces, the albedo decreases with increasing plant water stress and vegetation height while for bare soil surfaces, it typically decreases with increasing water content and roughness (Idso et al., 1975; Iziomon and Mayer 2002; Wang, 2005).

According to the Stefan-Boltzmann relationship, both the incoming and terrestrial long-wave components of the radiation budget vary as function of temperature and emissivity. In both cases, the temperature and emissivity are regulated by multiple factors. For example, while the atmospheric emissivity, thus the incoming long-wave radiation, is primarily controlled by humidity (Brunt 1932; Swinbank 1963; Brutsaert, 1975) under clear-sky conditions, the air temperature varies because energy is exchanged with the surface, that is, the sensible heat flux, or transported horizontally via advection. In turn, these processes are influenced by such factors as the surface temperature and roughness, wind speed, turbulent intensity, and heterogeneity of the landscape (Oke, 1987; Monteith and Unsworth, 2008).

At the same time, the magnitude of the terrestrial long-wave radiation depends on the surface emissivity and temperature. Since the surface emissivity integrates both soil and vegetation emissivity, it varies with the texture, composition and moisture content of the soil, and the type, density, architecture, and fractional cover of the vegetation cover (Van de Griend and Owe, 1993; Humes et al., 1994; Norman and Becker 1995; Mira et al., 2007). The processes controlling the surface temperature are similarly complex. Consider, for example, the influence of soil moisture on surface temperature. A large soil moisture content both facilitates evaporative cooling [energy is preferentially partitioned into evapotranspiration when water is available (Wetzel and Chang, 1987)] and enhances the thermal conductivity of the soil thereby increasing the soil heat flux. Over time, without the input of additional energy, the soil near the surface becomes both cooler and drier due to both evaporative water loss and the transport of heat to deeper soil layers; as a result, the influence of these processes on the surface temperature is diminished. A similar effect can be seen with vegetation. The presence of vegetation not only reduces the surface temperature through evaporative cooling and shading (Friedl and Davis, 1994; Santanello and Friedl, 2003; Herb et al., 2008), it also boosts turbulent mixing, facilitating heat transfer to the atmosphere (Cellier et al., 1996; Nemani and Running, 1997).

Finally, the soil heat flux, the non-radiative quantity needed to determine the available energy, is defined as the amount of heat conducted through the soil. While the soil heat flux is, quite expectedly, strongly influenced by the surface temperature, it is also controlled by the ability of the soil to conduct heat. In turn, the thermal properties of the soil, and particularly the thermal conductivity, are dependent on the soil texture, compaction, mineral composition, and moisture content (Van Bavel and Hillel, 1976; Hillel, 1998; Abu-Hamdeh and Reeder, 2000). For example, Abu-Hamdeh (2003) found that the thermal conductivity of the soil can double as the soil moisture content, which is the dominant control, increases from 0 to 25%.

Atmospheric Uptake of Water

The ability of the atmosphere to take up additional water is a function of numerous factors including air temperature, humidity, wind speed, and turbulent intensity

(Alfieri et al., 2007; Moene and van Dam, 2014). The first two factors determine the amount of water vapor that can be transferred to the atmosphere. The amount of water vapor that can be carried by the atmosphere increases exponentially with temperature. As a result, both the amount of water vapor that can be transported into the atmosphere and the rate of evapotranspiration increase with increasing temperature (Monteith and Unsworth, 2008). At the same time, this implies that the atmosphere can hold only a finite amount of water at any given time before becoming saturated. As the humidity approaches saturation, the atmosphere becomes less able to take up additional water vapor. This, in turn, impedes evapotranspiration.

The effects of wind speed and the intensity of turbulent mixing are similarly interconnected because both factors influence how quickly water vapor can be transferred to the atmosphere. In the case of wind speed, this is accomplished by moving moist air away from the surface and, thereby, maintaining a moisture gradient. By enhancing the amount of turbulent mixing, increased wind speed also increases the rate at which water vapor can be transported from the surface to the atmosphere (Moene and van Dam, 2014). Other factors that can increase the intensity of turbulent mixing, thus increasing the efficiency of evapotranspiration, include the surface roughness, which varies with distribution and density of the vegetation, and the temperature gradient between the surface and the atmosphere that drives convective mixing (Brutsaert, 1982).

Additional Controls on Transpiration

In addition to the environmental controls discussed previously, transpiration is controlled by the response of vegetation to environmental conditions. To understand the controls on evapotranspiration that are unique to plants, it is useful to review the physiology of leaves. The leaf surface is typically covered by a waxy film, the cuticle, which minimizes evaporative water loss except through pores called stomata. The plant uses these pores to uptake the carbon dioxide needed for photosynthesis (Hopkins, 1999). Since water loss via transpiration occurs through the stomata, then environmental factors that impact photosynthesis and the size of the stomata also influence water loss (Lambers et al., 1998).

There are multiple environmental factors that can influence water loss from plants due to transpiration. The first of these factors is the amount of available radiative energy. Since sunlight is necessary for photosynthesis, with the exception of the 8% of plant species that follow the crassulacean acid metabolism (CAM) photosynthetic pathway, the stomata tend to close as the level of incident solar radiation decreases. As a result, transpiration also decreases (Monteith and Unsworth, 2008; Jones, 2014). The stomata also close in response to decreasing humidity (increasing water vapor pressure deficit); this prevents excessive water loss when there is a high evaporative demand (Campbell and Norman, 1998). In contrast, plants tend to open their stomata as the ambient carbon dioxide level increases. This is because the assimilation rate of carbon dioxide is limited by the amount of energy available to drive photosynthesis; the excess carbon dioxide will not hasten the process (Moene and van Dam, 2014).

The effect of air temperature on transpiration is more complex. Photosynthesis tends to increase with increasing temperature until some optimal temperature is reached. As the temperature continues to increase, both photosynthetic activity and transpiration decline (Monteith and Unsworth, 2008; Jones, 2014). The optimal temperature, which varies between 20 °C and 40 °C depending on the plant species, is the temperature at which metabolic processes are maximized. If the temperature exceeds the optimal value, photosynthetic activity decreases to prevent cellular damage. In turn, the stomata close to block carbon dioxide uptake thereby also reducing transpiration (Moene and van Dam, 2014).

In-Situ Measurement Techniques

Numerous techniques have been developed to measure evapotranspiration. Examples of these methods range from mass balance-based approaches, such as lysimetry, to micrometeorological approaches, such as eddy covariance, that use measurements of turbulent flow to determine evapotranspiration. Each of the measurement approaches is built around its own unique set of theoretical and practical considerations. As a result, differing measurement techniques have differing strengths and weaknesses. Although a comprehensive analysis of all of the methods for measuring evapotranspiration is beyond the scope of this discussion, overviews of lysimetry and eddy covariance are given as representative approaches for mass balance-based and micrometeorological approaches, respectively.

Mass-Balance Techniques

As discussed above, the change in the amount of water in an agricultural field is equal to the difference in the water inputs and outputs including evapotranspiration. By taking steps, such as the installation of berms to prevent runoff, to eliminate the other output terms, evapotranspiration is directly proportional to the change in the storage term. The mass-balance techniques take advantage of this while using gravimetric, neutron scattering, or electrical conductance measurements to determine the change is the soil moisture content.

Weighing lysimetry, which is among the most well-established and regarded mass-balance techniques, can be used to illustrate the strength and limitations of the mass-balance approach. Beginning with the work of Pruitt and Angus (1960), significant advances have been made in the design of weighing lysimeters so that today the accuracy of evapotranspiration estimates can be better than 0.05 mm d^{-1} (Marek et al., 1988; Howell et al., 1991; Allen et al., 2011). Modern weighing lysimeters measure the change in soil moisture, thus evapotranspiration, based on the change in mass of a large isolated block of soil and associated vegetation using counterbalanced load cells or hydraulic scales. The block of soil can be either reconstructed to match the soil horizon of the surrounding field or, preferably, extracted as a soil monolith directly from the field. Typically, the depth of the soil block is between 1 m and 2 m to ensure that it extends beyond the rooting depth of the vegetation; similarly, the surface of the soil block generally ranges between 0.5 m^2 and 40 m^2 (Howell et al., 1985; Phene et al., 1989; Allen et al., 2011). Counterbalanced scales are necessary because soil water accounts for less than 20% of the total mass of the lysimeter (Howell et al., 1995).

The key strength of weighing lysimeters is their ability to very accurately measure evapotranspiration over a range of timescales from as little as 30 min to several days. As pointed out by Allen et al. (2011) among others, a properly calibrated and maintained weighing lysimeter can achieve very accurate measurements of evapotranspiration. Additionally, lysimeters require a much smaller footprint compared to micrometeorological techniques such as eddy covariance. Typically, the measurement area of a weighing lysimeter is on the order of few square meters; in contrast, the source area of an eddy covariance system is on the order of several hundred square meters.

The primary concern when using a lysimeter is its ability to properly represent the surrounding field. As pointed out by Allen et al. (2011), failure to represent the surrounding field conditions can result in errors in the field scale evapotranspiration estimates of 50% or more. For sparse or widely-spaced canopies, the relatively small measurement area may not adequately capture the variability of the surrounding field. Moreover, even in the visually homogenous monoculture of an agricultural field, the vegetation height and density within the lysimeter can differ significantly from the field as a whole (Kustas et al., 2015). Differences in phenology, density, and biomass of the plants within the lysimeter and the surrounding area can result significant errors when estimating evapotranspiration for the field as a whole (Martin et al., 2001; Allen et al., 2011; Alfieri et al., 2012). Discrepancies in soil structure and soil conditions within the lysimeter and the surrounding field can result in differences not only in crop phenology, vegetation density, and rooting structure, but also direct evaporation from the soil ((Allen and Fisher, 1990; Dugas and Bland, 1991; Allen et al., 2011). As a result, the evapotranspiration measurements within the lysimeter may not be representative.

Micrometeorological Techniques

Micrometeorological approaches, such as the Bowen ratio–energy balance and eddy covariance methods, take advantage of the principles of Monin–Obukhov similarity theory (Monin and Obukhov, 1954) and conservation of mass to measure evapotranspiration and other surface fluxes (Kaimal and Finnigan, 1994; Foken, 2008a). Monin–Obukhov similarity theory states that the vertical transport and turbulent properties within the surface boundary layer (the layer of air extending from the surface to a height of a few tens of meters where vertical turbulent fluxes are invariant with height) can be defined as universal functions of height above the surface (Stull, 1988; Ayra, 2001). As a result, measurements collected at some height above the surface within the surface boundary layer are identical to those at the surface.

Monin–Obukhov similarity theory, and thus the measurement techniques, is underpinned by a number of interconnected assumptions. The first of these is that the surface boundary layer is well-mixed and in equilibrium with the underlying surface. Similarly, these approaches are predicated on the assumption that transport is fully turbulent and in the vertical directions only. In turn, these assumptions imply that the measurements are collected over a uniform surface and there is no net horizontal (advective) transport of heat, water vapor, or other scalar quantities. These approaches are also predicated on the assumption of stationarity. In this context, stationarity implies that the bulk atmospheric conditions (for example, air temperature and atmospheric pressure) do not change during the measurement period.

Eddy covariance is the most prevalent of the micrometeorological techniques. It uses high-frequency ($\geq$ 10 Hz) measurements of the vertical wind velocity and the scalar quantity of interest to determine the turbulent flux. The flux is proportional to the covariance of the two quantities; for example, evapotranspiration, or more commonly the latent heat flux, its equivalent expressed in terms of energy, is determined according to:

$$\mathrm{ET}=1000\frac{\Delta t}{\rho_w}\overline{w'p'_v} \quad [3.1a]$$

$$\lambda E=\lambda_v\overline{w'p'_v} \quad [3.1b]$$

where ET is evapotranspiration (mm), Dt is length of the measurement period (s), r_w is the density of water (kg m^{-3}), w is the vertical wind velocity (m s^{-1}), r_v is the concentration of water vapor in the air (kg m^{-3}), λE is the latent heat flux (W m^{-2}), and l_v is the latent heat of vaporization (kg J^{-1}). The overbar denotes mean values while prime indicates deviations from the mean.

As with any measurement technique, eddy covariance has both advantages and shortcomings. The strength of the method is its ability to provide direct, independent, continuous, and nondestructive measurements of the turbulent fluxes. Additionally, the measurement is representative because it is area-integrated over a large source area, typically on the order of several hundred square meters. (Foken 2008a; Allen et al., 2011; Foken et al., 2012a). This does, however, suggest that upwind extent of the surface being measured, that is, the fetch, also has to be large to ensure the measurements represent the surface of interest; as a rule of thumb, the minimum fetch should one hundred times the measurement height (Leclerc and Thurtell, 1990). With an uncertainty on the order of 20 W m^{-2}, eddy covariance is the dominant method for collecting highly accurate measurements of surface fluxes (Meyers and Baldocchi, 2005; Foken, 2008a)

Nonetheless, because the eddy covariance method is underpinned by the continuity relationships and Monin–Obukhov similarity theory, it is highly sensitive to the underlying assumptions discussed above. Moreover, since those assumptions are rarely, if ever, fully satisfied, a suite of standard corrections and adjustments must be applied to the data during post-processing (Foken et al., 2004, 2012b; Foken 2008a). As a result, a thorough understanding of boundary–layer processes and micrometeorology is needed to implement eddy covariance and interpret the results.

Another issue with the eddy covariance method is the well-known closure problem. Although it can vary significantly from site to site, the energy available to drive the fluxes often exceeds the sum of the turbulent fluxes by 10% to 30% (Wilson et al., 2002; Foken, 2008b; Leuning et al., 2012). Although the causes of the closure problem are not fully understood and remain an area of active research, a range of reasons for the imperfect closure have been proposed. Some of these include a failure to fully account for the storage and other minor terms in the energy balance relationship, the development of advection and organized coherent structures as a result of surface heterogeneity, and imperfect sensor response (Mahrt, 1998; Foken, 2008b; Leuning et al., 2012; Kochendorfer et al., 2012; Frank et al., 2013). Regardless of the cause, closure is often forced by assuming the measurement of available energy is accurate and adjusting the turbulent fluxes maintaining a constant Bowen ratio (Twine et al., 2000). However, recent studies by Prueger et al. (2005), Alfieri et al. (2009), and others have suggested the underlying assumptions of this closure method might not be satisfied under all atmospheric conditions. As a result, caution is required when forcing closure with eddy covariance data.

Methods based on Reference Evapotranspiration

Another approach for estimating evapotranspiration from agricultural systems scales reference evapotranspiration derived from local meteorological data using empirically determined crop coefficients (Jensen, 1968; Allen et al., 1998). It is important to recognize that this approach does not estimate the actual evapotranspiration; instead, it provides an estimate of the maximum possible evapotranspiration from the crop given the environmental conditions and ample water. In brief, the crop evapotranspiration is estimated as the product of the reference evapotranspiration of a well-watered reference crop and a coefficient that accounts for the influence of crop-specific characteristics on evapotranspiration (Pereira et al., 2015). The reference evapotranspiration is determined using standardized forms of the Penman–Monteith relationship. The reference crop is a short cold-season grass for short crops less than 12 cm in height while it is alfalfa for tall crops near 50 cm in height (ASCE–EWRI, 2005). As such, this approach is both straight-forward to implement and able to provide a useful estimate of the crop evapotranspiration for practical applications.

However, other factors, such as local climate and soil conditions, crop varietal and phenological state, and irrigation and other agricultural practices, also influence evapotranspiration. Thus, the use of a generalized crop coefficient can introduce significant uncertainty into the evapotranspiration estimates obtained via this approach (Hunsaker et al., 2003; Allen et al., 2005; Gonzalez-Dugo et al., 2009). To overcome this uncertainty, either the generalized crop coefficients must be adjusted to account for site and crop-specific conditions or a local crop coefficient must be determined. However, the data needed to determine site-specific crop coefficients may not be readily available. This can be even more problematic for complex landscapes characterized by a mosaic of crop types or significant variability in vegetation cover where the utility of this approach can be particularly limited (Xia et al., 2016).

Remote Sensing of Evapotranspiration

A multitude of remote sensing-based approaches for estimating evapotranspiration have been developed in recent years (Kalma et al., 2008; Kustas and Anderson, 2009; Wang and Dickinson, 2012). These approaches take advantage of remote sensing data to estimate evapotranspiration across the continuum from sub-field to regional and global scales. For example, some models, such as the Surface Energy BALance (SEBAL; Bastiaanssen et al., 1998a,b) and Mapping Evapotranspiration with high Resolution and Internalized Calibration (METRIC; Allen et al., 2007a,b) models, use within-scene spatial variations in land-surface temperature to scale moisture flux between dry and wet endpoints. Other models, such as the Two-Source Energy Balance (TSEB; Norman et al., 1995; Kustas and Norman, 1999, 2000) model, along with the closely-related Atmosphere-Land Exchange Inverse (ALEXI) and flux disaggregation (DisALEXI) models (Anderson et al., 1997, 2005, 2007, 2011), use temporal change in land surface temperature to estimate evapotranspiration by simultaneously solving for the soil and canopy components of the surface energy budget. Although model performance varies somewhat depending on the timeframe and conditions considered, comparisons of the remote sensing-based estimates of evapotranspiration with in situ observations suggest that these models are typically accurate to within 20 to 30% (Kalma et al., 2008; Wang and Dickinson 2012).

To understand how remote sensing can be used to monitor evapotranspiration, the ALEXI/DisALEXI model can be used as an example. The ALEXI component of the model first uses multi-temporal land surface temperature data from geostationary and polar orbiting satellites to estimate the surface energy fluxes at a relatively coarse scale, typical near 5 km. Using higher resolution imagery from Landsat, MODIS, or airborne platforms, these fluxes are then disaggregated to finer resolution, which is typically 30 m to 60 m when using Landsat imagery, by the DisALEXI component of the model (Fig. 1). Moreover, recent advances that combine data from multiple sensor systems with data fusion techniques provide daily evapotranspiration estimates at a 30-m resolution (Cammalleri et al., 2014; Semmens et al., 2016; Gao et al., 2017).

More specifically, the ALEXI model uses the time-differenced surface temperature to explicitly estimate the surface energy fluxes from the canopy and soil. By using the difference between the temperature measured in the early morning, typically 1 h after local sunrise when the sensible heat flux is very close to zero, and the period of interest, the model's sensitivity to potential errors in the remote sensing based estimates of surface temperature is minimized (Anderson et al., 1997). To do this, the ALEXI model decomposes the surface temperature measured via remote sensing between the bare soil and canopy temperature by treating the measured value as the area-weighted average of the temperatures of the two surfaces. Then, using the surface temperature estimates for a given surface type, a family of equations relating the surface fluxes and temperature are solved simultaneously to determine the fluxes (Kustas and Norman, 1999). To account for the effects of plant stress, the original form of the model uses a

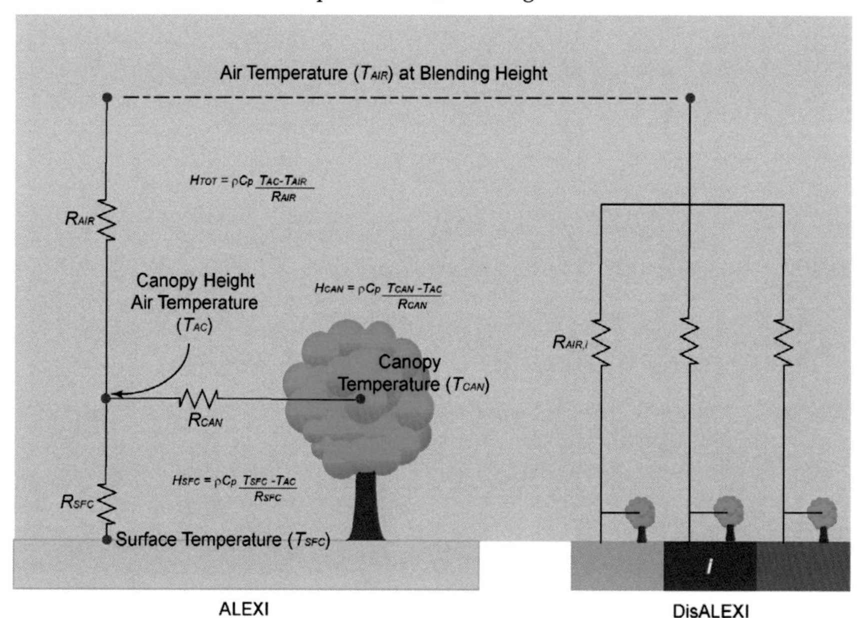

Fig. 1. Schematic of the thermal remote sensing-based ALEXI/DisALEXI model.

modified form of the Priestley–Taylor relationship that allows the Priestley–Taylor coefficient to vary in direct response to changes in canopy temperature. The modified Priestley–Taylor relationship used by the model is expressed as:

$$\lambda E_c = \alpha f_g \frac{\Delta}{\Delta + \gamma} R_{n_c} \qquad [4.1]$$

where λE_c is the latent heat flux from the canopy (transpiration), α is the Priestley–Taylor coefficient, f_g is the greenness fraction, Δ is the slope of the water vapor pressure-temperature curve, γ is the psychrometric constant, R_{nc} is the net radiation of the canopy. More recently, model formulations have been developed that use the Penman–Monteith relationship (Colaizzi et al., 2012, 2014) or a light use efficiency-based approach (Anderson et al., 2008) to determine transpiration.

The ALEXI model fluxes can be spatially disaggregated to finer spatial scales using higher-resolution thermal imagery from Landsat, MODIS, or another platform. Specifically, the DisALEXI component of the model system solves the same family of equations within each ALEXI pixel using higher resolution data while maintaining the same regional atmospheric forcings and total flux.

Regardless of the underlying modeling framework, the chief advantage of remote sensing-based approaches for estimating evapotranspiration is the ability to provide spatially distributed estimates across a range of scales from subfield to regional and continental scales. Unfortunately, platforms with a high spatial resolution tend to have a lower temporal resolution. In other words, the return interval for higher resolution satellites tends to be longer than for satellites with a relatively coarse spatial resolution. For example, while MODIS provides imagery on a daily basis, it is at a 1-km spatial resolution; in contrast, Landsat provides imagery less frequently, once every 16 d, but at a much higher spatial resolution, 30 m. Cloud cover further complicates matters by preventing the collection of useful data during a given overpass. Although interpolation methods have been developed to fill the gaps between image acquisitions, their utility is limited for lengthy gaps (Alfieri et al., 2017). However, data fusion approaches that can leverage remote sensing observations from multiple satellite platforms (e.g., Cammalleri et al., 2014) may provide reliable daily evapotranspiration estimates useful to agriculture.

Conclusion

As competing demands for limited water resources continue to grow, accurate estimates of evapotranspiration become increasingly important to manage water resources effectively. With regard to agriculture, information about evapotranspiration is critical to maximize crop yield, ensure agricultural sustainability, safeguard food security, and foster rural development (UN FAO, 2002). There are multiple techniques for collecting these data with each having its own strengths and weaknesses. Some examples of field-based techniques include lysimetery and eddy covariance. While these techniques can provide continuous measurements, their footprint is limited to field scales. Other techniques that combine crop coefficients with estimates of potential or reference evapotranspiration derived from local meteorological data can also provide continuous flux estimates. However, these techniques are semi-empirical and may be difficult to extend over heterogeneous agricultural landscapes with a patchwork of different crop types. In contrast, the remote sensing-based methods can provide estimates of evapotranspiration on much larger spatial scales with some satellite sensors providing these estimates at subfield resolutions. However, the temporal resolution of field-scale satellite data tends to be much lower. Data fusion approaches may hold some promise in generating field scale evapotranspiration estimates for agricultural areas.

References

Abu-Hamdeh, N.H. 2003. Thermal properties of soils as affected by density and water content. Biosystems Eng. 86:97–102. doi:10.1016/S1537-5110(03)00112-0

Abu-Hamdeh, N.H., and R.C. Reeder. 2000. Soil thermal conductivity: Effects of density, moisture, salt content, and organic matter. Soil Sci. Soc. Am. J. 64:1285–1290. doi:10.2136/sssaj2000.6441285x

Alexandratos, N., and J. Bruinsma. 2012. World agriculture towards 2030/2050: 2012 revision. Food and Agriculture Organization of the United Nations, Rome, Italy.

Alfieri, J.G., P.D. Blanken, D.N. Yates, and K. Steffen. 2007. Variability in the environmental factors driving evapotranspiration from a grazed rangeland during severe drought conditions. J. Hydrometeorol. 8:207–220. doi:10.1175/JHM569.1

Alfieri, J.G., P.D. Blanken, D. Smith, and J. Morgan. 2009. Concerning the measurement and magnitude of heat, water vapor, and carbon dioxide exchange from a semiarid grassland. J. Appl. Meteorol. Climatol. 48:982–996. doi:10.1175/2008JAMC1873.1

Alfieri, J.G., W.P. Kustas, J.H. Prueger, L.E. Hipps, S.R. Evett, J.B. Basara, C.M.U. Neale, A.N. French, P. Colaizzi, N. Agam, M.H. Cosh, J.L. Chaves, and T.H. Howell. 2012. On the discrepancy between eddy covariance and lysimetry-based surface flux measurements under strongly advective conditions. Adv. Water Resour. 50:62–78. doi:10.1016/j.advwatres.2012.07.008

Alfieri, J.G., M.C. Anderson, W.P. Kustas, and C. Cammalleri. 2017. Effect of the revisit interval and temporal upscaling methods on the accuracy of remotely sensed evapotranspiration estimates. Hydrol. Earth Syst. Sci. 21:83–98. doi:10.5194/hess-21-83-2017

Allen, R.G., and D.K. Fisher. 1990. Low-cost electronic weighing lysimeters. Trans. ASABE 33:1823–1833. doi:10.13031/2013.31546

Allen, R.G., L.S. Pereira, D. Raes, and M. Smith. 1998. Crop evapotranspiration (guidelines for computing crop water requirements). FAO irrigation and drainage paper no. 56, Food and Agriculture Organization of the United Nations, Rome, Italy.

Allen, R.G., A.J. Clemmens, C.M. Burt, K. Solomon, and T. O'Halloran. 2005. Prediction accuracy for projectwide evapotranspiration using crop coefficients and reference evapotranspiration. J. Irrig. Drain. Eng. 131:24–36. doi:10.1061/(ASCE)0733-9437(2005)131:1(24)

Allen, R.G., M. Tasumi, and R. Trezza. 2007a. Satellite-based energy balance for mapping evapotranspiration with internalized calibration (METRIC)—Model. J. Irrig. Drain. Eng. 133:380–394. doi:10.1061/(ASCE)0733-9437(2007)133:4(380)

Allen, R.G., M. Tasumi, A. Morse, R. Trezza, J.L. Wright, W. Bastiaanssen, W. Kramber, I. Lorite, and C.R. Robison. 2007b. Satellite-based energy balance for mapping evapotranspiration with internalized calibration (METRIC)—Applications. J. Irrig. Drain. Eng. 133:395–406. doi:10.1061/(ASCE)0733-9437(2007)133:4(395)

Allen, R.G., L.S. Pereira, T.A. Howell, and M.E. Jensen. 2011. Evapotranspiration information reporting: I. Factors governing measurement accuracy. Agric. Water Manage. 98:899–920. doi:10.1016/j.agwat.2010.12.015

Anderson, M.C., J.M. Norman, G.R. Diak, W.P. Kustas, and J.R. Mecikalski. 1997. A two-source time-integrated model for estimating surface fluxes using thermal infrared remote sensing. Remote Sens. Environ. 60:195–216. doi:10.1016/S0034-4257(96)00215-5

Anderson, M.C., J.M. Norman, J.R. Mecikalski, R.D. Torn, W.P. Kustas, and J.B. Basara. 2005. A multiscale remotes sensing model for disaggregated regional fluxes to micrometeorological scales. J. Hydrometeorol. 5:344–363.

Anderson, M.C., J.M. Norman, J.R. Mecikalski, J.A. Otkin, and W.P. Kustas. 2007. A climatological study of evapotranspiration and moisture stress across the continental United States based on thermal remote sensing: 1. Model formulation. J. Geophys. Res. 112. doi:10.1029/2006JD007506

Anderson, M.C., J.M. Norman, W.P. Kustas, R. Houborg, P.J. Starks, and N. Agam. 2008. A thermal-based remote sensing technique for routine mapping of land-surface carbon, water and energy fluxes from field to regional scales. Remote Sens. Environ. 112:4227–4241. doi:10.1016/j.rse.2008.07.009

Anderson, M.C., W.P. Kustas, J.M. Norman, C.R. Hain, J.R. Mecikalski, L. Schultz, M.P. Gonzalez-Dugo, C. Cammalleri, G. d'Urso, A. Pimstein, and F. Gao. 2011. Mapping daily evapotranspiration at field to continental scales using geostationary and polar orbiting satellite imagery. Hydrol. Earth Syst. Sci. 15:223–239. doi:10.5194/hess-15-223-2011

Anderson, M.C., R.G. Allen, A. Morse, and W.P. Kustas. 2012. Use of Landsat thermal imagery in monitoring evapotranspiration and managing water resources. Remote Sens. Environ. 122:50–65. doi:10.1016/j.rse.2011.08.025

ASCE-EWRI. 2005. The ASCE standardized reference evapotranspiration equation. Report of the Task Committee on Standardization of Reference Evapotranspiration. Environmental and Water Resources Institute, American Society of Civil Engineers, Reston, VA.

Ayra, P.S. 2001. Introduction to micrometeorology. Academic Press, San Diego, CA.

Basile, B., J. Marsal, M. Mata, X. Vallverd, J. Bellvert, and J. Girona. 2014. Phenological sensitivity of cabernet sauvignon to water stress: Vine physiology and berry composition. Am. J. Enol. Vitic. 67: 452-461.

Bastiaanssen, W.G.M., M. Menenti, R.A. Feddes, and A.A.M. Holtslag. 1998a. A remote sensing surface energy balance algorithm for land (SEBAL): 1. Formulation. J. Hydrol. 212-213:198–212. doi:10.1016/S0022-1694(98)00253-4

Bastiaanssen, W.G.M., H. Pelgrum, J. Wang, Y. Ma, J.F. Moreno, G. Roerink, and T. van der Wal. 1998b. A remote sensing surface energy balance algorithm for land (SEBAL): 2. Validation. J. Hydrol. 212-213:213–229. doi:10.1016/S0022-1694(98)00254-6

Bastiaanssen, W.G.M., R. Allen, P. Droogers, G. d'Urso, and P. Steduto. 2007. Twenty-five years modeling irrigated and drained soils: State of the art. Agric. Water Manage. 92:111–125. doi:10.1016/j.agwat.2007.05.013

Bravdo, B.A., Y. Hepner, C. Loinger, S. Cohen, and H. Tabacman. 1985. Effect of irrigation and crop level on growth, yield and wine quality of cabernet sauvignon. Am. J. Enol. Vitic. 36:132–139.

Brunt, D. 1932. Notes on radiation in the atmosphere. Q. J. R. Meteorol. Soc. 58:389–420. doi:10.1002/qj.49705824704

Brutsaert, W. 1975. On a derivable formula for long-wave radiation from clear skies. Water Resour. Res. 11:742–744. doi:10.1029/WR011i005p00742

Brutsaert, W. 1982. Evaporation into the atmosphere. D. Reidel Publishing, Dordrecht, The Netherlands. doi:10.1007/978-94-017-1497-6

Cammalleri, C., M.C. Anderson, F. Gao, C.R. Hain, and W.P. Kustas. 2014. Mapping daily evapotranspiration at field scales over rainfed and irrigated agricultural areas using remote sensing data fusion. Agric. For. Meteorol. 186:1–11. doi:10.1016/j.agrformet.2013.11.001

Campbell, G.S., and J.M. Norman. 1998. Introduction to environmental biophysics. Springer Verlag, New York. doi:10.1007/978-1-4612-1626-1

Cellier, P., G. Richard, and P. Robin. 1996. Partition of sensible heat fluxes into base soil and atmosphere. Agric. For. Meteorol. 82:245–265. doi:10.1016/0168-1923(95)02328-3

Colaizzi, P.D., W.P. Kustas, M.C. Anderson, N. Agam, J.A. Tolk, S.R. Evett, T.A. Howell, P.H. Gowda, and S.A. O'Shaughnessy. 2012. Two-source model estimates of evapotranspiration using component and composite surface temperatures. Adv. Water Resour. 50:134–151. doi:10.1016/j.advwatres.2012.06.004

Colaizzi, P.D., N. Agam, J.A. Tolk, S.R. Evett, T.A. Howell, P.H. Gowda, S.A. O'Shaughnessy, W.P. Kustas, and M.C. Anderson. 2014. Two-source energy balance model to calculate E, T, and ET: Comparison of priestley-taylor and penman-monteith formulations and two time scaling methods. Trans. ASABE 57:479–498.

Dingman, S.L. 2002. Physical hydrology. Prentice Hall Publishing, Upper Saddle Creek, NJ.

Dugas, W.A., and W.L. Bland. 1991. Springtime soil temperatures in lysimeters in Central Texas. Soil Sci. 152:87–91. doi:10.1097/00010694-199108000-00004

Erekul, O., and W. Kohn. 2006. Effect of weather and soil conditions on yield components and bread-making quality of winter wheat (Triticum aestivum L.) and winter triticale

(Triticosecale Wittm.) varieties in North-East Germany. J. Agron. Crop Sci. 192:452–464. doi:10.1111/j.1439-037X.2006.00234.x

Escalona, J.M., J. Flexas, and H. Medrano. 1999. Stomatal and nonstomatal limitations of photosynthesis under water stress in field-grown grapevines. Aust. J. Plant Physiol. 26:421–433. doi:10.1071/PP99019

Esteban, M.A., M.J. Villanueva, and J.R. Lissarrague. 2001. Effect of irrigation on changes in the anthocyanin composition of the skin of cv. Tempranillo (*Vitis vinifera* L.) grape berries during ripening. J. Sci. Food Agric. 81:409–420. doi:10.1002/1097-0010(200103)81:4<409::AID-JSFA830>3.0.CO;2-H

Eltbaakh, Y.A., M.H. Ruslan, M.A. Alghoul, M.Y. Othman, K. Sopian, and T.M. Razykov. 2012. Solar attenuation by aerosols: An overview. Renew. Sustain. Energy Rev. 16:4264–4276. doi:10.1016/j.rser.2012.03.053

Foken, T., M. Göckede, M. Mauder, L. Mahrt, B. Amiro, and W. Munger. 2004. Post-field data quality control. In: X. Lee, W. Massman, and B. Law, editors, Handbook of micrometeorology. Kluwer Academic Publishers, Dordrecht, The Netherlands. p. 181–208.

Foken, T. 2008a. Micrometeorology. Springer-Verlag, Berlin, The Germany.

Foken, T. 2008b. The energy balance closure problem: An overview. Ecol. Appl. 18:1351–1367. doi:10.1890/06-0922.1

Foken, T., M. Aubinet, and R. Leuning. 2012a. The eddy covariance method. In: M. Aubinet, T. Vesala, and D. Papela, editors, Eddy covariance. Springer-Verlag, Berlin. p. 1–19. doi:10.1007/978-94-007-2351-1_1

Foken, T., R. Leuning, S.P. Oncley, M. Mauder, and M. Aubinet. 2012b. Eddy corrections and data quality control. In: M. Aubinet, T. Vesala, and D. Papela, editors, Eddy covariance. Springer-Verlag, Berlin. p. 85–131. doi:10.1007/978-94-007-2351-1_4

Frank, J.M., W.J. Massman, and B.E. Ewers. 2013. Underestimates of sensible heat flux due to vertical velocity measurement errors in non-orthogonal sonic anemometers. Agric. For. Meteorol. 171-172:72–81. doi:10.1016/j.agrformet.2012.11.005

Franzluebbers, A.J. 2002. Water infiltration and soil structure related to organic matter and its stratification with depth. Soil Tillage Res. 66:197–205. doi:10.1016/S0167-1987(02)00027-2

Frenken, K., and V. Gillet. 2012. Irrigation water requirement and water withdrawal by country. Food and Agriculture Organization of the United Nations, Rome, Italy.

Friedl, M.A., and F.W. Davis. 1994. Sources of variation in radiometric surface temperature over a tallgrass prairie. Remote Sens. Environ. 48:1–17. doi:10.1016/0034-4257(94)90109-0

Gao, F., M.C. Anderson, X. Zhang, Z. Yang, J.G. Alfieri, W.P. Kustas, R. Mueller, D.M. Johnson, and J.H. Prueger. 2017. Toward mapping crop progress at field scales through fusion of Landsat and MODIS imagery. Remote Sens. Environ. 188:9–25. doi:10.1016/j.rse.2016.11.004

Gonzalez-Dugo, M.P., C.M.U. Neale, L. Mateos, W.P. Kustas, J.H. Prueger, M.C. Anderson, and F. Li. 2009. A comparison of operational remote sensing-based models for estimating crop evapotranspiration. Agric. For. Meteorol. 149:1843–1853. doi:10.1016/j.agrformet.2009.06.012

Guttieri, M.J., J.C. Stark, K. O'Brien, and E. Souza. 2001. Relative sensitivity of spring wheat grain yield and quality parameters to moisture deficit. Crop Sci. 41:327–335. doi:10.2135/cropsci2001.412327x

Hamza, M.A., and W.K. Anderson. 2005. Soil compaction in cropping systems: A review of the nature, causes and possible solutions. Soil Tillage Res. 82:121–145. doi:10.1016/j.still.2004.08.009

Hanasaki, N., S. Kanae, T. Oki, K. Masuda, K. Motoya, N. Shirakawa, Y. Shen, and K. Tanaka. 2008. An integrated model for the assessment of global water resources. Part 2: Applications and assessments. Hydrol. Earth Syst. Sci. 12:1027–1037. doi:10.5194/hess-12-1027-2008

Herb, W.R., B. Janke, O. Mohseni, and H.G. Stefan. 2008. Ground surface temperature simulation for different land covers. J. Hydrol. 356:327–343. doi:10.1016/j.jhydrol.2008.04.020

Hillel, D. 1998. Environmental soil physics. Academic Press, San Diego.

Hopkins, W.G. 1999. Introduction to plant physiology. John Wiley & Sons, New York.

Howell, T.A., R.L. McCormick, and C.J. Phene. 1985. Design and installation of large weighing lysimeters. Trans. ASABE 28:106–112. doi:10.13031/2013.32212

Howell, T.A., A.D. Schneider, and M.E. Jensen. 1991. History of lysimeter design and use for evapotranspiration measurements. In: R.G. Allen, T.A. Howell, W.O. Pruitt, L. Walter, and M.E. Jensen, editors, Lysimeters for evapotranspiration and environmental measurements. American Society of Civil Engineers, New York. p. 1–9.

Howell, T.A., A.D. Schneider, D.D. Dusek, T.H. Marek, and J.L. Steiner. 1995. Calibration and scale performance of the Bushland weighing lysimeters. Trans. ASABE 38:1019–1024. doi:10.13031/2013.27918

Howell, T.A. 2001. Enhancing water use efficiency in irrigated agriculture. Agron. J. 93:281–289. doi:10.2134/agronj2001.932281x

Humes, K.S., W.P. Kustas, M.S. Moran, W.D. Nicols, and M.A. Weltz. 1994. Variability of emissivity and surface temperature over a sparsely vegetated surface. Water Resour. Res. 30:1299–1310. doi:10.1029/93WR03065

Hunsaker, D.J., P.J. Pinter, E. Barnes, and B.A. Kimball. 2003. Estimating cotton evapotranspiration crop coefficients with a multispectral vegetation index. Irrig. Sci. 22:95–104. doi:10.1007/s00271-003-0074-6

Idso, S.B., R.D. Jackson, R.J. Reginato, B.A. Kimball, and F.S. Nakayama. 1975. The dependence of bare soil albedo on soil water content. J. Appl. Meteorol. 14:109–113. doi:10.1175/1520-0450(1975)014<0109:TDOBSA>2.0.CO;2

Iglesias, A., and L. Garrote. 2015. Adaptation strategies for agricultural water management under climate change in Europe. Agric. Water Manage. 155:113–124. doi:10.1016/j.agwat.2015.03.014

Iziomon, M.G., and H. Mayer. 2002. On the variability and modelling of surface albedo and long-wave radiation components. Agric. For. Meteorol. 111:141–152. doi:10.1016/S0168-1923(02)00013-8

Jensen, M.E. 1968. Water consumption by agricultural plants. In: T.T. Kozlowski, editor, Water deficits and plant growth: Development, control, and measurement. Academic Press, New York. p. 235–386.

Jones, H.G. 2014. Plants and microclimates: A quantitative approach to environmental plant physiology. Cambridge Univ. Press, Cambridge, U.K.

Kaimal, J.C., and J.J. Finnigan. 1994. Atmospheric boundary layer flows: Their structure and measurement. Oxford Univ. Press, Oxford, U.K.

Kalma, J.D., T.R. McVicar, and M.F. McCabe. 2008. Estimating land surface evaporation: A review of methods using remotely sensed surface temperature data. Surv. Geophys. 29:421–469. doi:10.1007/s10712-008-9037-z

Kim, D., and V. Ramanathan. 2008. Solar radiation budget and radiative forcing due to aerosols and clouds. J. Geophys. Res. 113. doi:10.1029/2007JD008434

Kochendorfer, J., T.P. Meyers, J. Frank, W.J. Massman, and M.W. Heuer. 2012. How well can we measure the vertical wind speed? Implications for fluxes of energy and mass. Boundary-Layer Meteorol. 145:383–398. doi:10.1007/s10546-012-9738-1

Kopp, G., and J.L. Lean. 2011. A new, lower value of total solar irradiance: Evidence and climate significance. Geophys. Res. Lett. 38. doi:10.1029/2010GL045777

Kustas, W.P., and J.M. Norman. 1999. Evaluation of soil and vegetation heat flux predictions using a simple two-source model with radiometric temperatures for partial canopy cover. Agric. For. Meteorol. 94:13–29. doi:10.1016/S0168-1923(99)00005-2

Kustas, W.P., and J.M. Norman. 2000. A two-source energy balance approach using directional radiometric temperature observations for sparse canopy covered surfaces. Agron. J. 92:847–854. doi:10.2134/agronj2000.925847x

Kustas, W.P., and M.C. Anderson. 2009. Advances in thermal infrared remote sensing for land surface modeling. Agric. For. Meteorol. 149:2071–2081. doi:10.1016/j.agrformet.2009.05.016

Kustas, W.P., J.G. Alfieri, and N. Agam. 2015. Reliable estimation of water use at field scale in an irrigated agricultural region with strong advection. Irrig. Sci. 33:325–338. doi:10.1007/s00271-015-0469-1

Lambers, H., F.S. Chapin, and T.L. Pons. 1998. Plant physiological ecology. Springer Verlag, New York. doi:10.1007/978-1-4757-2855-2

Leclerc, M.Y., and G.W. Thurtell. 1990. Footprint predictions of scalar fluxes using a Markovian analysis. Boundary-Layer Meteorol. 52:247–258. doi:10.1007/BF00122089

Leuning, R., E. van Gorsel, W.J. Massman, and P.R. Isaac. 2012. Reflections on the surface energy imbalance problem. Agric. For. Meteorol. 156:65–74. doi:10.1016/j.agrformet.2011.12.002

Mahrt, L. 1998. Flux sampling errors for aircraft and towers. J. Atmos. Ocean. Technol. 15:416–429. doi:10.1175/1520-0426(1998)015<0416:FSEFAA>2.0.CO;2

Marek, T.H., A.D. Schneider, T.A. Howell, and L.L. Ebeling. 1988. Design and construction of large weighing monolithic lysimeters. Trans. ASABE 31:477–484. doi:10.13031/2013.30734

Martin, E.C., A.S. de Oliveira, A.D. Folta, E.J. Pegelow, and D.C. Slack. 2001. Development and testing of a small weighable lysimeter system to assess water sue by shallow-rooted crops. Trans. ASABE 44:71–78. doi:10.13031/2013.2309

Meyers, T.P., and D.D. Baldocchi. 2005. Current micrometerological flux methodologies with applications in agriculture. In: J.L. Hatfield and J.M. Baker, editors, Micrometeorology in agricultural systems. Agronomy Society of America, Madison, Wisconsin. p. 381–396.

Meyers, T.P., and S.E. Hollinger. 2004. An assessment of storage terms in the surface energy balance of maize and soybean. Agric. For. Meteorol. 125:105–115. doi:10.1016/j.agrformet.2004.03.001

Mira, M., E. Valor, R. Boluda, V. Caselles, and C. Coll. 2007. Influence of soil water content on the thermal infrared emissivity of bare soils: Implication for land surface temperature determination. J. Geophys. Res. 112. doi:10.1029/2007JF000749

Moene, A.F., and J.C. van Dam. 2014. Transport in the atmosphere-vegetation-soil continuum. Cambridge Univ. Press, Cambridge, U.K.

Monin, A.S., and A.M. Obukhov. 1954. Basic laws of turbulent mixing in the ground layer of the atmosphere. Tr. Akad. Nauk SSSR Geophiz. Inst. 151:163–187.

Monteith, J.L., and M.H. Unsworth. 2008. Principles of environmental physics. Academic Press, San Diego.

Nemani, R., and S. Running. 1997. Land cover characterization using multitemporal red, near-IR data from NOAA/AVHRR. Ecol. Appl. 7:79–90. doi:10.1890/1051-0761(1997)007[0079:LCCUMR]2.0.CO;2

Norman, J.M., and F. Becker. 1995. Terminology in thermal infrared remote sensing of natural surfaces. Agric. For. Meteorol. 77:153–166. doi:10.1016/0168-1923(95)02259-Z

Norman, J.M., W.P. Kustas, and K.S. Humes. 1995. A two-source approach for estimating soil and vegetation energy fluxes from observations of directional radiometric surface temperature. Agric. For. Meteorol. 77:263–293. doi:10.1016/0168-1923(95)02265-Y

Oke, T.R. 1987. Boundary layer climates. Routledge, London.

Oweis, T.Y., H.J. Farahani, and A.Y. Hachum. 2011. Evapotranspiration and water use of full and deficit irrigated cotton in the Mediterranean environment in northern Syria. Agric. Water Manage. 98:1239–1248. doi:10.1016/j.agwat.2011.02.009

Ozturk, A., and F. Aydin. 2004. Effect of water stress at various growth stages on some quality characteristics of winter wheat. J. Agron. Crop Sci. 190:93–99. doi:10.1046/j.1439-037X.2003.00080.x

Panozzo, J.F., and H.A. Eagles. 2000. Cultivar and environmental effects on quality characters in wheat. II. Protein. Aust. J. Agric. Res. 51:629–636. doi:10.1071/AR99137

Pereira, L.S., R.G. Allen, M. Smith, and D. Raes. 2015. Crop evapotranspiration estimation with FAO56: Past and future. Agric. Water Manage. 147:4–20. doi:10.1016/j.agwat.2014.07.031

Perry, C., P. Steduto, R.G. Allen, and C.M. Burt. 2009. Increasing productivity in irrigated agriculture: Agronomic constraints and hydrological realities. Agric. Water Manage. 96:1517–1524. doi:10.1016/j.agwat.2009.05.005

Phene, C.J., R.L. McCormick, K.R. Davis, J.D. Pierre, and D.W. Meek. 1989. A lysimeter feedback irrigation controller system for evapotranspiration measurements and real time irrigation scheduling. Trans. ASABE 32:477–484. doi:10.13031/2013.31029

Pietola, L., R. Horn, and M. Yu-Halla. 2005. Effects of trampling by cattle on the hydraulic and mechanical properties of soil. Soil Tillage Res. 82:99–108. doi:10.1016/j.still.2004.08.004

Prueger, J.H., J.L. Hatfield, W.P. Kustas, L.E. Hipps, J.I. MacPherson, and T.B. Parkin. 2005. Tower and aircraft eddy covariance measurements of water vapor, energy and carbon dioxide fluxes during SMACEX. J. Hydrometeorol. 6:954–960. doi:10.1175/JHM457.1

Pruitt, W.O., and D.F. Angus. 1960. Large weighing lysimeter for measuring evapotranspiration. Trans. ASABE 3:13–18. doi:10.13031/2013.41105

Qadir, M., T.M. Boers, S. Schubert, A. Ghafoor, and G. Murtaza. 2003. Agricultural water management in water-starved countries: Challenges and opportunities. Agric. Water Manage. 62:165–185. doi:10.1016/S0378-3774(03)00146-X

Radford, B.J., D.F. Yule, D. McGarry, and C. Playford. 2001. Crop responses to applied soil compaction and to compaction repair treatments. Soil Tillage Res. 61:157–166. doi:10.1016/S0167-1987(01)00194-5

Santanello, J.A., and M.A. Friedl. 2003. Diurnal covariation in soil heat flux and net radiation. J. Appl. Meteorol. 42:851–862. doi:10.1175/1520-0450(2003)042<0851:DCISHF>2.0.CO;2

Santos, T.P., C.M. Lopes, M.L. Rodriguez, C.R. de Souza, J.P. Maroco, J.S. Pereira, J.R. Silva, and M.M. Chaves. 2003. Partial rootzone drying: Effects on growth and fruit quality of field-grown grapevines (Vitis vinifera). Funct. Plant Biol. 30:663–671. doi:10.1071/FP02180

Saxton, K.E., and W.J. Rawls. 2006. Soil water characteristic estimates by texture and organic matter for hydrologic solutions. Soil Sci. Soc. Am. J. 70:1569–1578. doi:10.2136/sssaj2005.0117

Scanlon, B.R., R.C. Reedy, J.B. Gates, and P.H. Gowda. 2010. Impacts of agroecosytems on groundwater resources in the Central High Plains, USA. Agric. Ecosyst. Environ. 139:700–713. doi:10.1016/j.agee.2010.10.017

Schultz, B., and D. De Wrachien. 2002. Irrigation and drainage systems research and development in the 21st century. Irrigation and Drainage 51:311–327. doi:10.1002/ird.67

Semmens, K.A., M.C. Anderson, W.P. Kustas, F. Gao, J.G. Alfieri, L.G. McKee, J.H. Prueger, C.R. Hain, and C. Cammalleri. 2016. Monitoring daily evapotranspiration over two California vineyards using Landsat 8 in a multi-sensor data fusion approach. Remote Sens. Environ. 185:155–170. doi:10.1016/j.rse.2015.10.025

Song, J. 1998. Diurnal asymmetry in surface albedo. Agric. For. Meteorol. 92:181–189. doi:10.1016/S0168-1923(98)00095-1

Stull, R.B. 1988. An introduction to boundary layer meteorology. Kluwer Academic Publishers, Dordrecht, The Netherlands. doi:10.1007/978-94-009-3027-8

Swinbank, W.C. 1963. Long-wave radiation from clear skies. Q. J. R. Meteorol. Soc. 89:339–348. doi:10.1002/qj.49708938105

Tarjuelo, J.M., J.A. Rodriguez-Diaz, R. Abadia, and E. Camaco. 2015. Efficient water and energy use in irrigation modernization: Lessons from Spanish case studies. Agric. Water Manage. 162:67–77. doi:10.1016/j.agwat.2015.08.009

Thenkabail, P.S., M.A. Hanjra, V. Dheeravath, and M. Gumma. 2010. Global croplands and their water use from remote sensing and non-remote sensing perspectives. In: Q. Weng, editor, Advances in environmental remote sensing: Sensors, algorithms and applications. Taylor and Francis, CRC Press, Boca Raton, FL. p. 383–419.

Turral, H., M. Svendsen, and J.M. Faures. 2010. Investing in irrigation: Reviewing the past and looking to the future. Agric. Water Manage. 97:551–560. doi:10.1016/j.agwat.2009.07.012

Twine, T.E., W.P. Kustas, J.M. Norman, D.R. Cook, P.R. Houser, T.P. Meyers, J.H. Prueger, P.J. Starks, and M.L. Wesley. 2000. Correcting eddy–covariance flux underestimates over a grassland. Agric. For. Meteorol. 103:279–300. doi:10.1016/S0168-1923(00)00123-4

UN ESA. 2011. World population prospects: 2010 revision. Department of Economic and Social Affairs, Population Division of the United Nations, New York.

UN FAO. 2002. Crops and drops: Making the best use of water for agriculture. Food and Agriculture Organization of the United Nations, Rome, Italy.

UN FAO. 2003. Unlocking the water potential of agriculture. Food and Agriculture Organization of the United Nations, Rome, Italy.

UN FAO. 2009. How to feed the world in 2050. Food and Agriculture Organization of the United Nations, Rome, Italy.

Van Bavel, C.H.M., and D. Hillel. 1976. Calculating potential and actual evaporation from a bare soil surface by simulations of concurrent flow of water and heat. Agric. Meteorol. 17:453–476. doi:10.1016/0002-1571(76)90022-4

Van de Griend, A.A., and M. Owe. 1993. On the relationship between thermal emissivity and the normalized difference vegetation index for nature surfaces. Int. J. Remote Sens. 14:1119–1131. doi:10.1080/01431169308904400

Wang, K., and R.E. Dickinson. 2012. A review of global terrestrial evapotranspiration: Observation, modeling, climatology, and variability. Rev. Geophys. 50:RG2005.

Wang, S. 2005. Dynamics of surface albedo of a boreal forest and its simulation. Ecol. Modell. 183:477–494. doi:10.1016/j.ecolmodel.2004.10.001

Wallace, J.S. 2000. Increasing agricultural water use efficiency to meet future food production. Agric. Ecosyst. Environ. 82:105–119. doi:10.1016/S0167-8809(00)00220-6

Wetzel, P.J., and J.-T. Chang. 1987. Concerning the relationship between evapotranspiration and soil moisture. J. Clim. Appl. Meteorol. 26:18–27. doi:10.1175/1520-0450(1987)026<0018:CTRBEA>2.0.CO;2

Wilson, K., A. Goldstein, E. Falge, M. Aubinet, D. Baldocchi, P. Berbigier, C. Bernhofer, R. Ceulemans, H. Dolman, C. Field, A. Grelle, A. Ibrom, B.E. Law, A. Kowalski, T Meyers, J. Moncrieff, R. Monson, W. Oechel, J. Tenhunen, R. Valentini, and S. Verma. 2002. Energy balance closure at Fluxnet sites. Agric. For. Meteorol. 113:223–243. doi:10.1016/S0168-1923(02)00109-0

Xia, T., W.P. Kustas, M.C. Anderson, J.G. Alfieri, F. Gao, L.G. McKee, J.H. Prueger, H.M.E. Geli, C.M.U. Neale, L. Sanchez, M.M. Alsina, and Z. Wang. 2016. Mapping evapotranspiration with high resolution aircraft imagery over vineyards using one- and two-source modeling schemes. Hydrol. Earth Syst. Sci. 20:1523–1545. doi:10.5194/hess-20-1523-2016

Photosynthesis

Elizabeth A. Ainsworth*

Abstract

The process by which crops harvest solar energy, convert it to chemical energy, and use it to fix carbon dioxide (CO_2) for growth is photosynthesis. This process has evolved over 2000 million yr and ultimately provides the food, feed, fiber, and fuel for human life. Photosynthesis evolved to be extremely sensitive to environmental factors including light, temperature, CO_2 concentration ([CO_2]) and water availability. While many of the protein complexes and enzymes involved in harvesting light energy and fixing CO_2 are conserved across higher plants, distinct pathways of photosynthesis have evolved in response to environmental factors. This chapter describes the photosynthetic process, the sensitivity of photosynthesis to environmental conditions, and ongoing efforts to improve photosynthesis in crops.

Photosynthesis is the primary source of energy for all of life on Earth and is the process by which plants harvest sunlight energy, convert it to chemical energy, and use it to fix atmospheric CO_2 to produce sugars. In doing so, plants emit oxygen (O_2) and water vapor to the atmosphere. Since the process of photosynthesis evolved over 2000 million years ago, it has shaped the atmospheric composition and climate of the planet, and today photosynthesis is ultimately responsible for the vast majority of humankind's food, feed, fiber, and fuel. Across the globe, photosynthesis or net primary productivity, defined as the net flux of C into plants per unit time, can be limited by a number of factors including the availability of light, temperature, and water (Fig. 1). The total productivity of a crop field per unit land area is largely determined by the amount of solar radiation incident on the land and the efficiency by which the crop canopy intercepts light and converts it into biomass (as formulated by Monteith [1977]). This chapter first provides a description of the photosynthetic components, processes, and pathways used by crops then discusses the sensitivity of photosynthesis to climatic and environmental factors including radiation, atmospheric [CO_2], temperature, and water deficit. With the advanced mechanistic and genetic understanding of photosynthesis, research

---

Abbreviations: [CO_2], CO_2 concentration; 2PG, 2-phosphoglycolate; 3-PGA, phosphoglycerate; ABA, abscisic acid; ATP, adenosine triphosphate; CAM, crassulacean acid metabolism; LHC, light-harvesting complex; NADPH, nicotinamide adenine dinucleotide phosphate hydrogen; NPQ, nonphotochemical quenching; PAR, photosynthetically active radiation; PEPCase, phosphoenolpyruvate carboxylase; PPFD, photosynthetic photon flux density; PQ, plastoquinone; PQH_2, plastohydroquinone; PSI, Photosystem I; PSII, Photosystem II; Rubisco, ribulose-1,5-biphosphate carboxylase; RuBP, ribulose 1,5-bisphosphate; VPD, vapor pressure deficit.

USDA–ARS Global Change and Photosynthesis Research Unit, 1201 W. Gregory Drive, Urbana, IL. (Lisa.Ainsworth@ars.usda.gov).

doi:10.2134/agronmonogr60.2014.0035

© ASA, CSSA, and SSSA, 5585 Guilford Road, Madison, WI 53711, USA.
Agroclimatology: Linking Agriculture to Climate, Agronomy Monograph 60.
Jerry L. Hatfield, Mannava V.K. Sivakumar, John H. Prueger, editors.

effort and resources have been devoted to improving the photosynthetic process to enhance crop yields for a growing global population (Horton 2000; Long et al., 2006, 2015; Murchie et al., 2008; Zhu et al., 2010; Parry et al., 2011; Ort et al., 2015). In the final section of this chapter, the theoretical basis for those efforts to improve photosynthesis and enhance crop productivity in the future is discussed.

Photosynthetic Energy Conversion: Thylakoid Membrane Components and Processes

Photosynthesis in all higher plants occurs in the chloroplasts, specialized organelles that evolved following an endosymbiotic event between a eukaryote and a cyanobacterium approximately 1 billion years ago. The chloroplast contains an inner membrane system of flattened vesicles called thylakoids, which house the protein complexes needed to perform light-driven reactions of photosynthesis and energy transduction. The thylakoids differentiate into two interconnected domains: cylindrical stacked structures called grana and unstacked regions called stroma laemellae (Fig. 2). Although the protein complexes contained in the thylakoids are often drawn in a linear fashion representing electron transfer processes, the membranes have three-dimensional architecture that provides helical connections between the sheet-like stroma lamellae and the granal thylakoid stacks (Austin and Staehelin, 2011). The tightly stacked arrangement of granum thylakoid membranes ensures a large surface-area-to-volume ratio and a high stability of the ultrastructure. Additionally, there is flexibility in the thylakoid membrane enabling response to dynamic environmental conditions such as rapidly fluctuating light levels.

The primary reactions of oxygenic photosynthesis are mediated by four transmembrane proteins and two mobile carriers located in the thylakoid membrane (Rochaix, 2014). These proteins are highly conserved in higher plants (Hohmann-Marriott and Blankenship, 2011) and collectively capture energy from the sun and facilitate electron transfer reactions to convert free energy of absorbed light into chemical forms (adenosine triphosphate [ATP] and nicotinamide adenine dinucleotide phosphate hydrogen [NADPH]) that can be used in metabolism (Fig. 3). The chlorophyll-containing photosystems (PSI and PSII)

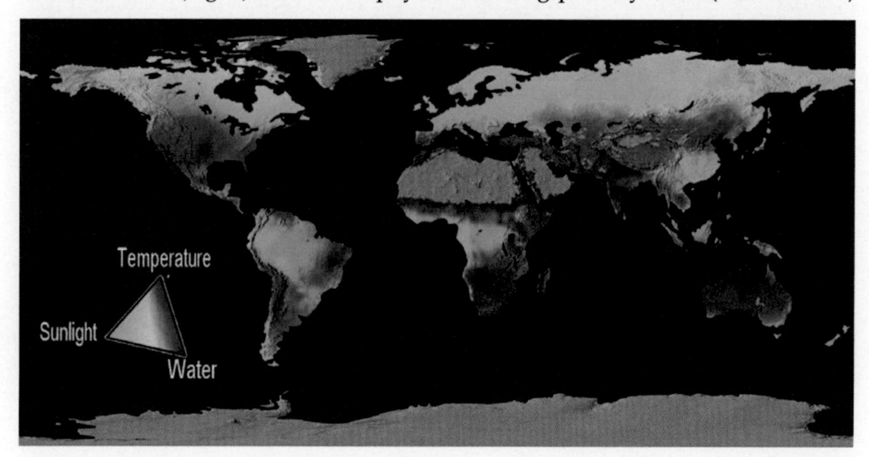

Fig. 1. Geographical variation in availability of sunlight, temperature, and water that pertain to limitations of net primary productivity across the planet. Figure used with permission from Running et al. (2004).

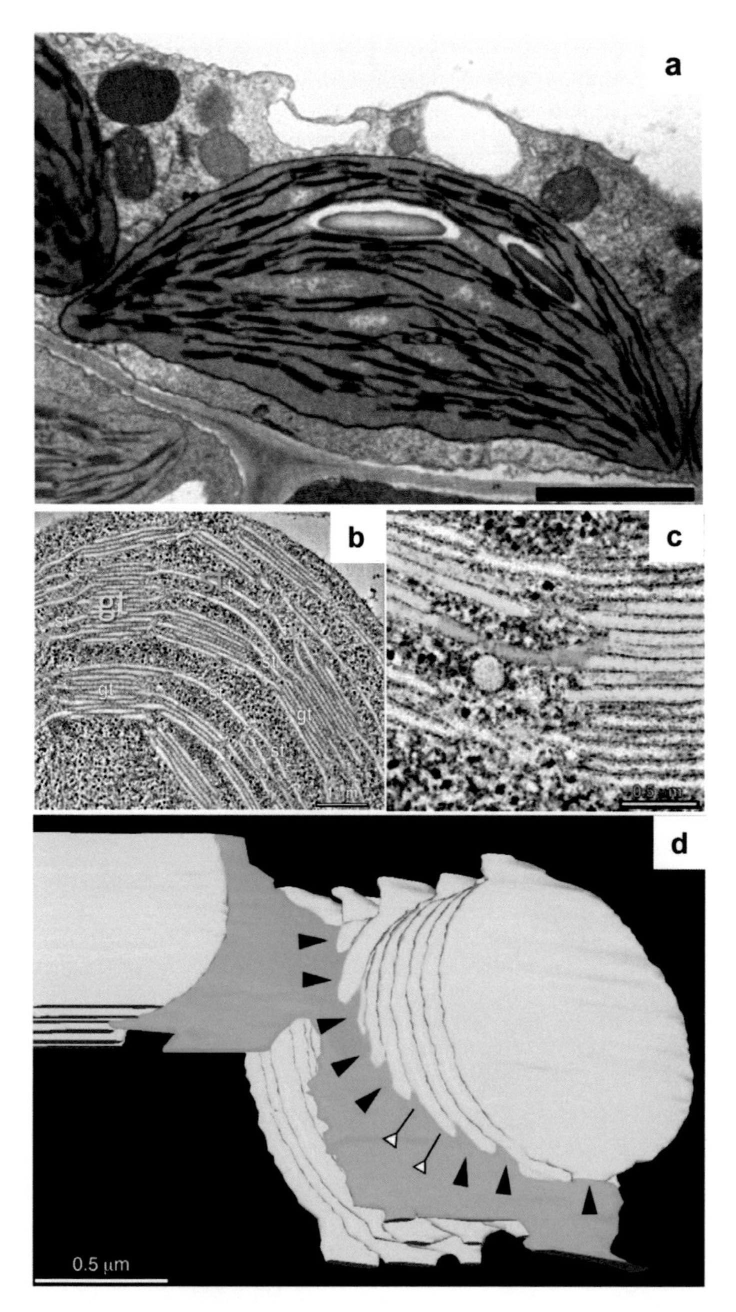

Fig. 2. (a) Transmission electron micrograph of *Arabidopsis thaliana* (L.) Heynh. chloroplast from 10-d-old seeding. Scale bar is 2 mm. (b) Electron tomogram showing the interconnecting grana stacks (gt) and the stroma thylakoids (st). (c) Superimposed tomographic slices with the grana stacks colored yellow and stroma thylakoid colored green. (d) Tomographic model of the grana stack and stroma thylakoid with arrowheads indicating the slit-like stroma-grana connections. (a) Image from Hyman and Jarvis (2011); (b–d) adapted from Austin and Staehelin (2011).

have large, light-harvesting antenna or complexes (LHCs) that absorb photons and transfer energy to reaction centers where stable charge separation across the thylakoid membrane occurs. Another transmembrane protein, the Cytochrome b_6f complex, provides the electron transfer connection between the two reaction center complexes, and the final transmembrane protein is ATP synthase. The four protein complexes of the thylakoid membrane are heterogeneously arranged in different regions of the thylakoid membrane (Fig. 3) and in different cells in C4 plants. Photosystem II is located predominantly in the stacked regions of the thylakoid membrane where the associated LHCII complexes form extensive protein–protein interactions across the thylakoids, while PSI and ATP synthase are found in the unstacked regions (Nelson and Yocum, 2006). Cytochrome b_6f complexes are more evenly distributed in the thylakoid stacks and stroma lamellae. The smaller molecules plastohydroquinone (PQH_2) and plastocyanin serve as the mobile carriers to transfer electrons between the large transmembrane proteins.

In higher plants, the major mode of electron transport is noncyclic or linear electron transfer in which water is oxidized to O_2 and $NADP^+$ is reduced to NADPH. This conversion of light energy into chemical energy requires PSII and PSI to work in close collaboration. Photosystem II preferentially absorbs red light (680 nm) and oxidizes water to molecular O_2 in the thylakoid lumen. In the process, it releases protons in the lumen and transfers electrons to plastoquinone (PQ) to ultimately form PQH_2. This mobile carrier then moves to the Cytochrome b_6f complex where PQH_2 is oxidized, and electrons are transferred to a Rieske Fe-S center to hemes and finally to another mobile Cu protein carrier, plastocyanin. Oxidation of PQH_2 in the Cytochrome b_6f complex is also coupled to proton transfer to the lumen, adding to the proton motive force. Plastocyanin moves to PSI where it reduces oxidized $P700^+$. Photosystem I preferentially absorbs light at 700 nm and facilitates electron transfer from $NAPD^+$ to NADPH in the stroma by the action of ferredoxin and ferredoxin-$NADP^+$ reductase. The collective action of

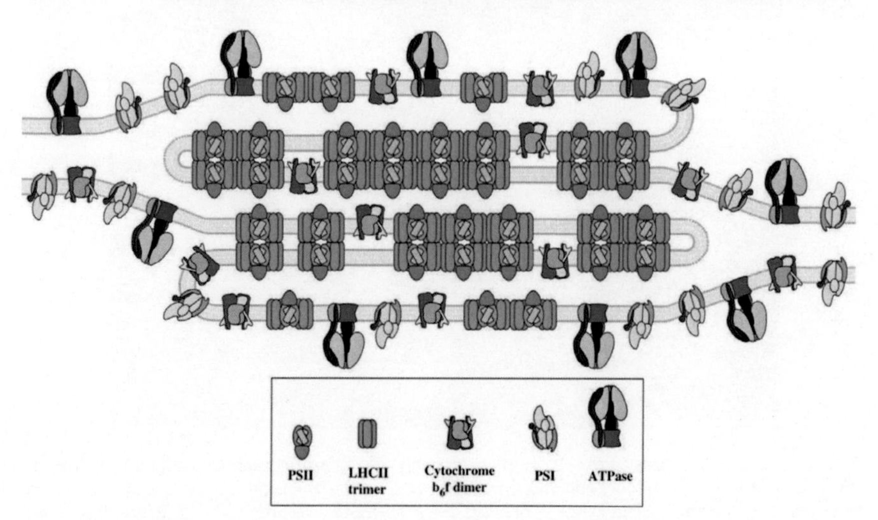

Fig. 3. Heterogeneity of the distribution of protein complexes in the thylakoid membrane in higher plants. Used with permission from Allen and Fosberg (2001) and Blankenship (2014). PSII, photosystem II; LHCII, light harvesting complex II; PSI, Photosystem I; ATPase, ATP synthase.

these reactions moves protons to the thylakoid lumen where ATP synthase uses the proton gradient to produce ATP (Fig. 4). Adenosine triphosphate and NADPH are then consumed in the enzymatic C fixation reactions in the stroma.

Plants are also capable of alternative electron transport processes including cyclic electron flow and the Mehler reaction. Cyclic electron flow involves electron flow around PSI that results in ATP synthesis but does not involve a terminal electron acceptor. Electrons are passed from NADPH or ferredoxin to PQ, increasing the ΔpH. This process adjusts the ATP/NADPH production ratio and is thought to be important for situations when linear electron flow is inhibited, for example during abrupt light transitions (Murchie and Niyogi, 2011). The Mehler reaction, or water–water cycle, occurs by photoreduction of O_2 to water at the reducing side of PSI using electrons generated from water in PSII. The sequence of reactions involves many antioxidant enzymes and functions to scavenge superoxide and hydrogen peroxide in the chloroplast where high concentrations of those reactive oxygen molecules can cause damage. Like cyclic electron flow, the Mehler reaction can also dissipate excess energy in the thylakoid membrane and is thought to be active during light transitions and stress conditions when the electron transport chain has the potential to be highly reduced.

Photosynthetic Carbon Metabolism

The C3 Photosynthetic Carbon reduction cycle

The vast majority of plants, including major food crops such as wheat (*Triticum aestivum* L.), rice (*Oryza sativa* L.), and soybean [*Glycine max* (L.) Merr.], perform C3 photosynthesis to fix atmospheric CO_2 into C skeletons that subsequently drive plant growth and metabolism. The C3 photosynthetic C reduction cycle uses ATP and NADPH generated from the electron transport chain to synthesize carbohydrates from atmospheric CO_2 and water. The net reaction of the C3 cycle is as follows:

$$3CO_2 + 9ATP + 6NADPH \rightarrow \text{Triose phosphate} + 9ATP + 8P_{\text{inorganic}} + 6NADP^+$$

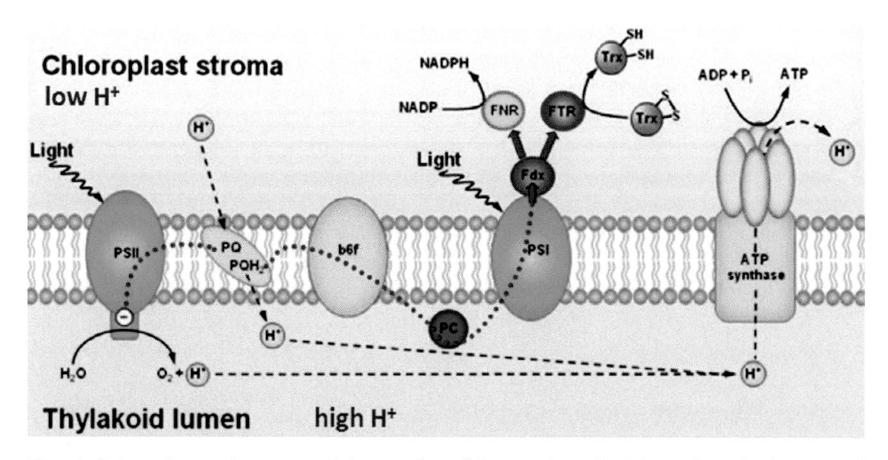

Fig. 4. Integral membrane proteins and mobile carriers that transfer electrons and protons across the thylakoid membrane in plants. Modified from Meyer et al. (2009). PSII, Photosystem II; PQ, plastoquinone; PQH$_2$, plastohydroquinone; b6f, cytochrome *b*$_6$f complex; PC, plastocyanin; PSI, Photosystem I; Fdx, ferredoxin; FNR, ferredoxin-NADPH reductase; FTR, ferredoxin-dependent thioredoxin (Trx).

Therefore, the synthesis of a 6-C sugar requires 18 ATP molecules and 12 NADPH molecules. Nearly 90% of the energy originally available in the ATP and NADPH molecules is ultimately available in the carbohydrate products, making the C3 cycle energetically highly efficient.

The C3 photosynthetic cycle starts with the addition of CO_2 to the five-carbon sugar, ribulose 1,5-bisphosphate (RuBP). This reaction is catalyzed by the enzyme RuBP carboxylase/oxygenase (Rubisco), a dual functioning, competitive enzyme that also catalyzes the reaction of O_2 and RuBP. The first stable product of the carboxylation reaction is a three-carbon compound, phosphoglycerate (3-PGA), and for this reason, the process is referred to as the C3 cycle. In addition to the carboxylation phase of the C3 cycle described above, there are two other phases of the cycle: reduction and regeneration. The reductive phase of the C3 converts 3-PGA to glyceraldehyde phosphate and is catalyzed by two enzymes: 3-PGA kinase and glyceraldehyde 3-phosphate dehydrogenase (Fig. 5). The regenerative phase of the cycle converts triose phosphates back to RuBP, the sugar acceptor for CO_2. Most of the triose phosphates produced in the Calvin cycle are used to regenerate RuBP, but approximately one-sixth of the C exits the cycle for biosynthesis of carbohydrates as well as a broad range of other compounds required for plant metabolism, growth, and defense (Raines, 2011; Fig. 5).

The thylakoid reactions that capture solar energy and produce ATP and NADPH must be coordinated with the C fixation reactions, and this coordination is accomplished in a number of ways. Many enzymes in the C3 cycle are regulated by light via thioredoxin. Electrons transported through the thylakoid membrane reduce ferredoxin, which in turn interacts with the enzyme ferredoxin:thioredoxin reductase to convert a light-activated signal into a thiol signal. Reduced thioredoxins interact with specific disulfide sites on C3-cycle proteins, thereby linking enzyme activity with light availability. Rubisco activity is also modulated by pH and Mg^{2+} concentration in the stroma, both of which are light dependent, and Rubisco activase, the catalytic chaperone for Rubisco, is also redox regulated. These regulations ensure that the catalytic activity of C3-cycle enzymes is tightly coordinated to activity in the thylakoid membrane.

The C2 Photorespiratory Cycle

The competitive enzyme kinetics of the carboxylation phase of the C3 cycle can be substituted with oxygenation of RuBP, which produces the toxic compound 2-phosphoglycolate (2PG). This compound is then recycled back to 3-PGA in a series of enzymatic steps that shuttle metabolites between multiple cellular compartments. The photorespiratory cycle recovers one molecule of 3PGA from two molecules of 2PG, with one out of four 2PG C atoms released as CO_2. The cycle involves C and N recycling, has a direct impact on cell energetics and reduces the maximum quantum yield of photosynthesis. The photorespiratory pathway is also tightly linked to cellular redox processes, energy metabolism, and signaling processes that regulate growth, environmental responses, and programmed cell death (Foyer et al., 2009). Rates of photorespiration in C3 plants increase with temperature because the enzymatic properties of Rubisco change and stimulate RuBP oxygenation more than carboxylation and also because the solubility of CO_2 in water decreases with temperature more than the solubility of O_2 (Foyer et al., 2009). Thus, for plants to maximize C economy, there was strong selection pressure to reduce photorespiration rates under conditions favoring that reaction including heat, drought, salinity, or low atmospheric [CO_2]. Not surprisingly, under those conditions, C4 photosynthesis evolved (Sage, 2004).

The C4 Cycle

Plants with C4 photosynthesis have a series of anatomical and biochemical modifications to the C3 pathway that concentrate CO_2 around Rubisco. C4 photosynthesis occurs in only ~3% of land plant species but evolved independently over 60 times and is the dominant photosynthetic strategy of the world's most productive crops including maize (*Zea mays* L.) and sugarcane (*Saccharum officinarum* L.) (Sage, 2004). The initial step of C4 photosynthesis is fixation of inorganic C by phosphoenolpyruvate carboxylase (PEPCase) into a four-carbon organic acid, hence the name C4 photosynthesis. This reaction occurs in mesophyll cells, and

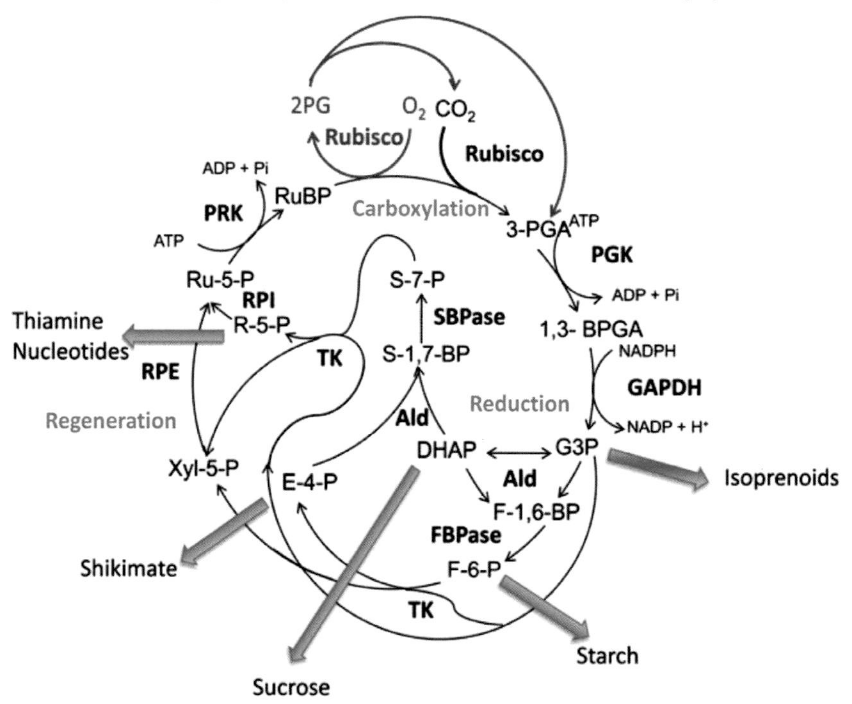

Fig. 5. C3 carbon fixation reactions. The first phase of the C3 cycle is carboxylation of RuBP by Rubisco, resulting in 3-PGA. 3-PGA is reduced to the triose phosphates, glyceraldehyde phosphate (G3P) and dihydroxyacetone phosphate (DHAP) by 3-PGA kinase (PGK) and glyceraldehyde 3-phosphate dehydrogenase (GAPDH) in the reductive phase of the cycle. The regeneration phase of the cycle converts triose phosphates to RuBP. It is catalyzed by aldolase (Ald) and either fructose-1,6-bisphosphatase (FBPase) or sedoheptulose-1,7-bisphosphatase (SBPase), producing fructose-6-phosphate (F-6-P) and sedoheptulose-7-P (S-7-P). F-6-P and sedoheptulose-7-P are then used in reactions catalyzed by transketolase (TK), R-5-P isomerase (RPI), and ribulose-5-P (Ru-5-P) epimerase (RPE) to produce Ru-5-P. The final step in regeneration converts Ru-5-P to RuBP, catalyzed by phosphoribulokinase (PRK). Rubisco is a dual-functioning enzyme and can also fix O_2, forming PGA and 2-phosphoglycolate (2PG), and the subsequent process of photorespiration to recycle the 2PG (shown in red) releases CO_2 and PGA. Export points of C backbones from the pathway for production of isoprenoids, starch, sucrose, shikimate, and thiamine nucleotides are shown with blue arrows. Adapted from Raines (2011).

in most C4 plants, the four-carbon acid then moves to bundle sheath cells where Rubisco is localized. The C4 acid is then decarboxylated, and the CO_2 released is fixed by Rubisco (Fig. 6). The [CO_2] in the bundle sheath cells can be many times higher than atmospheric [CO_2], therefore, photorespiration is suppressed.

Many C4 species have dimorphic chloroplasts that reflect the different biochemical functions of the mesophyll and bundle sheath cells. Mesophyll chloroplasts contain all the transmembrane complexes required for the light reactions and linear electron flow and produce ATP and NADPH. Synthesis of tetrapyrroles, the core of chlorophyll molecules, is localized to mesophyll chloroplasts as is lipid and hormone synthesis. Bundle sheath chloroplasts contain very little PSII and lack stacked thylakoids but house Rubisco and the enzymes needed for starch synthesis (Majeran et al., 2005).

In addition to CO_2 saturated C fixation, C4 plants have greater water and N-use efficiencies than C3 plants. Efficiency of water use is improved by differences in stomatal aperture and the competitive enzyme kinetic properties of the carboxylase. In C4 species, PEPCase fixes C at a greater rate than Rubisco in C3 species, which allows for a steeper air-to-leaf CO_2 gradient in C4 species and greater net CO_2 assimilation rate (*A*) for a given stomatal conductance (g_s) (Osborne and Sack, 2012). Because C4 plants saturate Rubisco with CO_2, they require 50 to 80% less Rubisco for a given photosynthetic rate thereby enabling

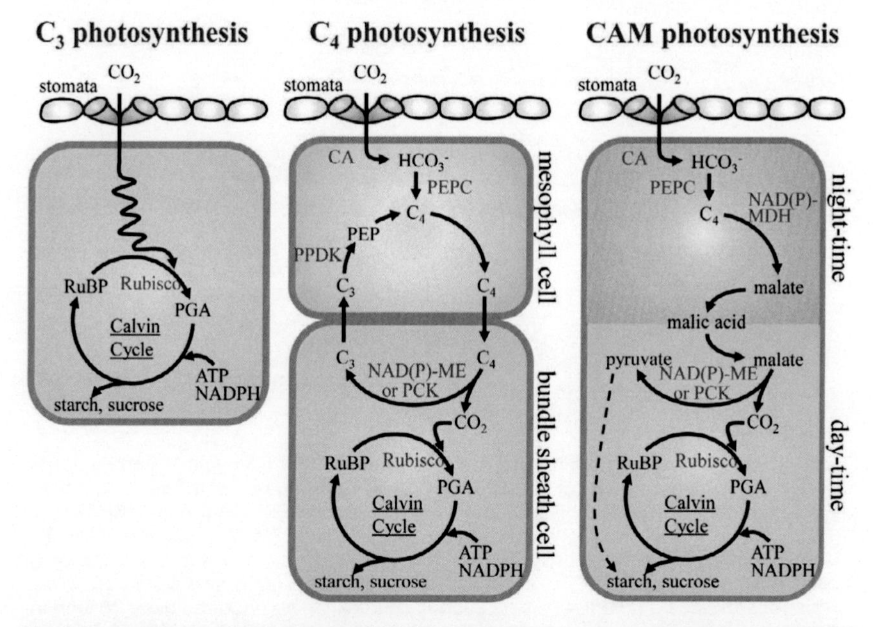

Fig. 6. Comparison of key features of C3, C4, and crassulacean acid metabolism (CAM) photosynthesis. C4 and CAM photosynthesis concentrate CO_2 around Rubisco either using spatial (C4) or temporal (CAM) separation of the initial carboxylation reaction by PEP carboxylase and the ultimate carboxylation by Rubisco. Adapted with permission from Yamori et al. (2014). CA, carbonic anhydrase; PGA, phosphoglyceric acid; RuBP ribulose-1;5-bisphosphate; PEP, phosphoenolpyruvate; Rubisco, ribulose-1,5-bisphosphate carboxylase/oxygenase; PEPC, phosphoenolpyruvate carboxylase; NAD(P)-ME, NAD(P)-malic enzyme; PCK, phosphoenolpyruvate carboxykinase; PPDK, pyruvate phosphate dikinase; NAD(P)-MDH, NAD(P)-malate dehydrogenase.

them to allocate less N to Rubisco and to photorespiratory proteins. Comparisons of maize and rice showed that only 8.5% of leaf N was partitioned to Rubisco in maize in contrast to 27% of leaf N partitioned to Rubisco in rice (Makino et al., 2003). Because crops with C4 photosynthesis maintain greater C assimilation, water and N-use efficiencies, engineering the pathway into C3 crops has become a major research endeavor.

Crassulacean Acid Metabolism

Another modification to C3 photosynthesis that minimizes water loss during photosynthesis is crassulacean acid metabolism (CAM). Crassulacean acid metabolism evolved multiple times and is most commonly found in desert species and epiphytes with stem succulents and small woody growth forms. It is estimated that 6% of plant species have CAM, with pineapple [*Ananas comosus* (L.) Merr. var. *comosus*] and agave (*Agave* spp.) being two of the most important agricultural CAM species (Borland et al., 2011). Like C4 plants, CAM plants use PEPcase to incorporate inorganic C into a C4 acid, but they do this at night when the water vapor pressure deficit between the leaf and the atmospheric is minimized, which greatly reduces transpiration. In the light, CAM plants close their stomata to prevent water loss and then decarboxylate malate, releasing CO_2 for Rubisco to fix in the C3 cycle during the day (Yamori et al., 2014; Fig. 6). CAM plants temporally separate initial C capture by PEPcase and ultimate CO_2 fixation by Rubisco, whereby C4 plants spatially separate these processes. The photosynthetic process—whether C3, C4 or CAM—is sensitive to environmental factors, including light, temperature, and atmospheric [CO_2] (Table 1), and those responses will be covered in the following sections.

Leaf and Canopy Light Interception and Acclimation

Leaf-Level Response of Photosynthesis to Light

For all plants, regardless of photosynthetic type, light is a variable resource in time and space, and photosynthesis is highly responsive to irradiance. Of total solar output, only ~50% is used by plants to drive photosynthesis (wavebands between 400 and 740 nm, termed photosynthetically active radiation [PAR]). Intercepted PAR has three fates in leaves: absorption, transmittance, or reflectance. Chlorophylls absorbs strongly in the blue and red wavelengths, so transmitted and reflected light is enriched in green.

Table 1. Physiological features of C3, C4, and crassulacean acid metabolism (CAM) photosynthesis.

	C3	C4	CAM
Initial carboxylating enzyme	Rubisco	PEPcase	PEPcase
Specialized leaf anatomy	None	Kranz	Large vacuoles
Chloroplasts	One type	Dimorphic	One type
Chlorophyll a/b ratio (dimensionless)	2.8	3.9	2.5–3.0
Energy requirement (CO_2/ATP/NADPH)	1:3:2	1:5:2	1:6.5:2
Transpiration ratio (mol H_2O mol^{-1} CO_2)	700–1300	400–600	18–125
CO_2 compensation point (mmol CO_2 m^{-2} s^{-1})	30–70	0–10	0–5 (in the dark)
Photorespiration	Yes	Minimal, only in bundle sheath cells	Yes (in the p.m. when C3 photosynthesis occurs)

At the leaf level, A increases linearly with photosynthetic photon flux density, and the response of A to photosynthetic photon flux density (PPFD) is well-described by a nonrectangular hyperbola (Fig. 7):

$$A = \frac{\Phi_{CO_2} I + A_{sat} - \sqrt{(\Phi_{CO2} I + A_{sat})^2 - 4\theta\theta_{CO2} I A_{sat}}}{2\theta} - R$$

where φ_{CO2} is the maximal apparent quantum efficiency of photosynthetic CO_2 fixation, that is, the initial slope of the light response curve relating the number of CO_2 molecules fixed per photon absorbed; θ is the convexity of the hyperbola; A_{sat} is the light-saturated rate of photosynthetic CO_2 uptake; I is the PPFD; and R is the rate of CO_2 efflux from mitochondrial respiration (Long et al., 1994). As shown in Fig. 7 for soybean, A increases linearly with PPFD then approaches a plateau. The Φ_{CO2} value is determined by the efficiency with which absorbed photons are used in the reduction of NADP and phosphorylation of ATP and the enzymatic reactions that use reductant and ATP.

Acclimation to the Light Environment

Both chloroplasts and leaves acclimate in the short and long term to the light environment. Chloroplasts alter the content of thylakoid proteins, pigments, and Calvin cycle enzymes to take maximum advantage of available light. Chloroplasts also move around in mesophyll cells to maximize absorption in low-light environments or avoid light if levels are too high. These dynamic chloroplast-level changes are reversible and thought to be controlled by signals within the

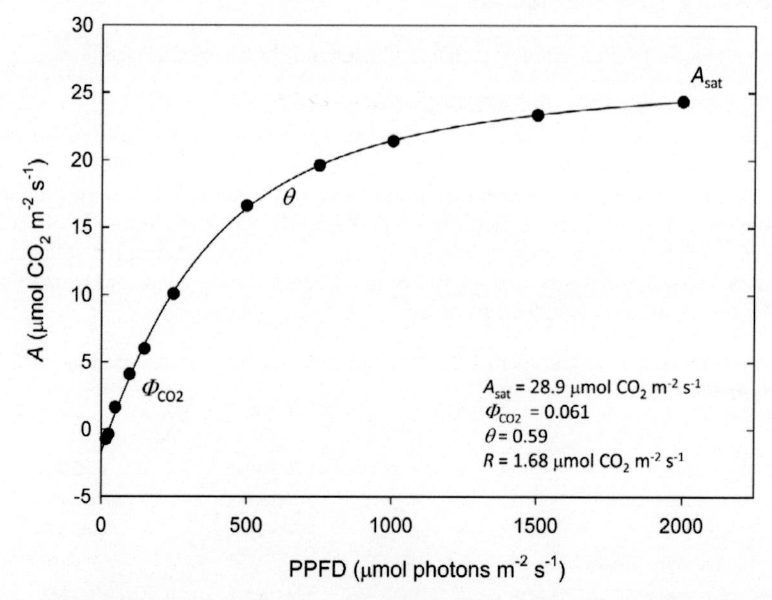

Fig. 7. The hyperbolic response of net CO_2 assimilation (*A*) to photosynthetic photon flux density (PPFD) of a field-grown soybean leaf. The response was measured at 25°C and 400 mmol mol^{-1} CO_2. F_{CO2} is the maximum apparent quantum efficiency of photosynthetic CO_2 fixation, q is the convexity of the hyperbola, A_{sat} is the light-saturated rate of photosynthetic CO_2 uptake, and *R* is the rate of CO_2 efflux from mitochondrial respiration.

chloroplasts (Murchie et al., 2008). Leaf-level acclimation to the light environment, which occurs during early leaf expansion, is generally not reversible. Sun leaves are typically thicker and may have more palisade mesophyll cell layers than shade leaves. The total number of chloroplasts per cell, the content of total protein, and Rubisco are also strongly associated with leaf-level acclimation to the light environment. Other leaf-level acclimation processes include heliotrophic leaf movement, common in many crop species including sunflower (*Helianthus annuus* L.), alfalfa (*Medicago sativa* L.), cotton (*Gossypium hirsutum* L.), soybean, and other bean species. Thus, the canopy architecture and the composition of leaves and chloroplasts enhance light absorption, and photoacclimation sets the maximum photosynthetic capacity of the leaf.

Thermal Dissipation of Excess Light

Apparent from the light response curve of photosynthesis (Fig. 7) is that leaves at the top of a crop canopy experience saturating or near-saturating light levels on clear-sky days. In fact, photosynthesis in C3 crops, including wheat, soybean, and rice, saturates at light levels that are below the maximum intensity of sunlight (Murchie et al., 2008). Thus, plants have evolved dynamic processes for thermal dissipation of excess absorbed light (Ort, 2001; Li et al., 2009). When light-harvesting complexes within a leaf absorb more energy than is required for photosynthesis, the excess excitation energy can be damaging to the chloroplast and leaf, so excess energy is dissipated through either photochemical or nonphotochemical processes to avoid photooxidative damage. Nonphotochemical quenching (NPQ) refers to diverse processes that increase thermal dissipation in the LHCs and includes short-term (minutes) and long-term (hours to days) processes. The mechanisms governing sensing and responding to excess light have been reviewed (Ort, 2001; Li et al., 2009; Murchie and Niyogi, 2011), and a key outcome of NPQ is that it changes the thylakoid membrane to a photoprotected state (Ort, 2001), thereby reducing the photosynthetic efficiency and Φ_{CO2} of leaves (Murchie et al., 2008). Because of the slow relaxation of NPQ, it may limit photosynthesis in canopies where light intensity is highly variable. It has been modeled that crop canopies could lose up to 30% of canopy C gain as a result of the slow relaxation rate of NPQ (Zhu et al., 2004). Therefore, accelerating the recovery from NPQ is a current target for enhancing canopy photosynthesis (Murchie and Niyogi, 2011).

Canopy Light Absorption and Photosynthesis

At the canopy scale, the efficiency of light absorption and the efficiency of converting it to biomass set the maximum theoretical limitation on crop yield. Total photosynthetic C gain of a crop canopy is the summation of photosynthesis in sunlit and shaded leaves. For a modern maize canopy, >80% of the total seasonal C fixation occurs in sunlit leaves (Dohleman and Long, 2009). Photosynthetic productivity of a shaded leaf will be compromised as new leaves develop above it. Photosynthetic capacity falls in older leaves because the older leaves acclimate to the shade environment and also because nutrients are translocated to younger, developing leaves (Murchie et al., 2008).

Improvements in capture and conversion of light energy have contributed greatly to yield gains over the past century and crop canopies with high leaf area indices (i.e., leaf area per unit ground area) are extremely efficient at absorbing radiation. When a crop canopy is closed, 90 to 95% of the light is intercepted by the canopy (Dohleman and Long, 2009; Koester et al., 2014). Early in the growing season, the efficiency of light interception is much lower, but increased plant density, optimal canopy architecture, and rapid canopy development in crops can minimize this period of time. Maize plant density in the United States has increased at an average rate of 1000 plants ha^{-1} over the past 50 yr, and along with that change in density, modern maize hybrids have developed more efficient photosynthesis and enhanced capacity to recover photosynthetic rates following stress events (Duvick, 2005). There is opportunity to further enhance canopy light capture by increasing the length of the growing season, and there is strong evidence in soybean that later maturation dates have been a target of selection over the past 80 yr (Rinker et al., 2014).

Response of Photosynthesis to Carbon Dioxide Concentration

The availability of CO_2 at the site of Rubisco can limit the rate of photosynthesis. Our understanding of this limitation and the biochemical and environmental parameters associated with it have been greatly advanced by the development of robust leaf-level models of C3 and C4 photosynthesis (Farquhar et al., 1980; Farquhar and von Caemmerer, 1982; Sharkey, 1985; von Caemmerer, 2000), which have been widely incorporated into canopy, ecosystem, landscape, and earth-system models (Bernacchi et al., 2013). According to the model, in C3 plants, A can be limited by one of three processes: (i) the maximum Rubisco carboxylation capacity ($V_{c,max}$), (ii) the rate in which the light reactions generate ATP and NADPH for use in the photosynthetic reduction cycle (J_{max}), or (iii) the rate in which inorganic phosphate is released during triose phosphate utilization (Fig. 8). The C4 model (von Caemmerer, 2000) predicts photosynthetic response to CO_2 is limited initially by PEPcase activity (V_{pmax}) and by Rubisco activity at higher $[CO_2]$ (V_{max}) (Fig. 8).

How will Rising Atmospheric CO_2 Concentration impact Crop Photosynthesis?

Current atmospheric $[CO_2]$ is 400 μmol mol^{-1} and has increased from 280 mmol mol^{-1} in the early 1800s (Le Quéré et al., 2015). As atmospheric $[CO_2]$ continues to rise, it will directly increase photosynthesis in C3 crops because of two properties of Rubisco. First, because of the competitive enzyme kinetics, increasing $[CO_2]$ will competitively inhibit the oxygenation reaction, which produces glycolate and leads to photorespiration. Second, the Michaelis-Mentin constant (K_m) for CO_2 is close to atmospheric $[CO_2]$, so increasing CO_2 supply will increase the velocity of carboxylation. C4 crops, in contrast to C3 crops, use PEPcase as the initial carboxylase, and therefore, O_2 is not a competing substrate, so they are typically CO_2–saturated at current atmospheric $[CO_2]$ (Fig. 8). Therefore, in the absence of a stress such as drought, which results in stomatal closure and decreasing intercellular $[CO_2]$, the rise in atmospheric $[CO_2]$ anticipated over the next few decades is not likely to directly stimulate C4 photosynthesis in crops (Leakey, 2009).

By the middle of this century, atmospheric $[CO_2]$ will likely exceed 550 μmol mol^{-1}. Many experiments have investigated how crops will respond to this anticipated midcentury atmospheric $[CO_2]$ in open field environments using either open-top chambers or free-air CO_2 enrichment (FACE) technology. These experiments demonstrate that photosynthesis is stimulated by growth at elevated atmospheric $[CO_2]$ in major food crops, which leads to greater biomass production and crop yield (Bishop et al., 2014). These experiments also revealed that investment in photosynthetic capacity decreases at elevated atmospheric [CO_2], which can be measured as lower $V_{c,max}$ and J_{max} (Ainsworth and Rogers, 2007). This response can be interpreted as an optimization response because at elevated atmospheric [CO_2] Rubisco is operating more efficiently, and therefore, there is less need for the plant to invest as heavily in that enzyme, freeing N reserves for growth or other processes (Drake et al., 1997). Environmental and genetic factors influence the degree of acclimation of photosynthesis to elevated atmospheric [CO_2]. For example, under low N availability or conditions of sink limitation, photosynthetic acclimation is more pronounced (Ainsworth and Long, 2005; Leakey et al., 2009), and in wheat, it has been observed at the biochemical (Adam et al., 2000), leaf (Wall et al., 2000), and whole-canopy (Brooks et al., 2000) levels.

A second physiological response of both C3 and C4 crops to rising atmospheric [CO_2] is a reduction in stomatal conductance (g_s) (Ainsworth and Rogers, 2007). Decreased g_s at elevated atmospheric [CO_2] at the leaf level generally leads to less evaporative cooling and an increase in canopy temperature. However, despite a small increase in canopy temperature, results from free-air CO_2 enrichment experiments have shown that the leaf-level response of decreased g_s is consistent with less evapotranspiration at the canopy level and greater soil moisture availability (Hussain et al., 2013; Bernacchi and VanLoocke, 2015). Of course, climate change is leading to more than just rising atmospheric [CO_2], and crops in the future will face warmer temperatures, more variability in precipitation, and in some regions, greater air pollution. Understanding how photosynthesis

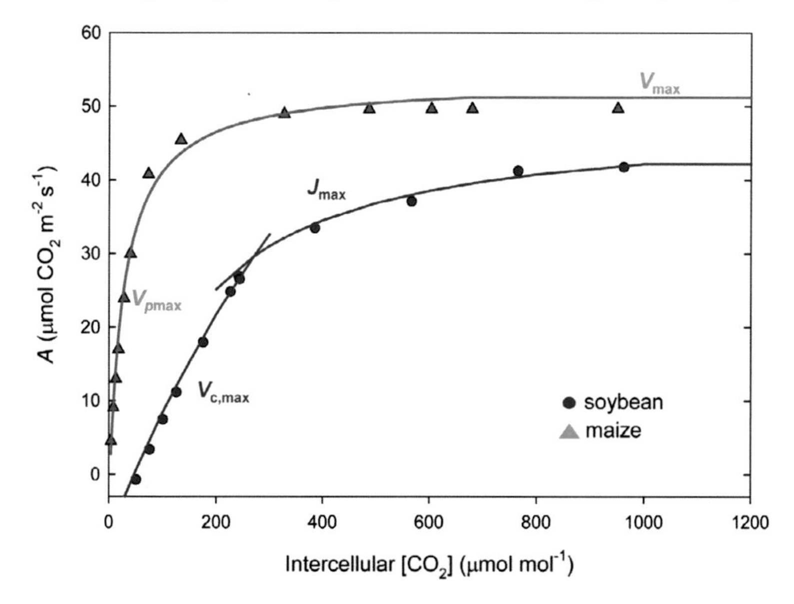

Fig. 8. The response of net CO_2 assimilation rate (*A*) to intercellular [CO_2] for a C3 crop, soybean, and a C4 crop, maize. Plants were grown in the field in Champaign, IL, and a mature leaf from the top of the canopy was measured for CO_2 response at 25°C and 1750 mmol m^{-2} s^{-1} PPFD. In soybean, *A* is limited by the maximum Rubisco carboxylation capacity ($V_{c,max}$) at low intercellular [CO_2] (<280 mmol mol^{-1}) and limited by J_{max} at higher [CO_2] (>280 mmol mol^{-1}). In maize, *A* is limited by PEP carboxylase activity (V_{pmax}) at low [CO_2] (>80 mmol mol^{-1}) and limited by Rubisco activity at higher [CO_2] (V_{max}).

responds to all of these changing environmental conditions in concert is fundamental for refining crop productivity predictions for the future, and ignoring photosynthetic acclimation to rising [CO_2] (i.e., reductions in $V_{c,max}$ and J_{max} in C3 species) can lead to errors in modeled estimates of crop C gain (Bagley et al., 2015).

Temperature Response of Photosynthesis

Temperature is a major determinant of the geographical distribution of crops, the length of the crop growing season, and ultimately, crop productivity. Photosynthesis functions between 0 and 30°C in cold-adapted plants active in winter and early spring and between 7 and 40°C in plants from hot environments (Sage and Kubien, 2007). The temperature optimum for photosynthesis generally varies with photosynthetic type and corresponds to the midpoint of the range of nonharmful temperatures for which the species is adapted (Yamori et al., 2014). Crops have a high capacity for thermal acclimation of photosynthesis in which they modify the photosynthetic proteins and enzymes to improve performance in a new growth environment (Berry and Björkman, 1980). For example, wheat grown at cooler temperatures shows a lower thermal optimum of photosynthesis than wheat grown at warmer temperatures (Fig. 9). Following a 5 to 10°C change in temperature, the thermal optimum of photosynthesis will shift in the direction of the new temperature. For indeterminate crops that continue to grow new leaves over a growing season, the optimum temperature for photosynthesis can vary over the growing season (Rosenthal et al., 2014). However, across a long growing season, crops are often exposed to temperatures that are outside of the optimal range for photosynthesis. For example, the mean maximum air temperature in July in Champaign, IL, over the past 30 yr is 29.4°C, which is approximately the optimum temperature for photosynthesis of soybean grown in that area and measured in July (Rosenthal et al., 2014). However, in 2014 and 2015, maximum air temperatures ranged from 19.4 to 33.3°C, which would have reduced maximum photosynthetic rate on those days by ~20%.

Effects of Low Temperature on Photosynthesis

Photosynthesis in warm-climate crops is substantially reduced following exposure to low-temperature stress especially in crops with tropical evolutionary histories (Allen and Ort, 2001). Both sustained growth at suboptimal temperatures and exposure to occasional short chilling episodes challenge crops in the field and reduce photosynthetic C gain. At low temperatures, the catalytic turnover number of enzymes is slowed, and so a key feature of acclimation to sustained exposure to cold temperatures is increased expression of proteins and enzymes to overcome the reduction in activity. Increased expression of cold-stable isozymes is also a common response. Similarly, temperature impacts membrane fluidity and increasing the ratio of unsaturated to saturated fatty acids is a common acclimation response to low-temperature exposure (Yamori et al., 2014).

The mechanisms of damage from short-term chilling differ from acclimation responses to prolonged exposure to cold stress and also differ if chilling occurs in the light vs. in the dark. Exposure to chilling temperatures in the light results in photodamage to PSII, while this is not common with chilling in the dark. Photodamage occurs because there are limited sinks for absorbed light energy, as low temperature decreases CO_2 fixation and photorespiration rates, and subsequently there is greater oxidative damage to PSII. By contrast, chilling in the dark interferes with carbohydrate metabolism, inhibits Rubisco activity, causes stomatal closure, and results in increased energy dissipation as heat (Allen and Ort, 2001).

Effects of High Temperature on Photosynthesis

Continued emissions of greenhouse gases is projected to increase annual temperatures by 2.5 to 4.3°C by 2080 to 2099 with growing season temperatures expected to warm more than the annual averages (Christensen et al., 2007). The effects of warmer temperatures on photosynthesis will be one of the most important determinants of the impact of global warming on crop yield (Ainsworth and Ort, 2010). As temperature rises, $V_{c,max}$ increases, but the ability of Rubisco to discriminate against O_2 declines, and there is increased solubility of O_2 relative to CO_2. This results in an initial increase in photosynthetic C assimilation with increasing temperature up to an optimum and then a decrease in photosynthetic C assimilation above that optimum as photorespiration increases (Fig. 9). Above the thermal optimum for photosynthesis, electron flow through PSII is reduced, cyclic electron transport at PSI is activated, and thylakoid membrane leakiness increases, which reduces ATP production and limits RuBP regeneration (Sage and Kubien, 2007). Rubisco activase has also been shown to be heat labile at temperatures where photosynthesis and the activation state of Rubisco decline (Crafts-Brandner and Salvucci, 2000).

Crops in the field are exposed to variation in temperature throughout the day and throughout the growing season. Field observations of crops exposed to high-temperature stress have shown that the response of photosynthesis to heat stress can be closely linked to water availability (Hatfield et al., 2011). This

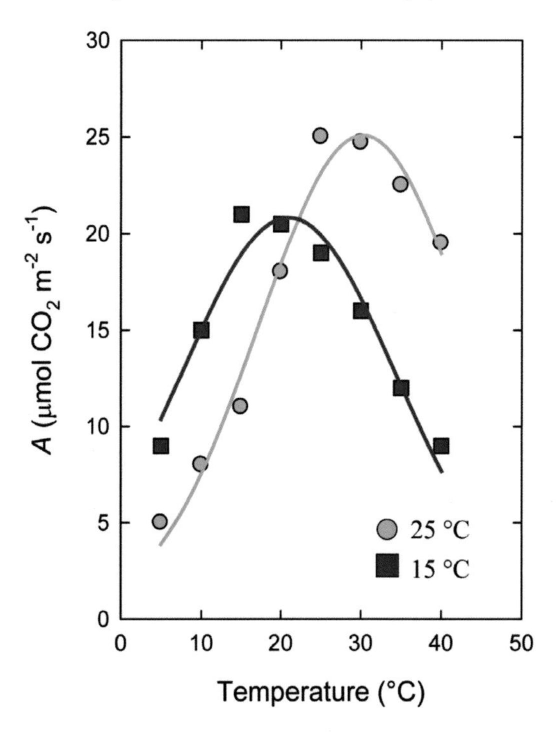

Fig. 9. Contrasting temperature responses of net CO_2 assimilation rate (*A*) of winter wheat exposed to 15 or 25°C. Redrawn using data from Yamasaki et al. (2002).

is especially pronounced in semiarid environments where a large vapor pressure gradient exists between the leaf and the low humidity atmosphere. A study of cotton cultivars in Arizona showed that, if provided with well-watered conditions, modern cotton cultivars can use evaporative cooling to decrease leaf temperature by 6°C or more and maintain high rates of photosynthesis even if air temperatures exceed 40°C. However, under water-limiting conditions, photosynthetic rates plummeted as transpiration is limited and leaf cooling was minimal (Carmo-Silva et al., 2012). Even in the absence of water stress, high temperatures can reduce photosynthetic C gain. Both season-long warming (+3.5°C) or short-term heat waves (+6°C) imposed with infrared heaters to soybean canopies significantly decreased daily C gain and subsequent soybean yields in the US Midwest (Ruiz-Vera et al., 2013; Siebers et al., 2015). The combination of elevated atmospheric [CO_2] and season-long warming decreased photosynthesis and yield in soybean in a warmer growing season, providing experimental evidence that rising atmospheric [CO_2] may not protect from the detrimental impacts of rising temperature on crop yields in temperate growing regions (Ruiz-Vera et al., 2013).

Response of Photosynthesis to Water Limitation

Limitation to adequate water resources is a major environmental constraint to crop yield and net primary productivity around the globe (Fig. 1; Boyer, 1982), and droughts are projected to become more intensive as a result of global climate change over this century (Trenberth et al., 2013). The impact of drought on crop yields will depend on the timing, intensity, and duration of the stress, and most crops will experience water deficit at some period during the growing season. In semiarid regions, crops experience water deficit daily as temperatures and the vapor pressure deficit (VPD) between the leaf and atmosphere substantially increase during the day, resulting in a midday depression of g_s and A (Brodribb et al., 2015). Plants employ a number of signaling pathways that enable rapid responses to short-term water deficit in addition to long-term growth and physiological adjustments that improve plant tolerance to drought (Chaves et al., 2003). In the short term, osmotic changes are perceived by membrane receptors, which initiate signal cascades that alter cellular growth, gene expression patterns, and improve osmoregulation, macromolecular stabilization, and membrane protection (Chaves et al., 2003). In the long term, exposure to drought leads to physiological adjustments to minimize water loss and maximize water uptake including sustained root growth and reduced investment in leaf area to limit transpiration (Chaves et al., 2003).

All plants face a fundamental trade-off between C gain and water loss, and not surprisingly, stomatal closure is one of the earliest responses to moderate drought stress. Stomatal closure protects water transport capability, thus avoiding cellular dehydration, xylem cavitation, and ultimately plant death. Closure of stomata limits water loss and also transpirational cooling, resulting in greater canopy temperature. Under field conditions, the decrease in photosynthesis under moderate drought stress is thought to primarily be due to diffusional limitations from both reduced stomatal and mesophyll conductance (Chaves and Oliveira 2004; Flexas et al., 2006). Stomatal closure in response to moderate soil water deficit is promoted by abscisic acid (ABA), a chemical signal that is synthesized in roots exposed to dry soils (as well as stems and leaves) and carried through the transpiration stream to leaves where it regulates stomatal and growth responses to drought (Wilkinson and Davies, 2002). Abscisic acid transport is modulated by other hormones and xylem pH, and the sensitivity of guard cells to ABA varies among species and environmental conditions (Wilkinson and Davies, 2002). In addition to a role in promoting stomatal closure, there is evidence that ABA acts indirectly to control growth under water stress by inhibiting ethylene synthesis (Sharp, 2002) and inducing gene expression to enable drought acclimation (Mickelbart et al., 2015). Both the closure of stomata and inhibition of growth under water deficit restricts CO_2 diffusion into the leaf and constrains photosynthetic C assimilation (Chaves et al., 2003).

Metabolic limitation of photosynthesis can also occur in plants subjected to increasing levels of drought stress (Lawlor and Cornic, 2002). As stomatal and mesophyll conductance decrease with increasing drought stress, intercellular and chloroplastic [CO_2] drop and *A* is reduced. Initially, electron transport capacity can remain high during drought stress, resulting in an imbalance between light energy harvested and its use in the photosynthetic C reduction cycle. Thus, photoprotective mechanisms to use the excess energy are induced including primarily increased photorespiration but also Mehler reaction activity and NPQ (Lawlor and Cornic, 2002; Medrano et al., 2002). Carbohydrate and N metabolism are subsequently impacted by drought stress, resulting in altered leaf carbohydrate and amino acid profiles (Lawlor and Cornic, 2002).

There is a linear relationship between the amount of water used by a crop for transpiration and crop C assimilation and biomass production resulting from the dependence of photosynthesis and expansion growth of plants on transpiration rates (Brodribb et al., 2015). Over the past 50 to 80 yr, g_s has inadvertently increased in many crops selected for high yields (Roche, 2015). Thus, maximizing CO_2 entry to leaves via high g_s is an important feature for high crop yields. On the other hand, when there is not ample water supply, then high g_s and transpiration early in the season may reduce soil water availability during later reproductive stages and negatively impact production. Thus, efforts to improve crop tolerance to water deficit have focused on strategies to limit the transpiration rate of crops when VPD is high (Sinclair, 2012; Messina et al., 2015). The partial stomatal closure at high VPD would result in greater water conservation that would support growth later in a growing season, and simulation studies with sorghum, soybean, and maize predict that in most years there would be improvements in production (Sinclair, 2012; Messina et al., 2015). However, because limiting g_s would also limit potential C assimilation in years with ample water availability, this strategy may decrease maximum yield potential (Brodribb et al., 2015). Thus, there may be a different suite of hydraulic traits that would maximize crop productivity in years with ample vs. limiting water supply (Brodribb et al., 2015).

Targets for Improving Photosynthesis to Enhance Crop Yields

Monteith (1977) described the limitations to total biomass production by the following equation:

$$\text{Total biomass} = \sum_{\text{planting}}^{\text{maturity}} QIE$$

where *Q* is the solar radiation incident on the canopy over the duration of the crop period, *I* is the interception of solar radiation by the crop canopy, and *E* is the radiation-use efficiency or total dry matter content produced per unit radiation interception. This equation has been extended to understand the theoretical limits of yield potential (Y_p) by the following equation:

$$Y_p = Q\varepsilon_i\varepsilon_c\varepsilon_p$$

where ε_i is the efficiency with which radiation is intercepted by the crop canopy, ε_c is the efficiency with which intercepted radiation is converted to biomass, and ε_p is the partitioning of biomass to harvested product, commonly called harvest index. Breeding for seed yield gains has improved ε_p in modern crops to 50 to 60%, and it has been argued that there is little room for further improvement (Hay, 1995; Horton, 2000; Long et al., 2006). Likewise, breeders have effectively selected for longer growing seasons to maximize light interception, and modern crop cultivars have rapidly closing canopies, so ε_i can exceed 80 to 90% in some growing seasons. Thus, it is thought that there is little room from improvement of ε_i (Long et al., 2006; Zhu et al., 2010). Hence it has been suggested that improvements in yield must come from greater photosynthesis or ε_c (Horton, 2000; Long et al., 2006).

A number of diverse targets have been discussed to improve photosynthetic efficiency in C3 crops, many of which may also improve efficiency in C4 crops (Long et al., 2006, 2015; Zhu et al., 2010; Ort et al., 2015). These include targets for expanding and optimizing canopy light capture by extending the usable spectrum of crop light harvesting, allowing more light to transmit through the canopy, and altering canopy architecture (Ort et al., 2011, 2015; Blankenship and Chen, 2013); more rapid relaxation of NPQ and heat dissipation at PSII (Murchie and Niyogi, 2011); improving Rubisco to minimize photorespiration (Whitney et al., 2011); enhancing the regenerative capacity of the C3 cycle (Raines, 2011); converting C3 crops to C4 (von Caemmerer et al., 2012); adding components of cyanobacterial or algal systems to pump CO_2 and compartmentalize Rubisco (Long et al., 2015); and synthetic engineering of a photorespiratory bypass (Peterhansel et al., 2012; Ort et al., 2015). These improvements have estimated efficiency gains of 5 to 60%, and many have additional benefits of improving water and N-use efficiency (Long et al., 2015). While the timeline for some of these improvements is 10 to 30 yr, for other improvements, including more rapid relaxation of heat dissipation at PSII and optimization of regeneration of the Calvin cycle, transgenic plants are already available and being field tested. The possibility of improving crop photosynthesis for improved yields is bolstered by tremendous fundamental understanding of the process enabling computational modeling to predict the outcome of photosynthetic alterations and ever-more rapid and sophisticated technologies for genetic transformation (Ort et al., 2015; Long et al., 2015).

References

Adam, N.R., G.W. Wall, B.A. Kimball, P.J. Pinter, Jr., R.L. LaMorte, D.J. Hunsaker, F.J. Adamsen, T. Thompson, A. Matthias, S. Leavitt, and A.N. Webber. 2000. Acclimation response of spring wheat in a free-air CO_2 enrichment (FACE) with variable soil nitrogen regimes. 1. Leaf position and phenology determine acclimation response. Photosynth. Res. 66:65–77. doi:10.1023/A:1010629407970

Ainsworth, E.A., and S.P. Long. 2005. What have we learned from 15 years of free-air CO_2 enrichment (FACE)? A meta-analytic review of the responses of photosynthesis, canopy properties and plant production to rising CO_2. New Phytol. 165:351–372. doi:10.1111/j.1469-8137.2004.01224.x

Ainsworth, E.A., and D.R. Ort. 2010. How do we improve crop production in a warming world? Plant Physiol. 154:526–530. doi:10.1104/pp.110.161349

Ainsworth, E.A., and A. Rogers. 2007. The response of photosynthesis and stomatal conductance to rising [CO_2]: Mechanisms and environmental interactions. Plant Cell Environ. 30:258–270. doi:10.1111/j.1365-3040.2007.01641.x

Allen, D.J., and D.R. Ort. 2001. Impacts of chilling temperatures on photosynthesis in warm-climate plants. Trends Plant Sci. 6:36–42. doi:10.1016/S1360-1385(00)01808-2

Allen, F.J., and J. Fosberg. 2001. Molecular recognition in thylakoid structure and function. Trends Plant Sci. 6:317–326. doi:10.1016/S1360-1385(01)02010-6

Austin, J.R., and L.A. Staehelin. 2011. Three-dimensional architecture of grana and stroma thylakoids of higher plants as determined by electron tomography. Plant Physiol. 155:1601–1611. doi:10.1104/pp.110.170647

Bagley, J., D.M. Rosenthal, U.M. Ruiz-Vera, M.H. Siebers, P. Kumar, D.R. Ort, and C.J. Bernacchi. 2015. The influence of photosynthetic acclimation to rising CO_2 and warmer temperatures on leaf and canopy photosynthesis models. Global Biogeochem. Cycles 29:194–206. doi:10.1002/2014GB004848

Bernacchi, C.J., J.E. Bagley, S.P. Serbin, U.M. Ruiz-Vera, D.M. Rosenthal, and A. VanLoocke. 2013. Modelling C3 photosynthesis from the chloroplast to the ecosystem. Plant Cell Environ. 36:1641–1657. doi:10.1111/pce.12118

Bernacchi, C.J., and A. VanLoocke. 2015. Terrestrial ecosystems in a changing environment: A dominant role for water. Annu. Rev. Plant Biol. 66:599–622. doi:10.1146/annurev-arplant-043014-114834

Berry, J.A., and O. Björkman. 1980. Photosynthetic response and adaptation to temperature in higher plants. Annu. Rev. Plant Physiol. 31:491–543. doi:10.1146/annurev.pp.31.060180.002423

Bishop, K.A., A.D.B. Leakey, and E.A. Ainsworth. 2014. How seasonal temperature or water inputs affect the relative response of C3 crops to elevated [CO_2]: A global analysis of open top chamber and free air CO_2 enrichment studies. Food Energy Security 3:33–45. doi:10.1002/fes3.44

Blankenship, R.E. 2014. Molecular mechanisms of photosynthesis. 2nd ed. Wiley-Blackwell, Hoboken, NJ.

Blankenship, R.E., and C.M. Chen. 2013. Spectral expansion and antenna reduction can enhance photosynthesis for energy production. Curr. Opin. Chem. Biol. 17:457–461. doi:10.1016/j.cbpa.2013.03.031

Borland, A.M., A.B. Zambrano, J. Ceusters, and K. Shorrock. 2011. The photosynthetic plasticity of crassulacean acid metabolism: An evolutionary innovation for sustainable productivity in a changing world. New Phytol. 191:619–633. doi:10.1111/j.1469-8137.2011.03781.x

Boyer, J.S. 1982. Plant productivity and environment. Science 218:443–448. doi:10.1126/science.218.4571.443

Brodribb, T.J., M.M. Holloway-Phillips, and H. Bramley. 2015. Improving water transport for carbon gain in crops. In: V.O. Sadras and D.F. Calderini, editors, Crop physiology: Applications for genetic improvement and agronomy. Elsevier, Philadelphia, PA. p. 251–281. doi:10.1016/B978-0-12-417104-6.00011-X

Brooks, T.J., G.W. Wall, P.J. Pinter, Jr., B.A. Kimball, R.L. LaMorte, S.W. Leavitt, A.D. Matthias, F.J. Adamsen, D.J. Hunsaker, and A.N. Webber. 2000. Acclimation response of spring wheat in a free-air CO_2 enrichment (FACE) atmosphere with variable soil nitrogen regimes. 3. Canopy architecture and gas exchange. Photosynth. Res. 66:97–108. doi:10.1023/A:1010634521467

Carmo-Silva, A.E., M.A. Gore, P. Andrade-Sanchez, A.N. French, D.J. Hunsaker, and M.E. Salvucci. 2012. Decreased CO_2 availability and inactivation of Rubisco limit photosynthesis in cotton plants under heat and drought stress in the field. Environ. Exp. Bot. 83:1–11. doi:10.1016/j.envexpbot.2012.04.001

Chaves, M.M., J.P. Maroco, and J.S. Pereira. 2003. Understanding plant responses to drought: From genes to the whole plant. Funct. Plant Biol. 30:239–264. doi:10.1071/FP02076

Chaves, M.M., and M.M. Oliveira. 2004. Mechanisms underlying plant resilience to water deficits: Prospects for water-saving agriculture. J. Exp. Bot. 55:2365–2384. doi:10.1093/jxb/erh269

Christensen, J.H., B. Hewitson, A. Busuioc, A. Chen, X. Gao, I. Held, et al. 2007. Regional climate projections. In: S. Solomon, D. Qin, M. Manning, Z. Chen, M. Marquis, K.B. Averyt, M. Tignor, H.L. Miller, editors, Climate Change 2007: The physical science basis. Contribution of Working Group I to the Fourth Assessment Report of the Intergovernmental Panel on Climate Change. Cambridge Univ. Press, Cambridge, UK, and New York. p. 847–940.

Crafts-Brandner, S.J., and M.E. Salvucci. 2000. Rubisco activase constrains the photosynthetic potential of leaves at high temperature and CO_2. Proc. Natl. Acad. Sci. USA 97:13430–13435. doi:10.1073/pnas.230451497

Dohleman, F.G., and S.P. Long. 2009. More productive than maize in the Midwest: How does Miscanthus do it? Plant Physiol. 150:2104–2115. doi:10.1104/pp.109.139162

Drake, B.G., M.A. Gonzalez-Meler, and S.P. Long. 1997. More efficient plants: A consequence of rising atmospheric CO_2? Annu. Rev. Plant Physiol. Plant Mol. Biol. 48:609–639. doi:10.1146/annurev.arplant.48.1.609

Duvick, D.N. 2005. The contribution of breeding to yield advances in maize (*Zea mays* L.). Adv. Agron. 86:83–145. doi:10.1016/S0065-2113(05)86002-X

Farquhar, G.D., and S. von Caemmerer. 1982. Modeling of photosynthetic response to environmental conditions. In: O.L. Lange, P.S. Nobel, C.B. Osmond, and H. Ziegler, editors, Physiological plant ecology I. Encyclopedia of plant physiology new series. Vol. 12B. Springer–Verlag, Berlin. p. 549–587.

Farquhar, G.D., S. von Caemmerer, and J.A. Berry. 1980. A biochemical model of photosynthetic CO_2 fixation in C_3 species. Planta 149:78–90. doi:10.1007/BF00386231

Flexas, J., J. Bota, J. Galmés, H. Medrano, and M. Ribas-Carbó. 2006. Keeping a positive carbon balance under adverse conditions: Responses of photosynthesis and respiration to water stress. Physiologia Plantarum 127:343–352. doi:10.1111/j.1399-3054.2006.00621.x

Foyer, C.H., A.J. Bloom, G. Queval, and G. Noctor. 2009. Photorespiratory metabolism: Genes, mutants, energetics and redox signaling. Annu. Rev. Plant Biol. 60:455–484. doi:10.1146/annurev.arplant.043008.091948

Hatfield, J.L., K.J. Boote, B.A. Kimball, L.H. Ziska, R.C. Izaurralde, D. Ort, A.M. Thomson, and D. Wolfe. 2011. Climate impacts on agriculture: Implications for crop production. Agron. J. 103:351–370. doi:10.2134/agronj2010.0303

Hay, R.K.M. 1995. Harvest index: A review of its use in plant breeding and crop physiology. Ann. Appl. Biol. 126:197–216. doi:10.1111/j.1744-7348.1995.tb05015.x

Hohmann-Marriott, M.F., and R.E. Blankenship. 2011. Evolution of photosynthesis. Annu. Rev. Plant Biol. 62:515–548. doi:10.1146/annurev-arplant-042110-103811

Horton, P. 2000. Prospects for crop improvement through the genetic manipulation of photosynthesis: Morphological and biochemical aspects of light capture. J. Exp. Bot. 51:475–485. doi:10.1093/jexbot/51.suppl_1.475

Hussain, M.Z., A. VanLoocke, M.H. Siebers, U.M. Ruiz-Vera, R.J.C. Markelz, A.D.B. Leakey, D.R. Ort, and C.J. Bernacchi. 2013. Future carbon dioxide concentration decreases canopy evapotranspiration and soil water depletion by field-grown maize. Glob. Change Biol. 19:1572–1584. doi:10.1111/gcb.12155

Hyman, S., and R.P. Jarvis. 2011. Studying Arabidopsis chloroplast structural organisation using transmission electron microscopy. In: Chloroplast Research in Arabidopsis: Methods and Protocols, Vol. 1. Methods Mol. Biol. 774:113–132. doi:10.1007/978-1-61779-234-2_8

Koester, R.P., J.A. Skoneczka, T.R. Cary, B.W. Diers, and E.A. Ainsworth. 2014. Historical gains in soybean (*Glycine max* Merr.) seed yield are driven by linear increases in light interception, energy conversion, and partitioning efficiencies. J. Exp. Bot. 65:3311–3321. doi:10.1093/jxb/eru187

Lawlor, D.W., and G. Cornic. 2002. Photosynthetic carbon assimilation and associated metabolism in relation to water deficits in higher plants. Plant Cell Environ. 25:275–294. doi:10.1046/j.0016-8025.2001.00814.x

Le Quéré, C., R. Moriarty, R.M. Andrew, J.G. Canadell, S. Sitch, J.I. Korsbakken, et al. 2015. Global carbon budget 2015. Earth System Sci. Data 7:349–396. doi:10.5194/essd-7-349-2015

Leakey, A.D.B. 2009. Rising atmospheric carbon dioxide concentration and the future of C_4 crops for food and fuel. Proc. Biol. Sci. 276:2333–2343. doi:10.1098/rspb.2008.1517

Leakey, A.D.B., E.A. Ainsworth, C.J. Bernacchi, A. Rogers, S.P. Long, and D.R. Ort. 2009. Elevated CO_2 effects on plant carbon, nitrogen and water relations: Six important lessons from FACE. J. Exp. Bot. 60:2859–2876. doi:10.1093/jxb/erp096

Li, Z., S. Wakao, B.B. Fischer, and K.K. Niyogi. 2009. Sensing and responding to excess light. Annu. Rev. Plant Biol. 60:239–260. doi:10.1146/annurev.arplant.58.032806.103844

Long, S.P., S. Humphries, and P.G. Falkowski. 1994. Photoinhibition of photosynthesis in nature. Annu. Rev. Plant Physiol. Plant Mol. Biol. 45:633–662. doi:10.1146/annurev.pp.45.060194.003221

Long, S.P., A. Marshall-Colon, and X.G. Zhu. 2015. Meeting the global food demand of the future by engineering crop photosynthesis and yield potential. Cell 161:56–66. doi:10.1016/j.cell.2015.03.019

Long, S.P., Z.G. Zhu, S.L. Naidu, and D.R. Ort. 2006. Can improvement in photosynthesis increase crop yields? Plant Cell Environ. 29:315–330. doi:10.1111/j.1365-3040.2005.01493.x

Majeran, W., Y. Cai, Q. Sun, and K.J. van Wijk. 2005. Functional differentiation of bundle sheath and mesophyll maize chloroplasts determined by comparative proteomics. Plant Cell 17:3111–3140. doi:10.1105/tpc.105.035519

Makino, A., H. Sakuma, E. Sudo, and T. Mae. 2003. Differences between maize and rice in N-use efficiency for photosynthesis and protein allocation. Plant Cell Physiol. 44:952–956. doi:10.1093/pcp/pcg113

Medrano, H., J.M. Escalona, J. Bota, J. Gulías, and J. Flexas. 2002. Regulation of photosynthesis of C_3 plants in response to progressive drought: Stomatal conductance as a reference parameter. Ann. Bot. (Lond.) 89:895–905. doi:10.1093/aob/mcf079

Messina, C.D., T.R. Sinclair, G.L. Hammer, D. Curan, J. Thompson, Z. Oler, C. Gho, and M. Cooper. 2015. Limited-transpiration trait may increase maize drought tolerance in the US corn belt. Agron. J. 107:1978–1986. doi:10.2134/agronj15.0016

Meyer, Y., B.B. Buchanan, F. Vignols, and J.P. Reichheld. 2009. Thioredoxins and glutaredoxins: Unifying elements in redox biology. Annu. Rev. Genet. 43:335–367. doi:10.1146/annurev-genet-102108-134201

Mickelbart, M.V., P.M. Hasegawa, and J. Bailey-Serres. 2015. Genetic mechanisms of abiotic stress tolerance that translate to crop yield stability. Nat. Rev. Genet. 16:237–251. doi:10.1038/nrg3901

Monteith, J.L. 1977. Climate and the efficiency of crop production in Britain. Philos. Trans. R. Soc. Lond. 281:277–294. doi:10.1098/rstb.1977.0140

Murchie, E.H., and K.K. Niyogi. 2011. Manipulation of photoprotection to improve plant photosynthesis. Plant Physiol. 155:86–92. doi:10.1104/pp.110.168831

Murchie, E.H., M. Pinto, and P. Horton. 2008. Agriculture and the new challenges for photosynthesis research. New Phytol. 181:532–552. doi:10.1111/j.1469-8137.2008.02705.x

Nelson, N., and C.F. Yocum. 2006. Structure and function of Photosystems I and II. Annu. Rev. Plant Biol. 57:521–565. doi:10.1146/annurev.arplant.57.032905.105350

Ort, D.R. 2001. When there is too much light. Plant Physiol. 125:29–32. doi:10.1104/pp.125.1.29

Ort, D.R., S.S. Merchant, J. Alric, A. Barkin, R.E. Blankenship, R. Bock, et al. 2015. Redesigning photosynthesis to sustainably meet global food and bioenergy demand. Proc. Natl. Acad. Sci. USA 112:8529–8536. doi:10.1073/pnas.1424031112</jr

Ort, D.R., X.G. Zhu, and A. Melis. 2011. Optimizing antenna size to maximize photosynthetic efficiency. Plant Physiol. 155:79–85. doi:10.1104/pp.110.165886

Osborne, C.P., and L. Sack. 2012. Evolution of C_4 plants: A new hypothesis for an interaction of CO_2 and water relations mediate by plant hydraulics. Philos. Trans. R. Soc. Lond. B Biol. Sci. 367:583–600. doi:10.1098/rstb.2011.0261

Parry, M.A.J., M. Reynolds, M.E. Salvucci, C. Raines, P.J. Andralojc, X.G. Zhu, G.D. Price, A.G. Condon, and R.T. Furbank. 2011. Raising yield potential of wheat. II. Increasing photosynthetic capacity and efficiency. J. Exp. Bot. 62:453–467. doi:10.1093/jxb/erq304

Peterhansel, C., C. Blume, and S. Offermann. 2012. Photorespiratory bypasses: How can they work? J. Exp. Bot. 64:709–715. doi:10.1093/jxb/ers247</jrn

Raines, C.A. 2011. Increasing photosynthetic carbon assimilation in C_3 plants to improve crop yield: Current and future strategies. Plant Physiol. 155:36–42. doi:10.1104/pp.110.168559

Rinker, K., R. Nelson, J. Specht, D. Sleper, T. Cary, S.R. Cianzio, et al. 2014. Genetic improvement of U.S. soybean in maturity groups II, III and IV. Crop Sci. 54:1–14. doi:10.2135/cropsci2013.10.0665

Rochaix, J.D. 2014. Regulation and dynamics of the light-harvesting system. Annu. Rev. Plant Biol. 65:287–309. doi:10.1146/annurev-arplant-050213-040226

Roche, D. 2015. Stomatal conductance is essential for higher yield potential of C_3 crops. Crit. Rev. Plant Sci. 34:429–453. doi:10.1080/07352689.2015.1023677

Rosenthal, D.M., U.M. Ruiz-Vera, M.H. Siebers, S.B. Gray, C.J. Bernacchi, and D.R. Ort. 2014. Biochemical acclimation, stomatal limitation and precipitation patterns underlie decreases in photosynthetic stimulation of soybean (*Glycine max*) at elevated [CO_2] and temperatures under fully open air field conditions. Plant Sci. 226:136–146. doi:10.1016/j.plantsci.2014.06.013

Ruiz-Vera, U.M., M. Siebers, S.B. Gray, D.W. Drag, D.M. Rosenthal, B.A. Kimball, D.R. Ort, and C.J. Bernacchi. 2013. Global warming can negate the expected CO_2 stimulation in photosynthesis and productivity for soybean grown in the Midwestern United States. Plant Physiol. 162:410–423. doi:10.1104/pp.112.211938

Running, S.W., R.R. Nemani, F.A. Heinsch, M. Zhao, M. Reeves, and H. Hashimoto. 2004. A continuous satellite-derived measure of global terrestrial primary production. Bioscience 54:547–560. doi:10.1641/0006-3568(2004)054[0547:ACSMOG]2.0.CO;2

Sage, R.F. 2004. The evolution of C_4 photosynthesis. New Phytol. 161:341–370. doi:10.1111/j.1469-8137.2004.00974.x</jrn

Sage, R.F., and D.S. Kubien. 2007. The temperature response of C_3 and C_4 photosynthesis. Plant Cell Environ. 30:1086–1106. doi:10.1111/j.1365-3040.2007.01682.x

Sharkey, T.D. 1985. Photosynthesis in intact leaves of C_3 plants: Physics, physiology and rate limitations. Bot. Rev. 51:53–105. doi:10.1007/BF02861058

Sharp, R.E. 2002. Interaction with ethylene: Changing views on the role of abscisic acid in root and shoot growth responses to water stress. Plant Cell Environ. 25:211–222. doi:10.1046/j.1365-3040.2002.00798.x

Siebers, M.H., C.R. Yendrek, D. Drag, A.M. Locke, L. Rios Acosta, A.D.B. Leakey, E.A. Ainsworth, C.J. Bernacchi, and D.R. Ort. 2015. Heat waves imposed during early pod development in soybean (*Glycine max*) cause significant yield loss despite a rapid recovery from oxidative stress. Glob. Change Biol. 21:3114–3125. doi:10.1111/gcb.12935

Sinclair, T.R. 2012. Is transpiration efficiency a viable plant trait in breeding for crop improvement? Funct. Plant Biol. 39:359–365. doi:10.1071/FP11198

Trenberth, K.E., A. Dai, G. van der Schrier, P.D. Jones, J. Barichivich, K.R. Briffa, and J. Sheffield. 2013. Global warming and changes in drought. Nat. Clim. Chang. 4:17–22. doi:10.1038/nclimate2067

von Caemmerer, S. 2000. Biochemical models of leaf photosynthesis. CSIRO Publishing, Collingwood, Australia.

von Caemmerer, S., W.P. Quick, and R.T. Furbank. 2012. The development of C_4 rice: Current progress and future challenges. Science 336:1671–1672. doi:10.1126/science.1220177

Wall, G.W., N.R. Adam, T.J. Brooks, B.A. Kimball, P.J. Pinter, Jr., R.L. LaMorte, F.J. Adamsen, D.J. Hunsaker, G. Wechsung, F. Wechsung, S. Grossman-Clarke, S. Leavitt, A.D. Matthias, and A.N. Webber. 2000. Acclimation response of spring wheat in a free-air CO_2 enrichment (FACE) atmosphere with variable soil nitrogen regimes. 2. Net assimilation and stomatal conductance of leaves. Photosynth. Res. 66:79–95. doi:10.1023/A:1010646225929

Whitney, S.M., R.L. Houtz, and H. Alonso. 2011. Advancing our understanding and capacity to engineer nature's CO_2–sequestering enzyme, Rubisco. Plant Physiol. 155:27–35. doi:10.1104/pp.110.164814

Wilkinson, S., and W.J. Davies. 2002. ABA-based chemical signalling: The co-ordination of responses to stress in plants. Plant Cell Environ. 25:195–210. doi:10.1046/j.0016-8025.2001.00824.x

Yamasaki, T., T. Yamakawa, Y. Yamane, H. Koike, K. Satoh, and S. Katoh. 2002. Temperature acclimation of photosynthesis and related changes in photosystem II electron transport in winter wheat. Plant Physiol. 128:1087–1097. doi:10.1104/pp.010919

Yamori, W., K. Hikosaki, and D.A. Way. 2014. Temperature response of photosynthesis in C_3, C_4 and CAM plants: Temperature acclimation and temperature adaptation. Photosynth. Res. 119:101–117. doi:10.1007/s11120-013-9874-6

Zhu, X.G., S.P. Long, and D.R. Ort. 2010. Improving photosynthetic efficiency for greater yield. Annu. Rev. Plant Biol. 61:235–261. doi:10.1146/annurev-arplant-042809-112206

Zhu, X.G., D.R. Ort, J. Whitmarsh, and S.P. Long. 2004. The slow reversibility of photosystem II thermal energy dissipation on transfer from high to low light may cause large losses in carbon gain by crop canopies: A theoretical analysis. J. Exp. Bot. 55:1167–1175. doi:10.1093/jxb/erh141

Biological Linkages to Climatology

Jerry L. Hatfield,* Erica Kistner-Thomas, and Christian Dold

Abstract

Crops as part of agroecosystems are dependent on natural resources for water, carbon dioxide, and a temperature regime in which they experience no harmful effects. How the plant responds to this environment determines the capability of a particular species to thrive. Temperature and precipitation are the two primary climate variables that determine the range of plants along with the productivity of plants. The development and utilization of agroclimatic indices incorporating climate variables have been used extensively to estimate where plants can be grown. These indices utilize a combination of temperature to define the optimum environment for a plant and precipitation to define whether the water resources are sufficient to allow for plant growth. The temperature response in these indices utilizes the fact that each species has a unique temperature range with a lower and upper threshold and an optimum temperature. For example, cool-season species with optimums near 15 to 20 °C will exist in different areas or at different times of the year than warm season species with optimum temperatures between 27 and 32 °C. Defining the proper environment may not be only a function of the optimum temperatures but for some species the exposure to temperatures below a given threshold before they can flower and set fruit. The linkage between plants and the climate centers on the energy balance of the plant which determines its temperature and rate of water use in a given environment. Understanding these linkages provides the framework needed to evaluate all agricultural systems and their response to a changing climate.

Plant responses to the climate determines the adaptability of plants to a given area. The complexities of the interactions between plants and climate focus on the temperature being within the unique temperature range of a given plant species and the availability of water. Temperature and precipitation as climate factors determine the suitability of any environment; however, what is required is a framework for the quantitative assessment of these factors and how they interact with biological processes. The foundational process in developing this framework is the temperature requirements of the biological components of agricultural systems. All organisms have a lower threshold, an upper threshold, and an optimal temperature which links the organism with its surrounding environment and in subsequent sections, we will detail differences among plants in

---

Abbreviations: CWSI, crop water stress index; GDD, growing degree days; ET, evapotranspiration; FF, probability of frost free growing period; PACSI, Poone AgroClimatic Suitability Index PMSD, potential soil moisture deficit; SW, soil water content; Tbase, base temperature; Tmax, temperature maximum; Tmin, minimum temperature; WHC, water holding capacity; WRSI, water requirements satisfaction index.

National Laboratory for Agriculture and the Environment, 1015 N University Blvd., Ames, Iowa 50011. *Corresponding author (Jerry.hatfield@ars.usda.gov)

doi:10.2134/agronmonogr60.2016.0004

© ASA, CSSA, and SSSA, 5585 Guilford Road, Madison, WI 53711, USA.
Agroclimatology: Linking Agriculture to Climate, Agronomy Monograph 60.
Jerry L. Hatfield, Mannava V.K. Sivakumar, John H. Prueger, editors.

their temperature response and incorporate information on insects to demonstrate how the climate dictates plant and insect responses and how these factors determine the potential overlap between economically viable plants and potential insect or disease damage.

Temperature Response of Agricultural Systems

Each species has a specific temperature range; these have been summarized in Hatfield et al. (2011). The lower threshold or base temperature is the temperature below which a plant ceases growth, while the upper threshold is the boundary at which growth ceases because temperatures exceed physiological thresholds. A good example of the differences between species is the distinction between cool-season and warm-season turfgrasses; cool-season have an optimum range between 15 and 24 °C, whereas warm-season turfgrasses have an optimum between 27 to 35 °C (DiPaola and Beard, 1992). Temperature responses for turfgrasses show differences in temperature ranges for root growth responses, which are lower than temperature limits for shoot growth with 10 to 18 °C for cool-season species and 24 to 30 °C for warm-season species (DaCosta and Huang, 2013). The differences in temperature responses for roots and shoots separate plants into those better suited to cool environments or warm environments, as illustrated by the differences between broccoli (*Brassica oleracea* L. var. *italic* Plenck) and maize (*Zea mays* L.). Broccoli is considered a cool-season vegetable that will not grow when exposed to temperatures above 22 °C, while maize will not grow efficiently at temperatures where broccoli thrives. Each species has a unique temperature response which will determine the reaction to changing temperatures. Weeds exhibit a similar response to turfgrasses; summer annuals [*Amaranthus albus* L., *A. palmeri* S. Watson, *Digitaria sanguinalis* (L.) Scop., *Echinochloa crus-galli* (L.) P. Beauv., *Portulaca oleracea* L., and *Setaria glauca* (L.) P. Beauv.] have a base temperature of 13.8 °C while winter annuals [*Hirschfeldia incana* (L.) Lagr.-Foss. and *Sonchus oleraceus* L.] showed base temperatures of 8.3 °C (Steinmaus et al., 2000). For grain crops, there are lower optimum temperatures during the reproductive period than during vegetative growth, indicating that exposure to high temperatures during the reproductive phases will have a negative impact on production (Hatfield et al., 2011). Exposure to temperatures will cause different responses depending on the stage of development of the plant (Hatfield and Prueger, 2015).

Another unique feature of many perennial plants is the requirement for exposure to temperatures below the lower limit but above 0 °C before flowering can occur. Fruit trees show a different requirement for chilling exposure (Hatfield et al., 2011) with cherries (*Prunus avium* (L.) L.) requiring over 700 h of exposure below 7 °C, and many apples (*Malus pumila* Mill.) require only 400 h while other apple varieties require nearly 1000 h. An analogous factor for crops is vernalization in small grains where crops like wheat (*Triticum aestivum* L.) require exposure to temperatures below 5 °C for a number of hours before flowering can commence. The increasing temperatures over the winter experienced in many areas of the world are creating conditions in which exposures to the cool temperatures may not be achievable during the winter period, as shown in the projections for the central Valley of California by Luedeling et al. (2009). Their analysis showed that with the continual warming, many perennial crops would not be exposed to adequate chilling hours by 2050. Plant response to temperature is species-specific and exposure to temperature within a given species' unique temperature range will directly affect productivity and growth rates.

Temperature has always been found to be important as a determinant for insect pest dynamics. Porter et al. (1991) listed the following direct effects of temperature on insects: limiting geographical range, overwintering, population growth rates, number of generations per annum, dispersal, and migration. Temperature also indirectly influences host plant and refugia availability, host plant quality, and crop–pest synchrony (Patterson et al., 1999). Insects are poikilothermic and are directly under the control of temperature for their growth. As long as the ambient temperature aligns with an insect's optimal growth range, then populations will grow quickly and higher pest densities will result. Conversely, temperatures outside the upper and lower thresholds halt insect growth altogether, comparable to the physiological thresholds exhibited by their host plants.

Two key effects of temperature in temperate regions, where winter temperatures seasonally fall below freezing, are to influence insect winter survival and range limitations (Bale et al., 2002). Temperate insect species have evolved behavioral and physiological mechanisms for surviving cold temperatures. They may overwinter as eggs in a diapause state, move to protected insulated habitats, or migrate to overwintering sites in southern latitudes. Regardless, winter mortality is often very high, which in turn influences crop pest densities in subsequent growing seasons. For example, the northern and western corn rootworms, (*Diabrotica barberi* and *Diabrotica virgifera virgifera*), are key pests of corn in the Great Plains and Midwestern United States, whose outbreak potential is directly linked with the overwintering survival of the egg stage. Under future climate projections, corn rootworms outbreaks may increase in severity and frequency as rising winter temperatures could reduce overwintering mortality, enhance development rates, and enable northward range expansions (Diffenbaugh et al., 2008). Rising temperatures, particularly in the winter, are also influencing northward range shifts observed in several major orders of insect pests (Diptera, Hemiptera, Coleoptera, Lepidoptera, and Isoptera) over the last fifty years (Bebber et al., 2013).

While insect responses to temperature are species-specific, there is a general census that insects with strong dispersal capabilities, multiple host plants, and the ability to produce multiple generations per year adapt better to both short- and long-term changes in temperature (Bale et al., 2002; Tobin et al., 2008). Since rising global temperatures are reducing cold limitations while simultaneously enhancing the developmental rates for many species, climate change has the potential to significantly alter the global pest management landscape. For instance, aphids are important agricultural pests throughout the world that are expected to benefit from rising temperatures due to their low developmental threshold, short generation time, and strong dispersal capabilities (Sutherst et al., 2007). Unfortunately, current crop–pest outcome models are limited due to the complexities of interacting abiotic and biotic factors driving insect outbreaks and subsequent yield losses in agroecosystems. To date, the potential impact of rising temperatures on pest–crop synchrony is poorly understood and additional research is greatly needed to improve crop models in the context of climate change.

Table 1. Agroclimatic indices and their application.†

Abbreviation	Agroclimatic variables	Description	Application	Limitations	References
GDD	T_a, T_{base}	Plant growth in relation to temperature	Hybrid phenology evaluation in response to climate	Crop specific T_{base} needed; Different GDD calculations	Neild and Richman (1981)
PSMD	ET_o, P	Soil water variability assessment for potato production	Evaluate current and future production areas, Irrigation management	Monthly time periods, Limited to specific region and crop	Daccache et al. (2012)
PACSI	O, FF, WRSI	Climatic limitations of corn production	Evaluate current and future production areas	Limited to specific region and crop	Moeletsi and Walker (2012)
WRSI	ET_p, k_c, P, SW, WHC	Water stress quantification	Drought monitoring and yield forecasting	Many agroclimatic variables needed	Frere and Popov (1979)
CWSI	T_a, T_c, T_{dmin}, T_{dmax}	Water stress quantification	Irrigation management, Drought monitoring	Many agroclimatic variables needed, difficult to apply in field	Idso et al. (1981); Jackson et al. (1981)

†ET_p, potential evapotranspiration; ET_o, reference evapotranspiration; FF, probability of frost free growing period; k_c, crop coefficient; O, Probability of planting conditions; P, Precipitation; SW, soil water content; T_a, air temperature; T_b, crop specific base temperature; T_c, canopy temperature; T_{dmin}, lower temperature boundary; T_{dmax}, upper temperature boundary; WHC, water holding capacity.

Agroclimatic Indices

The linkage between crops and climate has been quantified in many different types of indices (Table 1). The most recognized form of an index is growing degree days (GDD) used by Neild and Richman (1981) where they combined GDD and seasonal precipitation patterns to evaluate the differences among maize hybrids in phenology at various locations around the world. Growing degree days for a given species is calculated as $GDD = (T_{max} + T_{min})/2 - T_{base}$, where T_{max} is the maximum temperature (°C), T_{min} the minimum temperature (°C), and T_{base} the base temperature (°C). There have been several different forms of GDD calculations to incorporate the effect of temperatures above the upper threshold on plant growth rates. The use of a GDD approach has taken many different forms and is related to phenological development of crops. Maize has been the most studied crop in terms of GDD models and the recent analysis by Kumudini et al. (2014) revealed that models incorporating an upper threshold temperature response were more robust across a range of environments for estimating temperature effects on phenology. Growing degree days have been extensively used by producers to determine suitable hybrids to plant in different areas or to adjust planting dates in vegetables to ensure timely harvest. Thus, the GDD parameter remains widely used in management of crops.

Development of tools to define where crops can be produced is critical to understand crop distribution and productivity (van Wart et al., 2013). Estimation of crop distribution within arable areas is necessary to determine whether a species can thrive in an agroclimatic zone. Zomer et al. (2008) extended this concept to demonstrate how climate zones could be used to evaluate technologies that would enhance the ability of management practices to offset the impacts of climate change on crop production. There have continued to be advances in the development of agroclimatic indices to evaluate the suitability of a location for a particular crop since the efforts of Neild and Richman (1981). Siddons et al. (1994) cautioned that development of robust agroclimatic indices requires observations collected over long time periods and extensive observations from experimental locations. They applied this approach to the development of agroclimatic zones for production of white lupin (*Lupinus albus*) in England and Wales. Illustrative of the evolution of agroclimatic indices is provided by Holzkämper et al. (2013) where they included six factors into a suitability index for crops. These factors are: average daily minimum temperatures below 0 °C for frost impacts; daily mean temperature to determine plant growth; average daily maximum temperature above 35 °C for heat stress; average daily soil water availability [precipitation–reference evapotranspiration (ET)]; and length of the phenological period (days) to account for the effects of changing phenological development on biomass accumulation and crop yield. They related their suitability index to maize yields for locations around the world to demonstrate a positive relationship between productivity and the suitability index. This approach is a refinement of the effort by Neild and Richman (1981) to add more factors into their index to more closely match crop physiological responses.

Agroclimatic zones represent the combination of factors affecting plant growth to evaluate the potential to produce a grain or forage crop (e.g., Neild and Richman, 1981; Simane and Struik, 2003; Araya et al., 2010; Daccache et al., 2012; Falasca et al., 2012; Moeletsi and Walker, 2012; van Wart et al., 2013). The form of the index depends on the assumption of the factors limiting growth. For example, Araya et al. (2010) evaluated a combination of factors for the semiarid areas in Ethiopia to determine the suitability of this region for growing barley (*Hordeum vulgare* L.) and teff [*Eragrostis tef* (Zuccagni) Trotter]. Because water is a primary limitation in this region and the precipitation pattern during the summer rainfall period is extremely variable, their index used a combination of precipitation and reference ET to determine the suitability to grow these crops after 15 June. Their index was based on the cumulative five-day total of precipitation relative to cumulative ET over this same period to determine whether adequate soil moisture would be present in the seed zone for crop establishment. They defined the length of the growing period as the time between when adequate soil water was available for initial crop growth and the cessation of growth when ET exceeded precipitation for five cumulative days that placed a water limitation on growth (Araya et al., 2010). The dynamics of soil water use or ET can be altered by conservation practices that place plant residue on the soil surface to decrease soil water evaporation from soils and change soil water status in the root zone (Hatfield et al., 2001). In semiarid regions, where water is the primary limitation, the availability of soil water becomes the dominant factor determining crop suitability. Araya et al. (2010) found incorporating soil water was superior to using temperature and altitude as the foundation for agroclimatic zones and were able to identify eight regions compared to the traditional five regions and demonstrated this increase in spatial resolution helped producers better manage drought.

Soil water availability is often the determining factor in crop production in all ecosystems and the application has ranged from determination of irrigation water requirements to potential impacts on production caused by water deficits. Daccache et al. (2012) incorporated soil water variability to evaluate the need for irrigation for viable potato (*Solanum tuberosum* L.) production in England and Wales. Their index was based on the potential soil moisture deficit (PSMD) index defined as

$$\mathrm{PSMD}_i = \mathrm{PSMD}_{i-1} + \mathrm{ET}_{oi} - P_i \quad [1]$$

where PSMD_i is the value in month i and PSMD_{i-1} is the value for the previous month, $\mathrm{ET}_{\mathrm{oi}}$ is the reference ET for the current month calculated with the Penman–Monteith equation formulated by Allen et al. (1994), and P_i the precipitation in the current month. They utilized this approach to demonstrate that increased variation in precipitation would decrease in potato production in an area currently suited for production, unless supplemental water could be supplied via irrigation. This type of analysis could be utilized to determine the need for supplemental irrigation to ensure crop production in regions with highly variable precipitation.

Identification of potential new areas for crop production is one application of agroclimatic analysis. One example of this methodology is illustrated by Falasca et al. (2012) to determine potential production areas for castor bean (*Ricinus communis* L.) in the semiarid regions of Argentina. Suitable growing regions were defined relative to temperature ranges, water requirements, and length of the growing season. The classification scheme for castor bean derived from this method is described as:

- optimal (> 750 mm; temperature 24.0 to 27.0 ◦C; > –8 °C; > 180 frost-free days);
- very suitable (> 750 mm; temperature 21.0 to 23.9 °C; > –8 °C; > 180 frost-free days);
- suitable with humid regime (> 750 mm; temperature 16.0 to 20.9 °C; > –8 °C; > 180 frost-free days);
- suitable 1 with sub-humid regime (450 to 750 mm; temperature 24.0 to 27.0 °C; > –8 °C; > 180 frost-free days);
- suitable 2 with sub-humid regime (450 to 750 mm; temperature 21.0 to 23.9 °C; > –8 °C; > 180 frost-free days);
- suitable 3 with sub-humid regime (450 to 750 mm; temperature 16.0 to 20.9 ◦C; > –8 °C; > 180 frost-free days);
- marginal due to humidity (200 to 450 mm); marginal due to temperature (< 16.0 °C);
- marginal due to frosts (< 180 frost free days or < –8 °C);
- and not suitable areas (combination of two or more of the following variables: < 200 mm; < 16.0 °C;

This approach could identify areas suitable for production and serve as a potential framework to quantify the impacts of climate change by incorporating changes in temperature and precipitation to examine shifts in suitable areas. Another form of this type of framework was developed by Moeletsi and Walker (2012) to quantify climate risk index for maize in South Africa based on three climatic parameters; onset of rains, frost risk, and drought risk. These parameters were incorporated into the Poone AgroClimatic Suitability Index (PACSI) and utilized a weighed distribution of climate parameters as:

$$\mathrm{PACSI} = O \times 0.3 + FF \times 0.3 \times WRSI \times 0.4 \quad [2]$$

where O is the probability planting conditions are met, FF is the probability of a frost-free growing period, and the water requirements satisfaction index (WRSI). These indices require sufficient data to develop the probability of the different indices including the length of record sufficient to develop robust probability assessments (Siddons et al., 1994). An aspect of this index is the assessment of drought risk which is a complex interaction by soil water holding capacity and any change in the soil affecting water availability (Eq. [2]).

Precipitation effects on crop productivity are defined by the occurrence of the water deficits in the soil profile which fail to meet the evaporative demand. Agroclimatic indices for arid and semiarid regions are often based on precipitation amounts adequate to exceed the ET rate at the time of planting to ensure crop establishment (Neild and Richman, 1981; Araya et al., 2010; Daccache et al., 2012; Moeletsi and Walker, 2012; Holzkämper et al., 2013). Moeletsi and Walker (2012) provided a framework for the evaluation of soil water dynamics based on the WRSI to determine the potential to meet crop water requirements at any growth stage as:

$$\mathrm{WR}_i = \mathrm{ET}_{\mathrm{pi}} \times k_{ci} \qquad [3]$$

where WR_i is the water requirements for a decadal period during the growth of the crop, $\mathrm{ET}_{\mathrm{pi}}$ is the potential ET during this decadal period, and k_{ci} the crop coefficient for this corresponding period of growth. They developed a soil water balance to determine the dynamics during the season using the plant available water (WA_i) for a given decadal period as

$$\mathrm{WA}_i = P_i - \mathrm{SW}_{i-1} \qquad [4]$$

and P_i is the precipitation in a given decadal period and $\mathrm{SW}_{i\text{-}1}$ is the soil water in the profile for the previous decadal period. Soil water holding capacity (WHC) becomes a critical component of this approach because SW is a function of WHC. Frere and Popov (1979) computed the WRSI as:

$$\mathrm{WRSI}_i = \mathrm{WRSI}_{i-1} - \frac{\mathrm{WD}_i}{\sum_{i=1}^{\mathrm{end}} \mathrm{WR}} \qquad [5]$$

with WD_i the water deficit for decadal period I, defined as

$$\mathrm{WD}_i = \mathrm{WR}_i - P_i - \mathrm{SW}_{i-1} \text{ when } \mathrm{WR}_i > P_i + \mathrm{SW}_{i-1} \qquad [6]$$

Or

$$\mathrm{WD}_i = 0 \text{ when } \mathrm{WR}_i = P_i + \mathrm{SW}_{i-1} \qquad [7]$$

In this process soil water in the profile is quantified as:

$$\mathrm{SW}_i = P_i + \mathrm{SW}_{i-1} - \mathrm{WR}_i \qquad [8]$$

$$SW_i = WHC \text{ when } SW_i = WHC \quad [9]$$

$$SW_i = 0 \text{ when } SW_i = 0 \quad [10]$$

Using this methodology, Moeletsi and Walker (2012) were able to evaluate the suitability for maize production for various planting dates with a correlation of 0.8 between the PACSI and grain yields.

As a result of climate change, precipitation is changing in intensity and frequency and these changes directly affect WA_i (Eq. [3]). Precipitation patterns are projected to increase in annual totals, with decreasing summer precipitation amounts in the United States (Collins et al., 2013; Walsh et al., 2014). If we link these precipitation patterns with the PACSI (Eq. [2]), then maize production could become more variable among years because of soil water availability. Egli and Hatfield (2014a; 2014b) found that county-level maize and soybean [*Glycine max* (L.) Merr.] yield was directly related to soil water holding capacity. Counties across the Midwest with better soils exhibited a higher average grain yield across 50 yr of data (Egli and Hatfield, 2014a; 2014b). At the field scale, Hatfield (2012) found the largest effect on maize yields in the central U.S. was the lack of sufficient soil water during the grain-filling period to meet the evaporative demand. The inability of the soil to supply water impacts all scales from fields to ecosystems. The effect on ecosystems was evaluated by Porporato et al. (2004), who found that a soil water balance model incorporating hydroclimatic variability (i.e., frequency and intensity of precipitation events) determined vegetative conditions. Soil water is the dominant factor affecting vegetative productivity in both cultivated and natural systems, and the ability of the soil to infiltrate and store precipitation is a critical factor to offset the impact of increasing variability in precipitation. The linkage between crop distribution and productivity is linked to both temperature and precipitation with the availability of water through the soil being the most critical factor.

Linkages of the Atmosphere to Plants

Leaf Scale

Some agroclimatic indices incorporate soil water status and the ET rate of the crop. The ET rate of a crop is determined by the energy balance of the crop. To understand this linkage, it is illustrative to use a leaf model and then extend it to the canopy level. The rate of water lost via transpiration from an individual leaf is a function of net radiation balance of the leaf, the water supply to the leaf, the conductance of the air for the transfer of water vapor, which is a function of leaf shape and windspeed, and the stomatal conductance which determines how rapidly the leaf releases water vapor into the atmosphere. These relationships are described by the following equation.

$$S_t(1-\alpha_1)+L_d-\epsilon\sigma T_a^4=\frac{\rho C_p(T_l-T_a)}{r_a}+\frac{\rho C_p}{\gamma^*}\frac{(e_o-e_a)}{r_s+r_a} \quad [11]$$

Where S_t is the incoming solar radiation (W m^{-2}), a_l is the albedo of the leaf, L_d is the incoming longwave radiation (W m^{-2}), esT^4 is the longwave radiation emitted by the leaf at the leaf temperature (T_l), ρ is the density of air (kg m^{-3}), C_p is the specific heat of air (J kg^{-1} C^{-1}), T_l is leaf temperature (°C), T_a is the air temperature (°C), r_a is the aerodynamic conductance to heat transfer (s m^{-1}), γ^* is the psychrometric constant (kPa C^{-1}), e_o is the saturation vapor pressure at T_l (kPa), e_a is the actual vapor pressure of the air (kPa), r_s is the stomatal conductance (s m^{-1}). One of the critical components in the leaf-scale energy balance is the actual temperature of the leaf (T_l) because it provides a direct linkage between air temperature, T_a, and the energy balance and water supply to the leaf. Leaf temperature determines the physiological reactions including the respiration rate. A leaf with maximum stomatal conductance will be below air temperature and the interactions among the components in Eq. [11] have been described in Hatfield and Prueger (2011). The changes in leaf relative to air temperature was originally observed by Tanner (1963) who showed that leaf temperature varied from air temperature and could be measured with thermocouples attached to plant leaves. Thermocouples have been replaced by infrared thermometers which measure emittance in the 8 to 14 mm range and have become small and relatively inexpensive. Following Tanner's observations, there has been a set of observations that linked leaf temperature with water stress, solar radiation, air temperature, and water vapor pressure (Wiegand and Namken, 1966; Ehrler et al., 1978). This initial research and the development of relationships between leaf temperature and plant reaction to water stress has prompted a series of studies about the linkage between plants and their environment. However, the majority of the efforts in the recent progress has been directed more toward the canopy level processes rather than leaf level and the ability to routinely measure surface temperatures with infrared thermometers without physically attaching an instrument to the leaf has expanded the capabilities of obtaining these observations.

Canopy-Level Feedbacks

Evaluation of linkages between biological processes and climate for plants is more easily understood at the canopy level because we grow plants in different configurations and typically observe them at this level rather than at the leaf level. Extension from the leaf to canopy level requires an incorporation of the soil interaction with the plant canopy and the atmosphere to accomplish this transition, the conductance terms in Eq. [11] represent values associated with canopy level processes rather than leaf level. However, when we transition from the leaf to the canopy, there is an additional component that has be considered in the energy exchanges, canopy level responses have to account for the energy exchanges between the soil and the atmosphere. These are often considered in two-layer energy balance models, with one layer for the canopy and the second layer for the soil surface. Two factors that become critical in this approach is the spatial arrangement of the canopy and the temporal dynamics of the canopy size. Perennial crops, such as trees and vineyards, will have a more consistent canopy architecture over the growing season with a variation in the amount of leaf cover and shape induced by pruning. Perennial grasses or pastures, are more consistent in their cover; however, there may be a variation in the spatial arrangement of the grasses creating a clumping pattern with potential bare soil areas between the clumps. Annual crops grown in rows can be treated as an expanding volume over the course of the growing season, and the interactions between the soil surface and plant canopy will change throughout the growing season in response to temperature, windspeed, vapor pressure, and soil surface wetness. These dynamics are complex because the micro-environment that canopy leaves are exposed to is dependent on the soil surface conditions. For example, when the canopy is small, the microclimate within the canopy is determined by the soil surface wetness and the resultant cooling of the air via evaporation. If the soil surface is dry, then there is little soil water evaporation and the energy reaching the soil surface increases the sensible heat component of the energy balance. This warms the air in the canopy, increases the water vapor deficit and increases the potential ET rate for the canopy. Plants exposed to these conditions will have a greater chance of water stress which limits their growth rate. Conversely, a wet soil surface increases the vapor pressure of the air surrounding the canopy which lowers the evaporative demand and reduces the water use by the canopy, with the net result being a negative impact on canopy water use and also canopy temperatures.

As the canopy develops and completely covers the soil surface, the energy balance is dominated by canopy vegetation and the soil water availability to meet the atmospheric demand. There are interactions among the parameters Eq. [11], for example, Lhomme and Monteny (2000) described the theoretical relationship between stomatal conductance (r_s) and canopy temperatures (T_c) to demonstrate the sensitivity of canopy temperatures to r_s. As the conductance decreases due to stomatal closure, canopy temperatures will increase because of the reduced supply of water for evaporative cooling.

These relationships between water status of a canopy and canopy temperature has prompted the development of a series of indices to quantify crop water status using canopy and air temperature differences (Maes and Steppe, 2012), of which the Crop Water Stress Index (CWSI) is widely applied (Idso et al., 1981; Jackson et al., 1981):

$$\mathrm{CWSI} = \frac{\left(dT - dT_{\mathrm{min}}\right)}{\left(dT_{\mathrm{max}} - dT_{\mathrm{min}}\right)} \qquad [12]$$

Where dT (°C) is the difference between air temperature (T_a) and actual observed canopy temperatures (T_c), often observed with infrared thermocouple sensors or infrared thermography, dT_{min} (°C) is the non-water limited, lower baseline of dT, dT_{max} (°C) is the upper baseline of dT and represents the canopy in a virtually non-transpiring state.

The CWSI is an approach to normalize measured dT to acknowledge changes of T_c by atmospheric conditions, especially net radiation, vapor pressure deficit (VPD), and windspeed. The lower baseline, dT_{min}, is linearly related to VPD. Transpiration increases with increasing VPD, which declines dT under well-watered conditions owing to a reduction of T_c. The dT_{min} is either directly measured as a series of dT of well water crops and VPD (Idso et al., 1981; Idso 1982), or is derived theoretically using the energy balance concept (Jackson et al., 1981; 1988). The upper baseline, dT_{max}, is often applied as a constant dT value at which plants are severely water stressed (Gardner et al., 1992). The CWSI ranges between 0 and 1, although field measurements may exceed these boundaries. Although the CWSI has been successfully applied to remotely quantify water stress and to schedule irrigation, it has not reached widespread adaption in agricultural production. The main challenge is the requirement of location, crop species and plant development specific dT_{min} and dT_{max} baselines. In addition, measurements are restricted to certain time periods (typically around noon) with high plant foliage cover and constant (and high) net radiation, which makes the application cumbersome for growers.

One of the major differences between a leaf and the canopy is the albedo of the surface. Albedo of a leaf remains relatively constant during the growing season until the leaf begins to senesce; however, albedo of a canopy is dependent on the amount of ground cover, the albedo of the soil, and the soil surface wetness. At the canopy level there is a separation of the evaporation from the soil surface and the plant canopy. These relationships between soil water evaporation (E) from plant transpiration (T) as part of the canopy ET have been described by Ritchie (1971). He found that soil water evaporation is affected by soil water content of the surface and degree of plant cover on the surface. The interactions between the soil surface and plant canopies has been described in a series of papers by Ritchie and Burnett (1971), Ritchie et al. (1972), and Ritchie and Jordan (1972). These papers provide the foundation for many of the simulation models on the response of plants to soil water and partitioning of evaporation and transpiration. As we begin to evaluate the linkages between crops and climate, the canopy dynamics relative to stage of growth and the microclimate become important factors in determining crop response.

Responses of Plants to Carbon Dioxide

Typically, carbon dioxide (CO_2) is not considered as part of the climate; however, given the changes in CO_2 concentrations over the past decades, this component becomes an integral part of the environment affecting plant response. The CO_2 effect on plant growth and development has been evaluated for a number of species (Kimball, 1983; Kimball and Idso, 1983; Ainsworth et al., 2002; Hatfield et al., 2011; Izaurralde et al., 2011) and show a positive impact, while temperature and precipitation effects show mixed effects by themselves and in combination with increased CO_2. The interaction among these variables introduces a new level of complexity in our understanding of the linkage between plants and climate. Carbon dioxide is one of the fundamental building blocks for plants in the incorporation of sugars via the photosynthetic process and the subsequent transformation of these sugars into a wide range of metabolic products. The changes in concentrations of CO_2 in the atmosphere have increased the concern about the potential impacts from rising CO_2 levels. Concentrations of CO_2 have increased from an average of 325 μmol mol^{-1} from the early 1970s to 400 μmol mol^{1} at the present time with projections of 550 μmol mol^{-1} by mid-century and over 700 μmol mol^{-1} by 2100 leading to increased questions about the effect of these concentrations on plants. These changes in CO_2 concentration have many positive impacts on plant growth and productivity. Over the past 40 yr there have been numerous studies conducted in chambers where CO_2 concentrations could be controlled relative to the current ambient condition. The most popular of these chambers has been the free air CO_2 experiment (FACE) that has increased our understanding of the changing CO_2 concentration on growth, water use, and yield for a variety of plants (Kimball, 1983; Kimball and Idso, 1983; Kimball and Mauney, 1993; Kimball et al., 1999, 2001, 2002; Ainsworth and Long, 2005; Ainsworth and Rogers, 2007; Kimball, 2010). The impacts of increasing CO_2 vary among plants with C_4 plants [maize and sorghum, *Sorghum biocolor* L. Moench] showing a smaller response than C_3 plants (barley, *Hordeum vulgare* L.; beans, *Phaseolus vulgaris* L.; cotton, *Gossypium hirsutum* L.; peanut, *Arachis hypogaea* L.; rice; soybean; sugarbeet, *Beta vulgaris* L.; wheat) for a number of physiological parameters including grain quality. Plants respond positively to increasing CO_2 in the absence of temperature stress and exhibit a more positive response when water stress occurs because of the impact on water use efficiency (Wu and Wang, 2000; Bernacchi et al., 2007; Chun et al., 2011; Abebe et al., 2016).

The positive effect on growth and yield occurs concurrently with a decrease in stomatal conductance and crop ET as summarized by Hatfield et al. (2011). In C_3 plants, photosynthesis is limited by three distinct physiological processes: i) maximum Rubisco carboxylation capacity ($V_{c,max}$); ii) the rate in which the light reactions generate ATP and NADPH for use in the photosynthetic reduction cycle (J_{max}); and iii) the rate in which inorganic phosphate is released during triose phosphate utilization (Bernacchi et al., 2013). For C_4 plants, photosynthetic response to CO_2 concentrations is limited by PEP carboxylase activity (V_{pmax}) and by Rubisco activity at higher CO_2 (V_{max}) (von Caemmerer, 2000). Increasing CO_2 exhibits a greater benefit on C_3 plants because these limiting factors are overcome by the increased internal concentration of CO_2. In contrast, C_4 plants show a saturation of CO_2 at the reaction sites and consequently no or limited photosynthetic response to increasing CO_2. The net effect of increasing CO_2 increases all physiological parameters with the largest effect on stomatal conductance which reduces water use from the leaf with no loss in photosynthetic uptake translating into enhanced water use efficiency. Leakey (2009) conceptualized that improved photosynthesis in C_4 plants occurs under water stress conditions because decreased stomatal conductance reduces leaf transpiration and improves the internal water status of the plant. The increased gradient of CO_2, because of the enhanced CO_2 in the atmosphere, maintains the intercellular CO_2 concentration. This conceptual model has been confirmed for maize and grain sorghum where relatively high canopy photosynthetic rates were observed concurrently with decreased transpiration rates, resulting in enhanced water use efficiency when both plants were grown at elevated CO_2 of 720 μmol mol^{-1}, but not at 360 μmol mol^{-1} (Allen et al., 2011). Differences between the two CO_2 concentrations were related to the ability of the corn and grain sorghum plants to maintain a higher leaf water potential under the higher CO_2 concentration as a result of the lower transpiration rate with the resultant effect being more rapid leaf area expansion leading to increased photosynthesis (Allen et al., 2011; Chun et al., 2011). There is a potential advantage of C_4 plants under limited soil water conditions or variable soil water supply (Bunce, 2010).

Elevated CO_2 in C_3 plants, increases photosynthetic rates as a result of overcoming the limitation of $V_{c\ max}$ under high light conditions (Bunce, 2014). Increased photosynthesis under higher CO_2 concentrations led to positive impacts on growth parameters. In spring wheat, elevated CO_2 increased the number of tillers and ears per plant by 13% at the beginning of stem elongation and leaf area increased throughout the growth cycle of the wheat (Ewert and Pleijel, 1999). The effects of increased CO_2 from 330 to 660 μmol mol^{-1} coupled with changes in temperature using different temperature regimes, 29/21, 33/25, and 37/29°C, increased tiller number and total dry weight across all temperature regimes in rice (Manalo et al., 1994).

Elevated CO_2 effects are not universal across all plants, Franzaring et al. (2008) found for oilseed rape (*Brassica napus* var. Campino) there was an early transition from vegetative to reproductive stages and earlier senescence. Vegetative biomass increased at the cessation of flowering but at harvest there was no difference between CO_2 concentrations on seed yield due to the decrease in the length of the grain-filing period. Increased CO_2 was associated with an increase in the pod walls, but seed size was not affected nor was seed oil content or total oil yields (Franzaring et al., 2008). This is in contrast to the observations on oilseed rape by Frenck et al. (2011) where increased CO_2 did not compensate for the negative yield impacts induced by high temperatures and O_3. There is an interaction between the effect of water stress which both $V_{c\,max}$ and J_{max} and increased CO_2 offsets this effect (Aljazairi and Nogués, 2015). They proposed that changes in the Rubisco protein content at both low and high CO_2 concentrations under increased CO_2, causes a reallocation of nitrogen from the leaf to other plant parts; however, there was no concurrent increase in sink strength for C. Accumulation of photosynthate in the leaves due to higher CO_2 may not be effectively translocated to the sinks for grain development leading to a limitation in grain production (Aljazairi and Nogués, 2015). In soybean, the interaction of the rising temperatures with increasing CO_2 showed that higher CO_2 produced more vigorous growth at 20 to 15 °C but at higher temperatures (30 to 25°C) there was a reduction in seed yield even at 700 μmol mol^{-1} due to a reduction in seed size (Heinemann et al., 2006). There are interactions between CO_2 and water stress, for example, soybean grown at 740 μmol mol^{-1} compared to 380 μmol mol^{-1} showed plant height increased by 25.4%, leaf area by 15.8%, shoot dry weight by 33.4%, and seed yield by 25.3% under normal water and high CO_2 at the seed-filling stage. However, under drought conditions, elevated CO_2 had no significant effect on plant height, leaf area, and seed yield, although shoot dry weight was increased by 56% (Li et al., 2013). This effect was attributed to greater biomass allocation toward the stems. In this experiment, photosynthetic rates were increased by 21.7 to 43.3% in the elevated CO_2 concentrations compared to ambient CO_2 levels under adequate soil water (Li et al., 2013).

Most of the effort on increasing CO_2 has focused on shoot growth and yield; however, the increasing CO_2 concentrations will have an effect on root growth and function. One effect of increasing CO_2 on root growth across a number of species was the increased allocation of C to roots stimulating lateral rooting and root branching (Madhu and Hatfield, 2013). Improved root density and proliferation to explore the soil to extract more water and nutrients will benefit plants. Observations by Pacholski et al. (2015) revealed barley, sugar beet, and wheat grown under ambient and 550 μmol mol^{-1} increased root biomass by 54% during the early vegetative period; however, there was an accompanying change in the root to shoot ratio during the growing season with differences among the species. Increasing the allocation of C to the roots increased the biomass with greater partitioning into roots early in the growing season with a decrease for the remainder of the growing season (Pacholski et al., 2015). Understanding the root system response and the interactions with water and nutrients in the soil profile could help provide strategies to quantify plant response to climate. Roots are a critical part of understanding the linkages of biology to climate and expanding our knowledge base of how plants are responding to variation in climates that will help improve our understanding of agroclimatology.

Challenges in Increasing Our Knowledge Base on Biological Linkages

Agroclimatic indices developed over decades of experience in quantifying the interactions among temperature, precipitation, and crop response provide a framework for assessing the impacts of climate change on plant distribution and productivity. Application of agroclimatic indices focus on the capability of the given environment to allow for a given species the capability to grow but does not focus on the level of productivity. For example, the region for growing perennial trees, with a specific cooling temperature requirement before flowering, can be defined if the winter temperatures meet this requirement; however, this does not define the level of productivity. Current agroclimatic indices incorporate both temperature and precipitation to define whether the climate can meet the temperature requirements and the water availability so the plant has the capacity to meet the atmospheric demand. Until recently the effect of CO_2 was not considered in agroclimatic indices; however, the rapid increase in atmospheric concentrations of CO_2 demonstrate that we need to begin to incorporate the effect of this variable on plant response. The major challenge for agroclimatology and plant response is to increase our understanding of the interaction among temperature, water, and CO_2 regimes for a given region. Utilizing the energy balance as a framework for the energy exchanges between a leaf or a canopy and the soil and atmosphere provides a more robust framework for these analyses and utilizing this type of physical approach enhances our understanding.

As we go into the future, our current understanding of agroclimatology must broaden beyond the effects on plants and begin to take into account that these same climate variables also drive insect, weed, and disease responses to truly approach a more holistic and ecological view of plants in their environment. The changing temperature regime will affect the ranges of all pests and will become a factor that limits productivity of economic plants. There have been noticeable changes in crop distribution in North America with a northward expansion of corn and soybean production. One of the challenges will be to improve the genetics of these plants to increase their productivity in these shorter seasons. An additional challenge will be to initiate the development of a more ecological approach for how we address the interactions among crops and pests. We have the framework for quantifying these interactions and now the challenge will be to expand our knowledge of how the biological components of agricultural systems and climatology link together. If we address this challenge, then ensuring our capability to provide the food and feed resources for a growing population will be a reality.

References

Abebe, A., H. Pathak, S.D. Singh, A. Bhatia, R.C. Harit, Vinod Kumar. 2016. Growth, yield and quality of maize with elevated atmospheric carbon dioxide and temperature in north–west India. Agric. Ecosyst. Environ. 218:66–72. doi:10.1016/j.agee.2015.11.014

Ainsworth, E.A., and S.P. Long. 2005. What have we learned from 15 years of free-air CO_2 enrichment (FACE)? A meta-analytic review of the responses of photosynthesis, canopy properties and plant production to rising CO_2. New Phytol. 165:351–372.

Ainsworth, E.A., and A. Rogers. 2007. The response of photosynthesis and stomatal conductance to rising [CO_2]: Mechanisms and environmental interactions. Plant Cell Environ. 30:258–270. doi:10.1111/j.1365-3040.2007.01641.x

Ainsworth, E.A., P.A. Davey, C.J. Bernacchi, O.C. Dermody, E.A. Heaton, D.J. Moore, P.B. Morgan, S.L. Naidu, H.Y. Ra, X. Zhu, P.S. Curtis, and S.P. Long. 2002. A meta-analysis of elevated [CO_2] effects on soybean (*Glycine max*) physiology, growth and yield. Glob. Change Biol. 8:695–709. doi:10.1046/j.1365-2486.2002.00498.x

Aljazairi, S., and S. Nogués. 2015. The effects of depleted, current and elevated growth [CO_2] in wheat are modulated by water availability. Environ. Exp. Bot. 112:55–66. doi:10.1016/j.envexpbot.2014.12.002

Allen, L.H., Jr., V.G. Kakani, J.C.V. Vu, and K.J. Boote. 2011. Elevated CO_2 increases water use efficiency by sustaining photosynthesis of water-limited maize and sorghum. J. Plant Physiol. 168:1909–1918. doi:10.1016/j.jplph.2011.05.005

Allen, R.G., M. Smith, A. Perrier, and L.S. Pereira. 1994. An update for the definition of reference evapotranspiration. ICID Bulletin 43:1–34.

Araya, A., S.D. Keesstra, and L. Stroosnijder. 2010. A new agro-climatic classification for crop suitability zoning in northern semi-arid Ethiopia. Agric. Meteorol. 150:1057–1064. doi:10.1016/j.agrformet.2010.04.003

Bale, J.S., G.J. Masters, I.D. Hodkinson, C. Awmack, T.M. Bezemer, V.K. Brown, J. Butterfield, A. Buse, J.C. Coulson, J. Farrar, J.E. Good, R. Harrington, S. Hartley, T.H. Jones, R.L. Lindroth, M.C. Press, I. Symrnioudis, A.D. Watt, and J.B. Whittaker. 2002. Herbivory in global climate change research: Direct effects of rising temperature on insect herbivores. Glob. Change Biol. 8:1–16.

Bebber, D.P., M.A. Ramotowski, and S.J. Gurr. 2013. Crop pests and pathogens move polewards in a warming world. Nat. Clim. Chang. 3:985–988. doi:10.1038/nclimate1990

Bernacchi, C.J., B.A. Kimball, D.R. Quarles, S.P. Long, and D.R. Ort. 2007. Decreases in stomatal conductance of soybean under open-air elevation of CO_2 are closely coupled with decreases in ecosystem evapotranspiration. Plant Physiol. 143:134–144. doi:10.1104/pp.106.089557

Bernacchi, C.J., J.E. Bagley, S.P. Serbin, U.M. Ruiz-Vera, D.M. Rosenthal, and A. VanLoocke. 2013. Modelling C_3 photosynthesis from the chloroplast to the ecosystem. Plant Cell Environ. 36:1641–1657.

Bunce, J.A. 2010. Leaf transpiration efficiency of some drought-resistant maize lines. Crop Sci. 50:1409–1413. doi:10.2135/cropsci2009.11.0650

Bunce, J.A. 2014. Limitations to soybean photosynthesis at elevated carbon dioxide in free-air enrichment and open top chamber systems. Plant Sci. 226:131–135. doi:10.1016/j.plantsci.2014.01.002

Chun, J.A., Q. Wang, D. Timlin, D. Fleisher, and V.R. Reddy. 2011. Effect of elevated carbon dioxide and water stress on gas exchange and water use efficiency in corn. Agric. Meteorol. 151:378–384. doi:10.1016/j.agrformet.2010.11.015

Collins, M., R. Knutti, J. Arblaster, J.-L. Dufresne, T. Fichefet, P. Friedlingstein, X. Gao, W.J. Gutowski, T. Johns, G. Krinner, M. Shongwe, C. Tebaldi, A.J. Weaver, and M. Wehner. 2013. Long-term climate change: Projections, commitments and irreversibility. In: T.F. Stocker, D. Qin, G.–K. Plattner, M. Tignor, S.K. Allen, J. Boschung, A. Nauels, Y. Xia, V. Bex, and P.M. Midgley, editors, Climate change 2013: The physical science basis. Contribution of Working Group I to the Fifth Assessment Report of the Intergovernmental Panel on Climate Change. Cambridge Univ. Press, Cambridge.

Daccache, A., C. Keay, R.J.A. Jones, E.K. Weatherhead, M.A. Stalham, and J.W. Knox. 2012. Climate change and land suitability for potato production in England and Wales: Impacts and adaptation. J. Agric. Sci. 150:161–177. doi:10.1017/S0021859611000839

DaCosta, M., and B. Huang. 2013. Heat-stress physiology and management. In: J.C. Stier, B.P. Horgan, and S.A. Bonos, editors, Turfgrass: Biology, use, and management. Agronomy Monograph 56. ASA, CSSA, SSSA, Madison, WI. p. 249–278. Doi:10.2134/agronmonogr56.c.

Diffenbaugh, N.S., C.H. Krupke, M.A. White, and C.E. Alexander. 2008. Global warming presents new challenges for maize pest management. Environ. Res. Lett. 3:044007. doi:10.1088/1748-9326/3/4/044007

DiPaola, J.M., and J.B. Beard. 1992. Physiological effects of temperature stress. In: D.V. Waddington, R.N. Carrow, and R.C. Shearman, editors, Turfgrass. Agronomy Monograph 32. ASA, CSSA, SSSA, Madison, WI 53711.

Egli, D.B., and J.L. Hatfield. 2014a. Yield gaps and yield relationships in central U.S. soybean production systems. Agron. J. 106:560–566. doi:10.2134/agronj2013.0364

Egli, D.B., and J.L. Hatfield. 2014b. Yield gaps and yield relationships in central U.S. maize production systems. Agron. J. 106:2248–2256. doi:10.2134/agronj14.0348

Ehrler, W.L., S.B. Idso, R.D. Jackson, and R.J. Reginato. 1978. Wheat canopy temperatures: Relation to plant water potential. Agron. J. 70:251–256. doi:10.2134/agronj1978.00021962007000020010x

Ewert, F., and H. Pleijel. 1999. Phenological development, leaf emergence, tillering and leaf area index, and duration of spring wheat across Europe in response to CO_2 and ozone. Eur. J. Agron. 10:171–184. doi:10.1016/S1161-0301(99)00008-8

Falasca, S.L., A.C. Ulberich, and E. Ulberich. 2012. Developing an agro-climatic zoning model to determine potential production areas for castor bean (*Ricinus communis* L.). Ind. Crops Prod. 40:185–191. doi:10.1016/j.indcrop.2012.02.044

Franzaring, J., P. Högy, and A. Fangmeier. 2008. Effects of free-air CO_2 enrichment on the growth of summer oilseed rape (*Brassica napus* cv. Campino). Agric. Ecosyst. Environ. 128:127–134. doi:10.1016/j.agee.2008.05.011

Frenck, G., L. van der Linden, T.N. Mikkelsen, H. Brix, and R.B. Jørgensen. 2011. Increased [CO_2] does not compensate for negative effects on yield caused by higher temperature and [O_3]. *In: Brassica napus* L. Eur. J. Agron. 35:127–134. doi:10.1016/j.eja.2011.05.004

Frere, M., and G.F. Popov. 1979. Agrometeorological crop monitoring and forecasting. FAO Plant Production and Protection Paper 17:66.

Gardner, B., D. Nielsen, and C. Shock. 1992. Infrared thermometry and the crop water stress index. I. History, theory, and baselines. J. Prod. Agric. 5:462–466. doi:10.2134/jpa1992.0462

Hatfield, J.L. 2012. Spatial patterns of water and nitrogen response within corn production fields. In: G. Aflakpui, editor, Agric Sci. Intech Publishers, London. p. 73–96.

Hatfield, J.L., and J.H. Prueger. 2011. Agroecology: Implications for plant response to climate change. In: S.S. Yadav, R.J. Redden, J.L. Hatfield, H. Lotze-Campen, and A.E. Hall, editors, Crop adaptation to climate change. Wiley-Blackwell, West Sussex, United Kingdom. p. 27–43. doi:10.1002/9780470960929.ch3

Hatfield, J.L., and J.H. Prueger. 2015. Temperature extremes: Effects on plant growth and development. Weather and Climate Extremes 10:4–10. doi:10.1016/j.wace.2015.08.001

Hatfield, J.L., T.J. Sauer, and J.H. Prueger. 2001. Managing soils for greater water use efficiency: A Review. Agron. J. 93:271–280. doi:10.2134/agronj2001.932271x

Hatfield, J.L., K.J. Boote, B.A. Kimball, L.H. Ziska, R.C. Izaurralde, D. Ort, A.M. Thomson, and D.W. Wolfe. 2011. Climate impacts on agriculture: Implications for crop production. Agron. J. 103:351–370. doi:10.2134/agronj2010.0303

Heinemann, A.B., A. de H.N. Maia, D. Dourado-Neto, K.T. Ingram, and G. Hoogenboom. 2006. Soybean (*Glycine max* (L.) Merr.) growth and development response to CO_2 enrichment under different temperature regimes. Eur. J. Agron. 24:52–61. doi:10.1016/j.eja.2005.04.005

Holzkämper, A., P. Calanca, and J. Fuhrer. 2013. Identifying climatic limitations to grain maize yield potentials using a suitability evaluation approach. Agric. Meteorol. 168:149–159. doi:10.1016/j.agrformet.2012.09.004

Idso, S., R. Jackson, P. Pinter, R. Reginato, and J. Hatfield. 1981. Normalizing the stress-degree-day parameter for environmental variability. Agric. Meteorol. 24:45–55. doi:10.1016/0002-1571(81)90032-7

Idso, S.B. 1982. Non-water-stressed baselines: A key to measuring and interpreting plant water stress. Agric. Meteorol. 27:59–70. doi:10.1016/0002-1571(82)90020-6

Izaurralde, R.C., A.M. Thomson, J.A. Morgan, P.A. Fay, H.W. Polley, and J.L. Hatfield. 2011. Climate impacts on agriculture: Implications for forage and rangeland production. Agron. J. 103:371–380. doi:10.2134/agronj2010.0304

Jackson, R.D., S.B. Idso, R.J. Reginato, and P.J. Pinter. 1981. Canopy temperature as a crop water stress indicator. Water Resour. Res. 17:1133–1138. doi:10.1029/WR017i004p01133

Jackson, R.D., W.P. Kustas, and B.J. Choudhury. 1988. A reexamination of the crop water stress index. Irrig. Sci. 9:309–317. doi:10.1007/BF00296705

Kimball, B.A. 1983. Carbon dioxide and agricultural yield. An assemblage of 430 prior observations. Agron. J. 75:779–788. doi:10.2134/agronj1983.00021962007500050014x

Kimball, B.A. 2010. Lessons from FACE: CO2 effects and interactions with water, nitrogen, and temperature. In: D. Hillel and C. Rosenzweig, editors, Handbook of climate change and agroecosystems: Impacts, adaptation, and mitigation. Imperial College Press, London, UK. p. 87–107. doi:10.1142/9781848166561_0006

Kimball, B.A., and S.B. Idso. 1983. Increasing atmospheric CO_2: Effects on crop yield, water use, and climate. Agric. Water Manage. 7:55–72. doi:10.1016/0378-3774(83)90075-6

Kimball, B.A., and J.R. Mauney. 1993. Response of cotton to varying CO_2, irrigation, and nitrogen: Yield and growth. Agron. J. 85:706–712. doi:10.2134/agronj1993.00021962008500030035x

Kimball, B.A., K. Kobayashi, and M. Bindi. 2002. Responses of agricultural crops to free-air CO_2 enrichment. Adv. Agron. 77:293–368. doi:10.1016/S0065-2113(02)77017-X

Kimball, B.A., R.L. LaMorte, P.J. Pinter, Jr., G.W. Wall, D.J. Hunsaker, F.J. Adamsen, S.W. Leavitt, T.L. Thompson, A.D. Matthias, and T.J. Brooks. 1999. Free-air CO_2 enrichment (FACE) and soil nitrogen effects on energy balance and evapotranspiration of wheat. Water Resour. Res. 35:1179–1190. doi:10.1029/1998WR900115

Kimball, B.A., C.F. Morris, P.J. Pinter, Jr., G.W. Wall, D.J. Hunsaker, F.J. Adamsen, R.L. LaMorte, S.W. Leavitt, T.L. Thompson, A.D. Matthias, and T.J. Brooks. 2001. Elevated CO_2, drought and soil nitrogen effects on wheat grain quality. New Phytol. 150:295–303. doi:10.1046/j.1469-8137.2001.00107.x

Kumudini, S., F.H. Andrade, K.J. Boote, G.A. Brown, K.A. Dzotsi, G.O. Edmeades, T. Gocken, M. Goodwin, A.L. Halter, G.L. Hammer, J.L. Hatfield, J.W. Jones, A.R. Kemanian, S.-H. Kim, J. Kiniry, J.I. Lizaso, C. Nendel, R.L. Nielsen, B. Parent, C.O. St ckle, F. Tardieu, P.R. Thomison, D.J. Timlin, T.J. Vyn, D. Wallach, H.S. Yang, and M. Tollenaar. 2014. Predicting Maize Phenology: Intercomparison of Functions for Developmental Response to Temperature. Agron. J. 106:2087–2097. doi:10.2134/agronj14.0200

Leakey, A.D.B. 2009. Rising atmospheric carbon dioxide concentration and the future of C4 crops for food and fuel. Proc. Biol. Sci. 276:2333–2343. doi:10.1098/rspb.2008.1517

Lhomme, J.P., and B. Monteny. 2000. Theoretical relationship between stomatal resistance and surface temperatures in sparse vegetation. Agric. Meteorol. 104:119–131. doi:10.1016/S0168-1923(00)00155-6

Li, D., H. Liu, Y. Qiao, Y. Wang, Z. Cai, B. Dong, C. Shi, Y. Liu, X. Li, and M. Liu. 2013. Effects of elevated CO_2 on the growth, seed yield, and water use efficiency of soybean (*Glycine max* (L.) Merr.) under drought stress. Agric. Water Manage. 129:105–112. doi:10.1016/j.agwat.2013.07.014

Luedeling, E., M. Zhang, and E.H. Girvetz. 2009. Climatic changes lead to declining winter chill for fruit and nut trees in California during 1950–2099. PLoS One 4(7):E6166. doi:10.1371/journal.pone.0006166

Maes, W., and K. Steppe. 2012. Estimating evapotranspiration and drought stress with ground-based thermal remote sensing in agriculture: A review. J. Exp. Bot. 63:4671–4712.

Madhu, M., and J.L. Hatfield. 2013. Dynamics of plant root growth under increased atmospheric carbon dioxide. Agron. J. 105:657–669. doi:10.2134/agronj2013.0018

Manalo, P.A., K.T. Ingram, R.R. Pamplona, and A.O. Egeh. 1994. Atmospheric CO_2 and temperature effects on development and growth of rice. Agric. Ecosyst. Environ. 51:339–347. doi:10.1016/0167-8809(94)90145-7

Moeletsi, M.E., and S. Walker. 2012. A simple agroclimatic index to delineate suitable growing areas for rainfed maize production in the Free State Province of South Africa. Agric. For. Meteorol. 162-163:63–70. doi:10.1016/j.agrformet.2012.04.009

Neild, R.E., and N.H. Richman. 1981. Agroclimatic normals for maize. Agric. Meteorol. 24:83–95. doi:10.1016/0002-1571(81)90035-2

Pacholski, A., R. Manderscheid, and H.-J. Weigel. 2015. Effects of free air CO_2 enrichment on root growth of barley, sugarbeet and wheat grown in a rotation under different nitrogen supply. Eur. J. Agron. 63:36–46. doi:10.1016/j.eja.2014.10.005

Patterson, D.T., J.K. Westbrook, R.J.V. Joyce, P.D. Lingren, and J. Rogasik. 1999. Weeds, insects, and diseases. Clim. Change 43:711–727. doi:10.1023/A:1005549400875

Porter, J.H., M.L. Parry, and T.R. Carter. 1991. The potential effects of climatic change on agricultural insect pests. Agric. Meteorol. 57:221–240. doi:10.1016/0168-1923(91)90088-8

Porporato, A., E. Daly, and I. Rodriguez-Iturbe. 2004. Soil water balance and ecosystem response to climate change. Am. Nat. 164:625–632. doi:10.1086/424970

Ritchie, J.T. 1971. Dryland evaporative flux in a subhumid climate. I. Micrometeorological influences. Agron. J. 70:723–728.

Ritchie, J.T., and E. Burnett. 1971. Dryland evaporative flux in a subhumid climate: II. Plant influences. Agron. J. 63:56–62. doi:10.2134/agronj1971.00021962006300010019x

Ritchie, J.T., E. Burnett, and R.C. Henderson. 1972. Dryland evaporative flux in a subhumid climate: III. Soil water influences. Agron. J. 64:168–173. doi:10.2134/agronj1972.00021962006400020013x

Ritchie, J.T., and W.R. Jordan. 1972. Dryland evaporative flux in a subhumid climate: IV. Relation to plant water status. Agron. J. 64:173–176. doi:10.2134/agronj1972.00021962006400020014x

Siddons, P.A., R.J.A. Jones, J.M. Hollis, S.M. Hallett, C. Huyghe, J.M. Day, T. Scott, and G.F.J. Milford. 1994. The use of a land suitability model to predict where autumn-sown determinate genotypes of the white lupin (*Lupinus albus*) might be grown in England and Wales. J. Agric Sci. 123:199–205. doi:10.1017/S0021859600068465

Simane, B., and P.C. Struik. 2003. Agroclimatic analysis: A tool for planning sustainable durum wheat (*Triticum turgidum* var. durum) production in Ethiopia. Agric. Ecosyst. Environ. 47:31–46. doi:10.1016/0167-8809(93)90134-B

Steinmaus, S.J., T.S. Prather, and J.S. Holt. 2000. Estimation of base temperatures for nine weed species. J. Exp. Bot. 51:275–286. doi:10.1093/jexbot/51.343.275

Sutherst, R., R.H.A. Baker, S.M. Coakely, R. Harrington, D.J. Kriticos, and H. Scherm. 2007. Pests under global change: Meeting your future landlords? In: J.G. Canadell, editor, Terrestrial ecosystems in a changing world. Springer, Berling, Germany. p. 211–226. doi:10.1007/978-3-540-32730-1_17

Tanner, C.B. 1963. Plant temperature. Agron. J. 55:210–211. doi:10.2134/agronj1963.00021962005500020043x

Tobin, P.C., S. Nagarkatti, G. Loeb, and M.C. Saunders. 2008. Historical and projected interactions between climate change and insect voltinism in a multivoltine species. Glob. Change Biol. 14:951–957. doi:10.1111/j.1365-2486.2008.01561.x

van Wart, J., L.G.J. van Bussel, J. Wolf, R. Licker, P. Grassini, A. Nelson, H. Boogaard, J. Gerber, N.I.D. Mueller, L. Claessens, M.K. van Ittersum, and K.G. Cassman. 2013. Use of agro-climatic zones to upscale simulated crop yield potential. Field Crops Res. 143:44–55. doi:10.1016/j.fcr.2012.11.023

von Caemmerer, S. 2000. Biochemical models of leaf photosynthesis. CSIRO Publishing. Collingwood, Australia.

Walsh, J., D. Wuebbles, K. Hayhoe, J. Kossin, K. Kunkel, G. Stephens, P. Thorne, R. Vose, M. Wehner, J. Willis, D. Anderson, S. Doney, R. Feely, P. Hennon, V. Kharin, T. Knutson, F. Landerer, T. Lenton, J. Kennedy, and R. Somerville. 2014. Ch. 2: Our changing climate. In: J.M. Melillo, T.C. Richmond, and G.W. Yohe, editors, Climate change impacts in the United States: The Third National Climate Assessment. U.S. Global Change Research Program, Washington, D.C. p. 19-67. doi:10.7930/J0KW5CXT.

Wiegand, C.L., and L.N. Namken. 1966. Influences of plant moisture stress, solar radiation, and air temperature on cotton leaf temperature. Agron. J. 58:582–586. doi:10.2134/agronj1966.00021962005800060009x

Wu, D., and G. Wang. 2000. Interaction of CO_2 enrichment and drought on growth, water use, and yield of broad bean (*Vicia faba*). Environ. Exp. Bot. 43:131–139. doi:10.1016/S0098-8472(99)00053-2

Zomer, R.J., A. Trabucco, D.A. Bossio, and L.V. Verchot. 2008. Climate change mitigation: A spatial analysis of global land suitability for clean development mechanism afforestation and reforestation. Agric. Ecosyst. Environ. 126:67–80. doi:10.1016/j.agee.2008.01.014

Modeling the Effects of Genotypic and Environmental Variation on Maize Phenology: The Phenology Subroutine of the AgMaize Crop Model

Matthijs Tollenaar,* Kofikuma Dzotsi, Saratha Kumudini, Kenneth Boote, Keru Chen, Jerry Hatfield, James W. Jones, Jon I. Lizaso, R.L. Nielsen, Peter Thomison, Dennis J. Timlin, Oscar Valentinuz, Tony J. Vyn and Haishan Yang

The predictive ability of process-based maize models is predicated on accurate prediction of phenology in terms of maize hybrids (genetics), abiotic factors affecting maize development (environment), and genotype × environment interactions. Future climate change could result in periods with relatively high temperatures, and some have concluded that this warming will result in a reduction of U.S. maize yield (e.g., Schlenker and Roberts, 2009). However, assessment of effects of temperature change on maize yield without a comprehensive understanding of the impact of temperature on maize phenology would be highly questionable. For instance, changes in crop management and genetics of maize hybrids grown in the U.S. Corn Belt since the early 1980s have included earlier maize planting and a longer duration of the post-flowering period (Sacks and Kucharik, 2011; Tollenaar et al., 2017). The combination of the effects of management and genetic changes on maize phenology has contributed substantially to the improvement in U.S. maize production from the 1980s to the 2010s (Tollenaar

---

Abbreviations: ASI, anthesis to silking interval; CRM, comparative relative maturity; CRMAT, Comparative relative maturity rating of a hybrid; DSSAT, decision support system for agrotechnology transfer; GDD, growing degree days; GFP, grain filling period; GTI, general thermal index; LOO, leave-one-out; PhaseDur, phase duration; Photp, photoperiod sensitivity; RLA, rate of leaf appearance; RM, relative maturity; RMSEP, root mean square errors of prediction; TI, tassel initiation; TLU, thermal leaf units; TnLeaf, total number of initiated leaves.

M. Tollenaar, Ojai, CA; K. Dzotsi and J.W. Jones, University of Florida, Agricultural and Biological engineering Dept., Gainesville, FL 32611; K. Dzotsi, The Climate Corporation, St. Louis, MO 63141; S. Kumudini, Ojai, CA; K. Boote, University of Florida, Dept. of Agronomy, Gainesville, FL 32611; K. Chen, R.L. Nielsen, and T.J. Vyn, Purdue University, Dept. of Agronomy, West Lafayette, IN 47907; J. Hatfield, National Lab for Agriculture and the Environment, Ames, IA 50011; J.I. Lizaso, Universidad Politecnica de Madrid, Madrid, Spain; P. Thomison, Ohio State University, Department of Horticulture and Crop Sciences, Columbus, OH 43210; D.J. Timlin, USDA-ARS, Beltsville, MD 20705; O. Valentinuz, Parana, Argentina; H. Yang, University of Nebraska – Lincoln, Department of Horticulture and Agronomy, Lincoln, NE 68583. *Corresponding author (mtollena@uoguelph.ca)

doi:10.2134/agronmonogr60.2017.0038

© ASA, CSSA, and SSSA, 5585 Guilford Road, Madison, WI 53711, USA.
Agroclimatology: Linking Agriculture to Climate, Agronomy Monograph 60.
Jerry L. Hatfield, Mannava V.K. Sivakumar, John H. Prueger, editors.

et al., 2017). Hence, changes in phenology can strongly influence any effects of future changes in temperature on crop yield and, crop models that are used to evaluate the impact of climate change on crop yield should have a robust phenology framework.

Temperature is the most important factor that influences rate of maize development and all important aspects of the temperature-dependent rate of development should be addressed in a maize phenology model. First, the effect of temperature on rate of development throughout the life cycle is in many models captured by a broken stick model comprised of two linear segments that meet at the optimum temperature for development (e.g., Ritchie and NeSmith, 1991). However, the temperature-dependent rate of development has an enzyme-like response (Parent et al., 2010) and we showed recently that nonlinear empirical functions and process-based functions are better predictors of maize development than the linear-empirical functions (Kumudini et al., 2014). Second, the temperature-dependent rate of development is not constant throughout the life cycle as is assumed in most maize models. Maize development is much less temperature responsive during the post-flowering than during the pre-flowering phase (Stewart et al., 1998; Kumudini et al., 2014). Third, the temperature-dependent rate of development in maize is controlled by the temperature of the growing point (Brouwer et al., 1973). Vinocur and Ritchie (2001) showed the necessity of using growing-point temperature rather than air temperature to predict development of field-grown maize during early phases of development when the growing point is below the soil surface (i.e., up to about V6 ~ 9 leaf tips).

While temperature is the most important factor, photoperiod, incident solar radiation, and crop stressors with or without effects on the source/sink ratio, may also influence rate of maize development. Although these factors should be included in a maize phenology model, this effort is challenged by the paucity of quantitative information on the effect of these factors on rate of development in commercial maize hybrids. When maize is exposed to photoperiods > 13 h per day, flowering may be delayed (e.g., Warrington and Kanemasu, 1983). The increase in the planting to flowering interval is associated with an increase in leaf number, and leaf number is sensitive to photoperiod during a short period prior to tassel initiation (Tollenaar and Hunter, 1983; Kiniry et al., 1983). Whether duration of the interval from tassel initiation to flowering and/or the post-flowering period is influenced by photoperiod has not been established, although one published report indicated that duration of the tassel-initiation-to-flowering period was also influenced by photoperiod (Ellis et al., 1992a). To the best of our knowledge, photoperiod sensitivity of current temperate maize germplasm (i.e., released after 1990) has not been reported. Photoperiod sensitivity increases from temperate to tropical germplasm (Coles et al., 2010; Gouesnard et al., 2002), and the sensitivity of tropical germplasm may have been reduced since the last decades of the 20th century due to the introgression of less sensitive temperate germplasm, usually performed to boost the yield potential of the tropical germplasm. Although the impact of various stresses on the anthesis-silking interval (ASI) has been fairly well documented (e.g., Lafitte and Edmeades, 1994; Edmeades et al., 2000), little information is available on the influence of factors such as N stress (McCullough et al., 1994), water stress (Muchow and Carberry, 1989), incident radiation (Birch et al., 1998), and source/sink ratio (Tollenaar and Daynard, 1982) on maize development.

The phenology of maize hybrids or genotypes may differ in respect to (i) the temperature-dependent rate of development, (ii) the duration of the life cycle (planting to black layer), and (iii) the duration of various phases of development as a proportion of the life cycle. Genotypic variation in temperature-dependent rate of development is relatively small (Tollenaar et al., 1984; Van Esbroeck et al., 2008), although modification of this trait is frequently used to "calibrate" phenology in maize models. However, a few outliers such as the low (20–22°C) optimum temperature for rate of progress to tassel initiation (TI) in highland tropical germplasm vs. temperate genotypes (28–30°C) have been documented (Ellis et al., 1992b). The duration of the life cycle (i.e., the planting to black layer interval) of temperate North American maize hybrids is related to their relative maturity (RM), also called comparative relative maturity (CRM); RM is particularly useful in large scale assessments based on crop models. Duration of the total and component-parts of life cycle have not remained constant during the past 80 yr due to changes in grain drying (Cavalieri and Smith, 1985) and planting dates (Sacks and Kucharik, 2011). Although the period of grain dry down between black layer and grain harvest is not considered part of the life cycle of maize, the period is important in that it may influence the RM classification of a maize hybrid. No formal definition of maize-hybrid RM exists despite the extensive use of RM classification of US maize hybrids. In general, RM is assigned to new and improved maize hybrids in breeder's yield trials based on their grain percent moisture at grain harvest in comparison to that of a set of maize hybrids with known RMs. Also, the relationship between RM and thermal accumulation during the planting to black layer interval varies among seed companies (Yang et al., 2004). The duration of the grain-filling period as a proportion of the life cycle has increased in U.S. maize hybrids from the 1980s to the 2010s (Sacks and Kucharik, 2011; Tollenaar et al., 2017) and variation in this important trait should be incorporate in maize models.

In this chapter, we describe the phenology routines of a new process-based maize model (AgMaize) being implemented in the Decision Support System for Agrotechnology Transfer (DSSAT), and we evaluate the model using a large number of diverse datasets, and compare the outcome of the evaluation to predictions by the existing maize phenology model in DSSAT: the CERES-Maize phenology model. The phenology routines in established maize models such as CERES-Maize and APSIM, which are similar in their approaches to modeling crop development, were developed 20 to 30 yr ago (Jones and Kiniry, 1986; Wilson et al., 1995) and these models have not kept up with technological changes in maize phenology that have occurred since their inception. Some current maize models require calibration of genotype coefficients for duration of the life cycle and its component phases, as well as for other genotype-specific crop processes, but this calibration is frequently not possible or practical. For instance, the use of gridded models for the prediction of regional and global maize productivity (e.g., Lobell et al., 2014) requires a phenology model that encompasses genotypic variation across a wide range of maize hybrid relative maturities. To address the need for a maize model that can evaluate production across a range of spatial and temporal scales, effects of relative maturity and genetic improvement on maize phenology should be incorporated. The phenology module of AgMaize has been designed to overcome some of the limitations of the phenology routines in current process-based maize models. The objectives of this chapter are to (i) present

Table 1. Definition of phenological stages in AgMaize (see Table 2 for definition of variables).

Growth Stage	Definition	Leaf stage (TLU)
0	Planting	0
1	Germination	-
2	Emergence	2
3	End of juvenile phase	4
4	Tassel initiation	0.63 * TnLeaf - 2.185
5	Appearance of topmost leaf	TnLeaf
6	Anthesis	TnLeaf + TAInt
7	Silking (*SILK*)	TnLeaf + TAInt + ASI
8	Onset of linear grain filling	SILK + 0.25 × GFP
9	50% milk line	SILK + 0.82 × GFP
10	Black layer	SILK + GFP

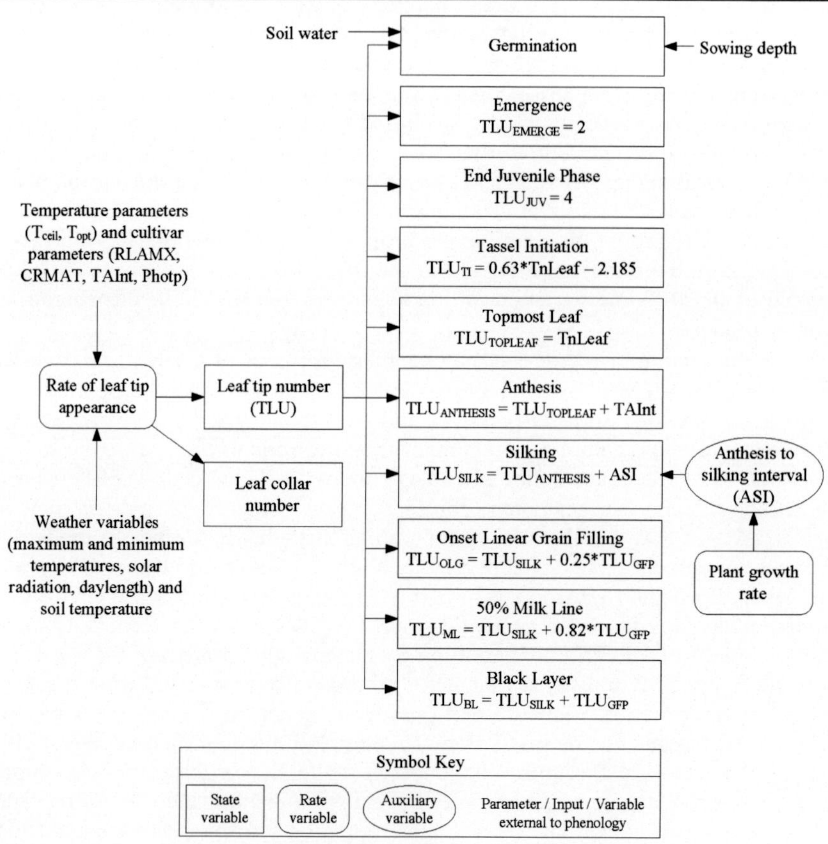

Fig. 1. Overview of the simulation of phenological stages in AgMaize.

a detailed description of the phenology module of AgMaize, and (ii) to compare results of the evaluation and testing of AgMaize across a range of temperate environments and genotypes with those of CERES-Maize.

Material and Methods

AgMaize Phenology

The life cycle of maize in AgMaize is divided into three phases: the pre-flowering, the flowering, and the post-flowering or grain-filling period. An overview of the representation of the different phenological stages in AgMaize is depicted in Table 1 and Fig. 1. Definition and units of variables used in the phenology module of AgMaize are presented in Table 2.

Leaf number is central to the phenology framework of AgMaize during the pre-flowering period. Rate of development during the pre-flowering period is based on the relationship between rate of leaf-tip appearance and temperature (Tollenaar et al., 1979); rate of leaf appearance is the inverse of the phyllochron (i.e., duration between two successive leaf-tip appearances). In AgMaize, measured or observed rate of leaf appearance (RLA; leaves per day) is distinguished from temperature-generated rate of leaf appearance [RLA (°C); leaves (°C) d^{-1}]. Phase duration in thermal time (PhaseDur) is computed as:

$$\mathrm{PhaseDur} = \int \mathrm{RLA}(^{\circ}\mathrm{C})\,dt \quad [1]$$

where phase duration is expressed in thermal leaf units (TLU; 1 TLU = 1 leaf (°C)]. For instance, RLA (°C) is approximately 0.5 leaves (°C) d^{-1} at the optimum temperature for development (31°C; Tollenaar et al., 1979) and, consequently, thermal accumulation after 2 d at 31°C will be approximately 1 TLU. There are a number of advantages to using thermal leaf units to account for thermal accumulation in maize phenology: (i) the thermal leaf unit methodology is among the most precise published thermal prediction approaches of maize phasic development (Kumudini et al., 2014); (ii) there is frequently a direct relationship between observed and predicted values: from the third to the topmost leaf, thermal accumulation from planting to any leaf-tip stage is equal to the leaf-tip stage (i.e., $TLU_n = n$), where the nth leaf-tip stage is defined as the stage of development when the tip of the nth leaf is first visible from a horizontal plane at the level of whorl. As number of leaves is related to the duration from planting to flowering, phase duration of the planting to silking interval can be expressed in terms of TLU; (iii) Thermal accumulation of a phyllochron is 1 TLU (i.e., PHINT in CERES-Maize), which enables a seamless conversion between thermal accumulation and predicted leaf number.

The response of maize development to temperature during the post-flowering or grain-filling period (i.e., silking to physiological maturity) exhibits a different temperature profile than that of the pre-flowering period. Rate of development is less responsive to temperature during this phase than during the pre-flowering period (Stewart et al., 1998; Kumudini et al., 2014) and AgMaize employs the GTI, General Thermal Index (Stewart et al., 1998), to quantify the effect of temperature on rate of development during the post-flowering period. The response to temperature during grain-filling period have been described in detail in Section 2.1.3.

Table 2. Definition and units of (i) species, (ii) genotype, and (iii) general variables used in AgMaize and genotype variables used in CERES-Maize.

Parameter	Unit	Definition
		AgMaize species parameters
T_{min} (0)	°C	Minimal temperature for leaf tip appearance
T_{ceil} (43.7)	°C	Ceiling temperature for leaf tip appearance
T_{opt} (32.1)	°C	Optimum temperature for leaf tip appearance
		AgMaize genotype parameters
CRMAT	d	Comparative relative maturity rating of a hybrid
Photp	leaves h^{-1}	Photoperiod sensitivity for day length values greater than 12.5 h
RLAMX	leaves d^{-1}	Rate of leaf appearance at optimum temperature
GFPYR	GTI	Change in duration of grain filling period (GFP) due to the year of commercial release of the hybrid
TAInt	TLU	Interval between emergence of topmost leaf tip and anthesis
		AgMaize variables
PGR	g $plant^{-1}$ d^{-1}	Average plant growth rate from Growth Stages 5 to 7
ASI	TLU	Anthesis to silking interval
TLU_{BL}	TLU	Accumulated TLU from planting to black layer formation
DayLength	hour	Day length duration (civil twilight)
GFP	TLU	Duration of grain-filling or post-flowering period
GTI	°Cd	General thermal index
GTI_{GFP}	°Cd	Duration of GFP in GTIs (°Cd)
MeanT	°C	Average air temperature between the end of juvenile phase and tassel initiation
mrad	MJ m^{-2}	Mean solar radiation during the previous week
RLA	leaves d^{-1}	Rate of leaf appearance (measured)
RLA(°C)	Leaves (°C) d^{-1}	Computed rate of leaf appearance based on RLA-temperature relationship
$RLAF_{rad}$	-	Effect of solar radiation on the rate of leaf appearance
TLU	Leaf (°C)	Thermal leaf unit
TLU_{AnthRM}	TLU	Thermal time from planting to anthesis of hybrid with a RM relative maturity grown under reference conditions
$TLU_{Anthesis}$	TLU	Thermal time from planting to anthesis
TLU_{Silk}	TLU	Thermal time from planting to silking
TLU_{TI}	TLU	Thermal time from planting to tassel initiation
TnLeaf	TLU	Computed total number of initiated leaves
$TnLeaf_{RM}$	TLU	TnLeaf of hybrid of RM relative maturity grown under reference conditions during the photoperiod/temperature-sensitive phase
$\Delta TnLeaf_{Phot}$	TLU	Change in initiated leaf number due to deviation of photoperiod from reference conditions during the photoperiod/temperature-sensitive phase
$\Delta TnLeaf_{Temp}$	TLU	Change in initiated leaf number due to deviation of temperature from reference conditions during the photoperiod/temperature-sensitive phase
		CERES-Maize genotype variables
P1	°Cd	Thermal time from emergence to the end of juvenile phase
P2	d	Photoperiod sensitivity for day length values larger than 12.5 hours
P5	°Cd	Thermal time from silking to physiological maturity
PHINT	°Cd	Thermal time between successive leaf-tip appearances

Pre-flowering Period

Rate of Development

Except for germination, rate of development during the pre-flowering phase is based on the relationship between RLA and temperature (Tollenaar et al., 1979). Germination under non-limiting conditions is estimated to occur 1 d after planting, which is similar to the estimation of germination in CERES-Maize (Jones and Kiniry, 1986). The simplified β function (Yan and Hunt, 1999; Kim et al., 2012) is used to express the RLA-temperature relationship:

$$\text{RLA}(^{\circ}\text{C}) = \text{RLAF} \cdot \text{RLA}_{\text{max}} \left(\frac{T_{\text{ceil}} - T}{T_{\text{ceil}} - T_{\text{opt}}} \right) \qquad [2]$$

where RLA(°C) is temperature-generated rate of leaf appearance [leaves (°C) d^{-1}], RLA_{max} is the maximum value of RLA (°C) [RLA_{max} = 0.53 leaves (°C) d^{-1}], T is temperature used to compute RLA (°C), T_{opt} is the temperature when RLA (°C) is maximum (T_{opt} = 31.2°C), T_{ceil} is the lowest temperature above T_{opt} when RLA(°C) = 0 (T_{ceil} = 43.7°C), and *RLAF* (rate of leaf appearance factor) is the effect of incidence solar irradiance on RLA(°C) (see below for more detail). The temperature response of RLA is depicted in Fig. 2a in Kumudini et al. (2014). In AgMaize, the daily value of RLA (°C) is calculated as the mean of 24 hourly values, each obtained from applying Eq. [2] to 24 hourly temperatures T resulting from interpolation based on DSSAT inputs of daily maximum and minimum air temperatures. When using simulated soil temperature to compute RLA (°C) during early-season development, Eq. [2] is applied directly to the simulated mean soil temperature. The temperature response of RLA (°C) is an outcome of Eq. [2] and base temperature is not an input in this equation: rates computed from Eq. [2] rise from 0 to 0.01 leaves (°C) d^{-1} between 0 and 5°C, and to 0.06 leaves (°C) d^{-1} at 10°C. Yan and Hunt (1999) showed that this simplified β function was highly predictive when model parameters that were estimated from six constant temperatures were used to predict rates of 16 varying day-night temperature regimes in the Tollenaar et al.

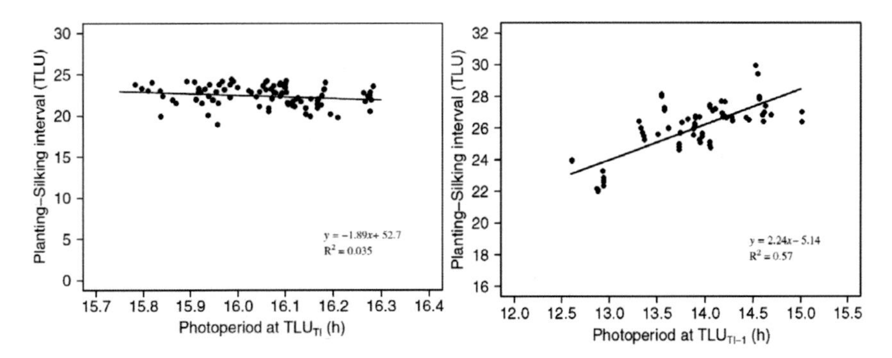

Fig. 2. Association between duration of the planting–silking interval and photoperiod (a) at the estimated leaf-tip stage at which tassel initiation (TLUTI) is attained in studies reported by Nielsen et al. (2002) in which each datum represent the mean of three replications for each of three hybrids grown in 48 location/years, and (b) at the estimated leaf-tip stage at 1 TLU prior to the leaf stage at which tassel initiation is attained (TLUTI-1) in studies carried out by Monsanto in 2012 and 2013 with tropical germplasm consisting of five 5-RM groups in the 120-144 RM range in which each of the 73 RM groups/location/years represent the mean of about 28 hybrids.

(1979) dataset. However, incident short wave radiation was low (~10 MJ m^{-2} d^{-1}) in the Tollenaar et al. (1979) study and it has been our experience that Eq. [2] slightly underestimates RLA(°C) under field conditions. We have introduced modifying factors to account for relatively minor effects of incident solar radiation, nitrogen and water status of the crop on rate of leaf-tip appearance. Effect of incident radiation on the rate of leaf tip appearance ($RLAF_{rad}$) is estimated using data reported by Tollenaar (1999). Between the 3- and the 12-leaftip stage, $RLAF_{rad}$ increases linearly with incident daily short-wave radiation between 10 and 20 MJ m^{-2}:

$$RLAF_{rad,t} = 0.1 \times \min(\max(mrad_t, 10), 20) + 0.9 \qquad [3]$$

where mrad is the mean incident solar radiation during the previous week (MJ m^{-2} d^{-1}). $RLAF_{rad}$ ranges from 1 to 1.1 during this phase of development and $RLAF_{rad}$ is 1.1 beyond the 12-leaf tip stage. Nitrogen supply (e.g., McCullough et al., 1994) and water status of the crop (Muchow and Carberry, 1989) can influence rate of leaf appearance, but mechanisms by which these factors influence rate of development have not been incorporated yet in AgMaize.

Rate of leaf-tip appearance is influenced by temperature of the growing point (Brouwer et al., 1973). Soil temperature influences growing–point temperature when the growing point is below the soil surface (Vinocur and Ritchie, 2001), from planting until approximately the eight-leaftip stage, and air temperature influences growing-point temperature beyond the eight-leaftip stage (in both cases the actual growing point temperature may vary due to the plant's energy balance). Soil temperature is simulated following the procedure utilized by the EPIC model (Potter and Williams, 1994), and growing-point temperature prior to the eight-leaf stage is set equal to soil temperature at 5 cm below the soil surface.

Stages of Development

The pre-flowering period extends from planting to the appearance of the topmost leaf tip and this phase includes five phenological events: germination (Growth Stage 1), plant emergence (Growth Stage 2), end of juvenile phase (Growth Stage 3), tassel initiation (Growth Stage 4), and appearance of topmost leaf tip (Growth Stage 5). Thermal accumulation under non-limiting conditions starts in AgMaize at 1 d after planting and plant emergence occurs at 2 TLU (Wu, 1998). The period from the end of the juvenile phase to tassel initiation (TI) is the photoperiod- and temperature-sensitive phase of development in maize. The end of the juvenile phase has been shown to occur at the four-leaftip stage in a short-season maize hybrid (Tollenaar and Hunter, 1983) and Padilla and Otegui (2005) showed that the leaf-tip stage at which TI is attained is related to total number of initiated leaves (TnLeaf). Using the relationship reported by Padilla and Otegui (2005) and considering that thermal leaf unit accumulation of a phyllochron is 1 leaf °C per leaf, thermal leaf accumulation between planting and the leaf stage at which TI is attained (TLU_{TI}) can be estimated as:

$$TLU_{TI} = 0.63 \times TnLeaf - 2.185 \qquad [4]$$

Total number of initiated leaves (TnLeaf) is determined by maize genotype and varies with environmental conditions. Total leaf number in AgMaize is computed as the sum of the genotype effect ($TnLeaf_{RM}$), the temperature effect ($\Delta TnLeaf_{Temp}$), and the photoperiod effect ($\Delta TnLeaf_{Phot}$) on total leaf number:

$$TnLeaf = TnLeaf_{RM} + \Delta TnLeaf_{Temp} + \Delta TnLeaf_{Phot} \qquad [5]$$

where $\Delta TnLeaf_{Temp}$ and $\Delta TnLeaf_{Phot}$ represent changes in total leaf number due to temperature and photoperiod deviating from the reference conditions of 20 °C for $\Delta TnLeaf_{Temp}$ and 16 h for $\Delta TnLeaf_{Pho}$.

The impact of temperature on number of initiated leaves (ΔTnLeaf) is estimated by assuming that temperature response of total leaf number can be quantified by a broken stick model comprised of two linear segments that meet at 15 °C, rising for temperatures greater than and less than 15°C (Tollenaar and Hunter, 1983; Warrington and Kanemasu, 1983):

$$\Delta \mathrm{TnLeaf}_{\mathrm{Temp}} = \begin{cases} 0.1 \times (\mathrm{Mean}T_t - 15) - 0.5, & \text{if } \mathrm{Mean}T_t > 15 \\ 0.2 \times (15 - \mathrm{Mean}T_t) - 0.5, & \text{if } \mathrm{Mean}T_t \leq 15 \end{cases} \quad [6]$$

where MeanT_t is the average air temperature during the temperature- and photoperiod-sensitive phase (i.e., between the end of juvenile phase and TI).

Effect of photoperiod on the number of initated leaves (ΔTnLeaf) is dependent on genotype-specific photoperiod sensitivity (Photp, leaves h^{-1}). Number of leaves increases linearly with an increase in daylength between 12.5 and 16 h; leaf number at daylengths < 12.5 h is equal to that at 12.5 h and leaf number at daylengths > 16 h is equal to that at 16 h.

$$\Delta \mathrm{TnLeaf}_{\mathrm{Phot}} = \begin{cases} -3.5 \times \mathrm{Photp} & \text{For Daylength} \leq 12.5\ h \\ (\mathrm{Daylength} - 16) \times \mathrm{Photp} & \text{For } 12.5\ h < \text{Daylength} \leq 16\ h \\ 0, & \text{For Daylength} > 16\ h \end{cases} \quad [7]$$

where Daylength is the value of daylength (i.e., civil twilight) during the temperature- and photoperiod-sensitive phase. Photoperiod sensitivity varies among temperate maize genotypes and it is greater in tropical than in temperate maize germplasm. Photoperiod sensitivity of North American maize hybrids and inbred lines has been shown to vary between 0.2 and 0.7 leaves h^{-1} (Warrington and Kanemasu, 1983; Tollenaar and Hunter, 1983; Birch et al., 1998; Coles et al., 2010), and Photp = 1.3 leaves h^{-1} was recently reported for Chinese maize hybrids (Liu et al., 2013). The reported photoperiod sensitivity of North American hybrids involved pre-1985 germplasm and results of a planting-date study reported by (Nielsen et al., 2002) indicates that more-recent Corn Belt maize hybrids may be photoperiod insensitive. In their study, three commercial Corn Belt maize hybrids (106, 111, and 115 RM) were grown at four locations in Indiana and Ohio from 1991 to 1994 at three planting dates ranging from April to June. The leaf-tip stage at TI (TLU_{TI}) in this study was calculated using Eq. [4], where TnLeaf was estimated using Eq. [10] and the simplifying assumptions that TAInt = 2.5 and anthesis date = silking date: TLU_{TI} ranged from 8.1 to 11.8 using these assumptions. The duration of the planting–silking interval was neither significantly associated with photoperiod at TLU_{TI} (Fig. 2a), nor with photoperiod at four leaf-tip stages after tassel initiation (i.e., TLU_{TI+4}): $y = 0.82 \times \mathrm{Photoperiod} + 9.4$; $r^2 = 0.018$). In contrast, the planting–silking interval showed a negative response to photoperiod at four leaf-tip stages prior to TI (i.e., TLU_{TI-4}): $y = -4.83 \times \mathrm{Photoperiod} + 99.9$; $r^2 = 0.253$). The slope of the regression between the planting–silking interval and photoperiod was similar for the three hybrids. Although the regression was statistically significant, we presume that the response was associated with one or more other factors (e.g., temperature) and we assume that the Corn Belt maize hybrids in the Nielsen et al. (2002) study were photoperiod insensitive. These results support speculations that full-season Corn Belt hybrids (i.e., hybrids > 95 RM) have become photoperiod insensitive by 1990. Unless the photoperiod sensitivity of a maize hybrid is known, it is assumed that the photoperiod sensitivity of full season, pre-1990, North American hybrids is 0.65 leaves h^{-1}, and Photp = 0.0 leaves h^{-1} for North American, post-1990, commercial germplasm. Photoperiod sensitivity of tropical maize germplasm is much greater than that of temperate germplasm. Photoperiod sensitivity of commercial tropical germplasm was evaluated in trials performed by Monsanto Co. in India, Brazil, Philippines, Thailand, and El Salvador during 2012 and 2013. The methodology to determine TLU_{TI} and photoperiod sensitivity was the same as described above for the Indiana-Ohio dataset. Results showed a significant response of the planting-silking interval to photoperiod at 1 TLU prior to TLU_{TI} (i.e., TLU_{TI-1}) and apparent photoperiod sensitivity was 2.24 leaves per hour increase in photoperiod (Fig. 2b), which is similar to the mean photoperiod sensitivity of two tropical inbred lines reported by Coles et al. (2010).

Under reference conditions, TnLeaf = $TnLeaf_{RM}$ and $TnLeaf_{RM}$ can be estimated as:

$$TnLeaf_{RM}=TLU_{AntRM}-TAInt \quad [8]$$

where $TnLeaf_{RM}$ is total initiated leaf number of a hybrid with a relative maturity of RM and TAInt is the interval between the emergence of the topmost leaf tip and anthesis, TLU_{AntRM} is thermal accumulation from planting to anthesis in a large set of North American hybrids that were grown at a Corn Belt location from 2007 to 2012 (Kumudini et al., 2014). Hybrid RM in this dataset ranged from 75 to 120 RM and mean temperature and photoperiod during the temperature- and photoperiod-sensitive interval were approximately 20 °C and 16 h, respectively. The linear relationship between TLU_{AntRM} and RM in this dataset was:

$$TLU_{antRM}=0.12 \times RM+10.5 \quad [9]$$

The Flowering Period

The flowering period in AgMaize extends from the time of appearance of the tip of the topmost leaf (i.e., TnLeaf) until the first day that silks emerge from husks and includes appearance of the topmost leaf (Growth Stage 5), anthesis (Growth Stage 6), and silking (Growth Stage 7). Thermal accumulation from planting to anthesis ($TLU_{Anthesis}$) is estimated as:

$$TLU_{Anthesis}=TnLeaf+TAInt \quad [10]$$

where TAInt is the interval between the appearance of the topmost leaf tip and anthesis. TAInt is a genetic coefficient with a default value of 2.5 TLU. Thermal accumulation from planting to silking (TLU_{Silk}) is affected by stress during the anthesis-to-silking interval (ASI) and by genotypic variation in the response of ASI to stress.

$$TLU_{Silk} = TLU_{Anthesis} + ASI \quad [11]$$

The effect of stress on duration of the ASI is quantified by a relationship between ASI and above-ground plant growth rate (PGR, g $plant^{-1}d^{-1}$) during the critical period for kernel set between Growth Stages 5 and 7. Based on unpublished data, silking date equals anthesis date when PGR > 3 g $plant^{-1}d^{-1}$ and ASI increases exponentially when PGR declines below 3 g $plant^{-1}d^{-1}$ and silking date equals anthesis date when PGR > 3 g $plant^{-1}d^{-1}$:

$$ASI = \begin{cases} 0.3 \times (2^{3-PGR}-1), & \text{if } PGR \leq 3 \\ 0, & \text{if } PGR > 3 \end{cases} \quad [12]$$

Although duration of the ASI can sometimes be negative (i.e., silking can occur prior to anthesis), the model assumes that ASI > 0. The relationship between ASI and PGR varies among maize hybrids that are more or less stress tolerant, which is associated with dry matter partitioning to the ear and kernels. Historical yield improvement of temperate maize hybrids has been associated with increased stress tolerance resulting in a relatively higher rate of dry matter accumulation under stress, and newer, more stress-tolerant hybrids exhibit higher kernel-set efficiency (cf., Tollenaar and Lee, 2011).

The Grain-Filling Period

The grain-filling period extends from the time when silks are first visible until black layer formation, that is, physiological maturity (Daynard and Duncan, 1969). The stages of developmental predicted by the model during the grain-filling period are the onset of linear grain filling (Growth Stage 8), 50% milk line (Growth Stage 9), and black layer (Growth Stage 10). Approximately 80% of grain dry matter at maturity is accumulated between Growth Stages 8 and 9: kernel dry matter as a proportion of kernel dry matter at maturity is approximately 10% at Growth Stage 8 (Tollenaar and Daynard, 1978) and 90% at growth Stage 9 (Tollenaar, 1999).

The General Thermal Index (GTI; °C d^{-1}) is used to quantify the temperature dependent rate of development during the grain-filling (Stewart et al., 1998). An analysis of results of a planting-date study reported by Nielsen et al. (2002) showed that GTI during the post-flowering period predicts development more accurately than other thermal accumulators under relatively low temperatures (Kumudini et al., 2014). The GTI for the post-flowering period has a quadratic dependency on temperature:

$$\frac{dGTI_t}{dt} = 5.3581 + 0.11178 \times MeanT_t^2 \qquad [13]$$

where $\frac{dGTI_t}{dt}$ is rate of development in GTI at the mean temperature of day t for temperatures < 29°C is computed. Mean daily temperatures of more than 95%

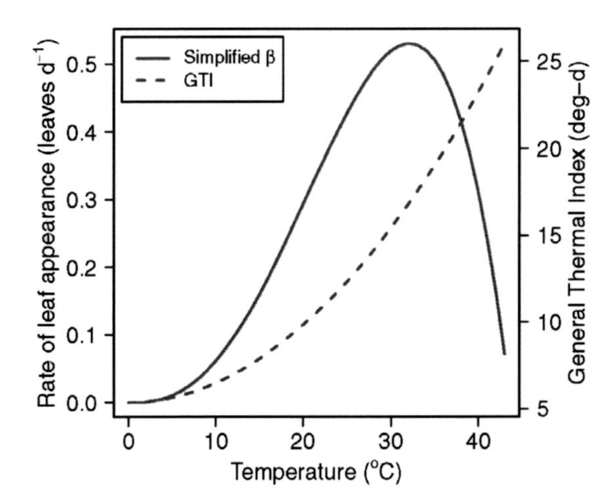

Fig. 3. Temperature response of GTI (General Thermal Index) and β function that are used in modeling the rate of development in AgMaize.

of data used in developing the rate of development vs. temperature relationship during the grain-filling period (GTI) were < 25°C (Stewart et al., 1998). To the best of our knowledge, response of development during the post-flowering period to mean temperature > 29°C is unknown and we assume that the response will follow the RLA (°C) response for temperatures > 29 °C. Relative changes in rate of development in terms of GTI and RLA (°C) are the same in the 25 to 29°C range (data not shown), but GTI continues to increase (see Fig. 3) and RLA (°C) levels off and declines beyond the optimum temperature (cf., Kumudini et al., 2014). To estimate rate of development in terms of RLA (°C) for temperatures > 29°C, rate of development is computed as:

$$\frac{dGTI_t}{dt} = \frac{dGTI_{29}}{dt} / RLA_{29} \times RLA_t \quad [13a]$$

where RLA_{29} is RLA(°C) at 29°C, $\frac{dGTI_{29}}{dt}/RLA_{29} = 28.5$, and $RLAt$ is RLA (°C) when temperature is t °C. The duration of the grain-filling period in terms of GTI accumulation (GTI_{GFP}) is related to hybrid Relative Maturity (RM) as:

$$GTI_{GFP} = 2.23 \times RM + 417 + GFPYR \quad [14]$$

where GFPYR is a genotype coefficient characterizing the change in the grain-filling period duration due to the year of commercial release of the hybrid. Eq. [14] is based on data from > 3000 hybrids in the range 75 to 119 RM grown at a location in the U.S. Corn Belt from 2007 to 2012 (Kumudini et al., 2014). The duration of the grain-filling period has increased by about 2 GTI thermal units per year between 1984 and 2013 in the U.S. Corn Belt (Tollenaar et al., 2017) due to technological changes in maize production, and for maize hybrids in this RM range. *GFPYR* can be estimated as:

$$GFPYR = 2 \times (HybridRelease - 2010) \quad [15]$$

where HybridRelease is the year when the hybrid was first released commercially between 1980 and 2010.

CERES-Maize Phenology

The phenology model in CERES-Maize recognizes six major stages, end of juvenile phase, tassel initiation, end of leaf growth (silking), beginning of grain filling, end of effective grain filling period and physiological maturity (Jones and Kiniry, 1986). The occurrence of these events is governed by the accumulation of growing degree days (GDD; °Cd) based on an optimum temperature of 34 °C and a base temperature of 8 °C, and using soil temperature before the appearance of the 10th leaf tip and air temperature afterward. Duration in °C to emergence is predicted as a linear function of sowing depth. A genotype parameter (P1) determines the duration of the emergence-end of juvenile phase. Tassel initiation is assumed to occur 4 d after the end of the juvenile phase but can be delayed by photoperiod. Variations in photoperiod sensitivity among genotypes are quantified by the number of days tassel initiation is delayed for each hour above a daylength of 12.5 h (genotype parameter P2). Silking to physiological maturity is determined by the genotype parameter P5. The duration of the effective grain filling phase accounts for 95% of P5. A lag phase of 170 °C d^{-1} separates silking from the beginning of grain filling. The genotype parameter PHINT defines the interval that separates the appearance of successive leaf tips.

Table 3. Characterization of datasets used for evaluating the phenology models.

Hybrid	Location and year †	Variables	Phenology observation	Data points per hybrid	Cross-validation subsets (*k*) per hybrid	Reference
Pioneer 33P67	Lincoln, NE† (2001) Manchester, IA (2002)	Location and year	-Silking date -Maturity date	2	2	Yang et al. (2004)
Pioneer 33A14	Lincoln, NE (1999-2000)	Year	-Silking date -Maturity date	2	2	Yang et al. (2004)
Avicta Complete Corn	Ames, IA (2011-2012)	Year	-Anthesis date -Maturity date	2	2	
Pioneer 3902	Ontario, Canada (1999-2001)	Year	-Silking date	3	3	Valentinuz and Tollenaar (2004)
Pioneer 34M91	Queenstown, MD (2006-2007)	Year	-Total leaf number -Leaf tips -Leaf ligules	2	2	Kim et al. (2012)
Pioneer 33B53	Georgetown, DE (2006-2007)	Year	-Total leaf number -Leaf tips -Leaf ligules	2	2	Kim et al. (2012)
XL72AA and DKC61-69	West Lafayette, IN (2012-2013) Wanatah, IN (2012) North Platte, NE (2012-2013)	Location and year	-Total leaf number -Leaf tips -Leaf ligules -Anthesis date -Silking date	5	5	
Pioneer 3245, Pioneer 3394 and Pioneer 3527	West Lafayette, IN (1991-1994) Butlerville, IN (1993-1994) South Charlestown, OH (1991-1994) Hoytville, OH (1993-1994)	Location, year, planting date	-Silking date -Maturity date	36	6	Nielsen et al. (2002)

† Definition of US States Abbreviations: IA Iowa; IN Indiana; MD Maryland; NE Nebraska; OH Ohio.

Table 4. Geographic coordinates and mean values of weather variables (DAYL, day length; SRAD, solar radiation; T_{MAX}, maximum temperature; and T_{MIN}, minimum temperature) from planting to AgMaize-simulated maturity at the locations of the datasets. Mid planting dates were used for the multiple-planting datasets.

Location	Longitude and latitude	Year	Average DAYL	Average SRAD	Average T_{MAX}	Average T_{MIN}
			(d)	(MJ m^{-2})	(°C)	(°C)
Manchester, IA	91.45 W, 42.47 N	2002	14.1	20	26	14
Ames, IA	93.77 W, 42.02 N	2011	14.1	19	25	15
		2012	14.3	22	29	16
Lincoln, NE	96.65 W, 40.82 N	1999	14.1	19	28	17
		2000	14.3	22	28	16
		2001	14.3	22	29	17
Ontario, Canada	80.42 W, 43.65 N	1999	14.3	20	26	13
		2000	13.6	17	23	11
		2001	14.1	19	23	12
Queenstown, MD	76.15 W, 38.91 N	2006	13.9	19	27	18
		2007	14.0	20	28	18
Georgetown, DE	76.65 W, 38.52 N	2006	13.6	17	26	17
		2007	13.8	20	28	16
West Lafayette, IN	87.033 W, 40.483 N	1991	14.3	21	30	17
		1992	13.8	19	25	13
		1993	13.9	18	26	15
		1994	14.2	21	26	14
		2012	14.3	23	30	16
		2013	14.1	18	27	15
Wanatah, IN	86.929 W, 41.443 N	2012	14.2	20	28	15
North Platte, NE	100.774 W, 41.09 N	2012	14.3	23	32	14
		2013	14.3	22	30	14
Butlerville, IN	85.483W, 39.05 N	1993	14.0	20	28	15
		1994	13.7	19	28	15
Hoytville, OH	83.75W, 41.2 N	1991	13.9	20	28	15
		1992	14.0	20	28	15
		1993	14.3	20	30	17
		1994	13.9	18	26	13
South Charlestown, OH	83.667W, 39.85 N	1993	14.1	20	29	16
		1994	14.1	20	29	15

Performance of Phenology Models Based on Non-Calibrated Genotypic Parameters

To assess the ability of the two phenology models to adequately predict leaf number, anthesis, silking and maturity dates of various hybrids when little or no information is available to estimate genotypic parameters, an analysis of simulation results without model calibration was conducted first. Maize hybrids from a number of datasets (Tables 3 and 4) were either characterized using the values of relative maturity ratings (RM) or the corresponding thermal time from emergence to physiological maturity (GDD) determined during their commercial release. The relative maturity values were used in AgMaize as the CRMAT genotypic parameter. The remaining AgMaize genotypic parameters were set to constant values as follows: RLAMX was 0.53 leaves d^{-1}; TAint was 2.5 leaves; GFPYR was calculated using Eq. [15]; PHOTP was 0.65 leaves h^{-1} for XL72AA and 0.0 leaves h^{-1} for other hybrids. For CERES-Maize simulations, the hybrids were assigned (based on their maturity GDD ratings) to one of the six generic maturity classes present in DSSAT v4.6 and varying in season length from 1389 to 1556 °C d^{-1}. CERES-Maize genotypic parameters for these generic hybrids varied with their season length ratings (Table 5a). Genotypic parameters for these generic maturity classes were used for predicting phenology characteristics of the hybrids without model calibration. In addition to cultivar characteristics, both models were executed within DSSAT v4.6 using standard DSSAT inputs, that is, management, soil physical and chemical properties and weather (daily minimum and maximum temperature, solar radiation and precipitation). For AgMaize, an option is available to the user to use simulated topsoil or air temperatures during early-season development. In this study, the modified EPIC soil temperature model (Potter and Williams, 1994) was used.

Estimating Genotypic Parameters

Phenology parameters were estimated for both crop models using data collected in optimum or near optimum conditions, allowing our study to concentrate on the effects of temperature, solar radiation, and photoperiod on development stages. For each model, genotypic parameters characterizing the hybrids tested were adjusted to obtain the smallest deviations between simulations and observations using the Nelder–Mead simplex algorithm (Nelder and Mead, 1965). For each hybrid, all available data were used for the parameter estimation. The estimation of the parameters followed a sequential approach that employed the structural relationships between the parameters and the model predictions (total leaf number, anthesis, silking and maturity dates), which was motivated by the need to reduce compensation of errors that often compromise parameter estimation procedures (Wallach, 2011).

Of the five AgMaize phenology parameters (CRMAT, TAInt, RLAMX, Photp and GFPYR, Table 2), three were estimated (CRMAT, TAInt and GFPYR). Variability in the rate of leaf-tip appearance across maize hybrids appears to be low (e.g., Tollenaar et al., 1984) and the value of RLAMX was set at 0.53 leaves d^{-1} (Yan and Hunt, 1999; Kim et al., 2012). The photoperiod sensitivity (Photp) was estimated as 0.65 leaves h^{-1} (Tollenaar and Hunter, 1983) for older hybrids (e.g., XL72AA) and assumed to be 0.0 leaves h^{-1} for newer hybrids (i.e., released after 1990). The sequential procedure for AgMaize can be summarized as follows:

1. The GFPYR parameter was estimated based on the year of commercial release of the hybrid using Eq. [15];
2. The CRMAT parameter was estimated using observed anthesis, silking, and maturity dates, assuming that the actual RM value for a hybrid may vary about the reported value by three units;
3. With GFPYR and CRMAT fixed at their values estimated from steps 1 and 2, TAInt (interval between the appearance of the topmost leaf and the occurrence of anthesis) was estimated using observed data on total leaf number. Evidence that this interval may be greater for an older hybrid than for a

Table 5. (a) Values of estimated AgMaize and CERES-Maize phenology parameters using all data (with non-calibrated values of parameters in parentheses), and (b) coefficient of variability of parameters estimated based on a cross-validation approach. Definition of variable names is in Table 2.

	AgMaize		CERES-Maize		
	CRMAT	TAInt	P1	P5	PHINT
	d	TLU	———————°Cd———————		
AvictaComplete	113 (111)	(2.5)	249 (160)	942 (780)	39 (49)
DKC61-69	109 (111)	3.8 (2.5)	206 (260)	(850)	39 (49)
Pioneer3245	117 (115)	(2.5)	245 (260)	870 (850)	39 (49)
Pioneer3394	110 (111)	(2.5)	224 (260)	831 (850)	39 (49)
Pioneer33A14	112 (113)	(2.5)	247 (240)	967 (850)	39 (49)
Pioneer33B53	109 (112)	4.5 (2.5)	(185)	(850)	39 (49)
Pioneer33P67	116 (114)	(2.5)	194 (240)	956 (850)	39 (49)
Pioneer34M91	112 (109)	4.5 (2.5)	(240)	(850)	39 (49)
Pioneer3527	108 (106)	(2.5)	224 (212)	816 (850)	39 (49)
Pioneer3902	89 (91)	(2.5)	224 (110)	(680)	39 (38.9)
XL72AA	116 (115)	3.9 (2.5)	187 (160)	(780)	39 (49)

(b)

	AgMaize		CERES-Maize	
	CRMAT	TAInt	P1	P5
	d	TLU	———————°Cd———————	
AvictaComplete	3	-	5	9
DKC61-69	0	4	4	-
Pioneer3245	1	-	1	1
Pioneer3394	0	-	3	1
Pioneer33A14	2	-	6	2
Pioneer33B53	0	0	-	-
Pioneer33P67	0	-	20	1
Pioneer34M91	0	1	-	-
Pioneer3527	1	-	1	1
Pioneer3902	2	-	7	-
XL72AA	1	3	4	-
MeanCV	1	2	6	2

newer hybrid (Chen and Vyn, unpublished data) suggests genotypic variations that can be captured by the TAInt genotypic parameter. A range of 2 to 4.5 leaves was assumed when estimating TAInt from observed data. The value of TAInt cannot be ascertained without data on final leaf number. Therefore, when final leaf number was not measured (which was the case for several datasets) a value of 2.5 leaves was used for TAInt. If anthesis, silking or maturity dates were not known and only final leaf number was observed, CRMAT and TAInt were simultaneously estimated.

Of the four genotypic parameters related to phenology in CERES-Maize (P1, P2, P5 and PHINT, Table 2), three were estimated (P1, P5 and PHINT) following a sequential approach similar to that of AgMaize. A photoperiod sensitivity of

0.3 d was used for XL72AA (older hybrid) due to the lack of reliable data at different latitudes for accurate estimation of this parameter. Newer hybrids were assumed to exhibit no photoperiod sensitivity. The interval between successive leaf-tip appearances (PHINT) was estimated from leaf-tip appearance data in the Indiana dataset (2012–2013, Table 3). The hybrids XL72AA and DKC 61–69 in this dataset had a PHINT value of 39 °C d^{-1} which is the same PHINT value used in DSSAT v4.6 for simulating the cultivar McCurdy 84AA in the Gainesville, Florida dataset. Based on this preliminary assessment a PHINT value of 39 °C d^{-1} was used for all hybrids. Using a known PHINT value also has the advantage of minimizing errors that could arise from the estimation of P1 (thermal time from seedling emergence to the end of juvenile phase) and PHINT both of which co-determine the final leaf number and silking date in CERES-Maize. Additionally, the rate of leaf tip appearance in AgMaize is modeled as a function of temperature and does not depend on specific hybrid traits, which supports the use of the same PHINT value for simulating all hybrids in CERES-Maize. Using a fixed value of PHINT, P1 was adjusted to minimize discrepancies between observed and CERES-Maize predicted anthesis or silking dates depending on the dataset. Next, the genotypic parameter P5 (thermal time from silking to physiological maturity) was estimated for accurate prediction of observed maturity dates if these data are available for the hybrid.

Estimating Model Prediction Errors

The number of data values available per hybrid for parameter estimation and independent testing of the models was limited for most hybrids. To avoid biases inherent to arbitrary separation of the data into calibration and validation sets and to maximize the use of the limited data available, a *k*-fold cross-validation approach was adopted to estimate errors associated with the parameter estimates (Jones and Carberry, 1994; Wallach, 2006). For each hybrid, all data available were subdivided into *k* subsets (Table 3, each subset may contain one or more treatments). Using *k*–1 subsets, parameters were estimated following the sequential procedure described earlier and these parameters were used to predict the omitted subset. The analysis was repeated until each subset is used for both parameter estimation and model testing exactly once. At the end of the procedure, *k* sets of parameter estimates and model predictions were obtained, where *k* is the total number of data subsets per hybrid (Table 3). For the special case where $k = n$, the total number of data points, a Leave-One-Out (LOO) cross validation was performed in which one data point was omitted and predicted during each parameter estimation-model testing cycle. For the LOO special case, the estimator of the parameter vector $\hat{\theta}$'s Root Mean Square Errors of Prediction (RMSEP) was calculated in the following way (considering a particular phenology event and a given hybrid):

$$\mathrm{RMSEP}(\hat{\theta})=\sqrt{n^{-1}\sum_{i=1}^{n}[y_i-f(\hat{\theta}_{-i})]^2} \quad [16]$$

where y_i is the phenology observation corresponding to a treatment i, f represents the crop model, and $\hat{\theta}_{-i}$ denotes the vector of parameters estimated using all the data with observations for treatment i omitted. The $\mathrm{RMSEP}(\hat{\theta})$ is to be interpreted as a measure of the average error of the model in predicting a phenology event using all data available. Spearman rank correlations (R) were also calculated using Eq. [17] to assess the strength of the relationship between the model predictions and observations, and model biases (BIAS) were calculated using Eq. [18] to evaluate the tendency of the models to overpredict (negative bias) or underpredict (positive bias) the observations.

$$R=\frac{\sum_{i=1}^{n}(x_i-\bar{x})(y_i-\bar{y})}{[\sum_{i=1}^{n}(x_i-\bar{x})]^{1/2}\cdot\left[\sum_{i=1}^{n}(yi-\bar{y})\right]^{1/2}} \quad [17]$$

where R is the general correlation coefficient computed on the rank of the two series x (observations) and y (model predictions) with respective means $\bar{x}$ and $\bar{y}$, n is the total number of data pairs.

$$\text{BIAS}=\frac{1}{n}\sum_{i=1}^{n}(y_i-y_i) \quad [18]$$

where y_i is the observation and $\hat{y}_i$ is the corresponding model prediction for observation i. The value of the bias can be interpreted as the quantity by which the model overpredicts or underpredicts the observations on average.

Maize Datasets

Data for a total of 11 maize hybrids corresponding to 131 unique combinations of location, year, hybrid, and planting date were used for the parameter estimation and testing of the phenology models (Table 3). The data were taken from field experiments in which measurements of leaf appearance rates and/or dates of occurrence of key development stages were recorded. The experiments were conducted in a variety of agro-ecological conditions (Table 4) and involved older and newer hybrids, which allowed assessment of the models' predictive ability of interactions between genotypes and the environment. Treatments used showed limited effects of water and nutrient deficits, which are not modeled in this study. It is worth noting that experiments that included different plant densities did not demonstrate an effect of varying densities on crop development. Daily weather data (maximum and minimum temperatures, precipitation and solar radiation) were collected at the sites where the field experiments took place. Detailed descriptions of the datasets, including the phenology observations taken, can be found in the references provided in Table 3. Aspects of the experiments relevant to our study are briefly described in this paper for each hybrid.

Pioneer 34M91: the dataset that used the hybrid Pioneer 34M91 was obtained from a study conducted at the University of Maryland's Wye Agricultural Research Center in 2006 and 2007 (Kim et al., 2012). The hybrid was planted on two fields with a row spacing of 0.76 m and a plant density of 6.9 plants m^{-2} on 8 May 2006 and 18 May 2007. The soils are classified as Mattapex silt loam and Whitemarsh silt loam (Kim et al., 2012). Leaf appearance rates were measured by monitoring the number of leaf tips and leaf ligules on five plants at biweekly intervals.

Pioneer 33B53: an experiment conducted on two commercial fields located in Georgetown, DE provided data for the maize hybrid Pioneer 33B53 planted on 10 April 2006 and 30 April 2007 in 0.76-m spaced rows, with a plant density of 7.3 plants m^{-2}. The soils at the two fields are classified as Pepperbox–Rosedale and Glassboro of textures loamy sand and sandy loam, respectively (Kim et al., 2012). Leaf appearance rates were measured in the same manner as in the University of Maryland experiment.

Pioneer 33A14 and 33P67: data for maize hybrid Pioneer 33A14 came from an experiment conducted in Lincoln, NE during 1999 to 2001 on a deep Kennebec silt loam (fine-silty, mixed, superactive, mesic Cumulic Hapludoll) and managed for near optimal growth conditions (Yang et al., 2004). Planting was on 13 May 1999 and 21 April 2000 with a row spacing of 0.76 m. In 2001, a different hybrid, Pioneer 33P67, was sown on 26 April. Three plant densities ranging from 6.9 to 11.3 plant m^{-2} were tested each year. Phenological observations included dates of silking and maturity. An additional treatment used for calibrating and testing Pioneer 33P67 came from a single-year experiment conducted in Manchester, IA and shared similar objectives with the Nebraska experiment (Yang et al., 2004). Planting date for the Manchester experiment was on 8 May 2002 with a row spacing of 0.51 m and a plant density of 8.4 plants m^{-2}. The soil in Manchester is a Kenyon loam (fine-loamy, mixed, superactive, mesic Typic Hapludoll).

Avicta Complete Corn: data for Avicta Complete Corn was obtained from a field experiment in Ames, IA on maize response to tillage methods and input intensity conducted during the period 2011 to 2012. The hybrid was planted on 3 May 2011 and 9 May 2012 at a row spacing of 0.76 m and a plant density of 8.6 plants m^{-2}. Dates of anthesis, beginning of grain filling and physiological maturity observed in the conventional tillage and high input treatment was used for evaluating the phenology models.

Pioneer 3902: a field experiment conducted at the Elora Research Station in Ontario, Canada from 1999 to 2001 and designed to compare the leaf senescence profiles of older and newer hybrids provided the data for Pioneer 3902 (Valentinuz and Tollenaar, 2004). The soil was a London loam soil (Aquic Hapludalf). The hybrids were planted on 13 May 1999, 26 May 2000 and 9 May 2001 in 0.76-m spaced rows. Plant densities were 1, 3.5 and 12 plants m^{-2}. Observed silking date from this experiment was used in our model evaluation.

XL72AA and DKC61–69: data for these two hybrids were obtained from experiments conducted at two research stations in Indiana (ACRE, Agronomy Center for Research and Education near West Lafayette, IN, and PPAC, Pinney Purdue Agricultural Center near Wanatah, IN) and one location in North Platte, NE during the 2012 to 2013 growing seasons, on the combined effects of nitrogen, cultivar, and plant density on maize performance. Soil characteristics used in the simulations were from a Xenia silt loam (fine-silty, mesic Aquic Hapludalfs) at ACRE, a Tracy silt loam (coarse-loamy, mixed, superactive, mesic Ultic Hapludalf) at PPAC and a well-drained silt loam Mollisols at North Platte. The hybrids were planted at ACRE on 17 May 2012 and 14 May 2013, at PPAC on 12 May 2012, and at North Platte on 9 May 2012 and 11 May 2013. Plant densities at ACRE and PPAC during the two years were 5.4, 7.9, and 10.4 plants m^{-2}. Plant density in North Platte was 12.2 plants m^{-2} in 2012 and 11 plants m^{-2} in 2013. Leaf tip and collar counts were taken two times during the 2012 season and three times during the 2013 season, and the total number of leaves produced by the plant was recorded. Anthesis, silking and milk line were monitored on selected plants allowing for the dates of occurrence of these events in 50% of the samples to be recorded.

Pioneer 3245, 3394, and 3527: data for these three hybrids were produced by field experiments conducted at four research stations, ACRE (near West Lafayette, Indiana), SEPAC (Southeast Purdue Agricultural Center near Butlerville, Indiana), NWBRF (Northwest Branch Research Farm near Hoytville, Ohio) and SWBRF (Southwest Branch Research Farm near South Charleston, Ohio) during the period 1991 to 1994. The experiments were designed to study the interactive effects of planting date and environment (location and year) on the development

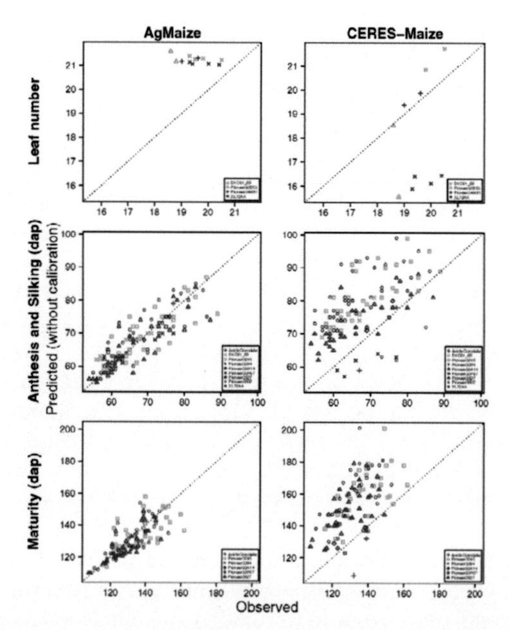

Fig. 4. Comparison of observed and simulated total leaf number, flowering day, and maturity day based on non-calibrated genotypic parameters.

of hybrids of different maturity ratings (Nielsen et al., 2002). Soils were a Drummer silty clay loam (Fine-silty, mixed, mesic Typic Haplaquolls) at ACRE, a Clermont silt loam (Typic Ochraqualf) at SEPAC, a Hoytville silty clay loam (Fine, illitic, mesic Mollic Ochraqualf) at NWBRF and a Kokomo silty clay loam (Fine, mixed, mesic, Typic Argiaquoll) at SWBRF. The hybrids were planted in 76-cm rows at a seeding rate of 67,000 and 74,000 seeds ha^{-1}, respectively at the Indiana and Ohio locations. At each location three planting dates (early, mid and late) were selected during the period April to June. The combinations of locations (4), years (4 in ACRE and SWBRF and 2 in SEPAC and NWBRF), hybrids (3) and planting dates (3) yielded a total of 108 treatments. Data collected included dates of silking and black layer formation.

Results and Discussion

The prediction of phenology for a number of diverse North American datasets by the uncalibrated phenology model of AgMaize was relatively accurate (Tables 5a and 6a; Fig. 4). The uncalibrated biases for silking date (-0.1 d) and maturity date (1.6 d) were low, whereas the bias for anthesis date (2.1 d) was greater (Table 6). The prediction of phenology in the same datasets by the uncalibrated phenology model of CERES-Maize was much less accurate than that for AgMaize (Tables 5a and 7a). The variability in the values of the parameters estimated using different subsets of the data (cross-validation) was smaller for AgMaize than for CERES-Maize (Table 5b). Such low variability may suggest smaller parameter uncertainties and an adequate modeling of the crop response to environmental factors. Considering the importance of the accurate prediction of phenology in crop growth models, the uncalibrated biases for silking date (-10 d) and maturity date (-22 d) by CERES-Maize (Table 7) as well as the RMSEs were high. The relatively high degree of accuracy in predicting phenology by AgMaize could have been the result of a high degree of commonality between the datasets used for model development and model evaluation. However, the dataset used for development of the relationships between RM and duration of development in AgMaize, *t* data from > 3000 hybrids grown at one location from 2007 to 2012 (Kumudini et al., 2014), was unrelated to the datasets used in this evaluation. The ability of the AgMaize phenology model to predict development stages over a range of environments and hybrids based on RM ratings and without calibration of other genotypic parameters also demonstrates the adequacy and stability of the temperature response functions used in the model as recently found in Kumudini et al. (2014). Note that this methodology can be used also when the RM rating of the hybrid is unknown by converting duration of the hybrid's life cycle in thermal accumulation into RM, although this methodology will be more variable (cf., Yang et al., 2004).

Accuracy of prediction of silking date by CERES-Maize was considerably improved after calibration of the genotype coefficients, whereas predictions by AgMaize were little affected by calibration (Table 5–7; Fig. 4–5). After calibration,

Table 6. Spearman correlation (R), root mean square error of prediction (RMSEP) and bias for evaluating the ability of AgMaize to predict different phenology variables in different hybrids: (a) without model calibration and (b) after model calibration.

(a)

	Total leaf number			Anthesis day			Silking day			Maturity day		
Hybrid	R	RMSE	Bias	R	RMSE	Bias	R	RMSE	Bias	R	RMSE	Bias
AvictaComplete				1	1.58	-1.5				1	4.74	1.5
DKC61-69	-0.8	1.61	-1.53	1	4.69	-4	0.95	3.97	-3			
Pioneer3245							0.84	4.96	0.22	0.71	9.76	2.17
Pioneer3394							0.84	4.54	-0.89	0.8	7.26	0.94
Pioneer33A14							1	1	1	-1	4.12	-1
Pioneer33B53	-1	2.71	-2.69									
Pioneer33P67								2.12	1.5	1	5.1	5
Pioneer34M91	1	1.97	-1.96									
Pioneer3527							0.86	4.56	0.39	0.86	6.11	1.78
Pioneer3902							0.5	5.1	0			
XL72AA	-0.8	1.4	-1.32	0.8	3.24	-0.5	0.9	3.85	2			
All	-0.32	1.84	-1.72	0.9	3.67	-2.1	0.85	4.58	-0.08	0.79	7.72	1.64

(b)

	Total leaf number			Anthesis day			Silking day			Maturity day		
Hybrid	R	RMSEP	Bias	R	RMSEP	Bias	R	RMSEP	Bias	R	RMSEP	Bias
AvictaComplete				1	1.41	-1				1	10.51	9.5
DKC61-69	-1	0.65	0	1	4.21	-3.25	0.97	3.63	-2.4			
Pioneer3245							0.84	4.97	-0.08	0.7	10.65	4.83
Pioneer3394							0.85	4.41	-0.69	0.81	8.28	4.75
Pioneer33A14							1	1.58	1.5	-1	5.83	3
Pioneer33B53	-1	0.49	-0.32									
Pioneer33P67								2.12	1.5	1	8.51	8.5
Pioneer34M91	-1	0.42	-0.27									
Pioneer3527							0.86	4.67	-0.06	0.86	7.05	3.97
Pioneer3902							0.5	5.45	0.33			
XL72AA	-1	0.62	0	0.8	3.57	-0.25	0.9	4.22	2.2			
All	0.02	0.58	-0.1	0.93	3.55	-1.6	0.85	4.59	-0.19	0.79	8.77	4.65

Table 7.(a) Spearman correlation (R), root mean square error of prediction (RMSEP) and bias for evaluating the ability of CERES-Maize to predict different phenology variables in different hybrids: (a) without model calibration and (b) after model calibration.

	Total leaf number			Silking day			Maturity day		
Hybrid	R	RMSE	Bias	R	RMSE	Bias	R	RMSE	Bias
AvictaComplete				1	5.7	3.5	1	17	15
DKC61-69	1	1.15	-1.15	1	9.88	-9.67			
Pioneer3245				0.76	14.29	-12.86	0.63	24.63	-21.17
Pioneer3394				0.78	16.33	-15.25	0.58	28.7	-25.39
Pioneer33A14				1	3.16	-3	-1	6.04	-5.5
Pioneer33P67					0	0		4	4
Pioneer3527				0.83	9	-7.44	0.53	23.77	-20.83
Pioneer3902				0.5	12.83	12.67			
Pioneer33B53	-1	2.28	1.66						
Pioneer34M91	1	0.34	-0.33						
XL72AA	0.8	3.57	3.55	0.8	7.95	6.8			
All hybrids	0.42	2.54	1.45	0.58	13.03	-9.97	0.59	25.33	-21.27

(b)

	Total leaf number			Silking day			Maturity day		
Hybrid	R	RMSEP	Bias	R	RMSEP	Bias	R	RMSEP	Bias
AvictaComplete				1	2.92	1.5	1	16.51	-0.5
DKC61-69	-0.4	0.97	-0.34	0.67	4.54	2.6			
Pioneer3245				0.81	5.25	0.67	0.45	12.77	0.86
Pioneer3394				0.83	4.68	0.78	0.51	10.72	0.17
Pioneer33A14				1	2.55	2.5	-1	5.1	1
Pioneer33P67					10	10		11	11
Pioneer3527				0.86	4.65	1.03	0.61	10.03	0.81
Pioneer3902				0.5	5.74	0.33			
XL72AA	-0.8	1.16	-0.62	0.56	8.12	6.4			
All hybrids	-0.63	1.07	-0.48	0.82	5.06	1.21	0.55	11.27	0.69

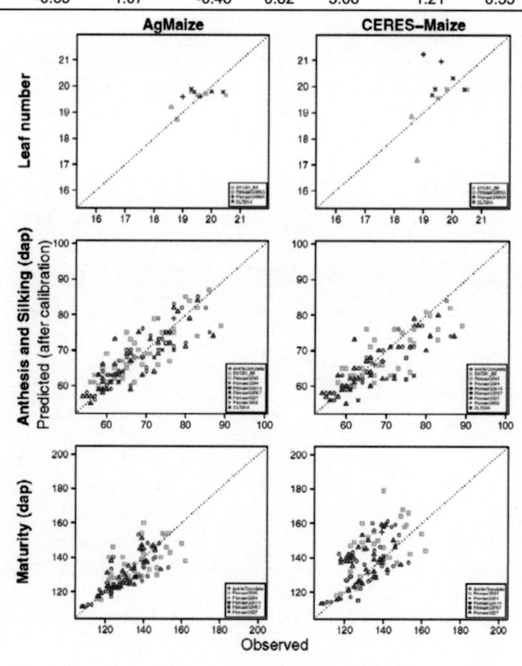

Fig. 5. Comparison of observed and simulated total leaf number, flowering day, and maturity day based on calibrated genotypic parameters.

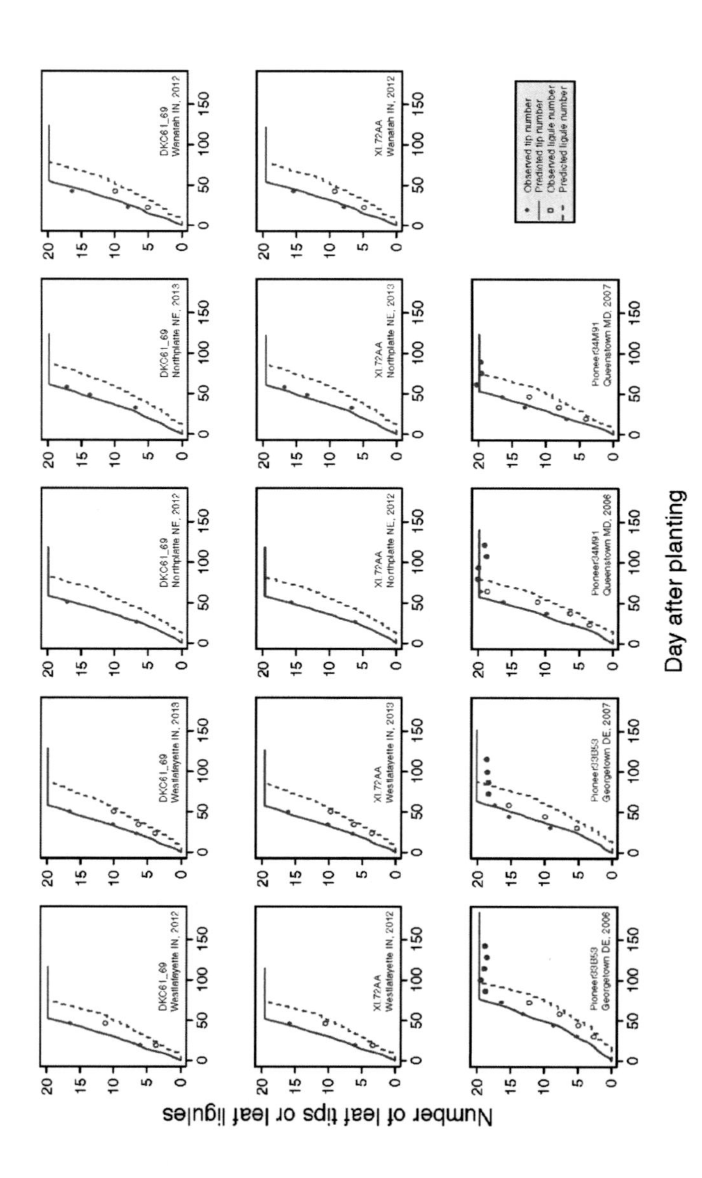

Fig. 6. Comparison of observed and AgMaize-predicted numbers of leaf tips and leaf ligules for several combinations of hybrids, locations and years.

Table 8. Errors statistics describing the ability of AgMaize to predict numbers of leaf tips and leaf ligules based on the simplified β function and using RLAMX (maximum rate of leaf tip appearance) = 0.53 leaves d^{-1}.

	Leaf tip number			Leaf ligule number		
Hybrid and environment	R	RMSE	Bias	R	RMSE	Bias
DKC61-69, North Platte NE, 2012	1	0.78	0.73			
DKC61-69, North Platte NE, 2013	1	1.42	-1.4			
DKC61-69, Wanatah IN, 2012	1	1.89	1.85	1	2.49	2.45
DKC61-69, West Lafayette IN, 2012	1	0.24	-0.24	1	1.61	1.4
DKC61-69, West Lafayette IN, 2013	1	0.31	-0.2	1	1.05	1.03
Pioneer 33B53, Georgetown DE, 2006	0.91	0.84	-0.55	0.91	2.03	0.77
Pioneer 33B53, Georgetown DE, 2007	0.91	1.74	-0.47	0.98	3.42	1.84
Pioneer 34M91, Queenstown MD, 2006	0.87	0.68	-0.46	0.94	2.59	1.21
Pioneer 34M91, Queenstown MD, 2007	0.94	0.68	0.21	0.81	3.38	2.16
XL72AA, North Platte NE, 2012	1	0.24	-0.23			
XL72AA, North Platte NE, 2013	1	1.67	-1.67			
XL72AA, Wanatah IN, 2012	1	1.3	1.3	1	2.01	2
XL72AA, West Lafayette IN, 2012	1	0.69	-0.59	1	1.01	0.85
XL72AA, West Lafayette IN, 2013	1	0.51	-0.44	1	0.85	0.8
All	0.95	1.08	-0.3	0.94	2.5	1.39

the bias for silking date and maturity date in CERES-Maize were 1.2 and 0.7 d, respectively (Table 7b). Variability of parameter during cross-validation was generally low (Table 5b). The difference between the calibrated and uncalibrated estimates of CRMAT in this evaluation varied seldom by more than 2 RM units for AgMaize, which may be in part attributable to the inherent constraints of the relationship between hybrid RM and phase duration. In AgMaize, phase duration of the pre- and post-flowering period are influenced both by hybrid relative maturity (CRMAT), but the relationship between phase duration and relative maturity differs for the two phases (i.e., Eqs. [6] and [14]). Calibrated values of TAInt (i.e., the interval between the emergence of the topmost leaf tip and anthesis) was on average 1.7 TLU greater than that estimated in the uncalibrated model (i.e., *TAInt* = 2.5 TLU), which may have been attributable to the failure of the AgMaize model to accurately estimate total leaf number (*TnLeaf*), that is, the difference between the calibrated and uncalibrated bias for TnLeaf was 1.6 TLU (Table 6).

The ability of AgMaize to predict the rate of leaf-tip appearance (RLA) and the rate of leaf-ligule appearance is depicted in Table 8 and Fig. 6. Whereas leaf tips were slightly overpredicted (i.e., the bias was -0.3 leaves), leaf ligules or V-stage were underpredicted (i.e., the bias was +1.4 leaf ligules). Overall, AgMaize predictions of leaf tips were associated with a RMSE smaller than that of leaf ligules. Hybrid-specific error measures showed that although AgMaize predictions of leaf ligules were biased, the model captured the observed patterns of variation of leaf tips and leaf ligules as demonstrated by high correlations (Table 8). For most locations and years, Spearman correlations were 1.0 (Fig. 6); the smallest correlation value was 0.87 for leaf tip (in Queenstown, MD, 2006) and 0.81 for leaf ligule (Queenstown, MD, 2007). It is worth noting that measurement error due to field variability is a possible reason for the lack of agreement between observations and simulations, because phenology observations are generally taken on several plants across experimental replications.

These predictions were based on a maximum leaf appearance rate (RLAMax) of 0.53 leaves d^{-1}, a ceiling temperature of 43.7 °C, and an optimum temperature of 32.1 °C considered as species parameters and not adjusted for specific hybrids. The results suggested that in datasets with no leaf tip or leaf ligule observations, an RLAMax value of 0.53 leaves d^{-1} would be adequate. The cardinal temperatures and the maximum leaf tip appearance rate used to describe the simplified β function in AgMaize were estimated using growth chamber data on Pioneer hybrid 3733 from Beltsville, MD (Kim et al., 2007) and used in the MAIZSIM crop model (Kim et al., 2012). Our ability to predict leaf tip and leaf ligule appearance using the growth chamber-estimated parameters from Kim et al. (2007) showed that the parameters describing the simplified β function (cardinal temperatures and RLAMax) can be considered stable across hybrids.

Accurate estimation of leaf ligules is important because (i) the V-stage is used in the large majority of reports that deal with U.S. corn production, and (ii) the simulation of leaf-area expansion of each leaf depends on the interval between leaf-tip and leaf-ligule appearance. Few datasets are currently available that report both leaf tips and leaf ligules, but substantial improvement in the accuracy of leaf-ligule prediction in AgMaize can be achieved when more data become available.

It should be pointed out that the model has not been tested or validated in a large area of the tropics and subtropics, both in terms of environments and germplasm. The model, however, will provide a framework for future evaluation of tropical germplasm grown in tropical and sub-tropical environments.

Conclusion

The accuracy of AgMaize and CERES-Maize in predicting the response of maize phenology to major environmental and genotypic influences was evaluated in temperate environments. The major influences on phenology are (i) effect of temperature on rate of development that is accumulated as thermal units, and (ii) the genotype-specific duration of growth phases that are defined in accumulated thermal units. The effect of temperature on rate of leaf-tip appearance was predicted fairly well by AgMaize and RLA(°C) appeared to be conservative across maize hybrids (Table 8; Fig. 6). Phase duration for maize hybrids ranging from 89 to 117 RM (Table 5a) that were grown from 1991 to 2013 under experimental conditions in North America (Table 4) was predicted more accurately by AgMaize than by CERES-Maize under uncalibrated conditions. Predictions of phase duration by CERES-Maize were similar to those by AgMaize when the models were calibrated by modifying genotype coefficients. Genotype coefficients necessary for model calibration are not always readily available, especially when using gridded models or when crop growth models are used for predictions in future climate-change scenarios. Therefore, models such as AgMaize that perform well with little calibration data have an advantage in those conditions. The phenology routine of AgMaize would be well suited when the RM or the accumulated thermal accumulation during the life cycle of the hybrid is known and when little or no additional information is available for genotype calibration.

The evaluation of the phenology routine of AgMaize did not include detailed analyses of relatively minor environmental influences in temperate production environments on phenology such as (i) effects of temperature and photoperiod on total leaf number (TnLeaf), (ii) effects of growing-point vs. air temperature on rate of development, and (iii) effects of abiotic stresses on rate of development (e.g., water and N stress, *ASI*). Although the impact of the minor influences on maize phenology may be an order of a magnitude smaller than that of the major influences, the impact of the minor environmental influences on predicting maize phenology increases as the effects of the major influences are fully accounted for. We showed that genotype effects on phase duration can be removed by calibration (e.g., Table 7a vs. 7b). It is not clear, however, whether the genotype calibration also has the impact of curvefitting of the minor environmental effects due to confounding the minor environmental effects with the genotype effects. To disentangle genotype and the minor environmental effects on phenology, further research is required that is designed specifically to a capture the direct impact of these effects on maize phenology. The improved accuracy of AgMaize in uncalibrated conditions may be a consequence of its ability to account separately for these other influences on maize phenology. However, since the impact of these minor environmental effects on maize phenology were not directly evaluated in this study, the improved performance of AgMaize under uncalibrated conditions cannot be attributed to these factors without further research to evaluate the contributions of these variables to the prediction of maize phenology.

References

Birch, C.J., G.L. Hammer, and K.G. Rickert. 1998. Temperature and photoperiod sensitivity of development in five cultivars of maize (*Zea mays* L.) from emergence to tassel initiation. Field Crops Res. 55:93–107. doi:10.1016/S0378-4290(97)00062-2

Brouwer, R., A. Kleinendorst, and J.Th. Locker. 1973. Growth responses of maize plants to temperature. p. 169-174. In: Plant response to climatic factors. Proc. Uppsala Symp. 1970, UNESCO, Paris.

Cavalieri, A.J., and O.S. Smith. 1985. Grain filling and field drying of a set of maize hybrids released from 1930 to 1982. Crop Sci. 25:856–860. doi:10.2135/cropsci1985.0011183X002500050031x

Coles, N.D., M.D. McMullen, P.J. Balint-Kurti, R.C. Pratt, and J.B. Holland. 2010. Genetic control of photoperiod sensitivity in maize revealed by joint multiple population analysis. Genetics 184:799–812. doi:10.1534/genetics.109.110304

Daynard, T.B., and W.G. Duncan. 1969. The black layer and grain maturity in corn. Crop Sci. 9:473–476. doi:10.2135/cropsci1969.0011183X000900040026x

Ellis, R.H., R.J. Summerfield, G.O. Edmeades, and E.H. Roberts. 1992a. Photoperiod, leaf number, and interval from tassel initiation to emergence in diverse cultivars of maize. Crop Sci. 32:398–403. doi:10.2135/cropsci1992.0011183X003200020024x

Ellis, R.H., R.J. Summerfield, G.O. Edmeades, and E.H. Roberts. 1992b. Photoperiod, temperature, and the interval from sowing to tassel initiation in diverse cultivars of maize. Crop Sci. 32:1225–1232. doi:10.2135/cropsci1992.0011183X003200050033x

Edmeades, G.O., J. Bolaños, A. Elings, J.-M. Ribaut, M. Bänziger, and M. E. Westgate. 2000. The role and regulation of the anthesis-silking interval in maize. In: M. Westgate and K. Boote, editors, Physiology and modeling kernel set in maize. CSSA Spec. Publ. 29. CSSA, Madison, WI. p. 43-73.

Gouesnard, B., C. Rebourg, C. Welcker, and A. Charcosset. 2002. Analysis of photoperiod sensitivity within a collection of tropical maize populations. Genet. Resour. Crop Evol. 49:471–481. doi:10.1023/A:1020982827604

Jones, C.A., and J.R. Kiniry. 1986. CERES maize: A simulation model of maize growth and development. Texas A&M Univ. Press, College Station, TX.

Jones, P., and P. Carberry. 1994. A technique to develop and validate simulation models. Agric. Syst. 46:427–442. doi:10.1016/0308-521X(94)90105-O

Kim, S., D. Gitz, R. Sicher, J. Baker, D. Timlin, and V. Reddy. 2007. Temperature dependence of growth, development, and photosynthesis in maize under elevated CO2. Environ. Exp. Bot. 61:224–236. doi:10.1016/j.envexpbot.2007.06.005

Kim, S., Y. Yang, D. Timlin, D. Fleisher, A. Dathe, V. Reddy, and K. Staver. 2012. Modeling temperature responses of leaf growth, development, and biomass in maize with MAIZSIM. Agron. J. 104:1523–1537. doi:10.2134/agronj2011.0321

Kumudini, S., F.H. Andrade, K.J. Boote, G.A. Brown, K.A. Dzotsi, G.O. Edmeades, T. Gocken, M. Goodwin, A.L. Halter, G.L. Hammer, J.L. Hatfield, J.W. Jones; A.R. Kemanian, S.-H. Kim, J. Kiniry, J.I. Lizaso, C. Nendel, R.L. Nielsen, B. Parent, C.O. St ckle, F. Tardieu, P.R. Thomison, D.J. Timlin, T.J. Vyn, D. Wallach, H.S. Yang, and M. Tollenaar. 2014. Predicting maize phenology: Intercomparison of functions for developmental response to temperature. Agron. J. 106:2087–2097. doi:10.2134/agronj14.0200

Kiniry, J.R., J.T. Ritchie, R.L. Musser, E.P. Flint, and W.C. Iwig. 1983. The photoperiod sensitive interval in maize. Agron. J. 75:687–690. doi:10.2134/agronj1983.00021962007500040026x

Lafitte, H.R., and G.O. Edmeades. 1994. Improvement for tolerance to low nitrogen in tropical maize. II. Grain yield, biomass production, and N accumulation. Field Crops Res. 39:15–25. doi:10.1016/0378-4290(94)90067-1

Liu, Y., R. Xie, P. Hou, S. Li, H. Zhang, B. Ming, H. Long, and S. Liang. 2013. Phenological responses of maize to changes in environment when grown at different lattitudes in China. Field Crops Res. 144:192–199. doi:10.1016/j.fcr.2013.01.003

Lobell, D.B., M.J. Roberts, W. Schlenker, N. Braun, B.B. Little, R.M. Rejesus, and G.L. Hammer. 2014. Greater sensitivity to drought accompanies maize yield increase in the U.S. Midwest. Science 344:516–519. doi:10.1126/science.1251423

McCullough, D.E., Ph. Girardin, M. Mihajlovic, A. Aguilera, and M. Tollenaar. 1994. Influence of N supply on development and dry matter accumulation of an old an a new maize hybrid. Can. J. Plant Sci. 74:471–477. doi:10.4141/cjps94-087

Muchow, R.C., and P.S. Carberry. 1989. Environmental control of phenology and leaf growth in tropically adapted maize. Field Crops Res. 20:221–236. doi:10.1016/0378-4290(89)90081-6

Nielsen, R.L., P.R. Thomison, G.A. Brown, A.L. Halter, J. Wells, and K.L. Wuethrich. 2002. Delayed planting effects on flowering and grain maturation of dent corn. Agron. J. 94:549–558. doi:10.2134/agronj2002.5490

Nelder, J., and R. Mead. 1965. A simplex method for function minimization. Comput. J. 7:308–313. doi:10.1093/comjnl/7.4.308

Padilla, J.M., and M.E. Otegui. 2005. Co-ordination between leaf initiation and leaf appearance in field-grown maize (*Zea mays*): Genotypic differences in response of rates to temperature. Ann. Bot. (Lond.) 96:997–1007. doi:10.1093/aob/mci251

Parent, B., O. Turc, M. Gibon, M. Stitt, and F. Tardieu. 2010. Modelling temperature-compensated physiological rates, based on the coordination of responses to temperature of developmental processes. J. Exp. Bot. 61:2057–2069. doi:10.1093/jxb/erq003

Potter, K., and J. Williams. 1994. Predicting daily mean soil temperatures in the EPIC simulation model. Agron. J. 86:1006–1011. doi:10.2134/agronj1994.00021962008600060014x

Ritchie, J.T., and D.S. NeSmith. 1991. Temperature and crop development. p. 5-29. In: J. Hanks and J.T. Ritchie, editors, Modeling plant and soil systems. Agronomy Monograph 31, ASA, CSSA, SSSA, Madison, WI.

Sacks, W.J., and C.J. Kucharik. 2011. Crop management and phenology trends in the U.S. Corn Belt: Impacts on yields, evapotranspiration and energy balance. Agric. For. Meteorol. 151:882–894. doi:10.1016/j.agrformet.2011.02.010

Schlenker, W., and M.J. Roberts. 2009. Nonlinear temperature effects indicate severe damages to U.S. crop yields under climate change. Proc. Natl. Acad. Sci. USA 106:15594–15598. doi:10.1073/pnas.0906865106

Stewart, D., L. Dwyer, and L. Carrigan. 1998. Phenological temperature response of maize. Agron. J. 90:73–79. doi:10.2134/agronj1998.00021962009000010014x

Tollenaar, M. 1999. Duration of the grain-filling period in maize is not affected by photoperiod and incident PPFD during the vegetative phase. Field Crops Res. 62:15–21. doi:10.1016/S0378-4290(98)00170-1

Tollenaar, M., and T.B. Daynard. 1978. Kernel growth and development at two positions on the ear of maize (*Zea mays*). Can. J. Plant Sci. 58:189–197. doi:10.4141/cjps78-028

Tollenaar, M., and T.B. Daynard. 1982. Effect of source-sink ratio on dry matter accumulation and leaf senesence of maize. Can. J. Plant Sci. 62:855–860. doi:10.4141/cjps82-128

Tollenaar, M., T.B. Daynard, and R.B. Hunter. 1979. Effect of temperature on rate of leaf appearance and flowering date in maize. Crop Sci. 19:363–366. doi:10.2135/cropsci1979.0011183X001900030022x

Tollenaar, M., J. Fridgen, P. Tyagi, and S. Kumudini. 2017. The contribution of solar brightening to US maize yield trend. Nature Climate Change 7:275–278.

Tollenaar, M., and R. Hunter. 1983. A photopenod and temperature sensitive period for leaf number in maize. Crop Sci. 23:457–460. doi:10.2135/cropsci1983.0011183X002300030004x

Tollenaar, M., and E.A. Lee. 2011. Strategies for enhancing grain yield in maize. Plant Breed. Rev. 34:37–82. doi:10.1002/9780470880579.ch2

Tollenaar, M., J.F. Muldoon, and T.B. Daynard. 1984. Differences in rates of leaf appearance among maize hybrids and phases of development. Can. J. Plant Sci. 64:759–763. doi:10.4141/cjps84-104

Valentinuz, O., and M. Tollenaar. 2004. Vertical profile of leaf senescence during the grain-filling period in older and newer maize hybrids. Crop Sci. 44:827–834. doi:10.2135/cropsci2004.8270

Van Esbroeck, G.A., J.A. Ruiz Corral, J.J. Sanchez Gonzalez, and J.B. Holland. 2008. A comparison of leaf appearance rates among teosinte, maize landraces, and modern maize. Maydica 53:117–123.

Vinocur, M.G., and J.T. Ritchie. 2001. Maize leaf development biases caused by air-apex temperature differences. Agron. J. 93:767–772. doi:10.2134/agronj2001.934767x

Wallach, D. 2006. Evaluating crop models. In: D. Wallach, D. Makowski, J. Jones, editors, Working with dynamic crop models: Evaluation, analysis, parameterization, and applications. Elsevier Science, Amsterdam, The Netherlands. p. 11–50.

Wallach, D. 2011. Crop model calibration: A statistical perspective. Agron. J. 103: 1144–1151.

Warrington, I., and E. Kanemasu. 1983. Corn growth response to temperature and photoperiod. II. Leaf initiation and leaf appearance rates. Agron. J. 75:755–761. doi:10.2134/agronj1983.00021962007500050009x

Wilson, D., R. Muchow, and C. Murgatroyd. 1995. Model analysis of temperature and solar radiation limitations to maize potential productivity in a cool climate. Field Crops Res. 43:1–18. doi:10.1016/0378-4290(95)00037-Q

Wu, J.W. 1998. On the relationship between plant-to-plant variability and stress tolerance in maize (*Zea mays* L.) hybrids from different breeding eras. M.S. thesis, Univ. of Guelph, Guelph, ON.

Yan, W., and L. Hunt. 1999. An equation for modelling the temperature response of plants using only the cardinal temperatures. Ann. Bot. (Lond.) 84:607–614. doi:10.1006/anbo.1999.0955

Yang, H., A. Dobermann, J. Lindquist, D. Walters, T. Arkebauer, and K. Cassman. 2004. Hybrid-maize—a maize simulation model that combines two crop modeling approaches. Field Crops Research 87:131–154. doi:10.1016/j.fcr.2003.10.003

Agroclimatology of Maize, Sorghum, and Pearl Millet

P.V.V. Prasad,* M. Djanaguiraman, Z.P. Stewart, and I.A. Ciampitti

Abstract

Coarse grains are a staple food for large populations in the arid and semiarid regions of the world. This chapter presents the growth, development, and climatic requirements (particularly temperature, precipitation, and photoperiod) of three important coarse grains, namely, maize, sorghum, and pearl millet. The environmental requirements for optimum growth varies with crop species. Most of the progression across growth stages and development is dependent on temperature and photoperiod. All crop species and stages of crop development have cardinal temperatures. Among the three cereals, maize has relatively lower cardinal temperatures, followed by grain sorghum and pearl millet. Total water requirement, water productivity, and response to irrigation is relatively greater in maize than in sorghum and pearl millet. Maize is relatively more sensitive to high temperature and drought stress during sensitive stages of gametogenesis, flowering, and early grain filling stages of crop development. Sorghum and pearl millet can perform better under high temperature stress and drier environments. All three-crop species are short-day plants and respond to changes in photoperiod; however, new hybrids are day neutral. There is large genetic diversity for response to environmental conditions such as temperature, water, and photoperiod, and to biotic stresses. Genetic variability needs to be systematically utilized for breeding to enhance stress tolerance to obtain higher and more stable yields in multiple environments. Projected change in climate and management practices will require a better understanding of crop agroclimatology, so that we can match crop phenology to optimum environments and efficient crop management practices to exploit full genetic potential.

Maize

Maize or corn (*Zea mays* L.) is a coarse cereal grown in a range of agroecological environments. Maize is a preferred staple food for over 900 million poor

---

Abbreviations: ASI, anthesis silking interval; LAR, leaf area ratio; NAR, net assimilation rates; QTL, quantitative trait loci; RGR, relative growth rates;

P.V.V. Prasad, M. Djanaguiraman, and I.A. Ciampitti, Department of Agronomy, Kanas State University, Manhattan, Kansas 66506; P.V.V. Prasad and Z.P. Stewart, Feed the Future Innovation Lab for Collaborative Research on Sustainable Intensification, Kansas State University, Manhattan, Kansas 66506; M. Djanaguiraman, Department of Crop Physiology, Tamil Nadu Agricultural University, Coimbatore, India.
*Corresponding author (vara@ksu.edu)

doi:10.2134/agronmonogr60.2016.0005

© ASA, CSSA, and SSSA, 5585 Guilford Road, Madison, WI 53711, USA.
Agroclimatology: Linking Agriculture to Climate, Agronomy Monograph 60.
Jerry L. Hatfield, Mannava V.K. Sivakumar, John H. Prueger, editors.

consumers, 140 million poor farm families, and about one-third of malnourished children (CIMMYT, 2010). Today, maize is the most important food crop in sub-Saharan Africa, Asia, and Latin America. During 2016, maize is produced on nearly 187 million hectares, with a production and productivity of 1060 million tons and 5640 kg ha^{-1}, respectively (FAO, 2016). All parts of the maize crop are used for food and non-food products. In developed countries, maize is largely used as livestock feed and as a raw material for industrial products, including biofuel. As global population increases and more people begin to consume meat, poultry, and dairy, demand for maize is expected to increase. The United States is the largest consumer and producer of maize in the world with an annual production of over 300 million metric tons. The United States is the largest producer of maize, followed by China, Brazil, Argentina, and India (FAO, 2016).

Maize was first domesticated 7000 to 10,000 yr ago in southcentral or western Mexico. After domestication, maize spread quickly throughout the Americas before European colonization (Mangelsdorf, 1974). There are six different types of maize grown today, namely flint corn, dent corn, sweet corn, flour corn, popcorn, and waxy corn (Brown et al., 1985; Hamilton et al., 1951; Ranum et al., 2014). In the case of flint corn [*Zea mays* L. var. *indurata* (Sturtev.) L. H. Bailey], the entire outer portion of the kernel is hard starch. Flint comes in many colors such as white, yellow, red-blue or other variations. Dent corn [*Zea mays* L. var. *indentata* (Sturtev.) L. H. Bailey], kernels are indented at the tip the hard starch is confined to the kernel only. Dent corn is about 95% of the U.S. production. Kernels may be yellow, white, or red. Sweet corn [*Zea mays* L. var. *saccharata* (Sturtev.) L. H. Bailey] is grown for food and harvested at 70% moisture content. It is a good source of energy with about 20% of its dry matter as sugar compared to 3% in dent corn. Sweet corn is also a good source of vitamin C and A. Flour corn (*Zea mays* L. var. *amylacea*), kernels are largely composed of soft starch with little or no hard starch and are easy to grind. Flour corn was primarily used by natives of the Andean Highlands of South America. Popcorn (*Zea mays* L. var. *everta*) kernels are small and an extreme form of flint corn. When heated to 170 °C, the grain swells, and bursts, turning inside out. At this temperature, the water held in the starch turns to steam and the pressure causing the explosion. Waxy corn (*Zea mays* L. var. *ceretina*), gained its name due to the waxy appearance of the kernel. The starch is entirely amylopectin, whereas, dent contains 78% amylopectin and 22% amylose. Waxy corn hybrids are raw materials for the industrial production of textile, packing materials and oil. Across the six different types of maize, the international maize and wheat improvement center (CIMMYT) has one of the largest germplasm collections (> 150,000 from ~100 different countries) and has a global mandate for maize research and capacity building activities.

Growth and Development

Maize utilizes a C_4 photosynthetic pathway and thus uses CO_2, solar radiation, water, and nitrogen (N) more efficiently relative to plants utilizing a C_3 photosynthetic pathway such as rice (*Oryza sativa* L.) and wheat (*Triticum aestivum* L.). Maize is grown as an irrigated or dryland crop based on the adequacy of rainfall and the availability of water for irrigation. Growth is defined as the accumulation of dry matter and development is defined as the plant's progression from vegetative to reproductive stages and maturity. During the life cycle of a maize plant, there are several key identifiable stages of development. The most common system for defining maize growth stages divides the growth into vegetative and reproductive development [Abendroth et al. (2011) and references therein].

Vegetative growth stages are determined using the leaf collar method. A plant is assigned a growth stage depending on the number of visible leaf collars. Vegetative stages defined by this method are defined as 'V' stages. For example, a plant with four visible leaf collars would be at fourth-leaf 'V4'. As the maize plant progresses through its development, some of the earlier leaves may detach from the stem due to stem expansion and aging, making it challenging to identify the growth stage by simply counting leaf collars. Visible with a vertical transect of the base of the stem, the fifth leaf node is located above the first visibly elongated node. From this point, the V-stage is determined by counting the nodes upward until the highest leaf node is identified. The final V-stage is VT; all the branches of the tassel are fully emerged. This stage occurs when the tassel is completely extended, and silks are not yet visible. During this time, all the leaves that will be grown on the plant are visible. The tassel will usually be visible for two to three days before the silk emergence, depending on the cultivar and environmental conditions. The potential kernels per row are set, a number of ovules and potential ear size are also defined at this point (further details at Ciampitti et al., 2016).

Reproductive stages are defined using 'R-stages.' Growth stage 'R1' is defined as silking, marked by the emergence of silks beyond the tip of the ear husk. The rest of the R stages relate to the development of the kernels on the ear (R2, blister; R3, milk; R4, dough; and R5, dent; R6, physiological maturity). Seeds start developing soon after florets are fertilized. The first florets to be fertilized are at the base of the ear, and the last seeds to be fertilized are at the ear tip. In assessing the reproductive growth stage, only the seeds in the middle section of the ear are assessed. Physiological maturity is the stage at which the kernel has achieved its maximum dry matter accumulation. The hard starch layer or black abscission layer called the 'black layer' is a sign of physiological maturity.

Table 1. Description of growth and developmental stages of maize (Ciampitti, 2016).

Stage	Description
VE	The coleoptile emerges from the soil surface.
V1	The collar of the first leaf is visible.
V2	The collar of the second leaf is visible.
Vn	The collar of leaf number "n" is visible. The maximum value of "n" represents the final number of leaves, which is usually 16 to 23, but by flowering, the lower 4 to 7 leaves have senesced.
VT	The last branch of the tassel is completely visible.
R0	Anthesis. Pollen shed begins.
R1	Silks are visible.
R2	Blister stage. Kernels are filled with clear fluid and the embryo can be seen.
R3	Milk stage. Kernels are filled with a white, milky fluid.
R4	Dough stage. Kernels are filled with a white paste. The embryo is about half as wide as the kernel. The top parts of the kernels are filled with solid starch.
R5	Dent stage. If the genotype is a dent type, the grains are dented. The "milk line" is close to the base when the kernel is viewed from the side in both flint and dent types.
R6	Physiological maturity. The black layer is visible at the base of the grain. Grain moisture is usually about 35%.

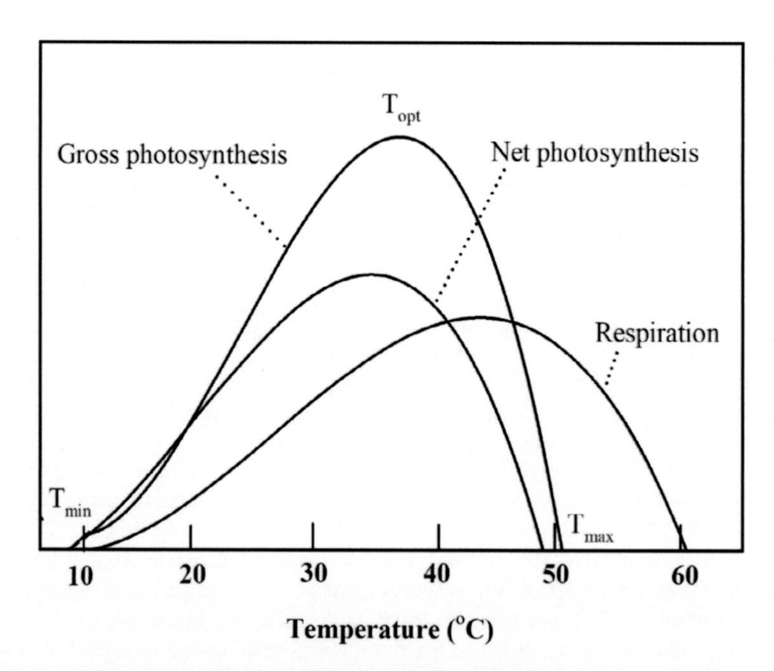

Fig. 1. Temperature effects on rates of gross photosynthesis, respiration, and net photosynthesis of maize. (Redrawn from Hopkins, 1999).

The various growth stages of maize are described in Table 1 along with visuals. Further details on corn growth and development can be found at https://www.bookstore.ksre.ksu.edu/pubs/MF3305.pdf (Accessed 20 Sept. 2018).

Climatic Requirements

Each crop has a certain set of climatic requirements for achieving maximum or optimum grain yields. These environmental conditions are described below. Deviation from optimum conditions will negatively influence yields. Maize plants are susceptible to drought and high temperatures; each year, an average of 15 to 20% of the potential world maize production is lost due to these stresses (Lobell et al., 2011). The total yield loss depends on when the stress occurs (plant growth stage), as well as the duration, intensity, and the severity of the stress. Maize is sensitive to photoperiod, any changes in photoperiod alter phenology influencing seed-set and grain yield.

Temperature

Maize is a warm weather crop and has high radiation and water use efficiency and can tolerate relatively warmer temperature compared with other temperate crops. Leaf photosynthesis of maize has an optimum temperature of 33 to 38 °C, and higher temperature inhibits the net photosynthetic rate (Crafts-Brandner and Salvucci, 2002). However, the photosynthetic rate of some maize cultivars was not reduced until 40°C (Massad et al., 2007). A typical leaf photosynthetic response to temperature is represented in Fig. 1. Gross and net photosynthetic rates are low at cooler temperatures, increase as temperature increases, and ceases when the temperature gets too high (Hopkins, 1999). Net photosynthesis is a difference between gross or total carbon assimilated through photosynthesis minus carbon used by respiration. The optimal temperature for net photosynthesis is generally lower than gross photosynthesis (Hopkins, 1999).

Germination is lower when the soil temperature is below 8 to 10 °C. Germination is faster and less variable at a soil temperature of 16 to 18 °C. At 25 °C soil temperature, maize emerges within four to 5 d, which is optimum for seedling establishment and for achieving early-season stand uniformity. Root and shoot dry weights decline by 10% for each degree increase between 25 and 35 °C (Walker, 1969). High temperature reduces both seedling establishment and growth (Weaich et al., 1996). The threshold temperature for leaf growth is about 35 °C (Dubey, 2005). Internode elongation and shoot growth are sensitive to high temperature during vegetative stages (Weaich et al., 1996). Maize plants become susceptible to high temperatures after reaching the eight-leaf stage (Chen et al., 2010). Extreme high temperature causes permanent tissue injury to develop leaves, dry quickly, a phenomenon called leaf firing. High temperature stress also induces leaf chlorosis (Karim et al., 1997).

The effects of high temperature stress are more prominent on reproductive than on vegetative growth (Stone, 2001). High temperature during the reproductive phase is associated with a decrease in yield due to a decrease in the number of grains and kernel weight. Maize yield is negatively correlated with accumulated degrees of daily maximum temperatures above 32 °C during grain fill (Dale, 1983). An increase in temperature of 6 °C during grain fill can cause a grain yield reduction of 10% in the U.S. corn belt (Thomson, 1975). Lobell and Burke (2010) suggested that an increase in temperature of 2 °C would result in a greater reduction in maize yields within sub-Saharan Africa than a decrease in precipitation by 20%. Lobell et al. (2011) reported that for each degree day spent above 30 °C the final maize yield would be reduced by 1% under optimal rainfed conditions and by 1.7% under drought conditions.

At reproductive stages, the ideal daytime temperature is from 26 to 30 °C under rainfed conditions, but maize can tolerate slightly higher temperatures under irrigated conditions. When temperatures reach 38 °C or higher water movement through the plant is affected even under irrigated condition leading to decreased growth and yield. Nighttime temperatures above 21 °C can result in increased wasteful respiration and lower dry matter accumulation. The reduced yield in maize under high temperature stress is attributed to negative impacts on sink (e.g., seed) activity rather than source (e.g., mature leaves) activity. It can also cause desiccation of tassel tissues, a phenomenon called tassel blasting, leading to decreased seed-set.

Moderate-high temperature stress occurring at early reproductive stages reduces pollen production, pollination, pollen tube growth, kernel set, and kernel weight, resulting in significant yield loss (Cheikh and Jones, 1994). It has been suggested that each 1 °C increase in temperature above optimum (25 °C) results in a reduction of 3 to 4% in grain yield (Shaw, 1983). Plants with severe leaf firing and tassel blasting lose considerable photosynthetic leaf area and produce small ears with fewer kernels and lower weight. The floral structure of maize, particularly the separation of male and female floral organs and the near-synchronous development of florets on a single ear, borne on a single stem, is extremely sensitive to temperature stress during anthesis (Johnson and Herrero, 1981). Anthesis Silking Interval (ASI) is critical for successful pollination and fertilization. Above optimal temperature can enhance ASI leading to lower

Table 2. Cardinal temperatures for different growth stages and developmental and yield processes of maize, sorghum, and pearl millet.

Crop	Stage	T_{min}	T_{opt}	T_{max}	Failure Temperature	Reference
		°C				
Maize	Up to Tasseling	9.3	28.3	39.2		Sanchez et al. (2014)
	Anthesis	7.7	30.5	37.3		Sanchez et al. (2014)
	Yield	–	20–25	35		
	Vegetative	–	25–37	–		Hatfield et al. (2008)
	Reproductive	–	18–22	–	35	Hatfield et al. (2008)
Sorghum	Vegetative	8	34	44		Alagarswamy and Ritchie (1991)
	Reproductive	–	32	40		Prasad et al. (2006)
	Yield	–	23–25	35		
	Vegetative	–	34	–		Hatfield et al. (2008)
	Reproductive	–	26–34	–	35	Hatfield et al. (2008)
Pearl millet	Germination	8–13.5	34	47–52		
	Vegetative stage	–	32–35	–		
	Reproductive stage	–	22–35	40–45	40	Djanaguiraman et al. (2018c)
	Grain or seed-set	–	22–25	> 25		
	Grain or seed growth	–	19–31	–		

fertilization. Extended exposure to a temperature beyond 32.5 °C reduces pollen germination (Herrero and Johnson, 1980). At temperatures above 38 °C, poor seed set in maize has been attributed to both a direct effect of high temperature (Johnson and Herrero, 1981) and pollen and silk desiccation due to its thin outer membranes (Schoper et al., 1986). High temperature stress during the phase of endosperm cell division and amyloplast biogenesis in maize kernels results in a reduction in the rate and duration of endosperm cell division, and thus in the number of cells formed (Jones et al., 1985). The cardinal temperatures of different growth stages of maize are presented in Table 2.

Maize is a frost sensitive plant, and frost damage happens by distorting the structural components of the cells by formation of ice crystals. Dry seeds in wet soil can withstand –5 and -10 °C for two days, but if the freezing temperature was preceded by exposure to 20 °C for one or two days, the seeds can die. After emergence, the aerial parts can be damaged or killed by frost. Later stages of growth and development are also sensitive to frost. Maize is moderately sensitive to chilling. Chilling injury is the physiological damage caused by temperatures between 0 and 12 °C. Imbibed seeds are killed by long-term exposure to chilling temperatures. The average seed or seedling mortality was 36 and 21% at 4 and 6 °C, respectively and no damage was observed at 8 and 10 °C (Miedama, 1982). Photosynthesis is decreased by low temperatures and damage increased with exposure time. Under cool and bright weather conditions, seedlings often develop chlorotic symptoms. This chlorosis is also the result of a combination of low temperature and high light intensity but differs from the chilling injury. At 13 to 14 °C maize seedlings show chlorotic symptoms and at 16 to 17 °C they gradually "re-greened". Chlorotic seedlings rapidly greened when the temperature was > 26 °C, indicating that the chlorophyll synthesizing system was not damaged at chilling temperatures. The inhibition of chlorophyll accumulation at low temperature and high light intensity was due to a higher rate of photooxidation of the chlorophyll compared to the rate of biosynthesis. Once chlorophyll is complexed with chloroplast lamellae, it is protected from photooxidation (Anderson and Robertson, 1960). Chlorotic plants may recover at warmer temperatures, but growth is somewhat inhibited in comparison with non-chlorotic plants (Miedama, 1982).

Heslop–Harrison (1961) reported that plants exposed to short photoperiods and cool nights (10 °C) in a greenhouse setting showed male sterility during flowering. The degree of male sterility is dependent on the stage of tassel development during cold temperatures. In the primary roots of maize seedlings, the rates of cell division and cell elongation were equally reduced when the temperature was decreased from 30 to 15 °C (Erickson, 1959). Water flows through the epidermis and cortex of the excised primary roots at 10°C was about one-fourth of that at 20 °C (Ginsburg and Ginzburg, 1971). This reduction could be due to an increase in the viscosity of water, which can decrease water flow and

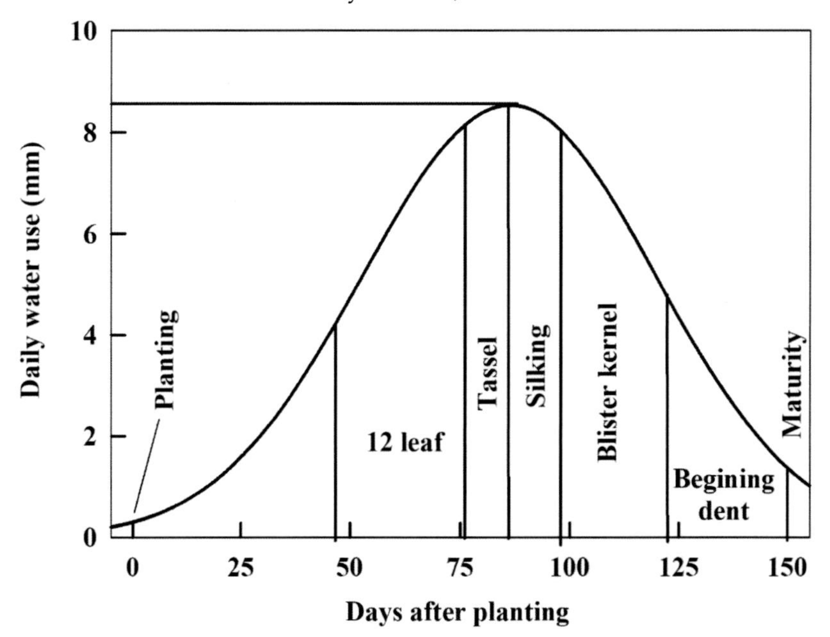

Fig. 2. Predicted daily water use (mm) of maize.

Table 3. Minimum temperature requirement for various physiological process in maize.

Process	Temperature Range (°C)
Chilling injury to seeds	0–5
Chilling injury to seedlings	0–5
Germination	6–10
Rate of emergence of leaf	8–15
Root growth	8–15
Water uptake	< 12
Net photosynthesis	10–15
Mobilization of carbohydrates	5–10
Shoot growth	8–15

can decrease water permeability of the roots. Lowering the temperature from 20 to 10 °C decreases relative growth rates (RGR) due to a higher decrease in net assimilation rates (NAR) compared with leaf area ratio (LAR) (Rajan et al., 1973). The RGR is the product of NAR and LAR. Minimum temperature requirement for various physiological processes in maize are shown in Table 3.

Rainfall

Water availability and drought episode are the important factors affecting maize yield. Drought stress is expected to worsen, according to optimistic climate change scenarios; and during the last few decades, major drought events have been recorded, and this event is projected to intensify in most parts of maize growing regions. Maize is grown mostly in regions having annual rainfall between 600 and 1100 mm but is also grown in areas having rainfall of about 400 mm. Maize requires 600 to 800 mm of water from sowing to maturity. Approximately 15 kg of grain is produced for every millimeter of water used, and a total of 250 L of water is consumed by a maize plant to maturity (Du-Plessis, 2003). A yield of 3100 kg ha^{-1} requires 350 to 450 mm of rain per season. The predicted water use pattern of maize is shown in Fig. 2. Soil moisture influences the speed of germination. Germination and emergence are rapid in moist soil. Water requirement of maize is low at early growth stages, then reaches its peak at reproductive growth stages, and again lowers at terminal growth stages. After germination, drought stress can significantly reduce the plumule and radicle growth, resulting in unusual seedling growth (Gharoobi et al., 2012). At the seedling stage, drought stress reduces shoot elongation more than root elongation (Khodarahmpour, 2011). Drought stress during vegetative stages, especially during V1 to V5, reduces leaf turgor resulting in leaf rolling, curling, or folding (Du-Plessis, 2003). Leaf folding reduces the leaf area exposed to sunlight, resulting in a reduced light interception which decreases photosynthetic activity. Cell division and cell elongation are reduced under drought stress which reduces leaf area. Reduction in leaf area under drought stress conditions is an adaptive strategy by maize plants. During reproductive growth stages, maize requires 8 to 9 mm of water d^{-1} $plant^{-1}$. The two weeks before and two weeks after pollination are the most crucial four weeks regarding maize water requirement. Pollen and silk development, pollination, embryo development, endosperm development and kernel development were the most sensitive growth stages to drought stress (Aslam et al., 2015). The amount of water available during silk growth substantially influences when silks emerge, their rate of growth, the duration of their receptivity, and their ability to supply water and nutrients to support pollen tube growth and fusion of the gametes. Maize plants grown under water stress during pollination produce ears with barren portions of the cob. Ill-filled or barren cob indicates mature ovules were not properly fertilized (Aslam et al., 2015). These unfertilized ovules begin to disintegrate and disappear before the ear reaches physiological maturity. Translocation of photosynthates to the reproductive parts rather than roots for their extensive elongation is the most probable reason for the increased susceptibility of maize during reproductive growth stages under drought stress (Setter et al., 2011). Apart from this, the monoecious nature and synchronized growth of floret in an ear make it highly susceptible to drought. If moisture stress occurs during their development, all will be affected resulting in decreased yield. Drought stress at flowering and pollination can result in unfilled kernels on the cob (Aslam et al., 2015). This can reduce grain yield by 6 to 8% each day the plant is water stressed. If the plant is water stressed after flowering, kernel size is reduced, and the risk of mycotoxin contamination increases.

About 18% of the total maize production area is affected by floods and water logging, resulting in a reduction of 25 to 30% of grain yield (Zaidi and Singh, 2001). Under waterlogged soil conditions, the diffusion rate of gases is estimated to be 100 times lower than that in the air, causing reduced gas exchange between root tissues and the atmosphere. Maize is especially sensitive to waterlogging during early seedling stages to tasseling (Zaidi et al., 2004). In maize, waterlogging was found to be associated with root decay and wilting. Under extended waterlogging of more than three days, the formation of lysigenous aerenchyma in the cortical region of roots and brace root development on above ground nodes has been observed in waterlogging tolerant genotypes (Mano and Omori, 2007). Production of adventitious roots with aerenchyma tissue under the waterlogging condition is not a constitutive trait but is an adaptive trait in maize. Aerenchyma is formed through ethylene-induced cell lysis (Vartapetian and Jackson, 1997), which provides a diffusion path of lower resistance for the transport of oxygen from aerial parts of the newly developed brace root to the roots present under severe anoxic conditions (Laan et al., 1989).

Photoperiod

Photoperiodism is the response to the length of the day, which enables a plant to adapt to seasonal changes in its environment. It is the duration rather than the quantity of light in a daily cycle that is more important in regulating flowering in certain plants. The length of the dark period rather than the comparative length of light and darkness is the decisive factor in flower induction. Maize originated in tropical regions but has developed into two major types of tropical and temperate maize germplasm through long-time artificial domestication. The supposed ancestor of maize, teosinte (*Zea mays* L. subsp. *parviglumis*), likely evolved photoperiod sensitivity to synchronize its reproductive phases to the wetter, short-day growing season (Campos et al., 2006). After post-domestication evolution of maize, it was spread from tropical to temperate regions, requiring adaptation to longer daylengths resulting in major genetic differentiation between temperate and tropical maize and substantially reduced photoperiod sensitivity of temperate maize (Gouesnard et al., 2002). The genetic homogeneity in those groups of maize has increased the vulnerability to biotic and abiotic stresses and limits future gains in productivity. Use of tropical maize, including inbred lines and landraces, has been promoted to increase genetic diversity in temperate breeding programs. Unfortunately, most tropical germplasm sources express numerous unfavorable phenotypes when grown in temperate locations. A major issue with the poor adaptability of tropical maize to temperate environments is the response of tropical maize to long day photoperiod. Integration of tropical and temperate maize germplasm into existing breeding programs is a promising way to improve maize agronomical traits, but many unfavorable traits can be associated. Tropical-adapted maize is more sensitive to photoperiod changes than temperate-adapted maize genotypes. In tropically-adapted maize lines, floral induction occurs in response to a developmental cue rather than daylength (Chen et al., 2014).

Maize is grown from 50 °N latitude through the temperate, subtropic, and tropical regions to 40 °S latitude. Effects of photoperiod changes on maize have been widely investigated such as anthesis-silking interval (ASI), plant height, total number of leaves, internode length, internode number, flowering time, identification of photoperiod sensitive chromosomal regions, and expression of photoperiod response related genes (Chen et al., 2014 and references therein). Tropical maize exhibits delayed flowering time, increased plant height, and a greater total leaf number when grown in temperate latitudes with daily dark periods < 11 h (Allison and Daynard, 1979). Short nights in temperate latitudes during summer induce late differentiation of tassels and tremendous vegetative growth in tropical maize. Late flowering delays pollination and seed-set and reduces the grain-filling period (before the first frost). Isolation of day-neutral genotypes and incorporation of insensitivity into diverse ranges of races, synthetics, and improved varieties would encourage more active utilization of new germplasm. Temperate maize populations that were introduced to long-day environments throughout North and South America during the spread of maize were selected to be insensitive to photoperiod (Coles et al., 2011). Lewis and Goodman (2003) suggested crossing improved exotic sources (tropical maize) with improved, adapted sources (temperate maize), and selection within this population can result in agronomically superior temperate inbred and hybrid lines. The other approach would be to identify photoperiod-insensitive exotic sources, crossing them with improved, adapted sources (temperate maize) and selection initiated from this population.

Crop Management

Maize grows best in a fertile and well-drained loamy soil. Maize is relatively well adapted to a wide range of soils with pH 5.0 to 8.0. Maize is not acid tolerant and is moderately sensitive to salinity. Sandy, gravelly and shallow soils increase the risk of drought; and good yields are unlikely on these soils unless there is supplementary irrigation or favorable rainfall distribution during the growing season. The choice of hybrid could have a significant impact on yield. Adequate soil moisture at the time of sowing is required. If the soil is too wet or too dry, or the maize seed is planted too deep (without soil moisture), seeds will be slow to emerge or fail to germinate. The seed should be sown at a depth of 3 to 5 cm in moist soil. Two soil preparation methods before sowing are commonly practiced in maize production (conventional tillage and no-tillage systems) with numerous variations. In conventional tilled systems, the land is commonly plowed, disked, or harrowed once or twice at a depth up to 30 cm before planting. In no-tillage or minimum tillage systems, stubble is retained from the previous crop, and maize is sown directly into the previous crop's stubble. The benefits of retaining crop residues or mulching are: reduced crusting of the soil surface; reduced surface evaporation; the reduced emergence of weeds; reduced soil erosion; reduced sandblasting damage to seedlings; improved rainfall infiltration; improved retention of soil carbon and increased production when implemented over a long period of time. Mulching has the potential to reduce the risk of crop failure as a result of drought (Verhulst et al., 2010). However, the application of conservation agriculture will vary with climate, biophysical soil characteristics, system management conditions, and farmer circumstances. Crop residue retention at the soil surface reduces soil evaporation compared with bare soil. The improved infiltration and reduced evaporation mean that more water is available for crops under conservation agriculture than for conventional tillage and no-tillage with residual removal. Gicheru (1994) showed that crop residue mulching resulted in more moisture in the soil profile (0–120 cm) throughout two crop periods over two years as compared to conventional tillage and tied ridges in a semiarid region. A high soil water content in conservation agriculture can buffer for short period of drought during the growing season.

Plant density and row spacing are critical factors to obtain maximum yield. The plant density factor varies with the target yield and the water deficit of each environment. Recent studies demonstrated a strong relationship between yield and plant density for maize, with plant density increasing as the maximum yield increases for maize (Assefa et al., 2016; 2017). Nitrogen (N) and phosphorus (P) and potassium (K) are commonly needed to be applied before planting if levels are low; however, secondary and micronutrients may also need to be applied if limiting. As far as possible, apply NPK fertilizers as per soil test recommendation and agro-ecologically appropriate fertilizer response functions. Total nutrients required to produce 12.0 Mg ha^{-1} of maize grain and 23.0 Mg ha^{-1} of total biomass include 286 kg N, 114 kg P_2O_5, 202 kg K_2O, 59 kg Mg, 26 kg S, 1.4 kg Fe, 0.5 kg Mn, 0.5 kg Zn, 0.1 kg Cu, and 0.08 kg B (Bender et al., 2013; Ciampitti et al., 2013a,b). Nitrogen fertilizer should be applied in split applications, matching crop demand, to avoid leaching, volatilization, and immobilization whereas full recommendations of P and K are often applied basally before sowing. Erosion, immobilization, and adsorption are the primary factors reducing labile forms of P and K for plant uptake. Proper soil nutrient status through increased root growth and yield improves water-use efficiency. Inversely, proper soil water status improves nutrient-use efficiency by improving mass flow, diffusion, and root interception. Finally, fertilizer application should be performed after weeding so that weeds do not benefit from applied fertilizers.

Pests and diseases incidence is significantly influenced by environmental conditions. The disease occurs when a virulent pathogen, susceptible host, and favorable environment are present. Global climate change has the potential to modify host physiology and resistance, and alter both stages and rates of pathogen development. Environmental conditions controlling disease development include rainfall, relative humidity, temperature, and sunlight. Changes in these factors under climate change are highly likely to affect the prevalence of diseases and the emergence of new diseases. The disease infection cycle includes inoculum survival, infection, latency period, production of new propagules, and dispersal, and all of these are highly influenced by environmental conditions. In general, fungi require high relative humidity or moist leaf surfaces for infection. Increased temperature reduces the latency period (generation time), resulting in a higher number of generations per season. The pathogens *Cercospora zeae-maydis* and *Cercospora zeina* that cause gray leaf spot in maize are highly sensitive to relative humidity. Under dry conditions (< 80% relative humidity) the pathogen growth stops, and hence, the infection stops (Thorson and Martinson, 1993). The prevalence of *Aspergillus flavus* in maize grain is higher in warmer environments (> 25 °C) compared to cooler environments (20 to 25 °C) (Shearer et al., 1992). As for the Fusariums, *F. graminearum* prevails in temperate maize-growing environments, whereas *F. verticillioides* and *F. proliferatum* are more widespread in tropical and subtropical environments (Miller, 1994). Increasing temperatures within maize-growing regions are highly likely to change the geographical distribution and predominance of *F. verticillioides*, particularly in cooler regions where it will replace *F. graminerum*. This shift in *Fusarium* species will result in a change in mycotoxins, from deoxynivalenol and zearalenone to fumonisin (Torres et al., 2007). Insect damage has been shown to follow *Fusarium* or *Aspergillus* ear rots (Miller, 2001). The incidence of the European maize borer increased *F. verticillioides* disease and fumonisin concentrations but not *F. graminierum* (Lew et al., 1991). Increased incidence of drought favors insect proliferation and herbivory, leading to increased incidence and severity of insect related damage as well as aflatoxin and fumonisin mycotoxins in maize. High temperatures have the potential to change the geographical distribution of crops, which in turn can lead to an expansion of the geographical distribution of pests and their associated pathogens, resulting in a change in the geographical distribution of diseases.

Maize can be attacked by a wide range of insects. However, maize stalk borer [*Chilo partellus* (Swinhoe)], pink stem borer [*Sesamia inferens* (Walker)], sugarcane leafhopper [*Pyrilla perpusilla* (Walker)], shoot bug [*Peregrinus maidis* (Ashmead)], armyworm [*Mythimna separate* (Walker)], fall armyworm (*Spodoptera frugiperda*), shoot fly (*Atherigona* spp.), corn leaf aphid [*Rhopalosiphum maidis* (Fitch)], cob borer [*Helicoverpa armigera* (Hübner)], and termites [*Macrotermes* spp. and *Odontotermes* spp.] are some of the major yield-reducing pests of maize from seedling emergence to harvest. Severe yield losses (35–80%) due to stem borer have been recorded in maize. A daytime temperature of > 13 °C and relative humidity of > 70% favors stem borer infestation, and every 1 °C reduction in nighttime minimum temperature decreases stem borer damage by 1.7% at 30 d after sowing. Shoot fly incidence is higher when daytime maximum temperatures are 30 to 34 °C, nighttime minimum temperatures are 24 to 25 °C, and relative humidity is 40 to 75%. However, low incidence has been observed under a daytime maximum temperature of 24 to 25 °C, a nighttime minimum temperature of 21 to 22 °C, and a higher relative humidity of 80 to 86% during the second and third week after sowing (Dhillon et al., 2013). Maize weevils, *Sitophilus oryzae* (L.) and *Sitophilus zeamais* Motschulsky, cause serious grain losses both under field and storage conditions in tropical countries. Most maize varieties and hybrids released for cultivation are susceptible to *C. partellus* during the vegetative stage and maize weevils, *Sitophilus* spp., under field and storage conditions (Hossain et al., 2007). Maize weevils can cause losses by direct consumption of grains or indirectly by creating a favorable environment for the establishment of other pests and/ or fungi during storage and by reducing quality (Tefera et al., 2011). Up to 80%

losses have been reported to occur for untreated maize grain stored in traditional structures depending on the period of storage (Boxall, 2002). In 2016, fall armyworm, a new pest in certain parts of Africa, caused significant yield losses, to the extent that several countries declared a state of emergency. Environmental conditions play a critical role in the spread of new pests into new regions. Thus, understanding the agroclimatology of crops and pests is important to develop best management practices.

Genomic Resources for Maize

The Maize Genetic and Genomics Database (MaizeGDB) (http://www.maizegdb.org) (Lawrence et al., 2008) is the online home for maize researchers because it has information on maize community news, genetic maps, mutations, and alleles, and direct access to requesting seeds. It also has a well-curated and annotated genome browser complete with *Robertson's Mutator* (*Mu*) and Activator/Dissociation transposon family (*Ac/Ds*) insertional mutations, multiple expression analysis tools, publications, and pathways databases. Gramene (http://www.gramene.org) (Monaco et al., 2014) is an online resource dedicated to genome-wide comparisons.

Sorghum

Sorghum [*Sorghum bicolor* (L.) Moench] is grown globally in warmer and slightly drier climatic conditions. World collection of sorghum consists of 235,711 accessions, housed in national and international gene banks. The international center ICRISAT has one of the largest germplasm collections in its gene bank (> 37,949 accessions, predominantly landraces from semiarid tropics) and has a global mandate for sorghum research and capacity building activities (Upadhyaya et al., 2009). The origin and early domestication of sorghum took place in northeastern Africa. The earliest known record of sorghum comes from an archaeological dig at Nabta Playa, near the Egyptian-Sudanese border, dated 8000 B.C. After domestication, sorghum spread from the highlands of Ethiopia to the semiarid Sahel, then to India, China, and Australia (Doggett, 1988). World cultivated sorghum has been classified into five basic races: bicolor, guinea, caudatum, kafir, and durra, and ten hybrid races which are combinations and stable intermediate hybrids of the basic races based on mature spikelet types (Harlan and Dewet, 1972). These include guinea-bicolor, guinea-caudatum, guinea-kafir, guinea-durra, caudatum-bicolor, kafir-bicolor, kafir-caudatum, kafir-durra, durra-bicolor, and durra-caudatum. It is believed that the four major forces of evolution (selection, mutation, random genetic drift, and migration) have promoted the wild sorghum complex *S. bicolor* subspecies *verticilliflorum*. This wild complex consisting of four races (*aethiopicum, arundinaceum, verticilliflorum,* and *virgatum*) is presumed to be the wild progenitor of cultivated sorghum (Doggett, 1988). According to Snowden (1936), the wild race *aethiopicum* gave rise to the races *bicolor* and *durra, arundinaceum* to *guinea,* and *verticilliflorum* to *kafir.* The race *caudatum* is believed to be a later domesticate, having been segregated out of the race *bicolor.*

Sorghum is the fifth largest cereal grain after wheat, maize, rice, and barley (*Hordeum vulgare* L.). During 2016, Africa and the Americas together contribute 75% of the total world sorghum production, while Asia contributes about 20%. India and Nigeria are the largest producers; each contributes ~13% of global

Table 4. Description of growth and developmental stages of grain sorghum (Ciampitti, 2015).

Growth stage		Description
Stage 0	Emergence	Occurs when the coleoptile is visible at the soil surface.
Stage 1	Three leaf stage	Occurs when the collars of three leaves can be seen without dissecting the plant.
Stage 2	Five leaf stage	Occurs when the collars of five leaves can be seen without dissecting the plant in about three weeks after emergence.
Stage 3	Growing point differentiation	At this stage, the growing point of the sorghum plant changes from vegetative to reproductive.
Stage 4	Final leaf visible in the whorl	At this point, all except the final three to four leaves are fully expanded representing approximately 80% of the total leaf area potential.
Stage 5	Boot stage	At this stage, all the leaves are fully expanded, providing maximum leaf area and light interception. The head has reached full size and is encompassed by the flag leaf sheath.
Stage 6	Half bloom	This stage is defined as when half of the plants in a field have started to bloom.
Stage 7	Soft dough	At this stage, the grain has a dough-like consistency and grain fill is occurring rapidly.
Stage 8	Hard dough	By this stage approximately three-fourths of the grain dry weight has been attained.
Stage 9	Physiological maturity	At this stage, the maximum total dry weight of the plant has been reached.

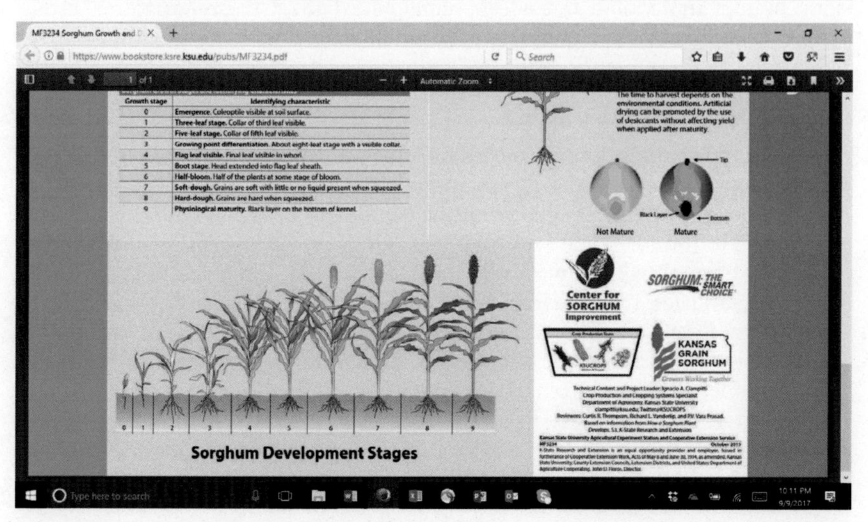

sorghum production. The average productivity of sorghum varies from 0.4 t ha^{-1} in Niger to 4.5 t ha^{-1} in Argentina (FAO, 2016). It is the only viable food grain crop for many of the world's most food insecure people and an important food source in Africa and India. Sorghum is moderately tolerant to drought and high temperature compared with other cereal crops. Over the past few years, sorghum production has increased steadily in Africa largely due to increases in land area under cultivation and slight increases in productivity. The average sorghum yield remains below 1 t ha^{-1} due to the use of traditional farming practices such as low level of inorganic fertilizer use, low or no pesticide use, and the use of traditional varieties or landraces (Waddington et al., 2010). However, where intensive agriculture is practiced, yields are much higher.

In the United States, sorghum is classified into four classes based on tannin content and color stipulated by the USDA Federal Grain Inspection Service (FGIS); namely sorghum, tannin sorghum, white sorghum and mix sorghum (Balota, 2012). The sorghum group is low in tannin content due to the absence of a pigmented subcoat and contains less than 98% white sorghum and not more than 3% tannin sorghum. The color of the seed coat in this class may appear, white, yellow, pink, orange, red or bronze. Tannin sorghum has high tannin content due to a pigmented subcoat and contains no more than 10% non-tannin sorghum. The color of the seed coat is usually brown but may be white, yellow, pink, orange, red, or bronze. The third group, white sorghum, has low tannin content due to the absence of a pigmented subcoat and contains no more than 2% of other classes. The color of the seed coat is white or translucent and includes sorghum containing spots that cover 25% or less of the kernel. The mix sorghum group does not meet the requirements for any of the other classes (Balota, 2012; Rooney and Miller, 1982).

Based on use, sorghum can be classified into different types: grain sorghum, sweet sorghum, forage sorghum, high biomass sorghum, and broom sorghum (Rooney, 2014). Grain sorghum in developing countries is primarily used as human food, while in developed countries it is used as animal feed. Sorghum is also used as raw materials for alcoholic beverages. Sweet sorghum is used for the production of sweetener syrup and biofuel production. Forage and high biomass sorghum are used for animal feed and bioenergy production. Broom sorghum is used as a material to make brooms. In recent years, there is an immerging focus on production of dual-purpose sorghum with focus on both grains for human consumption and biomass for animal feed. Use of grain sorghum as human food is critical and increasing, especially for gluten-free diets. Also, there are several benefits of sorghum grain such as high antioxidants, carotenoids, and low glycemic index.

Growth and Development

The developmental stage description and images are shown in Table 4. These stages are described from 0 to 9 (vegetative stages from 0 to 4; and reproductive stages from 5 to 9). The more detailed description on growth and development and physiology of sorghum is provided by Djanaguiraman et al. (2018a, b) and Roozeboom and Prasad (2018).

Seed germination (Stage 0) begins with the absorption of moisture. The radicle grows downward into the soil and forms the first primary seminal roots. The mesocotyl elongates from the seed, and a first node is formed at the base of the coleoptile just below ground level. The coleoptile grows and emerges above ground and remains as a sheath at the base of the seedling. Vegetative growth is from seedling (Stage 1) to flag leaf stage (Stage 4). Leaves and nodes develop at the rate of one per 3 to 6 d under optimum temperatures. Vegetative stages are described similar to those of maize with the appearance of leaf collars. The three-leaf stage (Stage 1) occurs when the collars of three fully developed leaves are visible, and the five-leaf stage (Stage 2) occurs when the collars of five fully developed leaves are visible. The sorghum root system consists of three types of roots: primary (seminal), secondary (adventitious), and nodal (crown) roots. Seminal roots develop directly from elongation of the radicle and are easily identified by

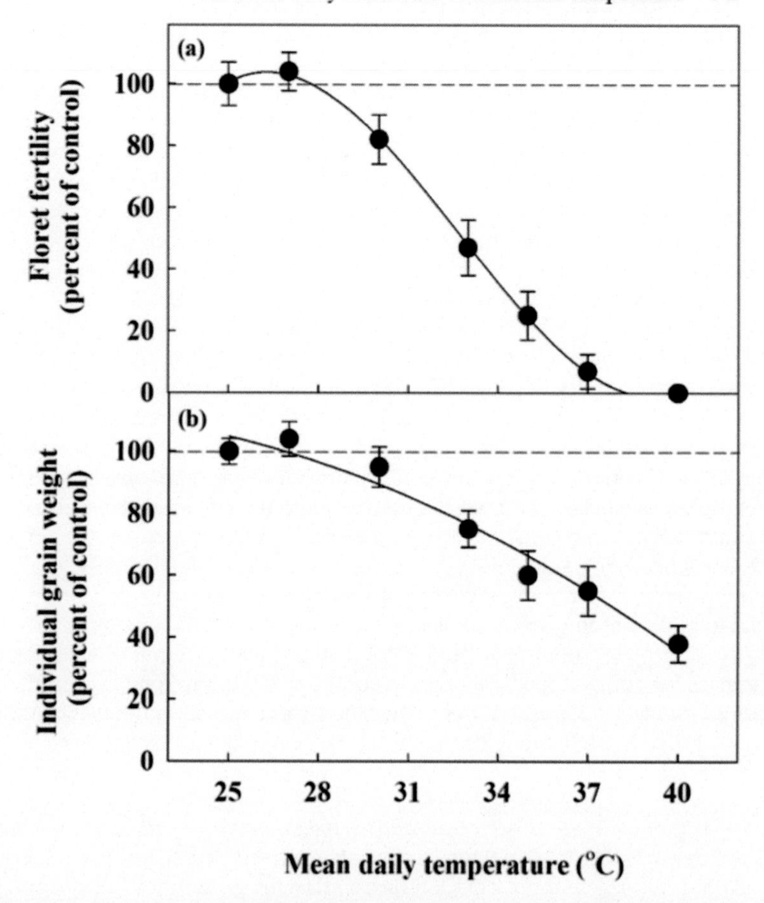

Fig. 3. Impact of different mean daily temperatures (°C) on (a) floret fertility; and (b) individual grain weight of grain sorghum, expressed as percent of control (25°C). (Adapted and redrawn from Prasad et al., 2015).

their continuity with the base of the culm. Seminal roots die subsequently after about 4 to 5 d. Secondary roots develop from the first and second internode on the mesocotyl and are branched and mainly used for absorption and uptake of water and nutrients. Nodal roots initiate from the base of the stem about 30 to 40 d after sowing. Nodal roots are thick above the ground level and thin in the soil, and produce many lateral roots.

The plant remains in a vegetative phase for about 30 to 40 d, during which the plant forms about 12 to 18 leaves. Each leaf consists of a sheath and a blade or lamina. The sheath is attached to the node and surrounds the internodes and sometimes the node above it. At stage 3, growing point differentiation occurs which is the time when the growing point comes off the soil and changes from producing leaves to forming heads. The top most leaf is called boot or flag leaf, which is short and broad. The stage at which the flag leaf is visible is termed Stage 4.

The sorghum inflorescence is called panicle and varies morphologically from compact to open, based on the length of the rachis, branch length, distance between whorls, and the angle of branching in the panicle. After a complete unfolding of the flag leaf, the peduncle elongates and forces the panicle out of the boot leaf. At Stage 5, the panicle exerts out of the boot leaf. Flowering (Stage 6) in sorghum occurs after 40 to 70 d depending on the cultivar and environment, i.e., mainly temperature. Anthesis (flower opening) starts about 2 d after the complete emergence of the inflorescence (panicle) from the boot leaf. It takes about 4 to 13 d for completion of flowering in the panicle depending on the cultivar, panicle size, and temperature. Maximum flowering generally occurs in the middle of this period about 3 to 6 d after the start of anthesis.

Anthesis in sorghum starts around midnight and proceeds up to 10 a.m. depending on the cultivar and environmental conditions (temperature and humidity). Maximum anthesis occurs between 6 and 8 a.m. The floret opens as a result of a swelling of the two lodicules at the base of the florets, which exerts pressure on the glumes forcing them to open in about 10 min. Anthers protrude first, followed by the stigma; a condition termed protoandry. Anthers shed pollen (dehiscence) just before or shortly after anthesis between 6 and 7 a.m., depending on cultivar and weather. On a dry and clear day, dehiscence occurs soon after sunrise, but on a cool and damp day, it may extend until 10 a.m. Pollen in the anthers remains alive for 3 to 6 h. After dehiscence, pollen grains are alive only 1 to 5 h, or even shorter depending on cultivar, temperature, and humidity. The stigma is receptive for about 10 d. Sorghum is a self-pollinating plant with cliestogamy. Pollination occurs with the shedding of pollen grains on the stigma. Pollen grains start to germinate immediately after contact with the stigma and develop pollen tubes through the style. Fertilization is complete in 2 h.

Organ differentiation and deposition of starch occur about 10 to 12 d after fertilization and embryo formation. The endosperm continues to grow until the seed reaches physiological maturity (about 30 d). The development of the seed (grain) follows a sequence of stages comprising a milk stage, early or soft dough stage (Stage 7), and a late or hard dough stage (Stage 8) to physiological maturity (Stage 9). At physiological maturity, a dark brown callus tissue is formed (black layer) at the base where the seed is attached to the spikelet (hilar region). This callus tissue stops the translocation of nutrients from the plant to the seed. Seeds typically contain 25 to 30% moisture and are fully viable at physiological maturity. For safe storage, seed moisture should be between 10 and 12%.

Climatic Requirements

Temperature

Sorghum is grown in a wide range of climates from the equator to over 50 °N lat. and 40 °S lat. and at altitudes from sea level to 1000 m. Sorghum is well adapted to hot, semiarid tropical environments and is relatively more drought and high-temperature tolerant compared with other cereal crops. The yield potential of sorghum is similar to other cereals; however, at present global yield averages are low as the crop is often grown during the hottest seasons of the year, in areas with limited rainfall, and in marginal or poor soils with low inputs applied to local landraces.

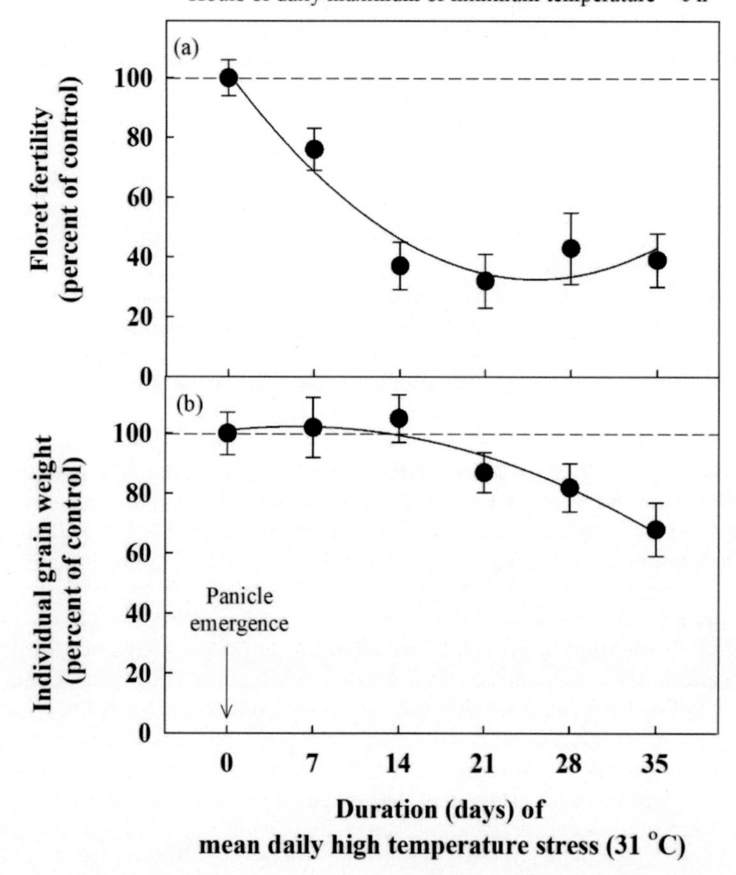

Fig. 4. Impact of high temperature stress (mean daily temperature = 31 °C) for different durations imposed at panicle emergence on (a) floret fertility; and (b) individual grain weight of grain sorghum, expressed as percent of control (25 °C). (Adapted and redrawn from Prasad et al., 2015).

Average soil temperature for seed germination ranges from 4.6 to 16.5 °C. Optimum soil temperature for germination range from 21 to 35 °C, with a maximum temperature of 40 to 48 °C, beyond which is lethal for the germination process. For leaf appearance, minimum temperature ranges from 7 to 12 °C, with an optimum between 26 and 34 °C. Temperatures above optimum decrease the rate of leaf appearance. Leaf elongation linearly increases from 13 to 32 °C and declines above 34°C. Minimum and maximum temperatures for leaf elongation are 15.5 and 43 °C, respectively. The base temperature for panicle initiation is 6.8 to 10.4 °C and the optimum are from 25.9 to 27.3 °C. Response to temperatures above or below the optimum is linear and delays panicle initiation. Eastin et al. (1976) observed that night temperatures of 5 °C above the optimum temperature reduced grain yield by

25 to 33% and 10 °C above the optimum reduced yield by 50%. Floral differentiation is the most sensitive phase to temperatures above optimum.

Grain yield is a product of grain number per unit area and grain weight. Grain number per unit area is determined by number of plants and number of grains per plant. Grain number per plant is determined during panicle initiation and flowering stage (Stages 3 through 5). High or low temperatures during these stages can significantly influence the viability of gametes and the ability of flowers to set seed. High temperatures above 32/22 °C (daytime maximum/nighttime minimum; daily mean = 27 °C) cause abortion of florets resulting in lower seed-set and grain numbers. A recent study by Prasad et al. (2015) has shown that daily mean temperature significantly affects floret fertility and individual seed weight (Fig. 3). Floret fertility decreased from about 100% at 25 °C mean daily temperature to 0% at 37.4 °C. High temperature stress (25 to 40 °C) during the grain filling stage for 10 d caused significant decreases in individual seed weight (Fig. 3). Individual grain weight decreased by 62% from 25 to 40 °C mean daily temperatures (100% to ~38% at 40 °C). Constant high temperature (36/26 °C; mean daily = 31 °C) during panicle emergence for the varied duration indicated that high temperature stress for 7 and 14 d periods decreased floret fertility to 76 and 37%, respectively. Individual grain weight under high temperature stress for 0, 7, and 14 d of high temperature stress were similar; however, a further increase in the duration of stress to 21, 28, and 35 d decreased individual grain weight to 87, 82, and 68%, respectively (Fig. 4). Flowering and 10 d before flowering are the two stages of grain sorghum reproductive development most sensitive to high temperature stress (Prasad et al., 2008; 2015; 2017). High temperature stress during these stages in the growth cycle causes a maximum reduction in seed-set, seed number, and seed yield. During seed filling, early seed-filling periods are relatively more sensitive to high temperature stress compared with later periods of seed filling. Seed yield losses during post-flowering stages are mainly due to decreases in seed size. The maximum potential grain number is set by the end of the flowering stage, and final grain yield is dependent on a number of filled grains and individual grain weight. Grain weight is a product of two factors: the rate of grain filling and the duration of the grain filling process. Both these processes are highly sensitive to temperature. High temperatures (> 36/26 °C) decrease both the rate of grain filling and grain filling duration, resulting in lower yields. The cardinal temperatures for various growth stages of sorghum are presented in Table 2.

In temperate zones, optimal growth can be constrained by low temperature stress in early spring. This constraint often determines the planting date in these areas. The germination and seedling establishment phase of sorghum is especially sensitive to cold temperature, which can negatively impact plant population and grain yield. Early-season low temperature stress can significantly reduce seedling growth and delay the time to flowering and maturity. Adams and Thompson (1973) observed that grain yield increased by 10% when they covered the soil with clear plastic, which kept the soil temperature about 2 °C higher during the seedling emergence stage. There is a significant genetic variation for cold tolerance during germination (Upadhyaya et al., 2016). There are significant focus and progress on enhancing cold tolerance during seed germination and early season vigor for sorghum genotypes. Increasing cold tolerance during germination and early seedling growth will allow sorghum to be planted early in the season in temperate zones (particularly in the mid-western region such

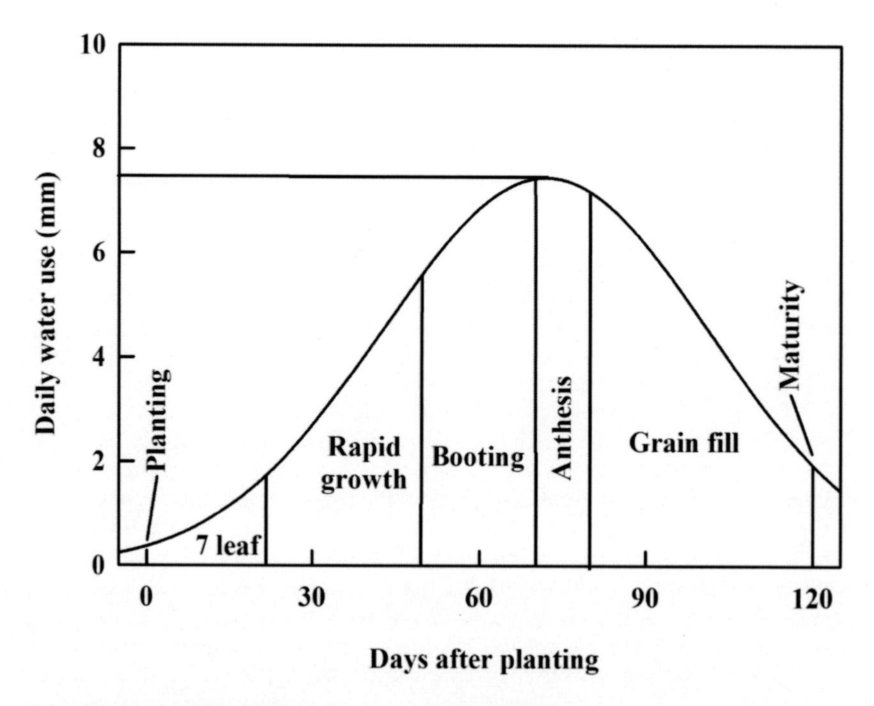

Fig. 5. Predicted daily water use (mm) of grain sorghum.

as Kansas, USA). Early planting dates would allow for the earlier establishment and faster growth to escape high temperatures and dry conditions during the flowering stage, permitting for a longer duration of the grain filling.

Paul (1990) showed that for sorghum a minimum temperature of 16 °C is necessary for all physiological processes to occur. Low temperature stress caused a significant decline in photosynthetic capacity and rate was more severely affected than the respiratory rate (Ercoli et al., 2004). After long exposure to low temperature (8 °C), plants were able to adapt by restoring photosynthesis; however, at lower temperatures, growth ceased. Low temperatures (< 10 °C) cause reduction of plant height, leaf area, and dry matter accumulation, possibly due to a reduction in chlorophyll synthesis and consequently photosynthesis. If average daily temperatures are below 20 °C, there is an extension of the growth period from 10 to 20 d for each 0.5 °C reduction in temperature. Low temperature during anthesis can affect panicle development and cause spike sterility through the effect on meiosis, resulting in pollen sterility. Structural and functional abnormalities in the reproductive organs of cold-stressed sorghum plants, as well as failed fertilization or premature abortion of the embryo, have been observed. Although both the anther and stigma have fully extended, low temperature may impact receptivity of the stigma, germination, and growth of the pollen tube or fertilization resulting in reduced seed-set and a lower number of seeds per panicle (Downes and Marshall, 1971). Similarly, cool temperatures ($< 27/22$ °C) during reproductive growth (early booting to maturity) significantly decrease seed numbers and grain yield as a result of lower seed-set. A minimum temperature below 10 °C during pre-flowering stages (23 to 27 d before flowering) significantly decreases pollen viability, leading to lower seed-set and yields. Similarly, exposure to cooler temperatures of 15 to 13 °C at the flowering stage for the duration of 10 d significantly decreased all components of yield, delayed maturity by about 10 d and was irreversible (Maulana and Tesso, 2013).

Rainfall

Maximum sorghum yield can be achieved if the soil moisture level is 600 to 700 mm during the crop cycle; however, a satisfactory yield can be obtained at a soil moisture level of 400 to 500 mm. Sorghum yield will be high if annual rainfall is > 800 mm and distributed during panicle initiation, flowering, and grain filling stages. Seed germination and seedling establishment require adequate soil moisture, critical to maintaining an optimum plant density. Water uptake by sorghum gradually increases from emergence, reaching a peak at flowering and gradually decreases until maturity. Sorghum plants require water use of about 3 to 5 mm ha^{-1} daily depending on growth stage. Maximum daily water use is highest during booting to seed-set (7 to 10 mm d^{-1}). Water use and requirement decline rapidly after the soft dough stage. The predicted daily water use of sorghum is shown in Fig. 5. About 90% of total water used by sorghum is extracted from a soil depth of 0 to 1.65 m. The rooting depth of sorghum, however, can extend to about 2.5 m. Water stored below 1 m deep is an important source of stored water at the end of growing season. Water use of sorghum is lower under low- than in high- evapotranspiration demand (Tolk and Howell, 2003). Even though sorghum is a C_4 plant, elevated atmospheric CO_2 has been reported to slightly reduce its water use due to partial closure of stomata. Soil management practices affect evapotranspiration by altering the heat balance at the soil surface and altering the temperature exchange rate between the soil and the atmosphere. Reduced tillage systems (minimum tillage or no tillage) decrease incoming heat energy, which is capable of evaporating water, changing the exchange rate between soil and atmosphere and trapping already vaporized water. Grain yield response to the water supply is greater with no tillage than conventional tillage. Planting date and density can also affect sorghum water use by altering canopy development. Optimum plant density can minimize the occurrence of drought stress. Sorghum yield is dependent on water supply (soil water at planting and in-season precipitation). Stone and Schlegel (2006) indicated that every millimeter of water above 100 mm resulted in an additional 16.6 kg of grain ha^{-1}. However, the relationship between grain yield and water is complex because yield is more sensitive to drought at certain growth stages. Grain yield is dependent on rainfall or irrigation distributed over the growing season meeting the demand of the plant at each growth stages.

Sorghum can tolerate drought stress for short periods; however, prolonged drought can result in decreased grain yield. Inuyama et al. (1976) reported 16 and 36% yield reduction due to 16 and 28 d of drought stress, respectively, during the vegetative stage of sorghum. Eck and Music (1979) reported that short period (13 to 15 d) of drought stress during early booting, heading and early grain filling stage did not have a significant effect on grain yield. However, drought stress of 27 to 28 d during early booting, heading, and early grain-filling reduced yield by 27 and 12%, respectively. Prolonged stress of 35 and 42 d at the beginning of the booting stage reduced grain yield by 43 and 54%, respectively. Sorghum is more sensitive to drought stress during reproductive stages compared to vegetative stages. For maximum yield, adequate soil moisture should be available during two critical stages of crop development. First, before booting and panicle exertion, during which potential seed number panicle^{-1} are determined and second at flowering to ensure seed-set and efficient early grain filling. Drought stress during reproductive stages can have an adverse effect on pollen and ovule development, fertilization, and cause premature abortion of fertilized ovules (Saini, 1997). Sorghum yield is a function of the number of harvested panicles, seeds per panicle and individual seed weight. Each of these factors is affected by the duration and severity of drought stress. Eck and Music (1979) reported that yield reduction due to drought stress at the early boot stage was due to both reduced seed size and seed number where as yield reduction due to drought stress at heading or later stages are due to reduced seed size. There is significant genetic variation among the sorghum germplasm collection to drought stress and traits associated with tolerance (Mutava et al., 2011) that include enhanced and efficient roots, increased water-use efficiency, slow-wilting and stay green (Prasad et al., 2018).

Sorghum grown in tropical and sub-tropical regions may suffer intermittent or long-term waterlogging due to heavy rains. In these areas, waterlogging causes deleterious effects due in large part to changes in plant metabolism and soil texture. Sorghum is moderately tolerant to short periods of water logging during the seedling stage; however, prolonged waterlogging damages seedlings. Even though both sorghum shoots and roots were susceptible to waterlogging, roots show quicker recovery than the shoots (Bhagwat et al., 1986). Waterlogging increases the resistance of roots to the radical movement of water, leading to a decrease in water potential and wilting. Sorghum plants ware most sensitive to flooding and respond with the highest reduction in growth and dry mass at the early vegetative and early reproductive stages. Long-term flooding causes a significant reduction in biomass production, increases the allocation of biomass to the roots and reduces leaf area. Waterlogging also decreases photosynthetic rate, stomatal conductance, and transpiration (Tari et al., 2012).

Photoperiod

Another environmental factor that has strong control of phenology, particularly time to flowering, is photoperiod or daylength. Sorghum is a short-day photoperiod sensitive crop. Progress toward flowering is accelerated when the daylength decreases below the critical photoperiod. Most sorghum accessions and landraces or cultivars are photoperiod sensitive. Major (1980) identified three genetic components to describe the response of the sorghum materials to photoperiod: (i) the basic vegetative phase, defined as the minimum thermal time required for panicle initiation under optimum daylength; (ii) the minimum/maximum optimal photoperiod (MOP), defined as the critical photoperiod beyond which the vegetative period is influenced by changes in daylength; and (iii) the photoperiod sensitivity slope, which, from the MOP, expresses the linear increase of time to flowering for individual varieties.

Substantial advancement has occurred in understanding how the duration from planting to panicle initiation, anthesis, and maturity in sorghum is modulated by photoperiod, temperature, and their interaction. In general, the duration for each growth stage is related to thermal time. The duration of panicle initiation comprises a juvenile or pre-inductive phase followed by an inductive photoperiod-sensitive phase. The rate of progress can be quantified by a linear response to mean temperature and photoperiod (Craufurd et al., 1999). Results have shown that the sensitivity of sorghum to photoperiod ranges from 0 to more than 40.5 d per 1-h increase in photoperiod, with a critical or threshold photoperiod that varies between 12 and 14 h (Major and Kiniry, 1991). Reciprocal transfer experiments between environments differing in the photo- and thermo-period revealed three distinct developmental phases leading to floral initiation in sorghum (Ellis et al., 1997). The duration of the initial juvenile or photoperiod insensitive phase was shown to be regulated by temperature. A relatively low temperature of 24°C was optimal for a rapid end of the juvenile phase. Following the juvenile phase, sorghum becomes sensitive to photoperiod but insensitive to temperature, with shorter photoperiods being more inductive to flowering. The photoperiod sensitive phase ends some time before floral differentiation, and stages after that are not sensitive to photoperiod (Ellis et al., 1997). Depending on planting date, sorghum that is sensitive to photoperiod can produce canopies that are between 1.5 and 5 m tall and can have 12 to > 30 leaves on the main stem.

Among the cereals, major genes regulating photoperiodism were first characterized and recognized in sorghum (Quinby, 1966). Paterson et al. (1995) connected a major photoperiod QTL in sorghum by mapping to corresponding regions in maize, rice, wheat, and barley genomes. Out of 30,000 accessions maintained by USDA, 25,000 were photoperiod sensitive. Photoperiod sensitive tropical sorghum varieties will not flower when grown in temperate regions because daylength during the summer never reaches the critical photoperiod, and by the time daylength becomes short enough for flowering, temperatures are too low for optimal grain yield. The exact photoperiod interval required for sorghum flowering is 11 to 12 h. There was significant variation among the accessions for photoperiod response. Tropical sorghum germplasm is an important source of dominant alleles for yield and plant height, while temperate germplasm has dominant alleles for earliness and/or maturity. The use of tropical by temperate crosses has produced several high-yielding varieties with desirable plant height (1.5 to 2.5 m) and maturity (100 to 120 d).

The sorghum conversion program started in the late 1950s to produce germplasm not limited by photoperiod sensitivity (Stephens et al., 1967). The purpose of this program was to convert tall and late photoperiod sensitive sorghum genotypes to short and early photoperiod insensitive sorghum. The impacts of this program are tremendous, resulting in numerous converted and partially converted lines which have been used by sorghum breeding programs to enhance a wide array of traits in modern sorghum cultivars and hybrids. As photoperiod changes with latitude (both north and south), as well as with time of year in temperate regions, care should be taken while selecting cultivars or hybrids based on photoperiod sensitivity and duration of critical photoperiod.

Crop Management

Sorghum has relatively high radiation and water-use efficiency compared to C_3 crops like rice and wheat. Dry matter accumulation depends on the difference between gross photosynthesis accumulation and respiration. Hence, factors like light, temperature, water, and nutrient availability will influence photosynthesis, respiration, dry matter production, and partitioning. Dry matter production is strongly dependent on leaf area up to panicle initiation and is dependent on temperature and radiation. Peacock and Heinrich (1984) showed that leaf emergence per day increased when temperature increased from 13 to 23 °C. Leaf expansion increased up to 34 °C and after that decreased. Similarly, below 15 °C, leaf expansion stops. Hence, management practices causing increased leaf growth and higher interception of solar radiation and the efficient use of resources results in increased grain yield.

Soil management practices have a positive relationship with yield. Well-drained soils with higher water holding capacity and well managed soils with better agronomic practices such as minimum tillage and crop rotation can improve resource-use efficiency and grain yield. Use of improved nutrient and fertilizer management practices will result in higher yields and a complementary increase in water-use efficiency. Adoption of high yielding varieties or hybrids, proper sowing time, proper crop protection management, and weed control can increase grain yield. An overview of crop management practices of sorghum is provided by Djanaguiraman et al. (2018a) and Ciampitti et al. (2018). No tillage systems conserve soil moisture and minimize adverse effects of drought stress. In no-till conditions, the average soil pore diameter increases leading to improved soil porosity and soil structure is thereby increasing the proportion of soil water available to the plant. Better water harvesting techniques and practices to enhance infiltration and minimize erosion will make water available to plants for biomass production and yield formation.

Insect pests and diseases are important biotic stressors that can reduce sorghum grain yield. The stem borers, aphids, green bugs and shoot flies are the most common pests of sorghum. Sorghum diseases, such as seedling and foliage diseases, root and stalk rot, head blights and molds, can also decrease grain yield. Diseases may cause leaf spots or leaf blights, wilts, and premature death of plants. Sorghum diseases can cause harvest losses, affect the quality of the harvested crop, and lead to losses in storage. To minimize losses due to sorghum pests and diseases, it is important to correctly identify the pest or disease present so that appropriate management steps can be taken (Fransmann, 2007; Sharma et al., 2015).

Sorghum aphid (*Melanaphis sacchari*) infestations are higher under warm and dry conditions. Populations of sorghum aphid can reach large sizes at the time of flowering, which triggers significant infestation. Most damage is caused when the temperature is very high, and the humidity is very low. Sorghum midge [*Stenodiplosis sorghicola* (Coquillett)] emergence requires a temperature between 20 and 30 °C. At 26 °C and humid conditions, more generations of the midge per year are possible. Shoot fly [*Atherigona soccata* (Rondani)] numbers are positively related to high humidity, and cool temperature, as high temperature does not favor fly abundance. Sorghum grain moth [*Sitotroga cerealella* (Olivier)] incidence is greatest under hot dry or hot wet conditions. The khapra beetle (*Trogoderma granarium*) requires a temperature of 35 °C for its development. If the temperature is more than 35 °C, the khapra beetle can be a serious storage pest. Results have indicated that the beetle's breeding is slow at 25 °C, very slow at 22.5 °C and populations decline at 20 °C and below. The sorghum anthracnose (*Colletotrichum graminicola*) disease most often develops during warm, humid conditions. Under humid conditions, gray colored spore masses are produced. In many instances, leaves can be entirely blighted, and when it attacks the stem, it is known as stalk rot. Leaf blight (*Helminthosporium turcicum* Pass.) causes seed rot and seedling blight, especially in cool and excessively moist soil. Under warm, humid conditions, the disease may cause serious damage by killing all leaves before plants have matured. Sorghum downy mildew (*Peronosclerospora sorghi*) requires warm temperatures and high humidity to thrive. Moisture in soils ensures suitable germination of oospores. Conidia can also be generated in moist environments, specifically when there has been rain as moisture is a key factor. Rain or high humidity causes leaf wetness, which is the optimal environment for the pathogen to produce the conidia. A normal temperature range for production is 13 to 24 °C. If these conditions are met, wind will disperse numerous conidia, which are the source of secondary inoculum.

Genomic Resources for Sorghum

With the publication of sorghum genome sequence (Paterson et al., 2009), sorghum genomics has taken a paradigm shift and generated enormous information that can be used for sorghum improvement. Sorghum genome is comparatively small (~730 Mb), making it an attractive model for functional genomics of Saccharinae and other C_4 grasses. Gramene (http://www.gramene.org/) is a curated, open-source, data resource for comparative genome analysis in the grasses (Ware et al., 2002). The gramene database has genome module, genetic diversity module, pathway module, protein module, genes or gene and allele module, ontologies module, markers module, maps module, QTL module, BLASTView module, gramene mart module, and species page. The other databases like PlantGDB, Phytozome, GreenPhylDB, CoGe, PLAZA, and SGRqtl, SorGSD are widely used as a resource for sequence and comparative genomic information.

Pearl Millet

Pearl millet [*Pennisetum glaucum* (L.) R. Br.] is a coarse cereal crop grown in tropical semiarid regions and sub-humid regions of the world, namely Africa and Asia. International Crops Research Institute for the Semi-Arid Tropics (ICRISAT) has the global mandate for research on pearl millet and has one of the largest germplasms collections in its gene bank (> 21,000 accessions) (Upadhyaya et al., 2011). Pearl millet is expected to have increased importance in the future adaptation of agriculture to climate change in sub-Saharan Africa, which will experience higher temperatures and drought stress conditions. Pearl millet is grown on more than 29 million hectares in the semiarid and arid climates of tropical and sub-tropical regions. Asia and Africa are the major producers with an acreage of 11 and 16 million hectares, respectively. Global production of pearl millet grain probably exceeds 10 million tons a year, to which India contributes nearly half. Pearl millet was likely domesticated along the southern margins of the Saharan central highlands some 4000 to 5000 yr Before Present. After that, it spread to India through trade, where it was domesticated due to its extreme tolerance to drought (Fuller, 2003). Pearl millet is one of the important cereals grown in the tropics after rice, wheat, maize, and sorghum. In Asia, pearl millet is mainly grown in India and China and is cultivated in small acreages in Myanmar, Nepal, and Pakistan. India is the largest producer of pearl millet in Asia. In Africa, Niger, Nigeria, and Sudan are the dominant pearl millet producing countries (FAO, 2016). In recent years, the crop has been cultivated in small areas in the United States. In the 3-yr period from 2008 to 2010, about 9 million hectares were planted, producing about 8.3 million metric tons a year. The average yield was 930 kg per hectare. Pearl millet is well adapted to production systems characterized by low rainfall, low soil fertility, low water availability, and high temperature, and thus can be grown in

Table 5. Growth stages and description of pearl millet.

Growth stage		Description
Stage 0	Coleoptile visible at soil surface	Emergence occurs when the coleoptile is visible at the soil surface.
Stage 1	Three leaf stage	Approximately 5 d after emergence of the coleoptile, the lamina of the third leaf can just be seen in the whorl of the second leaf without separating the first and second leaves.
Stage 2	Five leaf stage	About 13 to 15 d after emergence the lamina of the fifth leaf is visible. The first and second leaves ae fully expanded. The third leaf is still slightly rolled. Tiller leaves may be seen emerging from inside the sheaths of the basal leaves.
Stage 3	Panicle initiation	At this stage, the growing point changes from vegetative to reproductive (leaf primordia to spikelet primordia).
Stage 4	Final leaf visible in the whorl	The lamina of the final leaf is visible in the rolled lamina of the preceding leaf. The final leaf is easily distinguished form the preceding leaves as there are no other leaves within the rolled lamina of this leaf as it emerges from the whorl.
Stage 5	Panicle exertion or boot stage	The panicle at this stage is enclosed within the sheath of the flag leaf but has not yet emerged from the collar.
Stage 6	50% stigma emergence	Pearl millet is protogynous, hence the stigma begins to emerge about 3 to 5 d after panicle emergence, though this varies with genotype. Stigma emergence starts generally in the florets several centimeters below the tip of the panicle and then proceeds upward and downward simultaneously. 50% flowering is attained by the time the stigma emerges in the middle region of the panicle. It takes 2 to 3 d for the completion of stigma emergence and unpollinated stigmas may remain fresh for several days.
Stage 7	Milk stage	Within 6 to 7 d after fertilization the grains grow sufficiently to become visible within the floret. At this stage, they consist of the seed coat filled with first a watery and later a milky liquid.
Stage 8	Dough stage	This stage is identified by the changes in the endosperm from the liquid milk stage to a first semisolid and then a solid stage.
Stage 9	Black layer formation or physiological maturity	At this stage, formation of small black layer in the hilar region of the seed is evident. Black layer formation begins from the upper part of the panicle and proceeds down the panicle.

areas where other cereal crops, such as rice, wheat, or maize would not survive. Pearl millet is the basic staple food for households in the poorest countries and among the poorest people. It is a staple food for about 90 million people living in semiarid tropical regions of Africa and the Indian subcontinent. Nutritionally, pearl millet grains contain about 60 to 70% carbohydrate, 10 to 12% protein, 3 to 5% fat, 1.5 to 3% fiber and 1.5 to 2% ash (Burton et al., 1972).

Pearl millet is well adapted to dry climates and is mostly grown in areas with limited agronomic potential characterized by low rainfall, in the 200 to 500 mm range, and marginal soils. Pearl millet may grow from 50 cm to 4 m tall and may tiller profusely under favorable weather conditions (Krishna, 2013). Stems are pithy, tiller freely, and produce an inflorescence with a dense spike-like panicle. Pearl millet has long leaves that are slender and smooth or have hairy surfaces. The leaves may vary in color, from light yellowish green to deep purple. The leaves are long-pointed with a finely serrated margin. Pearl millet usually flowers from 40 to 55 d. The flowering structure (inflorescence) in pearl millet is called a panicle or head. The mature panicle is brownish in color. The seed begins developing after fertilization and matures 25 to 30 d later. The seeds are nearly white, yellow, brown, gray, slate blue, or purple in color. The size of the seed is about one-third that of sorghum and the weight about 8 mg on average.

Growth and Development

Pearl millet has three distinct growth phases: the vegetative phase, from emergence to panicle initiation of the main stem; the panicle development phase, from panicle initiation to flowering in the main stem; and the grain-filling phase, from flowering to physiological maturity (Maiti and Bidinger, 1981). Pearl millet has nine distinct growth stages which explain the three distinct growth phases as described in Table 5. The vegetative phase starts with the emergence of the seedlings and continues up to the point of panicle initiation. During this phase, the seedlings establish their primary root system (seminal roots) and produce adventitious roots. During this phase, all leaves and tillers are initiated by the end of this phase. There is little internode elongation, however, and the apical meristem remains at or below the soil surface. Dry matter accumulation is mostly confined to leaves and roots (Maiti and Bidinger, 1981). Panicle initiation is marked by the elongation of the apical dome and the formation of a constriction at the base of the apex. Floral initiation may occur at 50 to 80 d after sowing. During the panicle development phase, all leaves will fully expand, and the lower fully expanded leaves will start to senesce. Stem elongation occurs by sequential elongation of internodes beginning at the base of the stem. Tillers emerge, undergo floral initiation, and leaf expansion in patterns similar to that of the main stem. The first formed tillers follow the main stem closely in their development, whereas the development of the late tillers frequently ceases due to competition and/or suppression by the more advanced main stem and early tillers. Dry matter accumulation takes place in roots, leaves, and stems. The panicle undergoes a series of morphological and developmental changes during the stem elongation stage. The changes include the development of spikelets, florets, glumes, stigmas, anther, flowering, and pollination. Pearl millet is protogynous, which means that female reproductive organs emerge before male reproductive organs. The grain-filling phase begins with fertilization of florets in the panicle and continues to maturity of the plant. During this period, the increase in total plant dry weight is largely in the grain. Senescence of lower leaves continues, and by the end of the phase, normally only flag and next to the leaf will maintain greenness. This phase will end with the development of a small dark layer of tissue in the hilar region of the grain, which is indicative of physiological maturity.

Climatic Requirements

Temperature

Pearl millet can grow in a wide range of ecological conditions and can still yield well even under unfavorable conditions like high temperatures, water deficit, and poor soil fertility. It is generally grown between 40° N and 40° S of the equator, in warm and hot countries exhibiting characteristics of the semiarid environment. In the arid and semiarid regions, temperatures are high due to high solar radiation and scarce rainfall. Pearl millets grow better than sorghum under dry conditions due to their short growth cycle (60 to 75 d). Weather conditions best suited for pearl millet are bright sunny days with occasional light rain. Pearl millet is more tolerant to higher temperatures than probably any other cultivated cereal. For growth and development, temperature requirements of pearl millet

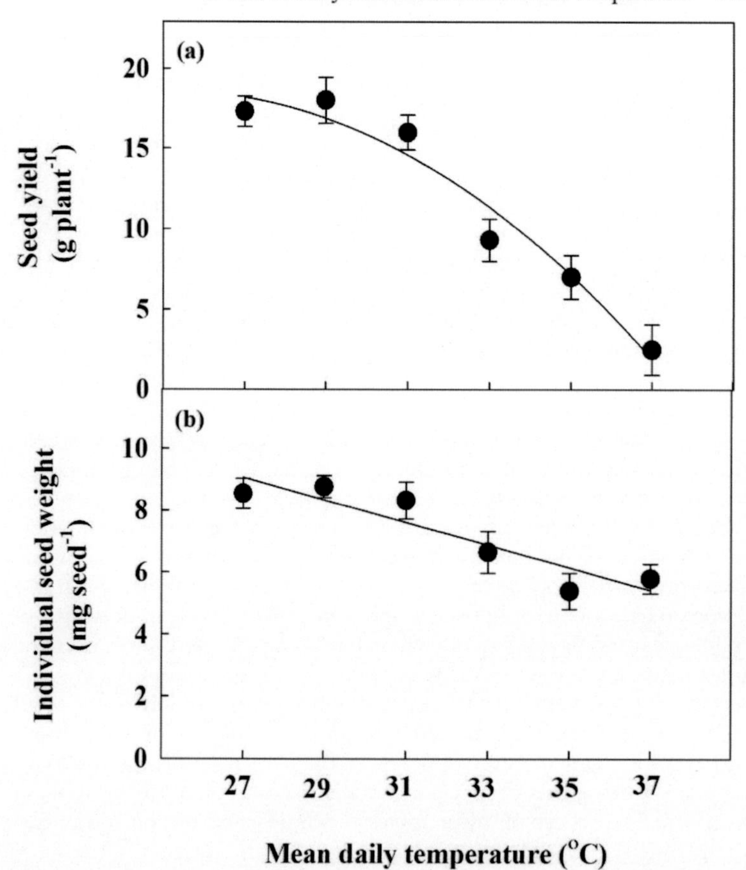

Fig. 6. Impact of high temperature stress (mean daily temperatures = 27, 29, 31, 33, 35, and 37 °C) from emergence to harvest (physiological maturity) on (a) seed yield (g plant^{-1}); and (b) individual seed weight (mg) of pearl millet. (Adapted and redrawn from Djanaguiraman et al., 2018c).

depend on the variety. The optimum temperature for plant growth ranges from 22 to 35 °C. The optimum mean daily temperature for growth and development is about 2 to 4 °C higher than for sorghum. The cardinal temperatures for different growth stages of pearl millet are shown in Table 2.

Pearl millet development begins at a base temperature of around 12 °C and a lethal temperature around 45 °C. Soil temperatures influence all aspects of early vegetative development including the emergence of seedlings, initiation, appearance, and a final number of leaves and tillers. For germination and emergence, soil temperature must reach 12 °C for germination to begin. Pearl millet seeds do not germinate under cool soil conditions. Poor emergence and seedling growth may result if planted before soil temperatures reach 23 °C. The caryopses require 18 °C from imbibition to the radical emergence and 29 °C from imbibition to seedling establishment. The optimum temperature for germination varies little across pearl millet genotypes averaging about 34 °C with a base temperature between 8 and 13.5 °C and an upper limit of between 47 and 52 °C (Mohamed et al., 1988). This wide range of temperature allows germination to proceed over a range of soil temperatures to which the caryopses are likely to be exposed. Maximum leaf growth occurs between 32 and 35 °C and are invariably linear between the base temperature and this optimum. The rate of leaf production accelerates at high temperatures (Pearson, 1975), although the number of leaf primordia on the main stem apex does not change from 18 to 30 °C. Ong (1983a) found a linear relationship between the rate of leaf extension and the temperature of the meristem. For a good photosynthetic response, the optimum temperature ranges from 31 to 35 °C. The temperature for root elongation ranges between 22 and 36 °C with an optimum of 32 °C. The optimum temperature for reproductive growth is 22 to 35 °C with a maximum of 40 to 45 °C. Each one-degree rise in temperature during the reproductive stage decreased the length of the growing period by about two days. Both the rate of spikelet production and the duration of the early reproductive phase are very sensitive to soil temperature since the meristem is at or close to the soil surface. Grain or seed-set has a temperature optimum of 22 to 25 °C and declines at temperatures below and above this range. Seed growth has a temperature optimum of 19 to 31 °C (Ong, 1983b). Increasing growth temperature from 36/26 °C (mean daily = 31 °C) to 44/34 °C (mean daily = 39 °C) resulted in decreased seed yield and no seed formation at about 42.3/33.2 °C (Djanaguiraman et al., 2018c; Fig. 6). Individual seed weight was decreased with increasing growth temperature; however, greater decrease was observed in seed yield compared to individual seed weight. (Djanaguiraman et al., 2018c). High temperature during flowering results in a loss of pollen viability and can reduce the receptivity of stigmas leading to reduced grain or seed set. Pearl millet is the most high-temperature tolerant cereal with the ceiling temperature of about 42 °C (Prasad et al., 2017). In contrast to other cereals, female reproductive organs are relatively more sensitive to high temperature stress compared to male reproductive organs (Djanaguiraman et al., 2018c).

Rainfall

Pearl millet grows on poor sandy soils and within a range of rainfall from 200 to 600 mm per annum. In regions with high rainfall, it can be used as a fodder crop. Heavy rains and high humidity are detrimental, particularly during flowering which can cause pollen to wash, prevent dehiscence and leads to poor seed-set. Prolonged spells of warm, rainless weather may be detrimental and may lead to reduced crop yields. At harvest time, dry, warm weather is most suitable. At sowing, poor soil moisture reduces seedling emergence, leading to poor crop establishment. During the vegetative growth period, the crop is well adapted to drought stress (Mahalakshmi et al., 1988) and can tolerate intermittent breaks in rainfall, which are a common feature of arid and semiarid regions of the world. Pearl millet has short developmental phases, rapid regrowth potential thus making use of asynchronous tillering under drought stress to make better use of short periods of water availability. Grain yield of the primary tiller will be compensated by the yield of secondary tillers because the early vegetative stage is less sensitive to drought stress. However, if the stress extends to the post-flowering period, yield reduction is more severe as the opportunity to recover is gradually lost (Lahiri and Kumar, 1966). The ability of pearl millet to grow in drier environments is due to a number of physiological and morphological characteristics, namely: rapid and deep root penetration and tillering capacity that compensates any reduction in yield contributing to components. such as a number of heads, length of the head, and grain weight. During the early flowering and grain-filling stages, pearl millet is most sensitive to drought (Mahalakshmi et al., 1988). Both timing of stress in relation to flowering and intensity of stress determine the reduction in grain yield (Mahalakshmi et al., 1988). Post flowering drought stress significantly decreases the grain yield of the main shoot panicle and the opportunity for yield compensation is less (Mahalakshmi et al., 1987). The development of tillers is arrested by drought stress during vegetative and early reproductive stages; however, pearl millet retains the potential to resume normal development when conditions again become favorable (Mahalakshmi and Bidinger, 1985). Drought stress during the reproductive stage is a serious limitation to grain yield, primarily when the total length of the period of available moisture is substantially reduced.

Photoperiod

Pearl millet is grown in a wide range of latitudes (11 to 29° N). Varieties grown in semiarid regions have short life cycles, less than 90 d to grain maturity, in a relationship with the short duration of the rainy season. Earliness is generally considered to be associated with the absence of photoperiod-sensitivity. The gene e_1e_1 imparts photoperiod insensitivity and makes most pearl millet genotypes carrying this gene mature 10 to 40 d earlier than other genotypes (Burton, 1981). Photoperiod-sensitive flowering is an example of phenotypic plasticity that can enhance adaptation to variable planting dates due to a scattered beginning of the rainy season, as typical for West and Central Africa. It enhances the simultaneous flowering of the cultivar in the target region, independent of the individual date of planting in different fields (Haussmann et al., 2007). The flowering date of early varieties would be independent of changes in daylength. However, experiments conducted in West and Central Africa have shown that early varieties varied about 10 d with different sowing dates. The sensitivity of pearl millet to photoperiod has evolved to trigger a drought escape mechanism. In other words, pearl millet flowers "on time" to ensure that it can complete its maturation cycle with the remaining soil moisture (Vadez et al., 2012). Any genotype with delayed flowering may be exposed to serious stress conditions during its reproduction phase. Burton (1965) classified 86% of the 250 accessions from Nigeria and Upper Volta as day neutral and the remainder as short-day plants. Subsequently, Ong and Everard (1979) reclassified them as facultative short-day and obligate short-day plants, respectively. Extended photoperiod imposed to short-day pearl millet lengthened the time taken to anthesis and increased plant height, a number of leaves and dry weight and reduced head length and head numbers per plant which contributed to decreased grain yield (Ong and Everard, 1979). Similarly, Carberry and Campbell (1985) found that by extending photoperiod from 13.5 to 14.5 and 15.5 h, the time taken to panicle initiation on the main stem increased from 16 to 23 and 34 d after emergence, respectively. Grain yield was not affected by photoperiod at higher plant populations, but yield was reduced by 35% when photoperiod was extended at low populations due to less contribution from the tillers. The number of tillers was unaffected by photoperiod; however, the number of productive tillers per plant was reduced up to 38% as photoperiod was extended. Plants under extended photoperiod have higher plant biomass and lower harvest index than plants under natural photoperiod.

Crop Management

In Africa, pearl millet is grown as a sole crop or intercropped with legumes, particularly cowpea. On sandy soils, pearl millet is planted either in a dry seedbed or immediately after rain. In the Sahel, rainfall can be sporadic causing poor crop stands or prolonged drought stress during the early vegetative stage. Drought and high temperature are detrimental to seedlings. Pearl millet is sown under no tillage conditions, and weeding is done after the emergence of seedlings. In sandy soil, the soil is dug over with a hoe and weeded before sowing. This facilitates breaking the soil crust and water infiltration. In some regions of the world, tractors are used for land preparation to bury weeds and stubble and to make a fine tilth. A major problem of rainfed agriculture in semiarid regions with short rainy seasons is determining the optimum sowing date, which is very critical for seedling establishment. Sowing is initiated after a good rain. Generally, farmers do not apply any inputs particularly inorganic fertilizer to pearl millet. However, pearl millet responds well to additional plant nutrition. Fertilizer applications should be supplied according to soil test results and agro-ecologically appropriate fertilizer response functions. Increased yield due to the application of fertilizer is accompanied by an increase in water-use efficiency. The beneficial effect of fertilizer could be attributed to the rapid early growth of leaves, which can contribute to a reduction in evaporative losses from the soil and increased water-use efficiency. Conservation of soil moisture is very critical in semiarid regions where pearl millet is grown. The use of organic mulch during the growing season would be a simple solution to conserve soil moisture. In West Africa, the probability of dry spells of 10 d or more is high, and crop water supply is often exhausted, requiring re-sowing after the next sufficiently large rain event. Delayed sowing can decrease grain yield. Hillel (1982) has suggested that one way to control evaporation during the first stage is to induce a temporarily higher evaporation rate, so the soil surface is rapidly desiccated. This hastens the end of the first stage and uses the hysteresis effect to help arrest or retard subsequent flow. Another approach is to optimize plant population and row and plant spacing to reduce evaporation and crop failure. Increased plant population from 5000 to 20,000 hills per hectare significantly increased grain yield by reducing evapotranspiration. Canopy cover can also be increased by the introduction of an intercrop. Pearl millet hybrids often give better grain yields than local open-pollinated cultivars. Genetically uniform single-cross hybrid cultivars are currently available which can improve grain yield. Most production in Asia (particularly in India) is hybrids, which yield much higher, compared with Africa where most pearl millet production is land races or cultivars with lower yields.

Insect pests can cause considerable economic damage to pearl millet. The major seedling pests of pearl millet are shoot flies, stem fly, chinch bugs, and spittle bug. Shoot flies (*Atherigona approximata*) can cause 12 to 46% yield loss in grain and 57% loss in dry fodder yield. This insect also causes damage to the ear head. Wild hosts such as *Cynodon dactylon* (L.) Pers. help the pest to survive during the off-season. Stem fly (*Agromyza* spp.) is occasionally reported on pearl millet. In the southeastern United States, the key insect pests on both forage and grain pearl millet are the chinch bug, *Blissus leucopterus leucopterus* (Say) (Heteroptera: Blissidae), and the false chinch bug, *Nysius raphanus* Howard (Heteroptera: Lygaeidae), which cause significant injury and loss of seedling stands. Spittle bug [*Poophilus costalis* (Walker)] infestation causes yellowing and wilting of mature pearl millet seedlings. The major foliage pests are hairy caterpillars, armyworms, aphids, and true bugs. High incidence of hairy caterpillars (*Amsacta albistriga* Walker, *Estigmene lactinea* Huebnerand, *Spilosoma obliqua* Walker) has been recorded on millets. Hairy caterpillars can cause complete defoliation because of their voracious feeding habits. The pest passes the hot summer as diapaused pupae in the soil. Moths emerge about 2 wk after the first rain. The pest can complete one or two generations depending on rainfall and its distribution. The other foliar pest is armyworms [*Mythimna separata* (Walker)]. Both the nymph and the adult of the sucking pests like aphids [*Rhopalosiphum maidis* (Fitch)], shoot bug [*Peregrinus maidis* (Ashmead)] and plant bug (*Aspavia armigera* Fab., *Callidea* spp.,) suck the sap from young leaves, causing distortion, yellowing, and wilting of the plants, leading to formation of the shriveled and chaffy grains. Some of the sucking pests act as vectors of plant viruses.

Stem borers are one of the major destructive groups of insects attacking millets. *Chilo partellus* Swinhoe and *Sesamia inferens* Walker are predominantly observed in millets. The pests bore into the stem and cause the dead heart symptom, resulting in the development of profuse additional tillers with unproductive spikes. Infestation of borers at later growth stages results in peduncle damage, leading to loss of grains and fodder. Pearl millet appears to be immune to borer attack at initial stages of crop growth, but it becomes susceptible to internode-injury later. Spike worms (*Heliocheilus albipunctella* De Joannis) and grain midges [*Geromyia penniseti* (Felt)] are also important pests, which can cause a yield reduction up to 85%. The female lays eggs on the spikelets or floral peduncles, the eggs hatch, and young larva feed on the flowers, leaving empty glumes. Older larvae cut the floral peduncles and eat between rachis and flowers, pushing out the destroyed flowers, and making a characteristic spiral trace on the spike. Management options include deep plowing before the rainy season to expose the diapausing pupae to predators and desiccation and delayed planting to avoid the coincidence of moth abundance at the flowering stage. Other common pests of pearl millet include head beetles [*Lytta tenuicollis* (Pallas), *Mylabris pustulata* (Thunb.), and *Psalydolytta rouxi* Laporte], head caterpillars (*Eublemma silicula* Swinh.), thrips [*Haplothrips ganglebauri* Schum. and *Thrips hawaiiensis* (Morgan)] and white grubs (*Holotrichia consanguinea* Blanch.). Among the pests of pearl millet, the major devastating pests of pearl millets are shoot fly, stem borer, and white grub. Information on economic injury levels, yield loss, the effectiveness of natural enemies on pearl millet pest are limited. Hence, need the based use of pesticide should be followed.

The most important diseases of pearl millet which can significantly decrease grain yield are downy mildew or green-ear disease [*Sclerospora graminicola* (Sacc.) Schroet.], ergot (*Claviceps fusiformis*), smut (*Tolyposporium penicillariae* Bref.), grain mold and blast (*Pyricularia grisea*). Downy mildew disease appears in the form of chlorosis at the base of the infected leaves followed by production of sporulation on the lower side of the leaves known as a half-leaf symptom. The white downy growth will be more at a moderate temperature (20–22 °C) and high relative humidity (> 95%). No sporulation will occur at a relative humidity of < 70%. At the time of panicle emergence, the green ear symptom will be visible. The malformed florets will be converted into leafy structures of diverse appearance. Ergot is a devastating disease. Infected ears show viscous, sugary exudates (honey-dew) out from the glumes. *Claviceps fusiformis* can grow over a wide temperature range, but the optimum is between 20 and 30 °C; however, below 10 °C the germination of fungal spores is highly retarded. Relative humidity of greater than 80% favors the disease development. In general, cool nighttime temperatures (< 15 °C) for two to three weeks before flowering is conducive for disease development (Miedaner and Geiger, 2015). Smut disease is also a floral disease. Immature, green sori, larger than the seed, develop on panicles during grain fill. A single sorus develops per floret. As grain matures, sori change in color from green to dark brown. Sori are filled with dark teliospores. Aerial populations of sporidia are greatest when minimum and maximum temperatures range between approximately 21 and 31 °C, and maximum relative humidity is greater than 80%. Grain mold of pearl millet is caused by various fungi. Grain mold on pearl millet tend to be more severe with humid conditions during grain fill and when grain harvest is delayed. The leaf blast of pearl millet appears as grayish, water-soaked lesions on the foliage that turns brown on drying. The symptom is more pronounced at a temperature of < 25 °C.

Genomic Resources for Pearl Millet

Pearl millet draft whole genome (~1.79 Gb) of reference genotype Tift23D_2B_1–P_1–P_5 indicating ~90% of the genome was assembled and contained 38,579 genes (Varshney et al., 2017). The pearl millet genome sequence provided a genetic blueprint of the species and presumed to facilitate the development of genomic tools that would expedite the development of improved cultivars through the application of marker-assisted breeding. Genomic resources like molecular markers, genetic linkage maps, and genes associated with specific traits and genetic mapping have been summarized by Dwivedi et al. (2012).

Conclusions

Maize, sorghum, and pearl millet are the major crops grown in arid and semiarid regions characterized by warmer temperatures and low and variable rainfall. Yield potential of maize is comparatively higher than sorghum and pearl millet. Maize has relatively lower cardinal temperatures, followed by grain sorghum and pearl millet. Optimum growth and yield for maize occur at relatively cooler temperature ranges compared to sorghum and pearl millet. The water requirement for maize is higher when compared to sorghum and pearl millet. Water requirement is much lower for pearl millet than sorghum and maize. The water productivity and response to irrigation is relatively greater in maize than in sorghum and pearl millet. However, maize is more sensitive to high temperature and drought stress resulting in lower yields, particularly when these stresses occur during sensitive stages of gametogenesis, flowering, and early seed filling stages of crop development. Sorghum and pearl millet can perform better under high temperature stress and drier environments. All three of these crop species are short-day plants and respond to changes in photoperiod or daylengths. However, some of the new hybrids or cultivars are day neutral and can be grown in a larger range of environments. There is significant genetic diversity for response to environmental conditions such as temperature, water, and photoperiod. Environmental conditions also influence the incidence of biotic pests (particularly diseases and pests) and their impact on host plants and crops. Genetic variability needs to be systematically utilized for breeding to enhance adaptation and stress tolerance to obtain higher yields in multiple environmental conditions. Projected changes in climate and management practices will require a better understanding of agroclimatology of crops so that we can match crop phenology and genotypes to optimum environments to take advantage and utilize the genetic potential. Similarly, a better understanding of climatic needs and understanding the environmental conditions of a location will allow us to develop and use management practices that will utilize available resources most efficiently to maximize yield and profitability.

Acknowledgments

We thank the following entities for supporting sorghum and millet research– Department of Agronomy; Great Plains Sorghum Improvement and Utilization Center at Kansas State University; Kansas Corn Commission; Kansas Grain Sorghum Commission; United Sorghum Checkoff Program; and United States Agency for International Development (Feed the Future Innovation Lab for Collaborative Research on Sustainable Intensification). Contribution no. 18-406-B of the Kansas Agricultural Experiment Station.

References

Abendroth, L.J., R.W. Elmore, M.J. Boyer, and S.K. Marlay. 2011. Corn growth and development. PMR 1009. Iowa State University, Extension. Ames, IA.

Adams, J.E., and D.O. Thompson. 1973. Soil temperature reduction during pollination and grain formation of corn and grain sorghum. Agron. J. 65:60–63. doi:10.2134/agronj1973.00021962006500010018x

Alagarswamy, G., and J.T. Ritchie. 1991. Phasic development in CERES sorghum model. In: T. Hodges, editor, Predicting crop phenology. CRC Press, Boca Raton, FL. p. 143–152.

Allison, J.C.S., and T.B. Daynard. 1979. Effect of change in time of flowering, induced by altering photoperiod or temperature, on attributes related to yield in maize. Crop Sci. 19:1–4. doi:10.2135/cropsci1979.0011183X001900010001x

Anderson, I.C., and D.S. Robertson. 1960. Role of carotenoids in protecting chlorophyll from photodestruction. Plant Physiol. 35:531–534. doi:10.1104/pp.35.4.531

Aslam, M., M.A. Maqbool, and R. Cengiz. 2015. Effects of drought on maize In: M. Aslam, M.A. Maqbool, and R. Cengiz, editors, Drought stress in maize (*Zea mays* L.) effects, resistance mechanisms, global achievements and biological strategies for improvement. Springer, New York. p. 5–17. doi:10.1007/978-3-319-25442-5_2

Assefa, Y., P.V.V. Prasad, P. Carter, M. Hinds, G. Bhalla, and R. Schon. 2016. Yield responses to planting density for US modern corn hybrids: A synthesis-analysis. Crop Sci. 56:2802–2817. doi:10.2135/cropsci2016.04.0215

Assefa, Y., P.V.V. Prasad, P. Carter, M. Hinds, G. Bhalla, and M. Jeschke. 2017. A new insight into corn yield: Trends from 1987 through 2015. Crop Sci. 57:2799–2811. doi:10.2135/cropsci2017.01.0066

Balota, M. 2012. Sorghum (*Sorghum vulgare* L.) marketability grain color and relationship to feed value. AREC-23NP, Virginia Cooperative Extension Bulletin, Blacksburg, VA. p. 1-3.

Bender, R.R., J.W. Haegele, M.L. Ruffo, and F.E. Below. 2013. Nutrient uptake, partitioning, and remobilization in modern, transgenic insect-protected maize hybrids. Agron. J. 105:161–170. doi:10.2134/agronj2012.0352

Bhagwat, K.A., S.R. Gore, and G. Banerjee. 1986. Waterlogging injury in sorghum seedlings and their kinetin effected rapid recovery. Plant Growth Regul. 4:23–31. doi:10.1007/BF00025346

Boxall, R. 2002. Damage and loss caused by the larger grain borer Prostephanus truncatus. Integrated Pest Management Reviews 7:105–121. doi:10.1023/A:1026397115946

Brown, W., M.S. Zuber, L.L. Darrah, and D.V. Glover. 1985. Origin, adaptation, and types of corn. National Corn Handbook. NCH-10, Cooperative Extension Service, Iowa State University, Ames, IA.

Burton, G.W. 1965. Photoperiodism in pearl millet, Pennisetum typhoides. Crop Sci. 5:333–335. doi:10.2135/cropsci1965.0011183X000500040014x

Burton, G.W., A.T. Wallace, and K.O. Rachie. 1972. Chemical composition and nutritive value of pearl millet (*Pennisetum typhoides* (Burm.) Stapf and E. C. Hubbard) grain. Crop Sci. 12:187–188. doi:10.2135/cropsci1972.0011183X001200020009x

Burton, G.W. 1981. A gene for early maturity and photoperiod insensitivity in pearl millet. Crop Sci. 21:317–318. doi:10.2135/cropsci1981.0011183X002100020027x

Campos, H., M. Cooper, G.O. Edmeades, C. Löffler, J.R. Schussler, and M. Ibañez. 2006. Changes in drought tolerance in maize associated with fifty years of breeding for yield in the US Corn Belt. Maydica 51:369–381.

Carberry, P.S., and L.C. Campbell. 1985. The growth and development of pearl millet as affected by photoperiod. Field Crops Res. 11:207–217. doi:10.1016/0378-4290(85)90103-0

Cheikh, N., and R.J. Jones. 1994. Disruption of maize kernel growth and development by heat stress. Plant Physiol. 106:45–51. doi:10.1104/pp.106.1.45

Chen, J., W. Xu, J.J. Burke, and Z. Xin. 2010. Role of phosphatidic acid in high temperature tolerance in maize. Crop Sci. 50:2506–2515. doi:10.2135/cropsci2009.12.0716

Chen, Q., H. Zhong, X.W. Fan, and Y.Z. Li. 2014. An insight into the sensitivity of maize to photoperiod changes under controlled conditions. Plant Cell Environ. 38:1479–1489. doi:10.1111/pce.12361

Ciampitti, I.A., S.T. Murrell, J.J. Camberato, and T.J. Vyn. 2013a. Maize nutrient accumulation and partitioning in response to plant density and nitrogen rate: I. Macronutrients. Agron. J. 105:783–795. doi:10.2134/agronj2012.0467

Ciampitti, I.A., S.T. Murrell, J.J. Camberato, and T.J. Vyn. 2013b. Maize nutrient accumulation and partitioning in response to plant density and nitrogen rate: I. Calcium, Magnesium, and Micronutrients. Agron. J. 105:1645–1657. doi:10.2134/agronj2013.0126

Ciampitti, I.A. 2015. Sorghum growth and development. Kansas State University, Manhattan, KS. https://www.bookstore.ksre.ksu.edu/pubs/MF3234.pdf (Accessed 17 Sept. 2018).

Ciampitti, I.A., R.W. Elmore, and J. Lauer. 2016. Corn growth and development. Kansas State University, Manhattan, KS. https://www.bookstore.ksre.ksu.edu/pubs/MF3305.pdf (Accessed 17 Sept. 2018).

Ciampitti, I.A., P.V.V. Prasad, A.J. Schlegel, L. Haag, R.W. Schnell, B. Arnall, and J. Lofton. 2018. Genotype × environment × management practices. US sorghum cropping systems. In: I.A. Ciampitti and P.V.V. Prasad, Sorghum: State of the art and future perspectives, ASA, Monograph 58, Madison, WI.

CIMMYT. 2010. Maize-Global alliance for improving food security and the livelihoods of the resource- poor in the developing world. CIMMYT and IITA to the CGIAR Comortium Board. El Batan, Mexico, p. 91.

Coles, N.D., C.T. Zila, and J.B. Holland. 2011. Allelic effect variation at key photoperiod response quantitative trait loci in maize. Crop Sci. 51:1036–1049. doi:10.2135/cropsci2010.08.0488

Crafts-Brandner, S.J., and M.E. Salvucci. 2002. Sensitivity of photosynthesis in a C4 plant, maize, to heat stress. Plant Physiol. 129:1773–1780. doi:10.1104/pp.002170

Craufurd, P.Q., V. Mahalakshmi, F.R. Bidinger, S.Z. Mukuru, J. Chantereau, P.A. Omanga, A. Qi, E.H. Roberts, R.H. Ellis, R.J. Summerfield, and G.L. Hammer. 1999. Adaptation of sorghum: Characterization of genotypic flowering responses to temperature and photoperiod. Theor. Appl. Genet. 99:900–911. doi:10.1007/s001220051311

Dale, R.F. 1983. Temperature perturbations in the Midwestern and South-eastern United States important for crop production. In: C.D. Raper and P.J. Kramer, editors, Crop reactions to water and temperature stresses in humid and temperate climates. Westview Press, Boulder, CO. p. 21–32.

Dhillon, M.K., V.K. Kalia, and G.T. Gujar. 2013. Insect-pest and their management: Current status and future need of research in quality maize. In: D.P. Chaudhar, S. Kumar, and S. Langyan, editors, Maize: Nutrition dynamics and novel uses. Springer India, p. 95-103.

Djanaguiraman, M., P.V.V. Prasad, and I.A. Ciampitti. 2018a. Improving sorghum crop management: overview. In: W.L. Rooney, editor, Achieving sustainable cultivation of sorghum. Burleigh Dodds Science Publishing, Cambridge, U.K. doi:10.19103/AS.2017.0015.13

Djanaguiraman, M., P.V.V. Prasad, and I.A. Ciampitti. 2018b. Sorghum crop physiology and development. In: W.L. Rooney, editor, Achieving sustainable cultivation of sorghum. Burleigh Dodds Science Publishing, Cambridge, U.K. doi:10.19103/AS.2017.0015.03

Djanaguiraman, M., R. Perumal, I.A. Ciampitti, S.K. Gupta, and P.V.V. Prasad. 2018c. Quantifying pearl millet response to high temperature stress: Thresholds, sensitive stages, genetic variability and relative sensitivity of pollen and pistil. Plant Cell Environ. 41:993–1007. doi:10.1111/pce.12931

Doggett, H. 1988. Sorghum, 2nd Edition, John Wiley & Sons, New York, NY.

Downes, R.W., and D.R. Marshall. 1971. Low temperature induced male sterility in Sorghum bicolor. Aust. J. Exp. Agric. Anim. Husb. 11:352–356. doi:10.1071/EA9710352

Dubey, R.S. 2005. Photosynthesis in plants under stressful conditions. In: M. Pessarakli, editor, Handbook of Photosynthesis. CRC Press, Boca Roton, FL. p. 717–737.

Du-Plessis, J. 2003. Maize production. Directorate agricultural information services, Department of Agriculture in Cooperation with ARC-Grain Crops Institute, Potchefshroom, South Africa.

Dwivedi, S., H. Upadhyaya, S. Senthilvel, C. Hash, K. Fukunaga, X. Diao, D. Santra, D. Baltensperger, and M. Prasad. 2012. Millets: Genetic and genomic resources. In: J. Janick, editor, Plant breeding reviews. Vol. 35. John Wiley & Sons. p. 247–375.

Eastin, J.D., I. Brooking, and S.O. Taylor. 1976. Influence of temperature on sorghum respiration and yield. Agronomy Abstracts. ASA, Madison, WI.

Eck, H.V., and J.C. Musick. 1979. Plant water stress effect on irrigated sorghum. I. Effect on yield. Crop Sci. 19:586–592.

Ellis, R.H., G.A. Guaufurd, R.J. Summerfield, and E.H. Robert. 1997. Effect of photoperiod, temperature and asynchrony between thermoperiod and photoperiod on development to panicle initiation in sorghum. Ann. Bot. (Lond.) 79:169–178. doi:10.1006/anbo.1996.0328

Ercoli, L., M. Mariotti, A. Masoni, and I. Arduini. 2004. Growth responses of sorghum plants to chilling temperature and duration of exposure. Eur. J. Agron. 21:93–103. doi:10.1016/S1161-0301(03)00093-5

Erickson, R.O. 1959. Integration of plant growth processes. Am. Nat. 93:225–235. doi:10.1086/282080

FAO. 2016. FAOSTAT. FAO, Rome. www.fao.org/faostat/en/#data/QC (Accessed 20 Sept. 2018).

Fransmann, B.A. 2007. Sorghum. In: P. Bailey, editor, Pests of field crops and pastures: Identification and control. CSIRO Publishing, Collingwood, Australia. p. 297–304.

Fuller, D.Q. 2003. African crops in prehistoric South Asia: A critical review. In: K. Neumann, S. Kahlheber, and E.A. Butler, editors, Food, fuel and fields: Progress in African archaeobotany. Heinrich-Barth Institut, Cologne, Germany, p. 239-271.

Gharoobi, B., M. Ghorbani, and M.G. Nezhad. 2012. Effects of different levels of osmotic potential on germination percentage and germination rate of barley, corn and canola. Iranian Journal of Plant Physiology 2:413–417.

Gicheru, P.T. 1994. Effects of residue mulch and tillage on soil-moisture conservation. Soil Technol. 7:209–220. doi:10.1016/0933-3630(94)90022-1

Ginsburg, H., and B.Z. Ginzburg. 1971. Radial water and solute flow in roots of Zea mays. III. Effect of temperature on THO and ion transport. J. Exp. Bot. 22:337–341. doi:10.1093/jxb/22.2.337

Gouesnard, B., C. Rebourg, C. Welcker, and A. Charcosset. 2002. Analysis of photoperiod sensitivity within a collection of tropical maize populations. Genet. Resour. Crop Evol. 49:471–481. doi:10.1023/A:1020982827604

Hamilton, T.S., B.C. Hamilton, B.C. Johnson, and H.H. Mitchell. 1951. The dependence of the physical and chemical composition of the corn kernel on soil fertility and cropping system. Cereal Chem. 28:163–176.

Harlan, J.R., and J.M.J. Dewet. 1972. A simplified classification of cultivated sorghum. Crop Sci. 12:172–176. doi:10.2135/cropsci1972.0011183X001200020005x

Hatfield, J., K.J. Boote, P.A. Fay, G.L. Hahn, R.C. Izaurralde, B.A. Kimball, T.L. Mader, J.A. Morgan, D.R. Ort, H.W. Polley, A.M. Thomson, and D.W. Wolfe. 2008. Agriculture In: The effects of climate change on agriculture, land resources, water resources, and biodiversity. A Report by the U.S. Climate Change Science Program and the Subcommittee on Global Change Research, Washington, D.C. p. 362.

Haussmann, B., Boureima, S.S., Kassari, I.A., Moumouni, K.H. and Boubacar, A. 2007. Mechanisms of adaptation to climate variability in West African pearl millet landraces - a preliminary assessment. SAT eJournal 3, p. 1–3.

Herrero, M.P., and R.R. Johnson. 1980. High temperature stress and pollen viability of maize. Crop Sci. 20:796–800. doi:10.2135/cropsci1980.0011183X002000060030x

Heslop-Harrison, J. 1961. The experimental control of sexuality and inflorescence structure in *Zea mays* L. Proc. Linn. Soc. London 172:108–123. doi:10.1111/j.1095-8312.1961.tb00875.x

Hillel, D. 1982. Introduction to soil physics. Academic Press, New York.

Hopkins, W.G. 1999. Introduction to plant physiology, Second ed. John Wiley & Sons, New York.

Hossain, F., B.M. Prasanna, R.K. Sharma, P. Kumar, and B.B. Singh. 2007. Evaluation of quality protein maize genotypes for resistance to stored grain weevil Sitophilus oryzae (Coleoptera: Curculionidae). Int. J. Trop. Insect Sci. 27:114–121. doi:10.1017/S1742758407814676

Inuyama, S., J.T. Musick, and D.A. Dusek. 1976. Effect of plant water deficit at various growth stages on growth, grain yield and leaf water potential of irrigated grain sorghum. Jpn. J. Crop Sci. 45:298–307. doi:10.1626/jcs.45.298

Johnson, R., and M.P. Herrero. 1981. Corn pollination under moisture and high temperature stress. Proceedings of the corn and sorghum industry research conference II, Chicago. American seed Trade Association, Washington, D.C., p. 66-77.

Jones, R.J., J.A. Roessler, and S. Ouattar. 1985. Thermal environment during endosperm cell division in maize: Effect on number of endosperm cells and starch granules. Crop Sci. 25:830–834. doi:10.2135/cropsci1985.0011183X002500050025x

Karim, M.A., Y. Fracheboud, and P. Stamp. 1997. Heat tolerance of maize with reference of some physiological characteristics. Annals of Bangladesh Agriculture 7:27–33.

Khodarahmpour, Z. 2011. Effect of drought stress induced by polyethylene glycol (PEG) on germination indices in corn (*Zea mays* L.) hybrids. Afr. J. Biotechnol. 10:18222–18227.

Krishna, K.R. 2013. Agroecosystems: Soils, climate, crops, nutrient dynamics and productivity. CRC Press, Boca Raton, FL. p. 100-104. doi:10.1201/b16300

Laan, P., M.J. Berrevoets, S. Lythe, W. Armstrong, and C.W.P.M. Blom. 1989. Root morphology and aerenchyma formation as indicators of the flood-tolerant of Rumex species. J. Ecol. 77:693–703. doi:10.2307/2260979

Lahiri, A.N., and V. Kumar. 1966. Studies on plant-water relationship III: Further studies on the drought mediated alterations in the performance of bulrush millet. Proc. Nat. Ins. Sci. India B 32:116.

Lawrence, C.J., L.C. Harper, M.L. Schaeffer, T.Z. Sen, T.E. Seigfried, and D.A. Campbell. 2008. MaizeGDB: The maize model organism database for basic, translational, and applied research. Int. J. Plant Genomics 2008:1–10. doi:10.1155/2008/496957

Lew, H., A. Adler, and W. Edinger. 1991. Moniliformin and the European corn borer (*Ostrinia nubilalis*). Mycotoxin Res. 7:71–76. doi:10.1007/BF03192189

Lewis, R.S., and M.M. Goodman. 2003. Incorporation of tropical maize germplasm into inbred lines derived from temperate x temperate adapted tropical line crosses: Agronomic and molecular assessment. Theor. Appl. Genet. 107:798–805. doi:10.1007/s00122-003-1341-x

Lobell, D.B., and M.B. Burke. 2010. On the use of statistical models to predict crop yield responses to climate change. Agric. For. Meteorol. 150:1443–1452. doi:10.1016/j.agrformet.2010.07.008

Lobell, D.B., M. Banziger, C. Magorokosho, and B. Vivek. 2011. Nonlinear heat effects on African maize as evidenced by historical yield trials. Nat. Clim. Chang. 1:42–45. doi:10.1038/nclimate1043

Mahalakshmi, V., and F.R. Bidinger. 1985. Flowering response of pearl millet to water stress during panicle development. Ann. Appl. Biol. 106:571–578. doi:10.1111/j.1744-7348.1985.tb03148.x

Mahalakshmi, V., F.R. Bidinger, and D.S. Raju. 1987. Effect of timing of water deficit on pearl millet (*Pennisetum americanum*). Field Crops Res. 15:327–339. doi:10.1016/0378-4290(87)90020-7

Mahalakshmi, V., F.R. Bidinger, and G.D.P. Rao. 1988. Timing and intensity of water deficits during flowering and grain filling in pearl millet. Agron. J. 80:130–135. doi:10.2134/agronj1988.00021962008000010028x

Maiti, R.K., and F.R. Bidinger. 1981. Growth and development of the pearl millet plant. Research Bulletin No 6. International Crops Research Institute for the Semi-Arid Tropics, Patancheru, India.

Major, D.J. 1980. Photoperiod response characteristics controlling flowering of nine crop species. Can. J. Plant Sci. 60:777–784. doi:10.4141/cjps80-115

Major, D.J., and J.R. Kiniry. 1991. Predicting day length effects on phenological processes. In: T. Hodges, editor, Predicting crop phenology. CRC Press, Boca Raton, FL. p. 15–28.

Mangelsdorf, P.C. 1974. Corn. Its origin, evolution, and improvement. Harvard Univ. Press, Cambridge, MA. p. 1-262. doi:10.4159/harvard.9780674421707

Mano, Y., and F. Omori. 2007. Breeding for flooding tolerant maize using teosinte as a germplasm resource. Plant Root 1:17–21. doi:10.3117/plantroot.1.17

Massad, R.S., A. Tuzet, and O. Bethenod. 2007. The effect of temperature on C4–type leaf photosynthesis parameters. Plant Cell Environ. 30:1191–1204.

Maulana, F., and T.T. Tesso. 2013. Cold temperature exposure at seedling and flowering stages reduces growth and yield component in sorghum. Crop Sci. 53:564–574. doi:10.2135/cropsci2011.12.0649

Miedaner, T., and H.H. Geiger. 2015. Biology, genetics, and management of ergot (Claviceps spp.) in rye, sorghum, and pearl millet. Toxins (Basel) 7:659–678. doi:10.3390/toxins7030659

Miedema, P. 1982. The effects of low temperature on *Zea mays*. Adv. Agron. 35:93–128. doi:10.1016/S0065-2113(08)60322-3

Miller, J.D. 1994. Epidemiology of Fusarim graminierum disease of wheat and corn. In: J.D. Miller and H.L. Trenholm, editors, Mycotoxins in grain: Compounds other than Aflatoxin. Eagan Press. St. Paul, MN. p. 19-36.

Miller, J.D. 2001. Factors that affect the occurrence of fumonisin. Environ. Health Perspect. 109:321–324.

Mohamed, H.A., J.A. Clark, and C.K. Ong. 1988. Genotypic differences in the temperature responses or tropical crops. II. Seedling emergence and leaf growth or groundnut (*Arachis hypogaea* L.) and pearl millet (*Pennisetum typhoides* S. & H.). J. Exp. Bot. 39:1129–1135. doi:10.1093/jxb/39.8.1129

Monaco, M.K., J. Stein, S. Naithani, S. Wei, P. Dharmawardhana, S. Kumari, V. Amarasinghe, et al. 2014. Gramene 2013: Comparative plant genomics resources. Nucleic Acids Res. 42:D1193–D1199. doi:10.1093/nar/gkt1110

Mutava, R.N., P.V.V. Prasad, M.R. Tuinstra, K.D. Kofoid, and J. Yu. 2011. Characterization of sorghum genotypes for traits related to drought tolerance. Field Crops Res. 123:10–18. doi:10.1016/j.fcr.2011.04.006

Ong, C.K., and A. Everard. 1979. Short day induction of flowering in pearl millet (Pennisetum typhoides) and its effect on plant morphology. Exp. Agric. 15:401–410. doi:10.1017/S0014479700013053

Ong, C.K. 1983a. Response to temperature in a stand of pearl millet (Pennisetium typhoides S. & H.): IV. Extension of individual leaves. J. Exp. Bot. 34:1731–1739. doi:10.1093/jxb/34.12.1731

Ong, C.K. 1983b. Response to temperature in a stand of pearl millet (Pennisetium typhoides S. & H.): II. Reproductive development. J. Exp. Bot. 34:337–348. doi:10.1093/jxb/34.3.337

Paterson, A.H., Y.R. Lin, Z. Li, K.F. Schertz, J.F. Doebley, S.R. Pinson, S.C. Liu, J.W. Stansel, and J.E. Irvine. 1995. Convergent domestication of cereal crops by independent mutations at corresponding genetic loci. Science 269:1714–1718. doi:10.1126/science.269.5231.1714

Paterson, A.H., J.E. Bowers, R. Bruggmann, I. Dubchak, J. Grimwood, H. Gundlach, G. Haberer, et al. 2009. The Sorghum bicolor genome and the diversification of grasses. Nature 457:551–556.

Paul, C. 1990. Sorghum agronomy. ICRISAT, Patancheru, India.

Peacock, J.M., and G.M. Heinrich. 1984. Light and temperature responses in sorghum. In: Agrometeorology of sorghum and millet in the semi-Arid tropics: Proc. Int. Symp., Patancheru, India. 15-20 Nov. 1982. ICRISAT, Patancheru, India, p. 143-158.

Pearson, C.J. 1975. Thermal adaptation of Pennisetum: Seedling development. Aust. J. Plant Physiol. 21:413–424.

Prasad, P.V.V., K.J. Boote, and L.H. Allen, Jr. 2006. Adverse high temperature effects on pollen viability, seed-set, seed yield and harvest index of grain-sorghum [Sorghum bicolor (L.) Moench] are more severe at elevated carbon dioxide due to high tissue temperature. Agric. For. Meteorol. 139:237–251. doi:10.1016/j.agrformet.2006.07.003

Prasad, P.V.V., S.R. Pisipati, R.N. Mutava, and M.R. Tuinstra. 2008. Sensitivity of grain sorghum to high temperature stress during reproductive development. Crop Sci. 48:1911–1917. doi:10.2135/cropsci2008.01.0036

Prasad, P.V.V., M. Djanaguiraman, R. Perumal, and I.A. Ciampitti. 2015. Impact of high temperature stress on floret fertility and individual grain weight of grain sorghum: Sensitive stages and thresholds for temperature and duration. Front. Plant Sci. 6:1–11. doi:10.3389/fpls.2015.00820

Prasad, P.V.V., R. Bheemanahalli, and S.V.K. Jagadish. 2017. Field crops and the fear of heat stress- opportunities, challenges and future direction. Field Crops Res. 200:114–121. doi:10.1016/j.fcr.2016.09.024

Prasad, P.V.V., M. Djanaguiraman, S.V.K. Jagadish, and I.A. Ciampitti. 2018. Drought and high temperature stress and traits associated with tolerance. In: I.A. Ciampitti and P.V.V. Prasad, editors, Sorghum: State of the art and future perspectives. Agronomy Monograph 58. ASA, Madison, WI.

Quinby, J.R. 1966. Fourth maturity gene locus in sorghum. Crop Sci. 6:516–518. doi:10.2135/cropsci1966.0011183X000600060005x

Rajan, A.K., B. Betteridge, and G.E. Blackman. 1973. Differences in the interacting effects of light and temperature on growth of four species in the vegetative phase. Ann. Bot. (Lond.) 37:287–316. doi:10.1093/oxfordjournals.aob.a084693

Ranum, P., J.P. Pena-Rosas, and M.N. Garcia-Casal. 2014. Global maize production, utilization, and consumption. Ann. N. Y. Acad. Sci. 1312:105–112. doi:10.1111/nyas.12396

Rooney, L.W., and F.R. Miller. 1982. Variation in the structure and kernel characteristics of sorghum. In: L.W. Roonery and D.S. Murty, editors, Proceedings of the International Symposium on Sorghum Grain Quality, 28-31 October 1981, ICRISAT, Patancheru, India. International Crops Research Institute for the Semi-Arid Tropics, Patancheru 502 324, Andhra Pradesh, India. p. 143.

Rooney, W.L. 2014. Sorghum. In: D.L. Karlen, editor, Cellulosic energy cropping systems. John Wiley & Sons, Ltd, Hoboken, NJ. p. 109–129. doi:10.1002/9781118676332.ch7

Roozeboom, K.L., and P.V.V. Prasad. 2018. Growth and development. In: I.A. Ciampitti and P.V.V. Prasad, editors, Sorghum: State of the art and future perspectives. Agronomy Monograph 58. ASA, Madison, WI.

Saini, H.S. 1997. Effects of water stress on male gametophyte development in plants. Sex. Plant Reprod. 10:67–73. doi:10.1007/s004970050069

Sánchez, B., A. Rasmussen, and J.R. Porter. 2014. Temperatures and the growth and development of maize and rice: A review. Glob. Change Biol. 20:408–417. doi:10.1111/gcb.12389

Schoper, J.B., R.J. Lambert, and B.L. Vasilas. 1986. Maize pollen viability and ear receptivity under water and high temperature stress. Crop Sci. 26:1029–1033. doi:10.2135/cropsci1986.0011183X002600050038x

Setter, T.L., J. Yan, M. Warburton, J.M. Ribaut, Y. Xu, M. Sawkins, E.S. Buckler, Z. Zhang, and M.A. Gore. 2011. Genetic association mapping identifies single nucleotide polymorphisms in genes that affect abscissic acid levels in maize floral tissues during drought. J. Exp. Bot. 62:701–716. doi:10.1093/jxb/erq308

Sharma, I., N. Kumari, and V. Sharma. 2015. In: E. Lichtfouse and A. Goyal, editors, Sustainable agriculture reviews – Cereals. Series 16. Springer International Publishing, CH-4052 Basel, Switzerland.

Shaw, R.H. 1983. Estimates of yield reductions in corn caused by water and temperature stress. In: C.D. Ruper, Jr. and P.J. Kramer, editors, Crop relations to water and temperature stress in humid temperate climates. Westview Press, Boulder, CO. p. 49–66.

Shearer, J.F., L.E. Sweets, N.K. Baker, and L.H. Tiffany. 1992. A study of Aspergillus flavus, Aspergillus parasiticus in Iowa crop field- 1988-1990. Plant Dis. 76:19–22. doi:10.1094/PD-76-0019

Singh, P.M., J.R. Gilley, and W.E. Splinter. 1976. Temperature thresholds for corn growth in a controlled environment. Trans. ASAE 19:1152–1155. doi:10.13031/2013.36192

Snowden, J.D. 1936. The cultivated races of sorghum. Adlard and Son, London.

Stephens, J.C., F.R. Miller, and D.T. Rosenow. 1967. Conversion of alien sorghums to early combine genotypes. Crop Sci. 7:396. doi:10.2135/cropsci1967.0011183X000700040036x

Stone, P. 2001. The effects of heat stress on cereal yield and quality. In: A.S. Basra, editor, Crop responses and adaptations to temperature stress. Food Products Press, Binghamton, NY. p. 243–291.

Stone, L.R., and A.J. Schlegel. 2006. Yield-water supply relationship of grain sorghum and winter wheat. Agron. J. 98:1359–1366. doi:10.2134/agronj2006.0042

Tari, I., G. Laskay, Z. Takacs, and P. Poor. 2012. Responses of sorghum to abiotic stresses: A Review. J. Agron. Crop Sci. 199:264–274. doi:10.1111/jac.12017

Tefera, T., F. Kanampiu, H.D. Groote, J. Hellin, S. Mugo, S. Kimenju, Y. Beyene, P.M. Boddupalli, B. Shiferaw, and M. Banziger. 2011. The metal silo: An effective grain storage technology for reducing post-harvest insect and pathogen losses in maize while improving small holders farmers food security in developing countries. Crop Prot. 30:240–245. doi:10.1016/j.cropro.2010.11.015

Thompson, L.M. 1975. Weather variability, climate change and grain production. Science 188:535-541.

Thorson, P.R., and C.A. Martinson. 1993. Development and survival of Cercospora zeae-maydis germlings in different relative humidity environments. Phytopathology 83:153–157. doi:10.1094/Phyto-83-153

Tolk, J.A., and T.A. Howell. 2003. Water use efficiencies of grain sorghum grown in the three USA southern Great Plains soils. Agric. Water Manage. 59:97–111. doi:10.1016/S0378-3774(02)00157-9

Torres, O.A., E. Palencia, L. Lopez de Pratdesaba, R. Grajeda, M. Fuentes, M.C. Speer, A.H. Merrill, Jr., K. O'Donnell, C.W. Bacon, A.E. Glenn, and R.T. Riley. 2007. Estimated fumonisin exposure in Guatemala is greatest in consumers of lowland maize. J. Nutr. 137:2723–2729. doi:10.1093/jn/137.12.2723

Upadhyaya, H.D., R.P.S. Pundir, S.L. Dwivedi, C.L.L. Gowda, V.G. Reddy, and S. Singh. 2009. Developing a mini core collection of sorghum for diversified utilization of germplasm. Crop Sci. 49:1769–1780. doi:10.2135/cropsci2009.01.0014

Upadhyaya, H.D., D. Yadav, K.N. Reddy, C.L.L. Gowda, and S. Singh. 2011. Development of pearl millet minicore collection for enhanced utilization of germplasm. Crop Sci. 51:217–233. doi:10.2135/cropsci2010.06.0336

Upadhyaya, H.D., Y.-H. Wang, D.V.S.S.R. Sastry, S.L. Dwivedi, P.V.V. Prasad, A.M. Burrell, R.R. Klein, G.P. Morris, and P.E. Klein. 2016. Association mapping of germinability and seedling vigor in sorghum under controlled low-temperature conditions. Genome 59:137–145. doi:10.1139/gen-2015-0122

Vadez, V., T. Hash, F.R. Bidinger, and J. Kholova. 2012. II.1.5 Phenotyping pearl millet for adaptation to drought. Front. Physiol. 3:1–12. doi:10.3389/fphys.2012.00386

Varshney, R.K., C. Shi, M. Thudi, C. Mariac, J. Wallace, P. Qi, H. Zhang, Y. Zhao, et al. 2017. Pearl millet genome sequence provides a resource to improve agronomic traits in arid environments. Nat. Biotechnol. 35:969–976. doi:10.1038/nbt.3943

Vartapetian, B.B., and M.B. Jackson. 1997. Plant adaptations to abiotic stress. Ann. Bot. (Lond.) 79:3–20. doi:10.1093/oxfordjournals.aob.a010303

Verhulst, N., B. Govaerts, E. Verachtert, A. Castellanos-Navarrete, M. Mezzalama, P. Wall, J. Deckers, and K.D. Sayre. 2010. Conservation agriculture, improving soil quality for sustainable production systems? In: R. Lal and B.A. Stewart, editors, Advances in soil science: Food security and soil quality. CRC Press, Boca Raton, FL. p. 137-208.

Waddington, S.R., X. Li, J. Dixon, G. Hyman, and M.C. de Vicente. 2010. Getting the focus right: Production constraints for six major food crops in Asian and African farming systems. Food Secur. 2:27–48. doi:10.1007/s12571-010-0053-8

Walker, J.M. 1969. One-degree increments in soil temperatures affect maize seedling behaviour. Soc. Soil Sci. Am. J. 33:729–736. doi:10.2136/sssaj1969.03615995003300050031x

Ware, D., P. Jaiswal, J. Ni, et al. 2002. Gramene: A resource for comparative grass genomics. Nucleic Acids Res. 30:103–105. doi:10.1093/nar/30.1.103

Weaich, K., K.L. Bristow, and A. Cass. 1996. Modelling pre-emergent maize shoot growth II. High temperature stress conditions. Agron. J. 88:398–403. doi:10.2134/agronj1996.00021962008800030007x

Zaidi, P.H., and N.N. Singh. 2001. Effect of waterlogging on growth, biochemical compositions and reproduction in maize. J. Plant Biol. 28:61–69.

Zaidi, P.H., S. Rafique, P.K. Rai, N.N. Singh, and G. Srinivasan. 2004. Tolerance to excess moisture in maize (*Zea mays* L.): Susceptible crop stages and identification of tolerant genotypes. Field Crops Res. 90:189–202. doi:10.1016/j.fcr.2004.03.002

Agroclimatology of Oats, Barley, and Minor Millets

M. Djanaguiraman, P.V.V. Prasad,* Z.P. Stewart, R. Perumal, D. Min, I. Djalovic, and I.A. Ciampitti

Abstract

Understanding the interactions of atmospheric variables and biological systems in agriculture, and applying this knowledge to improve productivity is critical. One of the most vulnerable sectors to changes in environmental conditions is rainfed agriculture because of limited and unreliable rainfall. Minor coarse grain crops such as oats, barley and minor millets (finger millet, fox tail millet, kodo millet, and proso millet) are primarily grown under rainfed conditions and in marginal lands of temperate, arid and semiarid regions. Yield potential of these crops is low compared to other cereals; however, they have unique nutritional qualities. Understanding growth and development stages and climatic requirements is important for development of best and efficient management practices. Oats are a cool climate crop and hence, are susceptible to hot and dry weather from heading until grain filling period. Barley can be grown during summer and winter seasons and is highly sensitive to frost at all growth stages. Minor millets are relatively more resilient to extreme climatic conditions, compared to other cereals. The area under cultivation of these coarse cereals is declining due to, non-availability of high yielding varieties, limited crop management options, high labor involved in their processing and negative perceptions of minor millets as a food for the poor relative to other crops; and limited support and procurement policies. The importance of minor millets for global food and nutritional security is increasing due to medicinal and nutritional properties and tolerance to abiotic and biotic stress.

Coarse cereals refer to grains other than wheat and rice and includes maize (*Zea mays* L.), sorghum (*Sorghum bicolor* L.), oats (*Avena sativa L.*), rye (*Secale cereale* L.), barley (*Hordeum vulgare* L.), pearl millet [*Pennisetum glaucum* (L.) R. Br.] and other minor millets such as finger millet [*Eleusine coracana* (L.) Gaertn.], kodo millet (*Paspalum scorbiculatum* L.), proso millet (*Panicum miliaceum* L. ssp. *miliaceum*),

M. Djanaguiraman, P.V.V. Prasad, D. Min, and I.A. Ciampitti, Department of Agronomy, Kansas State University, Manhattan, Kansas 66506; M. Djanaguiraman, Department of Crop Physiology, Tamil Nadu Agricultural University, Coimbatore, India; P.V.V. Prasad, Feed the Future Innovation Lab for Collaborative Research on Sustainable Intensification, Kansas State University, Kansas 66506; R. Perumal, Agricultural Research Center, Kansas State University, Hays, Kansas 67601; I. Djalovic, Institute of Field and Vegetable Crops, Novi Sad, Serbia. *Correspondence: P.V.V. Prasad (vara@ksu.edu)

doi:10.2134/agronmonogr60.2018.0020

© ASA, CSSA, and SSSA, 5585 Guilford Road, Madison, WI 53711, USA.
Agroclimatology: Linking Agriculture to Climate, Agronomy Monograph 60.
Jerry L. Hatfield, Mannava V.K. Sivakumar, John H. Prueger, editors.

foxtail millet (*Setaria italica* (L.) P. Beauv.), little millet (*Panicum sumatrense* Roth ex Roem. & Schult.) and barnyard millet [*Echinochloa frumentacea* Link/*Echinochloa utilis* Ohwi et Yabuno/*Echinochloa esculenta* (A. Braun) H. Scholz Except maize, sorghum and pearl millet, all other coarse grains in general are referred to as minor coarse grains or minor cereals. Across the globe, these crops are often cultivated on marginal lands of temperate, subtropical, and tropical regions. Even though the contribution of minor coarse grains to food production is very small (1.5%), it is essential for food security in many developing countries, especially in Asia and Africa. The minor coarse grains are warm season cereals, grown under rainfed farming systems with little to no external inputs and yield is often very low (less than one-ton per hectare) compared with other grain crops. Coarse grains are gaining importance because of their proximal constituents like dietary energy, vitamins, iron, zinc, insoluble dietary fiber, and phytochemicals with antioxidant properties (Bouis, 2000). Considering these nutritional properties, coarse grains have been designated as nutricereals. Minor coarse grains, namely oat, barley, and minor millets, are grown as alternate crops in areas where maize and sorghum fail to grow. Most of the minor coarse cereal crops have short life cycles; hence, are best suited in the regions where rainfall is low or erratic. Among the minor coarse grains cultivated in the world, barley leads the area under cultivation, followed by oats and others. In this chapter, growth stages, climatic requirements and management practices of oat, barley, and minor millets are discussed.

Oats

The cultivation of oats began about 4000 yr ago, and its center of origin is postulated to be in the Mediterranean basin or the Middle East. Domestication of oats occurred much later than wheat and barley. European colonists and immigrants brought oats to the United States, southern Canada, Australia, and New Zealand, where it became an important winter season crop. Oats occur in three ploidy levels, diploid, tetraploid, and hexaploid, with a base chromosome number of 7. The primary cultivated oat is hexaploid of $2n = 6x = 42$, originating as an aggregation of three diploid genomes (AA, CC, DD). The major hexaploid species include the winter habit weedy species, *A. sterilis,* the spring habit weedy species, *A. fatua,* and the cultivated species, *A. sativa.* The diploid oat is often referred to as "black oat". Oats are an annual grass grown in the temperate regions of Russia, Canada, United States, Nordic regions like Finland, Sweden, Norway, Northern British Isles, and cooler regions of the Central European Plains. Most oat seeds are characterized by the retention of the lemma and palea surrounding the caryopsis. However, there is a hulless variant called "naked oat".

Currently, oats are an important grain and forage crop in many parts of the world, grown as a monocrop or rotated with other cereals or legumes in 9.4 million hectares with a grain production of 22.9 million metric tons (FAOSTAT, 2016). Oats prefer cooler temperatures but can also grow in Mediterranean climate. In the Canadian Plains, oats are cultivated for grazing and silage. During 2016, the oat belt was 2.75 million hectares in Russian Federation, 0.93 million hectares in Canada, 0.83 million hectares in Australia, 0.48 million hectares in Poland, 0.47 million hectares in Spain, 0.40 million hectares in the United States of America and 0.34 million hectares in Brazil (Table 1; FAOSTAT, 2016). The average productivity of oats is 2.0 Mg of grain ha^{-1} (FAOSTAT, 2016). The inherent soil fertility of the geographic region, rate of fertilizer application, irrigation and agroclimate defines productivity. Oat grain has the highest protein content among the various cereals (12 to 20% in the dehulled kernel and 9 to 15% in the whole grain) and the content varies with genotype and growing environmental conditions (Peterson, 1992). Furthermore, it has relatively higher level of essential amino acids compared with wheat, maize and barley.

Growth and Development

Knowing how a crop develops is essential for comprehending the plant's response to environmental stresses and for providing knowledge that can inform management decisions. The duration of each growth stage depends on the genotype and the growing environment. Other factors, such as soil fertility, insect or disease damage, moisture stress, and weed competition, can also affect the duration of each growth stage. Oats belong to the monocotyledons and the seedlings have one cotyledon or seed-leaf and the root system is fibrous in nature. Oats produce branches, tillers at the base of the stem and the leaves emerge from the nodes with parallel venation. The stem or culm is solid and erect, which consists of hollow cylindrical tubes called internodes. The first stem is called the main stem, which produces tillers. Primary tillers arise in the axils of the lower leaves of the main stem at its base. Secondary tillers arise in the axils of primary tillers. Tillers arise during early growth stages between emergence of the third leaf and stem elongation. Not all tillers produced survive, and generally, the older tillers are most likely to survive.

The leaves are arranged in two rows alternating on opposite sides of the stem. The flat upper part of the leaf, the lamina, is joined at the sheath, the basal part where the leaves meet at the stem. The root system comprises of seminal and adventitious roots. Three to four seminal roots develop from the root primordia in the embryo. Adventitious roots arise at the basal nodes of the main stem and

Table 1. Top five countries for area under cultivation, total production and productivity of oats and barley.†

Area under cultivation (m ha)		Total production (Mg)		Productivity (kg ha^{-1})	
Countries	Value	Countries	Value	Countries	Value
			Oats		
Russia	2.75	Russia	4.76	Ireland	7901
Canada	0.93	Canada	3.02	United Kingdom	5787
Australia	0.82	Poland	1.36	New Zealand	5535
Poland	0.48	Australia	1.30	Denmark	5074
Spain	0.47	Finland	1.04	Chile	4945
			Barley		
Russia	8.13	Russia	17.99	United Arab Emirates	7976
Australia	4.11	Germany	10.73	Ireland	7822
Ukraine	2.86	France	10.31	New Zealand	7018
Spain	2.80	Ukraine	9.44	Netherlands	6862
Turkey	2.70	Australia	8.99	Saudi Arabia	6755

†Source: FAOSTAT (2016)

Fig. 1. Oats panicle from www.cropchatter.com/blast-in-oats. The inflorescence in compound, consisting of a series of flowering branches called spikelets (A). The spikelets is arranged in a panicle or directly on the main axis as a raceme. Each spikelet has one or more individual florets (B).

tillers which form the crown. The adventitious roots constitute the major part of the root system, these roots are produced continuously throughout the early stage of the life cycle up to anthesis. The inflorescence is compound, consisting of a series of flowering branches, namely spikelets. The spikelet may be arranged in a panicle or directly on the main axis as a raceme. Each spikelet has one or more individual flowers called florets (Fig. 1). The floret contains the female part (a superior ovary) and the male parts (the stamens) and is composed of three or a multiple of three. Sometimes, scale-like leftovers of the petals or lodicules may be found at the base of the floret. The floret is enclosed within two protective bracts or scales, the outer lemma and the inner palea. Following fertilization, the single ovary develops into a caryopsis, including an embryo and an endosperm.

The term growth refers to change in biomass or height or morphological changes, while, development refers to changes in stages and tissue or organ differentiation process at the apex. The growth and development of oats follows very similar patterns to wheat and matures in 6 to 11 months. Winter oat varieties take approximately 12 d from sowing to emergence, 267 d from sowing to anthesis, and 58 d from anthesis to harvest. However, spring oats take approximately 22 d, 100 d, and 58 d, respectively for sowing to emergence, sowing to anthesis, and anthesis to harvest, respectively. The distinction between autumn and spring sown types is the main determinant of maturation period. However, cultivar and environment also influence the maturation period (Table 2). Development is continuous and can be divided into vegetative and reproductive phases. Several scales for the description of macromorphological growth stages, visible to naked eye, have been proposed. Numerical scales like Zadoks, Feekes, and Haun have been devised to quantify the growth stages. The Zadoks system uses a two-digit code referring to the principal stage of growth from emergence (stage 0) to kernel ripening (stage 9) (Fig. 2). The second digit subdivides each principal growth stage (Zadoks et al., 1974). For example, '13' indicates principal stage 1 (seedling stage) subdivision 3 when leaves are at least 50% emerged from the main stem. Growth stages of oats can be classified into eight major stages, namely: germination, leaf production, tiller production and survival, stem elongation, preanthesis panicle development terminating in panicle emergence, anthesis, grain filling, and ripening.

Germination (Zadoks Growth Stage: ZGS 00–09)

This stage can be identified by emergence of the plumule and radical from the oat seed. Emergence of the seedling occurs in two to three weeks after sowing.

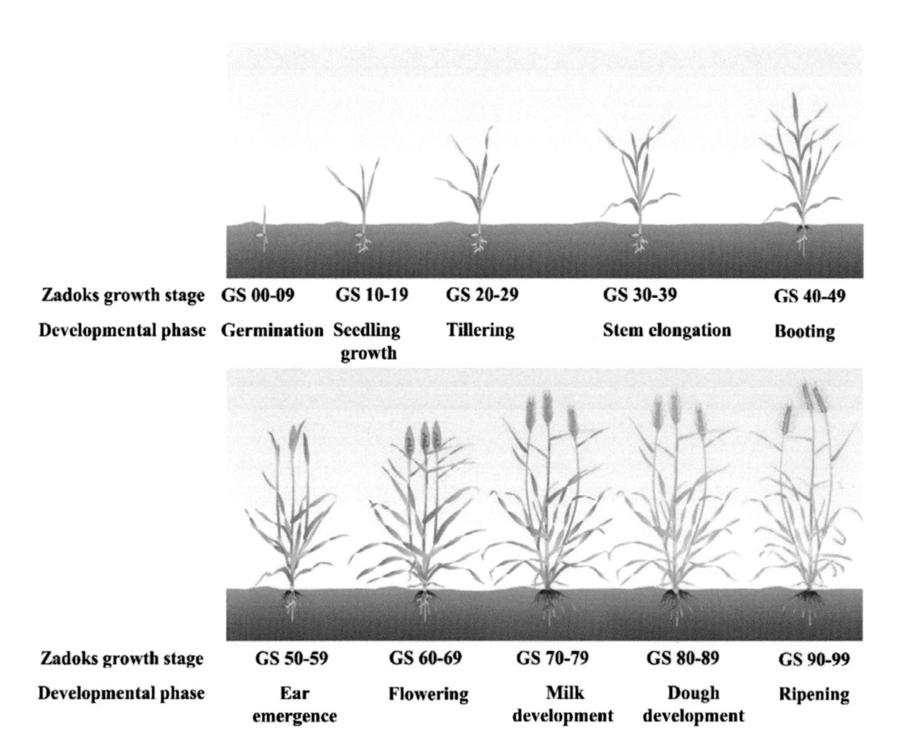

Fig. 2. Zadoks cereal growth stages ranging from 0 to 99 (Image from the GRDC cereal growth stages guide).

Table 2. Oat and barley phenology calculations using 0 °C as base temperature using universal growth staging scale (Miller et al., 2001).

Stage	Description	Growing degree days (°C)
	Oat	
Anthesis	Flowering commences	760–947
Seed fill	Seed fill begins. First grains have reached half of their final size.	1019–1229
Dough stage	Soft dough stage, grain contents soft but dry, fingernail impression does not hold.	1380–1625
Maturity complete	Grain is fully mature and dry down begins. Ready for harvest when dry.	1483–1738
	Barley	
Emergence	Leaf tip just emerging from above-ground coleoptile.	109–145
Leaf development	Two leaves unfolded.	145–184
Tillering	First tiller visible	308–360
Stem elongation	First node detectable	489–555
Anthesis	Flowering commences	738–936
Seed fill	Seed fill begins. First grains have reached half of their final size.	927–1145
Dough stage	Soft dough stage, grain contents soft but dry, fingernail impression does not hold.	1193–1438
Maturity complete	Grain is fully mature and dry down begins. Ready for harvest when dry.	1269–1522

Leaf Production (ZGS 10–19)

Leaves emerge on the main stem at regular intervals depending on temperature. The tips of the younger emerging leaves appear within the older leaves. As such, the oldest leaf is found at the base and on the outside of the plant. Leaf emergence happens from seedling emergence to panicle emergence and will vary from 9 to 12 leaves per plant. Initiation of leaves at the apical meristem (plastochron) occurs at a faster rate than the rate at which leaves emerge (phyllochron). Bunting and Drennan (1966) found that the plastochron at spring temperatures takes two or three days; however, the phyllochron takes from five to seven days. Cell division leads to an increase in size of the leaf primordium, followed by cell expansion in both the leaf lamina and sheath (White, 1995). The initiation of leaves continues until the transition phase of development, which occurs by slight elongation of the shoot apex prior to initiation of the panicle; the latter being influenced by seed rate, growing environment, vernalization, and photoperiod.

Tiller Production (ZGS 20–29)

The sequence of tiller production is synchronized with leaf production. Tiller production occurs during the two- or four-leaf stage. Primary tillers are produced in the axils of leaves on the main stem and secondary tillers in the axils of leaves of the primary tillers. Tiller growth depends on environmental conditions and older tillers are more likely to survive than younger tillers. Tiller initiation ceases when panicle development commences.

Stem Elongation (ZGS 30–39)

The stem elongation stage is characterized by sequential extension of the internodes of the four to seven younger leaves and of the peduncle, beginning with the oldest internode, which results in rapid increase in stem height. This stage occurs before panicle emergence. When oats are sown during autumn, overwintering during the vegetative stage causes slow growth of leaves and tillers, and stem elongation does not begin until temperature increases during spring.

Preanthesis Panicle Development Terminating in Panicle Emergence (ZGS 41–59)

The apical meristem first produces leaf primordia and then produces the primordia of the panicle branching. A change in shape of the meristem, namely a longer and more cylindrical elongation of the apical meristem, coincides with its transition from initiating leaf primordia to initiating panicle branch primordia. Formation of double ridges is the first step in the reproductive phase, during which pollen and pistil differentiate in a series of events. Spikelet differentiation begins at the tip of the main axis of the panicle and branches, proceeding basipetally so that the basal branches are the last to differentiate. Only single spikelets, as opposed to branched clusters of spikelets, are produced at the tips of the main axis and branches. Glumes will differentiate first followed by florets within the spikelet. The spikelet appears in an alternate fashion on the axis above the glume primordia and develops acropetally. Consequently, the basal floret is most advanced and will likely have larger grain than the florets above it. The second floret is usually fertile but the third may or may not produce a kernel depending on the position of the branch within the panicle and also depending on cultivar and environmental conditions. The floret parts develop acropetally in the following order: lemma, palea, stamens, lodicules, and pistil, and then the ovule primordium terminating the floral axis. As a result of the sequential development of spikelets and florets, a hierarchy of both the number and size produced is set up within the panicle.

Anthesis (ZGS 60–69)

At anthesis, the anther dehisces and ripe pollen is shed onto the feathery stigmas. The florets may open due to swelling of the lodicules at their base so that the anthers are exerted to hang outside the glumes. Oats are self-fertilized, so opening of the florets in this way is not essential to achieve pollination.

Grain Filling (ZGS 71–85)

Once pollinated, fertilization occurs within 24 h. The endosperm nucleus is formed by fusion of one or two male cell nuclei with the two polar nuclei in the embryo sac of the ovule. The zygote is formed by the fusion of the other male cell nucleus with the egg. The zygote will not begin to divide immediately; instead, several division cycles will be completed before the organs of the embryo are formed. The grains increase in size and weight as sugars are imported from photosynthesizing parts of the plant and converted into starch which is stored in the cells of the endosperm. Initially, water content will be high in the developing embryo and starch content will be low; however, during the developmental phase, accumulation of starch will occur and the water content will be reduced.

Ripening (ZGS 90–99)

Ripening occurs with senescence of older leaves. Following fertilization, senescence progresses as the leaves, internodes, and panicle lose their ability to photosynthesize. The whole plant dries out and the grain hardens as its water content decreases to less than 20%, at which stage it can be harvested.

Climatic Requirements

Oats are grown between 40 and 60 °N latitudes in North America, Europe, and Asia. The maritime climate also favors oat cultivation in Northern Europe. Oats are also grown in South America, Australia, and New Zealand. Oats grown at high altitudes or high latitudes, are more frequently damaged by autumn frosts prior to physiological maturity than by frosts at early growth stages. Oats are grown in cool and moist climates because it requires moisture to produce more dry matter. It is susceptible to hot and dry weather from heading until grain filling.

Temperature

Air temperature and photoperiod are meteorological elements that influence growth and development of oats. In general, winter oats are grown in areas with relatively mild winters and warm summers, while spring oats are grown in areas with relatively colder winters and cooler summers. Oat production in the winter–spring transition zone is risky because low temperature stress may cause winter injury and high temperature during post-juvenile growth and development may severely reduce grain yield of either winter or spring oat cultivars. Among the oat cultivars, the major difference in maturation period occurs in the length of the vegetative growth stage and not the grain-filling period. The earliness of early-maturing cultivars is due to a shortened vegetative growth stage rather than by a shortened in the grain-filling period. Soil and air temperatures are the most critical parameters for explaining the duration of seedling emergence and the vegetative to anthesis, respectively. Temperature has a profound effect on the rate of initiation and production of leaves and floral structures whereas daylength determines the final leaf number (Sonego, 2000). Winter oats require very low soil temperatures (3–4°C) for at least 10 to 56 d for vernalization. The seedling and vegetative stages are not sensitive to low temperatures and frost. However, the reproductive stages are sensitive to low temperature and frost. Freezing temperatures and frost can damage oat growth and development leading to reduction in yield. Seedling emergence, flowering and pollination are the most sensitive stages to chilling temperatures compared to other stages.

According to Leite et al. (2012), the maximum temperature for oat development is 35 °C and the minimum is 0 °C. The minimum, optimum, and maximum temperatures for oat development were 4, 22, and 30 °C from emergence to anthesis and 15, 25, and 35 °C from anthesis to maturation, respectively. High temperature during sporogenesis and flowering cause decreased seed-set; similarly, high temperature stress after anthesis causes reduced seed weight (Eagles et al., 1978). Winter oats cope extremely well with midwinter growth conditions; with only four hours of direct sunlight and daily mean temperatures of around 5 °C (Korner, 2008). Higher temperatures increased rate of growth from emergence to heading stages. At air temperatures of 24, 18.5 and 13 °C, oats took 50, 60, and 74 d, respectively, to reach the heading stage (Fulton, 1968). Grain yield decreased substantially as ambient temperature from emergence to the heading stage increased (Fig. 3a; Fulton, 1968). An increase in soil temperature up to 20 °C increases oat biomass and grain yield; however, further increase in temperature to 27 °C results in reduced grain yield (Nielsen et al., 1960). Hellewell et al. (1996) observed that eight hours of high nighttime temperature is more detrimental than 16 h of high daytime temperature as evident by decreased grain yield, kernel weight and length of the grain filling duration.

Rainfall

Oats can be classified into three botanical groups: (i) wild oats, (ii) the arid region red oats, and (iii) the cool or humid region white oats. The red oat varieties are best adapted to low desert valleys because they can endure winter frost, mature earlier, and suffer less damage from high temperature and drought stress than white oats. The white oats are suited to cool and high or low elevation. Sandhu

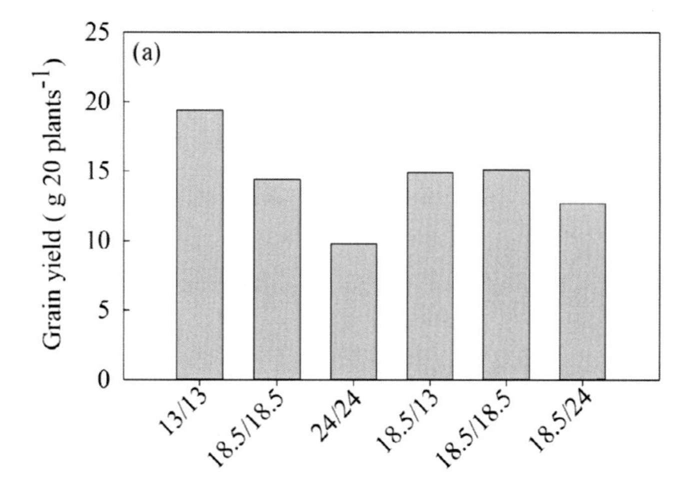

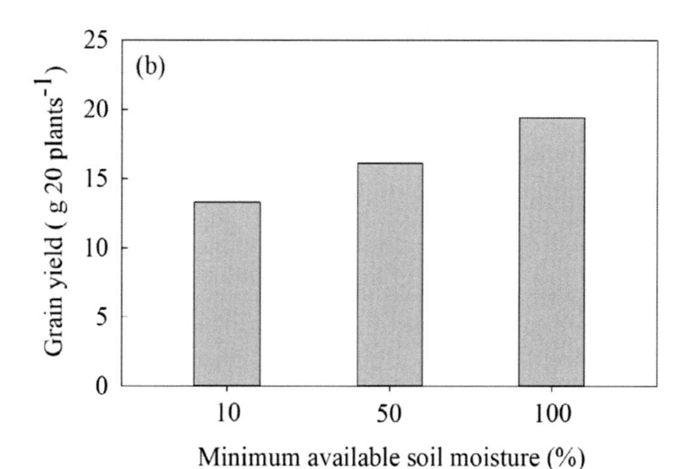

Fig. 3. Impact of season long (a) air/soil temperature and (b) moisture level on oat grain yield (Redrawn and adapted from Fulton, 1968).

and Horton (1977a) reported that severe drought stress for 10 d prior to panicle initiation and 10 d after panicle emergence caused 80% reduction in photosynthetic rate. However, seed yield was decreased by 20 and 58% by drought stress prior to panicle initiation and after panicle emergence, respectively. Yield loss due to drought stress prior to panicle initiation is associated with a reduction in the number of florets per panicle and drought stress after panicle emergence is associated with decreased numbers of panicles per plant and florets per panicle, coupled with an increase in floret sterility (Sandhu and Horton, 1977b). Chinnici and Peterson (1979) showed that all drought treatments imposed at four different stages of panicle development reduced yield, particularly at the booting stage. The black oat types have a better ability to withstand periods of early summer drought than modern white oats because of its superior root mass: shoot ratio during the seedling stage and an extended period of tillering. Oats grown in Mediterranean environments are highly susceptible to drought stress from germination to the early tillering and from anthesis to maturity. Oats are sensitive to waterlogging during early flower development or just before flowering. Oats are more affected by waterlogging than other cereals, with a comparatively greater reduction in tillering and number of grains per inflorescence. However, oats compensate better for such losses later in development, in particular by increased grain weight (Watson et al., 1976). Research by Fulton (1968) indicated that oat yield increased linearly with soil moisture availability (Fig. 3b).

Photoperiod

For genotypes that require vernalization, response to photoperiod sensitivity occurs only after the vernalization requirement is met. Response to photoperiod prior to heading is strong in northern European oat cultivars, whereas cultivars originating from southern areas are less responsive to daylength, especially at low temperatures. It is suggested that the overall duration of the growth period has remained unchanged, but the length of the vegetative and grain filling period has slightly shortened, so that the period for spikelet set is prolonged leading to increased grain yield. This is supported by the finding that oat cultivars tested in southern Finland had shorter vegetative period, lower grain-straw ratio and longer period for maximum floret production, higher panicle weight and the number of grains per panicle (Peltonen-Sainio, 1994).

In sub-tropical environments, oats are grown under no-till conditions rotated with soybean. Under these conditions, flowering time is more important, even as sowing time can be anytime from March to July. Oats are a long-day plant; thus, short night promotes flowering in most genotypes. Many cultivars do not produce seeds under short-daylengths and flowering is hastened under long-daylengths. Flowering is a response to the interaction of photoperiod and temperature. Cultivars grown in high latitudes respond more to photoperiod (delayed flowering in short-day conditions), and cultivars grown in low latitudes respond in an opposite way (Sorrels and Simmons, 1992). Daylength-sensitive cultivars may mature too late. White oats are sensitive to photoperiod and may not head until April. Winter oat varieties require a period of cold before they will produce heads. Genotypes vary in their responsiveness to vernalization, some genotypes require vernalization for panicle initiation, whereas in other genotypes vernalization reduces heading date. Even varieties grown in sub-tropical environments can respond to cold treatment. Vernalization here is defined as the acquisition or acceleration of the ability to flower by a chilling treatment (Chouard, 1960).

Production Practices

Oats are a cool climate crop and do not grow well in hot, dry conditions. They are grown in wide range of soils; however, they thrive best in well drained loamy soil with good organic matter. Optimum soil pH is between 5.0 and 6.5, but oats can tolerate high soil pH. The planting time for spring oats is from January to April and for winter oats from September to October. Planting is similar to wheat in terms of depth and row width. The seed is usually placed 2 to 4 cm below the soil surface under no-till conditions. Under irrigated conditions, seed depth is not critical, as moisture is available; however, under dryland conditions, seed placement is critical because the seed must be placed within the moist soil layer. Seeds placed above this layer or too deep will not germinate or will not reach the surface. In early, normal and late sowing conditions, a plant density of 175 to 200, 200 to 275 and 275 to 350 plants m^{-2} should be maintained to obtain optimum grain yield. Tillering in spring oats is lower than in winter types, and therefore much higher seeding rates (80 to 120 kg ha^{-1}) are needed.

Nutrient management of the oat crop is determined by soil and nutrient management strategies, including the previous crop, soil water and nutrient availability, yield potential, risk of lodging, timing of nutrient applications and nutrient sources available for use. Soil and/or plant tissue testing and regionally specific oat nutrient response functions are essential for determining adequate nutrient application rates. Oats grown in irrigated conditions require 90 kg of nitrogen (N), 20 to 25 kg of phosphorous (P) and 20 kg of potassium (K) for a grain yield potential of 4.5 Mg ha^{-1}. Under dryland conditions in high rainfall regions, 40 kg N, 10 kg P, and 10 kg K is recommended. Oats are generally grown in rainfed conditions; hence, irrigation may be supplied as needed. The current recommendation is five irrigations during the growing season (namely five-leaf, early stem elongation, flag-leaf, flowering, and grain-filling stages). Once the crop canopy is closed by the establishment of a solid crop stand, herbicide application is usually not required. However, in intercultural operations, hoeing once or twice, can be effective for early-season weed control. Broadleaf weeds can be controlled easily with recommended rates of appropriate herbicides. The important insect pests like aphids, bollworm, grain chinch bug, grain slug and mites and diseases like rust, mildew, spots and blotches, bunts and viral diseases should be diagnosed with the help of an expert. Recommendations based on their diagnosis may be adopted. Oats can be harvested at grain moisture content below 20%.

Barley

Barley belongs to the tribe *Triticeae,* and its cultivation has been traced back to 10,000 yr ago. It is postulated that cultivated barley was derived from *H. spontaneum* K. Koch. There are two different views about the origin of barley, namely, monophyletic and diphyletic. It is believed that barley was domesticated at least twice, over a similar period of time, at two different locations. Barley includes 32 species, and most of the cultivated barley species are diploids ($2n = 2x = 14$), but tetraploids ($2n = 4x = 28$) and hexaploids ($2n = 6x = 42$) are also present. Cultivated barley is classified into three major types: (i) six-rowed barleys, *H. vulgare*

L., (ii) two rowed barleys, *H. distichum* L., and (iii) irregular barleys, *H. irregulare* E. Aberg and G.A. Wiebe sp. nov. *H. vulgare* barley has three fertile flowers at each node of the rachis. However, in two rowed barley the median florets are fertile. In *H. irregulare*, the median florets are fertile, while the proportion of fertile and wanting lateral florets varies considerably. Barley is grown on around 47 million hectares worldwide, thereby representing the fourth most widely-grown cereal crop after wheat, maize, and rice. Barley is grown in parts of Europe, Asia, North America, and Australia. The top countries producing barley are Russia (18 million Mg), Germany (11 million Mg), France (10 million Mg), Ukraine (9 million Mg), Canada (9 million Mg), Australia (9 million Mg), Spain (8 million Mg), Turkey (7 million Mg), United Kingdom (4 million Mg) and United States of America (4 million Mg) (FAOSTAT, 2016).

The current global barley yield is 2.0 Mg ha^{-1} (FAOSTAT, 2016), although under favorable conditions, barley yield can be as much as 7 Mg ha^{-1}. The maximum reported on-farm yield of winter and spring barley was 13 and 10 Mg ha^{-1}, respectively. The majority of the world's barley production is from spring barley varieties, since it has wide adaptation to various environments and does not require vernalization. Winter barley varieties are sown in autumn and need vernalization. They can withstand a low temperature of -20 °C and yield 2 Mg ha^{-1} higher than spring varieties. Spring barley production predominates in Russia, Canada, Australia, South America, and Scandinavia. Spring barley matures in 90 to 120 d after planting and is mostly preferred to be grown as an alternate crop in spring wheat and maize crop rotation systems. In high latitudes, barley is sown during spring to avoid colder winters. Therefore, in these regions spring-type cultivars are grown. At low latitudes, air temperature is too high to induce vernalization in a winter type. In mid-latitudinal regions, including North Africa, southern Europe, Nepal, China and Japan, both spring and winter barley cultivars are generally sown in autumn. Barley is well adapted to different soils in temperate countries. Barley withstands alkalinity, but is susceptible to acid soils. It grows well on soils with low fertility and regions with relatively lower levels of precipitation, around 300 to 400 mm.

Growth and Development

After seed germination, shoot growth of barley is dependent on the activity of the shoot apical meristem, which produces the internode, leaf, and axillary bud. Barley is characterized by distichous arrangement of leaves and spikelets on the vegetative stems and inflorescence axes, respectively. The leaves are strap shaped with parallel venation and a prominent midrib. The leaf blade and proximal sheath are separated by the ligule. In barley, the tunica originates the epidermal layer and part of the mesophyll and the corpus gives rise to all other parts of the plant. Leaf growth is the result of the coordinated cell division and cell elongation process. Development of the vegetative internode involves both radial and vertical growth. The number of leaves varies by time of sowing and genotype. The last vegetative leaf formed on the culm is called the flag leaf, and the sheath of the flag leaf encloses the spike.

In barley, lateral shoots called tillers develop from the vegetative axillary meristems, which are present in the axils of the leaves at the base of the plant crown. Tiller development involves three stages: (i) initiation of vegetative axillary meristems, (ii) cell division and (iii) differentiation of vegetative axillary meristem cells into leaf primordia which produces an axillary bud and outgrowth of the axillary bud leading to tiller production. Primary tillers formed by the main culm can produce secondary tillers, which in turn may produce tertiary tillers. The mature barley inflorescence called spike, head, or ear consists of floral stem or rachis and floral units called spikelets. The spikelets consist of a floret and two subtending bracts called empty or outer glumes. Each rachis node bears three spikelets. In two-row barley cultivars, the central spikelet is fertile and the lateral spikelets are sterile or underdeveloped. In six-rowed barley genotype, all three florets mature to produce grains.

Two to four weeks after germination, transition from vegetative to the reproductive phase in spring barley will occur, provided that temperature and photoperiod are favorable. The double ridge stage marks the transition of the shoot apical meristem to an inflorescence meristem. During this stage, the apex will be elongated to 1 mm in length and each double ridge corresponds to a pair constituted by a leaf primordium and a lateral meristem. As inflorescence development progresses, leaf initiation fails to develop; however, the lateral meristems become the main growing points that will originate spikelet triplet meristems that in turn develop three spikelet meristems, namely, one central and two laterals. This is called the triple mound stage. Each spikelet meristem gives rise to a floral meristem. A detailed description of histogenesis and the sequence of floral organ differentiation was reported by Bossinger et al. (1992). Spikelet differentiation along the immature inflorescence is not uniform; spikelets in the central region develop earlier than the basal and apical spikelets. The apex continues to initiate new spikelet meristems until the awn primordium stage. After this stage, the spikelet undergoes growth and further differentiation followed by fertilization, caryopsis development, and grain filling.

After pollination, one of the two sperm cells released by the pollen tube enters the egg cell to produce the diploid zygote. The other sperm cell combines with the diploid nucleus of the central cell resulting in the primary triploid endosperm nucleus. This is called double fertilization, which initiates seed development. After fertilization, barley grains start to develop and enter the pre-storage phase when cell division and elongation takes place. During this stage, the lateral vascular bundles supply the photo-assimilates to the developing pericarp for growth and storage product accumulation. Caryopsis elongation concludes within 10 d after the flowering stage. After 10 d, the caryopsis strength is established, resulting in massive accumulation of starch and storage proteins in the endosperm and aleurone layers.

On average, barley genotypes spend about 60% of their life cycle in vegetative growth and about 40% of their life cycle in grain filling. The most distinct growth stages of barley are (i) germination and seedling establishment, (ii) stem extension, which includes jointing and shooting, (iii) heading and (iv) ripening. The growth stages of barley can be classified using Large–Feekes scale (Briggs, 1978) or decimal code system (Tottman et al., 1979). The Large–Feekes stage was originally developed to describe wheat and was modified for barley.

Large-Feekes scale of Barley Growth and Development

Germination and Seedling Establishment

Stage 0: Initial growth: Seed germination and growth occurs below the soil surface.

Stage 1: Coleoptile reaches the soil surface and leaf growth: The leaves appear and are numbered in sequence. Each leaf is visible when the auricle appears. During this stage, the crown is formed from which crown roots are formed.

Stage 2: Tillering: This stage is identified by the formation of new tillers. The main stem appears at ground level.

Stage 3: Secondary tillering: Secondary tillers grow from the crown. The number of secondary tillers vary between genotypes. Tertiary and even quaternary tillers may also arise from the crown. Some genotypes are more or less prostrate up to this stage.

Stage 4: Pseudo-stem: The pseudo-stem of rolled leaf sheaths begins to erect and the leaf sheaths begin to elongate.

Stage 5: Pseudo-stem erect: The pseudo-stem is erect.

Stem Extension (Jointing and Shooting)

Stage 6: First stem node visible. The first stem node is visible at the shoot base.

Stage 7: Second stem node visible. The second stem node is visible at this stage.

Stage 8: Appearance of flag leaf. The tip of the flag leaf, the last leaf, will be visible; however, the flag leaf is still rolled.

Stage 9: Appearance of the ligule of the flag leaf. The ligule of the flag leaf is visible. Basal leaves begin to senesce. Gametogenesis occurs at this stage.

Stage 10: Booting. The sheath of the flag leaf is fully visible and is swollen by the ear. The ear (head) is in boot.

Heading

10.1: The tips of the awns begin to emerge.

10.2: One quarter of the head is visible in the plant.

10.3: Half of the ear has emerged.

10.4: Three-quarters of the ear has emerged.

10.5. The ear has fully emerged.

Ripening

11.1: Kernels are milky.

11.2: Milk ripe: The kernels are soft and dry.

11.3: Kernel is hard to dent with thumbnail.

11.4: Dead-ripe: Grains start to shed and the straw starts to break.

Decimal Code Classification of Barley Growth and Development

The decimal code classification of barley growth stages includes eight different growth stages, namely: (i) seedling, (ii) tillering, (iii) stem elongation, (iv) booting, (v) ear emergence, (vi) flowering, (vii) milk development and (viii) dough development stages. The seedling stage includes: first leaf coleoptile (GS10), first leaf unfolded (GS11), three leaves unfolded (GS13), five leaves unfolded (GS15) and nine or more leaves unfolded (GS19). Similarly, the tillering stage has five stages: main shoot only (GS 20), main shoot and one tiller (GS21), main shoot and three tillers (GS23), main shoot and five tillers (GS25), and main shoot and nine or more tillers (GS29). The stem elongation stage consists of six substages, namely, ear at 1 cm (GS30), first node detectable (GS 31), third node detectable (GS33), fifth node detectable (GS35), flag leaf just visible (GS37), and flag leaf blade all visible (GS39).

The booting stage consist of four sub-stages: flag leaf sheath extending (GS41), flag leaf sheath just visibly swollen (GS43), flag leaf sheath swollen (GS45), and first awns visible (GS49). The ear emergence stage has three distinct substages, namely, first spikelet of the ear is just visible (GS51), half of the ear emerged (GS55) and ear completely emerged (GS59). Flowering stage consists of three distinct sub-stages: start of flowering (GS61), flowering half-way (GS65) and flowering complete (GS69). Milk development has four stages, namely, grain is watery ripe containing clear liquid and is approximately 3 mm long (GS71), early milk (GS73), medium milk (GS75), and late milk (GS77). Dough development has three stages: early dough (GS83), soft dough (GS85), and hard dough (GS87) and the ripening stage has two sub-stages, namely, grain is hard and difficult to divide (GS91) and grain is hard and cannot be dented by fingernail (GS92). The speed at which barley progresses through these growth stages is governed by temperature (warm conditions speed up development), vernalization (cool, not freezing, temperatures advance the start of flower initiation in young plants) and photoperiod (long day advances floral development). More detailed effects of temperature, vernalization, and photoperiod is provided below.

Climatic Requirements

Temperature

Barley can be grown as a summer or winter crop. The winter crop requires a temperature of 12 to 16 °C at vegetative and reproductive stages and 30 to 32 °C at maturity. Barley is very sensitive to frost at all growth stages. The emergence of new leaves in barley has a linear function with time of exposure to temperature. Leaf emergence increases with increasing temperature until an optimum temperature of 22.5 °C and decreases with further increase in temperature. The leaf growth rate per degree-day has a maximum at 17.5 °C. It has also been observed that leaf emergence and leaf growth under different temperatures but at the same degree-day are different (Tamaki et al., 2002). Karsai et al. (2008a) observed that barley growth and development was fastest under nonfluctuating environments; however, growth is delayed by longer photoperiods. Similar responses have been observed for temperature. Facultative barley is the most sensitive, followed by winter barleys, while spring barleys are the least sensitive to temperature. Crop yield is highly influenced by incidence of frost at the flowering stage. Flowering in barley is accelerated by an initial treatment with either short photoperiods or low temperature. In either case, treatment with long photoperiods was necessary for completion of flowering. In spring rye, short photoperiods result in the formation of increased number of leaves before heads were formed and heading was thereby delayed by such treatment. Hence, it seems that heading of both spring and winter varieties is accelerated by the application of long photoperiods, except during the very early stages of winter varieties when short photoperiods seem more effective (Purvis and Gregory, 1937). Barley is moderately tolerant to high temperature and drought stress. High temperatures during one or more growth phases will result in reduced yield. High temperature coupled with water stress is even more detrimental to yield, because photosynthesis is curtailed as stomata close to preserve water, thereby preventing plant leaves from exchanging oxygen and carbon dioxide. Temperature and/or water stress that hamper photosynthesis during the grain-fill period can result in decreased grain development and fewer kernels per spike. High temperature stress during grain filling decreases starch granule levels in the subaleurone layer. Formation of the endosperm cell wall and crushed cell layer are sensitive to high temperatures. Endosperm texture was generally more friable in high temperature treated grains than in control grains.

Vernalization is the induction of flowering by exposure to an extended period of low temperature. It is a characteristic feature of many temperate plants, including winter growth habit forms of barley. Barley plants that need a period of low temperature to induce the switch from vegetative to reproductive growth are referred to as winter barley. In general, winter varieties of barley are low temperature tolerant, photoperiod-sensitive, and strong-vernalization responsive, because they can flower without vernalization (Karsai et al., 2008b). In contrast, spring varieties are essentially the opposite and have minimal low temperature tolerance capacity, do not require vernalization, and are typically insensitive to short day photoperiod. In barley, a third growth habit class (facultative) occurs. These genotypes are low temperature tolerant like winter varieties but lack a vernalization requirement. Takahashi and Yasuda (1971) proposed a three-locus (*Sh2* = *VRN-H1* (chromosome 5H); *Sh* = *VRN-H2* (chromosome 4H); and *Sh3* = *VRN-H3* (chromosome 1H), where H indicates *Hordeum* genome) epistatic model for barley vernalization. The *VRN-H1* located on the long arm of chromosome 5H encodes a MADS-box transcription factor with a high similarity to the *Arabidopsis* meristem identity genes *APETALA1, CAULIFLOWER* and *FRUITFUL* (Yan et al., 2003). The recessive winter allele at *VRN-H1* is only expressed after exposure to cold which is important for the transition to reproductive growth. In barley, *VRN-H1* downregulates expression of *VRN-H2,* which is only expressed under long-day conditions (Shimada et al., 2009). Vernalization is mediated by three genes, a dominant allele *HvVrn2*, and two recessive alleles: *Hvvrn1* and *Hvvrn3*. Mutation of any of the vernalization alleles proved sufficient to abolish the vernalization requirement, leading to evolution of spring or facultative types. The spring growth habit trait was controlled by allelic variations in three vernalisation genes, *VRN1, VRN2* and *VRN3,* in barley. Presence or absence of the vernalization-critical region in intron 1 of the *VRN-H1* gene basically determines the growth habit of barley. Acquisition of the spring growth habit is alleged to be one of the driving forces for the expansion of the cultivable area.

Rainfall

The seasonal water requirement for barley depends on the variety, targeted yield and crop management. Barley is a drought-tolerant crop and requires 390 to 430 mm of rainfall for optimum yield. Maximum water use occurs for 21 to 28 d. Irrigation should be scheduled according to evapotranspiration and growth stage requirements, because barley is more sensitive to drought stress during jointing, booting, and heading. To optimize grain yield, soil moisture levels should remain above 50% of available moisture in the active root zone from the seedling to soft dough stage. Plant responses to drought stress are complex, and plants have evolved different strategies to alleviate the adverse effects of drought stress by altering their physiological, molecular, and cellular functions. A study on the effect of drought, high temperature, or the combination of both indicated that combined stress decreased grain weight by 30%, followed by drought (20%) and high temperature stress (5%). The reduction in duration of grain growth is the most important cause of reduced grain weight at maturity under drought alone or combined with high temperature

(Savin and Nicolas, 1996). It was discovered that landraces and wild species of barley have gene(s) for adaptation to drought tolerance.

Barley is well known for its low consumption of water per unit weight of dry matter produced compared with other cereals, and can be grown with limited irrigation. Studies on screening barley genotypes tolerant to drought stress showed a negative correlation between grain yield and days to heading and days to maturity under drought stress. However, 1000-grain weight, grains per spike, relative water potential and stay green were positively correlated with grain yield. Yield was also associated with osmotic adjustment and was least related to plant height (Vaeiz et al., 2010; Zare et al., 2011). Under drought stress, wide genotypic variability among barley genotypes was observed for carbon isotope discrimination, osmotic adjustments, transpiration efficiency, and water use efficiency (García del Moral et al., 1991). Several quantitative trait loci (QTLs) have been identified in barley associated with drought tolerance (Teulat et al., 2001, 2002; Tondelli et al., 2006). Four out of 12 drought tolerance QTLs of the consensus map are associated with regulatory candidate genes on chromosome 2H, 5H, and 7H, and two QTLs with effector genes on chromosome 5H and 6H (Tondelli et al., 2006). In another study, Teulat et al. (1998) found three QTLs for relative water content, four QTLs for osmotic potential, and one QTL for osmotic adjustment in barley under drought stress. Two chromosome regions [chromosome 1 (7H) and chromosome 6 (6H)] are involved in several osmotic adjustment related trait variations. The genomic regions involved in osmotic adjustments on chromosome 1 (7H) have also been reported in other cereals. Eight chromosome regions controlling carbon isotope discrimination have been observed in barley under drought stress. Carbon isotope discrimination provides an integrative measure of photosynthetic activity, mainly regulated by stomatal aperture. These chromosome regions also relate to plant water status and/or osmotic adjustment in barley. Six regions controlling agronomic traits were colocated with QTLs with carbon isotope discrimination (Teulat et al., 2002). A comparative analysis using differential gel electrophoresis of Egyptian barley landrace leaf proteome under drought stress revealed alterations in proteins related to the energy balance, transcription, protein synthesis, and proteins involved in metabolism and chaperones between the drought-tolerant and susceptible lines (Ashoub et al., 2013). In barley, a transcriptomic approach employed to study spike responses to drought stress revealed that different classes of homologous genes participating in late embryogenesis abundant protein biosynthesis, antioxidative pathways, and osmolyte synthesis are altered under drought stress (Abebe et al., 2010). This information can be used in marker-assisted breeding for improving drought tolerance of barley.

Photoperiod

Vernalization affects flowering time, predominantly by reducing the duration of the vegetative phase. In contrast, long photoperiods have little effect on the duration of the vegetative phase but strongly accelerate the late reproductive phase of inflorescence development. Barley is a facultative long-day species, requiring less than a certain number of hours of darkness in each 24 h period to trigger a switch from vegetative growth to reproductive growth. In some barley genotypes, short-daylengths accelerate flowering time in a similar way as cold treatment, so is therefore referred to as short-day vernalization. This concept of photothermal time applies only between the base and maximal temperatures as well as between the ceiling and critical photoperiods. Depending on the genotype, the ceiling and the critical photoperiods for barley as a long-day crop is $\leq$ 10 h d^{-1} and $\geq$ 13 h d^{-1}, respectively. Photoperiod and vernalization have a predominant impact on developmental rate only during certain parts of the pre-flowering period. During the vegetative phase, the crop initiates leaves until floral initiation; however, during the early reproductive phase, the spikelets are differentiated until the initiation of the terminal spikelet. Finally, during the late reproductive phase, when the stem internodes elongate, the floret primordia reaches their maximum number and then matures.

Photoperiod is governed by the dominant allele at the *Ppd-H1* gene on chromosome 2H. The basis of photoperiod insensitivity is a single nucleotide polymorphism (SNP79) (Turner et al., 2005). However, it has also been suggested that SNP48 is more strongly associated with the phenotype than SNP79 (Jones et al., 2008). The gene *Ppd-H1* exerted a pleiotropic effect on plant height, tiller number, spike length, number of grains produced per spike, and grain yield (Sameri et al., 2006). A second important gene is *Ppd-H2,* which affects flowering time when barley plants are exposed to a daylength under 10 h but has little effect under long day conditions (Szucs et al., 2006). The gene *Ppd-H2* probably encodes the FT-like gene *HvFT3* (Kikuchi et al., 2009). Photoperiod interacts with vernalisation to determine flowering time (Karsai et al., 2008b). This interaction operates via the integration of the low temperature and photoperiod response pathways to induce the transcription of *VRN3* via the induction of *VRN2* (Hemming et al., 2008).

In photoperiod-sensitive winter barley, *VRN-H2* represses *VRN-H3* to counteract the *Ppd-H1*-dependent long-day induction of *VRN-H3* prior to winter. *VRN-H2* expression is maintained at high levels, prior to vernalization, and downregulated by *VRN-H1* during exposure to cold. Upregulation of *VRN-H1* during vernalisation and the consequential downregulation of *VRN-H2* promotes inflorescence meristem identity at the shoot apex and accelerates inflorescence initiation. Downregulation of *VRN-H2* transcript levels in the leaves facilitates the upregulation of *VRN-H3* during long days mediated by *Ppd-H1* and possibly by *HvCO1* (Campoli et al., 2012). High levels of *VRN-H3* in turn upregulate *VRN-H1.*

In the case of late planting, the growth and development of barley would be negatively affected by short nights triggering reproductive growth too soon after the seedling emergence. In spring barley, long days are sensed by the phytochrome receptors in the leaf. The transition from a vegetative to reproductive crown meristem is controlled by the expression of photoperiod and *CONSTANS* genes in the leaf. In barley, there are two constans proteins, the first one is expressed during flower emergence until initiation, and the second one is highly expressed in late stages of terminal spike production and heading (Song et al., 2015).

Production Practices

The ideal pH for spring barley is 6.5. This ensures optimum availability of both macro- and micro- nutrients. Barley can grow in a wide range of soil including saline, sodic, and lighter soils. However, the crop thrives best in sandy to moderately heavy loam soils having neutral to saline pH. Acid soils are not suited for barley cultivation. Winter barley is sown during mid- to the end-September. Early September is appropriate at high altitude and in drought-prone areas, providing a good chance for establishment. The crop completes its tillering prior to winter and sowing early in September is encouraged for early high biomass production to withstand low temperature. Shallow, late sowing and low seed rates increase the risk of plant loss. Seeding rate depends on local conditions and soil type, but for two-rows, it should be 320 to 365 seeds m^{-2}. Six-row barleys have a higher ear weight and thus require a lower ear density for optimum yield. Hence, a seeding rate of 225 to 250 seeds m^{-2} is recommended. Barley cannot compensate for low plant density. A firm seedbed ensures better moisture and nutrient availability and reduces the risk of frost and plant losses.

Tiller emergence starts after the emergence of the third leaf. Maximum tiller number occur around the start of stem extension, but tillering can continue beyond this or stop and restart later under stress conditions. Shoot production completes before ear emergence; however, late tillers will be produced during recovery from stress. These late-formed tillers will not form an ear and will eventually die. Initially, canopy size increases slowly as successive leaves emerge and low light interception leads to low biomass. Biotic stress (disease) during early growth phases causes the plant to lose chlorophyll resulting in reduced light interception and canopy expansion. Canopy size reaches a plateau soon after flag leaf emergence. While the stem and ear continue to grow, the leaf area index decreases due to senescence of lower leaves. Once the ear and stem reach full growth, there is no further canopy expansion and total canopy starts to decline. Grain number is the key determinant of yield in spring barley and is dependent on ear number, which in turn is dependent on plant number and shoot number per plant. Shoot number per plant is determined by tiller production and tiller survival. Shoot survival is more important than shoot production in spring barley. Final ear number is decided at the anthesis stage. Crop with low ear number will yield less even under favorable grain filling conditions.

Nitrogen (N) is required in high amounts by barley and is frequently the most limiting nutrient for maximum crop growth. Fertilizer guidelines for N can be based on the results of the soil NO_3–N test or the consideration of expected yield, previous crop, soil organic matter content and agroecologically appropriate nutrient response functions. Split application of N increases N-use efficiency. Approximately 30% of total N is recommended to be applied at sowing or alternatively at crop emergence to meet early crop growth requirements. The remaining N should be applied around the mid-tillering stage of crop growth. Compound fertilizer containing P and K as well as N are commonly used in the seedbed. Urea is normally a less expensive source of N but is susceptible to loss of N by volatilization when applied to the soil surface. Use of urease inhibitors can minimize volatilization. The main application of N should be in advance of stem extension to drive canopy growth and tiller retention. Phosphorus and potassium are essential for establishment and growth and are also frequently limiting and should be applied according to plant available levels based on soil sampling and testing and crop response. Phosphorus requirement is highest at six weeks after sowing (root and tiller development) and plant K uptake is highest during the reproductive phase. Barley also requires secondary and micronutrients including sulfur (S), magnesium (Mg) and sometimes copper (Cu), manganese (Mn) and zinc (Zn); however, these elements are usually adequately supplied by the soil without additional application.

Herbicide application is recommended during the GS14-23 growth stage for maximum effect. Mostly sulfonylurea, hormone, or grass weed herbicide is recommended for barley. Glyphosate pre-drill is appropriate where perennial weeds are a problem. A mix of herbicide types will help prevent resistance (sulphonyl urea

+ hormone herbicide). To achieve the best control, apply the herbicide with three days of good growing conditions on either side of the herbicide application. Early (March) sown crops are less susceptible to pests resulting in higher yield. Barley is vulnerable to various pests (oats and wheat aphids, Russian aphids, and black sand mites) and diseases (leaf spot, rust, blotch and powdery mildew) that may lead to decreased yield and poor grain quality. High temperature, together with high humidity, encourages the incidence of rust. Utilizing different pest and disease control measures ranging from chemical, biological and other cultural practices aids in maintaining good grain yield and quality. Use of resistant cultivars and certified and disease-free seeds as control mechanisms for pests and diseases is encouraged.

Minor Millets

Millets are generally grown in semiarid regions of Asia and Africa and are resilient to extreme climatic conditions. They possess a C_4 photosynthesis system, and hence, avoids excessive photorespiration and have relatively high water-use efficiency and nutrient-use efficiency. Millets are a rich source of nutrients and possess high levels of protein compared to other cereals. As indicated above, millets are dryland crops and are cultivated in semiarid regions characterized by low and erratic rainfall and periodic drought. The important minor (small) millets are finger millet, fox tail millet, kodo mille and proso millet.

Finger millet is widely grown in 20 °N to 20 °S with an annual rainfall of 500 to 1000 mm. It is called finger millet because the inflorescence resembles the fingers of a human hand. The morphology of the inflorescence can be used to differentiate between the two subspecies, *africana* and *coracana*. Finger millet is an allotetraploid. Genomic donors of the A genome are most likely *Eleusine indica* and *Eleusine trisachya* (Liu et al., 2014). The B genome contributor has yet to be discovered. Domestication of cultivated finger millet, *E. coracana* started around 5000 yr ago in Western Uganda and the Ethiopian highlands and extended to the Western Ghats of India around 3000 BC (Hilu et al., 1979). Finger millet grain has high calcium, polyphenol, and fiber content. Finger millet also contains methionine and tryptophan amino acids which are often absent in starch-based diets of some subsistence farmers. The genus *Eleusine* includes eight species of annual and perennial herbs. Morphological, cytogenetic, and molecular evidence suggests that cultivated finger millet (subsp. *coracana*) were domesticated from wild populations of *E. coracana* subsp. *Africana* (Dida et al., 2008).

Finger millet is an important cereal crop in India, Ethiopia, Somalia, Kenya, Tanzania, Zaire, Zambia, and Uganda. Globally, 12% of the total millet area is under finger millet cultivation, covering more than 25 countries of Africa and Asia. In India, finger millet is cultivated over an area of 1.19 million hectares with a production of 1.98 million Mg, giving an average productivity of 1661 kg per ha (Sakamma et al., 2018). Finger millet is a preferred cereal in dry and semiarid tracts of India and Africa. Globally, finger millet is cultivated on an area of 4 to 4.5 million hectares with a production of 4.5 million Mg. Finger millet thrives well under tropical temperatures ranging from 25 to 38 °C and is a short-day crop. Finger millet seed can be stored for long periods of time with minimal insect damage and with little loss of viability. The genetic potential of finger millet genotypes with regard to its ability to withstand drought, disease, and soil fertility related constraints is important. The development of crosses between *Eleusine* strains from the Indian and African cropping belts significantly improved grain yield.

The International Crops Research Institute for the Semi-Arid Tropics (ICRISAT) conserved 6804 finger millet germplasm accessions originating from 25 different countries. From these large collections, ICRISAT has grouped all genotypes according to region of origin and other parameters. A subset of each group is selected as a representative of the genetic diversity of the crop. This group is termed the "core collection" and typically consists of ~10% of all available accessions. Core collections facilitate breeding by providing an efficient means to screen for desired traits from a large pool of genotypes. Mini-core collections, that represent ~1% of the total accessions, can be used by institutions to further streamline available genetic diversity.

The Indaf varieties (cross between Indian and African germplasm) are resistant to diseases, sturdy, withstand drought, and respond to fertilizers leading to yield increase from 1.5 or 2.5 Mg ha^{-1} to 4 to 5 Mg ha^{-1}. Finger millet can be directly sown or transplanted (2 to 3-wk-old seedlings) in the field. Transplanted crops had higher productivity than direct sown crops. Initial soil moisture is required for crop establishment. Overall, finger millet can be grown in all three seasons (Rabi, Kharif, and Summer) throughout the year in southern India. However, the growth and yield will be higher under Kharif season. Finger millet is a dryland crop, often grown as a rainfed crop without supplemental irrigation. Since it withstands intermittent drought, it is a staple cereal for subsistence farming communities in drought-prone areas. In relatively wetter areas, sorghum or pearl millet replaces finger millet. Finger millet monocrops grown under rainfed conditions are most common in drier areas of Eastern Africa; however, in tropical Central Africa, finger millet is commonly intercropped with a legume. Finger millet can also be grown intercropped with cassava, plantain, and vegetables. Finger millet is most commonly rotated with legumes such as cowpea [*Vigna unguiculata* (L.) Walp.] or pigeonpea [*Cajanus cajan* (L.) Huth.] in Eastern Africa. In India, finger millet is rotated with legumes like dolichos [*Lablab purpureus* (L.) Sweet], pigeonpea, blackgram, [*Vigna mungo* (L.) Hepper], cowpea and horse gram [*Macrotyloma uniflorum* (Lam.) Verdc.] or with oil seed crop like castor (*Ricinus communis* L.), Brassica species and soybean [*Glycine max* (L.) Merr.] or with cereals like maize, foxtail millet, sorghum, and little millet.

Foxtail millet is a C_4 grass and closely related to the hardy weed or green foxtail *Setaria viridis* (L.) P. Beauv., which is assumed to be its progenitor. Several hypotheses regarding the origin and domestication have been proposed and a multiple domestication hypothesis has been widely accepted. Li et al. (1995) suggest a multiple domestication hypothesis with three centers, that is, China, Europe, and Afghanistan-Lebanon. Foxtail millet or Italian millet is grown primarily in China, mostly for human consumption, but also for forage. It is named for the bushy, tail-like appearance of its immature panicles, and is considered one of the world's oldest crops. Currently, it is a minor cereal grown in Asia, Africa, Europe, and Americas. Foxtail millet is a short season crop that is well suited to fill the summer fallows that occur after harvest of wheat or maize. Foxtail millet is quick to mature, able to produce seed in 75 to 90 d, and is sometimes grown as a "catch-crop" in between the plantings of other species.

Globally, foxtail millet production is around 5.0 million Mg annually and productivity ranges from 0.8 to 1.8 t of grain per hectare depending on water availability. Foxtail millet is often grown in mountains or plains for seed and/or forage purpose. It is grown as a short-season cereal and harvested in 60 to 90 d. Foxtail is sown after the onset of the warm period and first rains. Late-maturing varieties with medium thick stems and profuse leaves are grown for forage. Foxtail millet reaches the vegetative stage during July to August. It puts forth leaves utilizing each rainfall and remains in the vegetative phase for longer durations. It is a shallow rooted cereal, yet it tolerates drought and erratic precipitation patterns. Foxtail millet grown in North America offers golden yellow grains, good for consumption by farm animals and birds. In the United States of America, foxtail millet is an important small grain cereal primarily grown for forage, but seeds are usually meant for birds. Like other crops, there is an abundance of foxtail millet germplasm, and these are available at the Chinese National Genebank in China, ICRISAT in India, National Institute of Agrobiological Sciences in Japan, and Plant Genetic Resources Conservation unit of United States Department of Agriculture. Foxtail millet has one of the largest collections of cultivated (46,070 accessions) as well as wild accessions (906 accessions) (http://www.fao.org/wiews).

Foxtail millet is grown in fallow or soils with constraints related to fertility and irrigation. Foxtail millet is grown as a monocrop or intercropped with legumes or oilseed crops. Foxtail millet is rotated with short-season legumes that support biological nitrogen fixation and is a preferred catch crop or short season crop in North Dakota, Nebraska, and Colorado. In West Asia, foxtail millet is sown immediately after the main cereal or lentil (*Lens culinaris* Medik.), with supplemental irrigation, to achieve rapid germination of seeds and to protect the crop from long dry periods. Foxtail millet is grown as a hardy drought-tolerant crop in South India in sequence with dryland cereals or legumes. Similarly, foxtail millet is grown in Northern dry regions of China. Landraces from the north of China are typically well-adapted to cold weather with short growing seasons and are highly sensitive to light and temperature changes, while those from southern regions grow better in high temperatures and humidity. Wheat–foxtail millet rotations actually improve fertilizer-use efficiency. In East Africa, foxtail millets are rotated with legumes such as pigeon pea or cowpea or is intercropped with other cereals such as sorghum or maize. Zhi et al. (2007) has reported a cytoplasmic male sterile material originating from a cross between green and cultivated foxtail millet, the products (hybrid and BC_1 plants) of which were all male sterile. In addition, genetic male sterility in foxtail millet has been exploited to produce hybrid cultivars, namely Suanxi 28 × Zhangnong 10 and Jigu 16 (Du and Wang, 1997).

Kodo millet belongs to the genus *Paspalum*, a diverse genus comprised of about 400 species, most of which are native to the tropical and subtropical regions of the Americas. The main center of origin and diversity of the genus is considered to be the South American tropics and subtropics. Kodo millet is a tetraploid, but diploids are also common. Almost all species have a basic chromosome number of 10; however, some species of *Paspalum* were reported whose basic chromosome number is 6 or 9. Kodo millet was domesticated roughly 3000 yr ago in India, and the grain contains a diverse range of high quality protein and antioxidant activities. Like finger millet, kodo millet is rich in fiber, and hence may be useful for diabetics. In Africa, kodo millet is referred to as black rice or bird's grass. Kodo millet is divided into three races (*regularis, irregularis,* and *variabilis*) based on panicle morphology. In southern India, there are two different varieties of kodo millets available, namely small (karu varagu) and large seeded (peru varagu).

Globally, 8000 accessions of kodo millet have been conserved, and the ICRISAT gene bank in India conserves 665 accessions of kodo millet under medium- and long-term storage. Kodo millet is commonly harvested in tropical Africa and India as a wild cereal and consumed as a famine food. It grows as a weed in rice fields, but farmers do not weed it out as it can be harvested as an alternative crop if the primary crop fails. In developed countries, it is considered to be a noxious weed. Kodo millet seed is light red or dark gray in color. The red grain is sweet and the gray is bitter in taste. It is mainly present in tropical and subtropical regions of the world. The grains are larger than foxtail and little millet grains. Kodo millet is primarily grown in India in the upland rice regions and in Indonesia, Philippines, Thailand, and Vietnam, Bangladesh, and Myanmar. It is grown as a monocrop or as a mixed culture. Kodo millet is also grown in the hot, arid regions of Asia, New Zealand, and the United States as a pasture crop. It is drought tolerant and can survive on marginal soils where other crops may not survive. The crop matures in 120 to 160 d with an average yield of 250 to 1000 kg per hectare and has a potential yield of 2000 kg per hectare. Kodo millet cultivation is confined to tribal areas and is grown under poor environments for food and fodder. Kodo millet grains possess excellent storage properties and can be stored for several years without pest damage at ambient storage conditions.

Proso millet is an annual herbaceous plant in the genera *Panicum* and was likely domesticated in China and Europe. The earliest records come from the Yellow River valley site of Cishan, China dating between 10,300 yr BP and 8700 yr BP. Evidence of proso millet also occurs at a number of pre-7,000 yr BP sites in Eastern Europe, in the form of charred grains and grain impressions in pottery (Hunt et al., 2008; Lu et al., 2009). Proso millet requires relatively less water and has a low transpiration ratio, since it is a C_4 cereal crop. Proso millet was domesticated by human tribes in Transcaucasia and Northern China during the Neolithic period and has been served as a cereal food for humans for 5000 yr. Currently, it is grown extensively in the Indo-Gangetic Plains, Russia, Ukraine, Romania, Turkey, and adjoining regions in Syria and Israel. Even today, it is the staple cereal for the local tribes in Korea and Mongolia. Proso millet produces whitish, gray, or brown grain.

Globally, more than 29,000 accessions of proso millet have been conserved and the ICRISAT gene bank in India conserves 849 accessions of proso millet under medium- and long-term storage. Proso millet cultivation is limited to low fertility and drought-prone regions of Russian plains, China, India, West Asia, and North America. Currently, the total cultivated area of proso millet in the U.S. is approximately 0.2 m ha (USDA-NASS, 2016), and most of this production is from Great Plains. Proso millet thrives well on neutral or slightly acidic soils with pH 6.0. It is a shallow-rooted crop and grows well intercropped with deep rooted crops such as cotton (*Gossypium hirsutum* L.), sunflower (*Helianthus annus* L.) or maize. Proso millet yields 10 Mg of forage per hectare and 1 to 1.5 Mg of grains per hectare. Proso millet is grown in areas receiving a rainfall of 200 to 450 mm annually. Proso millet is sown during warm periods, namely June or July in the Great Plains of the United States. Early planting is often conducted to avoid cold and frost damage. Proso millet requires 70 to 90 d to mature. Crop duration depends on number of degree-days and heat units received. Proso millet is susceptible to early frost or extreme temperatures at flowering. The short duration between the third and sixth leaf stage is critical for full expression of growth potential and yield formation. Proso millet is sown after wheat harvest to conserve soil moisture, in some cases it is sown after the harvest of maize. Specific rotations followed in the Great Plains are wheat-proso millet-fallow; wheat-corn-proso millet-fallow; and wheat-corn-proso millet-forage. Among the various rotations practiced that include proso millet, corn followed by wheat then proso millet seems to be popular in the region (Croissant et al., 2012). In parts of Florida, proso millet is sown first in June as a sole crop and is followed by a mixture of maize/proso millet and a legume during the next season. Proso millet tolerates drought and erratic precipitation patterns, but it needs protective irrigation during critical stages like flowering and grain filling (Croissant et al., 2012).

Growth and Development of Minor Millets

Minor millets mature in the range of 85 to 130 d after sowing. Generally, the growth stages of millets can be classified into three distinct stages, namely vegetative, pre- and post-flowering stages.

Vegetative Stages

The vegetative phase covers the period from germination to panicle initiation and depends on the cultivar used and climate conditions. The vegetative stage may be completed in 16 to 20 d after planting. An increase in number of leaves, tiller buds and plant height are characteristics of the vegetative phase.

Pre-flowering Stages

Rapid elongation of stem internodes and an increase in leaf area, accompanied by more tillers, is noticed during the pre-flowering stage. The reproductive phase occurs about 20 to 25 d from panicle differentiation to flowering of the main culm. This phase initiates when the panicle primordium is greater than 0.5 mm. Flower opening happens between sunrise and noon, as daytime temperature rise. The spikelets open and close in a short period of time (< 10 min), which favors self-pollination. Anthesis happens basipetally.

Post-flowering Stages

The post-flowering stage starts at flowering and continues to the end of physiological maturity, which covers a period of 20 to 30 d. During this period, the plant actively accumulates dry matter, particularly in the grain. Physiological maturity proceeds from top to bottom of the panicle. The ripening of the seed is not uniform throughout the panicle and delay in harvesting may cause losses due to shattering. Grain from the main panicle reach maximum dry weight and a small dark layer at the hilar region of the seed is formed. At maturity, the grain generally includes about 20% or less of moisture.

Detailed Classification of Minor Millet Growth Stages

In general, growth stages of minor millets are grouped into 11 distinct stages, namely, Stage 0: seedling emergence; Stage 1: 3-leaf stage; Stage 2: 5-leaf stage; Stage 3: crown root initiation stage; Stage 4: maximum tillering; Stage 5: panicle primordium initiation; Stage 6: flag leaf; Stage 7: booting; Stage 8: heading; Stage 9: milk; Stage 10: dough; and Stage 11: physiological maturity (Karim et al., 1993). In general, millets will take 5 to 8 d for seedling emergence, and 6 d to reach the three-leaf stage, 11 and 15 d to reach the five-leaf stage and crown root initiation stage. Maximum tillering will occur up to 30 d. Panicle primordium initiation will take place from 30 to 46 d. Appearance of the flag leaf occurs approximately 53 d after sowing.

Booting and heading will take place around 56 to 61 d after sowing. The milk stage can be identified around 68 d and the dough stage around 75 d.

Climatic Requirements of Minor Millets

In India, minor millets are grown during Kharif season. Although millets can be grown in low rainfall tracts receiving season rainfall of 350 to 400 mm, some can adapt to high rainfall up to 1000 mm. Minor millets are known for their drought-tolerance characteristics. Matsuura et al. (2012) studied the effects of drought stress before and after flowering in four millets, namely proso millet, little millet, foxtail millet and wild millet (*Seteria glauca* L.). The results indicated that drought stress prior to flowering significantly decreased grain yield in all four millets. However, drought stress after seed set to the maturity stage significantly decreased grain yield in proso and little millet, while the effect on foxtail and wild millet was negligible. The percent yield loss due to drought stress in various millets is presented in Table 3. In proso millet and little millet, the period immediately before and after the heading stage is very sensitive to drought stress. However, for foxtail and wild millet, only the period immediately before heading is the most sensitive stage for drought stress (Matsuura et al., 2012). The flowering stage in finger millet is the most sensitive to drought stress (Maqsood and Ali, 2007). In finger millet, drought stress significantly reduced leaf area, total dry matter accumulation, seed yield and harvest index (Maqsood and Ali, 2007). Drought stress increased the activity of superoxide dismutase, ascorbate peroxidase and glutathione reductase enzymes and decreased membrane

Table 3. Minor millet water requirement, duration, and magnitude of yield loss (%) due to drought stress at reproductive stages.

Crop	Water requirement (mm)	Temperature requirement (°C) from sowing to maturity†			Photoperiod requirement Heading to maturity	Duration (d) (Planting to maturity)‡ Vegetative to maturity		Yield loss (%) due to drought stress
		Min.	Opt.	Max.				
Oats	300	4 §	25 §	35 §	16–18 ¶	Spring: 150; Winter: 300	–	–
Barley	380–430	5 #	19 ††	28 #	16 #	Spring: 150; Winter: 210	–	–
Finger millet	350	–	25 ‡‡	–	8.5–11 ‡‡	90–120	–	53 §§
Foxtail millet	400	–	26–29 ‡‡	35–40 ‡‡	12 ¶¶	75–90	20 ##	–
Proso millet	350	–	–	–	Short day	60–90	64 ##	–
Little millet	350	–	–	–	Short day	45–60	80 ##	–

†Min, minimum temperature; Opt, Optimum temperature; Max, Maximum temperature.

‡Varies with genotype – data not available.

§ Leite et al. (2012)

¶ King and Bacon (1992)

Borthwick et al. (1941)

†† Garmash (2005)

‡‡ Anonymous (2014)

§§ Maqsood and Ali (2007)

¶¶ Doust et al. (2017)

Matsuura et al. (2012).

damage in tolerant genotypes compared to susceptible genotypes (Bhatt et al., 2011). Significant correlation between agronomic traits (plant height, panicle weight, grain weight panicle^{-1} and 1000-grain weight) and physiological traits (chlorophyll index, soluble protein, malondialdehyde and superoxide dismutase enzyme activity) with drought-resistant index was reported in foxtail millet (Zhang et al., 2012). A quick and simple screening tool for identification of drought tolerant foxtail millet genotypes is through germination in mannitol or polyethylene glycol medium. Lata et al. (2011) have also used lipid peroxidation measure to assess membrane integrity under drought stress for identification of drought tolerant genotypes.

The ideal temperature for germination is 24 °C (minimum 8 to 10 °C). A mean temperature of 26 to 29 °C is optimum for growth. Crop yields will be reduced if the temperature goes below 20°C. The thin stemmed minor millets, such as finger millet, foxtail millet, and proso millet, are often affected by lodging, especially under high soil fertility. Similarly, minor millets are sensitive to frost. The ideal photoperiod for flowering is short-daylengths. If daylength (length of light period) is increased, flowering is postponed. In high altitudes, due to low photoperiods, the duration of the vegetative phase is extended. Studies on cereals indicate that reproductive stages appear to be more vulnerable to high temperature stress than vegetative stages of crop development (Hatfield et al., 2011), and high temperature during reproductive stages can decrease the number of seeds panicle^{-1} and individual seed weight, resulting in lower grain yields (Djanaguiraman and Prasad, 2014). Similarly, in finger millet high temperature stress during reproductive stages decreases grain yield, and among reproductive stages, the booting, panicle emergence or flowering stages were found to be the most sensitive to high temperature than other stages (Opole et al., 2018; Fig. 4). A temperature threshold of $<$ 32/22 °C (daytime maximum/nighttime minimum temperature) was observed for finger millet (Opole et al., 2018; Fig. 5). Increasing temperature above the threshold ($<$ 32/22 °C) caused decreased number of fingers panicle^{-1}, number of seeds panicle^{-1}, 100-seed weight, grain yield plant^{-1} and harvest index. The decrease in grain yield plant^{-1} is associated with decreased number of seeds panicle^{-1} and individual seed weight (Opole et al., 2018; Fig. 5).

Grain numbers are a result of successful fertilization (seed-set), which mainly depends on the functionality of male (pollen) and female (ovule) gametes. Adverse environmental conditions such as drought and high temperatures during floral development and anthesis can negatively influence gamete viability and function leading to decreased floret fertility and consequently, seed-set (Prasad et al., 2008; Djanaguiraman and Prasad, 2014; Prasad et al., 2017). High temperature stress during the grain filling period can decrease individual grain size by decreasing grain filling duration and/or grain filling rate (Prasad et al., 2017). Apart from the yield components, high temperature stress decreases chlorophyll content, leaf temperature, photosystem II quantum yield and transpiration rate (Fig. 6) and thylakoid membrane damage increased. Reduced accumulation of chlorophyll in high temperature stressed plants may be attributed to impaired chlorophyll synthesis, its accelerated degradation or a combination of both. High temperature stress decreased PSII quantum yield indicating changes in energy allocation to the photosystems. The downregulation of PSII activity by high temperature stress is matched by the reduction in the capacity of downstream reactions. Decreased transpiration rate resulted in increased leaf temperature. Significant genetic variability has been observed for high temperature stress tolerance in the mini-core germplasm. Maximum genotypic variation was explained by panicle length, finger length and number of seeds panicle^{-1} at high temperatures. The accessions IE5201 and IE2312 were found to be high temperature tolerant by recording higher seed yield at high temperature than other genotypes.

Production Practices

There are several constraints to the production of minor millets, namely (i) millets are grown in poor, shallow, and marginal soils under rainfed conditions. Still, some of these millets are grown in hilly areas under shifting cultivation, (ii) seeds are often broadcast, which affects inter-cultivation operations and effective weed control; (iii) the minor millets are grown in mixed cropping practices, which is mostly suited to

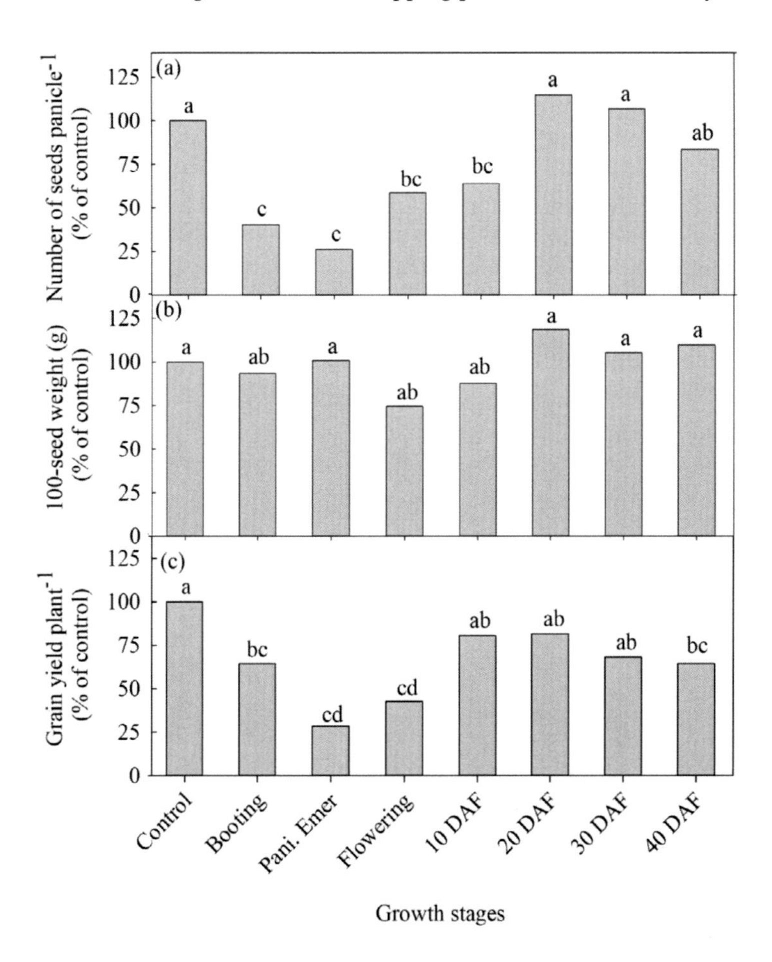

Fig. 4. Impact of short episodes of temperature stress [optimum (OT, 32/22 °C), high temperature (HT, 38/28 °C)] at different growth stages of finger millet on (a) number of seeds panicle-1; (b) 100-seed weight; and (c), and grain yield plant-1 (g). Abbreviations: control, continuous optimum temperature (27/18 °C), Pani. Emer., Panicle emergence; DAF, Days after flowering. Each datum is expressed as percentage of control. Means with the same letter are not significantly different at $p < 0.05$ (Redrawn and adapted from Opole et al., 2018).

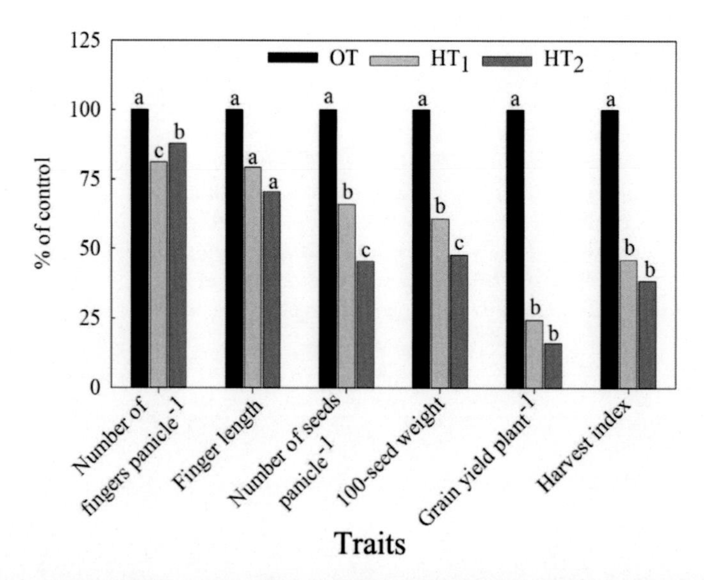

Fig. 5. Impact of season-long high temperature stress [optimum (OT, 32/22 °C), high temperature HT1, 36/26 °C and HT2, 38/28°C)] on number of fingers panicle−1; finger length (cm); number of seeds panicle−1; 100-seed weight (g); grain yield (g plant−1); and harvest index at maturity. Each datum is expressed as percentage of control. Means with the same letter are not significantly different at p < 0.05 (Redrawn and adapted from Opole et al., 2018).

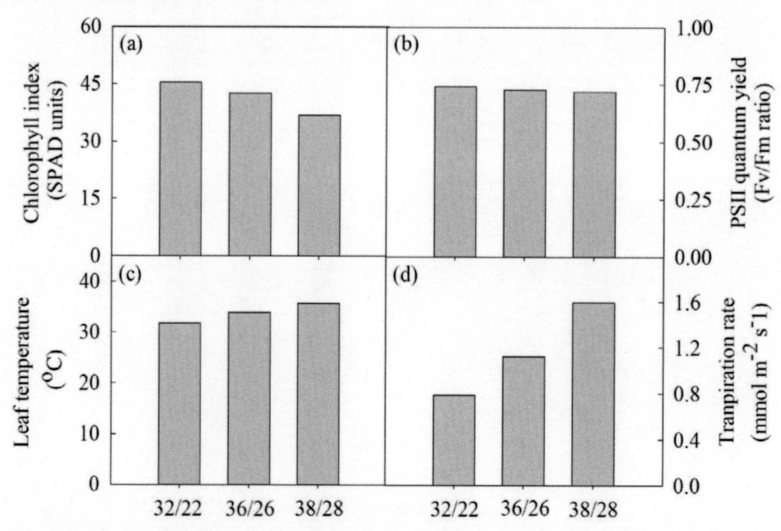

Fig. 6. Impact of season-long high temperature stress [optimum (32/22 °C), high temperature (36/26 °C and 38/28 °C)] on chlorophyll index, photosystem (PS) II quantum yield, leaf temperature (°C) and transpiration rate (mmol m^{-2} s^{-1}). Each datum is average of four measurements recorded at vegetative, booting, 50% flowering and 50% grain filling stage and expressed as percentage of control. (Redrawn and adapted from Opole et al., 2018).

subsistence agriculture and not remunerative, (iv) small millets are often cultivated under unfertilized conditions, and (v) non-adoption of improved varieties.

Finger millet is grown in different soils but mainly grown on red and laterite soils. Alluvial and black soils with good drainage are also suitable for finger millet cultivation. Under high elevations, early maturing varieties (90 to 100 d) are preferred, and in the plains, medium to late maturing varieties (110 to 120 d) are preferred. Finger millet is drought tolerant and the minimum water requirement for successful growing is 400 mm but can be grown in higher rainfall areas. In India, finger millet is grown mostly in the rainy season under dryland conditions. Land preparation for finger millet varies from little cultivation to intensive plowing followed by harrowing. Finger millet is cultivated on red and laterite soils that become hard and form a crust when dry; hence, to increase the infiltration and conserve moisture, tillage is essential. The finger millet seeds are very small and thus need to be sown at a shallow depth of 2 to 3 cm. In loosely prepared soils, the seeds are likely to sink to deeper layer when rain occurs. Therefore, soil compaction just below the seeding zone is recommended. To keep weeds under control and the soil surface loose, two to three intercultivations are essential.

Finger millet is sown during May to June or July depending on the duration of the cultivar. Shorter duration varieties can be sown later but their yield potential is usually lower. When long duration varieties are sown late in July, they head two to three weeks in advance leading to reduced yield. A medium duration (110 to 120 d) finger millet variety sown during the second or third week of July gave higher yield than when sown during August. Finger millet is drilled or broadcasted under dryland conditions, but under irrigated conditions it is usually transplanted. Transplanted finger millet has less weed pressure and usually yields higher. For transplanting, 3-wk-old seedlings raised in beds are used. For medium duration varieties (110 to 120 d), 25 to 30 d old seedlings are ideal for transplanting. In general, seedlings should be kept in a nursery for a week for every month of the crop's total life period. Thus, a variety of 120 d duration could be left in the nursery up to 4 wk. Finger millet is often broadcasted at a rate of 25 kg per hectare and thinned out later by cross cultivation in both directions using tine hoes. An optimum plant density to be maintained was 400 to 800 thousand plants per hectare. Yield will decline under low and high populations. In line sowing, row spacing of 25 cm and plant to plant spacing of 10 cm is adopted for about 400 thousand plants per hectare. Normally 25 to 30 cm row spacing is recommended for dryland finger millet to improve convenience for intercultivation and weeding. Finger millet responds well to nitrogen fertilizer. The general recommendation per hectare is 50:37.5:25 kg N, P_2O_5, and K_2O for dryland and 100:50:50 kg N, P_2O_5, and K_2O for irrigated conditions; however, this recommendation is highly dependent on soil nutrient availability, which should be assessed through soil analysis and agroecologically appropriate fertilizer response trials. Full P and K rates are recommended to be applied at sowing whereas, nitrogen should be applied in three or four split doses. It is essential that all of the N is applied before flowering. In general, finger millet is grown as a mixed crop under rainfed conditions during July which is very difficult for weed control using inter-cultivation.

In finger millet fields, it is very difficult to identify weeds during early stages because they are challenging to distinguish from the crop. However, under irrigated conditions, the situation is very different. Appropriate pre-emergence herbicide (Butachlor 50 EC, Oxyfluorfen 23.5 EC, and Bensulfuron-methyl + pretilachlor 6.6% G premix formulations) application of is recommended in sole finger millet crops. Similarly, selected post-emergence herbicide (Oxadiargyl 80 WP and Bispyribac-sodium 10% SC) application also provides effective weed control both in drill or transplanted finger millet. Along with chemical control, hand weeding once or twice is needed to control weeds. Finger millet requires 40 and 65 cm water under irrigated condition. Tillering and pre-flowering stages are the most critical stages for drought stress. Irrigation has to be provided at an interval of 8 to 12 d during summer months. However, during the rainy period three to four well-timed irrigations are enough to produce as much yield as when the crop is fully irrigated.

Foxtail millet is a 100-d duration grain crop suited to low and moderate rainfall ranging from 500 to 700 mm. In India, foxtail millet is mainly cultivated in South India as a mixed crop with cotton, castor, pigeon pea, sorghum, peanut and finger millet. It is also grown as a sole crop in black soils where it is followed by a winter crop like safflower (*Carthamus tinctorius* L.) or horse gram in a dry year. In hilly regions, foxtail millet is grown during the summer and matures in 2 months. Sowing during June is more advantageous than sowing soon after the receipt of monsoon in July. Early sowing in the monsoon produces higher yields than late sowings. The reduction in yield due to delayed planting is not compensated by increasing plant density. Optimum plant spacing is 30 cm between rows and 10 cm between plants within a row. The seeding rate for foxtail millet is 8 to 10 kg ha^{-1} for line sowing and 15 kg ha^{-1} for broadcasting. The response of foxtail millet to the application of nitrogen is positive up to 40 kg N ha^{-1}. Response to P application was not as marked as in the case of N. A combination of 60 kg N and 20 kg P_2O_5 per hectare gave the highest yield. Foxtail millet is generally grown as a mixed crop either with cotton or pigeon pea.

Kodo millet is a long-duration crop (110 to 130 d) compared to other small millets and grows well on shallow as well as deep soils. The kodo millet is often sown after the onset of monsoon, typically from the middle of June to the end of July in India. The seeding rate for kodo millet is 10 kg ha^{-1} for line sowing and 15 kg ha^{-1} for broadcasting. The optimum plant spacing is 22.5 cm between rows and 10 cm between plants within a row. Sowing 10 d before the onset of monsoon resulted in the highest grain and straw yield compared to sowing after the onset of monsoon. Delayed sowing increases the risk of drought stress at later stages. Early sowing also has comparatively less incidence of dead hearts caused by shootfly damage. Kodo millet is grown as a mixed crop with cereals (finger millet and maize) or oilseeds [Niger (*Guizotia abyssinica*), peanut (*Arachis hypogaea* L.), and soybean] in proportions of 1:1 and 2:1, respectively. A combination of kodo millet with peanut in a row proportion of 1:1 was found to be better than a sole crop. Kodo millet responds well to N and P fertilizer. A fertilizer rate of 40 and 20 kg of N and P ha^{-1}, respectively is recommended; however, plant-available nutrients determined through soil testing and agro-ecologically appropriate nutrient response trials should guide application rates.

Proso millet is a quick-growing short-duration cereal with a low moisture requirement and can be grown throughout the year. Two quick crops of proso millet per year are taken during March and June under irrigation. In South India (Tamil Nadu, Telangana, and Andhra Pradesh), proso millet is raised during rabi in black soils on stored soil moisture. Proso millet is usually broadcasted or occasionally sown in rows. Row to row spacing should be kept at 22 cm and plant to plant 10 cm with a seeding rate of 10 to 12 kg ha^{-1}. For better germination, seeds should be soaked in water for 24 h and should be planted less than 4 cm deep. In India, the Kharif crop is sown with onset of monsoon, preferably in the month July. However, in the states like Tamil Nadu and Andhra Pradesh, proso millet was sown during September through October. As an irrigated catch crop, proso millet is sown during mid-March to mid-May in the states of Bihar and Uttar Pradesh. Proso millet responds favorably to N fertilizer at low levels. Fertilizer application rates of 40:40 kg N and P ha^{-1} is recommended; however, application rates will vary by soil test results, previous crop and yield goals. For better effects, the full P rate should be applied basally, whereas, N should be applied in two equal splits at sowing and 35 to 40 d after sowing. Proso millet is generally grown as a summer crop though irrigation may increase yield. Usually a total of three irrigation timing is required for February, four for March and five for April planted crops. Flowering and grain filling stages are the most sensitive growth stages to environmental stress (particularly extreme temperature and water).

Conclusions

Crops such as oats, barley, and minor millets were domesticated and used as food across the globe. Most of these crops are grown by small farmers in Africa and Asia, where rainfall is uncertain and scanty, and soils are poor in fertility. Millet-based farming also enables farmers to suitably mix crops to minimize the risk of total crop failure during extreme environmental conditions. Minor millets have short life cycles and wide adaptation; hence, they play an important role in sustainable rainfed agriculture. The major reason for non-cultivation of millets is because it is not as remunerative as other major crops and lacks improved varieties, hybrids, production practices, and limited support and procurement policies. Millets are largely consumed in rural areas and are commonly known as poor man's grains. The global availability of large number of food products made from wheat and maize (e.g., bread, biscuits and other confectionaries) lead to the minor millets losing their value in the food chain. However, due to multiple health benefits and nutraceutical value such hypoglycemic properties, high phenolic contents, soluble fiber content, high antioxidants and starch-lipid-protein interactions in the minor millets, they are gaining in popularity or niche market. Owing to their potential medicinal and nutritional properties and tolerance to abiotic and biotic stress, these crops have great potential to address global food and nutritional security.

Acknowledgments

We thank the following entities for supporting millet research: Department of Agronomy, Kansas State University; and The Feed the Future Innovation Labs funded by United States Agency for International Development (Innovation Lab for Collaborative Research on Sustainable Intensification; and Innovation Lab for Collaborative Research on Sorghum and Millet). Contribution no. 18-490-B of the Kansas Agricultural Experiment Station.

References

Abebe, T., K. Melmaiee, V. Berg, and R.P. Wise. 2010. Drought response in the spikes of barley: Gene expression in the lemma, palea, awn, and seed. Funct. Integr. Genomics 10:191–205. doi:10.1007/s10142-009-0149-4

Anonymous. 2014, Status paper on coarse cereals, Directorate of Millet Development, Department of Agriculture and Co-operation, Ministry of Agriculture. Government of India, Jaipur, India. p. 87-114.

Ashoub, A., T. Beckhaus, T. Berberich, M. Karas, and W. Bruggemann. 2013. Comparative analysis of barley leaf proteome as affected by drought stress. Planta 237:771–781. doi:10.1007/s00425-012-1798-4

Bhatt, D., M. Negi, P. Sharma, S.C. Saxena, A.K. Dobriyal, and S. Arora. 2011. Responses to drought induced oxidative stress in five finger millet varieties differing in their geographical distribution. Physiol. Mol. Biol. Plants 17:347–353. doi:10.1007/s12298-011-0084-4

Borthwick, H.A., M.W. Parker, and P.H. Heinze. 1941. Effect of photoperiod and temperature on development of barley. Bot. Gaz. 103:326–341. doi:10.1086/335045

Bossinger, G., W. Rohde, U. Lundqvist, and F. Salamini. 1992. Genetics of barley development: Mutant phenotypes and molecular aspects. In: P.R. Shewry, editor, Barley: Genetics, biochemistry, molecular biology and biotechnology. CAB International, Wallingford. p. 231–263.

Bouis, H.E. 2000. Enrichment of food staples through plant breeding: A new strategy for fighting micronutrient malnutrition. Nutrition 16:701–704. doi:10.1016/S0899-9007(00)00266-5

Briggs, D.E. 1978. Barley. Chapman and Hall, London. p. 2-612. doi:10.1007/978-94-009-5715-2

Bunting, A.H., and D.S.H. Drennan. 1966. Some aspects of the morphology and physiology of cereals in the vegetative phase. In: F.L. Milthorpe and J.D. Ivins, editors, The growth of cereals and grasses. Proceedings of the Twelfth Easter School in Agricultural Science, Nottingham 1965, Butterworths, London, p. 20-38.

Campoli, C., B. Drosse, I. Searle, G. Coupland, and M. von Korff. 2012. Functional characterisation of HvCO1, the barley (*Hordeum vulgare*) flowering time ortholog of CONSTANS. Plant J. 69:868–880. doi:10.1111/j.1365-313X.2011.04839.x

Chinnici, M.F., and D.M. Peterson. 1979. Temperature and drought effects on blast and other characteristics in developing oats. Crop Sci. 19:893–897. doi:10.2135/cropsci1979.0011183X001900060035x

Chouard, P. 1960. Vernalization and its relations to dormancy. Annu. Rev. Plant Physiol. 11:191–238. doi:10.1146/annurev.pp.11.060160.001203

Croissant, R.L., G.A. Peterson, and D.G. Westfall. 2012. Dry land cropping systems. Colorado State University Extension Services Bulletin, Fort Collins, CO. p. 1-5.

Dida, M.M., N. Wanyera, M.L.H. Dunn, J.L. Bennetzen, and K.M. Devos. 2008. Population structure and diversity in finger millet (*Eleusine coracana*) germplasm. Trop. Plant Biol. 1:131–141.

Djanaguiraman, M., and P.V.V. Prasad. 2014. High temperature stress. In: M. Jackson, B. Ford-Lloyd, and M. Parry, editors, Plant genetic resources and climate change. CABI Publishers, Wallingford, UK. p. 201–220.

Doust, A.N., M. Mauro-Herrera, J.G. Hodge, and J. Stromski. 2017. The C4 model grass Setaria is a short-day plant with secondary long day genetic regulation. Front. Plant Sci. 8:1062. doi:10.3389/fpls.2017.01062

Du, R.-H., and T.-Y. Wang. 1997. Development and application of the two-line system for hybrid production in foxtail millet. In: Y. Li, editor, Foxtail millet breeding. Agricultural Press, Beijing, China. p. 591–616.

Eagles, H.A., R.M. Haslemore, and C.A. Stewart. 1978. Nitrogen utilization in Libyan strains of Avena sterilis L. with high groat protein and high straw nitrogen yield. N. Z. J. Agric. Res.. 21:65–72. doi:10.1080/00288233.1978.10427384

FAOSTAT. 2016. Food and agriculture data. Food and Agricultural Organization, Rome, Italy. www.fao.org/faostat

Fulton, J.M. 1968. Growth and yield of oats as influenced by soil temperature, ambient temperature and soil moisture supply. Can. J. Soil Sci. 48:1–5. doi:10.4141/cjss68-001

García del Moral, L.F., J.M. Ramos, M.B. García del Moral, and M.O. Jimenez-Tejada. 1991. Ontogenetic approach to grain production in spring barley based on path-coefficient analysis. Crop Sci. 31:1179–1185. doi:10.2135/cropsci1991.0011183X003100050021x

Garmash, E.V. 2005. Temperature controls a dependence of barley plant growth on mineral nutrition level. Russ. J. Plant Physiol. 52:338–344. doi:10.1007/s11183-005-0051-4

Hatfield, J.L., K.J. Boote, B.A. Kimball, L.H. Ziska, R.C. Izaurralde, D. Ort, A.M. Thomson, and D.W. Wolfe. 2011. Climate impacts on agriculture: Implications for crop production. Agron. J. 103:351–370. doi:10.2134/agronj2010.0303

Hellewell, K.B., D.D. Stuthman, A.H. Markhart, III, and J.E. Erwin. 1996. Day and night temperature effects during grain-filling in oat. Crop Sci. 36:624–628. doi:10.2135/cropsci1996.0011183X003600030017x

Hemming, M.N., W.J. Peacock, E.S. Dennis, and B. Trevaskis. 2008. Low-temperature and daylength cues are integrated to regulate FLOWERING LOCUS T in barley. Plant Physiol. 147:355–366. doi:10.1104/pp.108.116418

Hilu, K.W., J.M.J. de Wet, and J.R. Harlan. 1979. Archeobotany and the origin of finger millet. Am. J. Bot. 66:330–333. doi:10.1002/j.1537-2197.1979.tb06231.x

Hunt, H.V., M. Vander Linden, X. Liu, G. Motuzaite-Matuzevicuite, S. Colledge, and M.K. Jones. 2008. Millets across Eurasia: Chronology and context of early records of the genera Panicum and Setaria from archaeological sites in the Old World. Veg. Hist. Archaeobot. 17:S5–S18. doi:10.1007/s00334-008-0187-1

Jones, H., F.J. Leigh, I. Mackay, M.A. Bower, L.M.J. Smith, M.P. Charles, G. Jones, M.K. Jones, T.A. Brown, and P. Wayne. 2008. Population-based resequencing reveals that the flowering time adaptation of cultivated barley originated east of the Fertile Crescent. Mol. Biol. Evol. 25:2211–2219. doi:10.1093/molbev/msn167

Karim, M.A., S. Arabinda, M. Mohiuddin, A.K.M. Rahman, and A.B.M. Salahuddin. 1993. Study on the stages of development and agronomic parameters of foxtail millet (Setaria italica L. Beauv.) under Bangladesh conditions. Nettai Nogyo 37:28–31.

Karsai, I., B. Koszegi, G. Kovacs, P. Szucs, K. Meszaros, Z. Bedo, and O. Veisz. 2008a. Effects of temperature and light intensity on flowering of barley (*Hordeum vulgare* L.). Acta Biol. Hung. 59:205–215. doi:10.1556/ABiol.59.2008.2.7

Karsai, I., P. Szucs, B. Koszegi, P.M. Hayes, A. Casas, Z. Bedo, and O. Veisz. 2008b. Effects of photo and thermo cycles on flowering time in barley: A genetical phenomics approach. J. Exp. Bot. 59:2707–2715. doi:10.1093/jxb/ern131

Kikuchi, R., H. Kawahigashi, T. Ando, T. Tonooka, and H. Handa. 2009. Molecular and functional characterization of PEBP genes in barley reveal the diversification of their roles in flowering. Plant Physiol. 149:1341–1353. doi:10.1104/pp.108.132134

King, S.R., and R.K. Bacon. 1992. Vernalization requirement of winter and spring oat genotypes. Crop Sci. 32:677–680. doi:10.2135/cropsci1992.0011183X003200030019x

Körner, C. 2008. Winter crop growth at low temperature may hold the answer for alpine treeline formation. Plant Ecol. Divers. 1:3–11. doi:10.1080/17550870802273411

Lata, C., S. Bhutty, R.P. Bahadur, M. Majee, and M. Prasad. 2011. [Setaria italica (L.)] Association of an SNP in a novel DREB2-like gene SiDREB2 with stress tolerance in foxtail millet. J. Exp. Bot. 62:3387–3401. doi:10.1093/jxb/err016

Leite, J.G.D.B., L.C. Federizzi, and H. Bergamaschi. 2012. Potential impacts of climate change towards agricultural systems in South Brazil. Revista Brasileira de Ciencias Agrarias 7:337–343. doi:10.5039/agraria.v7i2a1239

Li, Y., S. Wu, and Y. Cao. 1995. Cluster analysis of an international collection of foxtail millet (Setaria italica (L.) P. Beauv.). Euphytica 83:79–85. doi:10.1007/BF01677864

Liu, Q., B. Jiang, J. Wen, and P.M. Peterson. 2014. Low-copy nuclear gene and McGISH resolves polyploid history of Eleusine coracana and morphological character evolution in Eleusine. Turk. J. Bot. 38:1–12. doi:10.3906/bot-1305-12

Lu, H., J. Zhang, K. Liu, N. Wu, Y. Li, K. Zhou, M. Ye, T. Zhang, H. Zhang, X. Yang, L. Shen, D. Xu, and Q. Li. 2009. Earliest domestication of common millet (Panicum miliaceum) in East Asia extended to 10,000 years ago. Proc. Natl. Acad. Sci. USA 106:7367–7372. doi:10.1073/pnas.0900158106

Maqsood, M., and S.N.A. Ali. 2007. Effects of environmental stress on growth, radiation use efficiency and yield of finger millet (*Eleusine coracana*). Pak. J. Bot. 39:463–474.

Matsuura, A., W. Tsuji, P. An, S. Inanaga, and K. Murata. 2012. Effect of pre- and post-heading water deficit on growth and grain yield of four millets. Plant Prod. Sci. 15:323–331. doi:10.1626/pps.15.323

Miller, P., W. Lanier, and S. Brandt. (2001). Using degree days to predict plant stages. Montguide 200103 AG 7/2001. Montana State University Extension Service, Bozeman, MT.

Nielsen, K.F., R.L. Halstead, A.J. MacLean, R.M. Holmes, and S.J. Bourget. 1960. The influence of soil temperature on the growth and mineral composition of oats. Can. J. Soil Sci. 40:255–263. doi:10.4141/cjss60-032

Opole, R.A., P.V.V. Prasad, M. Djanaguiraman, K. Vimala, M.B. Kirkham, and H.D. Upadhyaya. 2018. Thresholds, sensitive stages and genetic variability of finger millet to high temperature stress. J. Agro. Crop Sci. 204:477-492; doi:10.1111/jac.12279.

Peltonen-Sainio, P. 1994. Growth duration and above-ground dry-matter partitioning in oats. Agriculture and Food Science 3:195–198.

Peterson, D.M. 1992. Composition and nutritional characteristics of oat grain and products. In: H.G. Marshall and M.E. Sorrells, editors, Oat science and technology, Agron. Monogr. 33. ASA and CSSA, Madison, WI. p. 265–292.

Prasad, P.V.V., Staggenborg, S.A. and Ristic, Z. 2008. Impacts of drought and/or heat stress on physiological, developmental, growth and yield processes of crop plants. In: L.R. Ahuja, V.R. Reddy, S.A. Saseendran, and Q. Yu, editors, Response of crops to limited water: Understanding and modeling water stress effects on plant growth processes. Adv Agric. Syst. Modeling Series 1. ASA, Madison, WI. p. 301-355.

Prasad, P.V.V., R. Bheemanahalli, and S.V.K. Jagadish. 2017. Field crops and the fear of heat stress- Opportunities, challenges and future directions. Field Crops Res. 200:114–121. doi:10.1016/j.fcr.2016.09.024

Purvis, O.N., and Gregory, F.G. 1937. Studies in vernalisation of cereals. I. A comparative study of vernalisation of winter rye by low temperature and by short days. Ann. Bot. (Oxford, U.K.) 1:569-591.

Sakamma, S., K.B. Umesh, M.R. Girish, S.C. Ravi, M. Satishkumar, and V. Bellundagi. 2018. Finger millet (Eleusine coracana L. Gaertn.) production system: Status, potential, constraints and implications for improving small farmer's welfare. J. Agric. Sci. 10:162–179.

Sameri, M., K. Takeda, and T. Komatsuda. 2006. Quantitative trait loci controlling agronomic traits in recombinant inbred lines from a cross of oriental- and occidental-type barley cultivars. Breed. Sci. 56:243–252. doi:10.1270/jsbbs.56.243

Sandhu, B.S., and M.L. Horton. 1977a. Response of oats to water deficit. I. Physiological characteristics. Agron. J. 69:357–360. doi:10.2134/agronj1977.00021962006900030006x

Sandhu, B.S., and M.L. Horton. 1977b. Response of oats to water deficit. II. Growth and yield characteristics. Agron. J. 69:361–364. doi:10.2134/agronj1977.00021962006900030007x

Savin, R., and M.E. Nicolas. 1996. Effects of short periods of drought and high temperature on grain growth and starch accumulation of two malting barley cultivars. Aust. J. Plant Physiol. 23:201–210. doi:10.1071/PP9960201

Shimada, S., T. Ogawa, S. Kitagawa, T. Suzuki, C. Ikari, N. Shitsukawa, T. Abe, H. Kawahigashi, R. Kikuchi, H. Handa, and K. Murai. 2009. A genetic network of flowering-time genes in wheat leaves, in which an APETALA1/FRUITFUL-like gene, VRN1, is upstream of FLOWERING LOCUS T. Plant J. 58:668–681. doi:10.1111/j.1365-313X.2009.03806.x

Sonego, M. 2000. Effect of temperature and daylength on the phenological development of oats (*Avena sativa* L.). Ph. D thesis. Lincoln University, Lincoln, New Zealand.

Song, Y.H., J.S. Shim, H. Kinmonth-Schultz, and T. Imaizumi. 2015. Photoperiodic flowering: Time measurement mechanisms in leaves. Annu. Rev. Plant Biol. 66:441–464. doi:10.1146/annurev-arplant-043014-115555

Sorrels, M.E., and S.R. Simmons. 1992. Influence of environment on the development and adaptation of oat. In: H.G. Marshall and M.E. Sorrels, editors, Oat science and technology. ASA, Madison, WI. p. 115–163.

Szűcs, P., I. Karsai, J. von Zitzewitz, K. Meszaros, L.L. Cooper, Y.Q. Gu, T.H. Chen, P.M. Hayes, and J.S. Skinner. 2006. Positional relationships between photoperiod response QTL and photoreceptor and vernalization genes in barley. Theor. Appl. Genet. 112:1277–1285. doi:10.1007/s00122-006-0229-y

Takahashi, R., and S. Yasuda. 1971. Genetics of earliness and growth habit in barley. In: R.A. Nilan, editor, Barley genetics II. Proceedings of the second international barley genetics symposium. Washington State Univ. Press, Pullman, WA. p. 388-408.

Tamaki, M., S. Kondo, T. Itani, and Y. Goto. 2002. Temperature responses of leaf emergence and leaf growth in barley. J. Agric. Sci. 138:17–20. doi:10.1017/S0021859601001745

Teulat, B., D. This, M. Khairallah, C. Borries, C. Ragot, P. Sourdille, P. Leroy, P. Monneveux, and A. Charrier. 1998. Several QTLs involved in osmotic-adjustment trait variation in barley (*Hordeum vulgare* L.). Theor. Appl. Genet. 96:688–698. doi:10.1007/s001220050790

Teulat, B., C. Borries, and D. This. 2001. New QTLs identified for plant water status, water-soluble carbohydrate and osmotic adjustment in a barley population grown in a growth-chamber under two water regimes. Theor. Appl. Genet. 103:161–170. doi:10.1007/s001220000503

Teulat, B., O. Merah, X. Sirault, C. Borries, R. Waugh, and D. This. 2002. QTLs for grain carbon isotope discrimination in field-grown barley. Theor. Appl. Genet. 106:118–126. doi:10.1007/s00122-002-1028-8

Tondelli, A., E. Francia, D. Barabaschi, A. Aprile, J.S. Skinner, E.J. Stockinger, A.M. Stanca, and N. Pecchioni. 2006. Mapping regulatory genes as candidates for cold and drought stress tolerance in barley. Theor. Appl. Genet. 112:445–454. doi:10.1007/s00122-005-0144-7

Tottman, D.R., R.J. Makepeace, and M. Broad. 1979. An explanation of the decimal code for the growth stages of cereals, with illustrations. Ann. Appl. Biol. 93:221–234. doi:10.1111/j.1744-7348.1979.tb06534.x

Turner, A., J. Beales, S. Faure, R. Dunford, and D. Laurie. 2005. The pseudo-response regulator Ppd-H1 provides adaptation to photoperiod in barley. Science 310:1031–1034. doi:10.1126/science.1117619

USDA-NASS. 2016. Crop production 2015 summary. United States Department of Agriculture National and Agricultural Statistics Service, Washington, D.C.

Vaeiz, B., V. Bavei, and B. Shiran. 2010. Screening of barley genotypes for drought tolerance by agro-physiological traits in field condition. Afr. J. Agric. Res. 5:881–892.

Watson, E.R., P. Lapins, and R.J.W. Barron. 1976. Effect of waterlogging on the growth, grain and straw yield of wheat, barley and oats. Aust. J. Exp. Agric. Anim. Husb. 16:114–122. doi:10.1071/EA9760114

White, E.M. 1995. Structure and development of oats. In: R.W. Welch, editor, The oat crop, production and utilization. Chapman and Hall, London. p. 88–119. doi:10.1007/978-94-011-0015-1_4

Yan, L., A. Loukoianov, G. Tranquilli, M. Helguera, T. Fahima, and J. Dubcovsky. 2003. Positional cloning of the wheat vernalization gene VRN1. Proc. Natl. Acad. Sci. USA 100:6263–6268.

Zadoks, J.C., T.T. Chang, and C.F. Konzak. 1974. A decimal code for the growth stages of cereals. Weed Res. 14:415–421. doi:10.1111/j.1365-3180.1974.tb01084.x

Zare, M., M.H. Azizi, and F. Bazrafshan. 2011. Effect of drought stress on some agronomic traits in ten barley cultivars. Tech. J. Eng. Appl. Sci. 1:57–62.

Zhang, W., H. Zhi, B.H. Liu, J. Xie, J. Li, and W. Li. 2012. Screening of indexes for drought tolerance test at booting stage in foxtail millet. Plant Genet. Resour. 13:765–772.

Zhi, H., Y.-Q. Wang, W. Li, Y.-F. Wang, H.-Q. Li, P. Lu, and X.-M. Diao. 2007. Development of CMS material from intra-species hybridization between green foxtail and foxtail millet. Plant Genet. Resour. 8:261–264.

Modeling Soybean Phenology

C.F. Shaykewich and P.R. Bullock*

The Significance of Soybeans

Soybean [*Glycine max.* (L.) Merr.] is a grain legume grown in diverse environments throughout the world (Thies et al., 1995) and is a significant global agricultural crop. Soybean production has increased rapidly during the 21st century and currently exceeds 300 million metric tons per year (USDA, 2016a). This currently represents 60% of world oilseed production (USDA, 2016b). The United States is the largest producer of soybean (110+ million metric tons), followed closely by Brazil (100+ million metric tons) (USDA, 2016b).

Disruption of trade routes during World War II resulted in a rapid expansion of soybean acreage in the United States as the country looked for alternatives to imported edible fats and oils (Gibson and Benson, 2005). Soybeans are crushed for oil and meal. The oil is made into edible products including shortening, margarine, cooking oil, and salad dressings as well as industrial products including paint, varnishes, caulking compounds, linoleum, and printing inks (Gibson and Benson, 2005). Recently, several soybean oil-based lubricant and fuel products have been developed to replace non-renewable petroleum products (Gibson and Benson, 2005). After soybean oil extraction, the high protein meal remaining is processed either into soybean flour for human food or incorporated into animal feed.

In the United States, soybean production is largely in the temperate climate of the major corn-growing states where it is sown in the spring, typically May, and harvested in the fall. In South America, soybean production occurs over a range of climates from tropical in center-west Brazil to semitropical in southern Brazil (Flaskerud, 2003) and is linked to large-scale deforestation occurring in both the Humid Chaco and Semiarid Chaco regions of Argentina (Grau et al., 2005). Soybeans in South America are also a summer crop sown in October to November and harvested in April to May.

Abbreviations: CHU, corn heat unit; DID, daily increment of development; GDD, growing degree day; MG, soybean maturity group; PEUS, photoenergetic unit summations; RLA, rate of leaf appearance; RMSE, root mean square error; TH, threshold.

Department of Soil Science, University of Manitoba. *Corresponding author (paul_bullock@umanitoba.ca).

doi:10.2134/agronmonogr60.2018.0002

© ASA, CSSA, and SSSA, 5585 Guilford Road, Madison, WI 53711, USA.
Agroclimatology: Linking Agriculture to Climate, Agronomy Monograph 60.
Jerry L. Hatfield, Mannava V.K. Sivakumar, John H. Prueger, editors.

Table 1. Stages of Soybean Development (Fehr and Caviness, 1977).

Stage		
Abbreviated Stage	Title	Description
	Vegetative	
VE	Emergence	Cotyledons above soil surface
VC	Cotyledon	Unifoliate leaves unfolded sufficiently so that leaf edges are not touching
V1	First node	Fully developed leaves at unifoliate nodes
V2	Second node	Fully developed trifoliate leaf at node above unifoliate node
V3	Third node	Three nodes on main stem with fully developed. Leaves beginning with the unifoliate nodes.
Vn	nth node	n number of nodes on main stem with fully developed trifoliate leaves
	Reproductive	
R1	Beginning bloom	One open flower at any node on the stem
R2	Full bloom	Open flower at one of the two uppermost nodes on the main stem with a fully developed leaf
R3	Beginning pod	Pod 5 mm long at one end of the four uppermost nodes on the main stem with fully developed leaf
R4	Full Pod	Pod 2 cm long at one of the four uppermost nodes on the main stem with a fully developed upper leaf
R5	Beginning seed	Seed 3 mm long in a pod at one of the four uppermost nodes on the main stem with a fully developed leaf
R6	Full seed	Pod containing a green seed that fills the pod cavity at one of the four uppermost nodes on the main stem with a fully developed leaf
R7	Beginning maturity	One normal pod on the main stem which has reached its mature pod color
R8	Full Maturity	Ninety-five percent of pods have reached their mature pod color. Five to 10 days of drying weather are required after R8 before soybeans have less than 15% moisture

Soybean production is vulnerable to changing climate conditions. Significant increases in precipitation have led to an unprecedented expansion of the crop into the semiarid Chaco forests of Argentina, where previously the rainfall was insufficient for reliable production (Grau et al., 2005). Thus, any decline in precipitation in the major growing regions has potential to significantly shrink the area of production. Modeling of crop phenological development and the duration of time for critical phases is key to predicting crop productivity (Ritchie and Nesmith, 1991). Knowledge of the expected calendar date for specific phenological stages can be used to improve the timing of crop management operations to ensure that they occur at the developmental stages considered to be critical for enhancing seed yield or for mitigating pest-mediated yield loss (Torrion et al., 2011). Therefore, accurate phenological models of soybean development have an important role to play in the adaptation of soybeans to changing growing conditions. There has been considerable research conducted on the factors that impact soybean development and a number of models developed for the purpose of predicting the date at which the crop will reach specific stages.

Soybean Phenological Development

Fehr and Caviness (1977) prepared a description of soybean development stages which has become the standard in the literature (Table 1).

Temperature is the factor most highly correlated to the rate of phenological development in most crops (Gepts, 1987; Hodges, 1991; Johnson and Thornley, 1985; Morrison et al., 1989; Shaykewich, 1995; Wielgolaski, 1974). Temperature affects both vegetative and reproductive development in soybean. Vegetative development is highly sensitive to temperature (Hesketh et al., 1973; Thomas and Raper 1977). Hesketh et al. (1973) showed a linear relationship between node formation rate and temperature between about 8 and 30 °C. They also showed that the effect of temperature on reproductive development was in fact different from its effect on vegetative development. Sinclair (1984) found that the final number of main stem nodes is affected by photoperiod and temperature interactions.

Parker and Borthwick (1943) found that floral induction was optimal for foliage temperatures between 21 and 27 °C at night. Above 27 °C fewer primordia were formed. These results are similar to those later reported by Hesketh et al. (1973) for flowering and maturity. Their work showed that reproductive development was insensitive to temperatures between 22 and 28 °C, but that it slowed for temperatures outside this range. Temperature can also affect fruit set. The soybean pod consists of two halves of a single carpel. Thomas and Raper (1981) found that temperatures below 22 °C caused a decrease in carpel initiation and, therefore, fewer pods. Hesketh et al. (1973) found that no pods were formed at 14 or 18 °C. Grimm et al. (1994; p. 36) found "a reduction in temperature sensitivity as plants approach physiological maturity," and "that late maturity cultivars are less responsive to temperature than early maturity cultivars from R5 to R7 stage."

Soybean development is also affected by photoperiod. It is characterized as a quantitative, short-day plant, which means that most cultivars flower sooner under long nights than under short nights (Borthwick and Parker, 1938). Later-maturing cultivars are more sensitive to photoperiod than those that are early maturing, with some early maturing cultivars showing insensitivity to photoperiod (Criswell and Hume, 1972). According to Grimm et al. (1993), most soybean cultivars become sensitive to photoperiod at the time the unifoliates are mostly expanded.

Soybean response to photoperiod, however, is not restricted to flower development. Fisher (1963) showed that at least three short days were required after flowering for fruit to set. Johnson et al. (1960) found significant effects on the durations of phases from flowering, pod set, and end of flowering to maturity when photoperiod was increased at the beginning of each phase. Rates of seed filling may also be affected by photoperiod (Thomas and Raper, 1976, Cure et al., 1982).

Sensitivity to photoperiod is dependent on both cultivar and phase of development. Grimm et al. (1994, p. 34) found that "late-maturity cultivars require a longer night than early cultivars to attain a full rate of development." Also "soybean sensitivity to photoperiod changes as plants progress toward the R5 (beginning seed)... longer nights would be required to reach maximum rate of reproductive development from R1 (beginning bloom) to R5 than from V1 (first node) to R1". And further "longer nights would be required to reach maximum rate of reproductive development from R5 to R7 (beginning maturity) than from R1 to R5."

Early Growth of Soybeans

It is important to distinguish germination (i.e., the process that starts when the seed imbibes water and is complete when the radicle penetrates the testa) from emergence which is the penetrating of the soil surface by the stem. Covell et al. (1986) studied germination rate of soybeans as affected by temperature. They found threshold temperature (Tb) was 5 °C. Germination rate increased linearly with temperature to about 34 °C beyond which it declined. Cumulative germination was a function of thermal time, following a typical "S" shaped or sigmoidal growth function.

Pachepsky et al. (2002) reported that in most studies "just after emergence there was a pause in vegetative development, the length of which was cultivar-specific and temperature-dependent." In their own study they treated it as a "hidden" stage. Other authors (e.g., Hodges and French, 1985) refer to it as a "juvenile" stage.

Hesketh et al. (1973) found that rate of leaf appearance (RLA) increased linearly from 0.12 leaves d^{-1} at 14 °C to 0.46 leaves d^{-1} at 30 °C in their controlled environment experiments. Also in a controlled environment, Patterson (1992) observed a relatively rapid linear increase in RLA with mean temperatures from 8 to 22 °C, and then a slower rate of increase with mean temperature up to 36 °C. Rate of leaf appearance was about 0.18 leaves d^{-1} at 14 °C and about 0.36 leaves d^{-1} at 30 °C. Parker and Borthwick (1939) observed RLA to increase linearly between 13 and 24 °C and then level out to a constant value of about 0.35 leaves d^{-1}. In contrast, Jones and Laing (1978) found that RLA did not depend on temperature after the unrolling of the first trifoliate leaf and was about 0.55 leaves d^{-1} over the range of average daily temperatures between 12.5 and 27.5 °C. Hesketh et al. (1973) confirmed that rates of development of the first two nodes are significantly lower than for the later ones. Jones and Laing (1978) observed that RLA for the first two leaves was between 0.02 and 0.06 leaves d^{-1}, and increased linearly with mean daily temperature.

Models for Soybean Development

Growing Degree Days

The simplest estimate of thermal time is the growing degree day (GDD):

$$GDD = (Ta - Tb) \quad [1]$$

Where Ta is the mean daily temperature and Tb is a plant-specific base temperature.

"This calculation of thermal time is appropriate if:

1. The temperature response of development rate is linear over the range of temperature experienced.
2. Daily temperature does not fall below Tb for a significant part of the day.

3. Daily temperature does not exceed an upper threshold temperature for a significant part of the day. (Ritchie and Nesmith, 1991; p. 6-7)."

In these calculations it is assumed that plant temperature is the same as air temperature. "Even when temperature sensors are in the field where the crop is growing, air temperature at the shelter height (normally about 120 cm) may be different from temperature experienced at the plant height" (Ritchie and Nesmith, 1991). Also, because of absorption of solar energy, a plant leaf may be warmer than air temperature in daylight hours. Sometimes a well-watered plant may have a leaf temperature lower than air temperature if it is transpiring rapidly. Likewise, due to loss of "thermal" or long wave radiation emission, a plant may be cooler than air temperature at night.

Although there are no soybean phenological models based exclusively on GDD in the sections that follow, the GDD concept is applied as a part of some models.

Corn Heat Units

One of the earliest studies on soybean development as a function of maximum (Tmax) and minimum (Tmin) temperature was done at Iowa State University (Brown, 1962). It was later assumed that the relationship for corn was the same as for soybeans and the unit became known as the corn heat unit (CHU).

$$\text{If Tmin} > 4.4, \text{Ymin} = 1.8(\text{Tmin} - 4.4) \quad [2a]$$

$$\text{If Tmin} \leq 4.4, \text{Ymin} = 0 \quad [2b]$$

$$\text{If Tmax} > 10.0, \text{Ymax} = 3.33(\text{Tmax} - 10) - 0.084(\text{Tmax} - 10)^2 \quad [2c]$$

$$\text{If Tmax} \leq 10, \text{Ymax} = 0 \quad [2d]$$

$$\text{CHU} = (\text{Ymax} + \text{Ymin})/2 \quad [2e]$$

According to this equation, the nighttime minimum for development is 4.4 °C, the daytime minimum is 10 °C, and maximum development rate occurs at 30 °C. In this model, a given cultivar is characterized as needing a certain number of CHU to reach a given stage of development or maturity.

A weakness of the CHU is that the response function predicts an unrealistic rapid increase in development rate at temperatures just above 10 °C. In fact, it has been shown that this unit is not suitable for predicting phenological development of corn in Manitoba, Canada (Cutforth and Shaykewich, 1989).

Biometeorological Time Scale

Major et al. (1975) and Major and Johnson (1977) used Robertson's (1968) biometeorological time scale method to develop a biometeorological time scale for soybeans:

$$1 = M = \sum_{S_1}^{S_2} \{[a_1(L-a_0) + a_2(L-a_0)^2][b_1(T\max-b_0) + b_2(T\max-b_0)^2 + d_1(T\min-b_0) + d_2(T\min-b_0)^2]\} \quad [3]$$

in which a_0 is the base photoperiod, b_0 is the base temperature, and $a_{1...}d_2$ are "response" coefficients. The life of the crop is divided into phases with each phase having its own unique base photoperiod and temperature response coefficients. Using the equation above, the crop is assumed to have developed from stage S_1 to

Table 2. Regression coefficients for biometeorological time scale model for phenological development of Maple Presto and McCall soybeans (Burnett et al., 1985) (Photoperiod in hours, temperature in degrees Celsius).

	Phenological phase			
	Planting to emergence	Emergence to flowering	Flowering to begin pod fill	Begin pod fill to maturity
		Maple Presto		
a_0	Value of photoperiod effect = 1.0	5.664	0.9064	17.02
a_1		0.0116	0.0245	-0.6507
a_2		-0.0008381	0.0	-0.1516
b_0	13.0	6.326	12.15	-15.63
b_1	0.019	0.0659	0.016	0.001081
b_2	-0.0008129	0.0	0.0	-0.000001086
		McCall		
a_0	Value of photoperiod effect = 1.0	-5.227	26.71	11.79
a_1		1.197	-0.0028	0.0297
a_2		-0.004874	0.0	-0.006202
b_0	11.45	-0.1756	11.68	-332.5
b_1	0.014	0.00574	0.0765	0.001989
b_2	-0.0004485	-0.00007773	0.0	0.0

stage S_2 when the sum of the daily development rates equals unity. The equations developed by Major et al. (1975) predicted development more accurately than calendar days or "growing degree days" at all locations studied.

A similar approach has been used for soybean zonation in Manitoba, Canada (Burnett et al., 1985). Coefficients for calculating the daily rate of development as a function of temperature and photoperiod are given in Table 2.

Equations expressing development rate are obtained by regression analysis. This sometimes leads to some unrealistic threshold values such as the threshold photoperiods of –5.2 and 26.7 h in Table 2 above. Thus, this is a significant weakness to using this approach for modeling soybean phenology.

Energetic-Photo-Thermal Model

To incorporate more of the factors which might be significant to crop development, Sierra (1977) formulated a model describing the combined effects of solar radiation, photoperiod and air temperature on soybean development. Data were analyzed for four soybean cultivars over several years with several planting dates at each of two locations, Buenos Aires (34° 53′ S lat.) and Cerrillos (24° 54′ S lat.), Argentina. They used two phases of development were considered; emergence to flowering and flowering to ripeness.

Sierra (1977) took the approach of relating development to solar radiation summations, reasoning that solar radiation is a direct measure of thermal energy whereas temperature is a measure of translational kinetic energy. For the emergence to flowering phase, Sierra found that radiation sums required for phase completion (SRS) were more or less constant in the order of 754 to 837 MJ m^{-2} for photoperiods shorter than the photoperiodic threshold, that is, 14.5 h. From 14.5 to 15.1 h, the SRS requirement was a linear function of photoperiod, reaching a value of about 1674 MJ m^{-2} at 15.1 h. Beyond 15.1 h, SRS requirement was constant at about 1674 MJ m^{-2}. Thus, the contribution of a MJ m^{-2} to flowering progress when daily photoperiod was photoinductive ($\leq$ 14.5 h) was two times more than when daily photoperiod was non-photoinductive (> 15.1 h). At intermediate photoperiods, radiation sums were multiplied by an interpolated intermediate ratio. These ratios were used to convert SRS to photoenergetic unit summations (PEUS).

The variation in PEUS caused by variation in air temperature was corrected using a factor Q approximated by:

$$Q = 0.487 - 0.004059(t - 5)^2 + 0.000135(t - 5)^3 + 0.115(V + 1)^{1.371} + 0.018\,A \qquad [4]$$

in which t is the average phase temperature, V is the standard deviation of the average phase temperature, and A is the average phase temperature range. PEUS were then converted to energetic-photo-thermal unit summations (EPTUS) by:

$$\text{EPTUS} = \text{PEUS}/Q \qquad [5]$$

The coefficient of variation (CV) of the various corrected radiation summations required for completion of the emergence to flowering phase decreased in the order SRS > PEUS > EPTUS.

In this study, there was no effect of photoperiod on development in the flowering to ripeness phase.

This model requires a location-specific correction for the effectiveness of solar radiation in driving phenological development. The requirement for measured solar radiation data is a problem because historical data is available for very few locations. It is possible that modeled values of solar radiation (Kahimba et al., 2009) could be derived and utilized for soybean modeling with some measure of success.

Soyphen

Hodges and French (1985) assumed that soybean has a juvenile stage after emergence when it is primarily sensitive to temperature and insensitive to photoperiod. In subsequent phases, both temperature and photoperiod influence development. A daily increment of development (DID) was calculated for each day as a product of photoperiod [Dayf(DL)], temperature (Tempf) and water stress (Spif) functions:

$$\text{DID} = \text{Daylf(DL)} * \text{Spif(Spi)} * [\text{Tempf(Tx)} + \text{Tempf (Tn)}]/2 \qquad [6]$$

in which Tx and Tn are maximum and minimum temperatures, respectively. Thus the model is similar to the biometeorological time scale but with the addition of a water stress term (Spif).

Similar to the BMT time scale, the DID calculation is done for each phase of development. When $\sum$ DID over a period of time reaches 1.0, the phase is considered to have been completed.

Phases of development:

1. planting to emergence
2. emergence to the end of the juvenile stage

3. photoperiod-sensitive phase that ends when floral induction has occurred and floral buds have begun growing
4. floral development ending when first blossom opens
5. flowering
6. first pod growth that follows development of the earliest pod from first blossom to maturity
7. last pod growth that follows development of the last pod to final maturity

Development is modeled as a function of temperature and water stress to the end of the juvenile stage. Thereafter, the photoperiod sensitive phase lasts until sufficient photoperiod cycles have accumulated to trigger floral induction. Development of the first floral bud is modeled as a function of temperature and water stress until first bloom is reached. Then several plant organs and processes are modeled simultaneously as functions of temperature, photoperiod, and water stress. One function is the progress to first flower (R1 of Fehr and Caviness, 1977) through pod and bean growth to maturity. Another is the duration of the flowering period which lasts as long as new buds are forming and blooming. The last is the progress of the last pod to maturity. For each of these phases, a unique value for each of Daylf(DL), Spif(Spi), and [Tempf(Tx) + Tempf (Tn)]/2 was used.

For planting to emergence, daily soil temperature was estimated using a weighted average of air temperatures from the previous 14 d:

$$T_{soil} = \left[\sum_{i=7}^{13} (Tx_i + Tn_i)/2 + \sum_{i=0}^{6} \{ 2 \times (Tx_i + Tn_i)/2 \} \right] \Big/ 21 \qquad [7]$$

where Tx_i and Tn_i are maximum and minimum air temperatures (°C) for *i* days before the current day.

Photoperiod (DL) was calculated from latitude and day of the year:

Solar declination EI = 23.45 * Sin[360/365.25 * (JD - 80.6) * π/180] [8A]

JD = calendar day, for example, for 1 February, JD = 32.

C1 = - tan(Lat * π/180) * tan(EI * π/180), [8B]

C1 = -1 for C1 < -1, and [8C]

C1 = 1 for C1 > 1 [8D]

D2 = arcos(C1) * 24/π – period in which sun is above the horizon [8E]

D3 = [7 + abs(Lat)^2/300]/60 - twilight [8F]

DL = D2 + D3 [8G]

Preliminary coefficients and constants for the DID equation were subjectively determined for the five maturity groups for each growth stage. The model was repeatedly executed on the Columbia, MO meteorological data and model coefficients adjusted subjectively to improve estimation of stage dates. The following is an example of the calculation of DID for stage 6 (First pod growth stage), Maturity Group I:

DL = 15.0 h Daylf = 0.1 + [(16.95 – 15.0)/(16.95 – 13.55)]* (0.9 – 0.1) = 0.5588 [9A]

Tx = 25 C Spif = 1.2 + (50.0 – 50.0)/1.0 – 1.2)/(75.0 – 50.0) = 1.2 [9B]

Tn = 15 C Tempf = 0 + (0.05 – 0)*(25 – 0)/(34 – 0) = 0.02941 [9C]

Spi = 50.0 Tempf = 0 + (0.04 – 0) (15 – 0)/(34 – 0) = 0.01765 [9D]

DID = 0.5588 * 1.2 * (0.02941 + 0.01765)/2 = 0.0157 [9E]

In other words, Stage 6 was 1.57% completed during that day.

The equations developed from the Columbia data were independently tested on plantings at Spickard and Mt. Vernon, MO (Major et al., 1975). The model performance was excellent except for errors at Mt. Vernon resulting from the model estimating delays in emergence due to extreme dryness, which were not apparent in growth stage observations. However, there was poor agreement between reported and estimated maturity dates for 19 international sites of the International Soybean Program.

The Soyphen modeling approach requires the development of separate response functions for each phase of development for each maturity group. To determine production risk for a specific area using historical weather data, a separate model and analysis would be required for each individual cultivar. This would require a massive amount of analysis for each production region. Other research has shown that cultivar-specific parameters are not a necessity for obtaining acceptably accurate results from soybean phenology models (Setiyono et al., 2007)

Linear and Curvilinear Models

Sinclair et al. (1991) developed linear and logistic models for describing the rate of development from emergence to flowering. They defined the daily rate of development (*D*) as the reciprocal of the time between emergence and flowering. The linear equation was:

$$1/d = D = a + b\ \mathrm{Pav} + c\ \mathrm{Tav} \quad [10]$$

in which d is the number of days between emergence and flowering, P_{av} is the average photoperiod over the entire interval between emergence and flowering, T_{av} average temperature over the entire interval between emergence and flowering, and a, b and c are empirical coefficients.

The logistic model has the form:

$$D = D_{max} * f(T) * f(P) \quad [11]$$

in which D_{max} is the maximum developmental rate of individual cultivars which would occur under high temperature and short day conditions. In this case, T is daily average temperature, and P is daily photoperiod. The logistic function for temperature is:

$$\mathrm{f}(T) = 1/\{1 + \exp[-A(T - T_h)]\} \quad [12]$$

where A is a regression coefficient and T_h defines the temperature at which f(T) = 0.5. In this function, T_h is the midrange temperature in the flowering response and the value of A defines the range width; a small A describes a cultivar in which the flowering is responsive to temperature over a wide range.

During the juvenile phase, development rate is not affected by photoperiod and f(P) is set to 1.0. Thereafter the response to photoperiod is described by:

$$f(P) = 1 - \exp[B(P - P_c)\ (P \leq P_c) \quad [13A]$$

$$f(P) = 0.0\ (P > P_c) \quad [13B]$$

where B is a regression coefficient and P_c is the critical photoperiod beyond which there is no development toward flowering. As in the equation for temperature, a small B describes a cultivar in which the flowering is responsive to photoperiod over a wide range.

The study was conducted at Gainesville, FL and involved 13 genotypes and 10 sowing dates. Flowering date was defined as the date of first flower (R1 of Fehr and Caviness, 1977). Mean daily temperature was the average of daily maximum and minimum. In the regression, the average of the mean daily temperature over the growing period for each sowing date for each cultivar was calculated. Likewise, the average photoperiod was calculated. Then, multiple linear regression analysis was conducted to obtain the best fitting equation for each cultivar.

For the logistic model, an iterative procedure using successive approximations was applied. After each iteration step, the error was evaluated by:

$$e^2 = \sum_{j=1}^{n} (f_{\mathrm{j}} - f_j)^2 \quad [14]$$

where f_{j} and f_j are the observed and estimated number of days to flowering for each cultivar, respectively, and n is the number of observations. Iterations were continued with the objective of minimizing error.

Linear and logistic equations were derived for the 13 genotypes. The linear equation for "Stafford", a short season cultivar (Maturity Group IV), was:

$$1/d = D = 0.133 - 0.0088\ P_{\mathrm{av}} + 0.00065\ T_{\mathrm{av}} \quad r^2 = 0.74,\ S = 1.7\ \mathrm{d} \quad [15]$$

The logistic equation for "Stafford" was:

$$1/d = D = 0.0514\ \{1 - \exp[0.444(P - 15.57)]\}/\{1 + \exp[-0.204(T - 18.6)]\} \quad r^2 = 0.80,\ S = 1.6\ \mathrm{d} \quad [16]$$

Over all cultivars, the logistic model was superior to the linear model in all but one cultivar.

Caution is required in extrapolating for individual cultivar coefficients to environments, which vary substantially from the temperature and photoperiod regimes of Gainesville, FL.

In both the linear and logistic models, the coefficients b and B were smaller for later-maturing cultivars, indicating a greater inhibitory effect of longer photoperiods on development rate. The critical photoperiod (P_c) was smaller for later-maturing cultivars, indicating that flowering was more inhibited at shorter photoperiods in later-maturing cultivars than in shorter-season cultivars. Interestingly, no major variations in B were observed in the logistic model, indicating a near absence of variation among cultivars in the shape of the sensitivity to photoperiod below the individual values of P_c.

The temperature coefficient in the linear model tended to be larger for longer-season cultivars, indicating that development rate was more sensitive to temperature change in longer-season cultivars. In the logistic model, longer-season cultivars tended to have larger values of the midrange temperature, T_h, indicating the need for greater temperatures by these cultivars to achieve development rates and flowering dates equivalent to shorter season cultivars. Also, the coefficient A tended to be larger, meaning the sensitivity range to temperature was narrower. In the logistic model, cultivars with larger values of A were predicted to have greater development rates with increased temperatures around T_h.

Rodrigues et al. (2001), working in Brazil, used the same approach as Sinclair et al. (1991) to determine the best fitting linear regression equations estimating the time between emergence and flowering. They used nine cultivars, five seeding dates between the middle of September and the middle of January in 1995/1996 and 1996/1997. The time between emergence and flowering ranged from 33 to 52 d for the earliest cultivar to 54 to 88 d for the latest cultivar. Mean temperature (T) and mean photoperiod (F) over the emergence to flowering phase for each cultivar were regressed on $1/d$, the reciprocal of the time required for the phase to be completed. For Ocepar 6, a relatively short season cultivar, the equation obtained was:

$$1/d = 0.0925 - 0.00978\,F + 0.002892\,T,\ r^2 = 0.89\ S = 2.4\ \text{d} \quad [17]$$

The linear models in which daily development is modeled as the sum of a temperature and photoperiod effect do not seem to be theoretically sound. The logistic model is more acceptable because i) the logistic equation probably is the best representation of the response of a plant to the environmental parameters of temperature and photoperiod, and ii) modeling response as the product of a temperature and photoperiod effect seems to be more logical biologically. This is supported by Sinclair et al. (1991) who found the logistic model to be a better estimator of phenological development than a linear model. However, the method is still cultivar specific and thus would be difficult to adapt to specific regions.

Growing Photo-Thermal Days

SOYGRO

A model based on the concept of photo-thermal time is SOYGRO (Jones et al., 1991). It uses the relationship:

$$R(t) = F_r(T) * F(N) \quad [18]$$

where $R(t)$ is development rate on day t, F_r is a temperature function during phases sensitive to both temperature and photoperiod, and $F(N)$ is the night length function. Both T and N are functions of time at a specific location.

Progress from Stage 1 (S_1) to Stage 2 (S_2) is computed by integrating $R(t)$:

$$tp = \int_{S_1}^{t} R(t)\,\mathrm{d}t \quad [19]$$

where tp is photothermal time from S_1.

Table 3. Relationship between reproductive development of soybean and temperature (based on the work of Parker and Borthwick, 1943).

Temperature (°C)	7.8	10	15.5	21	27	32	37.5	40.5	41
$F_r(T)$	0.0	0.3	0.79	0.98	1.0	0.82	0.5	0.06	0.0

$F_r(T)$ is a piecewise linear function of nighttime temperature (Table 3). Thus, if $F_r(T)$ is 0.5 on a given day, development rate is one-half the maximum rate at the given photoperiod. Expressed in another way, this development, $R(t)$, is the number of photothermal days occurring during a real day. When tp reaches a threshold of f_{min}, the current phase is complete and the next phase begins.

$F_r(T)$ can also be expressed as:

$$F_r(T) = f_{min}/f \qquad [20]$$

where f_{min} is the minimum duration of a growth phase over all temperatures at optimal photoperiod, and f is the duration of the phase at temperature T. In the SOYGRO model, $F_r(T)$ is assumed the same for all growth phases sensitive to photoperiod and the threshold temperature is assumed to be 7.8 °C.

Relative duration of the reproductive phase in response to photoperiod is defined as:

$$f_N = f/f_{min} \qquad [21A]$$

defined as follows:

$$\text{If } N \geq 15, f_N = 1.0 \qquad [21B]$$

$$\text{If } 5 < N < 15, f_N = 1.0 + 0.45\ (15 - N) \qquad [21C]$$

$$\text{If } N \leq 5, f_N = 5.5 \qquad [21D]$$

and the night length function $F(N) = 1/f_N$

In the SOYGRO model, photothermal time is assumed to accumulate hourly during nighttime hours only. Numerically, it is computed for each hour in the night and summed until the threshold is reached for the stage S_1, and then S_2 occurs. Hourly temperatures are computed from daily maximum and minimum temperatures and photoperiod using a sine wave during daylight hours, starting 2 h after sunrise, and a linear decrease at night to the minimum temperature of the next morning.

Emergence and other early stages were predicted as a function of temperature only. The remaining growth stages, floral induction to physiological maturity, were modeled as a function of both temperature and photoperiod as described above.

Details of the model specific to stages, maturity groups, etc., can be found in Jones et al. (1991) or in the original reference (Jones et al., 1987).

This is a practical model that uses only one function to describe the response to temperature. However, since only nighttime temperature is used, some precision may be lost. These temperatures are estimated from daily maximum and minimum temperatures and the minimum temperature of the next day. Some calibration may be required to obtain values of minimum and optimum night length, but the original model provides a good starting point for developing a model to a specific region.

CROPGRO

Grimm et al. (1993) modeled the development of soybeans from sowing to flowering as a function of temperature and night length. For the juvenile phase, sowing to the unifoliate leaf stage, they assumed that development was a function of temperature only, and that this required 8 thermal days, that is, 8 d at optimal temperature under ideal moisture conditions. The rate of development is determined according to a spline function. The rate is zero at or below 6 °C, increases linearly to a maximum at 30 °C, remains at the maximum to 35 °C, and then decreases linearly to zero at 45 °C. A day in which temperature is held between 30 and 35 represents one thermal day. By integrating the rate of development over time, starting at sowing date, the end of the juvenile phase occurs when the threshold of 8 thermal days is reached.

From the end of the juvenile phase to flowering, development is assumed to be governed by a function of temperature and night length:

$$R(t) = F(N) \times F(T) \quad [22]$$

where $R(t)$ is development rate in photothermal days on day t, $F(N)$ is a night length function, and $F(T)$ is a temperature function. If both $F(N)$ and $F(T)$ are equal to unity, development progresses at its maximum rate and one photothermal day occurs during that calendar day. A linear plateau function was adopted for $F(N)$:

$$F(N) = 0 \text{ when } N \leq N_{min} \quad [23A]$$

$$= (N - N_{min})/(N_{opt} - N_{min}), \text{ when } N_{min} < N < N_{opt} \quad [23B]$$

$$= 1, \text{ when } N \geq N_{opt} \quad [23C]$$

where N is night length (h), N_{min} is the minimum night length below which there is no development, N_{opt} is the optimal night length.
For temperature:

$$F(T) = 0 \text{ when } T \leq T_{min} \quad [24A]$$

$$F(T) = (T - T_{min})/(T_{opt} - T_{min}) \text{ when } T_{min} < T < T_{opt} \quad [24B]$$

$$F(T) = 1 \text{ when } T \geq Topt \quad [24C]$$

where T is average hourly temperature, T_{min} is minimum temperature below which there is no development, and T_{opt} is the optimal temperature. Hourly temperatures were calculated from daily maximum and minimum temperatures and photoperiod using a sine wave during the daylight hours, starting 2 h after sunrise, and a linear decrease at night to the minimum temperature of the next morning.

Starting at the end of the juvenile phase, photothermal days are accumulated until flowering occurs. This defines the threshold for flowering (THf), that is, the number of photothermal days required from the end of the juvenile phase to flowering.

Data for this study were obtained from 12 cultivars grown at 18 locations from Isabela, PR (18° 03′ N lat.) to Ottawa, ON Canada (45° 25′ N lat.) over two years with only one sowing date at most locations. Computer programs were used to estimate values of the parameters in the model. Some of these results are shown in Table 4.

Table 4. Parameter estimates for four soybean cultivars, using linear plateau functions for night length and temperature, considering both day and night temperature.

Cultivar	MG†	N‡	RMSE§	N_{min}	N_{opt}	THf	T_{min}	T_{opt}
				h		photothermal d	°C	
Bragg	VII	89	5.28	8.70	11.25	21.00	2.44	26.73
Williams	III	115	3.45	7.64	9.91	18.45	2.66	25.59
Maple Arrow	00	40	4.01	2.73	10.00	15.15	2.43	25.06
Maple Presto	000	41	4.02	2.01	9.19	14.51	4.00	26.27

†MG = maturity group

‡N = number of observations

§RMSE = root mean square error between observed and simulated dates of flowering.

In addition to the above, Grimm et al. (1993) considered a linear plateau model using only night temperature, an inverse exponential night length model, a logistic model and a linear regression model. As a result of their study, Grimm et al. (1993) arrived at the following conclusions:

1."The linear plateau functions seem to be adequate for describing soybean cultivar sensitivity to both photoperiod and temperature to predict flowering.

2. A single equation can be used to describe temperature sensitivity of most cultivars to predict flowering date. In this case, T_{min} = 2.52 and T_{opt} = 25.28 °C.

3. There is a clear correlation between number of photothermal days (THf) from the end of the juvenile phase to flowering, and maturity groups of soybeans."

Grimm et al. (1994) used the same basic linear-plateau functions of temperature and night length to model development in each reproductive phase. In the case of R5 (beginning seed growth) accumulation begins on the date when R1 (flowering) occurs, while for predicting time to R7 (maturity) accumulation starts when R5 is observed. In each case, photothermal days were accumulated until a threshold (TH) was reached. They found that using the linear plateau model yielded less reliable predictions, that is, larger RMSE values, for R5 and R7 than for prediction of R1. Particularly in the case of development from R5 to R7, values of T_{min} take unrealistically low numbers, for example, –87.57 °C. For this reason, Grimm et al. (1994) tried an exponential function for temperature:

$$F(T) = 0 \text{ when } T \leq T_{min} \quad [25A]$$

$$= 1 - \exp[-B(T - T_{min})] \text{ when } T > T_{min} \quad [25B]$$

where B determines the shape of the curve. This model resulted in more realistic values of T_{min}, for example, –5.74 °C but only slightly better predictions than those of the original model. Grimm et al. (1994) summarized their study thus: for the reproductive phase of development, the results "do not yet permit conclusions about the shape of the response curves or the determination of base (minimum) values for night length and temperature for the four cultivars studied".

Most of the work of Grimm et al. (1993, 1994) was eventually incorporated into a model called CROPGRO (Piper et al., 1996).

The most appealing characteristic of this model is that a single equation can be used to describe temperature sensitivity of most cultivars to predict flowering date. In this case, T_{min} = 2.52 and T_{opt} = 25.28 °C. Also N_{min} and N_{opt} for cultivars Maple Arrow and Maple Presto, maturity groups 00 and 000, respectively, are quite similar at about 2.5 and 9 to 10 h, respectively. One calendar day with optimum night length and optimum temperature is assigned a value of one photothermal day. Relatively simple equations are used to calculate the actual number of photothermal days in any given calendar day. Predictions of flowering date were based on accumulation of a threshold number of photothermal days, and were relatively accurate. Although predictions of reproductive stages were less reliable, it would seem that this modeling approach is worthwhile.

Ottawa, on Canada Model

Cober et al. (2001) studied photoperiod and temperature responses in early maturing, near-isogenic soybean lines. They used six genotypes in a growth chamber study. Plants were grown from emergence to flowering at constant temperature treatments of either 18 or 28 °C and photoperiods of 10, 12, 14, 16, or 20 h. The model used for relating days from sowing to first day of flowering was:

$$1/f = b(T_A - T_B) + c(P_A - P_C) + d(T_A - T_D)(P_A - P_C) \quad [26]$$

in which b, c, and d are genetic coefficients, d is a temperature-photoperiod interaction coefficient

T_A = average daily temperature

T_B = base temperature

P_A = average photoperiod (h)

P_C = critical photoperiod (h), $(P_A - P_C)$ cannot be negative

T_D = set at 28 °C, corresponding to the high temperature treatment

An iterative procedure (Summerfield et al., 1993) was used to solve for f, that is, determine values for T_B, b, c, and d. P_C was determined by inserting a range of values into the equation and choosing the value which resulted in the lowest standard error of estimate. The resulting value was 13.5 h. Thus, the final relationship was:

$$1/f = 0.00173(T_A - 5.78) + c(P_A - 13.5) - 0.000338(T_A - 28)(P_A - 13.5) \quad R^2 = 0.96 \quad [27]$$

Cober et al. (2001) determined that the coefficient c was related (linear, positive) to the number of dominant, that is, late flowering and maturity, alleles. Using that relationship the equation became:

$$1/f = 0.00173(T_A - 5.78) - (0.00214 + 0.000812\,N)(P_A - 13.5) \quad [28A]$$

$$- 0.000338(T_A - 28)(P_A - 13.5) \quad R^2 = 0.89 \quad [28B]$$

in which N was the number of dominant alleles.

Stewart et al. (2003) defined phenological development as a variable D with a rate of change expressed as:

$$dD/dt = b[(T - T_B) - C_F(P - P_B) \quad [29A]$$

subject to:

$$(T - T_B) \geq 0 \text{ and } T \leq T_M \quad [29B]$$

$(P- P_B) \geq 0$ and $P \leq P_M$ [29C]

where: t = time

T = average daily temperature = (Tmax + Tmin)/2

T_B = base temperature below which T has no effect on rate of development (5.78 °C was assumed)

P = photoperiod (h)

P_B = base photoperiod below which photoperiod had no effect on development rate (13.5 h was used)

b = a genetic coefficient

T_M = an upper temperature limit above which there was no further effect on development rate (i.e., T was not allowed to exceed T_M) (30 °C assumed)

P_M = an upper photoperiod limit above which there was no further effect on development rate (i.e., P was not allowed to exceed P_M) (calculated to be 17.5 h)

The function included a cultivar specific photoperiod coefficient (c) and a temperature term that interacted with photoperiod:

$$C_F = c - d(T_D - T) \quad [30]$$

where d is a genetic coefficient and T_D is a temperature set to 28 °C; $c = C_F$ when $T = T_D$. The function was set to zero if negative and introduced an interaction between temperature and photoperiod.

In integral form, the stage of development (D) within a phase, for example, planting to first flower (R1 of Fehr and Caviness 1977), can be described by:

$$D = \int_{i=1}^{N} b[(T - T_B) - C_F(P - P_B)]\mathrm{d}t = 1 \quad [31]$$

where N is the number of days from planting to first flower (R1 of Fehr and Caviness, 1977).
Numerically:

$$D = \sum_{i=1}^{N} b[(T - T_B)\Delta t - C_F(P - P_B)\Delta t] = 1 \quad [32]$$

where Δt is a time step of 1 d.

This is compatible with the growing degree day concept; growing photodegree days (G_{PDD}) can be calculated by dividing the equation above by b:

$$G_{\mathrm{PDD}} = \sum_{i=1}^{N} [(T - T_B)\Delta t - C_F(P - P_B)\Delta t] = 1/b \quad [33]$$

This equation was used to calculate N by summing each day from planting given average daily temperature and photoperiod until the total equaled $1/b$, which was the time at first flower (R1 of Fehr and Caviness, 1977). The term "b" is the fraction of the phase, in this case from planting to first flower, that is completed per one growing photo-degree day. The values of constants and coefficients used for this calculation were:

$$T_B = 5.78\ °C,\ T_M = 30\ °C \text{ and } P_B = 13.5\ h \quad [34]$$

Values of other coefficients were determined by fitting the equation above to observed times from planting to first flower for 1998 and 1999 at Ottawa, ON Canada (45° 23′ N lat.) and Urbana, IL (40° 4′ N lat.) using a nonlinear least squares algorithm. Values for b and d were assumed to be constant for all isolines: that is, $b = 0.00214\ °C^{-1}\ d^{-1}$, and $d = 0.196\ h^{-1.}$ The value for b corresponds to 467 growing photo-degree days from planting to flowering. The photoperiod coefficient (c) was calculated for each of the 29 genetic isolines for both natural (C_N) and the 20 h daylength (C_{20}). As well, data for the 20 h daylength were used to solve for P_M, $P_M = 17.5$h.

The approach by Stewart et al. (2003) is somewhat more complex than that of Grimm et al. (1993, 1994). However, if a value can be assigned to the cultivar-specific coefficient that accounts for the interaction of temperature with photoperiod in the Stewart et al. (2003) model, then the approach is worth considering.

SOYDEV - Beta Function

Setiyono et al. (2007) created the soybean phenology model SOYDEV. In the phase from sowing to emergence, they assumed that 40 growing degree days above 5 °C were required for radicle appearance. Afterward, hypocotyl length, H_L, was calculated using accumulated thermal time until emergence:

$$H_L = H_{max}[1 - e^{(-a\ ATTra\ b)}] \quad [35]$$

where a and b are constants, H_{max} is the maximum hypocotyl length (cm), Tb is the base temperature (°C), and ATTra is accumulated thermal time since radicle appearance (°C d). The values of a, b, H_{max}, and Tb used in SOYDEV were 0.0128, 4.57, 5 cm, and 5 °C, respectively. Emergence is defined as when hypocotyl length H_L exceeds sowing depth by 0.6 cm.

For subsequent stages, daily development rate r was calculated from:

$$r = \text{Rmax}\ f(T)\ f(P) \quad [36]$$

where Rmax is the maximum development rate at optimum temperature and photoperiod, $f(T)$ is the temperature function, and $f(P)$ the photoperiod function.

They proposed a β function as a means of modeling daily development. For temperature, the equation for the model was:

$$\frac{f(T) = 2(T - T\min)^{\alpha}(T\text{opt} - T\min)^{\alpha} - (T - T\min)^{2a}}{(T\text{opt} - T\min)^{2\alpha}} \quad [37A]$$

$$\text{if Tmin} < T < \text{Tmax},\ \alpha = [\ln(2)]/\ln[(\text{Tmax} - \text{Tmin})/(\text{Topt} - \text{Tmin})] \quad [37B]$$

$$\text{and if } T \geq \text{Tmax, or } T \leq \text{Tmin},\ f(T) = 0 \quad [37C]$$

where for a given phase, Tmax is the temperature (°C) **above** which the development rate is zero, Tmin is the temperature **below** which the development rate is zero, Topt is the temperature at which the development rate is optimal, α is the β function shape factor, and T is the **mean** daily temperature. [It should be noted that the equations in the published journal paper (Setiyono et al., 2007) are not correct. The corrected equation given above has been obtained through direct communication with the authors.]

The β function for a particular phase of development is shown in Fig. 1b below. The linear spline function used in CROPGRO is given in Fig. 1a for comparison. For photoperiod, the equation was:

$$f(P)=\frac{(\{[(P-Popt)/m]+1\}(Pcrt-P)^{(Pcrt-Popt)/m})^{a}}{(Pcrt-Popt)} \quad [38A]$$

if Popt $\leq P \leq$ Pcrt, $\alpha = [\ln(2)]/\{\ln[((Pcrt-Popt)/m)+1]\}$ [38B]

and $f(P) = 0.0$ if $P >$ Pcrt, [38C]

$f(P) = 1.0$ if $P <$ Popt [38D]

Pcrt is the photoperiod (h) above which development rate is zero, Popt is the photoperiod **below** which development rate is optimal, P is photoperiod (based on −6.00° and – 0.83° solar elevation angle for floral induction and post-flowering photoperiod responses, respectively) and m is the constant (3.0 h).

Main stem node appearance rate (r_{NA}) after unifoliate stage was accumulated to simulate subsequent V stages and was calculated by:

$$r_{NA} = r_{NA,max} f(T)\ CF \quad [39A]$$

$$CF = -A(B - e^{-Cx}) \quad [39B]$$

where r_{NA} is the main stem node appearance rate (node d^{-1}), $r_{NA,max}$ is the maximum main stem node appearance rate, CF is the chronological function for reduction of main stem appearance rate, x is the days after V_{Δ} (onset of reduction in main stem node appearance rate), and A, B and C are the chronology coefficients. For indeterminate cultivars (cv. NE3001, MG 3.0), $A = 1.149$, $B = 0.1$, $C = 0.09$; for semi-determinate cultivars (cv. P93M11, MG 3.1) $A = 1.21$, $B = 0.175$ and $C = 0.187$.

With cultivar specific model calibration, root mean square errors (RMSE) of estimation of major phenological stages averaged 1.8 d in their long-term experiment and 3.3 d in the cultivar × sowing date experiment. Data from a cultivar × sowing date experiment were used to develop empirical equations for estimating key cultivar-specific model parameters for published soybean maturity (MG) group ratings. Compared to nine cultivar specific parameters derived from the full calibration, estimation of model parameters from readily available cultivar information only slightly decreased accuracy, resulting in RMSE (across stages and cultivars) values of 3.6 to 3.8 d. The authors claimed that SOYDEV may be particularly suitable for practical application because of the reduced need for cultivar-specific calibration.

Setiyono et al. (2010) later incorporated the "explicit floral induction and evocation processes in simulation of flowering, and use of nonlinear temperature and photoperiod functions for soybean development rates" into SoySim (UN, 2016). This is a crop simulation model ultimately used for estimating soybean growth and yield under near-optimal growth conditions, that is, no water or other stress. Among other components this model included phenology and temperature driven leaf expansion and senescence processes with logistic functions. One of the objectives in developing this model was to minimize the need for cultivar specific parameters. Data for development and validation of SoySim were obtained from field experiments in Nebraska, Indiana, and Iowa. Soybean cultivars ranged in

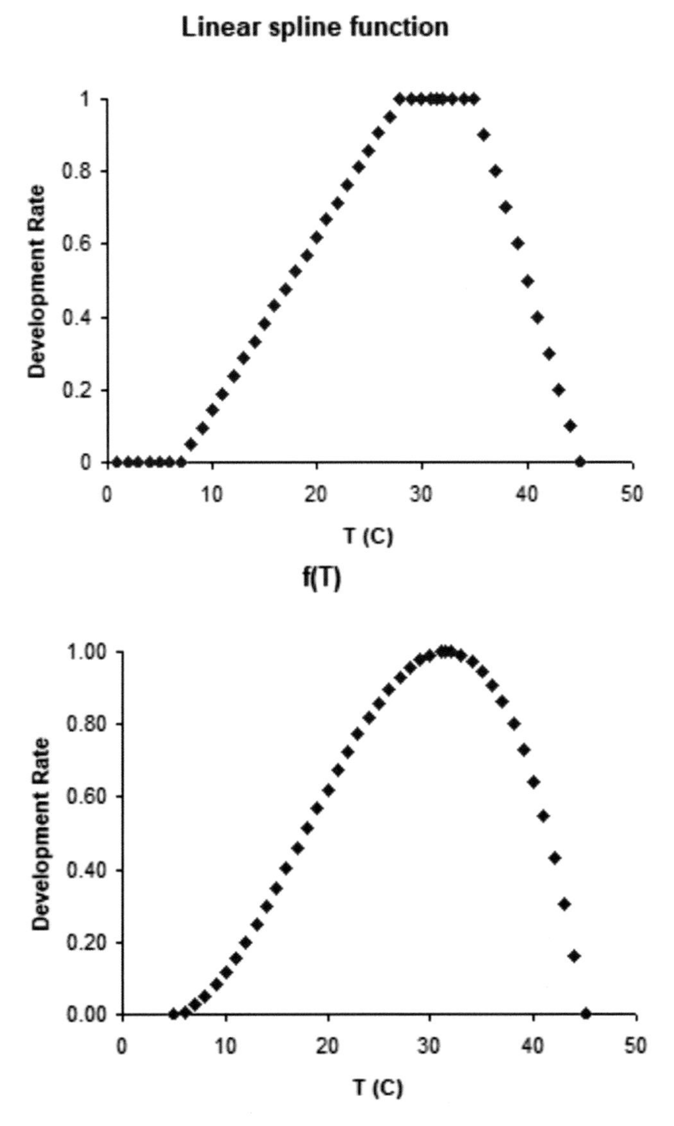

Fig. 1. Temperature functions describing soybean development rate towards the appearance of the unifoliate stage (V1). (a) Linear spline function (Jones et al., 2000) used in CROPGRO (Tmin = 7 °C, Topt1 = 28 °C, Topt2 = 35 °C, and Tmax = 45 °C). (b) Beta function (Wang and Engel, 1998) used in SOYDEV (Tmin = 5°C, Topt = 31.5 °C, and Tmax = 45 °C).

maturity group (MG) from 1.9 to 3.9. The resulting model requires only two cultivar specific parameters: "maturity group and stem termination growth habit."

Working in Nebraska, Indiana, and Iowa, Torrion et al. (2011, p. 1662) conducted a study "(i) to compare the performance of SoySim in predicting calendar dates of Vn and Rn stage occurrence with calendar dates of the observed occurrence of those stages in research plots or fields and. (ii) to identify possible reasons for any variance between the observed SoySim-predicted and researcher or producer-observed dates..". Their results showed RM SE values of 1.7 d, for V1, 3.2 d for R1, 4.2 d for R3.5, 5.3 d for R5, and 3.1 d for R7.

The β function has the distinct advantage that the response of development to an environmental variable seems to be "biologically correct", that is, small changes in response near the lower threshold and optimum values, a nearly linear response in the "mid range" and a somewhat rapid decrease in response approaching the maximum. The successful application of the β function in estimating leaf appearance rate in wheat (Jame et al., 1998) gives it additional credibility. The model seems rather complicated but the authors' claim that it has particular suitability in practical applications because of reduced need for cultivar specific calibration.

SIFESOJA

Sifesoja is a model simulating the phenology of soybeans developed in Argentina (Peltzer and Peltzer, 2013). It estimates dates of occurrence of principle reproductive stages: full flower (R2), initiation of grain filling (R5), maximum seed diameter (R6), and harvest maturity (R8) according to the scale of Fehr and Caviness (1977).

Beginning in 1995/1996, trials were conducted at EEE INTA Parana with soybeans of maturity groups III to VIII at seeding dates from 10 September to 15 February. Regression equations estimating the time from seeding to the reproductive phenological stages R2, R5, R6, R7, and R8 as a function of seeding date for each cultivar were derived. From these regression equations, software modeling the phenological development of cultivars was developed. The model is protected by patents and therefore a detailed description of the estimation equations is not available.

During 2008/2009 trials were replicated in the northern areas of La Paz and Gualeguay in the province of Entre Rios with the objective of incorporating the effects of latitude into the model.

In response to requests from professional agronomists and producers, it was decided to develop a version that could be used in the entire soybean growing area of Argentina. This was achieved by including data from various seeding dates and cultivars from experimental stations across the country.

The principle use of the model is in the selection of cultivars and the date of seeding to avoid critical periods of water stress, as well as programming the monitoring of diseases and the prediction of harvest date.

Validation consisted of comparing observed and simulated dates of R2, R5, R6, R7 and R8 for 6 seeding dates with 18 cultivars at Parana; and at Red Nacional de Evaluacion de cultivars de Soja (RECSO) (only at R2 and R8) at 22 locations at latitudes between 23° 48′ and 38° 20′ south lat.. There were 540 observations in the Parana data, and 290 in the RECSO data. At Parana 64.4% of simulated dates were within $\leq$ 3 d of the observed and 80.3% within $\leq$ 5 d. In the data from RECSO, 61.1% were within $\leq$ 3 d, and 79.2% within $\leq$ 5 d. In the total of 830 data points, the average difference between simulated and observed was 3.38 d.

Conclusion

The large amount of research conducted on modeling soybean development has provided several different approaches with potential to adapt to any region of the world. Every approach carries some limitations and a certain amount of "error" ranging from approximately 2 to 5 d in the occurrence of modeled and observed phenological development. This may be sufficient for purposes such as yield forecasting or determining the suitability of soybeans for specific growing regions. However, this level of accuracy is likely insufficient as a basis for crop stage specific management decisions on an individual farm. Torrion et al. (2011) noted that the SoySim model provided more accurate estimation of soybean stage calendar date in producer fields when it was started at the date of emergence rather than the date of seeding. However, accurate estimates of soybean emergence in producer fields are challenging.

The current state of knowledge on soybean phenological development models suggest that it is a worthwhile tool, especially in smaller research plots and could be usefully employed to assist with evaluating cultivar performance. These models also provide a mechanism to assess the suitability of specific growing regions for various cultivars using long-term historical weather records that are available for any particular location. Therefore, soybean phenology models make a useful contribution to the ongoing development and adaptation of this globally-significant agricultural crop.

References

Borthwick, H.A., and M.W. Parker. 1938. Photoperiodic perception in Biloxi soybeans. Bot. Gaz. 100:374–387. doi:10.1086/334792

Brown, D.M. 1962. Soybean ecology. I. Development temperature relationships from controlled environment studies. Agron. J. 52:493–495.

Burnett, R.B., G.W. Falk, and C.F. Shaykewich. 1985. Determination of climatically suitable areas for soybean [Glycine max (L.) Merr.] production in Manitoba. Can. J. Plant Sci. 65:511–522. doi:10.4141/cjps85-073

Cober, E.R., D.W. Stewart, and H.D. Voldeng. 2001. Photoperiod and temperature responses in early-maturing near-isogenic soybean lines. Crop Sci. 41:721–727. doi:10.2135/cropsci2001.413721x

Covell, S., R.H. Ellis, E.H. Roberts, and R.J. Summerfield. 1986. The influence of temperature on seed germination rate in grain legumes. J. Exp. Bot. 37:705–715. doi:10.1093/jxb/37.5.705

Criswell, J.G., and D.J. Hume. 1972. Variation in sensitivity to photoperiod among early maturing soybean strains. Crop Sci. 12:657–660. doi:10.2135/cropsci1972.0011183X001200050031x

Cure, J.D., C.D. Patterson, C.D. Raper, Jr., and W.A. Jackson. 1982. Assimilate distribution in soybeans as affected by photoperiod during seed development. Crop Sci. 22:1245–1250. doi:10.2135/cropsci1982.0011183X002200060039x

Cutforth, H.W., and C.F. Shaykewich. 1989. Relationship of development rates of corn from planting to silking to air and soil temperature and to accumulated thermal units in the Prairie environment. Can. J. Plant Sci. 69:121–132. doi:10.4141/cjps89-014

Fehr, W.R., and C.E. Caviness. 1977. Stages of soybean development. Iowa Coop. Ext. Serv. Spec. Rep. 80. Iowa State University, Ames, IA.

Fisher, J.E. 1963. The effects of short days on fruitset as distinct from flower formation in soybeans. J. Exp. Bot. 41:871–873.

Flaskerud, G. 2003. Brazil's soybean production and impact. Extension Service, North Dakota State University EB-79. North Dakota State University, Fargo, ND. https://library.ndsu.edu/repository/bitstream/handle/10365/4906/eb030079.pdf. (Accessed 1 August 2016).

Gepts, P. 1987. Characterizing plant phenology: Growth and development scales. In: K. Wisiol and J. D. Hesketh, editors, Plant growth modeling for resource management, Vol. II: Quantifying plant processes. CRC Press, Boca Raton, FL, p. 3-24.

Gibson, L., and G. Benson. 2005. Origin, history and uses of soybean (Glycine max). Department of Agronomy, Iowa State University, Ames, IA. http://agron-www.agron.iastate.edu/Courses/agron212/Readings/Soy_history.htm (Accessed 1 August 2016).

Grau, H.R., N.I. Gasparri, and T.M. Aide. 2005. Agriculture expansion and deforestation in seasonally dry forests of north-west Argentina. Environ. Conserv. 32:140–148. doi:10.1017/S0376892905002092

Grimm, S.S., J.W. Jones, K.J. Boote, and J.D. Hesketh. 1993. Parameter estimation for predicting flowering date of soybean cultivars. Crop Sci. 33:137–144. doi:10.2135/cropsci1993.0011183X003300010025x

Grimm, S.S., J.W. Jones, K.J. Boote, and D.C. Herzog. 1994. Modeling the occurrence of reproductive stages after flowering for four soybean cultivars. Agron. J. 86:31–38. doi:10.2134/agronj1994.00021962008600010007x

Hesketh, J.D., D.L. Myhre, and C.R. Willey. 1973. Temperature control of time interval between vegetative and reproductive events in soybeans. Crop Sci. 13:250–253. doi:10.2135/cropsci1973.0011183X001300020030x

Hodges, T., and V. French. 1985. Soyphen: Soybean growth stages modeled from temperature, daylength and water availability. Agron. J. 77:500–505. doi:10.2134/agronj1985.00021962007700030031x

Hodges, T. 1991. Temperature and water stress effects on phenology. In: T. Hodges, editor, Predicting crop phenology. CRC Press, Boca Raton. p. 7–13.

Jame, Y.W., H.W. Cutforth, and R.M. de Pauw. 1998. Testing leaf appearance models with 'Katepwa' spring wheat grown in a semiarid environment of the Canadian Prairie. Canadian Agricultural Engineering 40:249–256.

Johnson, H.W., H.A. Borthwick, and R.C. Leffel. 1960. Effects of photoperiod and time of planting on rates of development of the soybean in various stages of the life cycle. Bot. Gaz. 122:77–95. doi:10.1086/336090

Johnson, I.R., and J.H.M. Thornley. 1985. Temperature dependence of plant and crop processes. Ann. Bot. (Lond.) 55:1–24. doi:10.1093/oxfordjournals.aob.a086868

Jones, P.G., and D.R. Laing. 1978. Simulation of the phenology of soybeans. Agric. Syst. 3:295–311. doi:10.1016/0308-521X(78)90015-X

Jones, J.W., K.J. Boote, S.S. Jagtap, G.G. Wilkerson, G. Hoogenboom, and J.W. Mishoe. 1987. SOYGRO 5.4 Technical Documentation. Agric. Eng. Dep. Res., Rep., University of Florida, Gainesville, FL.

Jones, J.W., K.J. Boote, S.S. Jagtap, and J.W. Mishoe. 1991. Soybean Development. Chapter 5. In: J. Hanks and J.T. Ritchie, editors, Modeling plant and soil systems. Agronomy 31. American Society of Agronomy, Madison, WI. p. 72-90.

Jones, J.W., J. White, K. Boote, G. Hoogenboom, and C.H. Porter. 2000. Phenology module in DSSAT v 4.0. Documentation and Source Code Listing. Agricultural and Biological Engineering Department, University of Florida, Gainesville, FL.

Kahimba, F.C., P.R. Bullock, R. Sri Ranjan, and H.W. Cutforth. 2009. Evaluation of the SolarCalc model for simulating hourly and daily incoming solar radiation in the Northern Great Plains of Canada. Can. Biosystems Eng. 51: 1.11-1.21.

Major, D.J., D.R. Johnson, J.W. Tanner, and I.C. Anderson. 1975. Effects of daylength and temperature on soybean development. Crop Sci. 15:174–179. doi:10.2135/cropsci1975.0011183X001500020009x

Major, D.J., and D.R. Johnson. 1977. A review of the effect of daylength and temperature on development of soybeans [Glycine max. (L.) Merr.]. Lethbridge Research Station Mimeo Rep 2. Agriculture and Agri-Food Canada, Lethbridge, AB.

Morrison, M.J., P.B.E. McVetty, and C.F. Shaykewich. 1989. The determination and verification of a baseline temperature for the growth of Westar summer rape. Can. J. Plant Sci. 69:455–464. doi:10.4141/cjps89-057

Pachepsky, Y.A., V.R. Reddy, L.B. Pachepsky, F.D. Whisler, and B. Acock. 2002. Modeling soybean vegetative development in the Mississippi Valley. International Journal of Biotronics 31:11–24.

Parker, M.W., and H.A. Borthwick. 1939. Effect of variation in temperature during photoperiodic induction upon initiation of flower primordial in Biloxi soybean. Bot. Gaz. 101:145–167. doi:10.1086/334856

Parker, M.W., and H.A. Borthwick. 1943. Influence of temperature on photoperiodic reactions in leaf blades of Biloxi soybeans. Bot. Gaz. 104:612–619. doi:10.1086/335174

Patterson, D.T. 1992. Temperature and canopy development of velvetleaf (Abutilon theophrasti) and soybean (Glycine max). Weed Technol. 6:68–76. doi:10.1017/S0890037X0003431X

Peltzer, H.F., and N.G. Peltzer. 2013. Modelo de simulacion de fenologia de soja (Sifesoja): Una herramienta util para evitar el estros hidico durante el period critico. Congreso Argentino de Agroninformatica, CAI, Bueños Aires, Argentina. p. 155-160.

Piper, E.L., K.J. Boote, J.W. Jones, and S.S. Grimm. 1996. Comparison of two phenology models for predicting flowering and maturity date of soybean. Crop Sci. 36:1606–1614. doi:10.2135/cropsci1996.0011183X003600060033x

Ritchie, J.T., and D.S. Nesmith. 1991. Temperature and crop development. Chapter 2, p. 5-29. In: J. Hanks and J.T. Ritchie, editors, Modeling plant and soil systems. Agronomy Monograph no. 31. American Society of Agronomy, Madison WI doi:10.2134/agronmonogr31.c2

Robertson, G.W. 1968. A biometeorological time scale for a cereal crop involving day and night temperatures and photoperiod. Int. J. Biometeorol. 12:191–223. doi:10.1007/BF01553422

Rodrigues, O., A.D. Didonet, J.C.B. Lhamby, P.F. Bertagnolli, and J.S. da Luz. 2001. Resposta quantitativa do florescimento da soja à temperatura e ao fotoperíodo. Pesqi. Agropecu. Bras. 36:431–437. doi:10.1590/S0100-204X2001000300006

Setiyono, T.D., A. Weiss, J. Specht, A.M. Bastidas, K.G. Cassman, and A. Dobermann. 2007. Understanding and modeling the effect of temperature and daylength on soybean phenology under high-yield conditions. Field Crops Res. 100:257–271. doi:10.1016/j.fcr.2006.07.011

Setiyono, T.D., K.G. Cassman, J.E. Specht, A. Dobermann, A. Weiss, H.S. Yang, S.P. Conely, A.P. Robinson, P. Pederson, and J.L. Bruin. 2010. Simulation of soybean growth and yield in near-optimal growth conditions. Field Crops Res. 119:161–174. doi:10.1016/j.fcr.2010.07.007

Shaykewich, C.F. 1995. An appraisal of cereal crop phenology modeling. Can. J. Plant Sci. 75:329–341. doi:10.4141/cjps95-057

Sierra, E.M. 1977. Energetic-photo-thermal development model for medium late and late soybeans cultivars. Agric. Meteorol. 18:277–291. doi:10.1016/0002-1571(77)90019-X

Sinclair, T.R. 1984. Cessation of leaf emergence in indeterminate soybeans. Crop Sci. 24:483–486. doi:10.2135/cropsci1984.0011183X002400030012x

Sinclair, T.R., S. Kitani, K. Hinson, J. Bruniard, and T. Horie. 1991. Soybean flowering date: Linear and logistic models based on temperature and photoperiod. Crop Sci. 31:786–790. doi:10.2135/cropsci1991.0011183X003100030049x

Stewart, D.W., E.R. Cober, and R.L. Bernard. 2003. Modeling genetic effects on the photothermal response of soybean phenological development. Agron. J. 95:65–70. doi:10.2134/agronj2003.0065

Summerfield, R.J., R.J. Lawn, A. Qi, R.H. Ellis, E.H. Roberts, P.M. Chay, J.B. Brouwer, J.L. Rose, S. Shanmugasundaram, S.J. Yeates, and S. Sandover. 1993. Towards the reliable prediction of time to flowering in six annual crops: II Soybean (Glycine max). Exp. Agric. 29:253–289.

Thies, J.E., P.W. Singleton, and B.B. Bohlool. 1995. Phenology, growth and yield of field-grown Soybean and Bush Bean as a function of varying modes of N nutrition. Soil Biol. Biochem. 27:575–583. doi:10.1016/0038-0717(95)98634-Z

Thomas, J.F., and C.D. Raper, Jr. 1976. Photoperiodic control of seed filling for soybeans. Crop Sci. 16:667–672. doi:10.2135/cropsci1976.0011183X001600050017x

Thomas, J.F., and C.D. Raper, Jr. 1977. Morphological response of soybeans as governed by photoperiod, temperature, and age at treatment. Bot. Gaz. 138:321–328. doi:10.1086/336931

Thomas, J.F., and C.D. Raper, Jr. 1981. Day and night temperature influence on carpel initiation and growth in soybean. Bot. Gaz. 142:183–187. doi:10.1086/337210

Torrion, J., T.D. Setiyono, K. Cassman, and J. Specht. 2011. Soybean phenology simulation in the North-Central United States. Agron. J. 103:1661–1667. doi:10.2134/agronj2011.0141

United States Department of Agriculture (USDA), Foreign Agricultural Service. 2016a. Production, supply and distribution. http://apps.fas.usda.gov/psdonline/ (Accessed 26 June 2016).

United States Department of Agriculture (USDA), Foreign Agricultural Service. 2016b. Oilseeds: World markets and trade. http://apps.fas.usda.gov/psdonline/circulars/oilseeds.pdf (Accessed 26 June 2016).

University of Nebraska (UN), Institute of Agriculture and Natural Resources. 2016. Soysim– Soybean growth simulation model. http://soysim.unl.edu (Accessed 26 June 2016).

Wang, E., and T. Engel. 1998. Simulation of phenological development of wheat crops. Agric. Syst. 58:1–24. doi:10.1016/S0308-521X(98)00028-6

Wielgolaski, F.E. 1974. Phenology in agriculture. In: H. Lieth, editor, Phenology and seasonality modeling. Springer-Verlag, New York. p. 369–381. doi:10.1007/978-3-642-51863-8_31

Modeling Canola Phenology

C.F. Shaykewich and P.R. Bullock*

The Significance of Canola

The canola plant belongs to the *Brassica* genus (Canola Council of Canada 2016). In the history of agricultural crops, canola is relatively new. Researchers at Agriculture and Agri-Food Canada and the University of Manitoba developed *B. rapa* in the late 1970s as the first canola-quality crop cultivated in Canada, followed shortly by canola-quality *B. napus* (Gulden et al., 2008). Despite its recent origin, canola is now the world's second largest oil crop at 10 to 15% of world oil crop production between marketing years 1999 to 2000 and 2008 to 2009 (USDA Economic Research Service, 2016). In 2010, China had a total production of 12.6 Mt, about one third of the global production (Wang et al., 2012). In 2014 to 2015, the European Union was the largest global canola producer at 33% of global production, followed by Canada at 22% and China at 20% (Gervais, 2015).

Canola seeds are 44% oil, more than double the oil content of soybeans, with high-protein meal produced from the other 56% of the canola seed (Canola Council of Canada, 2016). Although canola was originally developed from rapeseed, the two plants have very different nutritional profiles. To use the name "canola", an oilseed plant must meet an internationally-regulated standard. "Seeds of the genus *Brassica* (*Brassica napus* L., *Brassica rapa* L., or *Brassica juncea* (L.) Czern.) from which the oil shall contain less than 2% erucic acid in its fatty acid profile and the solid component shall contain less than 30 μmol of any one or any mixture of 3-butenyl glucosinolate, 4-pentenyl glucosinolate, 2-hydroxy-3 butenyl glucosinolate, and 2-hydroxy- 4-pentenyl glucosinolate per gram of air-dry, oil-free solid" (Canola Council of Canada, 2016).

Canola production occurs at higher latitudes in areas with dry weather and shorter growing seasons (USDA Economic Research Service, 2016). Winter varieties are seeded before the winter begins in Europe, Ukraine, Russia, and parts of China. In these areas temperatures do not get cold enough to kill overwintering plants, which emerge quickly in spring and produce a yield 20 to 30% larger

Abbreviations: CD, calendar days; DB, days to bud; DEF, days from emergence to the commence of flowering; DTF, days from emergence to first flowering; DSE, days from sowing to emergence; DVS, development stages; E, emergence; ED, end of flowering; FF, first flower; FLN, final leaf number; GDD, growing degree day; LAR, leaf appearance rate; M, maturity; MF, physiological maturity; OF, onset of flowering; Pav, mean photoperiod in hours; PD, photothermal day; RMSD, root mean square deviation; SD, sowing date; Tb, base temperature; Tmax, max temperature; Topt1, lower optimum temperature; Topt2, upper optimum temperature.

Department of Soil Science, University of Manitoba, Winnepeg, MB R3T 2N2. *Corresponding author (Paul.Bullock@umanitoba.ca)

doi:10.2134/agronmonogr60.2018.0003

© ASA, CSSA, and SSSA, 5585 Guilford Road, Madison, WI 53711, USA.
Agroclimatology: Linking Agriculture to Climate, Agronomy Monograph 60.
Jerry L. Hatfield, Mannava V.K. Sivakumar, John H. Prueger, editors.

than spring sown varieties. The spring varieties are primarily planted in parts of China, India, Canada, and the United States (USDA Economic Research Service, 2016). Canola in Australia is usually sown in autumn, but with spring varieties that do not require vernalisation (winter chilling) to flower, and then ripens in late spring or early summer, after a 5 to 7 mo growing season (Walton et al., 1999). Because both winter and spring canola types are being produced in the world, the literature pertaining to canola phenology contains examples of both. In instances when the canola type has been clearly indicated in a publication, the type is specified in this review. However, some publications are not clear whether a spring or winter genotype was studied. In these cases, only the canola variety name is provided when describing the research.

Canola is not produced in warm subtropical or tropical environments as a result of its sensitivity to heat during the reproductive phase. Climate change could have a significant impact on canola production if the traditional growing regions experience higher mean temperatures and more frequent extreme heat events. Knowledge of canola response to growing season weather conditions will be fundamental to improving both the management of the crop and plant breeding results under a changing climate. Despite the relatively recent appearance of canola among global oilseed crops, there has been considerable research into phenological modeling of the crop, which promises to assist with our understanding of how best to maximize production.

Canola Phenological Development

Thomas (1995) proposed a method for describing the stages of canola phenological development that has been utilized by many others for canola phenology research (Table 1). Canola phenological development is most strongly impacted by temperature, as is the case for most crops (Gepts, 1987; Hodges, 1991; Johnson and Thornley, 1985; Morrison et al., 1989; Shaykewich, 1995; Wielgolaski, 1974). Plant growth begins at a baseline temperature below which plant growth and development are negligible. Plant growth and development rates increase progressively to an optimal temperature above which the rate of growth begins to decline and finally stops at some upper threshold. This relationship between plant growth and development with temperature is so unique that many experiments measuring various growth and development parameters all produce similar curves (Johnson and Thornley, 1985). In addition to temperature, it is well known that photoperiod has a significant effect on phenological development of many plants.

Plant Emergence

An overall estimated baseline soil temperature for emergence of three spring and two winter cultivars of canola of 0.9 °C was determined by Vigil et al. (1997). Nykiforuk and Johnson-Flanagan (1994) determined that low temperature has an injurious effect on the germination of canola. Soil temperatures for seeding canola should be between 15 and 20 °C (Anonymous, in Nykiforuk and Johnson-Flanagan, 1994). Canola can germinate at soil temperatures ranging from 2 to 25 °C (Thomas, 1995), but temperatures below 10 °C result in slow germination rates. Blackshaw (1991) concluded that soil temperatures between 10 and 25 °C resulted in greater than 90% emergence and that soil temperatures of 5 and 30 °C resulted in only slightly lower germination. Nykiforuk and Johnson-Flanagan (1994) used 22 °C as the optimum temperature for canola germination.

Plant Development

Photoperiod has been identified frequently as an important input for canola phenology models (Habekotté, 1997; Robertson and Lilley, 2016; Saseendran et al., 2010; Deligios et al., 2013; Husson and Leterme, 1997). King and Kondra (1986) quantified the response of several canola genotypes to photoperiod. They studied seven *B. napus* genotypes (Altex, Regent, Midas, Oro, 74G-1382, 75G-908, 81–58410K) and three *B. rapa* genotypes (Tobin, Torch, S76–4478). Canola plants were grown from seed at photoperiods of 12, 14, 16, 18, 19, and 20 h at a constant temperature of 20 °C. Days from emergence to first flower (DTF) were observed. In general, as photoperiod increased, DTF decreased for all genotypes, but the magnitude of the decrease varied widely with genotype (Fig. 1). The greatest reduction in DTF occurred between 12 and 16 h, with smaller reductions between 16 and 18 h, and very little change beyond 18 h. *Brassica napus* genotypes were more responsive to increasing photoperiod than *B. rapa* genotypes. King and Kondra (1986) concluded that the optimal photoperiod was less than 18 h for all genotypes.

Faraji (2010) created an illustration of the combined effects of temperature and photoperiod on canola development. Studies were conducted at the Agricultural Research Station of Gonbad, Iran (37° 15′ 18″ N, 55° 10′ 17″ E, 45 masl) during the 2002 through 2004 seasons. Four sowing dates (Nov 6, Nov 21, Dec 6, and Dec 21) and four genotypes (Hyola 401, S3, Quantum and Option500) were involved. Days from emergence to the commencement of flowering (DEF), and from commencement of flowering to physiological maturity (DFM) decreased significantly with a delay in sowing date (SD). In both seasons, there were strong linear relationships between SD and the number of days from sowing to emergence (DSE), SD and DEF and SD and DFM. When data from the four genotypes were combined, there were negative linear relationships between mean air temperature

Table 1. Canola Phenological Stages (Thomas, 1995).

Growth Stage	Description
0.0	Pre-emergenc=e
1.0	Seedling
2.1	1st true leaf expanded
2.2	2nd true leaf expanded
2.3, 2.4, etc.	3rd true leaf expanded, etc. for each additional leaf
3.1	Flower cluster visible at center of rosette
3.2	Flower cluster raised above level of rosette
3.3	Lower buds yellowing
4.1	1st flower open
4.2	Many flowers opened, lower pods elongating
4.3	Lower pods starting to fill
4.4	Flowering complete, seed enlarging in lower pods
5.1	Seed in lower pods full size, translucent
5.2	Seeds in lower pods green
5.3	Seeds in lower pods green-brown or green-yellow, mottled
5.4	Seeds in lower pods yellow or brown
5.5	Seeds in all pods brown, plant dead

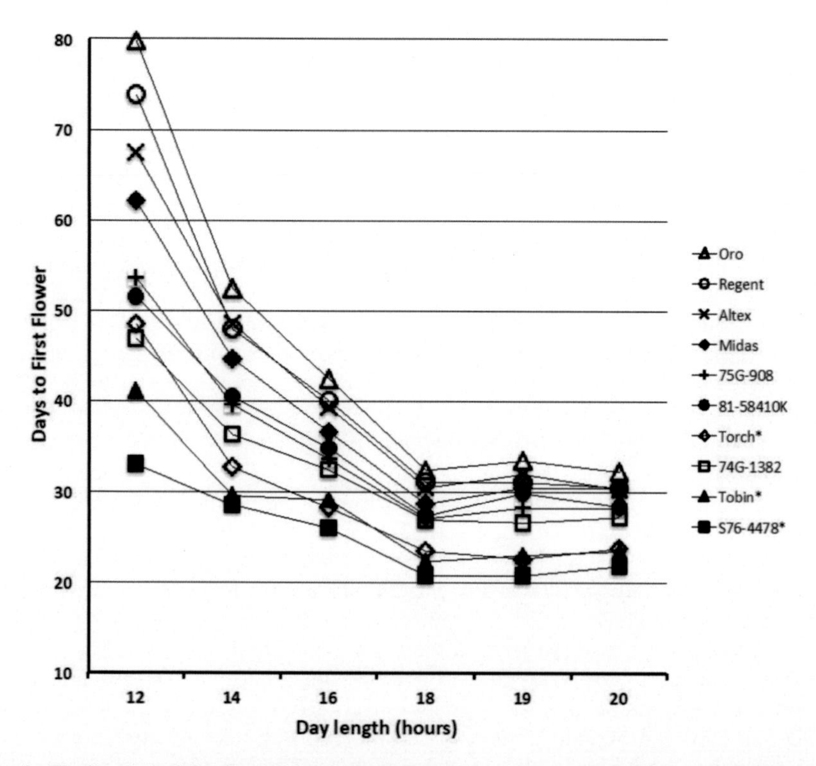

Fig. 1. The impact of day length on the number of days from emergence to first flower for several canola varieties. Varieties marked with * are *Brassica rapa* and all others are *B. napus*. Data from King and Kondra (1986).

Table 2. Relationship of phenological stage duration to mean air temperature (Faraji, 2010)

Days from sowing to emergence: both years	Y‡ = 34.0 -1.66 Tav†	$R^2 = 0.86$
Days from emergence to flowering 2002/2003:	Y = 405 – 34.9 Tav	$R^2 = 0.83$
Days from emergence to flowering 2003/2004:	Y = 418 – 24.7 Tav	$R^2 = 0.87$
Days from flowering to physiological maturity 2002/2003:	Y = 162 - 6.0 Tav	$R^2 = 0.95$
Days from flowering to physiological maturity 2003/2004:	Y = 175 - 6.9 Tav	$R^2 = 0.94$

† Duration (days) of the phenological stage

‡ Mean air temperature (°C) during the phenological stage

(Tav) at different developmental stages and the duration of those periods (Table 2). Faraji (2010) also developed regression equations relating days from emergence to flowering (DEF) to mean photoperiod in hours (Pav) for the 4 different genotypes in the study (Table 3). Faraji (2010) concluded that relationships of SD, temperature and photoperiod with phenology of canola genotypes can be used for developing growth and developmental models.

Vernalization

Phenological development of canola is affected by vernalization, which is the exposure of germinated seeds or seedlings to temperatures just above freezing. Seedlings exposed to these conditions exhibit phenological development that is more responsive to the effects of temperature and photoperiod than seedlings that are not vernalized. A study by Murphy and Scarth (1994) demonstrated the effects of vernalization. They worked with five genotypes with several countries of origin: Global (Sweden), Karat (Sweden), Marnoo (Australia), Regent (Canada), and Westar (Canada). Seeds were moistened in distilled water and subjected to a temperature of 4 °C for 14, 28, and 42 d under dark conditions. In a second experiment, seeds of Global and Marnoo received the same temperature treatment for 2, 4, 6, 8, 10, 12, and 14 d. In the non-vernalized control, seeds were germinated at 22 °C for 48 h prior to the end of each vernalization treatment so as to reach a growth stage similar to that of vernalized seeds. After treatment, seeds were grown under 22/16 °C day/night temperature and 20/4 h day/night photoperiod. Response was measured by observing days to bud (DB), first flower (FF) and final leaf number (FLN).

In the first experiment, all genotypes had reduced DB, FF, and FLN with increasing times of exposure to 4 °C. There were significant differences between genotypes with regard to time required for an initial response and the magnitude of the response. In the second experiment with Global and Marnoo at least 6 d of exposure to 4 °C was required to decrease length of each growth stage and FLN. Murphy and Scarth (1994, p. 277) concluded "that the response to vernalization was cumulative and that there were differences in response among the five cultivars".

Since vernalization seems to be common to all canola genotypes, whether they are spring or winter types, its effect on phenological development has been considered in many phenological models (for example, Robertson and Lilley, 2016; Habekotté, 1997; Wang et al., 2012). In some genotypes, vernalization may decrease the time to flowering, while in others, flowering will not occur without vernalization. Whenever specific studies specified these differences, they are mentioned in this review.

Thermal Time as a Predictor of Phenological Development

Degree Day Systems

One of the earliest methods of quantifying the effect of temperature on phenological development was the degree day system. In this system, a base temperature is assumed and development rate is a linear function of temperature above this temperature. Growing degree days (GDD) are calculated by averaging the maximum and minimum temperatures for the day and subtracting the base temperature. Growing degree days are accumulated throughout the growing season, usually from planting date, and using field observations, a GDD requirement for each stage of development is determined. An example of this approach is the work of Miller et al. (2001). They

Table 3. Relationship of emergence to flowering duration to mean photoperiod (Faraji, 2010)

Option 500	- 2002/2003:	DEF† = 537 – 40.3 Pav‡	$R^2 = 0.99$
	- 2003/2004:	DEF = 391 – 29.7 Pav	$R^2 = 0.98$
S3	- 2002/2003:	DEF = 258 – 34.0 Pav	$R^2 = 0.99$
	- 2003/2004:	DEF = 515 – 43.6 Pav	$R^2 = 0.98$
Hyola401	- 2002/2003:	DEF = 440 – 32.3 Pav	$R^2 = 0.99$
	- 2003/2004:	DEF = 464 – 38.4 Pav	$R^2 = 0.99$
Quantum	- 2002/2003:	DEF = 521 – 38.9 Pav	$R^2 = 0.99$
	- 2003/2004:	DEF = 380 – 28.8 Pav	$R^2 = 0.98$

† Duration (days) of the emergence to flowering stage

‡Mean photoperiod (hours) during the phenological stage

assumed a base temperature of 0 °C and compiled a GDD > 0 requirement for various stages of two *Brassica* types from field observations (Table 4).

Kiniry et al. (1995) proposed Environmental Policy Integrated Climate (EPIC) model parameters for a number of crops grown in the Northern Great Plains region. For both *B. napus* and *B. rapa* canola, they assumed a base temperature for

Table 4. Phenology (Universal Growth Stage Scale) compared to cumulative GDD (0 °C and 32 °F base temperatures) for canola grown in Montana (Miller et al., 2001).

CANOLA (*B. napus*)

Data source: Stu Brandt, Scott, SK 1993-97 and Perry Miller, Swift Current, SK 1995-98

Emergence	**Stage**	**GDD °C**	**GDD °F**
Cotyledons completely unfolded.	1.0	152-186	305-366
Leaf Stages			
Two leaves unfolded.	1.2	282-324	539-615
Four leaves unfolded	1.4	411-463	771-865
Flowering			
Flowering begins. At least one open floret on 50% or more plants.	6.0	582-666	1079-1230
Flowering 50% complete	6.5	759-852	1398-1565
Seed fill			
Seed fill begins. 10% of seeds have reached final size.	7.1	972-1074	1781-1965
Maturity			
Seed begins to mature. 10% of seed has changed color.	8.1	1326-1445	2418-2633
Swathing			
40% of seed on main stem has changed color.			
Swathing recommended at this stage.	8.4	1432-1557	2609-2834

CANOLA (*B. rapa*)

Data source: Stu Brandt, Scott, SK 1993-97 and Perry Miller, Swift Current, SK 1995-98

Emergence	**Stage**	**GDD °C**	**GDD °F**
Cotyledons completely unfolded.	1.0	102-143	215-289
Leaf Stages			
Two leaves unfolded.	1.2	201-254	393-489
Four leaves unfolded.	1.4	300-365	572-689
Flowering			
Flowering begins. At least one open floret on 50% or more plants.	6.0	467-554	872-1029
Flowering 50% complete.	6.5	630-726	1166-1338
Seed fill			
Seed fill begins. 10% of seeds have reached final size.	7.1	826-934	1518-1713
Maturity			
Seed begins to mature. 10% of seed has changed color.	8.1	1152-1279	2105-2334
Swathing			
40% of seed on main stem has changed color.			
Swathing recommended at this stage.	8.4	1249-1382	2280-2519

phenological development of 5 °C and an optimum temperature of 21 °C for GDD calculations. The planting to maturity values of GDD were chosen by assuming that spring wheat and *B. napus* canola required a similar number of days to mature, and that spring barley and *B. rapa* canola had similar days to maturity. In Saskatchewan, average time to reach maturity for spring wheat and spring barley was 97 to 103 d and 89 to 94 d, respectively. On that basis, they assigned a requirement of 1000 to 1200 and 910 to 1015 GDD for maturity for *B. napus* and *B. rapa* canola, respectively. They suggested that these values may be somewhat higher at Lethbridge, AB.

While some workers have assumed a base temperature, others have conducted experiments to actually determine the base temperature, that is, the temperature below which plant development stops. For canola, such a study was conducted by Morrison et al. (1989). Initially canola was grown in growth chambers at mean daily temperatures of 10, 13.5, 15, 17.5, 20, 22, and 25 °C. Mean temperatures were established by setting the minimum and maximum temperatures five degrees lower and higher, respectively, than the desired mean. Plants were grown under a 16 h photoperiod at 45 to 50 Wm^{-2} photosynthetically active radiation (PAR). The number of days required for plants to reach physiological maturity was recorded. Plants grown in the 22 and 25 °C mean temperature regimes did not produce fully developed pods or seeds. For each of the other temperature regimes a mean percent development to physiological maturity per day (% DPM day^{-1}) was calculated. The relationship between % DPM d^{-1} and mean temperature, Tm, was logarithmic in nature. When subjected to regression analysis it produced the following:

$$\% \text{ DPM } d^{-1} = 2.01 \log_{10} \text{Tm} - 1.36 \quad r^2 = 0.971 \qquad [1]$$

When extrapolated to zero daily development, this equation returns a value of 4.77 °C. This base temperature was rounded to 5 °C.

One of the earliest Canadian studies testing models of phenological development of canola was conducted by Morrison and McVetty (1991). Nine field sites of "Westar" canola were seeded between mid-May to early June during 1984 to 1986 at Winnipeg, MB (49.92°N lat., 97.23°W long.). In addition, canola was grown in growth chambers using the same conditions as in the earlier study (Morrison et al., 1989). Phenological development was assessed by observing leaf appearance rate (LAR). Calendar days (CD) and growing degree days above 5 °C (GDD > 5) summed from emergence were used as measures of thermal time.

Field data were analyzed by calculating the linear regression of leaf number on a given date after seeding as a function of thermal time.

$$\#\text{leaves} = 0.022 \text{ GDD} > 5 - 0.061 \quad r^2 = 0.91 \qquad [2]$$

$$\#\text{leaves} = 0.247 \text{ CD} - 0.096 \quad r^2 = 0.85 \qquad [3]$$

Thus, under field conditions GDD > 5 was a better estimator of leaf number than CD.

In growth chamber studies, leaf appearance rate, that is, number of leaves GDD^{-1} or number of leaves CD^{-1}, was found to approximate a linear function of mean temperature (Tmean):

$$\text{Leaves } (\text{GDD} > 5)^{-1} = 0.0358 - 0.0009 \text{ Tmean } (r^2 = 0.79) \qquad [4]$$

$$\text{Leaves CD}^{-1} = 0.0679 + 0.0093 \text{ Tmean } (r^2 = 0.71) \qquad [5]$$

Ideally, a measure of thermal time that accurately estimates phenological development would have a slope of zero in the above equations, that is, leaves $(GDD > 5)^{-1}$ should be a characteristic of the crop, and be the same at all temperatures. In the equation above, Leaves $(GDD > 5)^{-1}$ is 0.026 at 10 °C and 0.0133 at 25 °C. Thus, although better than calendar days, it is clear that GDD > 5 is not the best thermal time measure for canola development.

Sunday (2014) grew seven cultivars of canola in a glasshouse under two temperature regimes: 15/20 °C and 10/15 °C. The study was conducted at the University of Stellenbosch near Capetown, South Africa. Crops were planted on 11 Feb. 2014 and harvested on 14 July 2014. Daylength was 13.20 h at planting and 10.48 h at harvest. Dates of occurrence of a number of growth stages (emergence, budding, flowering, seed ripening, and physiological maturity) were recorded. In addition, rate of leaf appearance in the vegetative phase was observed. The number of days required to reach a given growth stage from planting was multiplied by the mean of the set night/day temperatures of 17.5 °C for the 15/20 °C and 12.5 °C for the 10/15 °C to calculate growing degree day above 0 °C (GDD). Photothermal units (PTU) were calculated as the product of GDD and the daily daylight length.

Plants grown at the warmer temperature regime had a greater plant height, greater number of leaves per plant at the budding stage, greater leaf area at budding, and a higher relative growth rate from planting to budding and from budding to flowering. The number of calendar days from planting to emergence as well as vegetative and reproductive stages were fewer at the higher temperature regime. By contrast, GDD and PTU requirements to reach specific growth stages were higher at the higher temperature. Thermal time and photothermal time units should account for the effects of temperature and photoperiod on the variation in time required to reach various stages of development. The fact that this was not achieved in this study suggests that the thermal time and photothermal time units used may not be appropriate for canola or that there is an interaction between temperature and daylength that was not apparent in this study.

De Marco et al. (2014) determined thermal requirements of two canola hybrids (Hyola-433 and Hyola-61) at Tangara da Serra, Mato Grosso State, Brazil (14°39′ S lat., 57°25′ W long., 321.5 m). They considered four phenological stages: i) emergence (E), ii) early flowering (IF), iii) end of flowering (ED), and iv) physiological maturity (MF). They defined thermal time as degree days above a base temperature,

$$GD = (Tmed - Tb) \quad [6]$$

in which Tmed is the average daily temperature and Tb is the base temperature. They used a different base temperature for each phase. For Hyola-61 base temperatures were -0.8 °C (E- IF), 10 °C (IF-ED), and 7.2 °C (ED-MF); for Hyola-433 base temperatures were 0.3 °C (E-IF), 9.9 °C and 7.9 °C (ED-MF) as determined by Luz et al. (2012).

Thermic sums required to complete the life cycle were 1778 GD for Hyola-433, and 1816 GD for Hyola-61. Thermic sums required for the sub periods E-IF, IF-ED, and ED-MF were 1179, 414, and 185 GD for Hyola-433; and 1203, 436, and 177 GD for Hyola-61, respectively.

Nanda et al. (1995) conducted studies on various aspects of leaf growth of Brassica species at the Indian Agricultural Research Institute (28.08° N lat., 77.12° E long., 229 m). *Brassica campestris* L. (cv. Pusa Kalyani), *B. juncea* L. (cv. Varuna), *B. napus* L. (cv. BO 706) and *B. carinata* A. Braun (cv. Tall-1) were seeded at five dates between 19 Oct. and 9 Dec. 1991. During this time, photoperiod varied between 11.3 to 10.2 h, and mean daily temperature decreased from 24 to 15 °C. On average across species, time from sowing to appearance of the first leaf increased from 10.8 to 22.0 d from the first to last sowing date. When the reciprocal of time to first leaf was regressed on mean temperature during that period, it was found that the base temperatures for *B. campestris, B. juncea, B. napus* and *B. carinata* were 5.7, 6.5, 4.6 and 5.3 °C, respectively.

When number of leaves was plotted as a function of time expressed as days after sowing, number of leaves at a given time decreased from the first to the last sowing date. However, when time was calculated as days from the appearance of the first leaf, leaf number at a given time was virtually the same for all planting dates. Similarly when leaf number was plotted as a function of thermal time from sowing, expressed as degree days above 0 °C, leaf number at a given thermal time was generally similar for all planting dates. However, thermal time from sowing was not a better estimator of leaf number than calendar days from first leaf appearance. This suggests that 0 °C may not be an appropriate base temperature for thermal time calculations. Furthermore, for a given cultivar the relationship between number of leaves as a function of days after first leaf appearance was the same for all sowing dates, that is, differences due to seeding date occurred before appearance of the first leaf.

Final leaf number was greatest for *B. carinata* and least for *B. campestris* and *B. juncea*, but decreased from early to late sowing date.

Physiological-days

Wilson (2002) grew two canola cultivars at eight test sites ranging from 49° 31′ to 51° 10′ N lat., 97 22′ to 101 22′ W long., and 266 to 555 m in elevation on clay loam or clay soils in the agricultural region of Manitoba, Canada during the 1999 and 2000 growing seasons. Wilson used planting and emergence dates as starting dates for calendar days and thermal time calculations. The number of calendar days and growing degree days above 5 °C (GDD > 5) accumulated to each observed stage of development at each site were determined for individual cultivars. GDD > 5 were calculated by averaging the maximum and minimum temperatures for a day and subtracting 5 °C.

The physiological day (*P*-days) model was used to calculate another type of heat unit.

The *P*-days equation developed by Sands et al. (1979) was originally used for potatoes. It was calculated from temperature T (°C) as follows:

$$P = 0 \text{ (When } T < 7) \quad [7A]$$

$$P = k(1-(T-21)^2/(21-7)^2) \text{ (When } 7 < T < 21) \quad [7B]$$

$$P = k(1-(T-21)^2/(30-21)^2) \text{ (When } 21 < T < 30) \quad [7C]$$

$$P = 0 \text{ (When } T > 30) \quad [7D]$$

in which T = 7, 21, and 30 °C are the lower, optimum and upper threshold temperatures, respectively. (Henceforth, these *P*-days will be designated as $P\text{-days}_{(7,21,30)}$). The constant k is a scale factor and has been set at 10.

P-days (ΔP) accumulation can be calculated from daily maximum (Tx) and minimum (Tn) temperatures:

$$\Delta P = 1/24\ [5P(\text{Tn}) + 8P(2\text{Tn} + \text{Tx})/3 + 8P(\text{Tn} + 2\text{Tx})/3 + 3P(\text{Tx})] \quad [8]$$

Table 5. Coefficients of variation for eleven heat unit systems for *B. napus* L. cv 2273 (Wilson, 2002).

Heat Unit	Thermal time Accumulation	Stage										Average
	Beginning at:	2.2	2.3	3.1	3.2	4.2	4.3	5.1	5.2	5.4	5.5	(3.1-5.4)
Calendar Days	Planting	15.6	19.7	8.9	3.4	3.6	3.3	6.6	6.5	6.2	5.9	5.5
	50% emergence	29.5	28.7	5.3	4.9	4.3	2.8	7.0	6.9	7.5	5.1	5.5
GDD > 5	Planting	9.4	16.4	11.2	14.9	10.5	11.0	10.6	10.4	6.3	7.4	10.7
	50% emergence	16.6	12.4	10.0	6.7	7.3	5.2	10.3	8.3	4.6	9.4	7.5
P-days (7,21,30)	Planting	6.5	6.8	10.7	7.4	7.5	2.8	8.3	6.4	4.5	4.7	6.8
	50% emergence	20.8	12.5	9.0	5.6	6.0	3.7	6.6	5.7	3.3	6.0	5.7
P-days (5,21,30)	Planting	5.9	6.0	9.5	5.3	5.6	1.7	7.4	6.1	3.3	3.4	5.6
	50% emergence	21.3	14.8	6.8	3.6	4.2	2.2	6.0	5.4	2.6	4.5	4.4
P-days (5,18,30)	Planting	6.7	8.1	9.7	4.2	4.7	1.5	7.3	6.0	3.1	3.2	5.2
	50% emergence	22.4	17.1	5.9	2.5	3.5	0.9	5.9	5.3	3.1	3.7	3.9
P-days (5,17,30)	Planting	6.9	9.8	9.1	1.4	2.0	1.9	5.2	5.7	3.5	3.4	4.1
	50% emergence	22.9	18.1	5.6	2.2	3.4	0.5	6.0	5.4	3.5	3.6	3.8
P-days (5,16,30)	Planting	7.7	10.2	9.9	3.7	4.4	2.3	7.5	6.2	3.3	3.8	5.3
	50% emergence	23.4	19.2	5.3	2.1	3.3	0.4	6.1	5.5	3.9	3.7	3.8
P-days (5,19,30)	Planting	6.4	7.2	9.6	4.5	5.0	1.4	7.3	6.0	3.1	3.1	5.3
	50% emergence	21.9	16.2	6.2	2.8	3.7	1.4	5.9	5.3	2.8	3.9	4.0
P-days (5,20,30)	Planting	6.1	6.5	9.6	4.9	5.3	1.4	7.3	6.0	3.2	3.2	5.4
	50% emergence	21.6	15.5	6.5	3.3	4.0	1.8	5.9	5.4	2.7	4.2	4.2
P-days (5,16,27)	Planting	8.5	12.5	10.5	3.9	4.9	3.5	8.3	6.6	3.7	4.8	5.9
	50% emergence	24.3	21.0	5.3	2.5	3.7	1.2	6.8	5.8	4.7	4.1	4.3
P-days (5,16,34)	Planting	6.9	9.2	8.9	1.4	1.7	1.7	5.1	5.7	3.3	3.4	4.0
	50% emergence	22.9	18.0	5.4	2.1	3.1	0.6	6.0	5.5	3.4	3.7	3.7

The equation above attempts to calculate the average temperature at four periods during the day and weights the calculation of *P*-days for the length of time during the day that each average temperature typically prevails. *P*-days with a lower threshold of 5 °C, optimum temperatures of 16, 17, 18, 19, and 21°C, and upper threshold temperatures of 27, 30, and 34 °C were calculated for each cultivar individually. Calculations were made from planting and 50% emergence for a number of growth stages according to the scale proposed by Thomas (1995, Table 1).

For calendar days and each heat unit system, the coefficient of variation (CV) of the number of heat units to reach a given development stage was calculated. The heat unit with the lowest CV was deemed to be the best estimator of phenological development. In general, coefficients of variation (CV) decrease with growth stage for heat unit accumulations from planting and 50% emergence for several stages of development for 11 heat unit systems for the cultivars 2273 (Table 5) and Quantum (Table 6). But an unexpected result occurred during the early stages of development (stages 2.2 and 2.3); CV was usually higher when thermal time was accumulated from 50% emergence than from planting. This may have occurred due to difficulties in accurately determining the date of 50% emergence.

For the comparison of heat units, the average of CVs from Stages 3.1 (flower cluster visible at the center of rosette) to 5.4 (seeds in lower pods yellow or brown)

Table 6. Coefficients of variation for eleven heat unit systems for *B. napus* L. cv Quantum (Wilson, 2002).

Heat Unit	Accumulation	Stage									Average
	Beginning at:	2.2	2.3	3.1	3.2	4.2	4.3	5.2	5.4	5.5	(3.1-5.4)
Calendar Days	Planting	17.23	24.3	7.4	5.5	10.3	9.4	6.6	5.6	7.1	7.4
	50% emergence	32.5	35.3	8.3	10.0	4.3	9.9	6.0	6.9	7.6	7.6
GDD>5	Planting	10.0	11.5	21.7	17.0	21.0	18.0	12.9	7.4	8.8	16.3
	50% emergence	20.2	18.0	20.4	45.3	6.7	15.8	10.5	5.7	10.5	17.9
P-days (7,21,30)	Planting	7.9	12.0	17.3	8.8	15.5	11.2	8.7	4.9	7.6	11.1
	50% emergence	24.3	21.4	17.4	10.8	6.0	10.9	8.1	3.4	9.0	9.4
P-days (5,21,30)	Planting	8.8	12.8	15.9	6.3	6.5	10.0	7.7	3.1	6.9	8.3
	50% emergence	25.2	23.2	14.4	9.6	4.1	9.9	7.0	2.8	8.2	8.0
P-days (5,18,30)	Planting	9.5	14.0	11.9	5.6	12.0	8.4	7.0	2.9	6.7	7.9
	50% emergence	26.5	24.8	12.2	8.9	3.3	8.6	6.2	3.2	7.8	7.0
P-days (5,17,30)	Planting	10.0	14.8	8.7	4.8	10.1	7.8	6.2	3.1	6.7	6.8
	50% emergence	27.1	25.5	11.3	8.6	3.0	8.1	5.9	3.5	7.8	6.7
P-days (5,16,30)	Planting	10.8	15.1	10.1	4.8	10.8	7.3	6.5	3.0	6.8	7.1
	50% emergence	27.7	26.2	10.4	8.4	2.6	7.7	5.7	4.0	7.7	6.5
P-days (5,19,30)	Planting	8.9	13.5	12.7	6.0	12.6	8.9	7.2	3.0	6.7	8.4
	50% emergence	26.0	24.2	13.0	9.1	3.6	9.0	6.5	2.9	7.9	7.4
P-days (5,20,30)	Planting	8.5	13.1	13.5	6.5	13.1	9.5	7.5	3.2	6.8	8.9
	50% emergence	25.6	23.7	13.7	9.4	3.9	9.5	6.7	2.8	8.1	7.7
P-days (5,16,27)	Planting	12.1	16.2	8.9	4.4	9.5	6.0	6.3	3.5	7.0	6.4
	50% emergence	28.6	27.4	9.1	8.1	2.6	6.5	5.4	4.8	7.8	6.1
P-days (5,16,34)	Planting	9.9	14.8	8.9	5.1	10.9	8.4	6.4	3.0	6.7	7.1
	50% emergence	27.2	25.6	11.3	8.6	2.8	8.9	6.0	3.4	7.7	6.8

Table 7. Coefficients of variation for seven heat unit systems for both *B. napus* L. cv 2273 and Quantum (Wilson, 2002).

Heat Unit	Thermal time Accumulation	Stage							Average
	Beginning at:	2.2	3.1	3.2	4.2	4.3	4.3	5.4	5.5	(3.1-5.4)
Calendar Days	Planting	15.0	4.4	3.4	4.0	3.5	4.8	6.2	7.6	4.4
	50% emergence	29.0	4.0	4.4	3.8	1.7	4.2	6.9	8.3	4.2
GDD>5	Planting	8.6	11.1	10.3	10.2	4.3	8.0	7.6	10.0	8.6
	50% emergence	21.1	8.7	7.1	7.1	5.9	6.7	5.7	1.5	6.9
P-days (5,17,30)	Planting	8.8	7.0	3.9	5.0	1.8	4.8	2.7	7.4	4.2
	50% emergence	23.3	4.9	2.5	3.5	0.8	3.8	3.3	8.6	3.1
P-days (5,16,30)	Planting	9.7	6.7	3.7	4.8	2.2	4.9	2.9	7.4	4.2
	50% emergence	25.6	3.9	2.2	3.0	0.6	3.8	4.0	8.5	2.9
P-days (5,18,30)	Planting	8.1	7.2	4.2	5.6	1.5	4.7	2.6	7.4	4.2
	50% emergence	24.4	4.4	2.4	3.2	1.1	3.7	3.2	8.7	3.0
P-days (5,17,34)	Planting	8.0	7.0	4.2	4.1	1.5	4.7	2.7	7.5	4.2
	50% emergence	24.5	4.2	2.3	3.1	1.3	3.6	3.1	8.7	2.9
P-days (5,17,32)	Planting	8.4	7.1	4.1	5.1	1.7	4.7	2.7	7.5	4.2
	50% emergence	24.7	4.2	2.3	3.0	1.1	3.6	3.3	8.6	2.9

Table 8. Mean P-Days(5,17,30) for several stages of development (Wilson 2002).

Stage	Thermal time accumulation beginning at:	
	Planting	50% emergence
2.2	140	86
3.1	299	245
3.2	369	304
4.2	419	364
4.3	479	421
5.1	529	476
5.2	583	529
5.4	758	708
5.5	836	778

was utilized because these are the most important phenological stages. At Stages 3.1 to 3.2, the crop reaches complete ground cover. At this stage, the crop is most susceptible to sclerotinia infection. Stages 4.2 and 4.3 coincide with 20 to 30% bloom, the ideal time for applying foliar fungicides. Stages 5.2 and 5.4 were used because physiological maturity (when the crop would be swathed) occurs at Stage 5.3. For these stages, average CV were lower when calculated from 50% emergence than from planting. The exception was Quantum which did not show any improvement for calendar days and GDD > 5.

Although GDD > 5 is a method commonly used for estimating phenological development, GDD > 5 had the highest average CV (stages 3.1 to 5.4) of all the heat unit systems tested, including calendar days. The CVs were lower for the physiological day used for potatoes, $P\text{-days}_{(7,21,30)}$, than for GDD > 5, but not lower than for calendar days. This suggested that fine-tuning of the base, optimum, and upper temperature thresholds was required. Thus, a number of combinations were investigated (Tables 5 and 6).

$P\text{-days}_{(5,17,30)}$ had the lowest coefficient of variation of all the cardinal temperature combinations investigated for the canola varieties Quantum and 2273 combined at each site (Table 7). The *P*-day requirement for several stages of development are shown in Table 8.

Dickson (2014) conducted a study in Manitoba with seven field sites ranging from 49.34° 0′ N to 50.22° N lat. and 97.26° to 98.58° W long. Two cultivars that were commonly grown in 2009 were used (5020 and 71–45RR). Sites were seeded between May 20 and May 30 and visited weekly for growth stage determination, according to Thomas (1995). Thermal time was calculated using the *P*-day formula developed by Wilson (2002), that is, 5 °C, 17 °C and 30 °C were used as the minimum, optimum, and maximum temperatures, respectively. Calculation of thermal time began at seeding and accumulated to each growth stage at each site. The two cultivars had similar thermal time response, so the average accumulation at a given growth stage was calculated using data from both cultivars and all sites (Table 9).

Thermal time accumulations for a given growth stage differed somewhat from those observed by Wilson (2002). The newer cultivars in the Dickson (2014) study reached Growth Stages 3.2 and 4.2 in fewer *P*-days than those used by Wilson (2002), but reached Growth Stage 4.3 in a similar number of *P*-days. However, the cultivars used by Dickson (2014) required more *P*-days to reach Growth

Stages 5.2 and 5.4. Thus, the newer varieties required less thermal time to reach early vegetative stages but more thermal time during reproductive stages and to reach complete maturity. Dickson (2014) suggested that plant breeding efforts may be inherently causing this shift in canola phenology and that a longer reproductive phase, may be responsible for increased canola yields. This emphasizes the importance of constantly updating thermal time characterization as new cultivars become available.

Modeling Examples

CROPGRO

Saseendran et al. (2010) adapted the generic CROPGRO model_for simulating spring canola growth. They grew *B. napus*, cvs. Westar and Hyola, near Akron, CO (40°9′ N lat., 103°9′ W long., 1384 m). Planting dates were: 20 Apr. 1993; 7 Apr. 1994; 20 Apr. 2004; and 8 Apr. 2005. For the purpose of irrigation scheduling, the growing season was divided into three periods: a five week vegetative period, five week reproductive period, and a five week grain filling period. Four irrigation treatments were used: (i) a total of 234 mm applied in 15 equal weekly applications, (ii) no water applied during the five week vegetative period followed by 10 equal weekly applications totalling 234 mm, (iii) no water applied in the five week reproductive period with 234 mm divided evenly among five week vegetative period and the five week grain filling period, and (iv) 234 mm divided evenly in the first 10 wk and no water in the grain filling period.

They used a spline function to describe the effect of temperature on plant processes. According to this function, a process begins at some minimum temperature, increases linearly with temperature to a lower optimum temperature, remains constant to an upper optimum temperature, and finally decreases linearly to zero at some maximum temperature. Various sources in the existing literature were used to establish initial values for these. For vegetative development, they used 1.0 °C, 16.0 °C, 25.0 °C and 40.0 °C as the four cardinal temperatures. A thermal day was defined as one in which the temperature remains in the optimum range for the entire day. A photothermal day (PD) is a thermal day that occurs at an optimum photoperiod. For maximum leaf photosynthesis, the values were 5.0 °C, 28.0 °C, 29.0 °C and 40.0 °C. For canopy photosynthesis, the values were 5.0 °C, 20.0 °C, 28.0 °C and 40.0 °C. Saseendran et al. (2010) did not comment or speculate about possible reasons for the differences in optimum temperatures between leaf photosynthesis and canopy photosynthesis.

Table 9. Mean P-Days (5,17, 30) for several stages of development (Dickson, 2014).

Stage	Description	P- days
3.2	Flower cluster raised above rosette	298
4.2	Many flowers opened, lower pods elongating	405
4.3	Lower pods starting to fill	479
4.4	Flowering complete, seed enlarging in lower pods	601
5.2	Seeds in lower pods green	735
5.4	Seeds in lower pods yellow or brown	815

Table 10. Base, optimum and maximum temperatures for canola temperature response in three different phenological phases (Deligios et al., 2013).

	Tb	Topt1	Topt2	Tmax
		---------------°C----------------		
Vegetative	5 †	22 †	25 §	35 §
Early reproductive	0 ‡	21 ‡	25 §	35 §
Late reproductive	0 ‡	21 ‡	25 §	35 §

†Copani et al. (1994) and Morrison et al. (1992)

‡Nanda et al. (1996) and Morrison et al. (1989)

§Robertson et al. (2002) and Morrison and Stewart (2002)

Data collected from four irrigation treatments in 2005 were chosen for model calibration. This calibration established photothermal requirements for various phases of development. Thus, planting to emergence required 5.0 thermal days, emergence to first flower 16.5 photothermal days, first flower to first seed 13.0 photothermal days, and first seed to physiological maturity 22.79 photothermal days. The model developed was ultimately tested on 2006 data as a continuation of the 2005 study. "Crop phenology was simulated reasonably well with deviations of days to emergence within 1 to 2 d, flowering within 1 to 3 d, first pod within 1 to 5 d, and harvest maturity within 2 to 4 d from measured data" (Saseendran et al., 2010, p. 1614). For rainout shelter experiments in 1993 and 1994 "deviations in simulated plant emergence were off by 1 to 3 d, flowering by 1 to 4 d, first pod by 1 to 5 d, and harvest maturity was off by 1 to 6 d" (Saseendran et al., 2010, p. 1617).

Saseendran et al. (2010, p. 1620) concluded: "Accurate simulations of growth and development (growth stages) of the crop showed that the model has potential as a tool for development of decision support systems for canola management and for evaluation of canola as a potential alternative crop across the central Great Plains region."

Deligios et al. (2013) grew the cultivar Kabel, a short, very early maturing type, at two sites in Italy: Ottana (40°16′ N lat., 8°58′ E long., 202 m asl) and Ottava (40°46′ N lat., 8°29′ E long., 81m asl) in 2007 to 2008. The Ottana site was seeded 9 Nov. 2007 and Ottava on 13 Nov. 2007. Adapting CROPGRO like Saseendran et al. (2010), they divided the life cycle into three phases: vegetative, early reproductive and late reproductive. They also used a spline function to describe the response of phenological development to temperature: a base temperature (Tb), a lower optimum temperature (Topt1), an upper optimum temperature (Topt2), and a maximum temperature (Tmax). The values they used (Table 10) were taken from the literature.

Because work by other researchers such as Robertson et al. (2002) had shown that early maturing genotypes like Kabel are least responsive to vernalization, this component was not in include in the model. A critical maximum long day effect of 16 h and an apparent sensitivity, that is, slope of relative response of development vs. photoperiod (PP-SEN) of- 0.0021 h^{-1}, were used. The daylength effect was used only up to anthesis. In a calibration procedure the duration of the period between germination or emergence and flower appearance was adjusted until flowering date was simulated correctly. Then, the period between first seed and physiological maturity was adjusted until the maturity date was correct. This calibration ultimately established photothermal day requirements for the various phases of development: Emergence to flowering, 35.0, First flower to beginning pod, 9.0, First flower to beginning seed, 25.0, and Beginning seed to physiological maturity, 33.6.

European Studies with Winter Canola

Husson and Leterme (1997) sought to develop a model to estimate the date of first flowering for winter canola in France. Experimental sites were located at Cher (47° 0′ N lat., 2° 35′ E long., 4 site-years), Haute-Garonne (43° 25′ N lat., 1° 30′ E long., 3 site-years), Indre (46° 46′ N lat., 1° 36′ E long., 8 site-years), and Meurthe et Moselle (48° 40′ N lat., 6° 10′ E long., 8 site-years) in the years 1985 to 1988. They calculated mean daily temperature as the mean of maximum and minimum temperature. Only those days on which the mean exceeded 5 °C were considered.

These temperatures were summed. The mean of the photoperiods on the days when temperature exceeded 5 °C was calculated. Their model began calculations on 15 Nov. When the summed temperatures accumulated to first flowering (Sum T) for the variety Goeland were plotted against mean photoperiod (P_m), a linear relationship was found:

$$\text{Sum T} = 4010 - 305.8\, P_m \qquad [9]$$

This showed a large effect of photoperiod, for example, in a 10.0 h photoperiod, a Sum *T* of 952 was required, while in a photoperiod of 11.5 h, a Sum *T* of only 800 was required.

Husson and Leterme showed equations for other genotypes. They assumed that the slope in the equation above, that is, 305.8, was the same for all genotypes but the intercept changed. The larger the intercept, the later flowering was the genotype. They separated the data in two seeding dates, before and after 15 Sept.. When the predicted dates of first flowering were plotted against those observed, there was an approximately equal scatter around the 1:1 line, that is, seeding dates did not influence the accuracy of the prediction.

Similarly, neither latitude nor the magnitude of Sum *T* caused any bias in the predictions.

Habekotté (1997) developed a model for phenological development of winter oilseed rape cultivar Jet Neuf grown in northern Europe. He used four development stages (DVS): (i) emergence (E), (ii) onset of flowering (OF), (iii) end of flowering (ED) and (iv) maturity (M).

Development rate (d DVS dt^{-1}) was assumed to be a linear function of effective daily temperature (T_{eff}) expressed by the parameter a_T.

$$\text{DVS}(t) = \int_{t_j}^{t_e} d\text{DVS}/dt \qquad [10]$$

in which t_i and t_e represent time at the beginning and end of a particular phase. Then $d\,\text{DVS}/dt = T_{eff}\, a_{T',\chi}\;\; 0 \le \text{DVS} \le 1$

$$d\,\text{DVS}/dt = T_{eff}\, a_{T',\chi}\, F_v\, F_{p'}\; 1 \le \text{DVS} \le 2;\; F_v \ge 0;\; F_p \ge 0\;\; d\,\text{DVS}/dt = T_{eff}\, a_{T',\chi}\; 2 \le \text{DVS} \le 4 \qquad [11]$$

where χ refers to the relevant development phase. T_{eff} is calculated from average daily temperature T_{day} (mean of maximum and minimum) and a base temperature for each phase, $^{T}b,\chi$:

$$T_{eff} = T_{day} - T_{b,\chi}\;\; T_{b,\chi} < T_{day}\;\; T_{eff} = 0,\; T_b \ge T_{day} \qquad [12]$$

Table 11. Winter oilseed rapeseed optimal model parameter values obtained from calibration (Habekotté, 1997).

Period†	Parameter	Value	Dimension
S-E	$T_{b,1}$	0.3024	°C
	$a_{T,1}$	7.7212	10^{-3} d^{-1}°C^{-1}
E-OF	$T_{b,2}$	0.5444	°C
	$a_{T,2}$	2.0083	10^{-3} d^{-1}°C^{-1}
	$T_{v,min}$	-3.7182	°C
	$T_{v.op.1}$	0.7260	°C
	$T_{v.op.2}$	5.3770	°C
	$T_{v,max}$	17.2022	°C
	$R_{v,max}$	0.014553	°C
	P_b	5.7416	h
	P_{sat}	14.8014	h
OF-EF	$T_{b,3}$	4.1963	°C
	$a_{T,3}$	5.1036	10^{-3} d^{-1}°C^{-1}
EF-M	$T_{b,4}$	0.6870	°C
	$a_{T,4}$	1.4651	10^{-3} d^{-1}°C^{-1}

† S, seeding; E, emergence; OF, onset of flowering; EF, end of flowering; M, maturity.

F_v is the degree of vernalization, with 0.0 not vernalized and 1.0 fully vernalized.

Vernalization rate, d F_v dt^{-1}, is calculated from emergence until onset of flowering or until full vernalization. The dF_v dt^{-1} rate is zero up to some minimum temperature $T_{v,min}$, increases linearly to a lower optimum temperature $T_{v,opt1}$, remains constant to the upper optimum temperature $T_{v,opt2}$, and finally decreases linearly to zero at a maximum temperature $T_{v,max}$.

F_p is the photoperiod factor. It increases from zero at some basal photoperiod, P_b, to 1.0 at a saturating period, P_{sat}.

A literature search provided a range of values for the various parameters, for example, $T_{b,\chi}$, $a_{T,\chi}$, $T_{v,min}$, P_b, etc. Data from experiments conducted at Wageningen, the Netherlands (51°58′ N lat., 5°40′ E long.), Kiel, Germany (54°19′ N lat., 10°12′ E long.) and Dijon, France (47°16′ N lat., 5°05′ E long.) were used for calibration, i.e., establishing the values of these parameters (Table 11). This was accomplished by trial and error using an optimization method.

Validation of the resulting model used 31 sowing dates from Lelystad (52°34′ N lat., 5°38′ E long.) and Dijon-high (47°16′ N lat., 5°05′ E long.) locations. When actual and modeled dates of occurrence were compared, average errors were 1.7 d for emergence, 3.0 d for onset of flowering, 2.6 d for end of flowering and 5.5 d for maturity. The latter three were 38 to 75% lower than mean deviations of estimates based on calendar days.

Agricultural Production Systems Simulator

In the late 1990s, an Agricultural Production Systems Simulator (APSIM) was developed in Australia. This was a framework crop growth simulation model from which models for individual crops were developed. APSIM-Canola was one such model. Using a daily time step, it simulates crop phenological development, growth (e.g., leaf area), yield and N accumulation driven by inputs of temperature, photoperiod, radiation, and soil water and nitrogen supply. A detailed description of the model from which the following was derived was given by Robertson and Lilley (2016). They referred to a number of research publications from which the specific components of the model were based. Only the phenological development section of the model is described here. It is used with both winter and spring canola types in Australia.

In APSIM, phenological development is divided into phases: (i) sowing to emergence, (ii) emergence to end of basic vegetative (or juvenile) period, (iii) end of vegetative to floral initiation (FI), (iv) flowering, and (v) grain filling, ending in maturity. All phases are responsive to temperature, for example, thermal time. In addition, phase (ii) is sensitive to vernalization, and phase (iii) is sensitive to photoperiod.

Thermal time is calculated assuming a base temperature of 0°C, and optimum of 20 °C, and a maximum of 35 °C. It is calculated as degree days, that is, mean daily temperature °C minus the base of 0 °C up to 20 °C. Above 20 °C there is a linear decrease to the maximum temperature of 35 °C, for example, at 27.5 °C, degree days = 10. Thermal time is calculated by the method of Jones et al. (1986) dividing each day into 3 h periods with inputs of daily maximum and minimum temperatures. The thermal time requirement for sowing to emergence is set at 110 degree days and adjusted for sowing depth.

The amount of thermal time required for completion of phase (ii), emergence to end of vegetative period, is modified by vernalization, that is, exposure to non-freezing cold temperatures. In APSIM it is assumed that the base temperature for vernalization is 0 °C, the optimum is 2 °C, and the maximum is 15 °C. A mean daily temperature $\leq$ 0 °C produces 0.0 vernal days, 2 °C produces 1.0 vernal day, and $\geq$ 15 °C yields 0.0 vernal days. There is a linear increase from 0 to 2 °C, for example, 1 °C produces 0.5 vernal days; and a linear decrease from 2 to 15 °C, for example, 8.5 °C gives 0.5 vernal days. The number of vernalization days required for the plant to be completely vernalized is genotype specific. Robertson et al. (2002) give an example for the genotype "Monty" in which the thermal time required to complete this phase decreased linearly from 229 at 0.0 vernal days to a minimum of 1 at 25 vernal days. Lilley et al. (2015) give examples of vernalization requirement for four genotypes with a range of maturity ratings, with later maturating genotypes requiring a larger number of vernalization days.

The next phase, (iii) end of vegetative to floral initiation, is photoperiod sensitive. For example, Robertson et al. (2002) concluded that among the 15 genotypes studied, thermal time to flowering varied from 533 to 621 degree days and that this requirement was reduced by 61 to 156 degree-days h^{-1} as photoperiod increased from 10.8 to 16.3 h, with the decrease being greatest in later flowering genotypes. Thus, in the model, 10.8 and 16.3 h were chosen as the base and ceiling photoperiods. The genotype specific thermal time target is revised daily as photoperiod changes from start to end of the phase. Studies by Robertson et al. (2002), Farre et al. (2002) and Wang et al. (2012) have shown that the model simulates the time from sowing to flowering with a root mean square deviation (RMSD) of approximately 5 d.

From flowering to maturity phenological progress responds only to temperature. Degree day requirements vary from 500 to 1000 among genotypes grown in Australia, Canada, China and Europe.

Wang et al. (2012) applied APSIM to winter canola production in the Yangtze River Basin. Here canola is planted as a winter crop in a double cropping rotation with rice. Canola is seeded from early September to the end of October. Harvest occurs from the end of April to early May. Sowing cannot take place until rice is harvested and is often delayed by late rice maturity. Wang et al. (2012) quote several studies outside China in which canola yields declined with delayed sowing. To further explore this issue, they collated experimental data on canola phenology, growth and grain yield from multiple sites in the Yangtze River Basin to calibrate and validate the APSIM– Canola model and then simulate yield potential and yield response of canola under different sowing dates and historical climate variability.

Sites were located in south-eastern China near Shimen county, Hunan province (29°35′ N lat., 111°20′ E long., alt 138 m), in 2002, Wuhan, Hubei province (30°28′17.32″ N lat., 114°21′29.68″ E long., alt 23 m) in 2006 and 2007, and Nanjing, Jiangsu province (32°03′ N lat., 118°48′ E long., alt 25m) in 2003. Cultivars grown were Xiangzayou 1 at Shimen, Zhongshuang 10 at Wuhan, and Ningza 3 at Nanjing. A wide range of sowing dates from early September and mid-November were used. The APSIM model version 7.3 was used simulate growth of canola in response to sowing date and growing season weather conditions. The model estimates the phenological stages of germination, emergence, juvenile stage, floral initiation, flowering, grain filling and maturity from thermal time and photoperiod.

Several modifications to the model were made. The number of vernalization days required was increased from 25 to 50. For each genotype, a specific thermal time requirement, as modified by photoperiod, was assigned for the end of juvenile to floral initiation phase. Finally, maximum temperature for thermal time calculations was decreased from 35 °C to 30 °C.

In all cases, crop duration decreased with later sowing dates. At Wuhan where seeding dates ranged from early September to 15 November, growth duration from sowing to maturity decreased from approximately 250 to about 180 d. The range in growing season at the other two locations was less, probably because the range in sowing dates was smaller. The model predicted the change in timing of FI and maturity with different sowing dates, and could explain 98% of the variation in phenological stages as affected by sowing date and climate. The model was calibrated only for the first sowing dates and was able to simulate changes in phenology caused by changes in sowing date. Based on the simulation results, the reduction in the length of the growing period was mainly a result of a shortened vegetative phase.

Heat Stress

A significant issue in canola production is heat stress. Some of the early work on this subject was conducted by Morrison (1988). Growth chambers were used to produce mean temperature regimes of 10.0, 13.5, 15.0, 17.0, 20.0, 22.0 and 25 °C. Daily temperatures were set to vary from a minimum of 5 °C lower than the mean to 5 °C higher than the mean. Plants grown at the 22 and 25 °C mean temperature regimes did not produce fully developed pods or seeds, while plants grown at 20 °C or lower mean temperatures did.

Field studies of heat stress on canola were conducted at Ottawa, Canada (45°23′ N lat., 75° 40′ W long.) from 1989 to 1991 by Morrison and Stewart (2002).

Using several cultivars, three seeding dates, spaced approximately two weeks apart were used to obtain different temperature regimes during flowering. They created a heat stress index Hi:

$$\mathrm{Hi}=\sum_{j=1}^{n}(T\max-Tf)\Delta t \qquad [13]$$

in which Tmax is the maximum daily temperature from the beginning of bolting to the end of flowering, with Tf being the threshold temperature, Δt the daily time step and n the number of days during flowering.

To determine Tf, observed yield from the three seeding dates across 3 yr were plotted as a function of Hi for a range of values for Tf. Regression analysis of each temperature produced a relationship of the form:

$$Y = \mathrm{Ymax} - b\,\mathrm{Hi} \qquad [14]$$

in which Ymax is the y axis intercept and b was a regression coefficient. Tf was defined as that temperature that resulted in the lowest sum of squares of differences between calculated and observed yields. Morrison and Stewart (2002) determined this to be 29.5 °C

Flower number in all cultivars was negatively correlated with mean Tmax during vegetative development. Seed yield was negatively correlated with mean Tmax during vegetative development and flowering. This likely occurred because high mean Tmax reduced the number of flowers.

Angadi et al. (2000) took a different approach to studying the effect of heat stress on canola. They recognized that while temperatures during most of the growing season are moderate, there are usually short periods of heat stress. They tried to mimic this in a growth chamber experiment with five temperature treatments: i) 20/15 °C day/night temperature from planting to maturity– control; ii) control to early flowering, then 7 d at 28/15 °C, then control to maturity; iii) control to early pod, then 7 d at 28/15 °C, then control to maturity; iv) control to early flowering, then 7 d at 35/15 °C, then control to maturity; and v) control to early pod, then 7 d at 35/15 °C, then control to maturity. Four Brassica genotypes were used in this study: i) J90–4316, a "canola-quality" *B. juncea* breeding line developed at AAFC Saskatoon, ii) Oriental mustard (*B. juncea*) cv. Cutlass, iii) Argentine canola (*B. napus*) cv. Quantum, and iv) Polish canola (*B. rapa*) cv. Maverick (1997) and cv. Parkland (1998).

Angadi et al. (2000) found that temperatures of 28/15 °C had no influence on dry matter accumulation. However, the 35/15 °C treatment decreased dry matter by 14%. Early flowering was more sensitive to 35/15 °C (21% dry matter decrease) compared to early pod (8% dry matter decrease). Pooled over genotypes and runs, 35/15 °C decreased seed yield per plant 35% compared to the control. The early flowering stage had a 53% decrease in yield compared to an 18% yield reduction at early pod.

The reductions in seed yield were greater than reductions in shoot dry matter, therefore, harvest index in the 35/15 °C regime was reduced. The early flowering stage was more sensitive than the early pod stage. Seed yield of all genotypes was reduced at early flowering relative to the control. At 28/15 °C at early flowering yield increased for Cutlass and J90–4316. Thus optimum daytime temperature for seed yield of Cutlass and J90–4316 is closer to 28 °C than either 20 °C or 35 °C.

Generally high temperature stress increased the total number of pods but obstructed their development. The 35/15 °C treatment did not affect main stem seed yield in *B. rapa* and *B. juncea* J90–4316, increased yield in *B. juncea* Cutlass by 98%, and reduced yield in *B. napus* by 48%.

Morrison (1993) also conducted growth chamber experiments studying the effect of heat stress at several phenological stages on sterility. He used two cultivars, Westar and Delta, and various combinations of temperature regimes. A hot regime was 27/17°C (light/dark) and cold 22/15 °C. A 16 h photoperiod at 525 μmol m^{-2} s^{-1} photosynthetically active flux density was used. Control treatments were grown from seeding to maturity at the two regimes. In other treatments, plants grown in the hot cabinets were transferred to the cold cabinet at growth stages before, during and after flowering. Similarly plants from cold cabinets were transferred to hot cabinets at these stages. Plants transferred from one cabinet to the other remained in the transfer cabinet until maturity. Main raceme fertility, seeds per pod and thousand seed weight were among the parameters measured on each treatment.

Delta and Westar plants grown from planting to maturity in the hot regime, 27/17 °C, were almost entirely sterile. 62% of these plants produced short plump pods that contained no seed. This occurred in only 4.2% of the plants in the cold cabinet. Plants transferred from the hot to cold cabinets from early flowering to late flowering had significantly lower raceme fertility and seeds per pod. Plants transferred from cold to hot cabinets did not have significantly greater raceme fertility until the late flowering stage.

Observations were made in 10 pod segments from bottom to the top of the raceme. "Cold" check plants had more than six fertile pods per segment up to the 60th pod. "Hot" check plants had fewer than two fertile pods in all segments. As plants transferred from hot to cold increased in age, the position of fertile pods increased up the raceme. "Plants transferred to the hot cabinet before first flower had fewer than two fertile pods per 10 pod segment up the entire raceme, whereas those plants transferred after first flower produced a higher number of pods at the first two 10 pod segments. Plants transferred to the hot cabinet at late flowering produced pods in a similar pattern as the cold check plants until the 50th pod." From these results, Morrison (1993) concluded that late bud to early seed development was the most sensitive to heat stress.

Conclusion

Research on canola phenology modeling has been conducted around the world and is very extensive despite the relatively late arrival of canola as an important agricultural crop in the world. The results of the research have revealed that there are some common characteristics shared by the countless varieties that have been or are currently under commercial cultivation. As expected, air temperature is the main factor driving canola phenological development rate but the crop also exhibits photoperiod sensitivity and vernalization sensitivity during some phases. Canola's preference for cool climates is evident in the low threshold values for heat stress. However, there is considerable variability in the minimum, optimum, and maximum temperature thresholds among varieties grown in the various production areas around the world. Further, there is evidence to suggest that breeding programs aimed at increasing crop yield may also shift the thermal requirements of the crop. Therefore, there will be an ongoing need to continually update these critical thresholds if phenological models are to be used successfully for this crop.

References

Angadi, S.V., H.W. Cutforth, P.R. Miller, B.G. McConeky, M.H. Entz, S.A. Brandt, and K.M. Volkmar. 2000. Response of three Brassica species to high temperature stress during reproductive growth. Can. J. Plant Sci. 80:693–701. doi:10.4141/P99-152

Blackshaw, R.E. 1991. Soil temperature and moisture effects on downy brome vs. winter canola, wheat and rye emergence. Crop Sci. 31:1034–1040. doi:10.2135/cropsci1991.001 1183X003100040038x

Canola Council of Canada. 2016. What is canola? Online (Bergh.) http://www.canolacouncil.org/oil-and-meal/what-is-canola/ (accessed 10 July 2016).

Copani, V., S. Cosentino, R. Tuttobene, and C. Patane. 1994. Relazioni tra temperature, fotoperiodo, e fenofasi nei colza coltivato in ambiente mediterrano. Agic. Ric. 154:55–64.

Deligios, P.A., R. Farci, L. Sulasb, G. Hoogenboom, and L. Ledda. 2013. Predicting growth and yield of winter rapeseed in a Mediterranean environment: Model adaptation at a field scale. Field Crops Res. 144:100–112. doi:10.1016/j.fcr.2013.01.017

De Marco, K., R. Dallaccort, A. Santi, R.S. Okumura, M.H. Inoue, J.D. Barbieri, D.V. de Araujo, R.A.S. Martinez, and W. Fenner. 2014. Thermic sum and crop coefficient of canola (*Brassica napus* L.) for the region of Tangara da Serra, Mato Grosso State, Brazil. J. Food Agric. Environ. 12:232–236.

Dickson, T.J. 2014. Growing season weather impacts on canola phenological development and quality. M.S. Thesis, Department of Soil Science, University of Manitoba. Winnipeg, MB.

Faraji, A. 2010. Determination of phenological response of spring canola (Brassica napus L.) genotypes to sowing date, temperature and photoperiod. Seed and Plant Production Journal 26:25–41.

Farré, I., M.J. Robertson, G.H. Walton, and S. Asseng. 2002. Simulating phenology and yield response of canola to sowing date in Western Australia using the APSIM model. Aust. J. Agric. Res. 53:1155–1164. doi:10.1071/AR02031

Gepts, P. 1987. Characterizing plant phenology: Growth and development scales. In: K. Wisiol, and J.D. Hesketh, editors, Plant growth modeling for resource management, Vol. II: Quantifying plant processes. CRC Press, Boca Raton, FL, p. 3-24.

Gervais, P. 2015. Supply and demand outlook for the canola industry. http://www.canolacouncil.org/media/565635/JP%20Gervais%20Canola%20council%20v3.pdf (Accessed 8 July 2016).

Gulden, R. H., Warwick, S. I. and Thomas, A. G. 2008. The Biology of Canadian Weeds. 137. Brassica napus L. and B.rapa. Can. J. Plant Sci. 88(5): 951-966.

Habekotté, B. 1997. A model of the phenological development of winter oilseed rape (*Brassica napus* L.). Field Crops Res. 54:127–136. doi:10.1016/S0378-4290(97)00043-9

Hodges, T. 1991. Temperature and water stress effects on phenology. In: T. Hodges, editor, Predicting crop phenology. CRC Press, Boca Raton, FL. p. 7–13.

Husson, F., and P. Leterme. 1997. Construction et validation d'un modele de prediction de la date de floraison du colza d'hiver. Ol. Corps Gras Lipides 4:379–384.

Johnson, I.R., and J.H.M. Thornley. 1985. Temperature dependence of plant and crop processes. Ann. Bot. (Lond.) 55:1–24. doi:10.1093/oxfordjournals.aob.a086868

Jones, C.A., J.T. Ritchie, J.R. Kiniry, and D.C. Goodwin. 1986. Subroutine structure. In: C.A. Jones and J.R. Kiniry, editors, Ceres- Maize: A simulation model of maize growth and development. Texas A & M Univ. Press, College Station, TX. p. 49–111.

King, J.R., and Z.F. Kondra. 1986. Photoperiod response of spring rape (*Brassica napus* L. and *B. campestris* L.). Field Crops Res. 13:367–373. doi:10.1016/0378-4290(86)90037-7

Kiniry, J.R., D.J. Major, R.C. Izaurralde, J.R. Williams, P.W. Gassamn, M. Morrison, R. Bergentine, and R.P. And Zentner. 1995. EPIC model parameters for cereal, oilseed, and forage crops in the northern Great Plains region. Can. J. Plant Sci. 75:679–688. doi:10.4141/cjps95-114

Lilley, J.M., L.W. Bell, and J.A. Kirkegaard. 2015. Optimizing grain yield and grazing potential of crops across Australia's high-rainfall zone: A simulation analysis. 2. Canola. Crop Pasture Sci. 66:349–364. doi:10.1071/CP14240

Luz, G.L., S.L.P. Medeiros, G.O. Tomm, A. Bialozor, A.D. Amaral, and D. Pivoto. 2012. Baseline temperature and cycle of canola hybrids. Cienc. Rural 42:1549–1555. doi:10.1590/S0103-84782012000900006

Miller, P., Lanier, W., and Brandt, S. 2001.Using growing degree days to predict plant stages. Montana State University Extension Service Montguide MT200103 Ag 7/2001. Montana State University, Bozeman, MT.

Morrison, M.J. 1988. Phenological and agronomic studies of Brassica napus L. Ph.D. Thesis, Department of Plant Science, The University of Manitoba, Winnipeg, MB.

Morrison, M.J. 1993. Heat stress during reproduction in summer rape. Can. J. Bot. 71:303–308. doi:10.1139/b93-031

Morrison, M.J., and P.B.E. McVetty. 1991. Leaf appearance rate of summer rape. Can. J. Plant Sci. 71:405–412. doi:10.4141/cjps91-056

Morrison, M.J., P.B.E. McVetty, and C.F. Shaykewich. 1989. The determination and verification of a baseline temperature for the growth of Westar summer rape. Can. J. Plant Sci. 69:455–464. doi:10.4141/cjps89-057

Morrison, M.J., D.W. Stewart, and P.B.E. McVetty. 1992. Maximum area, expansion rate and duration of summer rape leaves. Can. J. Plant Sci. 72:117–126. doi:10.4141/cjps92-012

Morrison, M.J., and D.W. Stewart. 2002. Heat stress during flowering in summer Brassica. Crop Sci. 42:797–803. doi:10.2135/cropsci2002.0797

Murphy, L.A., and R. Scarth. 1994. Vernalization response in spring oilseed rape (*Brassica napus* L.) cultivars. Can. J. Plant Sci. 74:275–277. doi:10.4141/cjps94-054

Nanda, R., S.C. Bharagava, and H.M. Rawson. 1995. Effect of sowing date on rates of leaf appearance, final leaf numbers and areas in *Brassica campestris, B. juncea, B. napus* and *B. carinata*. Field Crops Res. 42:125–134. doi:10.1016/0378-4290(95)00026-M

Nanda, R., S.C. Barghava, and H.M. Rawson. 1996. Phenological development of Brassica campestris, B. juncea, B. napus and B. carinata grown in controlled environments and from 14 sowing dates in the field. Field Crops Res. 46:93–103. doi:10.1016/0378-4290(95)00090-9

Nykiforuk, C.L., and A.M. Johnson-Flanagan. 1994. Germination and early seedling development under low temperature in canola. Crop Sci. 34:1047–1054. doi:10.2135/cropsci1994.0011183X003400040039x

Robertson, M.J., and J.M. Lilley. 2016. Simulation of growth, development and yield of canola (Brassica napus) in APSIM. Crop Pasture Sci. 67:332–344. doi:10.1071/CP15267

Robertson, M.J., A.R. Watkinson, J.A. Kirkegaard, J.F. Holland, T.D. Potter, W. Burton, G.H. Walton, D.J. Moot, N. Written, I. Farre, and S. Asseng. 2002. Environmental and genotypic control of time to flowering in canola and Indian mustard. Aust. J. Agric. Res. 53:793–809. doi:10.1071/AR01182

Sands, P.J., C. Hackett, and H.A. Nix. 1979. A model of the development and bulking of potatoes (*Solanum tuberosum* L.) I. Derivation from well-managed field crops. Field Crops Res. 2:309–331. doi:10.1016/0378-4290(79)90031-5

Saseendran, S.A., D.C. Nielson, L. Ma, and L.R. Ahuja. 2010. Adapting CROPGRO fro simulating spring canola growth with both RZWQM2 and DSSAT 4.0. Agron. J. 102:1606–1621. doi:10.2134/agronj2010.0277

Shaykewich, C.F. 1995. An appraisal of cereal crop phenology modeling. Can. J. Plant Sci. 75:329–341. doi:10.4141/cjps95-057

Sunday, N.J. 2014. The effect of temperature on phenological responses and growth of canola cultivars. M.S. Thesis, Department of Agronomy, University of Stellenbosch, Belleville, South Africa.

Thomas, P. 1995. Canola growers manual. Canola Council of Canada, Ottawa, ON.

USDA Economic Research Service. 2016. Canola. http://www.ers.usda.gov/topics/crops/soybeans-oil-crops/canola.aspx (Accessed 8 July 2016).

Vigil, M.F., R.L. Anderson, and W.E. Beard. 1997. Base temperature and growing-degree-hour requirements for the emergence of canola. Crop Sci. 37:844–849. doi:10.2135/cropsci1997.0011183X003700030025x

Walton, G., N.J. Mendham, M. Robertson, and T. Potter. 1999. Phenology, physiology, and agronomy. P 9-14. In: P.A. Salisbury, T.D. Potter, G. McDonald, and A.G. Green, editor, Canola in Australia: The first thirty years. 10th International Rapeseed Congress, Canberra, Australia.

Wang, S., E. Wang, F. Wang, and L. Tan. 2012. Phenological development and grain yield of canola as affected by sowing date and climate variation in the Yangtze River Basin of China. Crop Pasture Sci. 63:478–488. doi:10.1071/CP11332

Wielgolaski, F.E. 1974. Phenology in agriculture. In: H. Lieth, editor, Phenology and seasonality modeling. Springer-Verlag, New York. p. 369–381. doi:10.1007/978-3-642-51863-8_31

Wilson, J.L. 2002. Estimation of phenological development and fractional leaf area of canola (*Brassica napus* L.) from temperature. M.A. Thesis, Department of Geography, University of Manitoba. Winnipeg, MB.

Vegetables and Climate Change

M.R. McDonald* and J. Warland

Vegetables make an important contribution the human diet and provide protein and important vitamins, minerals and fiber (Peet and Wolfe, 2000). The wide range of flavors also adds to the enjoyment of food. The major vegetables in the world are staples that provide significant starch in the diet. In order of greatest production these are: potato, cassava, and sweet potato. The other most important vegetables worldwide, in order of production are: tomatoes, cabbage, onions, watermelons, cucumbers, carrots, and green peppers (Salunkhe and Kadam, 1998).

Global warming will undoubtedly affect vegetable production. Vegetables tend to be low acreage, high value crops, where quality has a major influence on price and saleability. For instance, forked or misshapen carrots are perfectly acceptable as food, but are not marketable in most North American and European markets. Another example is sweet corn, where high temperatures during ripening may result in lower sugars and poor tip-fill. These problems can make the sweet corn unmarketable, while similar problems in grain corn only reduces yield (Peet and Wolfe, 2000). An increase in average temperatures could have a positive or negative effect on vegetable crops, depending on the climate of the region where they are grown. However, in addition to an increase in average temperature, global warming is expected to result in an increase in the frequency of extreme weather events, including 'heat waves' (Mearns et al., 1984). These could result in severe plant damage or reduction in quality, even if the seasonal average temperatures are unchanged.

This chapter focuses on the effects of increasing global temperatures on yield and quality of vegetable crops, with some discussion of studies on the possible interaction of increasing carbon dioxide concentrations and increasing temperatures. A discussion of some of the methods to study the effects of climate change on vegetables crops is included. Approaches to adapting vegetable production for global warming are outlined.

Classification of Vegetables According to Optimum Temperature

Temperature is certainly one of the main factors affecting the growth, development and yield of all crops. "The temperatures favourable to plant growth in general lie between 5° and 40 °C, and the optimum will vary with plant and stage of development" (Kader et al., 1974, p. 523).

Abbreviations: USVL, U.S. Vegetable Laboratory.

University of Guelph Ontario Agricultural College, Guelph, ON N1G 2W1. *Corresponding author (mrmcdona@uoguelph.ca)

doi:10.2134/agronmonogr60.2018.0006

© ASA, CSSA, and SSSA, 5585 Guilford Road, Madison, WI 53711, USA.
Agroclimatology: Linking Agriculture to Climate, Agronomy Monograph 60.
Jerry L. Hatfield, Mannava V.K. Sivakumar, John H. Prueger, editors.

Many studies have been conducted to determine the optimum temperatures for seed germination, growth and yield of vegetable crops. Much of this research was conducted in the 1960s and was not related to climate change, but to site selection, production zone and calculation of growing degree days to predict when a crop would be ready to harvest. This information was used to classify vegetables as cool- and warm-season crops (Kader et al., 1974) and to make recommendations for seeding date and need for frost protection, among other production practices. However, the information on optimum and maximum temperatures can be very useful in determining potential effects of global warming.

The groupings of cool-season and warm-season vegetables can be divided further into hardy, half-hardy (grouped as cool-season crops) tender, and very tender (warm-season crops) based on tolerance to frost. Hardy and half-hardy crops demonstrate some frost tolerance, while tender and very tender crops do not (Kader et al., 1974; Maynard and Hochmuth 1997). Cool season crops have seed that germinates at cooler temperatures (some as low as 3 °C), grow and develop at cooler temperatures, are tolerant of light frost and should be stored at temperatures close to freezing. Warm season crops, in contrast, have seeds that require warmer temperatures for germination (minimum 10–15 °C), grow best at warmer temperatures and suffer from chilling injury if stored at cold temperatures postharvest. Examples of 'hardy' cool season crops are asparagus, carrots, cabbage, garlic and onions. Some warm season vegetables are tomatoes, cucurbits, sweet potato and sweet corn. The edible portion of cool season crops include roots, bulbs, leaves, stems, and immature flowers (i.e. cauliflower) while the edible potion of warm season crops is the immature or mature fruit (Maynard and Hochmuth, 1997). These categories, while general, are very useful for understanding differences in the response of vegetable crops to temperature differences at different physiological stages.

Methods of Evaluating the Effects of Varying Weather Parameters on Vegetable Production and Yields

Investigations of quantitative relationships between recorded weather and actual crop yield have a long history. Many studies have relied on simple correlation analysis (e.g., Wallace, 1920; Monteith, 1981; Lobell and Asner, 2003; Peng et al., 2004), as well as on single or multivariate regression (e.g., Robertson, 1974; Thompson, 1986; Chmielewski and Potts, 1995). Wallace (1920) was one of the earlier studies to attempt to correlate weather and yield. In that paper, he determined correlation coefficients between corn yields in eight US states, and monthly mean temperature and monthly total precipitation. His records extended for 29 yr (1891–1919). The greatest R values he reported were 0.77, -0.75 and -0.66 for Kansas yield and July precipitation, Kansas yield and July temperature and Missouri yield and July temperature, respectively. In the discussion at the end of Wallace (1920), Charles F. Brooks made several relevant points. Varying planting dates could affect the analysis so that it may be better to compare weather records with time normalized by planting date, and use of monthly averages may obscure important details such that the week may be a more appropriate unit. As an example, he offers a month possessing a normal mean temperature which contained a very hot period and a very cold period.

Some studies (Andresen et al., 2001; Lobell and Asner, 2003) have shown that gains in grain yield after World War II may have been due in part to the milder climate experienced between roughly 1955 and 1975. The importance of the relationships between weather and yield in the field (as opposed to in models) is emphasized by Peng et al. (2004), who showed a clear decrease in rice yield of 10% for each 1 °C increase in the minimum (night) temperature during the dry season. Peng et al. (2004) used correlation and regression analysis on a short (12 yr, 1992–2003) record of rice yields from experimental plots at the International Rice Research Institute in the Philippines. Their highest correlation (R = -0.87) was between grain yield and T_{min}, showing an approximate 10% decline in yield for each 1 °C increase in dry season T_{min}. As the authors discuss, studies using crop growth models fail to find such a significant effect from nighttime temperatures, indicating a failure in our fundamental understanding of the role of daily temperature cycles in crop development. Many model studies (e.g., Rosenzweig and Parry, 1994) have made use of the CO_2 fertilization effect seen in some laboratory studies, whereby increased concentration of atmospheric CO_2 induces greater plant productivity; however there is no evidence for such an effect when long-term agricultural experiments are examined (Barnett, 1994), and only limited evidence of a fertilization effect in forest systems (Gedalof and Berg, 2010).

The studies discussed so far have relied on either growing season or monthly weather values. Caprio (1966) developed the 'iterative chi-square method' to find correlations between daily weather records and annual crop yields. This technique avoids the problem of a long (relative to crop phenological development) averaging period, and is able to provide some quantitative information on the threshold values that define, for example, a wet or a cool year. In the initial study, poor yields of winter wheat were found to correlate most strongly with large T_{max} during the period corresponding to the heading stage of development.

The iterative chi-square method compares the daily weather records of high and low yielding years against the record of 'normal' years. In the original methodology, high and low years are defined as the top and bottom quartiles based on a seven-year running mean of yields. The normal years are then the remaining 50% of recorded years. Two chi-square values were generated for each day of the growing season for the two weather variables and for high and low years. The two chi-square variables are for number of days above a cardinal value and number of days below a cardinal value. These are determined by taking the total days in the three week period surrounding each day and incrementing or decrementing the threshold value for days below or days above a certain temperature or amount of rainfall. The maximum chi-square achieved is assigned to that run with the corresponding 'cardinal', or threshold, value. If the high or low yield year shows fewer days than expected, chi-square is reported as negative, indicating a deficit of days above or below the cardinal value.

The chi-square analysis uses a three-week running window. Within a window, the number of days of high (low) yield in the record will be $l_a = 21\ Y_a$, where subscript a indicates high, low, or normal years and Y_a is the number of years in the category within the total data record. For the variable under consideration, such as T_{max}, S indicates the total number of points in both high (low) and normal years above (below) a threshold value. Thus the expected number of occurrences for exceeding or below the threshold is

$$E_a = \frac{l_a}{l_a + l_{a'}} S \qquad [1]$$

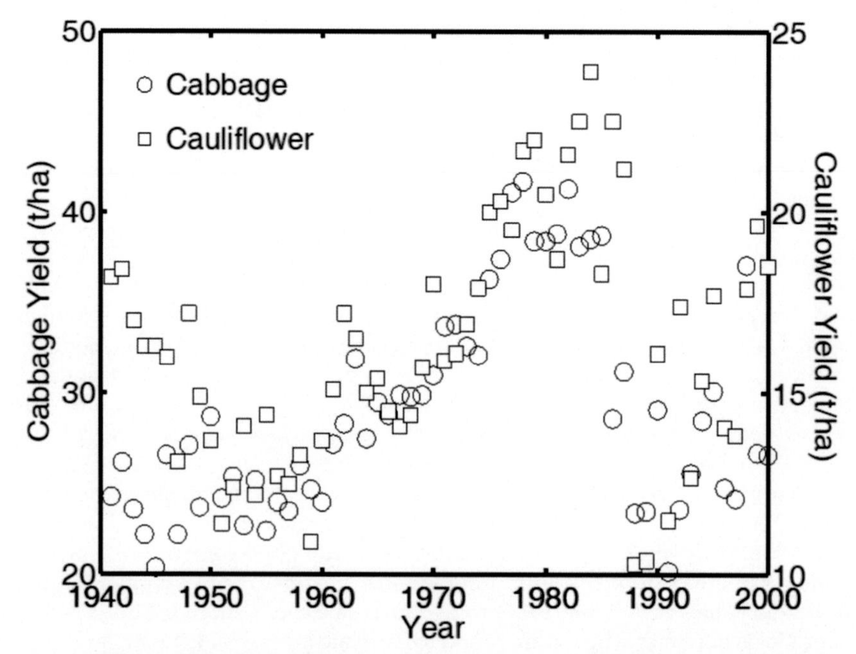

Fig. 1. Marketable yields, in tonnes per hectare (t ha-1), of cabbage and cauliflower in Ontario for the period 1940 to 2000.

The chi-square value is then computed as

$$\chi^2 = \frac{(x_o - E_o)^2}{E_o} + \frac{(x_n - E_n)^2}{E_n} \quad [2]$$

where x_a is actual number of occurrences of exceeding (below) the threshold value for high or low (*o*) yield years versus normal (*n*) years.

In applying the technique, the three-week window runs from the start of the year to the end of the year, and the threshold value for days above or below is incremented over a range of expected values, for example from 25 °C to 35 °C in one degree increments for T_{max}.

In defining high, low, and normal yield years, Caprio (1966) used a five-year running mean and set the lowest and highest quartiles as the low and high yield years, respectively, while setting the middle two quartiles as the normal years. The use of the running mean is intended to eliminate changes over time in management or other factors which may confound the climate analysis. However, given the accumulating evidence that a moderate climate may have made a larger contribution to increasing yields since the Second World War than previously thought (Lobell et al., 2005; McKeown et al., 2006), it may be appropriate, or even necessary, in some cases to use absolute yields in the analysis.

As a case in point, consider cabbage cauliflower yields in Ontario, shown in Fig. 1, for 1940 through 2000 (McKeown et al., 2005). In the period from approximately 1960 to 1980, one sees the linear increase in yield usually attributed to increased inputs and plant breeding, however, beginning about 1985, yields become very sporadic and never recover to values seen around 1980. This sharp change in production has been correlated with an increase in hot days in Ontario [days with regionally averaged maximum temperatures exceeding 30 °C (McKeown et al., 2005, 2006)]. The record of number of hot days per year in southern Ontario along the north shore of Lake Erie is shown in Fig. 2 (McKeown et al., 2005).

Effects of Weather Variables on Vegetables

There are well defined minimum, optimum, and maximum temperatures for most vegetable crops, although these will vary with cultivar and other environmental factors. Peet and Wolfe (2000) summarized many of the physiological disorders of vegetable crops that are associated with high (mostly) and low temperatures. Frost and freezing damage are common concerns where vegetables are grown in temperate regions, especially at the beginning and end of the growing season. Colder-than-optimum temperatures can also cause some crops to bolt, resulting in an unmarketable plant. This can occur in onions (Brewster, 2008) and carrots (Rubatsky et al., 1999), among others. Warmer growing seasons should reduce the risks from low temperature injury, but more variable weather patterns could still bring unseasonal frosts.

In general, longer growing seasons will be beneficial for vegetable production. For instance, warmer soil temperatures at seeding will improve the rate and uniformity of germination for most of these crops (Peet and Wolfe, 2000). Benefits are expected to be region specific, as some regions will become too warm for the production of some crops. The focus of the rest of this section will be on the detrimental effects of high temperatures on yield and quality.

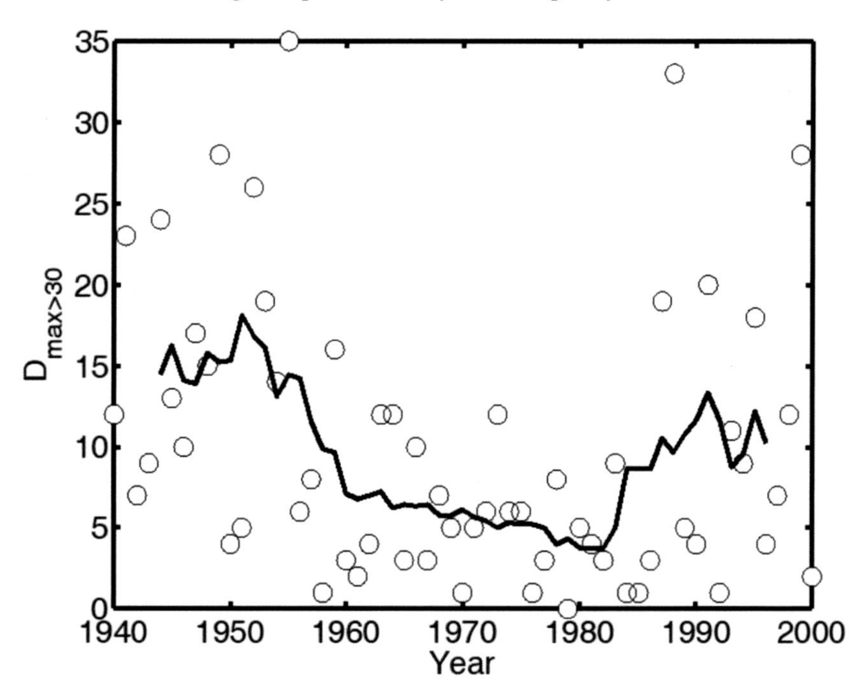

Fig. 2. Number of days with maximum temperature above 30 °C for the Lake Erie counties of southern Ontario. Line shows nine-year running mean.

There are many different negative effects of high temperatures on vegetable production. Lettuce seed goes in to dormancy at high soil temperatures (thermo-inhibition) (Wein, 1997; Yoong et al., 2016). High temperatures when vegetable seedlings are young can result in 'heat canker' when the soil surface heats to temperatures that kill the cells in the hypocotyl. Heat canker can kill a few or many seedlings. Irrigation (if available) may help to reduce the temperature of the soil surface, as can living mulches or mulches of other plant materials, such as straw. High temperatures, in combination with high relative humidity, contribute to 'black heart' in celery and tip burn in lettuce (Ryder, 1999), as a result of reduced translocation of calcium to the developing young tissue in the center of the plant. Lettuce and spinach may produce flower stalks (bolt) prematurely at high temperatures. Bolted plants are not marketable. High temperatures favor the production of more male flowers on cucumber plants, which results in lower yield (Wein, 1997). Tomatoes and peppers are warm-season crops, but yield and quality can be reduced by higher than optimum temperatures, especially at critical phases of development such as pollination (Wein, 1997).

Studies conducted in controlled environments can provide useful information on the changes in yield and quality in response to different growing conditions, but these have associated problems. For instance, there is a need for ventilation at high temperatures (Hand, 1984; Peet and Wolfe, 2000) and reduced light intensity, which does not mimic field conditions. Providing a realistic range of daily or weekly temperature fluctuations can also be challenging. An alternative approach is to evaluate how yields in a region vary with changes in temperature over many years, as discussed in the Methods section above.

Increases in yield over time (usually the 20th century) have been reported for grain corn in the five "corn belt' states of the United States (Thompson, 1986) and for grain corn, alfalfa, and the vegetables: potatoes, sweet corn, onions, carrots, table beet, peas, and snap (French) bean (Tiefenthaler et al., 2003) in Wisconsin. Thompson (1986) identified a yield increase for grain corn from 1930 to 1972, because of "improved weather for corn" especially below normal temperatures in July and August, accompanied with above normal precipitation for those months. Tiefenthaler et al. (2003) documented yield increases during the 1900s for all crops except alfalfa. The greatest increases were found for potatoes (7x) and grain corn (5x). However, examination of the graphs shows more variation from the 1980s onward. The authors attribute the yield increases for potatoes to improved nitrogen fertilization, since the same cultivar, Russet Burbank, has been grown for decades. Plant breeding effort was considered a major factor in the yield increase for the other crops, although with the observation that quality factors are important for alfalfa and vegetables, so yield per se is not usually the primary goal of the plant breeders. In contrast, a study of marketable yield of cool season Brassica crops in southern Ontario, Canada, from the 1940s to early 2000 found "no clear evidence of increased yield over the years due to improved management for any of the crops examined" (broccoli, cabbage, cauliflower, radish, and rutabaga) (Warland et al., 2006, p. 1213).

Studies in southern Ontario, Canada, over a time frame of 50 or 60 yr, investigated the association between marketable yield of several vegetable crops and the weather parameters: mean growing season temperature, number of days with maximum temperature at or above 30 °C (number of hot days), total growing season precipitation, and number of days with rain more than 2 mm (McKeown et al., 2005). Only the yields of field tomato (warm season crop) increased with increasing seasonal mean temperatures. Surprisingly, yields decreased with increasing number of days with rain, but this may be related to disease development in wet years rather than a direct effect on crop growth. The yields of cabbage, rutabaga, potato and onion were lower with increasing number of days with maximums over 30 °C. There was no relationship between yields and total precipitation, which may reflect the increase in irrigation for vegetable crops. However, yields of cabbage, onion and rutabaga increased with increasing number of days with rain.

This was refined further, with the focus on Brassica vegetables, also using the iterative approach discussed above (Warland et al., 2006). The study confirmed that high temperatures reduced the yield of these 'cool-season' crops. Low yields in broccoli were related to the number of days in August with temperatures over 29 °C. This is consistent with the findings of Bjorkman and Pearson (1998), which indicate that the negative effect of high temperatures on bud development and quality in broccoli means that production only takes place in regions where summer temperatures generally stay below 30 °C. In the Warland et al. (2006) study, yields of cabbage, cauliflower and rutabaga decreased about 10% for every 10 d when maximum air temperatures were higher than 30 °C.

The examination of marketable yield of vegetable crops over decades was extended to examine vegetables in Wisconsin, where vegetables are grown at a similar latitude as Ontario (Tesfaendrias et al., 2013). This work confirmed the negative effects of high temperatures on marketable yields. Yields of table beet (beetroot) for processing, green pea, onion, and sweet corn for processing were all lower as the number of 'hot' days increased. When the weather data were examined by month, a higher number of hot days during the mid-growth months of July (range 1–14 d) and/or August (range 1–13 d) was associated with decreased yields of all of the vegetables except carrot and potato. Carrot yields increased with increasing rainfall in June, a time of early crop establishment. Yields of potatoes were higher when minimum temperatures were higher in June (early growth) and lower with high nighttime temperatures in August (tuber filling).

A study to determine the optimum combination of nitrogen fertilizer and soil moisture for cabbage demonstrated the detrimental effects of high temperatures, specifically the number of hot days in a growing season, on cabbage yield. McKeown et al. (2010), conducted a trial over three years at the same site using the same cultivar. They applied a combination of five rates of N (0–400 kg ha^{-1}) and varying rates of water through drip irrigation (60–120 cm per season). There were 9, 1, and 29 d with maximum temperature over 30 °C in the 3 yr of the trial (2003-2005). The maximum yields attained in those years were 100, 108, and 57 tonnes ha^{-1}. The yields from the response surfaces for N and soil moisture indicated that yields were 13 and 49% lower in 2003 and 2005 than in 2004, which is an even greater reduction than that estimated by Warland et al. (2006). Nitrogen uptake in cabbage was shown to be maximum at 20 °C, lower at 15 °C and lowest at 25 °C (Hara and Sonoda, 1982). These results suggest that the effects of high temperatures on reducing yields cannot be ameliorated by sufficient water or nitrogen, which is a grave concern for the production of cabbage and other cool season vegetables under a warming climate.

Carbon Dioxide Levels and Interactions with Temperature

Increasing carbon dioxide levels in the atmosphere are a major contributor to global warming, but may have some positive effects on plant growth.

Carbon dioxide enrichment of the greenhouse environment for the production of vegetables and ornamentals has been studied extensively and increases of CO_2 concentration up to 1000 vpm are widely recommended (Hand, 1984). Keeping CO_2 levels high during times when greenhouses have to be ventilated remains a challenge. The response of a range of crops and wild plants to increased CO_2 levels was collated by Poorter (1993). He assembled the data from numerous reports where plants were grown at 600 to 720 ppm CO_2, as compared to the controls of 300 to 360 vpm CO_2 and calculated the ratio of yield at increased CO_2 over yield at 'normal' levels. Yield increases were found for all vegetables tested. The least responsive to increased CO_2 levels was pea (*Pisum sativum* L.) with a ratio of 1.05 and the greatest response was in carrot (*Daucus carota* L.) with a ratio of 210.

The interaction of CO_2 levels and temperature on yields of field vegetables has been investigated for cauliflower (Wheeler et al., 1995), onions (Daymond et al., 1997), carrots, beetroot, (Wurr et al., 1998), and French bean (Wurr and Fellows, 2000). All of these trials were conducted in the United Kingdom. There are some conflicting results although the overall conclusion is that yields of vegetable crops will increase with increasing temperature and CO_2 in the U.K. Interactions between temperature and CO_2 varied with crop and study.

A trial using a temperature gradient in polyethylene tunnels, with concentrations of 374 or 532 ppm CO_2, found that total dry weight of the plants decreased in a linear relationship with increasing temperatures from a mean of 12.3 °C to 18.6 °C, because of faster maturity (Daymond et al., 1997). There was no interaction with CO_2 concentration, but plant weight was much higher (29–37%) under high CO_2 conditions. Indeed the authors conclude that the positive effects of increasing CO_2 concentrations on yield were so great that "current estimates of climate change should be beneficial for bulb onion production" in the U.K.

Other studies came to similar conclusions, that climate change will have a positive effect on yields of most vegetable crops in the U.K, but for somewhat different reasons. Studies on onions, beetroot and carrots used day-lit controlled environment chambers with CO_2 concentrations from 350 to 750 ppm and temperatures from 12 to 18 °C (Wurr et al., 1998). Similar trials were conducted on French bean with the same range of CO_2 and temperatures from 14.5 to 18.5 °C (Wurr et al., 2000). These trials were also conducted in the U.K., but the use of controlled environment cabinets makes the location less of a factor. In this study, increasing temperature caused an increase in the fresh weight of onion bulbs. The authors conclude that increased temperature alone will not increase onion yields substantially, since the optimum temperature was 16.5 °C. However, with a concomitant increase in CO_2, there should be an increase in yield and quality. The optimum combination for yield of beetroot was 18 °C and 650 ppm CO_2, and for carrots the optimum was 16.5 °C and 700 ppm CO_2 (Wurr et al., 1998). The study on French beans (Wurr et al., 2000) demonstrated that increasing temperatures will greatly increase yields (39–118%); however, increasing CO_2 concentration had a small negative effect on the number and yield of beans.

The exception to increased yields with higher temperature in these studies is cauliflower. Studies on growth and yield of cauliflower were conducted in tunnels receiving either 328 or 531 ppm CO_2 and temperatures ranging from 14.0 to 17.1 °C (Wheeler et al., 1995). Curd mass (yield) was higher with high CO_2 at any temperature, with the average increase of 34%. There was no interaction between CO_2 concentration and temperature and the logarithm of curd weight decreased with increasing temperature.

A comparison of these studies indicates that there is no interaction between CO_2 concentration and temperature for yield of various vegetable crops. While increased CO_2 often resulted in a large increase in yields, this was not always the case. The temperature effects identified for French bean, a warm season or 'tender' crop (sensu Maynard and Hochmuth, 1997) are as expected. Maynard and Hochmuth (1997) indicate that the optimum temperature for snap (French) bean is 15 to 21 °C, while the optimum for cool season crops, cauliflower, onion, carrot and beetroot is 15 to 18 °C. These are mean temperatures, and daily fluctuations in temperature, including nighttime lows and daily highs, have an important influence on crop growth and crop quality, as discussed in the following section.

Adapting to Climate Change: Plant Breeding

High temperatures often have a strong detrimental effect on plants, which may be difficult to reverse or prevent through applying the optimum amount of water or fertilizer, as discussed above. Global warming "has compelled plant scientists to develop climate-change resilient crops, which can withstand broad spectrum stresses...Genomics appear to be a promising tool" (Kole et al., 2015, p. 1). The review also reports that public plant breeding programs have developed heat-tolerant cultivars of tomato, as well as cowpea, common bean and cotton. "Stress tolerant varieties offer the most sustainable, cost effective approach to combat the effects of increasing global temperature" (de la Pena et al., 2011, p. 231). Plant breeding is essential for maintaining yield and quality with variable weather conditions. However, breeding for resistance to heat stress can involve many different factors under different genetics (Peet and Wolfe, 2000). These authors offer the example of breeding for heat stress resistance in potato, which can be a problem because "a genotype possessing tolerance to one aspect of heat stress may not necessarily be tolerant to other stresses. High temperatures can also reduce yields by affecting ability of seed tubers to sprout, photosynthetic or dark respiration rates, [and] tuberization, each of which may be under separate genetic control" (Peet and Wolfe, 2000, p. 231). Despite these limitations, the AVDRC has used marker-assisted selection to transfer genes from wild relatives to tomato to pyramid genes for high yield and stress tolerance, producing varieties with higher yields under normal and stressful growing conditions (de la Pena et al., 2011).

Identifying genotypes of vegetables with a greater tolerance to high temperatures or a wider range of optimum temperatures for growth can be a challenge. Controlled environment trials that specifically look at a range of temperatures can be helpful, but these do not always mimic field conditions and the effects that the field environment has on yield or quality. One of the few breeding and selection programs that has targeted heat stress in vegetables is the program of the U.S. Vegetable Laboratory (USVL) in Charleston, South Carolina, to breed broccoli that is adapted to the southeastern United States (Farnham and Bjorkman, 2011). Early in the program, the USVL lines were tested in Geneva, New York, where days with high temperatures occurred at intervals during the summer. The quality of conventional hybrids such as 'Packman' was lower during the summer while the USVL lines produced consistent quality throughout the growing season. At the time of publication, the USVL lines did not have heads that were as high quality as the standard commercial cultivars. Instead the flower buds were larger and the heads were not as dense. This project has continued, with continuing success in breeding and selecting broccoli and by 2014 considerable improvements in broccoli cultivars were seen (http://www.hort.cornell.edu/bjorkman/lab/broccoli/ebreeding.php). Some of the best hybrid broccoli selected in 2011 have been commercialized.

Farnham and Bjorkman (2011) present this program to breed for heat stress tolerance as a case study, and recommend that breeding programs for stress tolerance deal with a number of questions prior to the start of the breeding program. These are: "1) What is the effect of the abiotic stress on the crop that is to be improved; 2) what will be the conditions of the selection environment; 3) what germplasm is available that contains the necessary genetic variation to initiate improvement; 4) what breeding scheme will be used to facilitate improvement; and 5) what will be the specific goals of the breeding effort" (Farnham and Bjorkman, 2011, p. 1093)?

The World Vegetable Center (Asian Vegetable Research and Development Center, AVRDC) has also been focused on breeding for tolerance to heat stress and adaptation to climate change. This organization had, as of 2011, a germplasm bank of over 57,000 accessions representing 161 genera and 407 species of vegetables. The goals of their breeding program are to address "the tropical constraints of high temperatures, water limitation and flooding" (de la Pena et al., 2011, p. 397). The authors point out that current levels of heat tolerance in tomato are not sufficient for optimum tomato production and have a program underway to identify heat tolerance genes in wild relatives of tomato. Their breeding program for sweet pepper has identified some lines from Central Europe that have high fruit set under high temperature conditions. They also have a program for breeding Chinese cabbage for tolerance to hot and humid conditions and have identified that heat tolerance is simply inherited. Finally, this research center is also collecting and evaluating indigenous vegetable crops from southern Asia and Africa and is selecting for nutritional quality and resistance to heat stress. They suggest focusing on indigenous crops of Sudan, Cameroon, and Nigeria, as these countries currently have climates that similar to those predicted for other areas of the world as global warming continues. Indigenous plants that are adapted to these climates may provide valuable germplasm for breeding programs (de la Pena et al., 2011).

Commercial seed companies routinely conduct cultivar evaluation trials in several locations over a number of years to determine how new cultivars respond in different environments. These are genotype by environment studies. The companies may be deliberately, or inadvertently, selecting for stress tolerance and consistent quality. Another approach to select for stress tolerance is to repeat the trials over many years, in an area where the weather changes considerably from year to year. In some situations data from long term cultivar trials, at a single site, can be used to identify cultivars that provide consistent yields over many years. This approach was used to assess cultivars of carrots and onions (Tesfaendrias et al., 2010). The onion and carrot cultivar trials in the Holland Marsh, southern Ontario, Canada, have been ongoing since the 1960s. The crops are grown on high organic matter (~70–80%) soil, within a 4 ha site (44° 5′ N lat., 79° 35′ W long.), and within 400 m of a weather station. Crop rotation and other standard production practices were used over the years. From 1989 on, a selection of five 'benchmark' cultivars were included in each trial, along with 15 to 30 trial cultivars, to provide a comparison with the new cultivars. Cultivar yield and quality was compared to the trial mean and to the benchmark cultivars. Variation in seasonal weather conditions were identified, as expected. There was a significant effect of weather (year) on the onions and for three of four data sets of carrots. The study found that yields of two of the five carrot cultivars, 'Cellobunch' and 'Sunrise,' were not affected by weather, which suggests that either they were not affected by abiotic stresses, or were able to compensate. Cellobunch had consistently high yields. There was a greater influence of weather on the yields of onions, as compared with carrots. Cultivars 'Corona' and 'Prince' tended to have the highest and most consistent yields. The general effects of weather parameters over time were also examined. The number of hot days in June (1–13 d) and August was related to lower yields in several of the onion cultivars, while the number of days with rain in June (3-12 d) was related to higher yields of most of the onions. Despite the trend for lower yields with more hot days in a growing season, the lowest onion yields were found in 1992, when cool temperatures prevailed following the eruption of Mount Pinatubo in 1991. Weather patterns that are substantially different from the norm are always potentially damaging, since growers normally choose cultivars that are adapted to the average conditions.

Other breeding programs are focused on specific qualities. For instance, the lettuce breeding program at Seed Biotechnology Center, University of California, Davis, has identified a "delay of germination" gene that controls high temperature dormancy and also flowering in lettuce (Huo et al., 2016)

Cultural Practices

There are a few cultural practices that can cool a crop or the soil, but more research in this area is warranted. There has been a great effort in horticulture to extend growing seasons and protect plants from cold temperatures, including the use of greenhouses, cold frames, floating row covers, and high tunnels. Little has been done to research methods to cool plants or soil. In some regions, seeding date can be adjusted to avoid the warmest part of the growing season. In other areas, choosing the growing region based on elevation can be used to choose the best climate for a specific crop. Evaporative cooling through overhead irrigation has been used in some fruit crops, such as apples and grapes. Overhead irrigation could provide some evaporative cooling of leaves or the soil of some vegetable crops, but this is not discussed in the literature, probably because so many factors go into decisions of when and how much to irrigate a crop. White or silver mulches are recommended for general vegetable production, to keep the soil cool during hot summers (Grubinger, 2004; https://www.uvm.edu/vtvegandberry/factsheets/plasticprimer.html). White mulch is recommended for the production of artichokes in warm areas of Virginia to reduce heat stress (Bratsch, 2014). However, more research is needed to determine the benefits of white and reflective mulches.

Alternative Crops and Crop Diversification

Many factors contribute to the selection of crops that will be produced in a certain area, in addition to climate. Important factors include markets, transportation, availability of labor, and food preferences. Air transportation of perishable food items around the world has expanded the palate of many people. There are competing pressures for crop specialization and crop diversification that are not related to climate. Many growers understand the need for the production of different crops and cultivars to "hedge their bets" through the normal variations in weather and market price from year to year. The emphasis on 'local' food has also contributed to the production of less common or 'ethnic' vegetables in many parts of North America and Europe. It is useful to remember that most vegetable crops, with the exception of (sweet) corn and squash, have been introduced to North America, so there are few indigenous crops. Thus there is greater crop diversification for vegetables than for most other categories of food, and this trend is expected to continue.

Vegetable crops that are indigenous to Africa and Asia could play an important role in adaptation to climate change in both developing countries and the western world. The Asian Vegetable Research and Development Center has an extensive germplasm collection of approximately 12,000 accessions of 200 species of indigenous vegetables collected from South and Southeast Asia and Africa (de la Pena et al., 2011). Many indigenous vegetables are easy to grow, are highly nutritious and are adapted to harsh environmental conditions. These crops can be developed for commercial production and the valuable traits can be used for breeding of more conventional vegetable crops. The plant families with many representatives of these useful indigenous vegetables are the Cucurbitaceae and Fabaceae (formerly Leguminosae) (de la Pena et al., 2011). Another example is vegetable amaranth (*Amaranthus* spp.). Leafy amaranth is described as hardy, drought tolerant, and able to grow on poor soils. It is interesting that this crop is currently imported into Canada from countries in the Carribbean, and is grown commercially, albeit in small quantities, in the province of Ontario, Canada, both in the field during the summer and in high tunnels in the fall and spring.

Summary and Conclusions

Vegetable crops must meet exacting quality standards in addition to producing acceptable yields. As such, the negative effects of increasing temperatures and more erratic weather patterns may influence vegetables more than some other groups of crops, such as field crops. However, the move of some retailers to sell "ugly" fruits and vegetables may provide a market for some produce that does not meet today's standards for quality.

Erratic weather patterns, as a result of global warming, will provide a challenge for all of agriculture. Higher CO_2 concentrations will not be a concern in vegetable production. The research that has been done suggests that yields will increase or not be affected.

Some regions of the world, for instance northern Europe, northern states of the United States, and parts of Canada, will benefit from higher temperatures during the growing season. However, the movement of vegetable production to different months, more northern latitudes or higher elevation, may not be possible for all countries.

Increasing temperatures will have the greatest negative effect on vegetable production and quality in many regions of the world. An increase in the daily or monthly mean temperature will not be as damaging as a 'heat wave', when there are several days with temperatures over 30 °C. The number of 'hot' days during the growing season has been shown in a number of studies to reduce the yield and quality of cool season vegetable crops. The negative effects of high daytime temperatures on cabbage yields were not ameliorated by optimum fertilization and soil moisture. This is probably the case for many cool-season vegetables.

The selection and adaptation of vegetable crops for tolerance to high temperature stress is an essential adaptation to global warming. Selection for vegetables that produce well at high temperatures and that provide consistent yields despite varying weather conditions is important. This will also provide more cultivars of temperate crops suitable for production in the tropics. Breeding and selection for heat tolerance, in addition to crop diversification, is the most promising approach to adaptation to climate change.

References

Andresen, J.A., G. Alagarswamy, C.A. Rotz, J.T. Ritchie, and A.W. LeBaron. 2001. Weather impacts on maize, soybean and alfalfa production in the Great Lakes Region, 1895–1996. Agron. J. 93:1059–1070. doi:10.2134/agronj2001.9351059x

Barnett, V. 1994. Statistics and the long-term experiments: Past achievements and future challenges. In: R. Leigh and A. Johnston, editors, Long-term experiments in agricultural and ecological sciences. CAB International, Wallingford, UK. p. 165–183.

Bjorkman, T., and K.J. Pearson. 1998. High temperature arrest of inflorescence development in broccoli (*Brassica oleracea* var. italica L.). J. Exp. Bot. 49:101–106. doi:10.1093/jxb/49.318.101

Bratsch, A. 2014. Specialty crop profile: Globe artichoke. Virginia Cooperative Extension. Blacksburg, VA. http://pubs.ext.vt.edu/438/438-108/438-108_pdf.pdf (accessed 28 July 2016).

Brewster, J, editor. 2008. Onions and other vegetable alliums. Crop Production Science in Horticulture 15. CABI Publishing, New York. doi:10.1079/9781845933999.0000

Caprio, J.M. 1966. A statistical procedure for determining the association between weather and non-measurement biological data. Agric. Meteorol. 3:55–72. doi:10.1016/0002-1571(66)90005-7

Chmielewski, F.-M., and J. Potts. 1995. 'The relationship between crop yields from an experiment in southern England and long-term climate variations'. Agric. For. Meteorol. 73:43–66. doi:10.1016/0168-1923(94)02174-I

Daymond, A.J, T.R. Wheeler, P. Hadley, R.H. Ellis, and J.I.L. Morison. 1997. The growth, development and yield of onion (*Allium cepa* L.) in response to temperature and CO2. J. Hort. Sci. 72:1, 135-145.

de la Pena, R.C., A.W. Evbert, P.A. Gniffke, P. Hanson, and R.C. Symonds. 2011. Genetic adjustment to changing climate:Vegetables. In: S.S. Yadav, R.J. Redden, J.L. Hatfield, H. Lotze-Campen, and A.E. Hall, editors, Crop adaptation to climate change. John Wiley & Sons Ltd., New York. p. 396-409.

Farnham, M.W., and T. Bjorkman. 2011. Breeding vegetables adapted to high temperatures: A case study with broccoli. HortScience 46:1093–1097.

Gedalof, Z., and A.A. Berg. 2010. Tree ring evidence for limited direct CO_2 fertilization of forests over the 20th century. Global Biogeochem. Cycles 24:GB3027. doi:10.1029/2009GB003699

Grubinger, V. 2004. Plastic mulch primer. University of Vermont, Burlington, VT. https://www.uvm.edu/vtvegandberry/factsheets/plasticprimer.html (accessed 5 Sept. 2016).

Hand, D.W. 1984. Crop responses to winter and summer CO_2 enrichment. Acta Hortic. 162:45–64. doi:10.17660/ActaHortic.1984.162.4

Hara, T., and Y. Sonoda. 1982. Cabbage-head development as affected by nitrogen and temperature. Soil Sci. Plant Nutr. 28:109–117. doi:10.1080/00380768.1982.10432376

Huo, H., I.M. Henry, E.R. Coppoolse, M. Verhoef-Post, J.W. Schut, H. de Rooij, A. Vogelaar, R.V.L. Joosen, L. Woudenberg, L. Comai, and K.J. Bradford. 2016. Rapid identification of lettuce seed germination mutants by bulked segregant analysis and whole genome sequencing. Plant J. 88:345–360. doi:10.1111/tpj.13267

Kader, A.A., J.M. Lyons, and L.L. Morris. 1974. Postharvest responses of vegetables to preharvest temperatures. HortScience 9:523–526.

Kole, C., M. Muthamilarasan, R. Henry, D. Edwards, R. Sharma, M. Abberton, J. Batley, A. Bentley, M. Blakeney, J. Bryant, H. Cai, M. Cakir, L.J. Cseke, J. Cockram, A. Costa de Oliveira, C. De Pace, H. Dempewolf, S. Ellison, P. Gepts, A. Greenland, A. Hall, K. Hori, S. Hughes, M.W. Humphreys, M. Iorizzo, A.M. Ismail, A. Marshall, S. Mayes, H.T. Nguyen, F.C. Ogbonnaya, R. Ortiz, A.H. Paterson, P.W. Simon, J. Tohme, R. Tuberosa, B. Valliyodan, R.K. Varshney, S.D. Wullschleger, M. Yano, and M. Prasad. 2015. Application of genomics-assisted breeding for generation for climate resiliant crops: Progress and prospects. Front. Plant Sci. 6:563.

Lobell, D., and G. Asner. 2003. Climate and management contributions to recent trends in U.S. agricultural yields. Science 299:1032. doi:10.1126/science.1078475

Lobell, D.B., J.I. Ortiz-Monasterio, G.P. Asner, P.A. Matson, R.L. Naylor, and W.P. Falcon. 2005. Analysis of wheat yield and climatic trends in Mexico. Field Crops Res. 94:250–256. doi:10.1016/j.fcr.2005.01.007

Maynard, D.N., and G.J. Hochmuth. 1997. Knott's handbook for vegetable growers. Fourth ed. John Wiley & Sons Inc., New York.

McKeown, A., Warland, J. and McDonald, M. R. 2005. Long-term marketable yields of horticultural crops in southern Ontario in relation to seasonal climate. Can. J. Plant Sci. 85: 431–438.

McKeown, A., J. Warland, and M.R. McDonald. 2006. Long-term climate and weather patterns in relation to crop yield: A mini review. Can. J. Bot. 84:1031–1036. doi:10.1139/b06-080

McKeown, A.W., S.M. Westerveld, and C.J. Bakker. 2010. Nitrogen and water requirements of fertigated cabbage in Ontario. Can. J. Plant Sci. 90:101–109. doi:10.4141/CJPS09028

Mearns, L.O., R.W. Katz, and S.H. Schneider. 1984. Extreme high temperature events: Changes in their probabilities with changes in mean temperature. J. Clim. Appl. Meteorol. 23:1601–1613. doi:10.1175/1520-0450(1984)023<1601:EHTECI>2.0.CO;2

Monteith, J.L. 1981. Climatic variation and the growth of crops. Q. J. R. Meteorol. Soc. 107:749–774. doi:10.1002/qj.49710745402

Peet, M.M., and D.W. Wolfe. 2000. Crop ecosystem responses to climate change: Vegetable crops. In: K.R. Reedy and H.F. Hodges, editors, Climate change and global crop productivity. CABI International, New York. p. 213

Peng, S., J. Huang, J. Sheehy, R. Laza, R. Visperas, X. Zhong, G. Centeno, G. Khush, and K. Cassman. 2004. Rice yields decline with higher night temperature from global warming. Proc. Natl. Acad. Sci. USA 101:9971–9975. doi:10.1073/pnas.0403720101

Poorter, H. 1993. Interspecific variation in the growth response of plants to an elevated ambient CO_2 concentration. Vegetatio 104-105:77–97. doi:10.1007/BF00048146

Robertson, G. 1974. Wheat yields for 50 years at Swift Current, Saskatchewan in relation to weather. Can. J. Plant Sci. 54:625–650. doi:10.4141/cjps74-112

Rosenzweig, C., and M. Parry. 1994. Potential impact of climate change on world food supply. Nature 367:133–138. doi:10.1038/367133a0

Rubatsky, V.E., C.F. Quiros, and P.W. Simon. 1999. Carrots and other Umbelliferae. Crop Production Science in Horticulture. CABI International, New York.

Ryder, E.J. 1999. Lettuce, endive and chicory. Crop Production Science in Horticulture 9. CABI International, New York.

Salunkhe, D.K. and S.S. Kadam. 1998. Handbook of vegetable science and technology: Production, compostion, storage, and processing. Marcel Dekker, Inc. New York.

Tesfaendrias, M., J. Warland, M.R. Mc, and M. Donald. 2010. Consistency of long-term marketable yield of carrot and onion cultivars in muck (organic) soil in relation to seasonal weather. Can. J. Plant Sci. 90:755–765. doi:10.4141/CJPS09175

Tesfaendrias, M., M.R. Mc, M. Donald, and J. Warland. 2013. Long-term yield of horticultural crops in Wisconsin in relation to seasonal climate in comparison with Southern Ontario, Canada. HortScience 48:863–869.

Thompson, L. 1986. Climatic change, weather variability, and corn production. Agron. J. 78:649–653. doi:10.2134/agronj1986.00021962007800040019x

Tiefenthaler, A.E., I. L. Goldman, and W. F. Tracey. 2003. Vegetable and corn yields in the United States, 1900-present. HortSience. 38: 10801082.

Wallace, H. 1920. Mathematical inquiry into the effect of weather on corn yield in the eight corn belt states. Mon. Weather Rev. 48:439–446. doi:10.1175/1520-0493(1920)48<439:MIITEO>2.0.CO;2

Warland, J., A.W. McKeown, and M.R. McDonald. 2006. Impact of high air temperatures on Brassicaceae crops in southern Ontario. Can. J. Plant Sci. 86:1209–1215. doi:10.4141/P05-067

Wein, H.C. 1997. The physiology of vegetable crops. CAB International, Wallingford, U.K. p. 662.

Wheeler, T.R., R.H.Ellis, P. Hadley and J.I.L. Morison. 1995. Effects of CO_2, temperature and their interaction on the growth, development and yield of cauliflower (*Brassica oleracea* L. botrytis). Sci. Hortic. (Amsterdam, Neth.) 60:181-197.

Wurr, D.C.E., D.W. Hand, R.N. Edmonston, J.R. Fellows, M.A. Hannah, and D.M. Cribb. 1998. Climate change: A response surface study of the effects of CO_2 and temperature on the growth of beetroot, carrots and onions. J. Agric. Sci. 313:125–133. doi:10.1017/S0021859698005681

Wurr, D.C.E. and J.R. Fellows. 2000. Temperature influences on the plant development of different maturity types of cauliflower. Acta Hortic. 539: 69-74

Yoong, F.-Y., L.K. O'Brien, M.J. Truco, H. Heqiang Huo, R. Sideman, R. Hayes, R.W. Michelmore, and J.K. Bradford. 2016. Genetic variation for Thermotolerance in lettuce seed germination is associated with temperature-sensitive regulation of ETHYLENE RESPONSE FACTOR1. Plant Physiol. 170:472–488. doi:10.1104/pp.15.01251

Cotton and Climate Change

Allyson A.J. Williams,* David McRae, Louis Kouadio, Shahbaz Mushtaq, and Peter Davis

Abstract

Cotton is grown in many climatically suitable countries across the world, in a diverse range of production systems. The levels of climate risk varies across these regions, as does the capacity to adapt to future changes in climate risk. In this chapter we discuss the general sensitivity of cotton to climate, recent climate changes that have influenced cotton in key producing countries, the likelihood of future relevant climate changes, and general adaptive strategies. These issues are considered in the case-study of Australia which has a high level of exposure to climate change.

Cotton, predominantly *Gossypium hirsutum* L., is grown commercially in more than 70 different countries, mostly in the longitudinal band between 37° N lat. and 32° S lat. (Fig. 1). The top major producers are China, the United States, and India, which together produced about 60% of the world production of cotton lint on average during 2000–2013 (FAO, 2014). Cotton represents about 44% of the global textile market and is the most widely produced natural fiber, with an estimated total international annual trade of AU$12 billion (Cotton Australia, 2015). The three largest exporters of cotton are the USA, India, and Australia (despite the last producing only 3% of the world's cotton). On average, 33–34 million hectares are planted to cotton annually around the world and produce about 26 million tons of lint (Fig. 2).

Although cotton is naturally a perennial shrub, commercially it is grown as an annual crop. With some resilience to high temperatures and moderate drought, cotton is ideally suited to arid and semiarid zones, where it is grown during spring and summer as either an irrigated or a rainfed crop, or as a combination of both. About 53% of the world's cotton-growing areas and 73% of all fiber-growing areas benefit from full or supplementary irrigation (ITC, 2011). The sensitivity of cotton to soil water availability has led to the development of large irrigation systems in many countries, especially in high-quality and high-value production systems such as in Australia.

Abbreviations: GCM, general circulation model; ENSO, El Niño–Southern Oscillation; IPM, integrated pest management; MDB, Murray-Darling Basin; SST, sea surface temperature; WUE, water-use efficiency.

International Centre for Applied Climate Sciences, Univ. of Southern Queensland, West St, Toowoomba 4350, Australia. *Corresponding author (Allyson.Williams@usq.edu.au).

doi:10.2134/agronmonogr60.2016.0009

© ASA, CSSA, and SSSA, 5585 Guilford Road, Madison, WI 53711, USA.
Agroclimatology: Linking Agriculture to Climate, Agronomy Monograph 60.
Jerry L. Hatfield, Mannava V.K. Sivakumar, John H. Prueger, editors.

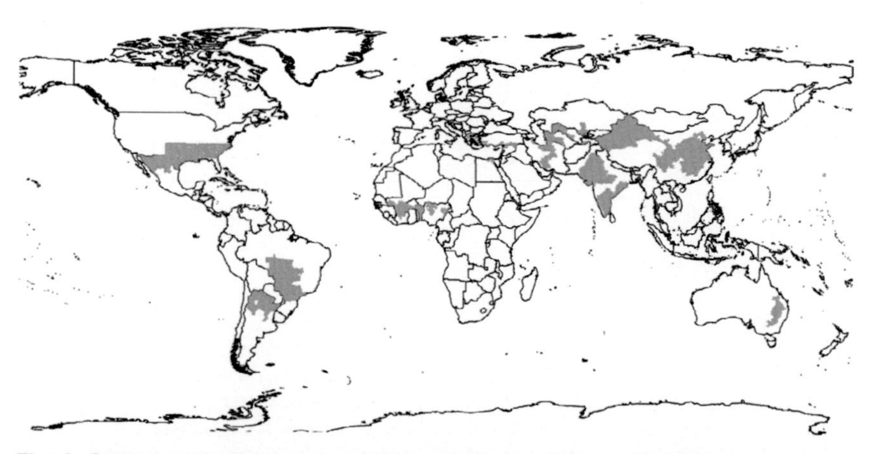

Fig. 1. Cotton production zones of top producer countries. Adapted from: China—Zhang, 2011; Argentina, Brazil, India, Pakistan, Turkey, USA, Uzbekistan—https://www.icac.org/cotton-economics-and-cotton-statistics/Cotton-Facts; Australia— http://cottonaustralia.com.au/australian-cotton/basics/where-is-it-grown; Egypt, Burkina Faso, Nigeria, Mali, Cote d'Ivoire, and Benin—OECD/SWAC, 2007; Iran—Iran Cotton Union (www.unicot.org); Tajikistan—World Bank, 2007.

Climatic Conditions for Cotton Growth

Adequate soil (i.e., excellent water-holding capacity, aeration, and good drainage) and climatic conditions are necessary for the growth and development of cotton. It requires a minimum daily air temperature of 15°C for germination, 21–27°C for vegetative growth, and 27–32°C during the fruiting period (Eaton, 1955; Freeland et al., 2010). Temperatures below 15°C slow the crop's growth and can have serious effects on the final yield, depending on the duration and phenological stages at which these temperatures occur (Waddle, 1984; Pettigrew et al., 2000; Singh et al., 2005). The total growing degree days from planting to harvest that are required for high yield vary between 1195 and 1275 degree-day heat units (base temperature 15.5°C). The resilience of cotton to high temperatures and drought is due to its vertical tap root. The crop, however, is sensitive to water availability, particularly at the height of flowering and boll formation. The amount of available water during the growing season ranges approximately from 500 to 1200 mm (Freeland et al., 2010). The optimum climate (i.e., temperature and crop water use) requirements for cotton are given in Table 1. The need for irrigation varies according to the region and environmental factors (soil and climate). For example, in Australia, cotton's average irrigation requirement was approximately 7.8 ML/ha during the 2012–2013 growing season (Australian Bureau of Statistics, 2014). To be efficient, the water supply for high production must therefore be adjusted to the specific requirements of each growth period.

Changes in weather conditions affect cotton yields by causing a variety of climatic stresses during the growth stages. Although cotton does have a natural resilience to drought and high temperature, water-deficit stresses can adversely affect the crop's growth and development, depending on the severity and duration of the stress, the growth stage at which the stress occurs, and the genotype (de Kock et al., 1990; Reddy et al., 2000; Loka and Oosterhuis, 2012). The reproductive period is the most sensitive to drought stress because water-deficit stress during

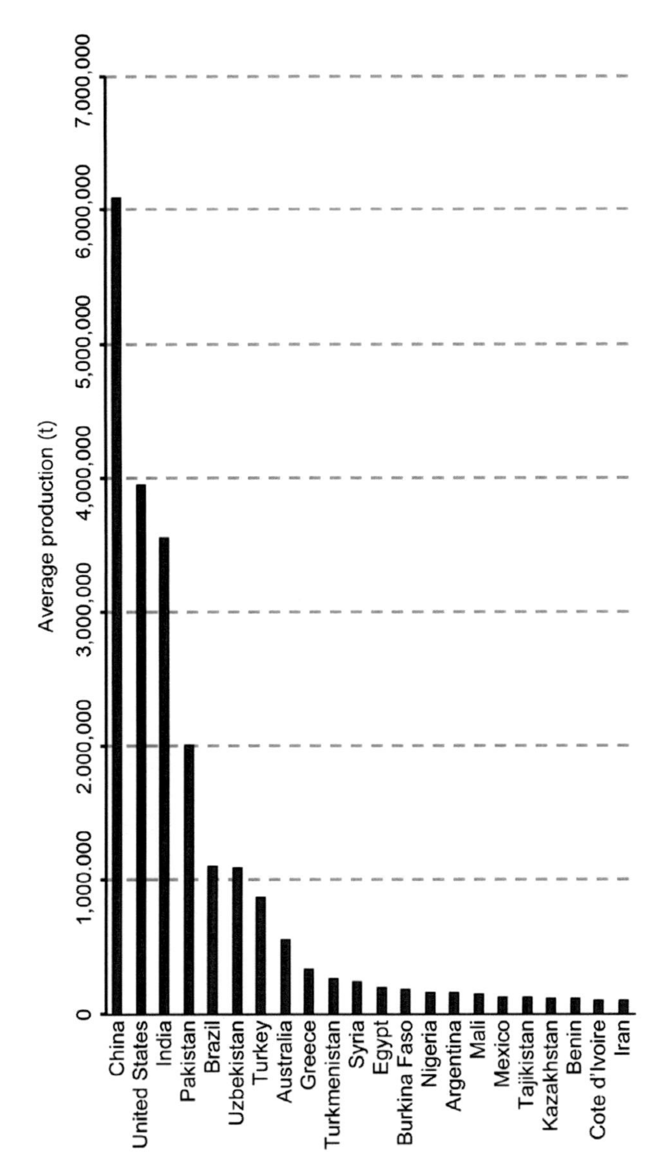

Fig. 2. Average cotton lint production over the 2000–2013 period. Countries with average production greater than 100,000 metric tons are represented. Source: FAO, 2014.

peak flowering can lead to significant yield losses compared with stress during the first squaring (unless the deficit is severe) or late boll maturation (Loka et al., 2011; Loka and Oosterhuis, 2012). Temperature stress can also lead to yield loss. Cold temperatures can stop plant growth or kill immature plants, and high temperatures can result in higher micronaire values (Wanjura and Barker, 1985; Bange et al., 2010a), which is an undesirable fiber characteristic. Moreover, the boll-maturation period, along with boll size, decreases as temperatures increase (Reddy et al., 1995a). Singh et al. (2005) reported that CO_2 assimilation and stomatal control

Table 1. Optimum daily temperature and water requirements for cotton.†

Growth stage	Average daily temperature	Daily crop water use
	°C	mm
Planting	>18 (soil)	>0
	>21 (air)	>0
Vegetative growth	21–27	1–2
First square	–	2–4
Reproductive growth	27–32	3–8
Peak bloom	–	8
First open boll	–	8–4
Maturation	21–32	4

† Adapted from Freeland et al. (2010).

of leaves may be negatively or positively affected by cool nights, depending on the length of exposure and the age of the leaves. Elevated CO_2 may be beneficial because photosynthesis is generally enhanced in cotton, that is, through an increased concentration of the substrate and a suppression of photorespiration, the water-use efficiency (WUE) is thus increased (Reddy et al., 2000, 2004).

Historical Climate Trends and Drivers in Key Cotton-Growing Regions

The key cotton-producing regions of China, the USA, and India (and the other regions shown in Fig. 1 and 2) have the climatic conditions and water supply that meet the discussed requirements for cotton production. However, all regions are subject to different levels of historical climate risk and climatic drivers and, consequently, different possible impacts resulting from climate change. The response of cotton production to the climate drivers varies depending on the specific production systems. Despite the difficulties in assessing specific climate risks to cotton production, the key risks from climate in each country can still be highlighted. The observed trends in climate at regional and global scales have been described in IPCC Assessment Report 5 (IPCC, 2014).

A major recognized driver of the year-to-year climate variability in the key cotton-producing countries is the El Niño–Southern Oscillation system (ENSO). ENSO is the abbreviation used to collectively describe the largest global interannual climate ocean and atmosphere circulation system. Changes in sea surface temperature (SST) anomalies in the central equatorial Pacific Ocean basin drive changes in global atmospheric patterns such as the Walker Circulation. El Niño refers to the oceanic component of ENSO. The Southern Oscillation (SO) refers to the atmospheric component: the shifting of atmospheric pressure between the central and eastern Pacific and the western Pacific that occurs between El Niño and La Niña events. (Allan 2000 provides a comprehensive review of the development of our understanding of the ENSO.) ENSO therefore is the coupled instability of the atmosphere ocean system driven by changes in SST patterns in the Pacific (Cane, 2000). The effect that ENSO has on the climates of various countries depends on the specific teleconnections, and there is often significant variation between countries. ENSO has major consequences for the climate globally, including that of the key cotton-producing regions (Ropelewski and Halpert, 1987). The following discussion details the historical climate trends and variability that affect the cotton industry in the key cotton-producing countries.

In Asia, Hijioka et al. (2014) reported that the mean annual temperature has risen across most regions during the past century. In East and South Asia, including China, India, and Pakistan, an increase of 0.5–2°C has been recorded during the 1901–2012 period. These changes were generally coupled with more hot days and warm nights and a decline in cool weather. The annual precipitation varied across these regions, with either increasing or decreasing trends observed. Particularly in India and Pakistan, there was a declining trend in the annual precipitation, which involved an increase in the number of monsoon-break days and a reduction in the number of monsoon depressions (Hijioka et al., 2014)

China

China is the largest cotton producer in the world. The three cotton-producing regions within China—the northern, the southern, and northwestern regions (entirely irrigated)—vary enormously in their climate, soil, and ecological characteristics (Hsu and Gale, 2001). Consequently, the different production systems have different levels of climate sensitivity. In the northwestern arid region of Xinjiang, cotton production is almost entirely irrigated, and consequently water availability is a key factor (Zhang, 2011). Irrigation depends on water from glacier melt and snowmelt (Fan et al., 2006). Streamflow has increased in recent decades because rising temperatures have resulted in more glacier melt. The southern region, which extends primarily across the Yangtze River basin, has seen enhanced mean and extreme precipitation since the 1970s, leading to a relatively sufficient water supply compared with other regions. However, the uneven temporal distribution of precipitation throughout the cropping season poses major problems for cotton growth, thereby having an adverse effect on yield (Zhao and Tisdell, 2009). In other cotton-growing areas of China, seasonal deficient mean precipitation has led to frequent droughts, namely during summer. Irrigation in such cases uses already-scarce surface and groundwater (Zhang, 2011).

The East Asian Monsoon, which affects both the immediate and neighboring regions (Lau and Li, 1984; Yasunari, 1991), is influenced by ENSO. The mature phase of ENSO often occurs in boreal winter and is normally accompanied by a weaker-than-normal winter monsoon along the East Asian coast (Zhang et al., 1996; Huang et al., 2003; Wang and Li, 2004). Consequently across southeastern China and Korea, warmer- and wetter-than-normal conditions are experienced during the ENSO winter and the ensuing spring (Zhang et al., 1999; Feng et al., 2011; Yuan et al., 2014). In the northwestern growing region, temperatures increase and rainfall increases in summer during a La Niña event (Tao et al., 2004).

India

The cotton industry in India is similarly disparate in the range of production systems used to grow the crop. Cotton is grown in three distinct zones across India (ITC, 2011): the semiarid central zone, where mostly rainfed cotton is grown; the southern zone, where both irrigated and rainfed systems are used; and the least climatically suitable, hot, humid northern zone, where all cotton is irrigated. The majority of these areas have a semiarid climate in which cotton is grown with water from the summer southwest monsoon (part of the southern zone is an exception: cotton is grown in winter because of the northeast monsoon). ENSO is a driver of climate variability across most of India (Joseph et al., 1994; Krishnaswamy et al., 2014), and this has been demonstrated to have a significant effect on many agricultural industries (Selvaraju, 2003; Patel et al., 2014; Mondal et al., 2015). The specific response of climate to ENSO shows spatial variation across India. The Indian summer monsoon (June–September) accounts for 75–90% of the all-India average annual rainfall. There are strong correlations between the Indian summer monsoon and ENSO (Parthasarathy and Pant, 1985). In general, an El Niño event is associated with a decrease in rainfall, and a La Niña event with an increase. However, there are spatial variations to this relationship, and as with most ENSO teleconnections, the effect is highly nonlinear. For example, Gujarat, which produces 31% of India's cotton, receives between 250 and 500 mm of rain in the northwest and more than 1500 mm in the south. During weak El Niño events, Gujarat's rainfall was higher than the long-term average; during moderate events, there was below-average rainfall; and in strong events, there was below-average rainfall, except in western Gujarat, which had above-average rainfall (Patel et al., 2014).

In conjunction with the decrease in summer monsoon rainfall during El Niño events, there has been a decrease in the total grain production in India (Selvaraju, 2003; Patel et al., 2014), especially in the summer (June–September season). Inversely, production increases during La Niña years, although in Mungari there were increased yields during strong El Niño events compared with La Niña or neutral years (Rao et al., 2007).

The United States of America

The United States of America is the third-largest producer country and the largest cotton exporter in the world. There are three main cotton-producing regions in the USA—the Southwest, the south-central area (Texas), and the Southeast—but there are also other smaller cotton-growing regions including western USA. Production in the Southwest and the West relies mostly on irrigation.

ENSO affects much of the USA (Legler et al., 1999; Garcia y Garcia et al., 2010). Throughout the cotton-growing regions of southeastern USA, a La Niña event is usually associated with a drier-than-average season. Baumhardt et al. (2014) found that in comparing the yield estimates for neutral and El Niño events, the limited rain during La Niña years reduced lint yields by 5–30%. In addition to this broader signal, ENSO signals are detected in the yields of cotton planted early or late but not in those crops planted mid-season (Paz et al., 2012) because the crop season straddles the transitional stage of ENSO. For crops planted early in the season, a La Niña correlates with higher yields.

Climate-related changes to agriculture have already been observed in the United States and globally. Although an increase in mean annual temperature has been observed during the past century and has been accompanied by more extremely hot seasons and a higher ratio of record-high to record-low daily temperatures (Romero-Lankao et al., 2014), some studies have shown that the warming has been less pronounced and less robust across areas of the central and southeastern USA (Alexander et al., 2006; Peterson et al., 2008). An increase in heavy precipitation across the USA and Mexico has occurred between the mid-twentieth and the early twenty-first centuries (DeGaetano, 2009; Peterson and Baringer, 2009). No clear trend has been found regarding changes in drought during the same period; Dai (2011) and Sheffield et al. (2012) reported mixed trends for dryness across the USA, Mexico, and Canada.

The Impact of Climate Change on Cotton Production

We have discussed the key climate variables that historically have directly affected cotton plant processes; namely, minimum and maximum temperatures, precipitation and streamflow for irrigated cotton, and evaporation. Much has been written about the effect of global warming on agricultural systems (IPCC, 2014). The cotton industry across the globe has been shown to be vulnerable to climate change, as demonstrated for China (Yang et al., 2014), India (Kranthi, 2009; Hebbar et al., 2013), the USA (Easterling et al., 2007), Israel (Haim et al., 2008), and Australia (Williams et al., 2014). In this section we discuss how changes in these key variables and atmospheric CO_2 levels will affect growth, and specifically how the main cotton-producing countries probably will be affected.

Changes in climatic conditions will affect cotton yields by causing various stresses during the growth stages (as detailed in Table 1) even though rising CO_2 levels will encourage increased biomass production (Easterling et al., 2007; Williams et al., 2014). The key variables of interest for future cotton production are shown in Table 2. It is not possible to present a single table documenting the projected changes to each of these key variables because there will be regional differences in the impact of climate change on these variables, and also because of differences in the scenarios produced by the different general circulation models (GCMs). "Model uncertainty" in the latter can be addressed by presenting a range of GCM climate-change futures or the "consensus" (the future scenario that most GCMs project) scenario. Here we discuss the "probable" future of some of the variables and the impact that a specific direction of change may have on cotton production.

Most of the global climate-change scenarios project increases in minimum and maximum temperatures, resulting in longer growing seasons and more heat waves. In terms of rainfall, the greatest certainty is an increase in the frequency of extreme events (both heavy rainfall and droughts); however, the change in the amount of annual and seasonal rainfall is one of the most uncertain variables. There is also agreement that the overall effect on cotton will mostly be adverse in the long term, and adaptation will be required.

Table 2. Projected changes in key variables of relevance to cotton production.

Climate variable	Implications for cotton farming system
Increased levels of atmospheric CO_2	Increased crop biomass Increased weed biomass Decrease in physiological water use Altered C/N ratio Altered N cycle Increase in pathogens Increase in insects numbers
Evaporation	Changes in evapotranspiration affect water use efficiency
Minimum temperatures	Change in physiological processes Change in weeds, pests, and diseases Change in irrigation needs
Maximum temperatures Days over 32°C	Heat stress
Droughts	Increased pressure on water supply for irrigation
Change in summer rainfall	Change in interannual variability affecting productivity Change in rainfall intensity affecting soil erosion and runoff

An increase in atmospheric CO_2 directly affects photosynthesis and may be beneficial since photosynthesis is generally enhanced in cotton through an increased concentration of the substrate, a suppression of photorespiration and thereby increasing WUE (Reddy et al., 2000, 2004). Despite the increase in growth rates in response to rising CO_2 concentrations (Kimball, 1983), there is no change in the duration of the phenological phases from emergence to reproductive initiation, the square period, and the boll-maturation period (Reddy et al., 2004). Cotton phenology is sensitive to other aspects of climate change (Reddy et al., 2000; Gérardeaux et al., 2013; Yang et al., 2014): as temperature increases, the development time decreases, especially during the boll-growth period. Although this effect may be moderated by CO_2 fertilization in the early twenty-first century, yields in some regions may significantly decline in the latter part of that century due to a shorter growing season, increases in temperature, and/or decreases in rainfall (Williams et al., 2014; Yang et al., 2014).

The potential adverse effects of water deficits and high temperatures have been discussed above, but it is noteworthy that rising temperatures may also lead to opportunities. These include growing cotton in regions that are currently too cold to grow cotton, having a more flexible growing season, or perhaps double-cropping cotton.

Although it is possible to identify the sensitivity of cotton to changes in specific climate variables and thereby infer the possible impacts of climate change, the actual effects that climate change could have on cotton production will vary depending on a multitude of factors, including a region's soil type, regional weather, and production system (e.g., Hebbar et al., 2013). It is therefore sometimes beneficial to use a crop-simulation model driven by future climate-change scenarios to specifically assess the impact (Rosenzweig and Iglesias, 1998, White et al., 2011, Hebbar et al., 2013; Williams et al., 2014). Such cotton growth models include CSM-CROPGRO (Jones et al., 2003), OZCOT (Hearn, 1994), COTCO2 (Wall et al., 1994), Cotton2K model (Marani, 2004), CropSyst (Stockle et al., 1994), and GOSSYM (Baker et al., 1983; Hodges et al., 1998). These crop models incorporate weather data, soil data, crop varieties, and crop management rules and therefore can quantify the effects of climatic conditions (and changes in other environmental components of a production system) and management options such as planting dates. However there are many caveats to be acknowledged when using crop models because, by definition, a crop model will simplify key processes, including plant growth, crop canopy development, soil management, and the effect of CO_2 on crop growth processes. Examples of their use in climate change applications include those of Gérardeaux et al. (2013), Cammarano et al. (2012), Wall et al. (1994), Haim et al. (2008), Buttar et al. (2012), Doherty et al. (2003), and Reddy et al. (2002).

Future Climate-Change Impacts on Key Regions of Cotton Production

Future climate scenarios have been developed using global-scale GCMs (IPCC, 2014). These climate projections are derived from a suite of GCMs rigorously selected through the Climate Model Intercomparison Project 5 (CMIP5) to provide representation of the world's climate (Taylor et al., 2011).

The maximum temperatures in cotton-growing regions are going to increase (Fig. 3). To assess the future change in actual maximum temperatures or the changes in the frequency of the temperatures above the optimum maximum temperature of 32°C requires the bias to be removed from a GCM. Such calibration requires a global-scale, gridded daily temperature database, which unfortunately does not exist. However, although these data limitations restrict analyses of changes in specific thresholds, it is still possible to assess changes in the extreme percentiles. An example of changes in the 95th percentile of daily maximum temperatures over the next few decades, as projected by the HADGEM GCM using the RCP 8.5 (Fig. 4), can be applied to local climate databases to identify the effect on optimal growing conditions. It is clear that there are hot spots in some countries that may present significant challenges for the cotton industry of those countries, which include China, southeast Pakistan, central Asia, and Brazil.

Of key importance in climate-change impact assessments is identifying the range of future scenarios as projected by different models (Christensen et al., 2007; Whetton et al., 2012). For example, in cotton-growing regions of USA, the HADGEM GCM projects a change of approximately 1.9–2.5°C in the southwestern region and 0.3–0.8°C in the southeastern region (Fig. 3). However, examining four other GCMs that are selected on the basis of their inclusion of atmospheric chemistry and ENSO-like behavior, it is clear that there is some uncertainty in the projections of the 95th percentiles of daily maximum temperature (Fig. 4). For example, in the southwestern USA, the difference in projections is between 2.5–3.8°C, and in the southeastern USA, the range is 0–1.3°C. Therefore, to account for this uncertainty in model output, risk assessments in the southwestern region will require a different approach than that in the southeast.

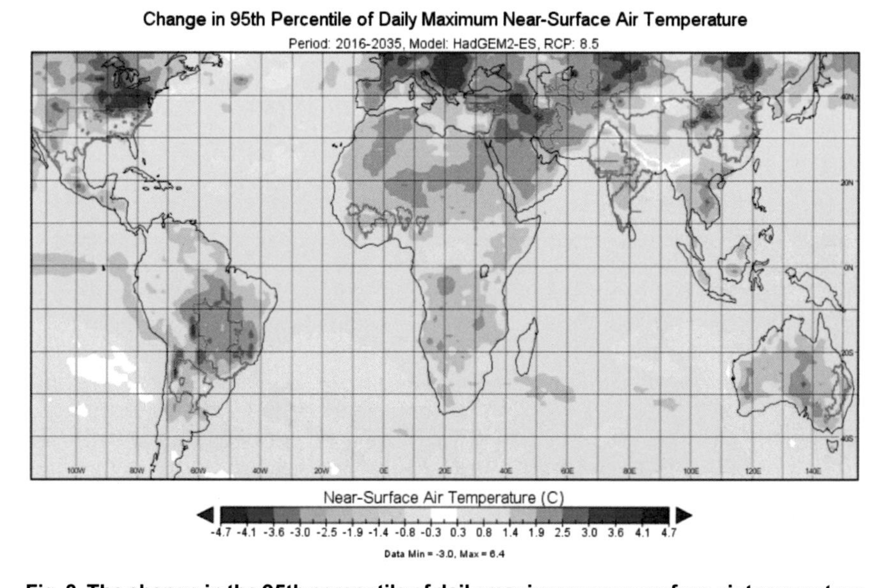

Fig. 3. The change in the 95th percentile of daily maximum near-surface air temperature as projected by the HadGEM2-ES general circulation model for 2016–2035 using the RCP8.5 emissions.

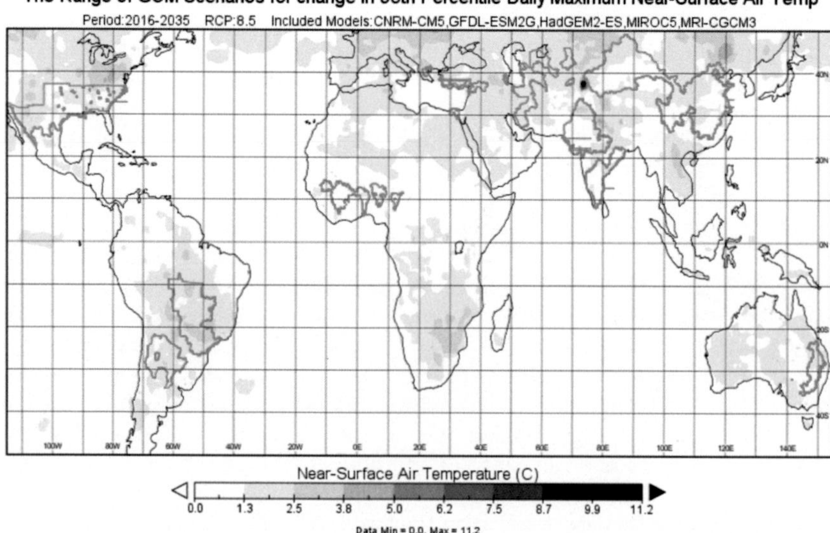

Fig. 4. The difference between the future projections of different general circulation models (GCMs; CNRM-CM5, GFDL-ESM2G, HadGEM2-ES, MIROC5 and MRI-CGCM3). Specifically, the difference between the GCM with the lowest change in 95th percentile for each pixel and the GCM with the greatest change in the 95th percentile for each pixel.

The following discussion focuses on the documented impacts that may result from climate change to cotton production in the key production countries of China, India, the USA, but potential impacts on Pakistan, and Uzbekistan are also considered. The impacts are not the same in each country because of the different levels of historical climate risk, the various changes to the climate, and the diverse production systems. Despite the difficulties in assessing specific impacts to cotton production, the key risks from climate change in each country can be highlighted.

China

Across most of China the most probable rainfall scenario for the next 100 years is for more rainfall in most seasons (IPCC, 2014), although less is expected in some regions in the near future. The increase will be accompanied by more frequent extremes in both temperature and precipitation (IPCC, 2014). In general, increases in temperatures are expected to have a positive effect on production, especially since in many regions the major constraint on production is the low night temperatures, which limit the growing season (People's Republic of China, 2004). The increase in frost-free conditions is expected to lengthen the growing season and the peak growing period. The proportion of bolls that open before frost is expected to rise by 5–10% for every 1°C increase in the annual average temperature (People's Republic of China, 2004).

In the northwestern arid region of Xinjiang, water availability is a key factor for the cotton production (Deng et al., 2006). Streamflow has increased in recent decades due to increased glacier melt resulting from warming. However, as global warming continues, it is expected that streamflow will decrease as the volume of glaciers decreases (Piao et al., 2010). Furthermore, the availability of water for irrigation is expected to decline due to both higher crop water requirements and increasing demand for water for nonagricultural use. This region is one of those in which low night temperatures are the principal limitation on the growing season.

The impacts of climate change vary across Xinjiang, where the changes are greater in the northern, colder region of Shihezi (Yang et al., 2014). The impacts of climate change include a shorter growing season and greater evapotranspiration. Yields are expected to increase in the early twenty-first century but to decrease in the later part of the century. Given the current growing varieties, increased water consumption will probably occur in warmer regions (Yang et al., 2014). Reducing the planting density will therefore be required as an adaptation strategy to cope with such increases.

India

As in most countries, temperatures in India are expected to increase. Greater variability in rainfall and more extreme events will have a larger impact on rainfed cotton than on irrigated cotton. The latter will be affected by reduced streamflow caused by a decline in snow and ice in the Himalayan glaciers and snowfields. Northern India will probably experience diminished precipitation, which, in combination with increased temperatures, will reduce yields. In central India (which encompasses more than 60% of the area of India's cotton-growing regions), cotton production is primarily a rainfed system and therefore is dependent on the monsoons. The most probable future climate scenario is for a slight increase in monsoon rainfall. In their study of southern Gujarat, Thakare et al. (2014) highlighted the dominance of rainfall as a production driver. Temperatures may rise less than in the northern Indian production zone and may overcome the production-limiting low winter temperatures and have a positive effect on boll growth (Hebbar et al., 2013).

The southern cotton-growing zone of India has regions influenced by the southwest monsoon (Kamataka and Andhra Pradesh) and northeast monsoon (Tamil Nadu). These regions have different impacts (Hebbar et al., 2013). The growing season in the southern zone has lower temperatures than the northern and central zones and therefore is less likely to be negatively influenced by the impacts of an increasing number of extreme high temperatures.

In addition, there are agroclimatic zones within these areas that also have variable impacts. For example, cotton production in the southern state of Tamil Nadu shows a variable response in 2050 climate projections (Arumugam et al., 2014): in the northern and western regions of the state, projections suggest that a decline in yield is most probable, whereas in the southern zone, a 15% increase in yield has been modeled. The impacts of climate change by 2070–2100 are projected to be greater in the northern and central zones, which will have higher temperatures than the southern zone. There is little change in precipitation across all zones. There is also little change in cotton yield within the 2030–2060 time frame; however, in the later part of the twenty-first century there may be a significant decline in yield that will be greater in the northern zone than in the southern.

The United States of America

Climate-change impacts on cotton production in the USA are expected to be regionally diverse (Mearns et al., 2000; Reddy et al., 2002; Doherty et al., 2003; Malcolm et al., 2012). In the primary growing regions of the central and southeastern USA, most projections for 2030 show increases in rainfall and extreme rain events (Karl et al., 2009). These will be significant for this mainly dryland cotton region. In the Southeast, approximately 80% of cotton is dryland. At the statewide level the simulated average changes in yield range from a 3% decrease to a 9% increase, with changes in individual states ranging from an approximately 10% decrease in Georgia to a nearly 40% increase in Tennessee (Doherty et al., 2003). The uncertainty in these projected yield changes reflect the differences in climate GCMs. If adaptation measures are implemented, there most probably will be an increase in yield across all states.

In the irrigated production systems in the Southeast, yields will most probably rise as a result of global warming, with projections ranging from a regional average increase of approximately 28% to individual-state average increases of 90% for Tennessee and 70% for Mississippi, but only 1–8% for Florida and Louisiana (Doherty et al., 2003). If farmers adapt their management practices (e.g., early planting dates, use of new cultivars, etc.) yield increases will probably be higher, with those in Florida and Louisiana increasing to approximately 12–20% and the regional average to 35%.

In Texas, cotton is grown in both irrigated and dryland conditions. Yield is projected to increase; however, there is more uncertainty in the eastern part of the state. Simulations using the GOSSYM model show that, with adaptation, yield may increase by more than 15% (Doherty et al., 2003).

The southwestern cotton production area is primarily irrigated. The impact of change in climate and atmospheric CO_2 levels on cotton yields are projected to increase by up to 20%, with increases being greatest in Arizona and New Mexico (Doherty et al., 2003). Water availability for irrigation is already under pressure from declining aquifer levels. In the coming decades, water will probably be further reduced as a result of increased use and decreased snowmelt from the Rocky Mountains.

A range of climate-change impact studies have included assessing yield changes on the basis of various adaptation measures, such as changes in planting date and planting (e.g., Doherty et al., 2003). In all studies across most regions of the USA, cotton yields increase, although only by a marginal amount in southern California according to the study by Doherty et al. (2003). Increases include 15–100% in dryland production in Texas, and between 5 and 20% in irrigated production systems (Doherty et al., 2003).

Pakistan and Uzbekistan

Although not in the top three cotton-producing countries, Pakistan and Uzbekistan are significant cotton producers and, in the context of climate change, are considered highly vulnerable (ITC, 2011; IPCC, 2014).

The majority of cotton in Pakistan is grown in the low-rainfall provinces of Punjab and Sindh and relies on irrigation from the Indus River. The high temperatures that occur during the summer growing season are the principal limitation on current production. The maximum temperature often exceeds 40°C, which is 10°C higher than the optimal range, and so it has negative effects on growth, maturity, and productivity. In addition, an increase in minimum temperatures may enhanced flower shedding and reduced boll size (Ashraf and Iftikhar, 2013).

Rainfall is generally expected to decline across Pakistan. Empirical relationships between cotton production and maximum temperature and precipitation indicate that cotton production may decline by 13% by 2030, accompanied by an increase in maximum temperatures of 1°C (Siddiqui et al., 2012). The potential decrease in rainfall and increase in temperatures will raise irrigation requirements. Furthermore, streamflows, and therefore the net availability of irrigation water, will possibly decrease due to glacier recession and reductions in rainfall. More important though, in a warming climate, as the glaciers and snowpacks melt, the availability of water will rise in early decades of the twenty-first century, but will decline.

Cotton in Uzbekistan is 100% irrigated. The projections for streamflows dependent on glacier melt are similar to those in Pakistan. However, much of the water available for irrigated agriculture in Uzbekistan is dependent on the water management policies of upstream neighboring countries (Abdullaev et al., 2007). Uzbekistan is projected to become warmer and most probably drier during the twenty-first century. Changes in temperature (increases) and rainfall (decreases) will probably be most significant in the summer (IPCC, 2014). Droughts are projected to increase in frequency and, combined with increasing aridity, will adversely affect cotton production systems (Lioubimtseva and Henebry, 2009). Given these climate projections, especially of the water available for irrigation, cotton yields will most probably decline. Yield decreases of 13 and 23% in the Syr Darya and Amu Darya catchments, respectively, have been modeled on the basis of increased evaporation and decreased water availability (Republic of Uzbekistan, 2008).

However, current cotton production is also temperature limited. Low temperatures prevent early planting and preclude the required number of degree days for optimum yields (ITC, 2011). Therefore, the projected increase in temperature will increase the length of the growing season, which may enable production to expand into new areas that were previously too cold.

Options for Adapting to Climate Change and Variability

Because of its indeterminate growth pattern, cotton, unlike many other crops, has a certain ability to compensate for changes in growing conditions (soil moisture, heat days etc.) resulting from climate variability. Despite this, it is generally considered difficult to make major adjustments to crop management after planting. It can therefore be argued that currently the best options to manage climate risk are to improve industry, farm and crop resilience, sustainability, and profitability (McRae et al., 2007; Traore et al., 2013). It can also be debated if changes in cotton management systems are as of the result of climate risk or a conscious means of ensuring ongoing economic viability, as long as suitable, practicable and workable options are developed given the parameters of the identified production system. The majority of the adaptation responses available for managing climate risk within a modern cotton production system include production-efficiency benefits and promote best management practices (McRae et al., 2007; Bange et al., 2010b). This is because they are focused on improving WUE, and improving general soil and crop management.

Water Use

As already discussed, irrigation is a key component of global cotton production. Bange et al. (2010b) identified a number of practices that growers could use to adapt to water availability constraints and improve WUEs. These options include a focus on more efficient irrigation systems (compared with furrow irrigation) such as center-pivot, drip, or lateral-move irrigation systems, as well as scheduling the timing of irrigation on the basis of crop soil-water use (through the use of technology such as moisture meters), crop stage (to identify plant demand), and weather and climatic conditions.

Managing nitrogen use, especially when water availability is constrained, through changes in fertilizer application, type, and timing has the positive potential to reduce extra vegetative growth and increase WUE. The use of a legume phase or cover crop in a cropping rotation plan would also have the positive benefit of improving soil health.

More accurately matching the area of cotton to be grown to the available irrigation water would reduce the crop failure risk, especially in years when the irrigation water supply is limited or when exceptionally dry or hot seasonal conditions are expected. The development and application of simulation models for cotton production have allowed growers, grower advisors, and researchers to tailor management decisions on the basis of crop water use and irrigation water management (Thorp et al., 2014). For example, Cotton2K (Marani, 2004) and OZCOT (Hearn, 1994), which incorporate hydrological tactical and strategic cotton irrigation-decision support systems (Richards et al., 2008; Thorp et al., 2014), have been used to assess the impact of different water allocations on yield while taking into account different climate scenarios.

Climate Risk Management

A climate-change risk assessment (such as in Cobon et al., 2009) for cotton production will be very regionally dependent. The benefit of such a process is that it helps industry and producers to estimate how vulnerable they are based on risks associated with climate change and to assess the possible options they should adopt.

Improved management of climate variability through the use of weather and seasonal climate forecasts may help growers optimize their immediate decision-making and longer-term tactical planning. Traore et al. (2013) identified a demand by African cotton growers for climatic information at an intraseasonal timescale and highlighted the need for effective agrometeorological services to assist agricultural communities in managing climate variability. Identified management options include adjustment of planting dates, fertilizer rates, and timing of application. Ritchie et al. (2004), in a review of risk management strategies using seasonal climate forecasting in irrigated cotton production, highlighted that minimizing risk by adjusting planting areas in response to seasonal climate forecasts and water allocation expectations can lead to significant gains in gross margin returns.

The protection of crop yields from a range of weed, insect, and mite attacks is an issue that all cotton growers face. The distribution and abundance of these pests and the damage caused by them is in part driven by changes in seasonal climate variability. Wetter- and warmer-than-normal years result in an increase in weed and insect loads, whereas drier- and hotter-than-normal years when irrigation water is readily available offer growers the opportunity to maximize production with minimal pest impact. The use of seasonal climate risk information to identify periods of the greatest threats from pests and weeds can allow growers to tailor treatment options. This process can occur at the farm level or on a wider industry basis through the use of integrated pest management (IPM) strategies such as those developed in Australia (Bange et al., 2010b).

The need to further develop tools and extension-based activities to assist growers in accessing climate risk information, interpreting the information to suit their production system, and developing alternative management options was also identified by Bange et al. (2010b) as a priority in a review of specific adaption options for the Australian cotton industry.

Case Study: The Impact of Climate Change on Cotton Production in Australia

In the selection of Australia as a case study, we aim to present a comprehensive analysis of the possible impact of climate change on cotton in Australia. The methods used here may then be used as a framework that can be used to assess impacts on other regions in the world. Previous sections in this chapter have already discussed the dependence of cotton production on climate and the global impact of climate change. The key effects of climate change result from CO_2 fertilization (Reddy et al., 1995b, 1996), water availability in the form of both rainfall and evaporative demand (Karl et al., 2009), and temperature (Hearn and Constable, 1984; Constable and Shaw, 1988; Reddy et al., 2002; Bange and Milroy, 2004; Schlenker and Roberts, 2009).

The Australian cotton industry is unique in a global context in that it exports more than 90% of its production. It is a large, rural, export earner for Australia: the gross value of Australian cotton production during the 2009–2010 El Niño event was AU$876.1 million, and AU$2.6 billion during the 2010–2011 La Niña event (ABARES, 2012). This shift in annual gross value highlights the role that climate variability plays as the primary driver of the volume of cotton produced seasonally in Australia (McRae et al., 2007; Cotton RD&E Strategy Working Group, 2011).

The majority (>95%) of cotton grown in Australia comes from the Murray-Darling Basin (MDB) catchment area (Murray Darling Basin Authority, 2014; Fig. 5). The climate characteristics of this region are relatively uniform in terms of temperature and rainfall seasonality. Consequently there are many similarities in production systems across the region. The 6-mo cotton-growing season starts in September/October (planting) and ends in March/April (picking). On a typical cotton-producing farm, cotton makes up the largest proportion of farm income (approximately 66%) in terms of the gross value of production, yet it composes only about 10% of the total farm area. About 80% of cotton-producing farms have irrigated land, and consequently overall production is sensitive to water availability. The availability of irrigation water is a limiting factor in cotton production. Water-use efficiency has increased by approximately 240% since the 1970s, and Australian cotton growers are now recognized as the most water-use efficient in the world, three times more efficient than the global average. In addition, Best Management Practices programs, IPM strategies, and the use of biotechnology has reduced pesticide use by more than 85% during 2000–2010 (http://www.agriculture.gov.au/ag-farm-food/crops/cotton).

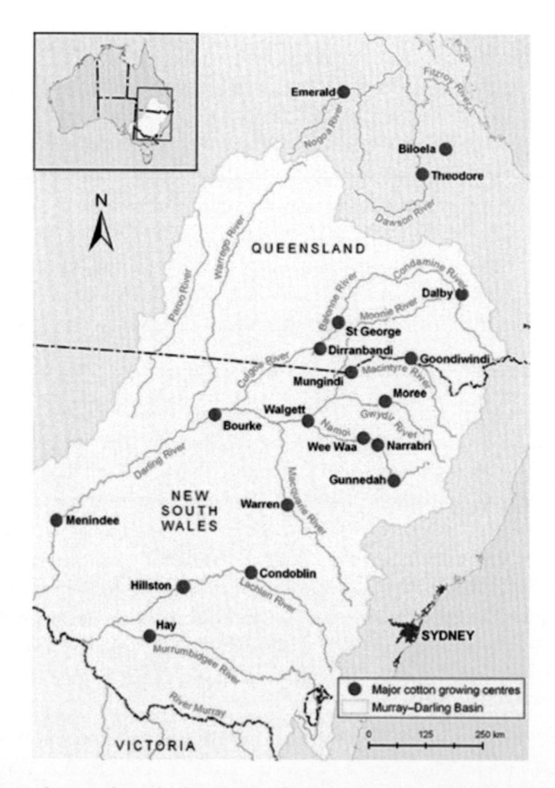

Fig. 5. Cotton growing regions in Australia. Source: National Water Commission (2011).

As the development of on-farm and off-farm infrastructure, such as irrigation schemes processing mills and shipping facilities, has continued to mature, the gross value of cotton produced in Australia has generally increased since 1985. However, the clear exceptions are the drought years: 1986–1987, 2002–2003, 2003–2004, 2006–2007, 2007–2008, and 2014–2015 (McRae et al., 2007; NLWRA, 2008; van Dijk et al., 2013; Cotton Australia, 2015). Because drought is a regular occurrence in the MDB, years with major streamflow or flood events are important for replenishing irrigation reserves (Maheshwari et al., 1995; Draper, 2007). Many of the drought years are associated with ENSO. Importantly, especially in the context of providing irrigation water for cotton systems, it is estimated that 86% of the MDB generates little runoff except during flood events (Prasad and Khan, 2002). With about 80% of Australian cotton-producing farms relying on irrigation, any lack of available irrigation water greatly reduces output, such as happened during the late 2000s, before the end of the Millennium Drought (2000–2010) (Heberger, 2011; Kirby et al., 2014; Williams et al., 2014).

The MDB will probably have future decreases in water availability due to reductions in rainfall, temperature increases, net evaporation, and decreased inflows as a result of climate change; government "buybacks" of environmental water under the Murray Darling Basin Plan (Murray Darling Basin Authority, 2010); and trading water for nonagricultural purposes. For Australian cotton producers to increase their resilience and respond most effectively to climate change, it is essential to understand the complexities of the key climate sensitivities of cotton-production variables and to quantify the possible effects on cotton yields (Bange et al., 2010b). One underpinning technique that reduces uncertainty is the quantification of production: this then enables determination of the expected value of the yield to provide the foundation for economic analysis of climate change impacts. Both empirical and biophysical simulation models (e.g., OzCot and APSIM) have been used to assess climate-change impacts on Australian cotton production (for example, Luo et al., 2013, 2014; Williams et al., 2014). The Agricultural Production Systems Simulator (APSIM; Keating et al., 2003) is used to simulate the yield of cotton in a typical farming enterprise to capture the complex interactions between the components of production.

The scenarios for future cotton yields include projections of both increasing yield by 2030 (Luo et al., 2014) and for decreasing yield in both 2030 and 2050 (Williams et al., 2014). Increases in minimum temperatures may result in a longer growing season, but increases in maximum temperatures will not be beneficial because the most probable response is a faster developing crop (Luo et al., 2014). The difference in the 2030 yield projections between the two studies is due to the differing methods. The key differences include the choice of GCMs for generating the future climate scenarios, GCM downscaling techniques, climate base periods, emissions scenarios, and cotton-model parameterizations. Such differences in methods are common and ought to be embraced as means of providing a range of equally likely possible futures that can provide useful information regarding climate sensitivity and climate adaptation.

In the worst-case rainfall-decreasing scenario (as per Williams et al., 2014), historical irrigation volumes will need to be increased by approximately 50% to maintain available soil water at the required 65%. This adaptation response will probably not be feasible due to costs and water availability. In the more favorable scenarios with increasing rainfall (as in studies by Luo et al., 2014, 2015), water use by dryland cotton will increase in some areas (Emerald and Carrabin) and decrease in Dolby and More. Water-use efficiency increased by between 0 and 24% (as per the above discussion on influences of elevated CO_2).

Adaptation

In Australia, cotton is grown in a region with the highest levels of climate variability in the country and so has required that farmers be able to diversify. Producers who can develop strategies to manage climate variability may be more readily equipped to consider adaptation options. Without adaptation to decreased water availability resulting from changes in climate and water policy, cotton enterprises will have to reduce the area of production. However, some research suggests that profitability need not suffer if there is appropriate adaptation (Challinor et al., 2009; Howden et al., 2010).

Specific adaptation options for the Australian cotton industry that may be applicable to other cotton-growing regions have been identified in a survey published by McRae et al. (2007) and further developed by Bange et al. (2010b). They focus on four key themes: policy and industry; crop and farm management; climate information and use; and management of pests, diseases, and weeds (Table 3). Many of the adaption responses identified in Table 3 are considered positive options for the industry because they offer immediate gains in production efficiency regardless of the magnitude of immediate and future climate risks. These themes and options also allow the entire cotton industry—from the grower to the researcher to industry and government—the opportunity to manage climate risk.

Water management through both industry policy development and crop and farm management has been identified as a key strategy in adapting to climate variability and change in cotton-growing regions (Bange et al., 2010b). But the increased risks to yield may not be easily overcome by a change in irrigation strategies because of the overall decrease in available water (Williams et al., 2014). It will be important to identify regionally specific management strategies. For example, for rainfed cotton across Australia's cotton-growing region, it is clear that in a scenario of increasing summer rainfall, regionally specific adaptation of planting configurations will create opportunities for increased production under climate change (Luo et al., 2014).

Information on farm-scale adaptation, as well as more tactical crop-level information, has been used for adaptation assessment (Rodriguez et al., 2011). Emphasizing the more strategic farm-level adaptation has allowed assessment of the impact of various management strategies on farm business profits, and on cost-benefit analyses (Rodriguez et al., 2014). Such analyses of gross margins over the long term suggest the capacity of cotton-production systems to adapt to climate change.

Overall, the differences in future scenarios of water availability and cotton production highlight the importance of understanding the uncertainties surrounding the generation of such projections. There is little certainty in future projections of summer rainfall in the MDB. Consequently, the benefit of using a range of potential climate scenarios is that they enable a more thorough understanding of the range of possible futures for the industry. Most important, the worst-case scenario of reduced rainfall and water allocation can be partially overcome through a variety of adaptation measures.

Conclusions

The multifaceted biophysical impacts of climate change on cotton production are complex, geographically varied, encumbered with varying degrees of uncertainty, and in some locations may be managed with a range of adaptations. The sensitivity of cotton to climate change has been assessed in many studies through the use of field experiments such as Free-Air Carbon Dioxide Enrichment (FACE), and also through biophysical and statistical crop-modeling techniques. Climate sensitivity analyses suggest that the effects of climate change on cotton production will be regionally specific. Cotton growth will probably be universally enhanced by limited increases in atmospheric CO_2. The irrigated production systems in China (Xinjiang), Pakistan, the western USA, and Australia will most probably be compromised by the reduction in the availability of water for irrigation. However, the dryland systems in areas of possible rainfall increase, such as India, the southeastern USA, and the Yellow River region of China would benefit, although rainfall is the most uncertain of variables in GCMs. The increase in temperatures could be beneficial to cotton growth in terms of lengthening the growing season, but in Pakistan, where heat is a major limiting factor to growth, heat stress could cause lower yields.

Although the use of biophysical models has enhanced our knowledge of physiological sensitivity of cotton to climate change, there are caveats that need

Table 3. Industry-identified adaptation options available.†

Policy/industry
Develop effective water trading systems
Support agricultural research for adaptation studies and plant breeding
Diversify enterprises (alternative uses of water)
Expand industry to other regions
Crop and farm management
Improve nitrogen use efficiency
Improve water use efficiency (alternative irrigation systems, row configurations etc.)
Select most appropriate cultivar for specific climate and environmental region
Improve cultivars to manage higher atmospheric CO_2 concentrations, temperatures, and related changes in environmental moisture
Shift optimal planting windows to take advantage of increased temperatures
Develop new crop rotation systems
Climate information and use
Downscale climate projections to regional or subcatchment scale
Improve management of climate variability through increased availability and use of seasonal climate risk information through development of appropriate decision support tools and extension
Managing pests, diseases, and weeds
Develop appropriate integrated pest and weed management systems
Develop pest predictive tools and indicators
Encourage industry to develop area wide management options

† Adapted from Bange et al. (2010b).

to be acknowledged when interpreting the results. These include the complexity of the models, the uncertainty about some of the processes, and the large number of parameters. By definition, a crop model will simplify key processes such as plant growth, crop canopy development, soil management, and the effect of CO_2 on crop growth processes. In addition, the sensitivity of the crop model to climate change will vary depending on the specific future climate scenario (different GCMs produce different future climate scenarios).

In addition to uncertainties within the crop models, there are also uncertainties in the GCMs that force the crop-model simulations. All GCMs have some inherent bias in their simulations. Additional uncertainties arise from the choice of emissions scenarios, the base time period from which changes are calculated, and the downscaling techniques, which vary in their ability to capture changes in daily events, which are important for both management and growth processes.

The capacity of the cotton industry to adapt to climate change differs markedly across the countries. Any adaptation strategies that are identified need to be considered in the context of other environmental, social, economic, and political frameworks. This consideration is especially relevant in regions that have little adaptive capacity and consequently have a cotton industry that is highly vulnerable to climate change, such as Pakistan. Adaptations will be most effective if they can integrate multiple risks from a range of interacting processes across spatial and temporal scales. Reduced water availability caused by climate change is the most common vulnerability across many countries with irrigated production systems, so besides benefiting cotton, adaptation will also include reductions in the risks of salinization and water loss through more modern irrigation systems.

References

Abdullaev, I., M. Giordano, and A. Rasulov. 2007. Cotton in Uzbekistan: Water and welfare. In: D. Kandiyoti, editor, The cotton sector in Central Asia: Economic policy and development challenges. Proceedings of a conference held at the School of Oriental and African Studies (SOAS), University of London, London, UK. p. 112–128.

Alexander, L.V., X. Zhang, T.C. Peterson, J. Caesar, B. Gleason, A.M.G. Klein Tank, et al. 2006. Global observed changes in daily climate extremes of temperature and precipitation. J. Geophys. Res. Atmos. 111:D05109. doi:10.1029/2005JD006290

Allan, R.J. 2000. ENSO and climatic variability in the last 150 years. In: H.F. Diaz and V. Markgraf, editors, El Niño and the Southern Oscillation: Multiscale variability, global and regional impacts. Cambridge Univ. Press, Cambridge, UK. p. 3–56.

Arumugam, S., K.R. Ashok, S.N. Kulshreshtha, I. Vellangany, and R. Govindasamy. 2014. Does climate variability influence yield of major crops? A case study of Tamil Nadu. Agric. Econ. Res. Rev. 27:61–71. doi:10.5958/j.0974-0279.27.1.005

Ashraf, S., and M. Iftikhar. 2013. Mitigation and adaptation strategies for climate variability: A case of cotton growers in the Punjab, Pakistan. Int. J. Agric. Ext. 1:30–35.

Australian Bureau of Agricultural and Resource Economics and Sciences (ABARES). 2012. Agricultural commodities: March quarter. Dept. of Agric. and Water Res., Australian Government, Canberra City, Australia. http://www.agriculture.gov.au/abares/publications/display?url=http://143.188.17.20/anrdl/DAFFService/display.php?fid=pe_agcomd9abcc004201203_12a.xml (accessed 6 Dec. 2016).

Australian Bureau of Statistics. 2014. Water use on Australian farms, 2012–13. Australian Bureau of Statistics, Beconnen, ACT, Australia. http://www.abs.gov.au/Ausstats/abs@.nsf/0/58229B62E45727E9CA257E5300138D69?OpenDocument (accessed 23 May 2016).

Baker, D.N., J.R. Lambert, and J.M. McKinion. 1983. GOSSYM: A simulator of cotton crop growth and yield. Tech. Bull. 1089. South Carol. Agric. Exp. Stn., Clemson.

Bange, M.P., G.A. Constable, D.B. Johnston, and D. Kelly. 2010a. A method to estimate the effects of temperature on cotton micronaire. J. Cotton Sci. 14:164–172.

Bange, M.P., G.A. Constable, D. McRae, and G. Roth. 2010b. Cotton. In: C. Stokes and M. Howden, editors, Adapting agriculture to climate change: Preparing Australian agriculture, forestry and fisheries for the future. CSIRO Publishing, Melbourne. p. 49–66.

Bange, M., and S. Milroy. 2004. Impact of short-term exposure to cold night temperatures on early development of cotton (*Gossypium hirsutum* L.). Crop Pasture Sci. 55:655–664. doi:10.1071/AR03221

Baumhardt, R.L., S.A. Mauget, P.H. Gowda, and D.K. Brauer. 2014. Modeling cotton lint yield response to irrigation management as influenced by El Niño–Southern Oscillation. Agron. J. 106:1559–1568. doi:10.2134/agronj13.0451

Buttar, G.S., S.K. Jalota, A. Sood, and B. Bhushan. 2012. Yield and water productivity of Bt cotton (*Gossypium hirsutum*) as influenced by temperature under semi-arid conditions of north-western India: Field and simulation study. Indian J. Agric. Sci. 82:44–49.

Cammarano, D., J. Payero, B. Basso, P. Wilkens, and P. Grace. 2012. Agronomic and economic evaluation of irrigation strategies on cotton lint yield in Australia. Crop Pasture Sci. 63:647–655. doi:10.1071/CP12024

Cane, M.A. 2000. Understanding and predicting the world's climate system. In: G.L. Hammer, editor, Applications of seasonal climate forecasting in agricultural and natural ecosystems: The Australian experience. Kluwer Academic, Netherlands. p. 29–50. doi:10.1007/978-94-015-9351-9_3

Challinor, A.J., F. Ewert, S. Arnold, E. Simelton, and E. Fraser. 2009. Crops and climate change: Progress, trends, and challenges in simulating impacts and informing adaptation. J. Exp. Bot. 60:2775–2789. doi:10.1093/jxb/erp062

Christensen, J., B. Hewitson, A. Busuioc, A. Chen, X. Gao, I. Held, et al. 2007. Regional climate projections. In: S. Solomon, D. Qin, M. Manning, Z. Chen, M. Marquis, K. Averyt, et al., editors, Climate change 2007: The physical science basis. Contribution of Working Group I to the Fourth Assessment Report of the Intergovernmental Panel on Climate Change. Cambridge Univ. Press, Cambridge, UK.

Cobon, D.H., G.S. Stone, J.O. Carter, J.C. Scanlan, N.R. Toombs, X. Zhang, et al. 2009. The climate change risk management matrix for the grazing industry of northern Australia. Rangeland J. 31:31–49. doi:10.1071/RJ08069

Constable, G., and A. Shaw. 1988. Temperature requirements for cotton. Agfact P5.3.5. Division of Plant Industries, New South Wales Dep. of Agric., NSW, Australia.

Cotton Australia. 2015. Statistics. Cotton Australia, Mascot, NSW, Australia. http://cottonaustralia.com.au/cotton-library/statistics (accessed 15 June 2015).

Cotton RD&E Strategy Working Group. 2011. Cotton sector research development and extension: Final strategy. Cotton Research and Development Corporation, Narrabri, NSW, Australia.

Dai, A. 2011. Drought under global warming: A review. Wiley Interdiscip. Rev.: Clim. Change 2:45–65. doi:10.1002/wcc.81

DeGaetano, A.T. 2009. Time-dependent changes in extreme-precipitation return-period amounts in the continental United States. J. Appl. Meteorol. Climatol. 48:2086–2099. doi:10.1175/2009JAMC2179.1

de Kock, J., L.P. de Bruyn, and J.J. Human. 1990. The relative sensitivity to plant water stress during the reproductive phase of upland cotton (*Gossypium hirsutum* L.). Irrig. Sci. 11:239–244. doi:10.1007/BF00190539

Deng, X.P., L. Shan, H. Zhang, and N.C. Turner. 2006. Improving agricultural water use efficiency in arid and semiarid areas of China. Agric. Water Manage. 80:23–40. doi:10.1016/j.agwat.2005.07.021

Doherty, R., L. Mearns, K.R. Reddy, M. Downton, and L. McDaniel. 2003. Spatial scale effects of climate scenarios on simulated cotton production in the Southeastern U.S.A. Clim. Change 60:99–129. doi:10.1023/A:1026030400826

Draper, C.S. 2007. The atmospheric water balance over the Murray-Darling Basin. BMRC Res. Rep. 127. Bureau of Meteorology, Melbourne, Vic., Australia.

Easterling, W.E., P.K. Aggrawal, P. Batima, K. Brander, L. Erda, S.M. Howden, et al. 2007. Food, fibre and forest products. In: M.L. Parry, O.F. Canziani, J.P. Palutikof, P.J. van der Linden, and C.E. Hanson, editors, Climate change 2007: Impacts, adaptation and vulnerability. Contribution of Working Group II to the Fourth Assessment Report of the Intergovernmental Panel on Climate Change. Cambridge Univ. Press, Cambridge, UK. p. 273–313.

Eaton, F.M. 1955. Physiology of the cotton plant. Annu. Rev. Plant Physiol. 6:299–328. doi:10.1146/annurev.pp.06.060155.001503

Fan, Z., X. Aili, Y. Wang, and Y. Chen. 2006. Formation, development and evolution of the artificially-irrigated oases in Xinjiang. Arid Zone Res. 23:410–418.

FAO. 2014. Crops, National production. FAOSTAT. FAO, Rome, Italy.

Feng, J., W. Chen, C.Y. Tam, and W. Zhou. 2011. Different impacts of El Niño and El Niño Modoki on China rainfall in the decaying phases. Int. J. Climatol. 31:2091–2101. doi:10.1002/joc.2217

Freeland, T.B., Jr., G. Andrews, B. Pettigrew, and P. Thaxton. 2010. Agrometeorology of some selected crops: Cotton. Guide to agricultural meteorological practices. World Meteorological Organ., Geneva, Switzerland. p. 10.11–10.19.

Garcia y Garcia, A., T. Persson, J.O. Paz, C. Fraisse, and G. Hoogenboom. 2010. ENSO-based climate variability affects water use efficiency of rainfed cotton grown in the southeastern USA. Agric. Ecosyst. Environ. 139:629–635. doi:10.1016/j.agee.2010.10.009

Gérardeaux, E., B. Sultan, O. Palai, C. Guiziou, P. Oettli, and K. Naudin. 2013. Positive effect of climate change on cotton in 2050 by CO_2 enrichment and conservation agriculture in Cameroon. Agron. Sustain. Dev. 33:485–495. doi:10.1007/s13593-012-0119-4

Haim, D., M. Shechter, and P. Berliner. 2008. Assessing the impact of climate change on representative field crops in Israeli agriculture: A case study of wheat and cotton. Clim. Change 86:425–440. doi:10.1007/s10584-007-9304-x

Hearn, A. 1994. OZCOT: A simulation model for cotton crop management. Agric. Syst. 44:257–299. doi:10.1016/0308-521X(94)90223-3

Hearn, A., and G. Constable. 1984. Irrigation for crops in a sub-humid environment VII. Evaluation of irrigation strategies for cotton. Irrig. Sci. 5:75–94. doi:10.1007/BF00272547

Hebbar, K., M. Venugopalan, A. Prakash, and P. Aggarwal. 2013. Simulating the impacts of climate change on cotton production in India. Clim. Change 118:701–713. doi:10.1007/s10584-012-0673-4

Heberger, M. 2011. Australia's millennium drought: Impacts and responses. In: P.H. Gleick, editor, The world's water: The biennial report on freshwater resources. New York, Springer. p. 97–125.

Hijioka, Y., E. Lin, J.J. Pereira, R.T. Corlett, X. Cui, G.E. Insarov, et al. 2014. Asia. In: V.R. Barros, et al., editors, Climate change 2014: Impacts, adaptation, and vulnerability. Part B: Regional aspects. Contribution of Working Group II to the Fifth Assessment Report of the Intergovernmental Panel on Climate Change. Cambridge Univ. Press, Cambridge, UK. p. 1327–1370.

Hodges, H.F., F.D. Whisler, S.M. Bridges, K.R. Reddy, and J.M. McKinion. 1998. Simulation in crop management: GOSSYM/COMAX. In: R.M. Peart and R.B. Curry, editors, Agricultural systems modeling and simulation. Marcel Dekker., New York. p. 235–281.

Howden, M., S. Crimp, R. Nelson, I. Jubb, P. Holper, and W. Cai. 2010. Australian agriculture in a climate of change. Paper presented at: Greenhouse 2009 Conference, Perth, Australia. 23–26 March 2009.

Hsu, H., and F. Gale. 2001. Regional shifts in China's cotton production and use In: L. King, editor, Cotton and wool: Situation and outlook yearbook. CWS-2001. Economic Res. Serv., USDA, p. 19–25.

Huang, R., L. Zhou, and W. Chen. 2003. The progresses of recent studies on the variabilities of the East Asian monsoon and their causes. Adv. Atmos. Sci. 20:55–69. doi:10.1007/BF03342050

Intergovernmental Panel on Climate Change (IPCC). 2014. Climate change 2014: Impacts, adaptation, and vulnerability. Part B: Regional Aspects. Contribution of Working Group II to the Fifth Assessment Report of the Intergovernmental Panel on Climate Change. Cambridge Univ. Press, Cambridge, UK.

International Trade Centre (ITC). 2011. Cotton and climate change: Impacts and options to mitigate and adapt. ITC, Geneva, Switzerland. http://www.intracen.org/Cotton-and-Climate-Change-Impacts-and-options-to-mitigate-and-adapt/ (accessed 6 Dec. 2016).

Jones, J.W., G. Hoogenboom, C.H. Porter, K.J. Boote, W.D. Batchelor, L.A. Hunt, et al. 2003. The DSSAT cropping system model. Eur. J. Agron. 18:235–265. doi:10.1016/S1161-0301(02)00107-7

Joseph, P.V., J.K. Eischeid, and R.J. Pyle. 1994. Interannual variability of the onset of the Indian summer monsoon and its association with atmospheric features, El Niño, and sea surface temperature anomalies. J. Clim. 7:81–105. doi:10.1175/1520-0442(1994)007<0081:IVOTOO>2.0.CO;2

Karl, T.R., J.M. Melillo, and T.C. Peterson. 2009. Global climate change impacts in the United States. Cambridge Univ. Press, New York.

Keating, B.A., P.S. Carberry, G.L. Hammer, M.E. Probert, M.J. Robertson, D. Holzworth, et al. 2003. An overview of APSIM, a model designed for farming systems simulation. Eur. J. Agron. 18:267–288. doi:10.1016/S1161-0301(02)00108-9

Kimball, B.A. 1983. Carbon dioxide and agricultural yield: An assemblage and analysis of 430 prior observations. Agron. J. 75:779–788. doi:10.2134/agronj1983.00021962007500050014x

Kirby, M., R. Bark, J. Connor, M.E. Qureshi, and S. Keyworth. 2014. Sustainable irrigation: How did irrigated agriculture in Australia's Murray-Darling Basin adapt in the Millennium Drought? Agric. Water Manage. 145:154–162. doi:10.1016/j.agwat.2014.02.013

Kranthi, K.R. 2009. Challenges and opportunities in cotton production research. Biosafety regulations, implementation and consumer acceptance. International Cotton Advisory Committee, Washington, DC. p. 16–20.

Krishnaswamy, J., S. Vaidyanathan, B. Rajagopalan, M. Bonell, M. Sankaran, R.S. Bhalla, et al. 2014. Non-stationary and non-linear influence of ENSO and Indian Ocean Dipole on the variability of Indian monsoon rainfall and extreme rain events. Clim. Dyn. 45:175–184. doi:10.1007/s00382-014-2288-0

Lau, K.M., and M.T. Li. 1984. The monsoon of East Asia and its global associations–A survey. Bull. Am. Meteorol. Soc. 65:114–125. doi:10.1175/1520-0477(1984)065<0114:TMOEAA>2.0.CO;2

Legler, D., K. Bryant, and J. O'Brien. 1999. Impact of ENSO-related climate anomalies on crop yields in the U.S. Clim. Change 42:351–375. doi:10.1023/A:1005401101129

Lioubimtseva, E., and G.M. Henebry. 2009. Climate and environmental change in arid Central Asia: Impacts, vulnerability, and adaptations. J. Arid Environ. 73:963–977. doi:10.1016/j.jaridenv.2009.04.022

Loka, D.A., and D.M. Oosterhuis. 2012. Water stress and reproductive development in cotton. In: D.M. Oosterhuis and J.T. Cothren, editors, Flowering and fruiting in cotton. Cotton Foundation, Memphis, TN. p. 51–58.

Loka, D.A., D.M. Oosterhuis, and G.L. Ritchie. 2011. Water-deficit stress in cotton. In: D.M. Oosterhuis, editor, Stress physiology in cotton. Cotton Foundation, Memphis, TN. p. 37–72.

Luo, Q., M. Bange, and L. Clancy. 2013. Temperature increase and cotton crop phenology. MODSIM2013, 20th International Congress on Modelling and Simulation, Adelaide, Australia. 1–6 Dec. 2013. Modelling and Simulation Society of Australia and New Zealand.

Luo, Q., M. Bange, and L. Clancy. 2014. Cotton crop phenology in a new temperature regime. Ecol. Modell. 285:22–29. doi:10.1016/j.ecolmodel.2014.04.018

Luo, Q., M. Bange, D. Johnston, and M. Braunack. 2015. Cotton crop water use and water use efficiency in a changing climate. Agric. Ecosyst. Environ. 202:126–134. doi:10.1016/j.agee.2015.01.006

Maheshwari, B.L., K.F. Walker, and T.A. McMahon. 1995. Effects of regulation on the flow regime of the River Murray, Australia. Regul. Rivers Res. Manage. 10:15–38. doi:10.1002/rrr.3450100103

Malcolm, S., E. Marshall, M. Aillery, P. Heisey, M. Livingston, and K. Day-Rubenstein. 2012. Agricultural adaptation to a changing climate: Economic and environmental implications vary by US region. USDA-ERS Economic Research Rep. ERR-136. USDA-ERS, Washington, DC. https://www.ers.usda.gov/publications/pub-details/?pubid=44989 (accessed 6 Dec. 2016).

Marani, A. 2004. Cotton2K Model version 4.0. http://departments.agri.huji.ac.il/plantscience/cotton/ (accessed 4 May 2015).

McRae, D., G. Roth, and M. Bange. 2007. Climate change in cotton catchment communities. A scoping study. Cotton Catchment Communities CRC.

Mearns, L.O., G. Carbone, W. Gao, L. McDaniel, E. Tsevtsinskaya, B. McCarl, et al. 2000. The importance of spatial scale of climate scenarios for regional climate change impacts analysis: Implications for regional climate modeling activities. Paper presented at: Tenth PSU/NCAR Mesoscale User's Workshop, Boulder, CO. 21–22 June 2000. Mesoscale and Microscale Division, National Center for Atmospheric Research, Boulder, CO, USA.

Mondal, P., M. Jain, R.S. DeFries, G.L. Galford, and C. Small. 2015. Sensitivity of crop cover to climate variability: Insights from two Indian agro-ecoregions. J. Environ. Manage. 148:21–30. doi:10.1016/j.jenvman.2014.02.026

Murray-Darling Basin Authority. 2010. Guide to the proposed Basin Plan. Murray-Darling Basin Authority, Canberra City, ACT, Australia.

Murray-Darling Basin Authority. 2014. Towards a healthy, working Murray–Darling Basin. Basin Plan annual report 2013–14. Murray Darling Basin Authority, Canberra City, ACT, Australia.

National Land & Water Resources Audit (NLWRA). 2008. Signposts for Australian agriculture framework: The Australian cotton industry/National Land & Water Resources Audit. Australian Government, National Land & Water Resources Audit, Canberra, Australia.

National Water Commission. 2011. Australian water markets: Trends and drivers 2007–08 to 2010–11. National Water Commission, Canberra, Australia.

Organization for Economic Cooperation and Development/Sahel and West Africa Club (OECD/SWAC). 2007. The strategic importance of cotton production and trade in West Africa. In: Cotton in West Africa: The economic and social stakes. OECD Publishing, Paris, France. p. 37–63. doi:10.1787/9789264025066-en.

Parthasarathy, B., and G.B. Pant. 1985. Seasonal relationships between Indian summer monsoon rainfall and the Southern Oscillation. J. Climatol. 5:369–378. doi:10.1002/joc.3370050404

Patel, H.R., M.M. Lunagaria, V. Pandey, P.K. Sharma, B. Bapuji Rao, and V.U.M. Rao. 2014. El Niño episodes and agricultural productivity in Gujarat. Tech. Bull.15. Department of Agricultural Meteorology, Anand Agricultural University, Anand, India.

Paz, J.O., P. Woli, A. Garcia y Garcia, and G. Hoogenboom. 2012. Cotton yields as influenced by ENSO at different planting dates and spatial aggregation levels. Agric. Syst. 111:45–52. doi:10.1016/j.agsy.2012.05.004

People's Republic of China. 2004. The People's Republic of China initial national communication on climate change. Under the aupices of the United Nations Framework Convention on Climate Change. UNFCCC, Switzerland.

Peterson, T.C., and M.O. Baringer, editors. 2009. State of the climate in 2008. Bull. Am. Meteorol. Soc. 90:S1–S196. doi:10.1175/BAMS-90-8-StateoftheClimate

Peterson, T.C., X. Zhang, M. Brunet-India, and J.L. Vázquez-Aguirre. 2008. Changes in North American extremes derived from daily weather data. J. Geophys. Res. Atmos. 113:D07113. doi:10.1029/2007JD009453

Pettigrew, W.T., J.C. McCarty, Jr., and K.C. Vaughn. 2000. Leaf senescence-like characteristics contribute to cotton's premature photosynthetic decline. Photosynth. Res. 65:187–195. doi:10.1023/A:1006455524955

Piao, S., P. Ciais, Y. Huang, Z. Shen, S. Peng, J. Li, et al. 2010. The impacts of climate change on water resources and agriculture in China. Nature 467:43–51. doi:10.1038/nature09364

Prasad, A., and S. Khan. 2002. Murray-Darling Basin dialogue on water and climate. A synthesis report of the Water and Climate Dialogue at the River Symposium, Brisbane, Australia. 6 Sept. 2002.

Rao, V.N., P. Singh, J. Hansen, T. Giridhara Krishna, and S.K. Krishna Murthy. 2007. Use of ENSO-base seasonal rainfall forecasting for informed cropping decisions by farmers in the SAT India. In: M.K. Sivakumar and J. Hansen, editors, Climate prediction and agriculture. Springer, Berlin. p. 165–180. doi:10.1007/978-3-540-44650-7_17

Reddy, K.R., P.R. Doma, L.O. Mearns, M.Y.L. Boone, H.F. Hodges, A.G. Richardson, et al. 2002. Simulating the impacts of climate change on cotton production in the Mississippi Delta. Clim. Res. 22:271–281. doi:10.3354/cr022271

Reddy, K.R., H.F. Hodges, and B.A. Kimball. 2000. Crop ecosystem responses to climatic change: Cotton. In: K.R. Reddy and H.F. Hodges, editors, Climate change and global crop productivity. CAB International, Wallingford, UK. p. 161–187. doi:10.1079/9780851994390.0161

Reddy, K., H. Hodges, and J. McKinion. 1995a. Carbon dioxide and temperature effects on pima cotton growth. Agric. Ecosyst. Environ. 54:17–29. doi:10.1016/0167-8809(95)00593-H

Reddy, K., H. Hodges, and J. McKinion. 1996. Can cotton crops be sustained in future climates? In: Proc. Beltwide Cotton Conf., Nashville, TN. 9–12 Jan. 1996. Natl. Cotton Counc. Am., Memphis, TN. p. 1189–1196.

Reddy, K.R., S. Koti, G.H. Davidonis, and V.R. Reddy. 2004. Interactive effects of carbon dioxide and nitrogen nutrition on cotton growth, development, yield, and fiber quality. Agron. J. 96:1148–1157. doi:10.2134/agronj2004.1148

Reddy, V.R., K.R. Reddy, and H.F. Hodges. 1995b. Carbon dioxide enrichment and temperature effects on cotton canopy photosynthesis, transpiration, and water-use efficiency. Field Crops Res. 41:13–23. doi:10.1016/0378-4290(94)00104-K

Republic of Uzbekistan. 2008. Second national communication of the Republic of Uzbekistan under the United Nations Framework Convention on Climate Change. UNFCCC, Switzerland.

Richards, Q.D., M.P. Bange, and S.B. Johnston. 2008. HydroLOGIC: An irrigation management system for Australian cotton. Agric. Syst. 98:40–49. doi:10.1016/j.agsy.2008.03.009

Ritchie, J.W., G.Y. Abawi, S.C. Dutta, T.R. Harris, and M. Bange. 2004. Risk management strategies using seasonal climate forecasting in irrigated cotton production: A tale of stochastic dominance. Aust. J. Agric. Resour. Econ. 48:65–93. doi:10.1111/j.1467-8489.2004.t01-1-00230.x

Rodriguez, D., H. Cox, P. deVoil, and B. Power. 2014. A participatory whole farm modelling approach to understand impacts and increase preparedness to climate change in Australia. Agric. Syst. 126:50–61. doi:10.1016/j.agsy.2013.04.003

Rodriguez, D., P. DeVoil, B. Power, and H. Cox. 2011. Adapting to change: More realistic quantification of impacts and better informed adaptation alternatives. Paper presented at: Fifth World Congress of Conservation Agriculture and Third Farming Systems Design Conference, Brisbane, Australia. 26–29 Sept. 2011.

Romero-Lankao, P., J.B. Smith, D.J. Davidson, N.S. Diffenbaugh, P.L. Kinney, P. Kirshen, et al. 2014. North America. In: V.R. Barros, C.B. Field, D.J. Dokken, M.D. Mastrandrea, K.J. Mach, T.E. Bilir, et al., editors, Climate change 2014: Impacts, adaptation, and vulnerability. Part B: Regional aspects. Contribution of Working Group II to the Fifth Assessment Report of the Intergovernmental Panel on Climate Change Cambridge Univ. Press., Cambridge, UK. p. 1439–1498.

Ropelewski, C.F., and M.S. Halpert. 1987. Global and regional scale precipitation patterns associated with the El Niño/Southern Oscillation. Mon. Weather Rev. 115:1606–1626. doi:10.1175/1520-0493(1987)115<1606:GARSPP>2.0.CO;2

Rosenzweig, C., and A. Iglesias. 1998. The use of crop models for international climate change impact assessment. In: G.Y. Tsuji, G. Hoogenboom, and P.K. Thornton, editors, Understanding options for agricultural production. Springer Netherlands, Dordrecht. p. 267–292. doi:10.1007/978-94-017-3624-4_13

Schlenker, W., and M.J. Roberts. 2009. Nonlinear temperature effects indicate severe damages to US crop yields under climate change. Proc. Natl. Acad. Sci. USA 106:15594–15598. doi:10.1073/pnas.0906865106

Selvaraju, R. 2003. Impact of El Niño–southern oscillation on Indian foodgrain production. Int. J. Climatol. 23:187–206. doi:10.1002/joc.869

Sheffield, J., E.F. Wood, and M.L. Roderick. 2012. Little change in global drought over the past 60 years. Nature 491:435–438. doi:10.1038/nature11575

Siddiqui, R., G. Samad, M. Nasir, and H.H. Jalil. 2012. The impact of climate change on major agricultural crops: Evidence from Punjab, Pakistan. Pak. Dev. Rev. 51:261–274.

Singh, B., L. Haley, J. Nightengale, W.H. Kang, C.H. Haigler, and A.S. Holaday. 2005. Long-term night chilling of cotton (*Gossypium hirsutum*) does not result in reduced CO_2 assimilation. Funct. Plant Biol. 32:655–666. doi:10.1071/FP05018

Stockle, C.O., S.A. Martin, and G.S. Campbell. 1994. CropSyst, a cropping systems simulation model: Water/nitrogen budgets and crop yield. Agric. Syst. 46:335–359. doi:10.1016/0308-521X(94)90006-2

Tao, F., M. Yokozawa, Z. Zhang, Y. Hayashi, H. Grassl, and C. Fu. 2004. Variability in climatology and agricultural production in China in association with the East Asian summer monsoon and El Niño southern oscillation. Clim. Res. 28:23–30. doi:10.3354/cr028023

Taylor, K.E., R.J. Stouffer, and G.A. Meehl. 2011. An overview of CMIP5 and the experiment design. Bull. Am. Meteorol. Soc. 93:485–498. doi:10.1175/BAMS-D-11-00094.1

Thakare, H.S., P.K. Shrivastava, and K. Bardhan. 2014. Impact of weather parameters on cotton productivity at Surat (Gujarat), India. J. Appl. Nat. Sci. 6:599–604.

Thorp, K.R., S. Ale, M.P. Bange, E.M. Barnes, G. Hoogenboom, R.J. Lascano, et al. 2014. Development and application of process-based simulation models for cotton production: A review of past, present, and future directions. J. Cotton Sci. 18:10–47.

Traore, B., M. Corbeels, M.T. van Wijk, M.C. Rufino, and K.E. Giller. 2013. Effects of climate variability and climate change on crop production in southern Mali. Eur. J. Agron. 49:115–125. doi:10.1016/j.eja.2013.04.004

van Dijk, A.I.J.M., H.E. Beck, R.S. Crosbie, R.A.M. de Jeu, Y.Y. Liu, G.M. Podger, et al. 2013. The Millennium Drought in southeast Australia (2001–2009): Natural and human causes and implications for water resources, ecosystems, economy, and society. Water Resour. Res. 49. doi:10.1002/wrcr.20123

Waddle, B.A. 1984. Crop growing practices. In: R.J. Kohel and C.F. Lewis, editors, Cotton. ASA–CSSA–SSSA, Madison, WI. p. 233–263.

Wall, G.W., J.S. Amthor, and B.A. Kimball. 1994. COTCO2: A cotton growth simulation model for global change. Agric. For. Meteorol. 70:289–342. doi:10.1016/0168-1923(94)90064-7

Wang, B., and T. Li. 2004. East Asian Monsoon-ENSO interactions. In: C.-P. Chang, editor, East Asian Monsoon. Vol. 2 of World Scientific Series on Asia-Pacific Weather and Climate. World Scientific, Singapore. p. 177–212. doi:10.1142/9789812701411_0005

Wanjura, D.F., and G.L. Barker. 1985. Cotton lint yield accumulation rate and quality development. Field Crops Res. 10:205–218. doi:10.1016/0378-4290(85)90027-9

Whetton, P., K. Hennessy, J. Clarke, K. McInnes, and D. Kent. 2012. Use of Representative Climate Futures in impact and adaptation assessment. Clim. Change 115:433–442. doi:10.1007/s10584-012-0471-z

White, J.W., G. Hoogenboom, B.A. Kimball, and G.W. Wall. 2011. Methodologies for simulating impacts of climate change on crop production. Field Crops Res. 124:357–368. doi:10.1016/j.fcr.2011.07.001

Williams, A., N. White, S. Mushtaq, G. Cockfield, B. Power, and L. Kouadio. 2014. Quantifying the response of cotton production in eastern Australia to climate change. Clim. Change 129:183–196.

World Bank. 2007. The cotton sector of Tajikistan —New opportunities for the international cotton trade. World Bank Group, Washington, DC. http://siteresources.worldbank.org/INTTAJIKISTAN/Resources/MB_300407_E.pdf (accessed 5 Mar. 2015).

Yang, Y., Y. Yang, S. Han, I. Macadam, and D.L. Liu. 2014. Prediction of cotton yield and water demand under climate change and future adaptation measures. Agric. Water Manage. 144:42–53. doi:10.1016/j.agwat.2014.06.001

Yasunari, T. 1991. The monsoon year-A new concept of the climatic year in the tropics. Bull. Am. Meteorol. Soc. 72:1331–1338. doi:10.1175/1520-0477(1991)072<1331:TMYNCO>2.0.CO;2

Yuan, Y., C. Li, and S. Yang. 2014. Decadal anomalies of winter precipitation over southern China in association with El Niño and La Niña. J. Meteorol. Res. 28:91–110.

Zhang, R., A. Sumi, and M. Kimoto. 1996. Impact of El Niño on the East Asian Monsoon. J. Meteorol. Soc. Jpn. 74:49–62.

Zhang, R., A. Sumi, and M. Kimoto. 1999. A diagnostic study of the impact of El Niño on the precipitation in China. Adv. Atmos. Sci. 16:229–241. doi:10.1007/BF02973084

Zhang, T. 2011. Country sector overview: The cotton sector in China. Sustainable Trade Initiative, Utrecht.

Zhao, X., and C. Tisdell. 2009. The sustainability of cotton production in China and in Australia. Working Paper 157. Working papers on Economics, Ecology and the Environment. School of Economics, Univ. of Queensland, Australia. p. 1–34.

Agroclimatology in Grasslands

David H. Cobon,* Walter E. Baethgen, Willem Landman, Allyson Williams, Emma Archer van Garderen, Peter Johnston, Johan Malherbe, Phumzile Maluleke, Ikalafeng Ben Kgakatsi, and Peter Davis

Abstract

Grasslands occupy nearly half the world's ice-free land and provide forage for livestock and native herbivores on almost one-third of the world's ice-free land. They are generally located in drier regions (arid and semiarid) with large diurnal temperature variations and high interannual rainfall variability. Rainfall is a key driver of pasture and livestock production, and managing drought is a common and challenging experience for pastoral managers. Here we examine the agroclimatic association of the worlds grasslands, looking specifically at grasslands in Australia, South America (Uruguay),

---

Abbreviations: ACS, annual cash surplus; AO, Antarctic Oscillation; AVHRR, advanced very high resolution radiometer; BoM, Bureau of Meteorology; CSIR, Council for Scientific and Industrial Research; CSIRO, Commonwealth Scientific and Industrial Research Organization; DAFF, Department of Agriculture, Forestry and Fisheries; ECMWF, European Centre for Medium Range Weather Forecasts; ENSO, El Niño–Southern Oscillation; GRAS, Climate, Environment, and Satellite Agriculture interdisciplinary team (GRAS for its Spanish acronym); IDSS, information and decision support system; IFDC, International Institute for Soil Fertility Management; INIA, National Institute for Agricultural Research; IOD, Indian Ocean Dipole; IRI, International Research Institute for Climate and Society; MCV, Managing Climate Variability; MJO, Madden Julian Oscillation; NDVI, normalized difference vegetation index; PDO, Pacific Decadal Oscillation; POAMA, Predictive Ocean Atmosphere Model for Australia; QBO, Quasi-Biennial Oscillation; SADC, Southern African Development Community; SAM, Southern Annular Mode; SAWS, South African Weather Service; SCF, seasonal climate forecasting; SESA, southeast South America; SOI, Southern Oscillation Index; SST, sea surface temperature; IDSS, information and decision support system; TWG, Technical Working Group.

D.H. Cobon, A. Williams (allyson.williams@usq.edu.au), and P. Davis (peterwdavis85@gmail.com), International Centre for Applied Climate Science, University of Southern Queensland, Toowoomba, Queensland, 4350, Australia; W.E. Baethgen (baethgen@iri.columbia.edu), International Research Institute for Climate and Society, The Earth Institute Columbia Univ., New York, NY 10964; W. Landman (willem.landman@up.ac.za), Dep. of Geography, Geoinformatics and Meteorology, Univ. of Pretoria, Pretoria, 0083, South Africa; E. Archer van Garderen (emma.archervangarderen@wits.ac.za), School of Geography, Archaeology and Environmental Studies, Univ. of the Witwatersrand, Johannesburg, South Africa; P. Johnston (johnston@csag.uct.ac.za), Climate Systems Analysis Group, Dep. Environment and Geographical Science, Univ. of Cape Town, 7701, South Africa; J. Malherbe (jmalherbe@csir.co.za) and P. Maluleke (MalulekeP@arc.agric.za), Agricultural Research Council, Institute for Soil, Climate and Water, 600 Belvedere St., Arcadia, Pretoria, South Africa; and I.B. Kgakatsi (ikalafengk@daff.gov.za), Climate Change and Disaster Management, Dep. of Agriculture, Forestry and Fisheries, 100 Hamilton St., Arcadia, Pretoria, South Africa. *Corresponding author (david.cobon@usq.edu.au).

doi:10.2134/agronmonogr60.2016.0013

© ASA, CSSA, and SSSA, 5585 Guilford Road, Madison, WI 53711, USA.
Agroclimatology: Linking Agriculture to Climate, Agronomy Monograph 60.
Jerry L. Hatfield, Mannava V.K. Sivakumar, John H. Prueger, editors.

and South Africa as examples of important grassland communities. These grazing regions are all affected by more than one climate driver, and the influence of the different drivers on rainfall varies geographically across the world. These climate drivers operate on scales from seasonal to interdecadal, but understanding of the El Niño–Southern Oscillation (ENSO) has provided the most widespread application of climate science into pastoral decision-making. Although ENSO is the most predictable climate driver, the reliability is limited to the austral spring and summer periods in years with strong ENSO anomalies, and even so, the reliability is moderate at best. Nonetheless, information on current conditions and seasonal forecasts have been generated and disseminated on national scales for decades (Australia late 1980s, Uruguay 1997, and South Africa early 1990s) but the uptake by agricultural decision-makers has been modest (one in three in Australia), and use by governments has largely been limited to crisis management during droughts. There has been little evidence and motivation by governments to manage the hydrological and hydro-illogical cycles by preparing pastoralists for drought by implementing strong policy platforms around early warning, preparedness, and national alerts. Despite the modest uptake of seasonal forecasts, there is evidence that their use in decision-making can increase productivity, profitability, and resource sustainability of pastoral enterprises in some parts of the world. These modest adoption rates have been attributed to poor presentation, lack of understanding of terminology, failure to show value and match forecast scale with decision-making at the value chain level, short lead times, poor reliability, and inappropriate dissemination methods. There is evidence of higher adoption rates at the regional scale where a combination of the following arrangements are most likely to contribute to more widespread use of climate forecasts in the future: (i) strong integration and support from institutions that generate and disseminate forecasts, (ii) information is customized for the region and industry, (iii) trust is built between the institutions and decision-makers, (iv) regional climate champions provide support and training in understanding and use of forecasts, and (v) institutions provide a climate service that delivers climate literacy and regionally relevant decision- and discussion-support tools. Incorporating climate science into climate-integrated agricultural models, decision support tools, training, and education and extension packages has made a valuable contribution to agricultural decision-making for governments, institutions, businesses, and pastoralists, but greater understanding is required to improve application and adoption for beneficial outcomes in both developing and developed world grassland communities.

Grasslands (including rangelands) occur naturally on all continents except Antarctica. They use nearly half the world's ice-free land and provide forage for livestock and native herbivores; habitat for native flora and fauna; and water, energy, mineral resources, and reserves for recreation and nature conservation. Rangelands are often distinguished from grasslands because they grow primarily native vegetation, and although grasslands contain native and nonnative pasture species, they are often occupied by improved pasture species that have been established by humans. Rangelands are also managed principally with extensive practices such as managed livestock grazing and prescribed fire rather than more intensive agricultural practices of seeding, irrigation, and the use of fertilizers. Here, when we use the word grassland we also include rangelands. Some of the world's largest expanses of grassland are found in tropical and subtropical Africa and Australia, and temperate and Montane grasslands in China and Tibet and in the Middle East and North and South America (Fig. 1). These grasslands are occupied by subsistence, nomadic, and commercial pastoralists and are grazed by wild and native herbivores and domestic animals, including cattle, sheep, and goats.

Livestock raised in combination with these ecosystems cover almost one-third of the world's ice-free land. Grasslands are a major source of nutrients for the production of livestock, particularly in areas where crop production is not feasible because of unsuitable climate or soil. In more favorable areas, mixed production systems use grassland, fodder crops, and grain feed crops as feed for ruminant (cattle, sheep, and goats) and monogastric (one stomach; mainly chickens and pigs) livestock. A major challenge for the grazing livestock industry is remaining competitive (economically, environmentally, and socially) with intensive or confined livestock industries, because its estimated that 75% of the growth in livestock production between 2003 and 2030 will be in intensive operations with much of this occurring in developing countries in Asia and to a lesser degree in Africa where changes in human population and sociocultural values are increasing the demand for meat as a food source (Thornton, 2010).

Grasslands, and particularly rangelands, often occur in drier areas with mean annual precipitation less than 800 mm and within the extreme ranges of temperature, with mean annual maximum temperatures above 33°C or mean annual minimum temperatures below −5°C (Fig. 2a, 2b, and 2c). As well as occupying areas of climatic extremes, grasslands experience high interannual variability in precipitation (discussed in more detail in the country sections below), which is a key driver of grass and animal production and resource deterioration and recovery (McKeon et al., 2004). In some areas, the understanding of climate science and the application of climate forecasts into agricultural decision-making has made significant improvements in agricultural production, profitability, and resource sustainability (Hammer et al., 2000a). For example, ENSO is a pattern of sea surface temperature (SST) and atmospheric conditions in the Pacific Ocean that affects climate (in different ways) around the world. For some locations and time periods, the ENSO influence on precipitation, temperature, and other climatic parameters is predicable at varying scales for up to 9 mo in advance. The use of

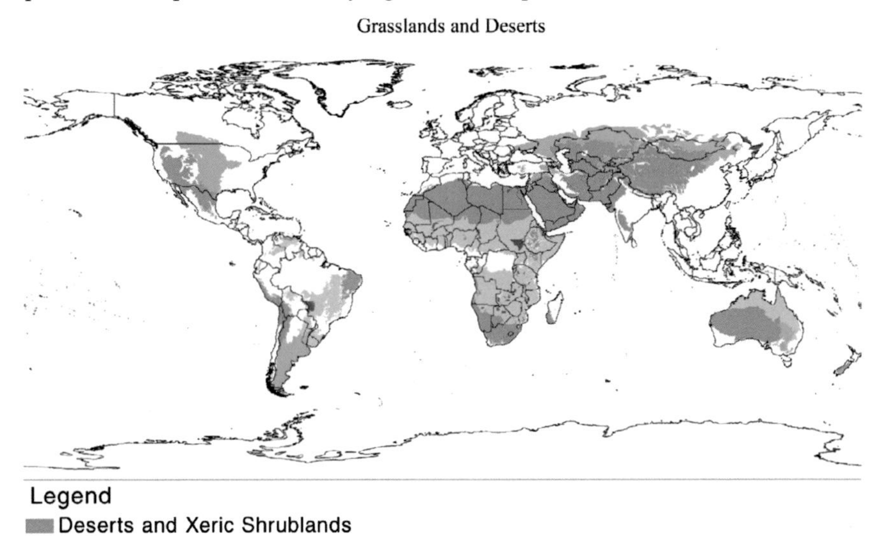

Fig. 1. Global distribution of grasslands.

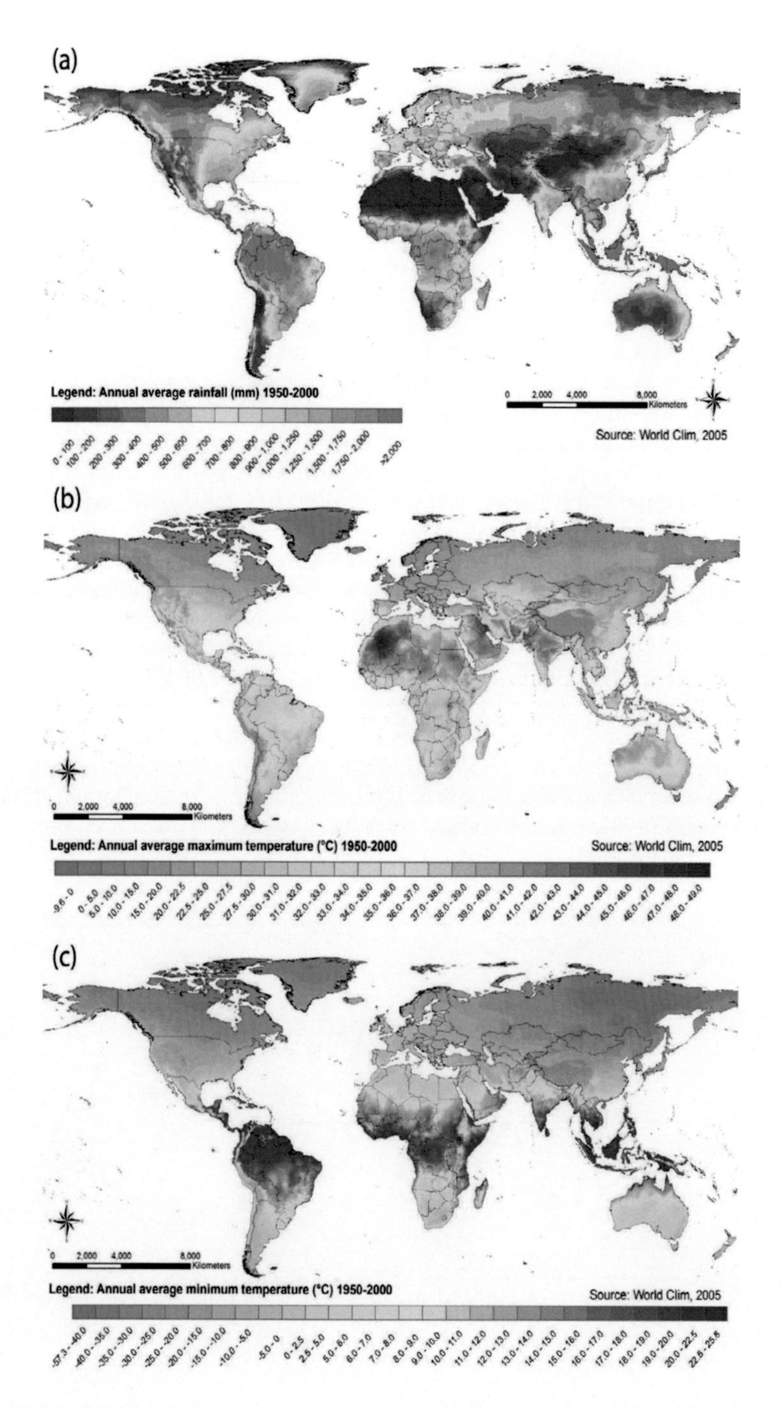

Fig. 2. World annual average rainfall (a), maximum temperature (b), and minimum temperature (c).

this knowledge of climate science and incorporating it into climate-integrated agricultural models, decision support tools, training, education, and extension packages has made a valuable contribution to agricultural decision-making for governments, institutions, businesses, and pastoralists.

In this chapter we focus on agroclimatology in grassland communities in Australia and also examine the use of climate information in Uruguay and South Africa. We discuss the history and problems associated with climate variability, the current forecast operating systems, how climate forecasts have been applied, the value of forecasts in decision-making, and provide some examples in use of forecasts and how the changing climate might influence things in the future. We examine common themes and discuss major lessons learned across these grassland communities to assist with future scenarios in other parts of the world.

Australia

Australia has been colloquially characterized as the land of "drought and flooding rains," yet imagery of a "wide brown land" reminiscent of the pastoral zone and being the driest inhabited continent in the world is deep rooted in the national culture. The pastoral industry covers about 70% of Australia (Australia State of the Environment Committee, 2001) (Fig. 3a), and areas north of about 30°S either have austral warm season dominant rainfall or are in arid regions with low and highly variable annual rainfall. The pastoral areas occupy major components

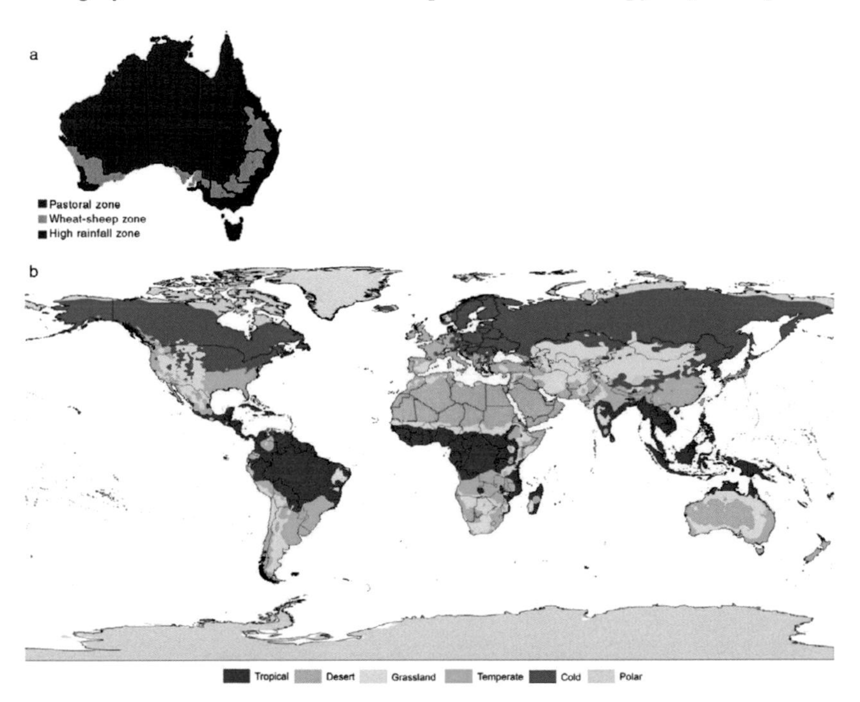

Fig. 3. Key agriculture regions in Australia (a) and the major climate classification groups defined by Köppen (1931) and modified by Geiger (b), seasonal rainfall zones (c), annual rainfall variability (d), and schematic representation of the main drivers of rainfall variability (e) (Australian Bureau of Meteorology).

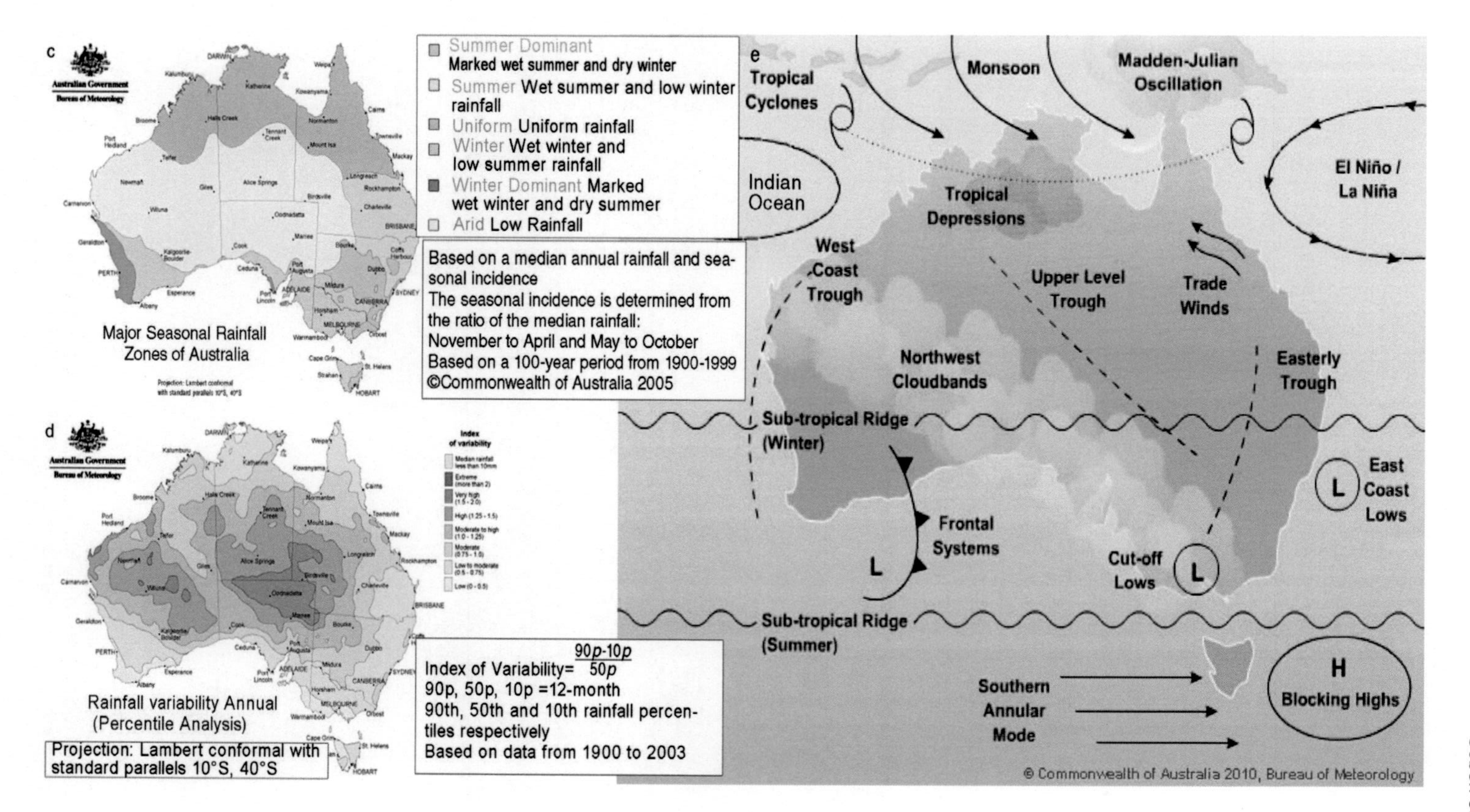
c
Australian Government
Bureau of Meteorology
Major Seasonal Rainfall Zones of Australia
Summer Dominant Marked wet summer and dry winter
Summer Wet summer and low winter rainfall
Uniform Uniform rainfall
Winter Wet winter and low summer rainfall
Winter Dominant Marked wet winter and dry summer
Arid Low Rainfall
Based on a median annual rainfall and seasonal incidence
The seasonal incidence is determined from the ratio of the median rainfall:
November to April and May to October
Based on a 100-year period from 1900-1999
©Commonwealth of Australia 2005
d
Australian Government
Bureau of Meteorology
Rainfall variability Annual (Percentile Analysis)
Projection: Lambert conformal with standard parallels 10°S, 40°S
Index of variability
Index of Variability= 90p-10p / 50p
90p, 50p, 10p =12-month
90th, 50th and 10th rainfall percentiles respectively
Based on data from 1900 to 2003
e
Tropical Cyclones
Monsoon
Madden-Julian Oscillation
El Niño / La Niña
Indian Ocean
Tropical Depressions
West Coast Trough
Upper Level Trough
Trade Winds
Easterly Trough
Northwest Cloudbands
Sub-tropical Ridge (Winter)
Frontal Systems
L
Cut-off Lows
L
East Coast Lows
L
Sub-tropical Ridge (Summer)
Southern Annular Mode
H
Blocking Highs
© Commonwealth of Australia 2010, Bureau of Meteorology

of the equatorial, tropical, grassland, and desert zones (Fig. 3b) (Köppen, 1931), whereas the wheat–sheep and high rainfall zones tend to be associated more with the subtropical and temperate Köppen classifications. The high year-to-year variability in rainfall has long been a major source of risk in pastoral management, financial performance, and resource degradation (White, 1978; Tothill and Gillies, 1992; McKeon et al., 2004).

Grasslands in Australia can be found across most of the continent. However, their productivity and use is highly dependent on the climate, soil type, etc. The following discussion of the climatology of Australia's grasslands is based on the three primary grazing regions: namely, the tropical zone (tropical and subtropical grasslands, savannahs, and shrublands), the arid center (deserts and xeric shrublands), and the temperate zone (temperate grasslands, savannahs, and shrublands). The rainfall characteristics of each of these three regions differs enormously and varies from a summer dominant rainfall zone in the north to a winter dominant zone in the southernmost regions (Fig. 3c). This discussion focuses on rainfall rather than other climate variables, as this is the primary driver of the interannual climate risk for the Australian grazing industry. The following sections discuss the synoptic and climate drivers of rainfall variability, the seasonal forecasting of rainfall, and possible changes driven by global warming in the future.

Climate Variability

The seasonality of the rainfall in these three zones (Fig. 3d) is caused by the occurrence of particular synoptic events.

- The high summer rainfall and low winter rainfall in the tropical region is caused by the position of the intertropical convergence zone as it moves southward toward the Australian land mass, causing the Australian monsoon trough (Nicholls et al., 1982) and associated mesoscale convection, trade winds caused by the subtropical ridge, heat lows, and other synoptic events such as tropical cyclones.
- Rainfall events in the arid region occur mostly in summer and are, as in themore tropical zone, caused by southerly incursions of the monsoon trough and, in addition, northwest cloud bands and upper-level low-pressure systems.
- In the temperate regions the key rain-bearing synoptic systems are the low-pressure and frontal systems, the east coast cyclones, and atmospheric-blocking and the resultant cutoff lows.

The interannual variability of rainfall in these three grazing zones is caused by variations in the strength and frequency of these synoptic-scale events (Fig. 3e). The variability as measured by the coefficient of variability is generally greatest in the more arid regions and also in the northeast of the continent. The interannual rainfall and synoptic patterns have been linked to recurring low-frequency larger scale processes in the ocean and atmosphere: the so-called "climate drivers." These climate drivers operate on scales from seasonal to interdecadal. They include ENSO (Allan et al., 1996; Philander, 1990), the Pacific Decadal Oscillation (PDO) (Power et al., 1999), the Indian Ocean Dipole (IOD) (Saji et al., 1999; Nicholls, 1989), the Southern Annular Mode (SAM) (Thompson and Wallace, 2000; Hall and Visbeck, 2002; Hendon et al., 2007), also known as the Antarctic Oscillation (AO), fluctuations in the longwave pattern (Risbey et al., 2009), fluctuations in the latitude and intensity of the subtropical ridge (Williams and Stone, 2009), and anthropogenic global warming (Dai, 2011; Nicholls, 2004). Other more subseasonal-scale systems include the Madden Julian Oscillation (Madden and Julian, 1971; Wheeler et al., 2009), which although it clearly has a strong influence on the onset and the active phase of the northern Australian wet season, tropical cyclone genesis, and extratropical synoptics, its influence is at intraseasonal time scales and therefore is not discussed further.

The effects of these large-scale drivers on rainfall vary depending on the specific region to be considered and also the specific season. The dominant driver of rainfall extremes, particularly in northern and eastern Australia, is ENSO, a quasi-periodic (between 2 and 7 yr cycles), coupled ocean–atmospheric system which is driven by variation in SSTs and atmospheric pressures in the central Pacific Ocean between the Asian and east Pacific regions (Nicholls, 1988; Stone et al., 1996). An El Niño (La Niña) event occurs when SSTs in the key region of the Pacific Ocean are anomalously warm (cool). A typical El Niño often results in below-average rainfall anomalies in Australia (Nicholls et al., 1996) with the reverse during La Niña (Allan, 2000; Lough, 2007). The ENSO has the strongest effect on winter and spring rainfall in eastern and northern Australia, extending through summer in the monsoonal north (Risbey et al., 2009). In these regions, ENSO accounts for between 30 and 50% of the variability. This is, however, a generalized relationship; there are many nuances to the rainfall–ENSO relationship. For example, large El Niño events have not always produced droughts and often produce quite different rainfall patterns to other El Niño events (Brown et al., 2009; Wang and Hendon, 2007). In addition to nonlinear responses of rainfall to ENSO (and other drivers), there are clearly complex interactions between ENSO and other drivers of drought across Australia (Nicholls, 1989, Meyers et al., 2007). The relationship between rainfall and ENSO varies on a multidecadal timescale too (e.g., Power et al., 2006; Nicholls et al., 1996), with greater coherency of Australian rainfall with ENSO apparent during cool phases of the IPO-PDO (Lough 2007, McKeon et al., 2004).

Australian rainfall is also affected by SSTs in the Indian Ocean (Ashok et al., 2003), as demonstrated by relationships between the IOD and rainfall (Cai et al., 2012, 2014a). A positive phase of the IOD (cooler SSTs in the north eastern Indian Ocean and warmer tropical SSTs in the west) is associated with lower than average rainfall (Meyers et al., 2007). The IOD primarily affects rainfall in the southern and western regions of the country (Ashok et al., 2003; Risbey et al., 2009); however, when combined with an ENSO event, these rainfall anomalies extend further east. Unlike the response of rainfall to the La Niña and El Niño extremes of the ENSO cycle, which is generally spatially similar, the spatial pattern is markedly different in positive and negative modes of the IOD. As is the case with other drivers, the relationship between climate and the IOD is nonlinear.

The three key grazing regions are all affected by more than one driver, and the influence of the different drivers on rainfall varies geographically across Australia. The drivers are not independent to each other, and the teleconnections between the drivers will intensify or dampen their individual impact on climate. The interrelationships among these climate modes are complex and the focus of many studies (e.g., Ashok et al., 2003; Meyers et al., 2007; Ummenhofer et al., 2009); however, several authors have proposed teleconnections among the IOD, ENSO, and SAM (e.g., Behera et al., 2006; Carvalho et al., 2005; England et al., 2006; L'Heureux and Thompson 2006; Ding et al., 2012), and ENSO, in particular, is especially highly correlated with the IOD (Cai et al., 2012; Risbey et al., 2009). The ENSO system operates in tandem with variations in the latitude of the subtropical ridge over eastern Australia with changes in the Southern Hemisphere Hadley cell and meridional circulation and the AO Index (Williams and Stone, 2009). Such interactions can make it difficult to identify the impact of individual drivers on rainfall variability and drought (e.g., Verdon-Kidd and Kiem 2009; Ummenhofer et al., 2009; Cai et al., 2014b).

It is important to understand the drivers of rainfall variability and rainfall because, for example, the climatic characteristics of a drought is partially determined by the climate driver forcing the drought (Verdon-Kidd and Kiem, 2009). Such characteristics include the seasonality of the rainfall deficits and the temperature changes, and the pattern of daily rainfall extremes. These will have impacts on patterns of subsequent runoff or pasture growth, for example, and hence the nature of the drought itself.

Current Climate Forecast Systems

The development of seasonal climate forecasting (SCF) in Australia has occurred primarily through the National Climate Centre in the Australian Bureau of Meteorology (BoM), the Commonwealth Scientific and Industrial Research Organization (CSIRO), various universities, and State Government departments. There are a range of SCF systems used in Australia (Stone et al., 1996; Alves et al., 2003), and these have developed significantly since the mid-1990s as our knowledge of climate systems has improved. In addition, the development of SCFs have been increasingly driven by agricultural industry bodies hungry for risk-reducing climate outlooks (Australian Government Department of Agriculture Fisheries and Forestry, 2004; Keogh et al., 2004a, 2004b, 2005; Hammer et al., 1996; Meinke and Hammer, 1997; Everingham et al., 2003; Abawi et al., 2000; Ash et al., 2000; Johnston et al., 2000; McKeon et al., 2000; Stafford Smith et al., 2000; McIntosh et al., 2005). Currently, all operational SCFs of rainfall are published with lead times of zero months and are probabilistic forecasts of total 3-mo rainfall amounts.

To skillfully forecast seasonal rainfall in Australia, a SCF model needs to successfully capture the influence of ENSO, which is the main influence on the rainfall in Australia. This has been achieved successfully by statistical and dynamical forecasting models. Here we discuss two operational models commonly used by the grazing industry to predict 3-mo rainfall totals with a lead time of zero to 1 mo.

The Australian BoM and the CSIRO have developed a dynamical SCF system, the Predictive Ocean Atmosphere Model for Australia (POAMA) (Alves et al., 2003). This is a global coupled ocean–atmosphere multimodel ensemble prediction system producing operational forecasts across Australia since 2002 (Lim et al., 2011; Wang et al., 2008; Hudson et al., 2013) and it is continuously undergoing improvements: as of 2013 the operational version is POAMA-2. POAMA has three key components: the coupling of the CSIRO Australian Community Ocean Model version 2 ocean model (Schiller et al., 2002) with the BoM's Atmospheric Model version 3.0 (Colman et al., 2005; Wang et al., 2005; Zhong et al., 2006); a data assimilation system for the initialization of the ocean, land, and atmosphere components (Hudson et al., 2011); and a system addressing forecast uncertainty via generating ensembles (Wang et al., 2011; Hudson et al., 2013). The model is continuously improving through developments in various components and data assimilation, and through incorporating POAMA with other operationally available models into a multimodel ensemble, which would increase the consistency and reliability of SCFs (Langford and Hendon 2013).

In addition to producing probabilistic forecasts of rainfall and temperature, POAMA is also used to predict SSTs in the key regions of the central and eastern tropical Pacific (Wang et al., 2008; Zhao and Hendon 2009), sea level and rainfall in the Pacific (Cottrill et al., 2013), and the onset of the North Australian wet season (Drosdowsky and Wheeler, 2014).

In conjunction with dynamical forecasting models, there are also a range of statistical SCF models utilized in Australia. The Southern Oscillation Index (SOI) was shown by McBride and Nicholls (1983) to be a predictor of spring and summer rainfall over much of the eastern half of Australia. The SOI–rainfall relationships have been assessed for forecasting skill in numerous studies identifying statistical methodologies for forecasting seasonal rainfall (e.g., Lo et al., 2007). As well as the simple linear, lagged relationships between SOI and rainfall and temperature, which formed the base of the BoMs seasonal outlooks since 1989, there are also more complex approaches, such as the SOI phase system (Stone et al., 1996; Stone and Auliciems, 1992), which identifies more subtle patterns of ENSO. The SOI phase system has been used operationally by the Queensland Government since 1997 to assess seasonal climate risk for a range of agricultural industries, including grain (Hammer et al., 1996), peanut (Meinke and Hammer, 1997), sugar (Everingham et al., 2003), water (Abawi et al., 2000), and pastoral industries (Cobon, 1999; Cobon and Toombs, 2013). This system has also led to improvements in forecast quality in a number of regions around the world affected by ENSO (Stone et al., 1996).

Both the statistical models and POAMA have the highest forecast skill in spring (September-October-November) (Langford and Hendon, 2013). This reflects the dominant influence of ENSO, the most predictable climate driver, especially in spring (McBride and Nicholls, 1983). The power of ENSO in spring is also reflected in the ability of POAMA to forecast skillfully with longer lead times of up to 4 mo (Langford and Hendon, 2013).

Major Problems

Forecast systems

Seasonal climate forecasts are used by 30 to 50% of agricultural producers in decision-making (Keogh et al., 2005; Keogh et al., 2004a; Australian Government Department of Agriculture Fisheries and Forestry, 2004). Desired levels of forecast reliability, forecast presentation style, use of terminology, proof of value, access to expertise, and perceptions regarding climate information have previously been identified as factors limiting the uptake and use of climate forecasts (Childs et al., 1991; Changnon et al., 1995; Nicholls 1999; Hartmann et al., 2002; McCrea et al., 2005).

Forecast reliability

Fluctuations in the Pacific Ocean SSTs at interannual and multiyear timescales have been identified as major contributors to year-to-year rainfall variability in the pastoral zone. This high variability in Australian rainfall is partly associated with ENSO, with El Niño referring to warmer than normal water in the central Pacific, resulting in drier conditions in eastern Australia. Alternatively, La Niña refers to cooler than normal water and wetter than normal conditions.

Australian rainfall in some seasons and areas is significantly correlated with SSTs and the SOI in the same or preceding seasons, but the correlations, although significant, are relatively low (Pittock, 1975; McBride and Nicholls, 1983; Chiew et al., 1998). To help manage this, calculations of forecast quality that show the skill of a forecast for a particular place and time are very useful, particularly for producers pondering high-risk decisions. Unfortunately, most of the operational forecasts are issued purely as national or state-scale maps showing chance of exceeding median rainfall without an accompanying map showing where the forecast has skill, which fails to alert the decision maker to a forecast that may have little useful application. These shortcomings in presentation of general forecast information to the public can lead to poor decision-making and lack of confidence in climate forecasting. On the other hand, targeted forecast campaigns that customize a forecast for the region and management system that are supported by professional and local champions (advocates and users) generate better understanding, greater confidence, and higher adoption rates by the decision-makers (Cobon et al., 2005; Cobon et al., 2008; Cliffe et al., 2016).

Since 1890 about 30% of the years have been El Niño and 30% La Niña, and the remainder are classed as neutral years. Unfortunately, the need to shorten and simplify our communication has led to a simplistic association of El Niño and drought and La Niña and flood, although history shows the opposite to occur in some years (e.g., 2009-2010, 1939-1940, and 1947-1948, 1964-1965). Extremes of rainfall have also occurred in neutral years with severe droughts (e.g., 8 yr between 1930 and 1938, 1984 to 1986, and 1959 to 1961) and floods (e.g., 1978-1979, 1983-1984, and 1990-1991) occurring during neutral years. Furthermore, in a study of nine degradation events in Australian rangelands between 1890 and 2002, 75% of the years in these extended drought periods were neutral years, five started in neutral years and seven finished with a neutral year type (McKeon et al., 2004). Forecasting of regional rainfall in neutral years with reasonable skill represents a major research challenge, and it may be that because the climate system is inherently chaotic by nature that events of this kind are unpredictable at a regional scale.

The current operational forecast systems vary spatially and temporally in their capacity to accurately predict forthcoming rainfall events. However, in general, ENSO-based systems are most useful in ENSO years (La Niña or El Niño), in the late austral winter or spring months (July to November), when issued with zero or short lead times, and in regions where the ENSO signal is strong. Knowing and understanding these limitations is important in decision-making, and tools that show the skill of a forecast are not widely used by forecasters nor are they understood by pastoralists.

Another limitation is that most forecasts are issued as probabilities that reflect the uncertain and nondeterministic nature of the climate system, but many pastoralists do not fully understand probabilities and other terms such as median (Coventry 2001; Cobon et al., 2008; Dalgleish et al., 2001; Keogh et al., 2004a, 2004b, 2005). For example, in regional western Australia (Gascoyne Murchison), 80% of pastoralists demonstrated they had a poor technical understanding of how to interpret the standard wording of a probabilistic median rainfall forecast (Keogh et al., 2005). This creates the potential for misunderstanding and misinterpretation, and highlights the need for providers of forecasts to simplify or provide further support to decision-makers. The challenge of educating producers exists because these terms and concepts are difficult to understand (and apply) (Tversky and Kahneman, 1974), and repeated tuition and support is often needed. Having local champions with climate science and applications training working in the regions has proved successful in increasing the use of seasonal forecasts in decision-making on pastoral enterprises (Cobon et al., 2008).

The operational rainfall forecasts are issued each month for the next 3-mo period (i.e., zero lead time). This rolling 3-mo forecast at zero lead time makes it difficult for pastoralists in Australia managing large properties to implement key decisions based on the forecast when the lag between the predictor and predictand is zero. For example, mustering on large northern beef properties may occur only twice a year, and on both occasions takes months to complete. Therefore to adjust stock numbers commensurate with a seasonal rainfall forecast requires some months of advanced warning. In addition, decisions related to stock numbers are made and implemented in the dry season, as access to the properties is limited during the wet season in the austral summer.

Several surveys of pastoralists in northern Australian pastoral regions showed they needed longer lead times and indicators of forecast quality for the forecasts to be useful (Keogh et al., 2005; Park et al., 2004; Paull, 2004; Ash et al., 2000). These surveys showed that forecasts for the austral warm season (November–March) issued first in June by using the April-May SOI phase and reissued each month for the same forecast period counting down from five to zero lead time would be most useful for application in management in these regions. These forecasts that target a particular period and are issued with lead times that countdown to the forecast period have been tested by using the SOI phase system and have been found to contain useful forecast quality in northern Australia and particularly northeastern Australia (Cobon and Toombs, 2013).

Information transfer

Seasonal climate predictions have been issued in Australia since the late 1980s, so the uptake of forecasts into agricultural decision-making on a national scale is relatively modest (30–40%), yet access to climate information is widespread on the internet and regularly reported on television news. On the other hand, in regions with access to local climate champions, adoption of seasonal forecasts into management decisions is much higher (75%) (Cobon et al., 2008; Cliffe et al., 2016), mainly because they deliver a climate service and not merely climate information. Local climate champions provide interaction between climate forecasters and pastoralists, they customize climate services for the local scale and to the local management and production cycles, they communicate face-to-face and provide support and understanding of climate terminology (climate literacy), and collectively the service they provide gives pastoralists the confidence to use climate forecasts in decision-making.

The remoteness and long distances to local towns in pastoral Australia can make traveling to workshops time-consuming and costly, so learning remotely can be more popular in these areas. Some of the reasons for this are discussed above. To facilitate wider use of climate information in remote regions, Australian CliMate was developed as a climate analysis tool for iOS devices. The App allows for interrogation and analysis of climate records (over the last 60 yr) relating to rainfall, temperature, and radiation. It has proved popular with agricultural decision-makers.

Decision-making

High interannual variability in rainfall is most likely to be responsible for extended periods of drought, some having made a particular impression in Australia (Lindesay, 2005).

Along with the notion of "drought and flooding rains" comes a more scientific description termed the "hydrological cycle" that outlines the cycle of wet to dry to wet. The timing of the cycle varies considerably, but unfortunately drought is too often considered to be an "uncommon visitor" or "rare and random event" (Wilhite, 2002, 2011, 2014); at some levels this has led to crisis rather than a risk management approach to drought management, akin to the hydro-illogical cycle (Wilhite, 1993). In the past when drought is declared, governments have responded by providing financial assistance to pastoralists, but little attention has been paid to preparedness, early warning, and national alerts when seasonal conditions are favorable.

The state of Queensland is currently (January 2017) 80% drought declared, and some pastoralists are now in their fourth year of drought. The economic, social, and resource impacts of events like this are devastating for rural communities, but unfortunately the crisis management approach of the governments are unlikely to be effective in reducing the risk associated with the next drought, since the recipients of the assistance are not expected to change behavior or enterprise management practices as a condition of the assistance. The reliance on the government for relief is contrary to encouraging self-reliance by producers investing in their own "drought-proofing" strategies. The drought risk management approach that the government has raised in the green paper (discussed below) through the application of preparedness and mitigation measures is an important step in increasing the self-reliance of pastoralists and reducing the reliance of government in drought relief.

Some pastoralists implement their own drought management strategies, such as (i) allocating livestock numbers to pastures at a safe (or low) carrying capacity so they consume only 15% of the pasture grown annually (Johnston et al., 1996), which leaves carryover pastures for use if a drought develops in subsequent years; (ii) having properties in different regions and transporting livestock from drought-affected areas to those less drought affected (iii); having a larger proportion of the herd–flock as castrate males (steers and wethers) compared with breeding females improves the ease and timeliness of forced livestock sales; (iv) having properties in the floodplain areas allows for pasture production from both rainfall and flooding from upstream; and (v) intensive feed lotting of cattle in or near cropping regions provides an option to destock pastures. These examples of drought management strategies used by pastoralists have varying levels of success in different droughts, but they lack the capacity to be effective during long multiyear droughts.

Under the current drought criteria for both declaration and revocation, 20% of Queensland has on average been officially drought declared since 1964 (Day et al., 2003). For those that live in rural areas, drought causes substantial economic and social pressures and, in some cases, deterioration of the resource that prolongs recovery (McKeon et al., 2004). In some circumstances, droughts can be part of a "triple whammy" where they align with low beef and wool prices at market (e.g., 1991–1995), which exacerbates the economic and social hardship of drought on pastoralists and rural communities.

Drought has been recognized as an important factor in divorce, suicide, and illness in pastoral areas (Munro and Lembit, 1997), and subsequently the Australian Government commissioned a review on the social impacts of drought, the first that exclusively investigated the issue (Drought Policy Review Expert Social Panel, 2008). The panel found that "people should be the priority (and not the farm property or the respective industry)," as they were "deeply concerned by the extent of distress in drought-affected communities in rural Australia. Too many farm decisions are made under stress and without adequate consideration of the needs of the family and in the absence of prior thought and planning. Family and business are intricately linked for the majority of farm families, but decision-making mostly occurs in separation and often at the expense of each other."

To recognize the importance of the social aspects in drought, we have developed a hydro-psychological cycle (Fig. 4) that shows the emotions, feelings, anxiety, and pressures of people that are associated with the hydrological and hydro-illogical cycles. Further recognition of the importance of people in drought was shown by the Australian Government by introducing the Farm Household Allowance in July 2014 to provide income support to families in financial hardship, including those suffering loss of income due to drought. Late in 2014, the Australian Government released a green paper that included increasing drought preparedness and in drought support as policy ideas. Initiatives under the

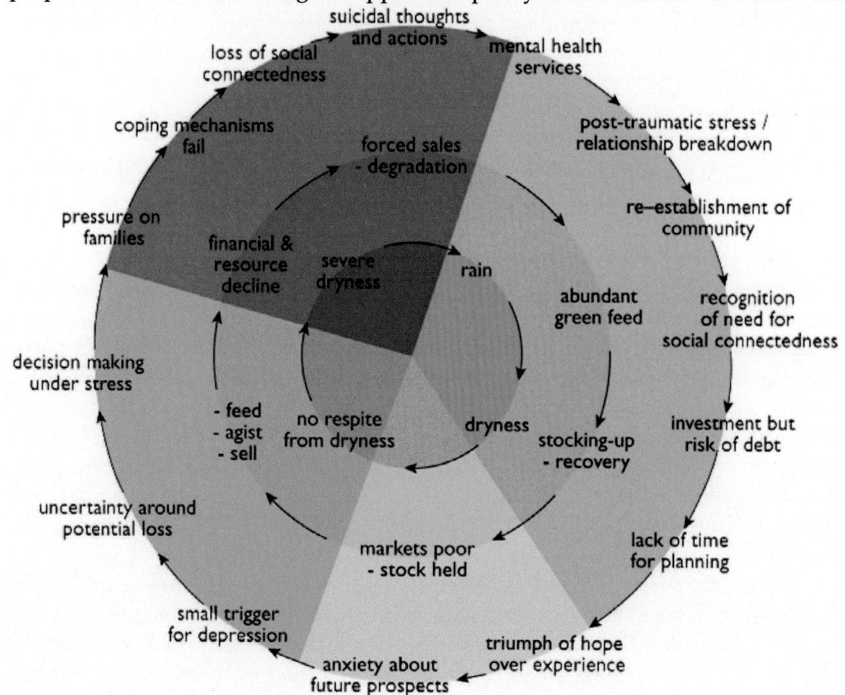

Fig. 4. The hydro cycles. Relationship between the hydrological (inner), hydro-illogical (middle), and hydro-psychological (outer) cycles and wet (green) and drying conditions (yellow-red).

drought preparedness for pastoralists included accelerated depreciation for new water and fodder infrastructure (e.g., hay sheds) and improving climate information. Providing incentives to pastoralists for fodder infrastructure in and near regions where crops are produced may be useful, but storing large amounts of native pasture hay as a fodder reserve for animals during drought on the large and extensive grazing properties that exist across vast areas in Australia is mostly unpractical, uneconomic, and in many cases will not provide more than survival feed for a small proportion of the normal herd–flock. In respect to improving climate information, the Bureau of Meteorology and other providers need to improve accuracy of seasonal forecasts and provide longer lead times and customized seasonal forecast systems as part of a climate service that tailors the climate information and decision support tools to the region and management system. Unfortunately the green paper says nothing about government-supported monitoring of key indicators, early warning systems, national drought alerts, or incentives to pastoralists for including drought in property planning and using sustainable stocking rate practices (e.g., low utilization of pasture or safe carrying capacity) (Johnston et al., 1996) that could reduce future impacts and lessen the need for government intervention in the future.

How Have Climate Forecasts Been Applied?

Climate forecasts are used to increase the chance of success or contribute to greater confidence in adopting new approaches to management, but their application in Australian pastoralism depends on many things, including location, enterprise (beef, sheep, and wool), and the nature of the management cycle. There are many decisions that can be changed based on a climate forecast (Table 1), but matching the timescale of decision with that of the forecast is important. For example, cutting and baling hay requires a forecast for the next few days, adjusting stocking rates for the next few months, but erecting infrastructure might require an outlook over the next few years. These three examples could be addressed by a short-term weather forecast, a seasonal forecast (e.g., SST and SOI), and a multiyear assessment (e.g., Quasi-Biennial Oscillation and PDO), respectively. Each forecast is characterized by different terminology, presentation methods, and

Table 1. Decisions pastoralists (beef, sheep, and wool) make using weather and climate forecasts.

Decisions		
Short-term (tactical); days, weeks, months; 7-10 days; MJO†	Long term (strategic); season, multiseason, multiyears; SOI, SST, QBO, PDO, POAMA, ECMWF†	Most useful climate forecasts
stocking rate management and adjustment (i.e., buy, breed, wean, sell, agistment), pasture burning, time of joining, animal supplementation, shearing, labor, helicopter hire, holidays	enterprise mix, water infrastructure, fencing, herd–flock dynamics, pasture improvement, woody weed and regrowth control, drought preparation (hay storage, feeding stations), herd reduction, markets–contracts, property, investment–divestment	Rainfall, model outputs of pasture growth

† MJO, Madden Julian Oscillation; SOI, Southern Oscillation Index; SST, sea surface temperature; QBO, Quasi-Biennial Oscillation; PDO, Pacific Decadal Oscillation; POAMA, Predictive Ocean Atmosphere Model for Australia; ECMWF, European Centre for Medium Range Weather Forecasts.

levels of accuracy, and it is important for the pastoralist to understand the pros and cons when making the decisions. Along with rainfall, stocking rate is an important driver of animal productivity in the pastoral zone, and because of this, making adjustments to stocking rate (e.g., buying, selling, and agistment–paying another landholder to run your livestock on their land) is a common means of using seasonal forecasts. An example of how pastoralists running large beef cattle and Merino sheep enterprises in northwestern Australia combine forecasts of rainfall in winter and summer with their annual management cycle is described by Keogh et al. (2005).

AussieGRASS is a tool developed in the early 1990s by the Queensland and Commonwealth governments to monitor key biophysical processes associated with pasture degradation and recovery, provide early warning of imbalances between livestock numbers and forage supply, an assessment of pasture curing and risk of fires and seasonal forecasts, and skill assessments of pasture growth (Fig. 5) (Carter et al., 2000, 2003).

AussieGRASS is a spatial implementation of the Grass Production model (GRASP) (Littleboy and McKeon, 1997). The data is run on a daily time-step and spatially interpolated to construct gridded data sets on a regular 0.05° by 0.05° grid (approximately 5 by 5 km) across Australia. The AussieGRASS spatial framework includes inputs of key climate variables (rainfall, evaporation, temperature, vapor pressure, and solar radiation), soil and pasture types, tree and shrub cover, domestic livestock, and other herbivore numbers. AussieGRASS continues to be a valuable tool for governments for drought assessment, rural fire authorities for bushfire hazard, and researchers and pastoralists for seasonal forecasts.

Value and adoption rates of climate forecasts

Across Australia seasonal climate forecasts are used by 30 to 50% of agricultural producers in decision-making (Keogh et al., 2005, 2004a; Australian Government Department of Agriculture Fisheries and Forestry, 2004; White 2001). Poor technical understanding (Keogh et al., 2005; Cobon et al., 2008) and social reasons (Marshall et al., 2011) have been given for low adoption rates in some regions. In regions where local climate champions provide a climate service, adoption of seasonal forecasts into management decisions can be much higher (75%) (Cobon et al., 2008; Cliffe et al., 2016). In regional western Australia (Gascoyne Murchison) more than 50% of pastoralists regularly access and use weather and climate forecasts and almost 75% considered climate information or tools useful (Keogh et al., 2005). However, in a shire in northeast Australia (Dalrymple Shire) the percentage of pastoralists using the SOI as a management tool was 8%, yet 70% of them would use the tool if greater reliability was available (Ash et al., 2000). As dynamic models of the climate and ocean system show promise for increasing the lead time and reliability of forecasts (Hunt and Hirst, 2000; Nicholls, 2000), there are greater prospects for higher adoption rates in future years.

The only grazing trial evidence of the value of seasonal forecasts has been conducted in northeastern Australia with beef cattle (O'Reagain et al., 2009, 2011). Here, stocking rates were adjusted over 15 yr based on an end-of-dry-season (October) standing pasture and SOI-based climate forecast for the forthcoming wet season (November–March). Over the 15 yr, the profitability was slightly higher and pasture condition significantly poorer relative to the constant strategy. The poor pasture condition occurred because of the carryover of high stocking

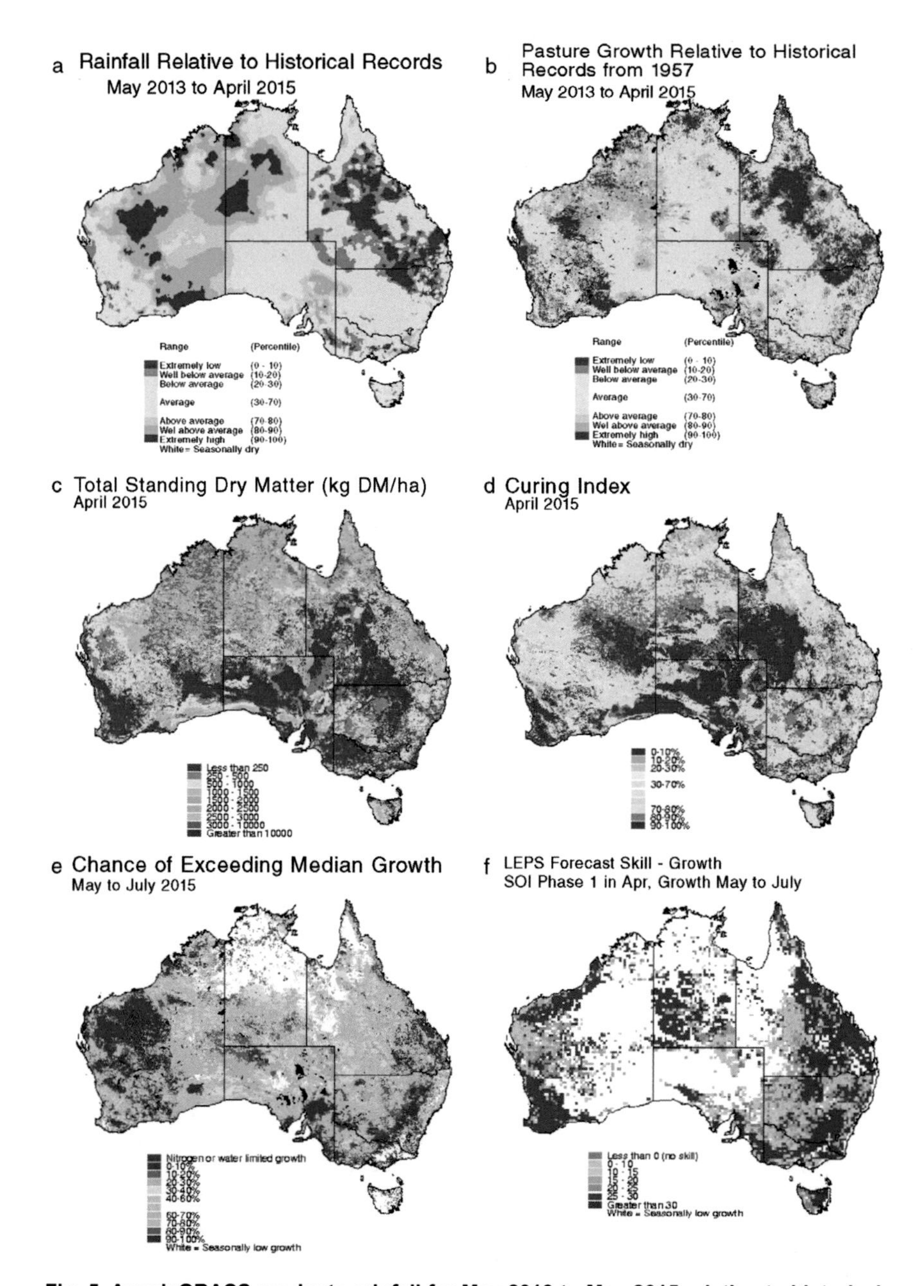

Fig. 5. AussieGRASS products rainfall for May 2013 to May 2015 relative to historical records (a) from 1890 pasture growth for May 2013 to May 2015 relative to historical records (b), from 1957 total standing dry matter from pastures in April 2015 (c) pasture curing index in April 2015 (d), probabilistic forecast provided in April 2015 of the chance of exceeding median pasture growth for May-July 2015 (e), and assessment of the skill of the forecast in d given as Linear Error in Probability Space (LEPS) for a negative SOI phase in April (f). (source: https://www.longpaddock.qld.gov.au/about/researchprojects/aussiegrass/index.html).

rates after a sequence of wet years. The interannual variability of stocking rates varied threefold in the 15 yr, which makes these large adjustments in stock numbers difficult on commercial properties.

The statistical climate forecast systems (Lag SOI, SOI phase, and SST phase) in Australia have now been integrated into pastoral simulation models to aid in the development of on-property decision-making (Hammer, 2000; Stone and deHoedt, 2000). Adjusting grazing pressure is a key grazing management decision that can have a major impact on resource condition, pasture yield and composition, soil loss, burning opportunities, growth of woody weeds, and profitability. There are many options for managing grazing pressure (Stafford Smith, 1992; Johnston et al., 2000), including using a seasonal forecast to adjust stocking rates, but no data is available to formally compare the effects of these options on production and resource status. Because of this simulation modeling, using linked models to simulate pasture growth, resource condition, soil loss, burning opportunities, liveweight change (GRASP), herd dynamics, and property economics (Herd-Econ) enables assessment of whole enterprise management of beef cattle (McKeon et al., 2000; Ash et al., 2000; Stafford Smith et al., 2000) and Merino sheep (Cobon and McKeon, 2002) in pastoral Queensland by using over a century of weather data.

McKeon et al. (2000) highlighted the potential value of achieving in June the skill from seasonal forecasting that was available in November by using average SOI in the August to October period as the indicator of season type from November to May. The simulation studies showed that a flexible beef cattle stocking rate strategy implemented in June using a forecast of next year's pasture growth would increase liveweight gain per hectare by 10%, reduce risk of liveweight loss by 57%, reduce risk of low pasture yield, but would slightly increase the risk of soil loss (4%).

Ash et al. (2000) found a modest but significant increase in beef production, but the responses varied with forecasting method; more benefit was gained if forecasts were available in June rather than November, relative value was greater for constant compared with flexible grazing strategies, increased animal production was not at the expense of the resource base, and if increased animal production was not the desired aim then significant reductions in soil loss can be achieved.

Stafford Smith et al. (2000) found that the level at which beef cattle stocking rate was adjusted in response to a forecast had more impact on cash flow than the forecast, more benefit in cash flow from a 12-mo forecast than a 6-mo forecast, the same level of cash flow could be achieved for a much lower risk of resource degradation (measured by soil loss) with forecasting, the benefits found in production per unit area (e.g., McKeon et al., 2000; Ash et al., 2000) do not translate to economic output at the whole enterprise level in a simple way, the outcomes were sensitive to market prices, trading strategies were relatively more favored over the constant stocking rate strategy as sale prices rose and especially as the margins between sale and purchase price increased, and higher stocking rates were favored at higher prices.

Cobon and McKeon (2002) found the annual cash surplus (ACS) in a wool growing enterprise of the best two forecast systems to be worth more than the constant ($1.40/ha) and flexible ($0.60/ha) grazing strategies, the forecast reduced the negative impact of degrading resource condition on ACS by about one-fifth compared with the constant strategy, increased the risk of making low ACS, and increased the frequency of high ACS compared with constant. The potential of forecasts was examined with perfect knowledge of forthcoming pasture growth. Compared with flexible strategy, perfect knowledge generated an extra $1.10/ha without causing degradation ($0.46/ha for responsive) and at lower risk (<$30,000 1 yr in 10 vs. 1 yr in 3). This demonstrates potential opportunities in seasonal forecasting as reliability of forecast methods improve.

The use of a forecast in decision-making can only be converted into value if the net benefits of using the forecast outweigh any costs arising from implementing the actions (e.g., losses from buying animals late) and errors caused by "incorrect" forecasts. These include missed opportunities from inaction and actions taken inappropriately. Indeed, the results from these modeling studies in northern Australia and one in southern Australia (Bowman et al., 1995) show only modest economic benefits over baseline strategies from using a forecast but real and significant benefits from a resource viewpoint.

The simulations provide a long-term view of benefits. In practice, pastoralists who experience a success early in their use of forecasts (or who fail to benefit by not acting) are likely to be oversupportive of the concept, while those that experience a failure will be unduly negative. It is likely that the benefits of forecasting are also strongly related to factors such as the degree of climate variability experienced and the resilience of the pasture type (Stafford Smith et al., 2000).

As the dynamic models improve and gain prominence, there are increasing opportunities for pastoralists to use climate forecasts in management of their enterprises. However, the results of the simulations highlight the need for extension about the economic and resource benefits (and costs) so that pastoralists are conscious of them when incorporating forecasts into their decisions and that they recognize they are just one more element to include in decision-making. Seasonal forecasting can provide value to pastoral enterprises, but ultimately this is only one small part of managing a profitable enterprise over the long term.

Affirmation from decision-makers

Some pastoralists have documented how they have learned from past experiences and altered management of the variable climate and complexity associated with pastoral enterprises (Landsberg et al., 1998; Mann, 1993; Fahey, 1997).

The Managing Climate Variability (MCV) program wants to enable Australian farmers and natural resource managers to manage the risks and exploit the opportunities that come from our variable and changing climate. The MCV Climate Champion program aims to help farmers manage climate risk by (i) giving farmers the best climate tools, products, practices and seasonal outlooks, and an understanding of how they might use that in their farm business; and (ii) giving climate researchers a chance to interact with farmers and get feedback about what regions and industries need from research.

Twenty farmers from around Australia, representing most major agricultural commodities, are taking part in the program. They have opportunities to (i) talk with researchers about the tools and information they need to help them manage climate risk; (ii) trial early research products and practices, and possibly influence the research; (iii) influence how research findings are communicated to farmers; and (iv) help farmers in their region and industry learn how to deal with the variable and changing climate (see http://www.managingclimate.gov.au/climate-champion-program/).

The Commonwealth Government announced in May 2015 significant funding in the R&D for Profit Program to improve the use of seasonal forecasts to increase farm profit across the cotton, grains, livestock, sugar, and rice industries.

Climate Change

Australian graziers are already conscious of potential impacts of climate change to their industry and their immediate livelihood (Bastin et al., 2014; McKeon et al., 2009). They are considered to be world leaders in managing seasonal climate risk through the use of seasonal climate forecasts (Cobon et al., 2009; Hall et al., 1998; Reeson et al., 2008). However, climate change will clearly have an impact on climate variability, in some regions more than others and in some seasons more than others, thereby influencing the choice of climate risk management strategies (Stokes et al., 2010; Crimp et al., 2010).

Because of the wide range in GCM future climate projections it is not known precisely what the changes will be, but there are some changes for which there is a high degree of confidence. On the timescale of the next 100 yr, the changes that are considered to most adversely affect the grazing industry, for which there is also a high degree of confidence, include increased maximum temperatures and heatwaves, more frequent droughts, and more intense rainfall (Bastin et al., 2014). Changes in rainfall are the most uncertain of all projections, especially over the next two decades when climate variability is still driving rainfall patterns. In southern Australia, winter and spring rainfall is most likely going to decline, but changes in other regions are uncertain. However, there is a high degree of consensus that by 2090 the time in drought and the frequency of extreme drought will increase across much of Australia. The increase in temperatures will increase the severity of drought. The resulting increase in evaporation exacerbates the effect of low rainfall amounts, causing droughts to be more severe even though the rainfall amounts may be no lower than in previous years. In addition to these direct effects on climate, there are also many indirect influences to the grazing industry through impacts on pasture productivity (Hall et al., 1998; Cullen et al., 2009; McKeon et al., 2009), forage quality (Wolfe et al., 1980; Howden et al., 1999b; Moore et al., 2009; McKeon et al., 2004, 2009; Salmon and Donnelly, 2007; Wand et al., 1999; Lilley et al., 2001), pests, diseases and weeds (Stokes et al., 2010), changes in pasture species composition (Orr et al., 2001; Park et al., 2003; Hill et al., 2004; McKeon et al., 2004; Stokes et al., 2010; Alcock et al., 2010), and animal husbandry and health (Howden and Turnpenny, 1997; Howden et al., 1999a).

The impact of climate change on the nature of the key drivers affecting Australia's grazing lands has been projected, albeit with some uncertainty. Global warming may cause changes to the characteristics of drivers. It may affect these drivers in terms of frequency, duration, amplitude, phase locking of the seasonal cycle, and the response of the hydrological cycle (Smith et al., 1997; Cane, 2005). Recent studies of CMIP5 future climate simulations provide guidance as to possible changes in ENSO resulting from climate change (Stevenson, 2012; Kim et al., 2014). There is model consensus for an increase in the frequency of extreme El Niño and La Niña events (Cai et al., 2014b; Power et al., 2013; Collins et al., 2010; Yeh et al., 2009). There is likely to be more extreme El Niño events such as those of 1997-1998 and 1982-1983 where the west Pacific warm pool moved significantly further east than in a moderate El Niño events (Lengaigne and Vecchi, 2010; McPhaden, 1999), but not necessarily an increase in the number of moderate

El Niño events. This increase in extreme El Niño events is suggested to be accompanied by a near doubling of extreme La Niña events from one in every 23 yr to one in every 13 yr (Cai et al., 2015). This increase in frequency of extremes of the ENSO cycle will pose adaptation and management challenges, as not only are the frequency of events increasing but the increase in extreme La Niña events are projected mostly to occur in the year immediately following an extreme El Niño, as occurred with the 1997-1998 El Niño and 1998-1999 La Niña.

Future changes in the IOD have also been suggested (Cai et al., 2014a; Ihara et al., 2008, Ummenhofer et al., 2009). Although the general consensus is for little change in the frequency of moderate positive and negative events (IPCC, 2013; Ihara et al., 2008; Cai et al., 2013), there is a suggestion of an increase in the frequency of extreme positive IOD events from one event every 17.3 yr in the instrumental record to one event every 6 yr in the 21st century (Cai et al., 2014a). There is, however, much less confidence in the stability of the teleconnections between regional climate and the key climate drivers (Cai et al., 2014a; Power and Smith, 2007; Jourdain et al., 2013; Chung et al., 2014).

Uruguay

Agricultural production in Uruguay is based on the highly fertile soils of the Pampas, an ecosystem in which native temperate and subtropical grasslands are used for livestock production or have been converted to improved pastures (grasses and legumes), croplands, or forest plantations. The country's economy largely depends on the agricultural sector. In recent years, about 80% of the total exports were agricultural commodities or agroindustrial products. Stimulated by good prices of agricultural commodities, the country's food exports increased by 400% in the last decade (OPYPA, 2014). The main agricultural exports are beef, soybean, rice, dairy products, and cellulose.

The vast majority of annual crops (with the exception of rice) are rain fed and, consequently, climate variability drastically affects both the agricultural productivity and the volume of export goods. Given the importance of agricultural production on agro-industrial activity and on the transportation–logistics sector, reductions in production have a multiplying effect on the general economy. For example, the impact of a recent drought (2008–2009) resulted in losses of about 300 million US dollars in the beef production sector. The associated losses in the national economy were about 1.0 billion US dollars, that is, a multiplying factor of more than 3.0 due to impacts on employment, logistics, and so forth. (Paolino et al., 2010).

Water excess can affect winter crop production (mainly through impacts on crop diseases and on grain industrial quality), and to a certain extent it can also impact livestock production (due to increased risk of diseases). However, the main climate-related threat to agricultural production in Uruguay is the occurrence of droughts.

One of the most critical rainfall periods for agriculture in Uruguay is late spring and summer (October–February). Average rainfall conditions during the summer months (90–130 mm/month, depending on the location) are typically not sufficient to compensate for evapotranspiration losses. Therefore, pasture and crop growth greatly depend on the soil's ability to store water. Natural grasslands in Uruguay, which are the key sustentation of livestock production, occupy about 70% of the total country's area and are mostly located in the northern and central regions of the country. Many soils in these regions are shallow (less than 30-cm depth) and therefore possess low water storing capacity. Pasture availability and consequent livestock production in these regions are thus highly dependent on the rainfall during the late spring and summer months.

On the other hand, summer crops (e.g., maize and soybeans) require adequate water supply during the critical flowering stage to attain good yields. These crops in Uruguay are grown in deep soils with relatively large water-holding capacity. However, the amount of stored water in these soils is typically insufficient to satisfy the crop water demand of more than 20 to 30 d, and yields of non-irrigated crops strongly depend on rainfall during the flowering months (late December to February, depending on the crop and planting date).

Climate Variability and Climate Outlooks

Uruguay is located in one of the regions of the world with the largest climate variability in the interannual time scale. Long-term mean values of precipitation are 60 to 130 mm/month throughout the year, but the coefficients of variation are 70 to 90% (Fig. 6).

Rainfall in Uruguay is strongly affected by ENSO: warm (cold) anomalies in the tropical Pacific increase the chances of positive (negative) rainfall anomalies (Cazes-Boezio et al., 2003; Diaz et al., 1998; Pisciottano et al., 1994). The impact of ENSO on rainfall is especially strong in the spring (October–November–December) and to a lesser degree in the early summer and fall. Also, the impacts of La Niña years are typically stronger and more consistent than those of El Niño years: the majority of La Niña years of the last decades have shown important rainfall deficits, especially in the spring and summer months which are critical for agricultural production (Baethgen and Magrin, 2000).

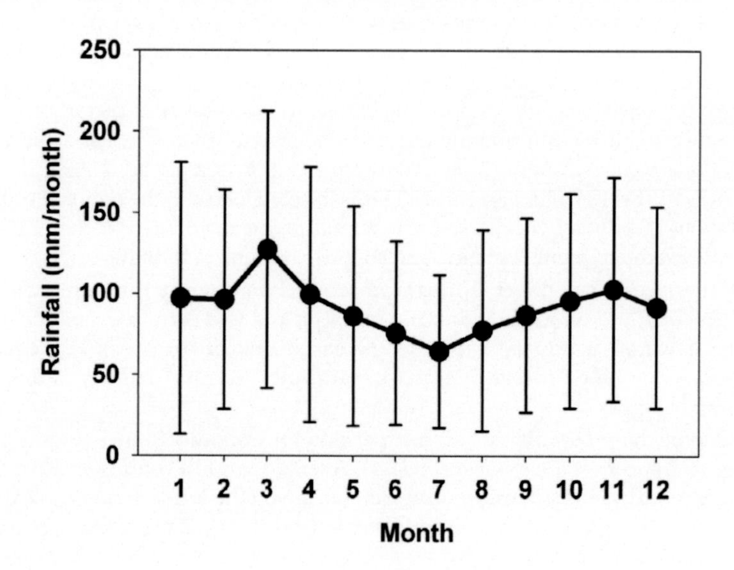

Fig. 6. Mean and standard deviation of monthly rainfall in southwest Uruguay (1915–2008) (source: INIA-GRAS, http://www.inia.uy/).

The strong relationship between ENSO phases and rainfall in southeast South America (SESA) allows for good skill of seasonal climate outlooks, especially in years with ENSO anomalies. In December 1997 during the strong El Niño event, Uruguay hosted the first Regional Climate Outlook Forum for the SESA region. Meteorologists from Argentina, Brazil, Paraguay, and Uruguay met with colleagues from the IRI (International Research Institute for Climate and Society, Columbia University), NOAA, and other international institutions and produced the first consensual seasonal climate forecast for the region. Since that time the SESA region has uninterruptedly organized Regional Climate Outlook Forums at least once per year. Also, in the last decade the National Meteorological Institute of Uruguay and a group of scientists from the University of Uruguay have been producing climate outlooks, but these outlooks include mainly qualitative information on the expected temperature and rainfall conditions for the following 3 months.

In 2012, the Ministry of Agriculture and Fisheries of Uruguay started a project funded with a loan from the World Bank to improve adaptation to climate change (World Bank, 2014). Within that project, the IRI of Columbia University has been collaborating with the National Meteorological Institute of Uruguay to produce improved climate outlooks. Within the next 2 years the project will produce seasonal forecasts with higher spatial resolution and will explore the predictability at subseasonal scale (i.e., less than 3 months) and the ability to predict "weather within climate" (i.e., probability of dry spells, extreme events, etc.).

Three strong La Niña episodes (1988-1989, 1999-1900, and 2010-2011) had large negative impacts on the Uruguayan economy. The episodes were characterized by extended periods with low rainfall during months that are critical for livestock and annual crop production, and strongly affected the agricultural sector.

Good years in Uruguay for natural grasslands in shallow soils and for annual summer crops in deeper soils are characterized by larger than normal rainfall during late spring and summer. The results presented in Fig. 7 show that in three recent La Niña episodes (1988-1989, 1999-1900, and 2010-2011), rainfall during this critical period was considerably below average.

Although Uruguay total land area is relatively small (approximately 190,000 km^2), large spatial variability is typically found in the spring and summer rainfall across the country's regions. For example, in the three studied years, rainfall in spring and summer was much lower in the northern region than in the southern or central regions. Also, the negative rainfall anomalies in 1999-2000 and 2010-2011 started earlier and lasted longer than in 1988-1989.

Drought of 1988-1989

In 1988 Uruguay had no institutions or special policies and programs in place to respond to droughts. At that time droughts were viewed as very low frequency phenomena which did not justify the development of special structures or programs. Consequently, governments had typically reacted to previous droughts with traditional "crisis management" responses such as special aid programs to affected regions.

In August 1989 when the drought had already shown important negative effects on agriculture, water resources, and hydroelectric dams, the government of Uruguay and the United Nations Development Program hired a consultant to assist in the development of reactive strategies to confront the drought. The

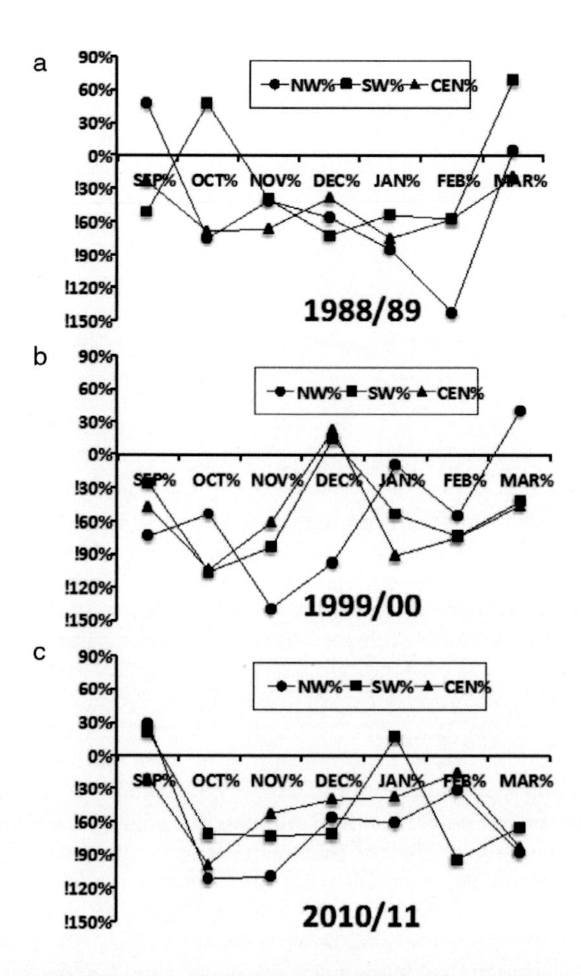

Fig. 7. Monthly rainfall in three locations of Uruguay (NW, northwestern region; SW, southwestern region; CEN, central region) in the spring and summer months during three recent droughts: 1988-1989 (a), 1999-1900 (b), and 2010-2011 (c).

consultant (Dr. Donald A. Wilhite, International Drought Information Centre, University of Nebraska) prepared a report to the government with a large list of recommendations for future droughts including the following:

1. Create a National Drought Commission to assist the government in the assessment, planning, management, and response activities.

2. Create a National Water Availability and Outlook Committee to develop a national drought plan to systematically prepare for the next major episodes. Among many other responsibilities, the National Water Availability and Outlook Committee should (i) continuously monitor water availability, (ii) distribute monthly summaries of national water availability, (iii) develop climatological tools (e.g., probability tables, indices, etc.), and (iv) explore the potential use of remote sensing data, such as advanced very high resolution radiometer (AVHRR), to provide a country-wide assessment of biomass activity.

In 1988 research on teleconnections and impacts of ENSO on rainfall in Uruguay was incipient. Ropelewski and Halpert (1987) had just published the first article that showed the correlation between ENSO anomalies and rainfall patterns in southeastern South America. Climate scientists from the University of Uruguay and the National Weather Service were starting the first research studies on ENSO impacts. Seasonal forecasts of rainfall were not available. Still, Dr. Wilhite's report to the government emphasized the need to consider ENSO in the climatic research national programs.

In summary, the 1988-89 drought found Uruguay with no institutional structures, with no capabilities to assess or monitor water availability, and with incipient research on the ENSO impacts on rainfall. Consequently, the government and the private sector could only respond to the drought with a crisis management approach. As a result, in the livestock sector 16% of the total cattle population was lost (Bartaburu et al., 2009), and the direct losses in the sector attributed to the drought were 300 million US dollars (approximately 630 million US dollars adjusted to 2015 values). However, the losses were much larger since the reduction in the population of breeding animals was felt for several years after 1988-1989. Other very important losses were reported in the forestry sector because of frequent fires and in the summer crops where yields at national level were reduced by more than 40%.

Drought of 1999-2000

Several changes had occurred in Uruguay during the period between both droughts. First, the government had created two institutions, the National Emergency System and the National Commission for Drought. The National Emergency System is appointed directly by the office of the President of Uruguay and had played a key role in crisis management activities during the 1997-1998 (El Niño) floods in Uruguay. On the other hand, the National Commission for Drought was created under the leadership of the Ministry of Agriculture and included representatives of the research community (agriculture and climate), a few governmental offices, and several organizations from the private sector.

Also, the National Institute for Agricultural Research (INIA) had established a Climate, Environment, and Satellite Agriculture interdisciplinary team (GRAS for its Spanish acronym), which had started collaborative research with the International Institute for Soil Fertility Management (IFDC–Uruguay Regional Office) and the IRI (Columbia University). The collaboration included research projects in the following areas:

(i) applications of seasonal climate forecasts in the agricultural sector (with climate scientists from the University of Uruguay), and

(ii) development of an information and decision support system (IDSS) for the agricultural sector of Uruguay (in collaboration with NASA's Goddard Institute for Space Studies; the Soils Department of Uruguay; and INTA, the National Agricultural Research Institute of Argentina).

These research projects started activities during the period when the first negative impacts of the 1999-2000 drought were starting to be felt in Uruguayan agriculture and proved to have major impacts on the government response. The INIA-IFDC-IRI research project that was establishing the IDSS included two activities that were extensively used by the public and private sectors responding to the 1999-2000 drought. First, the IDSS included calibrated and tested crop simulation models that were used to identify agronomic practices better adapted to the drought conditions. Also, they had started monitoring the vegetation status normalized difference vegetation index (NDVI) throughout the country by using satellite data that was updated every 15 d (using AVHRR at 1-km resolution).

The collaborative project also included the creation of a Technical Working Group (TWG) for improving the generation, dissemination, and applications of climate information (monitoring and seasonal forecasts). The TWG was constituted by researchers (agriculture and climate) working in the project mentioned above and by representatives of the major farmer organizations, agribusiness, and governmental offices. The TWG met periodically to discuss the most recent seasonal climate outlooks produced by the IRI (Fig. 8) and the current situation of vegetation (NDVI results; Fig. 9). During the TWG meetings the climate scientists presented the most recent climate outlooks and the results of their own climate research conducted at the national level. The agricultural scientists presented advances in tools for monitoring the vegetation status and for applying the available climate information to improve decisions. The stakeholders from the public and private sector discussed possible uses and shortcomings of the information they received. In addition to creating the adequate environment to improve the generation and applications of climate information (monitoring and forecasts), these meetings were also crucial for the dissemination of the climate outlooks in the agricultural sector.

All the information produced in these collaborative projects was published and continuously updated in the GRAS-INIA's Web page, which was visited by farmers, agribusiness representatives, agronomists, and government officers. In addition, IFDC and GRAS-INIA staff gave several live presentations and teleconferences in collaboration with the Ministry of Agriculture and Fisheries that reached all the major regions of the country. Researchers also appeared in several TV and radio programs and prepared numerous articles for the major newspapers and specialized magazines. In these communications, the researchers presented the evolution of the vegetation status monitored with remote sensing as well as the results of the latest IRI seasonal climate forecasts.

This continuous communication of researchers with public and private agricultural sectors had a major impact in providing stakeholders with the most updated, objective, and sound information on the status and evolution of the drought. Emergency situations are often characterized by the existence of an overflow of information from many different sources (national, international, and regional), with varying levels of objectivity and scientific soundness. This information overflow often causes confusion and hampers stakeholders from taking effective preventive or responsive actions. That was the situation in 1999-2000, and therefore identifying trustworthy sources of clear, relevant, and applicable information from the INIA-IFDC-IRI research team was crucial for making decisions at any level.

In this section we document the action of the Ministry of Agriculture and Fisheries that exemplify how the government of Uruguay used the information provided by the INIA-IFDC-IRI research group during the 1999-2000 drought. Several other similar examples exist in the country but were difficult to document (National Emergency System, agribusiness, and individual farmers).

During the early months of 2000 the impacts of the drought were already quite evident. Cattle herds were clearly suffering because of the lack of available

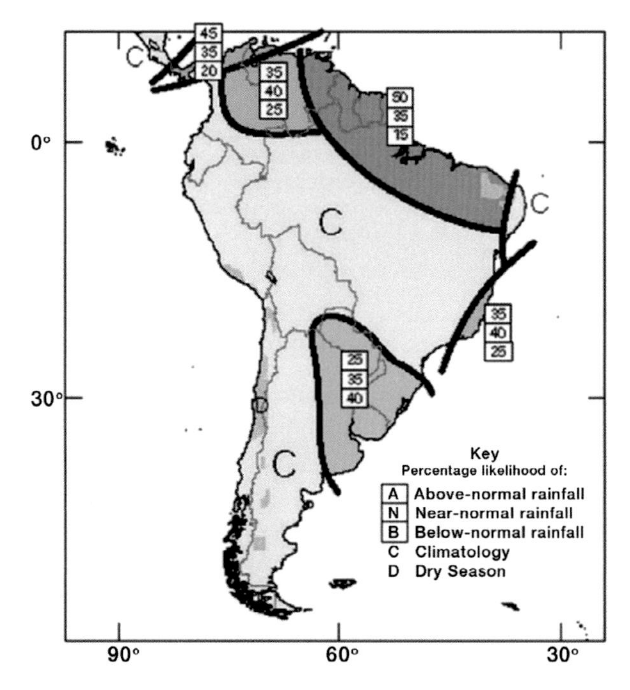

Fig. 8. Climate forecast issued by IRI in October 1999 for October–November–December.

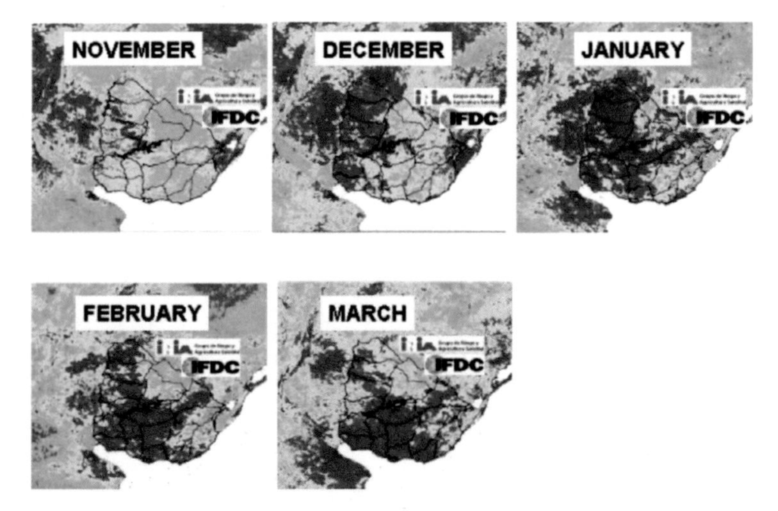

Fig. 9. Maps produced by the INIA-IFDC-IRI research team provided to the Ministry of Agriculture in 1999 showing normalized difference vegetation index values from NOAA's AVHRR.

forage and drinking water. The Ministry of Agriculture decided to establish an emergency plan and needed to set priorities for distributing aid to the different regions of the country. In the past, the definition of such priorities had mainly been based on reports prepared by the Ministry's staff working in the field. Since resources are always scarce, the Ministry field staff were usually not able to cover the entire country, and consequently several regions were left behind in the distribution of aid, even though they might have been in a critical situation. Also, similarly to what happens in the rest of the world, some regions in Uruguay can have more political influence than others. Confronted with the lack of objective and technically sound information, government decision-makers were often persuaded to prioritize such regions in the distribution of emergency aid.

When the impacts of the 1999-2000 drought started to fade away, the then Minister of Agriculture and Fisheries (Mr. Juan E. Notaro) sent a letter to the INIA-IFDC-IRI research group coordinators and wrote:

> "The results of your work during the recent drought were useful for making both operational and political decisions. From the operational standpoint, your work allowed us to concentrate our efforts in the regions highlighted as being the ones with the worst and longest water deficit [(1)]. We prioritized those identified regions for concentrating the use of our resources, both financial aid and machines for dams, water reservoirs, etc. We also established plans that were more flexible than usual (relaxing sanitary and commercial controls) to mobilize cattle from the most affected regions.
>
> "From the strictly political standpoint, your work provided us with objective information to defend our prioritization of regions, in a moment in which every governor, politician, and farmer in the country was asking for aid. We received no complaints in this respect. In the same line, your work also allowed to mitigate pressures since we provided the press and the general public with transparent, technically sound, and precise information.
>
> "In summary, the most important issue is that the celerity and precision of the information you provided allowed us to be effective in our decision-making and at the same time to publicly defend those decisions with technical solvency. We did not get complaints about politicizing our actions; on the contrary, our actions were complimented. And most importantly, the most feared threat of significant cattle deaths due to the drought never occurred thanks to the celerity of the actions taken, which would have been impossible without the information you provided to us."

Drought of 2010-2011

In March 2010 a newly elected government took office in Uruguay, and the new Minister of Agriculture and Fisheries defined five strategic lines for his administration. One of these strategic lines was oriented to "improve adaptation to climate change of the agricultural sector." The approach for implementing this line of work was based on an IRI paper (Baethgen, 2010) and proposed to improve adaptation to future climate change and increase resilience starting by improving adaptation to current climate variability. The Ministry of Agriculture and Fisheries implemented a project funded with a loan from the World Bank entitled "Development and Adaptation to Climate Change" (DACC) and asked the IRI to lead the efforts for establishing a National Agricultural Information System that would "assist the private sector to improve planning and decisions, and the government to elaborate public policy" (from the Ministry's resolution). The National Agricultural

Footnote (1) The Minister refers to the AVHRR images processed by the INIA-IFDC Research Group (both absolute and relative values). The data was presented in several public meetings and was published and updated in INIA's Web page.

Information System would take advantage of the work that had been developed in the previous years by INIA and collaborators, including the tools for incorporating climate monitoring and seasonal forecasts into agricultural decisions and plans that had successfully helped in the 1999-2000 drought. In addition, the IRI would work with the Uruguayan Meteorological Institute to improve the existing seasonal climate forecasts.

In September 2010 the seasonal climate forecast for October–November–December elaborated by the IRI showed high probabilities for lower than normal rainfall, and the ENSO outlook showed 98% probability of a La Niña year (Fig. 10). The IRI worked with staff of the Ministry of Agriculture, and this information was included in several press releases. In October 2010 the data produced by INIA collaborating with the IRI, started showing negative anomalies in the vegetation and low values of available water in the soils. This situation continued and worsened in the next 2 months. Normally, the soil–water balances are estimated by soil type (depending on the soil–water-holding capacity). However, given that decisions are made based on political boundaries (i.e., provinces and municipalities), the soil–water balances were also modified to estimate the values by the smallest political boundaries in Uruguay (i.e., "police districts"; Fig. 11).

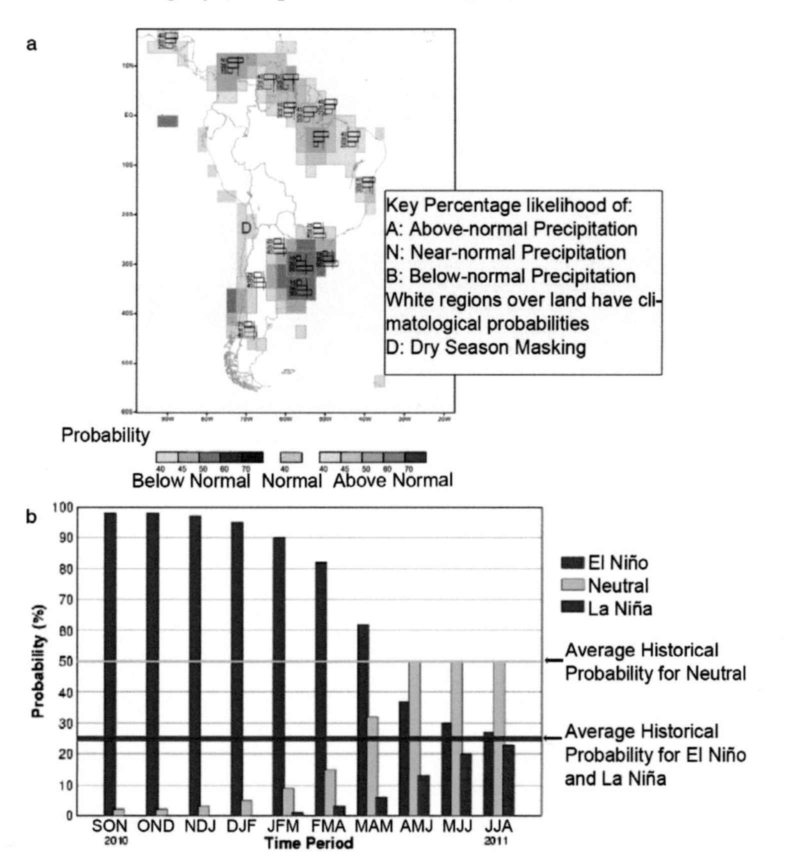

Fig. 10. IRI Climate Forecast issued in September 2010: rainfall seasonal forecast (a) and ENSO outlook (b).

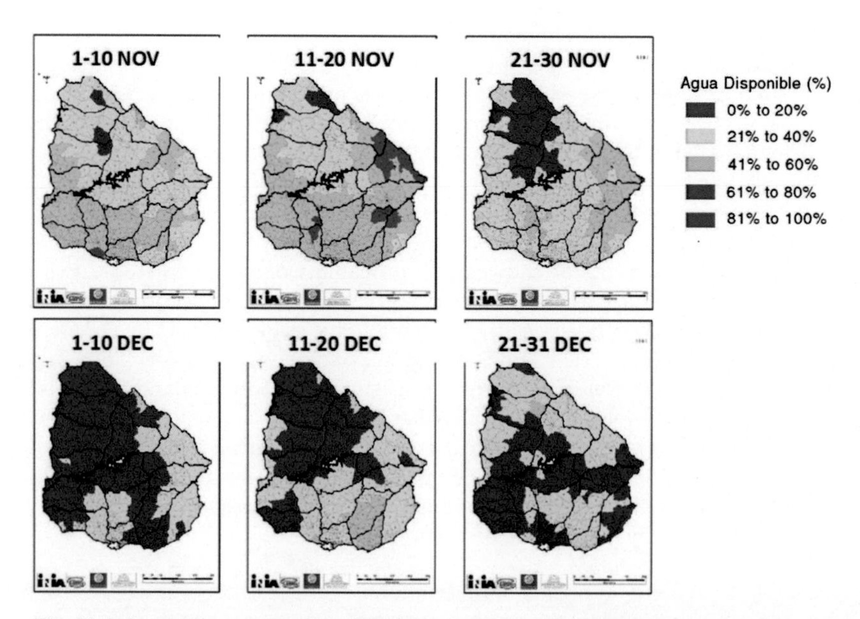

Fig. 11. Information provided by INIA-IRI in 2010 on soil–water balance at police district level. Based on this information an official emergency was declared for the northern region of Uruguay on 22 Dec. 2010.

Based on the information produced by the INIA-IRI team (Fig. 11), the Minister of Agriculture (Ing. Tabaré Aguerre) declared a national emergency in December 2010 for all police districts to the north of the Negro river (basically the northern half of the country). The declaration triggered a set of interventions including the exoneration of taxes to import animal feed and the establishment of a line of subsidized credit to help acquire animal feed and to implement "water solutions" for small farmers (wells, reservoirs, irrigation, etc.). The credit line was subsidized and implemented with the DACC project (World Bank loan). In the words of the Minister of Agriculture during the press conference when the credit lines were presented, those solutions were not intended "to avoid the impacts of the current drought but mainly to avoid impacts of future droughts."

In February 2011, when the rainfall was still below normal, especially in the north of the country (Fig. 11), the Minister of Agriculture was convened to the Parliament to provide a report of the current drought situation and to explain the concrete actions that his Ministry had taken to reduce the impacts on agricultural production. The minister had a long presentation in which he explained the approach that the Ministry was using to improve adaptation to climate change (based on IRI's approach; Baethgen, 2010). He also explained to the members of the Parliament that in August 2010, and then again in September 2010, his ministry had made public the seasonal forecast elaborated by the IRI showing the high chances of low rainfall in the spring related to the La Niña conditions. He explained that the main objective of sharing those climate forecasts was to induce the agricultural sector to take preventive actions. Furthermore, he explained that "the current drought had been anticipated and was confronted with financial tools that were specifically developed to reduce the impacts on small farmers" (3000 small farmers received some type of help with the subsidized credits mentioned above).

The minister defended his actions by explaining that in addition to making public IRI's seasonal climate forecast 2 months before the drought started, his ministry had also used objective information for monitoring the situation throughout the country. Consequently, the emergency was declared based on that technically sound and objective information. Minister Aguerre concluded his presentation explaining that the challenges of the current drought were not over. He showed a slide presenting the current situation of the vegetation and the soil–water content (INIA-IRI work) and the latest IRI's seasonal rainfall forecast for February–March–April 2011 which still showed some chances of below-normal rain.

Final Comments and Lessons Learned

Effective use of climate forecasts requires adequate institutions that can (i) produce the needed information and reports, and (ii) take advantage of the climate information for implementing actions to assist the agricultural sector. For example, no climate forecast information was available in the 1988-1989 drought, but even if information had been available, the lack of institutional arrangements would have prevented the Uruguayan government to take any preventive or responsive actions. In contrast, the droughts of 1999-20000 and 2010-2011 found Uruguay with adequate institutions to generate good and relevant climate information (INIA, IFDC, and IRI) and institutional arrangements that allowed the government to take effective action, including elaborating policy (National Emergency System, Ministry of Agriculture, and the DACC project).

Seasonal climate forecasts are especially useful when they are combined with good real-time information on the current situation of the agricultural sector (e.g., vegetation status and soil–water content). The combination of good climate forecast and monitoring systems results in actionable information for the institutions making decisions that can drastically reduce the socioeconomic impacts of droughts (e.g., declare emergencies, establish financial tools, prioritize aid, etc.).

Across the three Uruguay La Niña drought cases described above, a relationship of trust was achieved after more than a decade of collaborations between INIA and IRI with different agricultural stakeholders, including the national government. The successful integration of seasonal climate forecasts into actual decision-making required the trust of the stakeholders who were making decisions (farmers, governments, and agribusinesses) in the institutions generating and disseminating the information. Developing that kind of trust can be a lengthy process.

South Africa

South Africa is a predominantly semiarid country. Most of the country is a summer rainfall region. Average annual rainfall ranges between more than 1000 mm along the eastern coastal belt and northeastern escarpment, decreasing across the interior plateau westward to less than 200 mm over the northwestern interior and western coastal belt (Fig. 2a).

About 80% of the country is used for agriculture, but productivity is relatively low. Livestock farming constitutes the largest agricultural sector in South Africa, with 13.8 million cattle and 28.8 million sheep (Palmer and Ainslie, 2006). More than two-thirds of the country's surface is suitable for grazing. About 80% of the agricultural land is suitable for extensive livestock farming (Fig. 3b; Köppen [1931]), with rangelands playing an extremely important role for livestock production (Stroebel et al., 2008). Grazing capacity ranges from 1 Ha/LSU (hectares per livestock unit) over the subhumid grasslands of the Eastern Highveld to 70 Ha/LSU over the arid western interior (e.g., Morgenthal et al., 2005).

Beef produced in South Africa satisfies about 85% of the domestic need, with the rest imported from Botswana and Namibia, countries with extensive semiarid rangelands. In South Africa, cattle ranches dominate toward the semiarid savannah and grassland regions over the plateau, with more extensive practices further westward with decreasing rainfall. Dairy farming dominates toward the cooler southern and eastern coastal belt and Highveld. Sheep farming occurs primarily toward the more arid western and southern parts and to a lesser extent also over the cooler and high-lying eastern parts of the interior.

Climate variability has significant impacts on livestock farming in both marginal and high-potential areas. Livestock mortality in commercial and subsistence livestock farming is frequently associated with periods of drought (Ngaka, 2012). The importance of conservative stocking rates is stressed frequently by the National Agrometeorological Committee, realizing that decreasing potential of rangelands is mainly a result of unsustainable farming practices within the context of climate variability (Hoffman et al., unpublished final report, 1999; Archer, 2004; Wessels et al., 2012). Floods and drought can be catastrophic, resulting in soil–water erosion, thereby lowering the soil's capacity to support livestock and produce crops.

Projected change in rainfall and potential evapotranspiration may influence the suitability of different regions for the rain fed production of grazing grass and also the association between different species. Due mainly to projected changes in maximum temperatures, most of the interior is expected to become more arid (Engelbrecht and Engelbrecht, 2015).

Timely and accurate seasonal forecasts, distributed to an agricultural user community trained for the correct application of such forecasts, may play an important role in rangeland management (Haigh et al., 2015). Research on tailored forecasting, although still in its early stages, can add great value to livestock production and rangeland monitoring. The latter is driven by the ongoing struggle of food insecurity, which is projected to worsen with climate change across the world (Goddard et al., 2010).

History of Climate Variability

The ENSO phenomenon affects atmospheric circulation outside of the tropics in a relatively indirect way (Philander, 1990). For southern Africa south of about 15°S, however, interannual variations of rainfall and temperatures are affected such that during austral summer, drier and hotter than normal conditions are associated with El Niño events, whereas wetter and cooler conditions are associated with La Niña events (Mason and Jury, 1997; Lindesay, 1988; Van Heerden et al., 1988). Southern Africa tends to experience drier (wetter) conditions during approximately the same season when equatorial eastern Africa has a wetter (drier) than normal season (Ropelewski and Halpert, 1987). Over southern Africa, ENSO associations are best manifested in a northwest to southeast line across the region (Lindesay, 1988) and coincide with the preferred position of so-called tropical temperate trough systems, which are responsible for contributing significantly to the rainfall totals during wet seasons (Fauchereau et al., 2008; Washington and Todd, 1999).

ENSO affects southern African rainfall and temperature variations in the later, rather than in the early summer season. Tropical systems dominate the rain-producing systems during January and February, while during December, a shift from baroclinical systems occurs, and during March, a shift to baroclinical systems occurs (Van Heerden et al., 1988). The December to February season is often associated with the season of highest seasonal-to-interannual predictability over the region (Landman et al., 2012; Muchuru et al., 2014). In addition to ENSO, SST anomalies of the oceans adjacent to southern Africa are also related to southern African rainfall (Nicholson and Entekhabi, 1987; Walker, 1990; Jury and Pathack, 1991; Mason et al., 1994; Mason, 1995; Mason and Jury, 1997; Rocha and Simmonds, 1997). Sea surface temperatures over the central south Atlantic Ocean (Mason, 1995) and over the western tropical Indian Ocean have strong associations with southern African climate variability (Mason, 1990, 1995; Jury et al., 1993; Walker, 1990), but the seasonal rainfall over southern Africa also seems to be linked to a dipole in subtropical SST over the south Indian Ocean (Reason, 2001a, 2001b; Behera and Yamagata, 2001). The major ocean currents around southern Africa also play a role in its variability (Jury et al., 1993; Rouault et al., 2003).

Major Problems of Climate Variability

Severe droughts and floods are most often, although not exclusively, associated with ENSO. Critical sectors affected include the primary industries, settlements, and infrastructure, with rising concern in the region as to how to respond to increased frequency of extreme events (whether related to ENSO or not). In the Southern African Development Community (SADC) region, for example, disaster response takes the form of a number of different stakeholders, operating in different configurations. Key challenges exist in terms of national and regional early warning systems being effective, a situation currently of some concern at country and regional level (see, for example, Conway et al., 2015)

Costs of extreme events may be significant. In South Africa, for example, in January 2011 the government declared 33 disaster zones under conditions of extreme flooding (NDMC, 2011). Along with 88 deaths at the time of reporting, cost estimates of damage to infrastructure were set at around ZAR 160 billion (affecting 7 of the 9 provinces). Correctly costing impacts of extreme events remains an ongoing challenge in South Africa, as well as other SADC member states, one that is often a key area of attention in both climate services–focused programs, as well as climate change–focused reporting such as national communications.

Current Climate Forecast Operating Systems

In South Africa there are a number of institutions who are actively involved with the production of seasonal forecasts (Johnston et al., 2004). These comprise the Universities of Cape Town (UCT) and Pretoria (Rautenbach and Smith, 2001), the Council for Scientific and Industrial Research (CSIR), and the South African Weather Service (SAWS). The University of the Witwatersrand also contributed significantly to the seasonal forecasting effort during the late 1990s (Mason, 1998). Both UCT and SAWS have also been involved with seasonal forecast model development since the 1990s, ranging from purely statistical models (Jury et al., 1999; Landman and Mason, 1999), the use of general circulation models (Bartman et al., 2003; Landman et al., 2001; Rautenbach and Smith, 2001; Tennant and Hewitson, 2002), the development of multimodel ensemble systems (Landman and Beraki, 2012), and the development of a coupled ocean–atmosphere model (Beraki et al., 2014). A seasonal forecasting capability was also established at the CSIR in the last few years, with an additional focus on tailored forecasting for particular sectors (see, for example, the SASSCAL (http://www.sasscal.org/) projects in initiation, as well as Malherbe et al. (2014).

For approximately the last 12 yr, these three South African institutions have engaged in presenting their global climate model forecasts as maps on the website of the Global Forecasting Centre for Southern Africa (www.GFCSA.net). These institutions respectively administer the following global atmospheric general circulation models: SAWS the ECHAM4.5 (Roeckner, 1996), UCT the HadAM3P (Pope et al., 2000), and CSIR, the Conformal Cubic Atmospheric Model (McGregor, 2005). All of these are atmospheric models which use some form of prescribed SST anomalies, either persisted or predicted with various multimodels (Landman et al., 2011) as boundary forcing while keeping sea ice and the land surface constant. Although these three institutions issue seasonal forecasts routinely, the official seasonal forecasts are administered and issued by SAWS. The latter institution's forecasts are based on a multimodel system (Landman and Beraki, 2012) that combines statistically downscaled forecasts from global climate models run locally and internationally. Although nested regional climate models as seasonal forecasting tools have also been investigated (Kgatuke et al., 2008, Landman et al., 2005, 2009; Ratnam et al., 2011), such systems are currently not being used for operational forecast production in South Africa. It is possible that this may change in the near future, at least in terms of regional programs.

Seasonal forecast skill over southern Africa is modest. For the most part, forecast skill is restricted to the midsummer period, skill for temperatures is higher than for rainfall, and forecast models can really only reliably discriminate anomalously wet or dry summer seasons from the rest during El Niño and La Niña events (Landman and Beraki, 2012). Figure 12 (from Landman et al. [2012]) shows a typical verification result for December–January–February rainfall over South Africa. An important aspect of note is the overconfidence with which below-normal rainfall totals are usually predicted (the weighted regression line for the below-normal category is below the diagonal of perfect reliability). This result is mostly attributed to the tendency of models to predict drought conditions over the region during El Niño years (Landman and Beraki, 2012; Landman et al., 2014), although in the recent past El Niño events have not always been associated with droughts (e.g., 1997-1998 and 2009-2010). On the other hand, when wet summer seasons are predicted during La Niña seasons, forecasts tend to be reliable (e.g., 1999-2000 and 2010-2011) and may be applied for hydrological modeling on seasonal time scales (Muchuru et al., 2014, 2015).

How Have Climate Forecasts Been Applied?

Seasonal forecasts which are based on objective forecasting systems have been routinely issued in South Africa only since the early 1990s (Landman, 2014). These forecasts have at first mainly focused on the prediction of rainfall totals and average temperatures by using a variety of forecasting techniques and models, and very little work, if any, was done back then on model-based predictions of derivatives of rainfall or temperatures such as dry-land crop yield and flows in rivers. The development of application models for seasonal forecasting has been the focus of renewed energy and engagement (although not necessarily always with matching external funds), with the main focus falling on agricultural production (including livestock) and streamflows (including a specific focus on disaster management applications) (Malherbe et al., 2014; Muchuru et al., 2015). Notwithstanding the development of these so-called applications models, evidence that seasonal forecasts of rainfall, temperature, or any derived variable thereof are actually being incorporated into operational agricultural, or any other decision support systems, is lacking in South Africa. Even at times of critical decision-making, such as when the National Crop Estimates Committee (hosted by the Agricultural Research Council) meet each month to estimate crop yield in South Africa, the available seasonal forecasts play only a minor role in the decisions being made, despite the potential criticality of information provided. In fact, the committee for the most part makes use of analog years to estimate expected rainfall and temperature patterns (J. Malherbe, personal communication, 2016). Notwithstanding, seasonal forecasts produced by the three modeling groups in South Africa are being sent routinely to users. Such dissemination suggests that there may be users such as commercial farmers who make use of such forecasts in their risk assessments (Klopper et al., 2006). Whether or not there is only limited uptake in seasonal forecasts, one would have expected uptake to have steadily increased, owing to the fact that seasonal forecast skill has improved over the years as seasonal forecasting models for southern Africa have advanced from simple statistical models to state-of-the-art coupled climate models (Landman, 2014). In reality, however, forecast applications and their real utility may be more challenging to realize and require concerted effort, analysis, and attention.

The Department of Agriculture, Forestry and Fisheries (DAFF) implements an early warning system for disaster risk prevention, mitigation, and preparedness as per the Disaster Management Act 2002 (Act No. 57 of 2002). Providing information

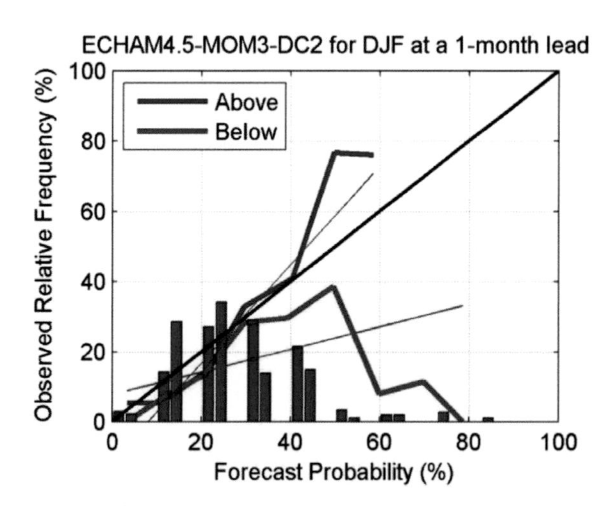

Fig. 12. Reliability diagram and frequency histogram for above- (75th percentile) and below- (25th percentile) normal December–January–February (DJF) rainfall 1-mo lead forecasts produced by downscaling a fully coupled ocean–atmosphere model to South African rainfall. The thick blue (red) curve and the blue (red) bars represent the wet (dry) category. The thin blue (red) line is the weighted least squares regression line of the wet (dry) reliability curve. After Landman et al. (2012).

to reduce the risk from adverse weather and climate conditions on agricultural productivity is a key aspect of the early warning system. The weather and climate conditions in early warning include the seasonal forecast information, which plays an important role as a tool that the farming community uses to assist them in planning their activities so as to reduce the impacts of disasters.

The DAFF, through the National Agrometeorological Committee, disseminates advisories on a monthly basis for disaster risk reduction. These advisories are distributed throughout the nine provinces in South Africa by email, websites (www.agis.agric.za), study group meetings, roving seminars, community radio, and farmer information days. Information in the advisories includes current climate conditions for each of the nine provinces (see Fig. 13 for rainfall current conditions May 2014 to March 2015), seasonal climate outlooks (see Fig. 14 for 3-mo forecast of rainfall May 2015 to September 2015), agricultural market conditions, strategies and tactics to help manage the climatic conditions forecast ahead, and information to help interpret the outlook. In addition, newsletters are distributed during May in preparation for the winter season and during September in preparation for the summer season. As an example, in the year up to May 2015, 24 roving seminars on weather and climate for farmers were conducted in all the districts in four provinces. In addition, capacity-building workshops are conducted in the provinces. These workshops include the interpretation of seasonal forecasts so that officials are able to assist farmers in using information. The aim of the roving seminars is to ensure that farmers become independent in the interpretation of seasonal forecasts and in using this information in the absence of officials.

The strategies and tactics for managing grazing and livestock associated with a drier than normal austral winter in 2015 reported in the May 2015 advisory were the following:

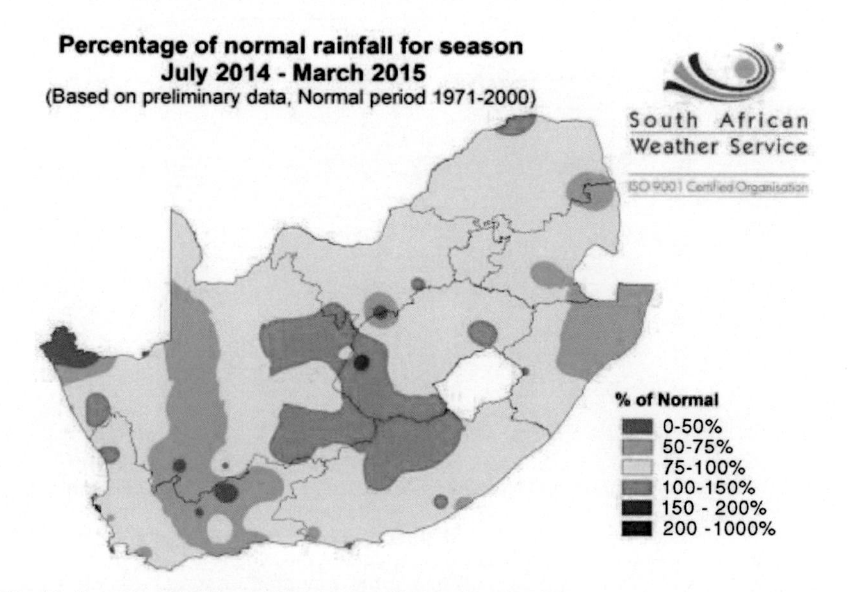

Fig. 13. Current rainfall conditions July 2014 to March 2015.

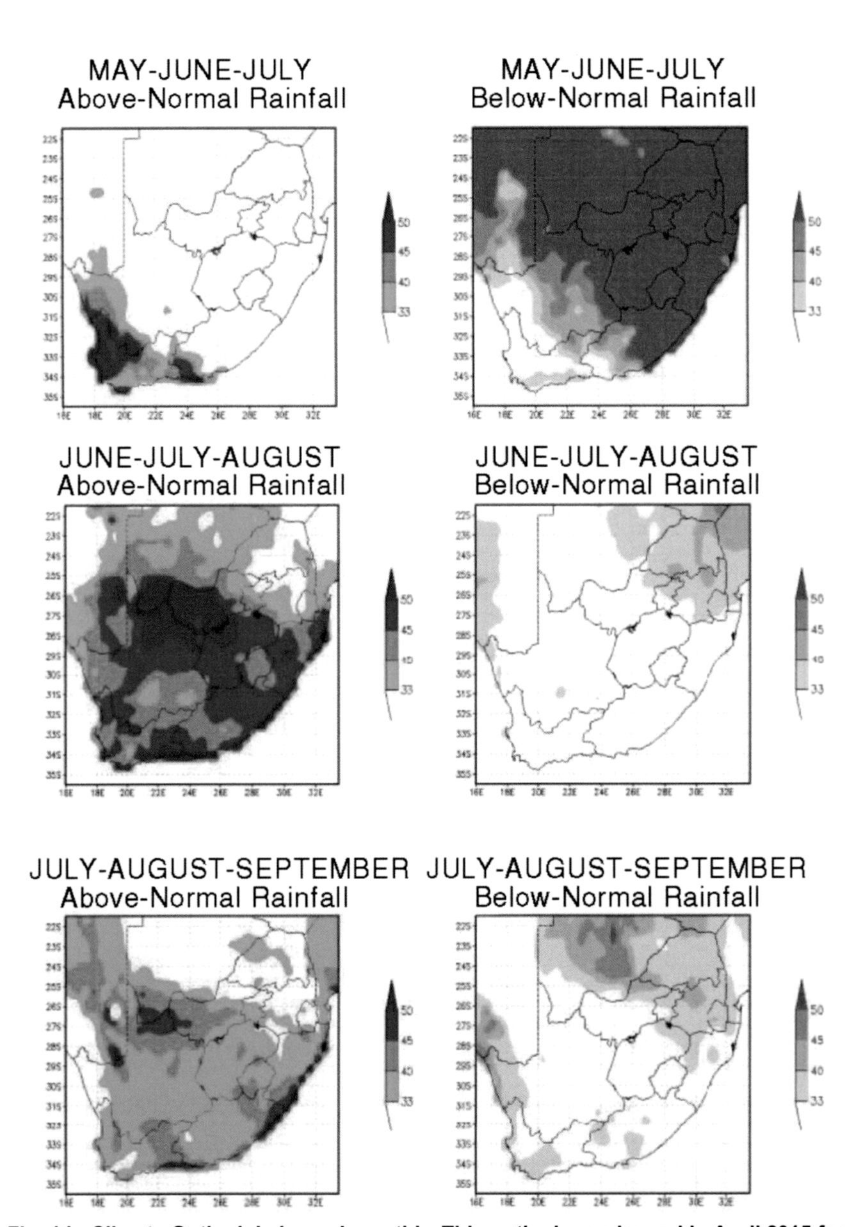

Fig. 14. Climate Outlook is issued monthly. This outlook was issued in April 2015 for May to September 2015, showing chance of above-normal rainfall (left-hand panel) and chance of below-normal rainfall (right-hand panel).

1) Keep stocking rates conservative and in balance with carrying capacity—avoid overgrazing.
2) Provide multiple drinking points for livestock.
3) Enhance nutritional value of dry grazing forage with licks. Phosphorous deficiency is a major problem. Licks should (in most cases) provide phosphorous, urea (to help with the break-down of dry vegetation), salt, and molasses.
4) Sell mature, marketable animals (to help prevent overstocking).
5) If available grazing forage is very low, herd animals into pens where different animals can be segregated and fed separately.
6) Subdivide grazing areas into camps of homogeneous units (in terms of species composition, slope, aspect, rainfall, temperature, soil, and other factors) to minimize selective grazing and provide for the application of animal and veld management practices such as resting and burning.
7) Determine the carrying capacity of different plant associations and calculate the stocking rate of each, and then decide the best ratios of large and small animals and of grazers or browsers.
8) Provide periodic full growing season rests (in certain grazing areas) to allow veld vigor recovery to maintain veld productivity at a high level as well as to maintain the vigor of the preferred species.
9) Eradicate invader plants.
10) Periodically reassess the grazing and feed available for the next few months and start planning in advance.
11) Wean calves early—lactating cows consume much more.
12) Maintain firebreaks in winter rainfall areas and construct the firebreaks in summer rainfall areas.

Advisories like these help livestock farmers reduce their stock numbers when conditions are not favorable and restock at a later stage. Unfortunately, these decisions can have a negative impact on the livelihood of subsistence and small-scale farmers as the commercial sector purchases their livestock at a much lower price. However, destocking minimizes the impact of overgrazing and other natural hazards in their operations and thereby helps to improve long-term productivity and livelihoods. The use of seasonal climate forecasts, however, is still a learning process for both subsistence and small-scale farmers in South Africa.

Value and Adoption Rates of Climate Forecasts and Affirmation of Use in Agricultural Decision-Making

Forecasting crop yields for southern Africa has been shown to be feasible (e.g., Cane et al., 1994; Malherbe et al., 2014), but it seems that the use of integrated climate and pasture models to predict pasture growth has not been applied in South Africa.

There is little specific evidence from the livestock and pastoral industries, but from a random sample of 75 maize farmers (Johnston, 2011), 73% acknowledged having received some sort of seasonal forecast, 33% said that they had confidence in the forecasts, but 55% indicated that forecast information was fairly or very relevant to their needs. All farmers agreed that the expected rainfall in terms of normal and/or below normal and/or above normal was of primary interest.

When asked to consider the usefulness of the limited spatial resolution of the forecasts, 63% stated very useful or useful, and 32% stated not useful. A number of farmers raised the proviso that the resolution was acceptable if the forecast was accurate. It was apparent that without any verification of the forecast available to farmers their confidence in the forecast was limited.

A question regarding the acceptable minimum accuracy for specific predictions of rainfall asked respondents to state the frequency with which a forecast would need to be right for each attribute of a forecast. The results for various climatic parameters, such as total seasonal rainfall or rainfall onset, showed that they expected a 60 to 80% strike rate of a forecast being within a specified forecast category (whether percentile or tercile).

In the maize example, the most popular forecast characteristics desired by users were given as (i) approximate timing and duration of dry spells, (ii) intraseasonal distribution of rainfall, (iii) prediction of monthly rainfall anomaly, (iv) approximate date of cessation of rainfall, and (v) approximate date of onset of rainfall.

Is the information in a seasonal forecast relevant and compatible? Can it then provide decision options depending on the variation of its content? Nicholls (2000) emphasized that the value of seasonal forecasts lies not only in accuracy but also in their ability to allow management options which farmers could use to take advantage of the forecasts. The effective interpretation by farmers would necessarily depend on the answers to these questions. A simple example was revealed when farmers were questioned about the profit–loss relationship with the actual rainfall in a prior year.

It was clear that farmers who used the forecast were, however, still skeptical—none were willing to credit the forecast with their success, though 78% said they "believed" it, with 23% stating it had been consistently accurate and 53% saying it was partially accurate. The balance of 24% said that it was not at all accurate.

Climate Change

Winsemius et al. (2014) show potential forecast utility of two types of tailored forecast, one addressing staple crops and another affecting livestock, as well as how such utility might change under conditions of projected climate change. In the case of livestock, potential utility under climate change is clear, as increased likelihood of heatwaves is a robust finding. As a result, not only would a currently tailored forecast for heat stress for livestock in the SADC region have potential significant utility (see, e.g., Archer van Garderen [2011] and Archer van Garderen et al. [2015]) but increased temperatures under climate change is likely to make such a forecast of even greater potential utility. It is important to note there that we are only considering in this example increased temperatures and not indirect drivers of stress for the agricultural sector, including effects on pests and pathogens.

Conclusion

Grasslands are important for livestock and native herbivores. Here we examine the agroclimatic association of the world's grasslands, looking specifically at grasslands in Australia, South America (Uruguay), and South Africa as examples of important grassland communities. They are generally located in drier regions (arid and semiarid) with large diurnal temperature variations and high interannual rainfall variability. Rainfall is a key driver of production and the way pastoralists manage the hydrological cycle can determine the difference between success and failure in relation to financial, resource, and social indicators. Drought is a recurring and challenging component of pastoral management. The hydro-illogical cycle describes disengaged cycles of hydrology with decision-making that leads to apathy when conditions are wet and favorable and panic during drought. The hydro-psychological cycle describes the social impacts of drought that have gained more notoriety in the last decade. Despite our understanding of these cycles there is little evidence from pastoralists, business, and governments for major investment in preparing for drought. National drought policies are more focused on crisis management during drought.

An understanding of the ENSO has provided the most widespread application of climate science into pastoral decision-making. Yet given that nearly three decades have passed, the global adoption of this technology in pastoral decision-making is relatively modest, the highest being in Australia where about one in three agricultural decision-makers use the information. Many reasons have been given for the modest adoption, but it has become evident that a mixture of institutional, dissemination, trust, customizing, and servicing arrangements are necessary for widespread adoption. In addition, relationships between rainfall and ENSO are nonlinear, the relationships are flipped in different parts of the world (e.g., in Australia and South Africa El Niños are aligned with lower than normal rainfall, whereas in Uruguay they align with higher than normal rainfall), interrelationships between climate drivers are complex, and the understanding of how teleconnections between different drivers intensify or weaken individual impacts on climate are not well understood. Nonlinearity in these relationships provides variation in forecast strength, lack of adequate lead times, and non-alignment of forecast strength at times when key decisions are made. For example, in Australia the period of best forecast strength is in spring, but for most of the northern Australian pastoral regions the summer period provides most of the rainfall needed for production of pastures. On the other hand, in South Africa the summer season is associated with the greater ENSO predictability, which aligns nicely with key decisions relating to allocating stock numbers based on production of pastures in summer.

Longer lead times provide more time for making decisions. Pastoralists in northern Australia rely on summer rainfall, and key management decisions can change depending on rainfall. The extensive nature of these grazing systems requires some weeks or months to make and implement a change to the standard management cycle. The SOI phases used at 0- to 2-mo lead time for the November–March period provided a forecast for the key rainfall period in northern Australia over a lead time long enough for key decisions to be made and implemented before November (Cobon and Toombs, 2013). Output from some GCMs can provide an experimental forecast with a 12-mo lead time (e.g., http://www.cpc.ncep.noaa.gov/products/predictions/90day/), but the level of forecast skill needs to be considered at these long lead times. In Uruguay, the warning of drought in December 2010 that was issued in August 2010 gave institutions and pastoralists 4 month's warning that enabled some preparation before the drought impacted. The forecasts were used in decision-making because of the trust that was built over a decade between the institutions issuing the forecasts and the users and decision-makers of the information. In South Africa, there is a lack of evidence to show that seasonal forecasts of rainfall are being incorporated into decision-making by farmers and institutions, although they are used by DAFF and the National Agrometeorological Committee in Disaster Management to reduce risk and provide an input into the early warning system for droughts and floods that can have catastrophic impacts. It is evident that more effort, analysis, and attention is required to improve the predictability, lead times, and trust that decision-makers have in the current operational (and experimental) forecast systems.

The moderate level of forecast skill referred to above was determined by analyzing across all seasons over years that represent a mixture of ENSO years. One of the major problems with using forecasts in decision-making is the lack of forecast skill in ENSO neutral years, which represent about 40% of all years. Currently we can only reliably discriminate anomalously wet or dry events from the rest during El Niño and La Niña events. Educating users of forecasts about these simple climate science principles can help users make better judgments about the use and value of climate forecasts.

The value of incorporating seasonal climate forecasts into decision-making of individual farmers can be substantial (see, e.g., Hammer et al. [2000b]). However, the use of climate forecasts in national government actions can affect the entire agricultural sector of a country (see, e.g., the Uruguay experience in the 2010 La Niña drought and in Disaster Risk Management in South Africa). Although assessing the value of such use is difficult, it is undoubtedly huge, especially in developing countries where those actions can determine the sustainability of small, vulnerable farmer communities. In Australia, the Centre for International Economics estimated of the value of a GCM-like forecast of rainfall of between $958 M and $1,930 M AUD for cropping and livestock combined (Centre for International Economics, 2014), or between 0.5 and 9.0% of the gross value added for the agricultural sector.

Meza et al. (2008) reviewed 33 papers across a number of countries on valuing seasonal climate forecasts for agriculture covering 58 different assessments. They showed that in terms of scale, the most common studies were at the individual crop or enterprise level. The vast majority were agronomic, single crop studies, with three horticultural and two livestock. The valuations observed ranged from zero to $700/ha, with horticultural studies generally being the highest, followed by agronomic studies, and then livestock. The values tended to be related to the intensity of the system and the value of the crop.

The extent to which seasonal climate forecast information is used by pastoralists in decision-making in South Africa and Uruguay is not known. However, it is likely that there is some use of seasonal climate forecast information in developing countries, because, for the most part, implementing coping strategies to manage climate variability requires only a macro view of the differences in mean annual or seasonal rainfall between climatology and El Niño and La Niña conditions (Kumar, 2010). The impact and coping strategies used during previous climate events is well known by farmers and institutions, but it is the responsibility of government, government agencies, and welfare and research organizations to use this predictable signal in the climate to warn the community about forthcoming climatic events that have caused catastrophic economic, social, and environmental hardship in the past.

A micro view of seasonal climate information requires detailed information about the climate history and the use of comprehensive application models (Kumar, 2010). This paradigm can provide site-specific information for any season about the climate means, probability distributions, strength, and predictability of the ENSO signal and forecast reliability, and when climate data is integrated with agricultural models, similar information can be provided about the predicted production of that commodity (Hammer et al., 1996; Potgieter et al., 2003). Using these climate-integrated agricultural production models to change decisions based on a climate driven prediction of modeled production has significantly increased profitability (Hammer, 2000; McKeon et al., 2000; Cobon and McKeon, 2002) and resource sustainability (McKeon et al., 2004) at the farm, local, and regional scales in Australia. The limitation for the application of these models in developing countries is that long-term (>30 yr) daily records of rainfall, temperature (max and min), radiation, evaporation, and humidity are often needed to run the models, and these records are either not always available or they have not been transferred from paper to digital format.

As is the case with the use of seasonal climate forecast information, neither do we know the extent to which agricultural decision-makers change decisions based on this information. Nor do we know the economic, social, or environmental implications of any changed decisions. These metrics are difficult to obtain, even in developed agricultural economies (Kumar, 2010). Surveys of climate forecast use in Australia suggest that between 30 and 50% of agricultural producers take seasonal climate forecasts into account when making farm management decisions, but how they used the information is open to interpretation. This level of use in Australia appears to be relatively high compared with farmers in other countries, particularly in the developing world. The shortcomings in the use and value of seasonal forecasts in agriculture are evident despite them being available to the public in various forms since the early 1990s. It would seem that targeted campaigns that customize a forecast for the region and management system that are supported by professional and local champions generate better understanding, greater confidence and trust, and produce higher adoption rates by the decision-makers (e.g., see Cobon et al. [2005] and [2008] and Cliffe et al. [2016] in Australia and the success of the targeted campaign in Uruguay in 2010). Local champions working in regional areas with a good understanding of climate science and its application in agriculture can develop the trust of farmers by providing links to the climate institutions issuing the forecasts, linking forecasts with on-farm decisions, and customizing forecasts for the industry at the regional scale. These activities ensure that the farmers are provided with a climate service and not merely climate information which is often the case when forecast are issued in the media and over the Web.

Despite all this, incorporating climate science into climate-integrated agricultural models, decision support tools, training, education, and extension packages has made a valuable contribution to agricultural decision-making for governments, institutions, business, and pastoralists. There remains an opportunity in the developed and developing grassland regions to use climate forecasts to improve food security, poverty, production, resource sustainability, and social outcomes.

Acknowledgments

The lead author acknowledges the Queensland Government through the Departments of Science Information Technology Innovation and the Arts, Environment and Resource Management and Primary Industries for their support in climate RD&E beginning in the late 1980s. The work has been funded by other State Governments, the Commonwealth Government, and many industry research development corporations over the years, including the Managing Climate Variability Program.

References

Abawi, G.Y., S.C. Dutta, T. Harris, J. Ritchie, D. Rattray, and A. Crane. 2000. The use of seasonal climate forecasts in water resources management. In: Proceedings of the Third International Hydrology and Water Resources Symposium of the Institution of Engineers. Institution of Engineers, Canberra, Australia. p. 20–23.

Alcock, D., P. Graham, A.D. Moore, J.M. Lilley, and E.J. Zurcher. 2010. GrassGro indicates that erosion risk drives adaptation of southern tablelands grazing farms to projected climate change. In: J. Palutikof, A. Ash, M. Stafford Smith, M. Waschka, M. Sparksman and K. Franzidis, editors, Climate adaptation futures—Preparing for the unavoidable impacts of climate change. Proceedings of the 2010 International Climate Change Adaptation Conference, Surfers Paradise, Australia. 29 June–1 July 2010. NACCRF: Brisbane, Australia. p. 67.

Allan, R.J. 2000. ENSO and climate variability in the last 150 years. In: H.F. Diaz and V. Markgraf, editors, El Niño and the Southern Oscillation: Multiscale variability and its impacts on the natural ecosystems and society. Cambridge Univ. Press, Cambridge, UK. p. 3–56.

Allan, R.J., J.A. Lindesay, and D.E. Parker. 1996. El Niño southern oscillation and climatic variability. CSIRO Publ., Melbourne, Australia.

Alves, O., G. Wang, A. Zhong, N. Smith, F. Tseitkin, G. Warren, and G. Meyers. 2003. POAMA: Bureau of Meteorology operational coupled model seasonal forecast system. In: Proceedings of national drought forum. Brisbane, Australia p. 49–56.

Archer, E.R.M. 2004. Beyond the "climate versus grazing" impasse: Using remote sensing to investigate the effects of grazing system choice on vegetation cover in the eastern Karoo. J. Arid Environ. 57:381–408. doi:10.1016/S0140-1963(03)00107-1

Archer van Garderen, E.R.M. 2011. Reconsidering cattle farming in Southern Africa under a changing climate. Weather Clim. Soc. 3(4):249–253. doi:10.1175/WCAS-D-11-00026.1

Archer van Garderen, E.R.M., C.L. Davis, and M.A. Tadross. 2015. A changing environment for livestock in South Africa. In: J. Emel and H. Neo, editors, The political ecologies of meat. Routledge, New York.

Ash, A.J., P. O'Reagain, G.M. McKeon, and M. Stafford Smith. 2000. Managing climatic variability in grazing enterprises: A case study for Dalrymple shire, north-eastern Australia. In: G.L. Hammer, N. Nicholls, and C. Mitchell, editors, Applications of seasonal climate forecasting in agricultural and natural ecosystems—The Australian experience. Kluwer Academic Press, Amsterdam, The Netherlands. p. 253–270. doi:10.1007/978-94-015-9351-9_16

Ashok, K., Z. Guan, and T. Yamagata. 2003. Influence of the Indian Ocean dipole on the Australian winter rainfall. Geophys. Res. Lett. 30(15). doi:10.1029/2003GL017926

Australian Government Department of Agriculture Fisheries and Forestry. 2004. Review of the Agriculture Advancing Australia Package 2000–2004. Australian Government Department of Agriculture Fisheries and Forestry, Canberra, Australia.

Australian State of the Environment Committee. 2001. Australia, state of the environment 2001: Independent report to the Commonwealth Minister for the Environment and Heritage. CSIRO, Publishing on behalf of the Department of the Environment and Heritage, Canberra, Australia.

Bastin, G., C. Stokes, D. Green, and K. Forrest. 2014. Australian rangelands and climate change – Pastoral production and adaptation. Ninti One Limited and CSIRO, Alice Springs, Australia.

Baethgen, W.E. 2010. Climate risk management for adaptation to climate variability and change. Crop Sci. 50(2):70–76. doi:10.2135/cropsci2009.09.0526

Baethgen, W.E., and G.O. Magrin. 2000. Applications of climate forecasts in the agricultural sector of south east South America. In: M.V.K. Sivakumar, editor, Climate prediction and agriculture, Proceedings of the START/WMO International Workshop. International START Secretariat, Geneva and Washington, DC. p. 248–266.

Bartaburu, D., E. Duarte, and E. Montes. 2009. Las sequías: Un evento que afecta la trayectoria de las empresas y su gente. In: M. Pereira, editor, Hermes Morales Grosskopf,

Familias y Campo: Rescatando estrategias de adaptación. Instituto Plan Agropecuario, Montevideo, Uruguay. p. 155–168.

Bartman, A.G., W.A. Landman, and C.J. de W, Rautenbach. 2003. Recalibration of general circulation model output to austral summer rainfall over southern Africa. Int. J. Climatol. 23:1407–1419. doi:10.1002/joc.954

Behera, S.K., J.J. Luo, S. Masson, S.A. Rao, H. Sakuma, and T. Yamagata. 2006. A CGCM study on the interaction between IOD and ENSO. J. Clim. 19(9):1688–1705. doi:10.1175/JCLI3797.1

Behera, S.K., and T. Yamagata. 2001. Subtropical SST dipole events in the southern Indian Ocean. Geophys. Res. Lett. 28:327–330. doi:10.1029/2000GL011451

Beraki, A.F., D. DeWitt, W.A. Landman, and C. Olivier. 2014. Dynamical seasonal climate prediction using an ocean-atmosphere coupled climate model developed in partnership between South Africa and the IRI. J. Clim. 27:1719–1741. doi:10.1175/JCLI-D-13-00275.1

Bowman, P.J., G.M. McKeon, and D.H. White. 1995. The impact of long range climate forecasting on the performance of sheep flocks in Victoria. Aust. J. Agric. Res. 46:687–702. doi:10.1071/AR9950687

Brown, J.N., P.C. McIntosh, M.J. Pook, and J.S. Risbey. 2009. An investigation of the links between ENSO flavors and rainfall processes in southeastern Australia. Mon. Weather Rev. 137(11):3786–3795. doi:10.1175/2009MWR3066.1

Cai, W., A. Santoso, G. Wang, E. Weller, L. Wu, K. Ashok, Y. Masumoto, and T. Yamagata. 2014a. Increased frequency of extreme Indian Ocean Dipole events due to greenhouse warming. Nature 510(7504):254–258. doi:10.1038/nature13327

Cai, W., S. Borlace, M. Lengaigne, P. Van Rensch, M. Collins, G. Vecchi, A. Timmermann, A. Santoso, M.J. McPhaden, L. Wu, M.H. England, G. Wang, E. Guilyardi, and F.F. Jin. 2014b. Increasing frequency of extreme El Niño events due to greenhouse warming. Nat. Clim. Chang. 4(2):111–116. doi:10.1038/nclimate2100

Cai, W., P. Van Rensch, T. Cowan, and H.H. Hendon. 2012. An asymmetry in the IOD and ENSO teleconnection pathway and its impact on Australian climate. J. Clim. 25(18):6318–6329. doi:10.1175/JCLI-D-11-00501.1

Cai, W., G. Wang, A. Santoso, M.J. McPhaden, L. Wu, F.F. Jin, and E. Guilyardi. 2015. Increased frequency of extreme La Niña events under greenhouse warming. Nat. Clim. Chang. 5(2):132–137. doi:10.1038/nclimate2492

Cai, W., X.T. Zheng, E. Weller, M. Collins, T. Cowan, M. Lengaigne, and T. Yamagata. 2013. Projected response of the Indian Ocean dipole to greenhouse warming. Nat. Geosci. 6(12):999–1007. doi:10.1038/ngeo2009

Cane, M.A. 2005. The evolution of El Niño, past and future. Earth Planet. Sci. Lett. 230:227–240. doi:10.1016/j.epsl.2004.12.003

Cane, M.A., Eshel, G. and Buckland, R.W. 1994. Forecasting Zimbabwean maize yield using eastern equatorial Pacific sea surface temperature. Nature. 370:204–205. doi:10.1038/370204a0.

Carter, J.O., D. Bruget, R. Hassett, B. Henry, D. Ahrens, K. Brook, K. Day, N. Flood, W. Hall, G. McKeon, and C. Paull. 2003. Australian grassland and rangeland assessment by spatial simulation (AussieGRASS). In: R. Stone and I. Partridge , editors, Science for drought: Proceedings of the National Drought Forum, April 2003. Brisbane, Australia. Queensland Department of Primary Industries. p. 152–159.

Carter, J.O., W.B. Hall, K.D. Brook, G.M. McKeon, K.A. Day, and C.J. Paull. 2000. AussieGRASS: Australian grassland and rangeland assessment by spatial simulation. In: G. Hammer, N. Nicholls, and C. Mitchell, editors, Applications of seasonal climate forecasting in agricultural and natural ecosystems External link icon—The Australian experience. Kluwer Academic Press, The Netherlands. p. 329–349. doi:10.1007/978-94-015-9351-9_20

Carvalho, L.M., C. Jones, and T. Ambrizzi. 2005. Opposite phases of the Antarctic Oscillation and relationships with intraseasonal to interannual activity in the tropics during the austral summer. J. Clim. 18(5):702–718. doi:10.1175/JCLI-3284.1

Cazes-Boezio, G., A.W. Roberton, and C.R. Mechoso. 2003. Seasonal dependence of ENSO teleconnections over South America and relationships with precipitation in Uruguay. J. Clim. 16:1159–1176. doi:10.1175/1520-0442(2003)16<1159:SDOETO>2.0.CO;2

Centre for International Economics. 2014. Analysis of the benefits of improved seasonal climate forecasting for agriculture, Report prepared for the Managing Climate Variability R&D Program. CIE, Canberra, Australia.

Changnon, S.A., J.M. Changnon, and D. Changnon. 1995. Uses and applications of climate forecasts for power utilities. Bull. Am. Meteorol. Soc. 76:711–720. doi:10.1175/1520-0477(1995)076<0711:UAAOCF>2.0.CO;2

Chiew, F.H.S., T.C. Piechota, J.A. Dracup, and T.A. McMahon. 1998. El Niño/Southern Oscillation and Australian rainfall, streamflow and drought: Links and potential forecasting. J. Hydrol. 204:138–149. doi:10.1016/S0022-1694(97)00121-2

Childs, I.R.W., P.A. Hastings, and A. Auliciems. 1991. The acceptance of long-range weather forecasts: A question of perception? Aust. Meteorol. Mag. 39:105–112.

Chung, C.T., S.B. Power, J.M. Arblaster, H.A. Rashid, and G.L. Roff. 2014. Nonlinear precipitation response to El Niño and global warming in the Indo-Pacific. Clim. Dyn. 42(7-8):1837–1856. doi:10.1007/s00382-013-1892-8

Cliffe, N., Stone, R., Coutts, J., Reardon Smith, K. and Mushtaq, S. 2016. Developing the capacity of farmers to understand and apply seasonal climate forecasts through collaborative learning processes. J. Agric. Educ. Ext. 22:311–325. doi:10.1080/1389224X.2016.1154473

Cobon, D.H. 1999. Use of seasonal climate forecasts for managing grazing systems in western Queensland. In: D. Eldridge and D. Freudenberger, editors, Proceedings of the VIth International Rangeland Congress. Townsville, Australia. Vol. 2. Aitkenvale, Australia. p. 855–857.

Cobon, D.H., K.L. Bell, J.N. Park, and D.U. Keogh. 2008. Summative evaluation of climate application activities with pastoralistists in western Queensland. Rangeland J. 30:361–374. doi:10.1071/RJ06030

Cobon, D.H., and G.M. McKeon. 2002. The value of seasonal forecasts in maintaining the resource and improving profitability in grazing systems—A case study in western Queensland. In: G.M. McKeon and W.B. Hall, editors, Learning from history—Can seasonal forecasting prevent land degradation of Australia's grazing lands? Technical report for the climate variability in agriculture program. Land and Water Australia, Canberra, Australia. p. 273–297.

Cobon, D.H., and J.N. Park., K.L. Bell, I.W. Watson, W. Fletcher, and M. Young. 2005. Targeted seasonal climate forecasts offer more to pastoralists. XX International Grassland Congress. Dublin, Australia. p. 556

Cobon, D.H., G.S. Stone, J.O. Carter, J. Scanlan, N.R. Toombs, X. Zhang, J. Willcocks, and G.M. McKeon. 2009. The climate change risk management matrix for the grazing industry of northern Australia. Rangeland J. 31(1):31–49. doi:10.1071/RJ08069

Cobon, D.H., and N.R. Toombs. 2013. Forecasting rainfall based on Southern Oscillation phases at longer lead-times in Australia. Rangeland J. 35(4):373–383. doi:10.1071/RJ12105

Collins, M., S.-I. An, W. Cai, A. Ganachaud, E. Guilyardi, F.-F. Jin, M. Jochum, M. Lengaigne, S. Power, A. Timmermann, G. Vecchi, and A. Wittenberg. 2010. The impact of global warming on the tropical Pacific Ocean and El Niño. Nat. Geosci. 3:391–397. doi:10.1038/ngeo868

Colman, R., L. Deschamps, M. Naughton, L. Rikus, A. Sulaiman, K. Puri, G. Roff, Z. Sun, and G. Embury, 2005. BMRC Atmospheric Model (BAM) version 3.0: comparison with mean climatology. BMRC Research Report No. 108, Bur. Met., Melbourne, Australia.

Conway, D., E. Archer van Garderen, D. Deryng, S. Dorling, T. Krueger, W.A. Landman, B. Lankford, K. Lebek, T. Osborn, C. Ringler, J. Thurlow, T. Zhu, and C. Dalin. 2015. Climate and southern Africa's water-energy-food nexus. Nat. Clim. Chang. 5:837–846. doi:10.1038/nclimate2735

Cottrill, A., H.H. Hendon, E.P. Lim, S. Langford, K. Shelton, A. Charles, and Y. Kuleshov. 2013. Seasonal forecasting in the Pacific using the coupled model POAMA-2. Weather Forecast. 28(3):668–680. doi:10.1175/WAF-D-12-00072.1

Coventry, W.L. 2001. Single event versus frequency formats for presenting climate forecast probabilities: Beyond reasoning abilities to judgement. Honours thesis, School of Psychology, University of Queensland, St. Lucia, Australia.

Crimp, S.J., C.J. Stokes, S.M. Howden, A.D. Moore, B. Jacobs, P.R. Brown, and P. Leith. 2010. Managing Murray–Darling Basin livestock systems in a variable and changing climate: Challenges and opportunities. Rangeland J. 32(3):293–304. doi:10.1071/RJ10039

Cullen, B.R., I.R. Johnson, R.J. Eckard, G.M. Lodge, R.G. Walker, R.P. Rawnsley, and M.R. McCaskill. 2009. Climate change effects on pasture systems in south-eastern Australia. Crop Pasture Sci. 60(10):933–942. doi:10.1071/CP09019

Dai, A. 2011. Drought under global warming: A review. Wiley Interdiscip. Rev. Clim. Chang. 2(1):45–65. doi:10.1002/wcc.81

Dalgleish, L., W. Coventry, and R. McCrea. 2001. Climate variability and farmers risk assessments and decision making. Final Report. Land and Water Australia, Canberra, Australia.

Day, K., D. Aherns, and G. McKeon. 2003. Simulating historical droughts: Some lessons for drought policy. In: R. Stone and I. Partridge, editors, Science for drought. Proceedings of the National Drought Forum, Brisbane. Department of Primary Industries, Queensland, Australia.

Diaz, A.F., C.D. Studzinski, and C.R. Mechoso. 1998. Relationship between precipitation anomalies in Uruguay and southern Brazil and sea surface temperature in the Pacific and Atlantic Oceans. J. Clim. 11:251–271. doi:10.1175/1520-0442(1998)011<0251:RBPAIU>2.0.CO;2

Ding, Q., E.J. Steig, D.S. Battisti, and J.M. Wallace. 2012. Influence of the tropics on the Southern Annular Mode. J. Clim. 25(18):6330–6348. doi:10.1175/JCLI-D-11-00523.1

Drosdowsky, W., and M.C. Wheeler. 2014. Predicting the onset of the north Australian wet season with the POAMA Dynamical Prediction System. Weather Forecast. 29(1):150–161. doi:10.1175/WAF-D-13-00091.1

Drought Policy Review Expert Social Panel. 2008. It's about people: Changing perspective. A report to government by an expert social panel on dryness. Report to the Minister for Agriculture, Fisheries and Forestry, Canberra, Australia.

Engelbrecht, C.J., and F.A. Engelbrecht. 2015. Shifts in Köppen-Geiger climate zones over southern Africa in relation to key global temperature goals. Theor. Appl. Climatol. 123:247–261. doi:10.1007/s00704-014-1354-1

England, M.H., C.C. Ummenhofer, and A. Santoso. 2006. Interannual rainfall extremes over southwest Western Australia linked to Indian Ocean climate variability. J. Clim. 19(10):1948–1969. doi:10.1175/JCLI3700.1

Everingham, Y.L., R.C. Muchow, R.C. Stone, and D.H. Coomans. 2003. Using Southern Oscillation Index phases to forecast sugarcane yields: A case study for north-eastern Australia. Int. J. Climatol. 23:1211–1218. doi:10.1002/joc.920

Fahey, D. 1997. Changing grazing management on "Keen-Gea": The strategies and findings—A producers experience. An introduction to management practices in the Dalrymple Shire. Grazing land management unit, Queensland Department of Primary Industries, Charters Towers, Australia. p 16–17.

Fauchereau, N., B. Pohl, C.J.C. Reason, M. Rouault, and Y. Richard. 2008. Recurrent daily OLR patterns in the Southern Africa/Southwest Indian Ocean region, implications for South African rainfall and teleconnections. Clim. Dyn. 32:575–591. doi:10.1007/s00382-008-0426-2

Goddard, L., Y. Aitchellouche, W. Baethgen, M. Dettinger, R. Graham, P. Hayman, M. Kadi, R. Martinez, H. Meinke, and E. Conrad. 2010. Providing seasonal-to-interannual climate information for risk management and decision-making. Procedia Environ. Sci. 1:81–101. doi:10.1016/j.proenv.2010.09.007

Haigh, T., E. Takle, J. Andresen, M. Widhalm, J.S. Carlton, and J. Angel. 2015. Mapping the decision points and climate information use of agricultural producers across the U.S. Corn Belt. Clim. Risk Manage. 7:20–30. doi:10.1016/j.crm.2015.01.004

Hall, A., and M. Visbeck. 2002. Synchronous variability in the Southern Hemisphere atmosphere, sea ice, and ocean resulting from the Annular Mode. J. Clim. 15(21):3043–3057. doi:10.1175/1520-0442(2002)015<3043:SVITSH>2.0.CO;2

Hall, W.B., G.M. McKeon, J.O. Carter, K.A. Day, S.M. Howden, J.C. Scanlan, P.W. Johnston, and W.H. Burrows. 1998. Climate change in Queensland's grazing lands: II. An assessment of the impact on animal production from native pastures. Rangeland J. 20:177–205. doi:10.1071/RJ9980177

Hammer, G. 2000. A systems approach to applying seasonal forecasts. In: G.L. Hammer, N. Nicholls, and C. Mitchell, editors, Applications of seasonal climate forecasting in agricultural and natural ecosystems—The Australian experience. Kluwer Academic Press, Amsterdam, The Netherlands. p. 51–65. doi:10.1007/978-94-015-9351-9_4

Hammer, G.L., D.P. Holzworth, and R.C. Stone. 1996. The value of skill in seasonal climate forecasting to wheat crop management in a region with high climatic variability. Aust. J. Agric. Res. 47:717–737. doi:10.1071/AR9960717

Hammer, G.L., N. Nicholls, and C. Mitchell. 2000a. Applications of seasonal climate forecasting in agricultural and natural ecosystems. In: Atmospheric and Oceanographic Sciences Library. Vol. 21. Springer, Berlin. doi:10.1007/978-94-015-9351-9

Hammer, G.L., N. Nicholls, and C. Mitchell, editors. 2000b. Applications of seasonal climate forecasting in agricultural and natural ecosystems—The Australian experience. Kluwer Academic Press, Amsterdam, The Netherlands.

Hartmann, H.C., T.C. Pagano, S. Sorooshian, and R. Bales. 2002. Confidence builders—Evaluating seasonal climate forecasts from user perspectives. Bull. Am. Meteorol. Soc. May:683–698. doi:10.1175/1520-0477(2002)083<0683:CBESCF>2.3.CO;2

Hendon, H.H., D.W. Thompson, and M.C. Wheeler. 2007. Australian rainfall and surface temperature variations associated with the Southern Hemisphere annular mode. J. Clim. 20(11):2452–2467. doi:10.1175/JCLI4134.1

Hill, J.O., R.J. Simpson, A.D. Moore, P. Graham, and D.F. Chapman. 2004. Impact of phosphorus application and sheep grazing on the botanical composition of sown pasture and naturalised, native grass pasture. Aust. J. Agric. Res. 55:1213–1225. doi:10.1071/AR04090

Howden, S.M., W.B. Hall, and D. Bruget. 1999a. Heat stress and beef cattle in Australian rangelands: Recent trends and climate change. In: D. Eldridge and D. Freudenberger, editors, People and rangelands: Building the future. Proceedings of the VI International Rangeland Congress, Townsville, Australia. p. 43–45.

Howden, S.M., G.M. McKeon, L. Walker, J.O. Carter, J.P. Conroy, K.A. Day, W.B. Hall, A.J. Ash, and O. Ghannoum. 1999b. Global change impacts on native pastures in south-east Queensland, Australia. Environ. Model. Softw. 14:307–316. doi:10.1016/S1364-8152(98)00082-6

Howden, S.M., and J. Turnpenny. 1997. Modelling heat stress and water loss of beef cattle in subtropical Queensland under current climates and climate change. In: D.A. McDonald and M. McAleer, editors, MODSIM 1997. International Congress on Modelling and Simulation Proceedings. University of Tasmania, Hobart. Modelling and Simulation Society of Australia, Canberra, Australia. p. 1103–1108.

Hudson, D., O. Alves, H.H. Hendon, and G. Wang. 2011. The impact of atmospheric initialisation on seasonal prediction of tropical Pacific SST. Clim. Dyn. 36:1155–1171. doi:10.1007/s00382-010-0763-9

Hudson, D., A.G. Marshall, Y. Yin, O. Alves, and H.H. Hendon. 2013. Improving intraseasonal prediction with a new ensemble generation strategy. Mon. Weather Rev. 141(12):4429–4449. doi:10.1175/MWR-D-13-00059.1

Hunt, B.G., and A.C. Hirst. 2000. Global climatic models and their potential for seasonal climatic forecasting. In: G.L. Hammer, N. Nicholls, and C. Mitchell, editors, Applications of seasonal climate forecasting in agricultural and natural ecosystems—The Australian experience. Kluwer Academic Press, Amsterdam, The Netherlands. p. 89–108. doi:10.1007/978-94-015-9351-9_7

Ihara, C., Y. Kushnir, and M.A. Cane. 2008. Warming trend of the Indian Ocean SST and Indian Ocean dipole from 1880 to 2004. J. Clim. 21(10):2035–2046. doi:10.1175/2007JCLI1945.1

IPCC. 2013. Climate Change 2013. The physical science basis. Contribution of Working Group I to the Fifth Assessment Report of the Intergovernmental Panel on Climate Change. IPCC, Bern Switzerland.

Johnston, P. 2011. The uptake and usefulness of seasonal forecasting products: A case study of commercial maize farmers in South Africa, Lambert Academic Publ., Saarbrükken, Germany

Johnston, P., G. McKeon, R. Buxton, D. Cobon, K. Day, W. Hall, and J. Scanlan. 2000. Managing climatic variability in Queensland's grazing lands—New approaches. In: G.L. Hammer, N. Nicholls, and C. Mitchell, editors, Applications of seasonal climate forecasting in agricultural and natural ecosystems—The Australian experience. Kluwer Academic Press, Amsterdam, The Netherlands. p. 197–226. doi:10.1007/978-94-015-9351-9_14

Johnston, P.A., E.R.M. Archer, C.H. Vogel, C.N. Bezuidenhout, W.J. Tennant, and R. Kuschke. 2004. Review of seasonal forecasting in South Africa: Producer to end-user. Clim. Res. 28(1):67–82. doi:10.3354/cr028067

Johnston, P.W., G.M. Mckeon, and K.A. Day. 1996. Objective safe grazing capacities for south-west Queensland Australia; development of a model for individual properties. Rangeland J. 18:244–258. doi:10.1071/RJ9960244

Jourdain, N.C., A.S. Gupta, A.S. Taschetto, C.C. Ummenhofer, A.F. Moise, and K. Ashok. 2013. The Indo-Australian monsoon and its relationship to ENSO and IOD in reanalysis data and the CMIP3/CMIP5 simulations. Clim. Dyn. 41(11-12):3073–3102. doi:10.1007/s00382-013-1676-1

Jury, M.R., H.M. Mulenga, and S.J. Mason. 1999. Exploratory longrange models to estimate summer climate variability over southern Africa. J. Clim. 12:1892–1899. doi:10.1175/1520-0442(1999)012<1892:ELRMTE>2.0.CO;2

Jury, M.R., and B. Pathack. 1991. A study of climate and weather variability over the tropical southwest Indian Ocean. Meteorol. Atmos. Phys. 47:37–48. doi:10.1007/BF01025825

Jury, M.R., H.R. Valentine, and J.R.E. Lutjeharms. 1993. Influence of the Agulhas current on summer rainfall along the southeast coast of South Africa. J. Appl. Meteorol. Climatol. 32:1282–1287. doi:10.1175/1520-0450(1993)032<1282:IOTACO>2.0.CO;2

Keogh, D.U., G.Y. Abawi, S.C. Dutta, A.J. Crane, J.W. Ritchie, T.R. Harris, and C.G. Wright. 2004b. Context evaluation: A profile of irrigator climate knowledge, needs and practices in the northern Murray–Darling Basin to aid development of climate based decision support tools and information and dissemination research. Aust. J. Exp. Agric. 44:247–257. doi:10.1071/EA02055

Keogh, D.U., K.L. Bell, J.N. Park, and D.H. Cobon. 2004a. Formative evaluation to benchmark and improve climate-based decision support for graziers in western Queensland. Aust. J. Exp. Agric. 44:233–246. doi:10.1071/EA01204

Keogh, D.U., I.W. Watson, K.L. Bell, D.H. Cobon, and S.C. Dutta. 2005. Climate information needs of Gascoyne Murchison pastoralists: A representative study of the Western Australian grazing industry. Aust. J. Exp. Agric. 45:1613–1625. doi:10.1071/EA04275

Kim, S.T., W. Cai, F.F. Jin, A. Santoso, L. Wu, E. Guilyardi, and S.I. An. 2014. Response of El Niño sea surface temperature variability to greenhouse warming. Nat. Clim. Chang 4:786–790. doi:10.1038/nclimate2326

Kgatuke, M.M., W.A. Landman, A. Beraki, and M. Mbedzi. 2008. The internal variability of the RegCM3 over South Africa. Int. J. Climatol. 28:505–520. doi:10.1002/joc.1550

Klopper, E., C.H. Vogel, and W.A. Landman. 2006. Seasonal climate forecasts– potential agricultural-risk management tools? Clim. Change 76:73–90. doi:10.1007/s10584-005-9019-9

Köppen, W. 1931. Klimakarte der Erde. Grundriss der Klimakunde, 2nd Ed. Walter de Gruyter & Company, Berlin and Leipzig.

Kumar, A. 2010. On the assessment of the value of the seasonal forecast information. Meteorological Applications 17: 385–392.

Landman, W.A. 2014. How the International Research Institute for Climate and Society has contributed towards seasonal climate forecast modelling and operations in South Africa. Earth Perspectives 1:22. doi:10.1186/2194-6434-1-22

Landman, W.A., and A. Beraki. 2012. Multi-model forecast skill for midsummer rainfall over southern Africa. Int. J. Climatol. 32:303–314. doi:10.1002/joc.2273

Landman, W.A., A. Beraki, D. DeWitt, and D. Lötter. 2014. SST prediction methodologies and verification considerations for dynamical mid-summer rainfall forecasts for South Africa. Water SA 40(4):615–622. doi:10.4314/wsa.v40i4.6

Landman, W.A., S. Botes, L. Goddard, and M. Shongwe. 2005. Assessing the predictability of extreme rainfall seasons over southern Africa. Geophys. Res. Lett. 32:L23818. doi:10.1029/2005GL023965

Landman, W.A., D. DeWitt, and D.-E. Lee. 2011. The high-resolution global SST forecast set of the CSIR. Proceeding of the 27th Annual Conference of South African Society for Atmospheric Sciences. Council for Scientific and Industrial Research, Hartbeespoort, North-West Province, South Africa, p. 39–40.

Landman, W.A., D. DeWitt, D.E. Lee, A. Beraki, and D. Lötter. 2012. Seasonal rainfall prediction skill over South Africa: 1- vs. 2-tiered forecasting systems. Weather Forecast. 27:489–501. doi:10.1175/WAF-D-11-00078.1

Landman, W.A., M.M. Kgatuke, M. Mbedzi, A. Beraki, A. Bartman, and A. du Piesanie. 2009. Performance comparison of some dynamical and empirical downscaling methods for South Africa from a seasonal climate modelling perspective. Int. J. Climatol. 29:1535–1549. doi:10.1002/joc.1766

Landman, W.A., and S.J. Mason. 1999. Operational long-lead prediction of South African rainfall using canonical correlation analysis. Int. J. Climatol. 19:1073–1090. doi:10.1002/(SICI)1097-0088(199908)19:10<1073::AID-JOC415>3.0.CO;2-J

Landman, W.A., S.J. Mason, P.D. Tyson, and W.J. Tennant. 2001. Retro-active skill of multi-tiered forecasts of summer rainfall over southern Africa. Int. J. Climatol. 21:1–19. doi:10.1002/joc.592

Landsberg, RG., Ash, A.J., Shepherd, R.K., and McKeon, G.M. 1998. Learning from history to survive in the future: Management evolution on Trafalgar Station, north-east Queensland. Rangeland J. 20:104–118.

Langford, S., and H.H. Hendon. 2013. Improving reliability of coupled model forecasts of Australian seasonal rainfall. Mon. Weather Rev. 141(2):728–741. doi:10.1175/MWR-D-11-00333.1

Lengaigne, M., and G.A. Vecchi. 2010. Contrasting the termination of moderate and extreme El Niño events in coupled general circulation models. Clim. Dyn. 35:299–313.

L'Heureux, M.L., and D.W. Thompson. 2006. Observed relationships between the El Niño-Southern Oscillation and the extratropical zonal-mean circulation. J. Clim. 19(2):276–287. doi:10.1175/JCLI3617.1

Lilley, J.M., T.P. Bolger, M.B. Peoples, and R.M. Gifford. 2001. Nutritive value and the nitrogen dynamics of Trifolium subterraneum and Phalaris aquatica under warmer, high CO2 conditions. New Phytol. 150:385–395. doi:10.1046/j.1469-8137.2001.00101.x

Lindesay, J.A. 1988. South African rainfall, the Southern Oscillation and a southern hemisphere semi-annual cycle. J. Climatol. 8:17–30. doi:10.1002/joc.3370080103

Lim, E.P., H.H. Hendon, D.L. Anderson, A. Charles, and O. Alves. 2011. Dynamical, statistical-dynamical, and multimodel ensemble forecasts of Australian spring season rainfall. Mon. Weather Rev. 139(3):958–975. doi:10.1175/2010MWR3399.1

Lindesay, J.A. 2005 Climate and drought in the subtropics—The Australian example. In: L.C. Botterill and D.A. Wilhite, editors, From disaster response to risk management. Australia's national drought policy. Advances in natural and technological hazards research. Vol. 22. Springer, Berlin. p. 15–36. doi:10.1007/1-4020-3124-6_3

Littleboy, M., and G.M. McKeon. 1997. Subroutine GRASP: Grass production model, Documentation of the Marcoola version of Subroutine GRASP. Appendix 2 of Evaluating the risks of pasture and land degradation in native pasture in Queensland. Final Project Report for Rural Industries and Research Development Corporation project DAQ124A. Canberra, Australia.

Lo, F., M.C. Wheeler, H. Meinke, and A. Donald. 2007. Probabilistic forecasts of the onset of the north Australian wet season. Mon. Weather Rev. 135(10):3506–3520. doi:10.1175/MWR3473.1

Lough, J.M. 2007. Tropical river flow and rainfall reconstructions from coral luminescence: Great Barrier Reef, Australia. Paleoceanography 22(2). doi:10.1029/2006PA001377

Madden, R.A., and P.R. Julian. 1971. Detection of a 40–50 day oscillation in the zonal wind in the tropical Pacific. J. Atmos. Sci. 28(5):702–708. doi:10.1175/1520-0469(1971)028<0702:DOA DOI>2.0.CO;2

Malherbe, J., W.A. Landman, C. Olivier, H. Sakuma, and J.-J. Luo. 2014. Seasonal forecasts of the SINTEX-F coupled model applied to maize yield and streamflow estimates over north-eastern South Africa. Meteorol. Appl. 21:733–742. doi:10.1002/met.1402

Mann, T.H. 1993. Flexibility—The key to managing a northern beef property. Proceedings of the XVII International Grasslands Congress. Vol. 3. p. 1961–1964. New Zealand Grassland Assoc., Palmerston North, New Zealand.

Marshall, N.A., I.J. Gordon, and A.J. Ash. 2011. The reluctance of resource users to adopt seasonal climate forecasts that can enhance their resilience to climate variability. Clim. Change 107:511–529. doi:10.1007/s10584-010-9962-y

Mason, S.J. 1990. Temporal variability of sea surface temperatures around Southern Africa: A possible forcing mechanism for the 18-year rainfall oscillation? S. Afr. J. Sci. 86:243–252.

Mason, S.J. 1995. Sea-surface temperature—South African rainfall associations, 1910–1989. Int. J. Climatol. 15:119–135. doi:10.1002/joc.3370150202

Mason, S.J. 1998. Seasonal forecasting of South African rainfall using a non-linear discriminant analysis model. Int. J. Climatol. 18:147–164. doi:10.1002/(SICI)1097-0088(199802)18:2<147::AID-JOC229>3.0.CO;2-6

Mason, S.J., and M.R. Jury. 1997. Climate variability and change over southern Africa: A reflection on underlying processes. Prog. Phys. Geogr. 21:23–50. doi:10.1177/030913339702100103

Mason, S.J., J.A. Lindesay, and P.D. Tyson. 1994. Simulating drought in southern Africa using sea surface temperature variations. Water S.A. 20:15–22.

McBride, J.L., and N. Nicholls. 1983. Seasonal relationships between Australian rainfall and the Southern Oscillation. Mon. Weather Rev. 111(10):1998–2004. doi:10.1175/1520-0493(1983)111<1998:SRBARA>2.0.CO;2

McCrea, R., L. Dalgleish, and W. Coventry. 2005. Encouraging the use of seasonal forecasts by farmers. Int. J. Climatol. 25:1127–1137. doi:10.1002/joc.1164

McGregor, J.L. 2005. C-CAM: Geometric aspects and dynamical formulation. CSIRO Atmos. Res. Tech. Paper No. 70:43.

McIntosh, P.C., A.J. Ash, and M. Stafford Smith. 2005. From oceans to farms: The value of a novel statistical climate forecast for agricultural management. J. Clim. 18:4287–4302. doi:10.1175/JCLI3515.1

McKeon, G.M., A.J. Ash, W. Hall, and M. Stafford Smith. 2000. Simulation of grazing strategies for beef production in north-east Queensland. In: G.L. Hammer, N. Nicholls, and C. Mitchell, editors, Applications of seasonal climate forecasting in agricultural and natural ecosystems—The Australian experience. Kluwer Academic Press, Amsterdam, The Netherlands. p. 227–252. doi:10.1007/978-94-015-9351-9_15

McKeon, G., W. Hall, B. Henry, G. Stone, and I. Watson. 2004. Pasture degradation and recovery in Australia's rangelands: Learning from history. Queensland Department of Natural Resources, Mines and Energy, Brisbane, Australia.

McKeon, G.M., G.S. Stone, J.I. Syktus, J.O. Carter, N.R. Flood, D.G. Ahrens, D.N. Bruget, C.R. Chilcott, D.H. Cobon, R.A. Cowley, S.J. Crimp, G.W. Fraser, S.M. Howden, P.W. Johnston, J.G. Ryan, C.J. Stokes, and K.A. Day. 2009. Climate change impacts on northern Australian rangeland livestock carrying capacity: A review of issues. A climate of change in Australian rangelands. Rangeland J. 31(1):1–30. doi:10.1071/RJ08068

McPhaden, M.J. 1999. El Niño: The child prodigy of 1997–98. Nature 398(6728):559–562. doi:10.1038/19193

Meinke, H., and G.L. Hammer. 1997. Forecasting regional crop production using SOI phases: An example for the Australian peanut industry. Aust. J. Agric. Res. 45:1557–1568.

Meyers, G., P. McIntosh, L. Pigot, and M. Pook. 2007. The years of El Niño, La Niña, and interactions with the tropical Indian Ocean. J. Clim. 20(13):2872–2880. doi:10.1175/JCLI4152.1

Meza, F.J., J.W. Hansen, and D. Osgood. 2008. Economic value of seasonal climate forecasts for agriculture: Review of ex-ante assessments and recommendations for future research. J. Appl. Meteorol. Climatol. 47:1269–1286. doi:10.1175/2007JAMC1540.1

Moore, A.D., L.W. Bell, and D.K. Revell. 2009. Feed gaps in mixed farming systems: Insights from the Grain & Graze program. Anim. Prod. Sci. 49:736–748. doi:10.1071/AN09010

Morgenthal, T.L., T. Newby, and D. Pretorius. 2005. A national long-term grazing capacity map for South Africa based on NOAA (AVHRR) satellite derived data. Poster presented at the Namaqualand Colloquium, Springbok, South Africa. 24–26 May 2005.

Muchuru, S., W.A. Landman, and D. DeWitt. 2015. Prediction of inflows into Lake Kariba using a combination of physical and empirical models. Int. J. Climatol. 36:2570–2580.

Muchuru, S., W.A. Landman, D. DeWitt, and D. Lötter. 2014. Seasonal rainfall predictability over the Lake Kariba catchment area. Water SA 40(3):461–469. doi:10.4314/wsa.v40i3.9

Munro, R.K., and M.J. Lembit. 1997. Managing climate variability in the national interest: Needs and objectives. Climate prediction for agricultural and resource management. Australian Academy of Science Conference, Canberra, Australia.

NDMC. 2011. NDMC Annual Report 2010/2011. Department of Cooperative Governance and Traditional Affairs, Pretoria, South Africa.

Ngaka, M.J. 2012. Drought preparedness, impact and response: A case of the Eastern Cape and Free State provinces of South Africa, Jàmbá J. Disaster Risk Stud. 4(1).

Nicholls, N., McBride J.L., and Ormerod, R.J. 1982. On predicting the onset of the Australian wet season at Darwin. Mon. Weather Rev. 110:14–17. doi.org/10.1175/1520-0493(1982)110%3C0014:OPTOOT%3E2.0.CO;2

Nicholls, N. 1988. El Niño–Southern Oscillation and rainfall variability. J Clim. 1:418–421. doi.org/10.1175/1520-0442(1988)001%3C0418:ENOARV%3E2.0.CO;2

Nicholls, N. 1989. Sea surface temperatures and Australian winter rainfall. J. Clim. 2(9):965–973. doi:10.1175/1520-0442(1989)002<0965:SSTAAW>2.0.CO;2

Nicholls, N. 1999. Cognitive illusions, heuristics and climate prediction. Bull. Am. Meteorol. Soc. 80:1385–1397. doi:10.1175/1520-0477(1999)080<1385:CIHACP>2.0.CO;2

Nicholls, N. 2000. Opportunities to improve the use of seasonal climate forecasts. In: G.L. Hammer, N. Nicholls, and C. Mitchell, editors, Applications of seasonal climate forecasting in agricultural and natural ecosystems—The Australian experience. Kluwer Academic Press, Amsterdam, The Netherlands. p. 309–327. doi:10.1007/978-94-015-9351-9_19

Nicholls, N. 2004. The changing nature of Australian droughts. Clim. Change 63(3): 323–336.

Nicholls, N., G.V. Gruza, J. Jouzel, T.R. Karl, L.A. Ogallo, and D.E. Parker. 1996. Observed climate variability and change. Cambridge Univ. Press, Cambridge, UK. pp. 133–192.

Nicholson, S.E., and D. Entekhabi. 1987. Rainfall variability in equatorial and southern Africa: Relationships with sea surface temperatures along the southwestern coast of Africa. J. Clim. Appl. Meteorol. 26:561–578. doi:10.1175/1520-0450(1987)026<0561:RVIEAS>2.0.CO;2

OPYPA. 2014. Oficina de Programación y Política Agropecuaria, Ministerio de Ganaderia, Agricultura y Pesca de Uruguay. MGAP, Montevideo, Uruguay.

O'Reagain, P.J., J.J. Bushell, C.H. Holloway, and A. Reid. 2009. Managing for rainfall variability: Effect of grazing strategy on cattle production in a dry tropical savanna. Anim. Prod. Sci. 49:85–99. doi:10.1071/EA07187

O'Reagain, P.J., J.J. Bushell, and W. Holmes. 2011. Managing for rainfall variability: Long-term profitability of different grazing strategies in a north Australian tropical savanna. Anim. Prod. Sci. 51:210–224. doi:10.1071/AN10106

Orr, D.M., W.H. Burrows, R.E. Hendricksen, R.L. Clem, M.T. Rutherford, M.J. Conway, and C.J. Paton. 2001. Pasture yield and composition changes in a Central Queensland black speargrass (Heteropogon contortus) pasture in relation to grazing management options. Anim. Prod. Sci. 41(4):477–485. doi:10.1071/EA00132

Palmer, T., and A. Ainslie. 2006. Country Pasture/Forage Resource Profiles South Africa. Prepared by Palmer and Ainslie in May 2002; livestock data updated by S.G. Reynolds in August 2006. http://www.fao.org/ag/AGP/agpc/doc/Counprof/southafrica/South-Africa.htm (accessed 2 Feb. 2017).

Paolino, C., M. Methol, and D. Quintans. 2010. Estimación del impacto de una eventual sequía en la ganadería nacional y bases para el diseño de políticas de seguros. In: Anuario OPYPA, 2010. Ministerio de Ganadería Agricultura y Pesca– Oficina de Programación y Política Agropecuaria. MGAP-OPYPA, Montevideo, Uruguay.

Park, J.N., D.H. Cobon, and K.L. Bell. 2004. Evaluating grazier knowledge of seasonal climate forecasting in the Mitchell grasslands of western Queensland. Proceedings of Australian Rangeland Society 13th Biennial Conference, Alice Springs, Australia. p. 275–276.

Park, J.N., D.H. Cobon, and D.G. Phelps. 2003. Modelling pasture growth in the Mitchell grasslands. In: MODSIM2003. International Congress on Modelling and Simulation Proceedings, 14–17 July. Townsville, Australia. Modelling and Simulation Society of Australia, Canberra, Australia. p. 519–524.

Paull, C. 2004. Processes for using climate risk management information in rangeland enterprises. Proceedings of Australian Rangeland Society 13th Biennial Conference, Alice Springs, Australia. 277–278.

Philander, S.G. 1990. El Niño, La Niña, and the Southern Oscillation, Academic Press, London, p. 293.

Pisciottano, G., A.F. Diaz, G. Cazes, and C.R. Mechoso. 1994. El Niño-Southern Oscillation impact on rainfall in Uruguay. J. Clim. 7:1286–1302. doi:10.1175/1520-0442(1994)007<1286:ENSOIO>2.0.CO;2

Pittock, A.B. 1975. Climatic change and the patterns of variation in Australian rainfall. Search 6:498–504.

Pope, V.D., M. Gallani, P.R. Rowntree, and R.A. Stratton. 2000. The impact of new physical parametrizations in the Hadley Centre climate model—HadAM3. Clim. Dyn. 16:123–146. doi:10.1007/s003820050009

Potgieter, A.B., Y.L. Everingham, and G.L. Hammer. 2003. On measuring quality of a probabilistic commodity forecast for a system that incorporates seasonal climate forecasts. Int. J. Climatol. 23:1195–1210. doi:10.1002/joc.932

Power, S., T. Casey, C. Folland, A. Colman, and V. Mehta. 1999. Inter-decadal modulation of the impact of ENSO on Australia. Clim. Dyn. 15(5):319–324. doi:10.1007/s003820050284

Power, S., F. Delage, C. Chung, G. Kociuba, and K. Keay. 2013. Robust twenty-first-century projections of El Niño and related precipitation variability. Nature 502(7472):541–545. doi:10.1038/nature12580

Power, S., M. Haylock, R. Colman, and X. Wang. 2006. The predictability of interdecadal changes in ENSO activity and ENSO teleconnections. J. Clim. 19(19):4755–4771. doi:10.1175/JCLI3868.1

Power, S.B., and I.N. Smith. 2007. Weakening of the Walker Circulation and apparent dominance of El Niño both reach record levels, but has ENSO really changed? Geophys. Res. Lett. 34(18). doi:10.1029/2007GL030854

Ratnam, J.V., S.K. Behera, Y. Masumoto, K. Takahashi, and T. Yamagata. 2011. A simple regional coupled model experiment for summer-time climate simulation over southern Africa. Clim. Dyn. 39:2207–2217. doi:10.1007/s00382-011-1190-2

Rautenbach, C.J.W., and I.N. Smith. 2001. Teleconnections between global sea-surface temperatures and the interannual variability of observed and model simulated rainfall over southern Africa. J. Hydrol. 254:1–15. doi:10.1016/S0022-1694(01)00454-1

Reason, C.J.C. 2001a. Subtropical Indian Ocean SST dipole events and southern African rainfall. Geophys. Res. Lett. 28:2225–2227. doi:10.1029/2000GL012735

Reason, C.J.C. 2001b. Evidence for the influence of the Agulhas Current on regional atmospheric circulation patterns. J. Clim. 14:2769–2778. doi:10.1175/1520-0442(2001)014<2769:EFTIOT>2.0.CO;2

Reeson, A.F., R.R.J. McAllister, S.M. Whitten, I.J. Gordon, M. Nicholas, and S.S. McDouall. 2008. The agistment market in the northern Australian rangelands: Failings and opportunities. Rangeland J. 30:283–289. doi:10.1071/RJ06042

Risbey, J.S., M.J. Pook, P.C. McIntosh, M.C. Wheeler, and H.H. Hendon. 2009. On the remote drivers of rainfall variability in Australia. Mon. Weather Rev. 137(10):3233–3253. doi:10.1175/2009MWR2861.1

Rocha, A., and I. Simmonds. 1997. Interannual variability of south-eastern African summer rainfall. Part I: Relationships with air-sea interaction processes. Int. J. Climatol. 17:235–265. doi:10.1002/(SICI)1097-0088(19970315)17:3<235::AID-JOC123>3.0.CO;2-N

Roeckner, E. 1996. Simulation of present-day climate with the ECHAM4 model: Impact of model physics and resolution. Max Planck Inst. Meteorol. Report 93.

Ropelewski, C.F., and M.S. Halpert. 1987. Global and regional scale precipitation patterns associated with El Niño/Southern Oscillation. Mon. Weather Rev. 115:1606–1626. doi:10.1175/1520-0493(1987)115<1606:GARSPP>2.0.CO;2

Rouault, M., P. Florenchie, N. Fauchereau, and C.J.C. Reason. 2003. South east tropical Atlantic warm events and southern African rainfall. Geophys. Res. Lett. 30(5). doi:10.1029/2002GL014840

Saji, N.H., B.N. Goswami, P.N. Vinayachandran, and T. Yamagata. 1999. A dipole mode in the tropical Indian Ocean. Nature 401(6751):360–363. doi:10.1038/43854

Salmon, L., and J. Donnelly. 2007. Variability in weather: What are the consequences for grazing enterprises. In: Proceedings of the 22nd Conference of the Grassland Society of NSW. Grassland Society of NSW, Queanbeyan, Australia.

Schiller, A., J.S. Godfrey, P.C. McIntosh, G. Meyers, N.R. Smith, O. Alves, G. Wang, and R. Fiedler. 2002. A new version of the Australian community ocean model for seasonal climate prediction. CSIRO, Canberra, Australia.

Smith, I.N., M. Dix, and R.J. Allan. 1997. The effect of greenhouse SSTs on ENSO simulations with an AGCM. J. Clim. 10(2):342–352. doi:10.1175/1520-0442(1997)010<0342:TEO GSO>2.0.CO;2

Stafford Smith, D.M. 1992. Stocking rate strategies across Australia. Range Manag. Newsl. 92:1–3.

Stafford Smith, M., R. Buxton, G.M. McKeon, and A.J. Ash. 2000. Seasonal climate forecasting and the management of rangelands: Do production benefits translate into enterprise profits? In: G.L. Hammer, N. Nicholls, and C. Mitchell, editors, Applications of seasonal climate forecasting in agricultural and natural ecosystems—The Australian experience. Kluwer Academic Press, Amsterdam, The Netherlands. p. 271–289. doi:10.1007/978-94-015-9351-9_17

Stevenson, S.L. 2012. Significant changes to ENSO strength and impacts in the twenty-first century: Results from CMIP5. Geophys. Res. Lett. 39(17). doi:10.1029/2012GL052759

Stokes, C. J., S. Crimp, R. Gifford, A.J. Ash and S.M. Howden 2010. "Broadacre grazing." Adapting agriculture to climate change: Preparing Australian agriculture, forestry and fisheries for the future. CSIRO, Collingwood, Australia. p. 153–170.

Stone, R., and A. Auliciems. 1992. SOI phase relationships with rainfall in eastern Australia. Int. J. Climatol. 12(6):625–636. doi:10.1002/joc.3370120608

Stone, R., and G. de Hoedt. 2000. The development and delivery of current seasonal forecasting capabilities in Australia. In: G.L. Hammer, N. Nicholls, and C. Mitchell, editors, Applications of seasonal climate forecasting in agricultural and natural ecosystems—The Australian experience. Kluwer Academic Press, Amsterdam, The Netherlands. p. 67–75. doi:10.1007/978-94-015-9351-9_5

Stone, R.C., G.L. Hammer, and T. Marcussen. 1996. Prediction of global rainfall probabilities using phases of the Southern Oscillation Index. Nature 384:252–255. doi:10.1038/384252a0

Stroebel, A., F.J.C. Swanepoel, N.D. Nthakheni, A.E. Nesamvuni, and G. Taylor. 2008. Benefits obtained from cattle by smallholder farmers: A case study of Limpopo Province, South Africa. Aust. J. Exp. Agric. 48:825–828. doi:10.1071/EA08058

Tennant, W.J., and B. Hewitson. 2002. Intra-seasonal rainfall characteristics and their importance to the seasonal prediction problem. Int. J. Climatol. 22:1033–1048. doi:10.1002/joc.778

Thompson, D.W., and J.M. Wallace. 2000. Annular modes in the extratropical circulation. Part I: Month-to-month variability. J. Clim. 13(5):1000–1016. doi:10.1175/1520-0442(2000)013<1000:AMITEC>2.0.CO;2

Thornton, P.K. 2010. Livestock production: Recent trends, future prospects. Philos. Trans. R. Soc. Lond. B Biol. Sci. 365(1554). doi:10.1098/rstb.2010.0134

Tothill, J.C. and C. Gillies. 1992. The pasture lands of northern Australia: Their condition, productivity and sustainability. Tropical Grasslands Society of Australia Occasional Publication No. 5. Meat Res. Corp., Brisbane, Australia.

Tversky, A., and D. Kahneman. 1974. Judgement under uncertainty: Heuristics and biases. Science 185:1124–1131. doi:10.1126/science.185.4157.1124

Ummenhofer, C.C., M.H. England, P.C. McIntosh, G.A. Meyers, M.J. Pook, J.S. Risbey, and A.S. Taschetto. 2009. What causes southeast Australia's worst droughts? Geophys. Res. Lett. 36(4). Doi:10.1029/2008GL036801

Ummenhofer, C.C., A. Sen Gupta, M.H. England, and C.J.C. Reason. 2009. Contributions of Indian Ocean sea surface temperatures to enhanced East African rainfall. J. Clim. 22:993–1013. doi.org/10.1175/2008JCLI2493.1

Van Heerden, J., D.E. Terblanche, and G.C. Schulze. 1988. The Southern Oscillation and South African summer rainfall. J. Climatol. 8:577–597. doi:10.1002/joc.3370080603

Verdon-Kidd, D.C., and A.S. Kiem. 2009. Nature and causes of protracted droughts in southeast Australia: Comparison between the Federation, WWII, and Big Dry droughts. Geophys. Res. Lett. 36(22). doi:10.1029/2009GL041067

Wand, S.J.E., G.F. Midgley, M.H. Jones, and P.S. Curtis. 1999. Responses of wild C4 and C3 grass (Poaceae) species to elevated atmospheric CO2 concentration: A meta-analytic test of current theories and perceptions. Glob. Change Biol. 5:723–741. doi:10.1046/j.1365-2486.1999.00265.x

Walker, N.D. 1990. Links between South African summer rainfall and temperature variability of the Agulhas and Benguela currents systems. J. Geophys. Res. 95:3297–3319. doi:10.1029/JC095iC03p03297

Wang, G., O. Alves, D. Hudson, H. Hendon, G. Liu, and F. Tseitkin. 2008. SST skill assessment from the new POAMA-1.5 system. BMRC Res. Lett. 8:2–6.

Wang, G., and H.H. Hendon. 2007. Sensitivity of Australian rainfall to inter-El Niño variations. J. Clim. 20(16):4211–4226. doi:10.1175/JCLI4228.1

Wang, G., D. Hudson, Y. Yin, O. Alves, H. Hendon, S. Langford, and F. Tseitkin. 2011. POAMA-2 SST skill assessment and beyond. CAWCR Res. Lett. 6:40–46.

Wang, G., N.R. Smith, and O. Alves. 2005. BAM3. 0 tropical surface flux simulation and its impact on SST drift in a coupled model. Bureau of Meteorology Research Centre. Melbourne, Australia.

Washington, R., and M. Todd. 1999. Tropical-temperate links in southern Africa and southwest Indian Ocean satellite-derived rainfall. Int. J. Climatol. 19:1601–1616. doi:10.1002/(SICI)1097-0088(19991130)19:14<1601::AID-JOC407>3.0.CO;2-0

Wessels, K.J., F. Van Den Bergh, and R.J. Scholes. 2012. Limits to detectability of land degradation by trend analysis of vegetation index data. Remote Sens. Environ. 125:10–22. doi:10.1016/j.rse.2012.06.022

Wheeler, M.C., H.H. Hendon, S. Cleland, H. Meinke, and A. Donald. 2009. Impacts of the Madden-Julian oscillation on Australian rainfall and circulation. J. Clim. 22(6):1482–1498. doi:10.1175/2008JCLI2595.1

White, B.J. 1978. A simulation based evaluation of Queensland's northern sheep industry. James Cook University, Department of Geography, Monograph Series No. 10. Townsville, Australia.

White, B.J. 2001. Survey shows forecasts use up. Climag Newsletter of the Climate Variability in Agriculture Program, Land and Water Australia, Canberra, Australia.

Wilhite, D.A. 1993. The enigma of drought. In: D.A. Wilhite, editor, Drought assessment management and planning: Theory and case studies. Kluwer Academic Publ., Boston, MA. p. 3–15. doi:10.1007/978-1-4615-3224-8_1

Wilhite, D.A. 2002. Combating drought through preparedness. Nat. Resour. Forum 26:275–285. doi:10.1111/1477-8947.00030

Wilhite, D.A. 2011. Breaking the hydro-illogical cycle: Progress or status quo for drought management in the United States. Eur. Water 34:5–18.

Wilhite, D.A. 2014. Changing the paradigm for drought management: Can we break the hydro-illogical cycle? Water Resources Impact 16:21–23.

Williams, A.A., and R.C. Stone. 2009. An assessment of relationships between the Australian subtropical ridge, rainfall variability, and high-latitude circulation patterns. Int. J. Climatol. 29(5):691–709. doi:10.1002/joc.1732

Winsemius, H.C., E. Dutra, F.A. Engelbrecht, E. Archer Van Garderen, F. Wetterhall, F. Pappenberger, and M.G.F. Werner. 2014. The potential value of seasonal forecasts in a changing climate in southern Africa. Hydrol. Earth Syst. Sci. 18(4):1525–1538. doi:10.5194/hess-18-1525-2014

Wolfe, E.C., R.D. FitzGerald, D.G. Hall, and O.R. Southwood. 1980. Beef production from lucerne and subterranean clover pastures. 1. The effects of pasture, stocking rate and supplementary feeding. Aust. J. Exp. Agric. Anim. Husb. 20:678–687. doi:10.1071/EA9800678

World Bank. 2014. With technology, rural producers in Uruguay adapt to climate change. http://www.worldbank.org/en/news/feature/2014/03/20/adaptacion-cambio-climatico (accessed 16 Jan. 2017).

Yeh, S.W., J.S. Kug, B. Dewitte, M.H. Kwon, B.P. Kirtman, and F.F. Jin. 2009. El Niño in a changing climate. Nature 461(7263):511–514. doi:10.1038/nature08316

Zhao, M., and H.H. Hendon. 2009. Representation and prediction of the Indian Ocean dipole in the POAMA seasonal forecast model. Q. J. R. Meteorol. Soc. 135(639):337–352. doi:10.1002/qj.370

Zhong, A., O. Alves, H. Hendon, and L. Rikus. 2006: On aspects of the mean climatology and tropical interannual variability in the BMRC atmospheric model (BAM 3.0). BMRC Research Rep. Bureau of Meteorology, Melbourne, Australia. p. 121.

Perennial Systems (Temperate Fruit Trees and Grapes)

Jeffrey A. Andresen* and William J. Baule

Abstract

Perennial fruit crops are produced across a wide variety of climates and geographical settings worldwide. As specialty crops with relatively high input costs, farmgate value, and multiple-year production cycles, they require high levels of management and are susceptible to a large number of weather- and climate-related impacts. This review considers eight economically important perennial crops grown that share similar environmental limitations: tree fruits apple, pear, sour (tart) cherry, sweet cherry, peach, apricot, plum, and vining fruit grapes. While the global area on which these crops is produced is limited with just under 2 million hectares or about 2.8% of total global non-forage related cropland (FAO, 2018), their economic impact is significant. For example, in the United States during 2016 these crops collectively had a net farmgate value of just under $11.9 billion on a production area of only 661,000 ha (NASS USDA, 2017). In contrast to some agricultural crops, crop quality is typically of greater importance in the production cycle than crop yield and quantity. Major weather and climate-related constraints include temperature fluctuations and extremes, especially spring freezes, insufficient or untimely rainfall, hail, and pressure from a variety of weather-dependent plant disease and insect pests. Recent climatic changes have resulted in a longer and warmer growing season in many production areas, with some regions becoming too warm to meet cool season chilling requirements. Impacts from projected future changes in climate include a continuation of recent trends, a potential poleward shift for some crop production areas, and the introduction of new and exotic pest species.

Geographic Origins and Current Production Areas

While these perennial crops have been bred and domesticated and by humans for centuries, their basic weather- and climate-related vulnerabilities are still linked with their ancestral origins. Geographically, most of these species originated in Asia or Europe in areas with a temperate climate in which part of the annual cycle is too cool to allow growth and development. As deciduous species which shed their leaves each season, they have developed a physiology and growth cycle which allows them to both survive both the cold season and complete their growth and development cycle during the warmer season. As such, attaining healthy

Dep. of Geography, Environment, and Spatial Sciences, Michigan State University, East Lansing, MI 48824. *Corresponding Author: J.A. Andresen (andresen@msu.edu)

doi:10.2134/agronmonogr60.2016.0016

© ASA, CSSA, and SSSA, 5585 Guilford Road, Madison, WI 53711, USA.
Agroclimatology: Linking Agriculture to Climate, Agronomy Monograph 60.
Jerry L. Hatfield, Mannava V.K. Sivakumar, John H. Prueger, editors.

Table 1. Perennial fruit crop total production (top), production area (middle), and yields (bottom) by country in 2016 (FAO, 2018).

	Apple		Pear		Cherry		Peach		Apricot		Plum		Grape	
Total Production 1000 kg	China	36,479,296	China	16,197,343	Turkey	455,667	China	11,459,217	Turkey	627,920	China	5841,097	China	455,667
	USA	4473,051	Argentina	802,238	USA	310,749	Italy	1510,203	Uzbekistan	421,720	Serbia	503,413	Italy	7663,722
	Iran	2847,867	USA	788,262	Iran	216,722	Spain	1350,980	Iran	345,248	Romania	502,135	USA	6908,261
	Turkey	2701,823	Italy	768,708	Italy	113,774	USA	1104,037	Algeria	233,237	USA	450,895	France	6030,992
	Poland	2679,868	Spain	451,783	Spain	92,833	Greece	796,507	Italy	227,393	Chile	291,675	Spain	6013,464
	Italy	2299,322	Turkey	416,507	Chile	76,671	Iran	699,335	Pakistan	192,887	Turkey	269,317	Turkey	4041,834
Production Area (ha)	China	2180,822	China	1092,244	Turkey	71,082	China	748,202	Turkey	111,754	China	1769,374	Spain	993,844
	Iran	202,477	Italy	36,523	USA	35,133	Spain	82,475	Iran	54,732	Serbia	144,150	France	776,343
	Poland	179,719	Spain	26,047	Italy	29,790	Italy	80,645	Uzbekistan	42,652	Romania	69,751	Italy	728,673
	Turkey	167,468	Argentina	24,033	Iran	26,820	USA	55,269	Algeria	40,669	USA	32,835	China	631,910
	USA	134,223	Turkey	22,446	Spain	24,889	Iran	54,455	Pakistan	28,703	Turkey	18,279	Turkey	469,212
	Italy	56,382	USA	21,404	Chile	15,072	Greece	43,669	Italy	18,576	Chile	17,459	USA	398,992
Yield (kg ha^{-1})	Italy	40,781	USA	36,828	USA	8845	USA	19,976	Italy	12,241	Chile	16,706	USA	17,314
	USA	33,326	Argentina	33,381	Iran	8081	Italy	18,727	Uzbekistan	9887	Turkey	14,734	China	16,217
	China	16,727	Italy	21,047	Turkey	6410	Greece	18,240	Pakistan	6720	USA	13,732	Italy	10,517
	Turkey	16,133	Turkey	18,556	Chile	5086	Spain	16,380	Iran	6308	Romania	7199	Turkey	8614
	Poland	14,911	Spain	17,345	Italy	3819	China	15,316	Algeria	5735	Serbia	3492	France	7768
	Iran	14,065	China	14,829	Spain	3730	Iran	12,842	Turkey	5619	China	3301	Spain	6051

dormancy and successful development of buds during the following growing season generally require a certain amount of time with environmental temperatures between 0 °C and approximately 5 to 10 °C during rest or dormant stages. These so-called chilling requirements generally prevent cultivation of such fruits in year round warm tropical climates without special management strategies.

Total production, production areas, and yields of major-producing countries for all seven of the tree fruit crops discussed are given in Table 1. All are members of the rose (Rosaceae) family. Apple (*Malus* × *domestica*) and pear are pome fruits and members of the genera *Malus* and *Pyrus*, respectively. Cherry, peach, apricot, and plum are all members of the genus *Prunus* and are categorized as stone fruits with single, stony pits. Grapes are a vining fruit from the genus *Vitus*. International apple production areas are illustrated in Fig. 1 and provide a representative example of the crops' collective geographical ranges. Current production areas generally range from 20 to 50° latitude in both hemispheres, with further poleward extension in regions with oceanic or maritime-moderated climates (e.g., western and northern Europe). Major Modified Köppen climate types include Cf Humid Subtropical and Marine West Coast, Cs Mediterranean, and Df Humid Continental. Some production also occurs in arid or semiarid (BW or BS) climates where irrigation is available and at high elevations in lower latitudes with cooler climates.

Apples

While most evidence suggests the geographic origin of wild apples is Kazakhstan in central Asia (Dzhangaliev, 2003), the true origins of cultivated apples remain unclear. Possible sources include the Caucasus Region of Southwest Asia, Central Asia, and Eastern Asia (Webster, 2005). Major current international producers (in order by size of production area) are China, the United States, Iran, Turkey,

Poland, and Italy (FAO, 2018). Among perennial specialty crops, apples tolerate a relatively large range of climates, ranging geographically from central Asia to North America, and high elevation sites where winter temperatures can fall as low as −40 °C, to subtropical areas where freezing temperatures are uncommon.

Pears

Pears (*Pyrus* spp) originated in western China. It is considered native to a large area from western Europe and North Africa to Asia Minor eastward to eastern Asia (Silva et al., 2014). Pears can be divided into two types that generally describe how they are consumed: the European pear (P. *communis* L.), with elongated and full-bodied texture which is usually consumed some time after harvest, and the Asian pear (P. *pyrifolia* (Burm. f.) Nakai and P. ussuriensis Maxim.) with sandy texture and rounded body which are consumed when fresh. Closely related to apples, pears are most commonly produced in cool, temperate climates, with climatic requirements similar to those of apples. Major current international producers (in order by size of production area) are China, Italy, Spain, Argentina, Turkey, and Spain (FAO, 2018).

Cherries

Cherries are divided into two major types: sweet (*Prunus avium* [L.] L.) and sour or tart (*Prunus cerasus* L.). The species developed over a large geographical area, originating in Central Asia, with secondary centers in Eastern Asia, Europe, and North America (Watkins, 1995). Cherries require a temperate climate with prolonged cold winter temperatures and an extended dormancy period at temperatures of 7 °C or below. During the growing season, they prefer regions with a cool spring and a mild summer. Major producers include Turkey, the United States, Italy, Iran, Spain, and Chile (FAO, 2018).

Peach

The peach (*Prunus persica* [L.] Batsch) is native to China and then gradually spread westward through Asia to southern Europe and later to other parts of Europe. It was introduced to the Americas in the 17th century along with European colonists and immigrants. Major fruit types include: freestone with melting flesh and

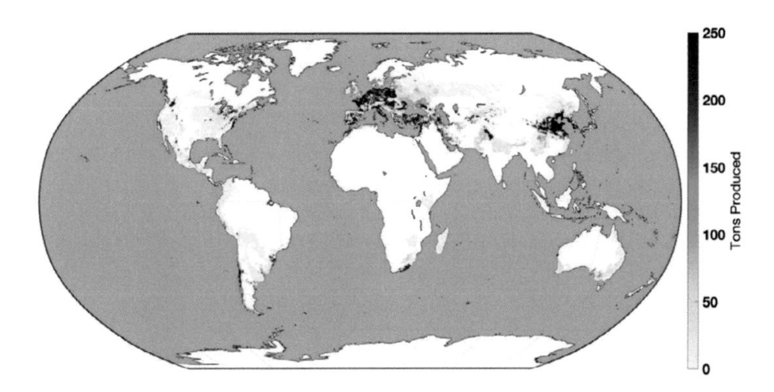

Fig. 1. Map of apple production (average percentage of land used for its production times average yield (t ha^{-1}) in each 10 km × 10 km grid cell) across the world compiled by the University of Minnesota Institute on the Environment with data from Monfreda et al. (2008).

white or yellow flesh, which are usually consumed fresh, and cling-stone, which are typically used in processing. The nectarine, a peach with a nonfuzzy skin, resulted from a mutation. While peaches require a cool winter to meet offseason chilling requirements, they are relatively more susceptible to damage from extreme cold than pome fruits. Major producing countries include China, Spain, Italy, the United States, Greece, and Iran (FAO, 2018).

Apricot

Apricot (*Prunus armeniaca* L.) is a native of central Asia and China. After gradual domestication in Asia, it was brought westward into the Middle East and eventually Europe. While it is slightly more cold tolerant than peach during full dormancy, it blooms very early, and is therefore subject to spring frost injury. It is also relatively more tolerant to drought than other tree fruit (Webster, 2005). Largest producing countries include Turkey, Iran, Uzbekistan, Algeria, Pakistan, and Italy.

Plums

Plums are members of the genus *Prunus domestica* L. and contain the two most commonly produced species, European plum (*Prunus domestica*) and Japanese plum *Prunus salicina* Lindl.). European plum is thought to have originated in southern Europe or southwestern Asia while Japanese plum originated in China (Webster, 2005). There are also plus species native to North America, but none are widely cultivated. Plums are consumed fresh or dried (prunes). Of the two major cultivated species, European plums are somewhat more tolerant of winter cold than the Japanese variety. Japanese plums also bloom earlier than the European variety and are more susceptible to freezing temperatures in the spring as a result. Largest producing countries in the world in 2016 were China, Serbia, Romania, the United States, Turkey, and Chile (FAO, 2018).

Grapes

On a yield per area basis, grape (species of *Vitis*, Vitaceae, or grape family) ranks among the most valuable crops in the world (FAO, 2018). Archeological records suggest that cultivation of the domesticated or European grape (*Vitis vinifera subsp. vinifera*), began 6000 to 8,000 yr ago in the Middle East from its wild ancestor *Vitis vinifera subsp. Sylvestris* (Myles et al., 2011). Cultivation of grape then gradually spread over time, especially to areas of Mediterranean Europe. Grapes derived from the European species have been utilized for the production of wine for centuries. Other types of grapes have also been developed for fresh consumption (table grapes), dried into raisins, or for fresh juice. A primary source for fresh juice is obtained from a North American native *Vitis labrusca* L., the fox grape. Another North American native is the muscadine grape (*Vitis rotundifolia* Michx.). Grape production has traditionally been associated with areas with warm, dry summers and mild winters typified by the Modified Köppen Cs (Mediterranean) climate category. However, grapes have also been produced successfully in many other climate types including areas both colder and wetter than the Mediterranean climate type. As of 2016, leading producing countries by area included Spain, France, Italy, China, Turkey, and the United States, although there has been a shift of production in recent decades from Europe to other areas, especially North and South America and Asia (FAO, 2018).

Meteorological and Climatological Constraints, Limitations, and Adaptations

Production of perennial tree fruits and grapes require significant investments of financial resources and time, given a relatively long production cycle of 20 to 30 years or longer. This necessitates careful strategic selection of cultivars and production site at the time of establishment as well as a consideration of long-term climatic conditions, including some measure of temporal variability. While more costly for initial establishment and early management than traditional row crop production systems, recent trends in commercial tree fruit production toward high density systems have resulted in significant increases in plant population densities of smaller trees to both increase potential yields, especially early in the production cycle, and to simplify harvest and pruning operations (Robinson et al., 2007). High density systems boost potential yields by increasing light interception rates in the canopy as well as increasing the harvest index (Hampson et al., 2002).

One critical environment-related factor often overlooked for perennial fruit crop production is crop quality, which on economic terms may be of equal or even greater importance to the producer than crop yield (Bates and Morris, 2009). Besides determining market value, crop quality relates directly to consumer-buying decisions and human nutrition and health. Quality parameters include the presence and concentrations of phytonutrients and secondary metabolites, bioactivity, as well as organoleptic properties such as color, visual appeal, aroma, taste, and texture (Ahmed et al., 2014).

There are a wide range of weather and climate-related factors that may impact perennial fruit crops during the course of a year. Associated variables include solar radiation, air temperatures, wind, precipitation, humidity, and soil water balance. Some influence crop yields, others crop quality, and some impact both. Some potential weather-related impacts on a perennial fruit crop during the course of a year are illustrated for sour cherry in a humid continental climate in Fig. 2. As shown, the type and timing of several impacts are directly related to the rate of development associated with changes in the annual temperature cycle. For example, abnormally wet weather during the growing season may lead to increased disease risk and to poor pollination, while abnormally dry weather may result in crop moisture stress. Extreme heat in late reproductive stages may lead to reduced fruit size and quality, and, if late enough, to soft fruit and further reductions in quality. Strong winds may result in wind whipped fruit and reductions in quality at a similar time of the season. In addition to longterm means and trends of these elements, their variability over time is a critical factor (Jones et al., 2012). In a comparative study of grape growing regions in California and Michigan, Schultze et al. (2014) found substantially higher interannual variability in seasonal growing degree day accumulations for the Michigan region in a temperate continental climate, which was in turn directly linked with variability in fruit quality. They found much of the seasonal variability in the Michigan production area associated with synoptic weather patterns during the spring season, with the observation that 71% of the seasons in which growing degree day total surpluses had developed by the first of July also had growing degree day surpluses at the end of the growing season.

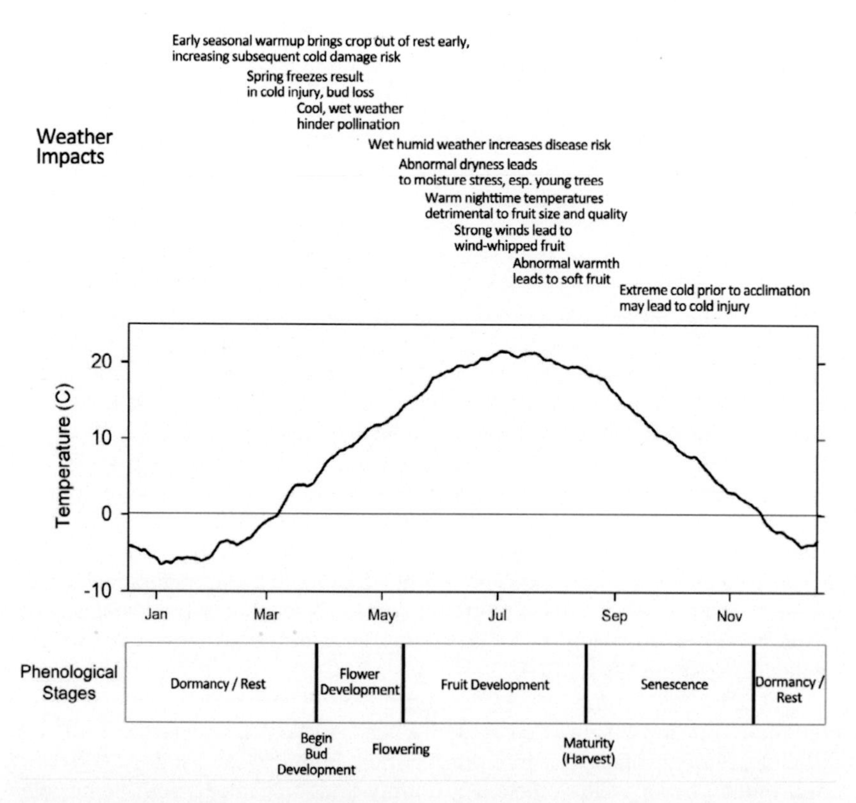

Fig. 2. Typical weather-related impacts for sour cherry as a function of calendar time, mean daily temperatures, and phenological stage in the northwestern Lower Peninsula of Michigan. Mean daily temperatures were derived from daily maximum and minimum temperature data at Traverse City, MI for the period 1981 to 2010.

Air Temperatures

Environmental temperatures play a critical role in determining where production of perennial temperate fruit crops is economically feasible. As noted earlier, the rate of growth and development is directly related to the amount of time ambient temperatures climb above a given temperature threshold (thermal time). Thus, one major climatological constraint is adequate heat accumulation for the crop to reach physiological maturity. There are some management strategies available to growers to address this issue. For example, physical removal of basal leaves near blooms affects grape leaf and fruit balance and improve fruit quality in cool climates, in which fruit maturity and phenolic ripening are frequent limitations (VanderWeide et al., 2018).

Another critical factor is a necessary period of rest or dormancy during the annual cycle. Most perennial fruit crops grown in temperate climates develop next year's buds during the summer. In the autumn, those buds go dormant. Lang et al. (1987) developed a classification scheme that identified three types of dormancy. Endodormancy is a state of rest that begins in the late fall season in response to cooling temperatures and shortening daylengths. It can only be reversed or broken by a sustained period of temperatures near to slightly above freezing. So-called chilling temperatures extend from 0 °C to approximately 10

°C or greater, depending on species. To quantify a given species and cultivar's chilling requirement, one chilling unit is defined as 1 h of exposure within the specified chilling temperature range. The hourly units are then summed up for a whole season to describe the amount of time a given cultivar needs to meet its chilling requirement. Failure to meet that minimum amount of exposure results in delayed and substandard foliation, flowering, and fruiting during the following growing season (Webster, 2005; Considine and Considine, 2016). Ecodormancy is a second, potentially reversible state of rest in which buds are dormant as a result of unfavorable environmental conditions such as in the early spring when temperatures remain below base growth threshold temperatures. A third type of dormancy, paradormancy, is a state of rest in one part of a plant in response to the inhibitory influence of another part of the same plant.

Perennial fruit production is strongly influenced by the occurrence of extreme temperature events, particularly cold temperatures, which can damage or kill exposed plant tissue. Plants can avoid freezing by supercooling. However, supercooled parts of buds can dehydrate, allowing the formation of ice and damage to the plant cell membranes (Pearce, 2001). The degree of vulnerability to cold damage depends on the phenological stage of the crop. Physically, this phenological dependence may be related to relative water content and air volume in the buds, which may change over time during development (Hewett, 1976). Maximum resilience to cold occurs during endodormancy during the cold season. As temperatures increase once again the following spring, the trees gradually lose their cold hardiness before becoming actively vegetative (Hewett and Hawkins, 1968). For example, 10% damage threshold temperatures for sour cherry increase from -34.4 °C at dormancy to −4.4 °C at side green stage to -3.3 °C at green tip to -2.2 °C at and following open cluster (Dennis and Howell, 1974). Depending on the magnitude of the seasonal warm-up, this transition can be rapid in some years. For example, beginning on a March first calendar date, only 120 base 4 °C growing degree day units are needed for the sour cherry crop to advance from dormancy to the side green stage (Zavalloni et al., 2006). In some crops such as grapes, there is a reserve system. Typically only the primary bud in the dormant bud becomes active during spring warmup while the secondary and tertiary buds remain inactive. However, those secondary and tertiary buds can become active if the primary bud is killed or severely damaged (Jackson, 2014).

Springtime freezes are a primary factor in describing interannual yield variability of perennial fruit crops. For example, during March, 2012 a persistent upper air jet stream ridging feature set up across much of eastern two thirds of North America and led to a three-week period of abnormally warm temperatures resulting in the warmest March on record (1895–2015 period of record) at many locations across the Midwest (Labe et al., 2017). Overwintering vegetation across the Great Lakes region broke dormancy weeks before normal, leaving it highly susceptible to frost or freeze damage. Near the end of March and throughout April 2012, climatologically normal temperatures returned to the region, including a series of frost and freeze events (normal last freezing temperatures in the region tend to occur in late April or early May). Fruit buds were severely damaged or lost due to the freezing temperatures. Michigan alone lost approximately 85% of its normal apple and 90% of its normal cherry production crop for the year, likely the worst single weather-related loss in the 150 years that tree fruit has been grown commercially in the region (Kistner et al., 2018). The total loss inclusive of

Table 2. Temperature-based thresholds for the crops at different stages of development. Values based on Santibanes (1994), Proebsting and Mills (1978), Ruml et al. (2010), and Dennis and Howell (1974).

		Apple	Pear	Sour Cherry	Sweet Cherry	Peach	Apricot	Plum	Grape
Base Temp. (°C)	Bud Dev.	0- 8	2–8	0–6	0–6	5–8	-2	3– 8	7–15
	Fruit Dev.	4 -10	8	4–10	4–10	10	3	8– 10	10
Optimum Temp. (°C)	Bud Dev.	10- 21	10–22	20–25	20–25	18–24	11–13	10–24	20–25
	Fruit Dev.	20- 23	21–25	18–25	18–25	12–26	25–32	18–22	20–30
SGDD	Bud Dev. (GDD_5)	190–380	170–280	200–350	200–350	150–220	N/A	130–200	250–450
	Fruit Dev. (GDD_{10})	900–1400	600–1100	300–450	300–450	400–800	N/A	570–1100	400–1200
Cold Injury Thresholds (°C)	Dormancy	-35- -30	-20	-20	-20	-25	-12- -8	-22	-18–15
	Bud Dev.	-5.6–2.2	-9–2.2	-8–2.2	-8–2.2	-3–2.8	-6.2–4.3	-9–2.5	-4–0
	Fruit Dev.	-2–1	-2–1.1	-2–1.1	-2–1.1	-1.5–1.1	-2.9–2.3	-2–1.1	-1–0
Chilling Hours (base 7 °C)		600–1500	600–1300	800–1200	800–1200	200–1000	320–920	500–1000	100–600

blueberries, grapes, peaches, sweet cherries, and other perennial crops was estimated at $209.8 million, and for some growers in the state it was the second major crop failure in 10 years following a similar type of event in 2002 (Knudson, 2012).

A number of temperature-related constraints for the perennial crops are given in Table 2. Ranges of various thresholds reflect variability in a number of cultural factors including cultivar, root stock, and pruning system. There are some pattern similarities in temperature sensitivity across the crop types, but there are also some differences. Base temperatures and temperatures of peak growth rate (optimal temperature) thresholds are generally warmer during bud development stages than fruit development later in the growing season. The lowest base temperatures for growth and development are observed in apricot, apple, and cherry, with the highest in grape. Similarly, the lowest optimal temperatures are generally observed in apple and cherries, with the highest in grape and apricots. Growing degree day (GDD) accumulations during bud development stages are lowest for plums and peaches and highest for grapes. Growing degree day thresholds for fruit development are lowest for cherries and highest for apple and grape, and are a direct reflection of the length of these stages of development. Base 7 °C chilling hour requirements are lowest for grapes (100–600) and peach (200–1000) and greatest for apple (600–1500). Cold injury threshold temperatures increase with increasing stages of development and are lowest for apple at dormancy (-35 °C) and highest for grape during fruit development (0 °C).

The selection of an orchard or vineyard site suitable for a specific crop and cultivar ranks as one of the most important strategic decisions made by a grower in the establishment of new production areas (Epstein et al., 2007; Jones and Hellman, 2003). Among the most critical criteria in that decision is the selection of a site in an area which is relatively less freeze prone (Batholic and Martsolf, 1979; Zabadal and Andresen, 1997). Given the overall tendency for cold air near the ground surface to run downhill due to differences in density during freeze events associated with nighttime radiational cooling of the surface (i.e., relatively clear, calm weather conditions), consideration of the landscape and associated topographical features during orchard or vineyard establishment is an important management decision. As a result, orchards and vineyards in most production areas are established on relatively higher topographical features including hilltops, ridges, or slopes just below the higher features. Exceptions would be in areas with high enough elevations to result in cooler (more frost prone) climates or in areas exposed to relatively high winds. Directional aspect should also be considered. Southern-facing slopes and surfaces warm up faster during the spring season, while the opposite is true of northern-facing surfaces. Heating rates on eastern- and western-facing surfaces are somewhere in between (Andresen et al., 2001; Aggelopoulou et al., 2010). An example of site selection in a temperate climate is given in Fig. 3. In this sour cherry orchard, trees were established on relatively higher areas of the landscape and the grower has deliberately not planted trees in the low-lying area in the foreground of the photo where cold air is more likely to accumulate, with more frequent and severe freezing temperature events during the spring season.

Another consideration is physical proximity to large bodies of water which may moderate air temperatures, reduce temperature extremes, and delay bloom in the spring thereby reducing the risk of freeze damage. For example, wine grape production has increased dramatically across the Great Lakes region of North America in recent decades (Che, 2010). Much of the increase in production area has taken place along the eastern shore of Lake Michigan where climate extremes are moderated by the proximity of the lake and in areas of higher topography where springtime freezes are relatively less common (Andresen and Winkler, 2009). However, extreme cold during the winter season is still a problem for *Vitis vinifera* L. cultivars, with a relatively mild -20 °C threshold for the beginning of cold injury during the dormant state (Schultze et al., 2014). The dependence of successful *vinifera* production in this region and the frequency of -20 °C minimum temperature events is illustrated in Fig. 4 with an overlay of current designated American viticultural areas (AVA). As can be seen in the figure, the frequency of such events in the relatively new "tip of the mitten" AVA along the northern edge of the lower peninsula is relatively high for *vinifera* production. As such, growers have relied more heavily on production of hybrid grape cultivars which are more resilient to extreme winter cold (Zabadal et al., 2009).

Fig. 3. Example of orchard site selection for sour cherry in Ionia County, MI. The grower has intentionally not planted trees in the low-lying area, relatively frost-prone area in the foreground (photo: J. Andresen).

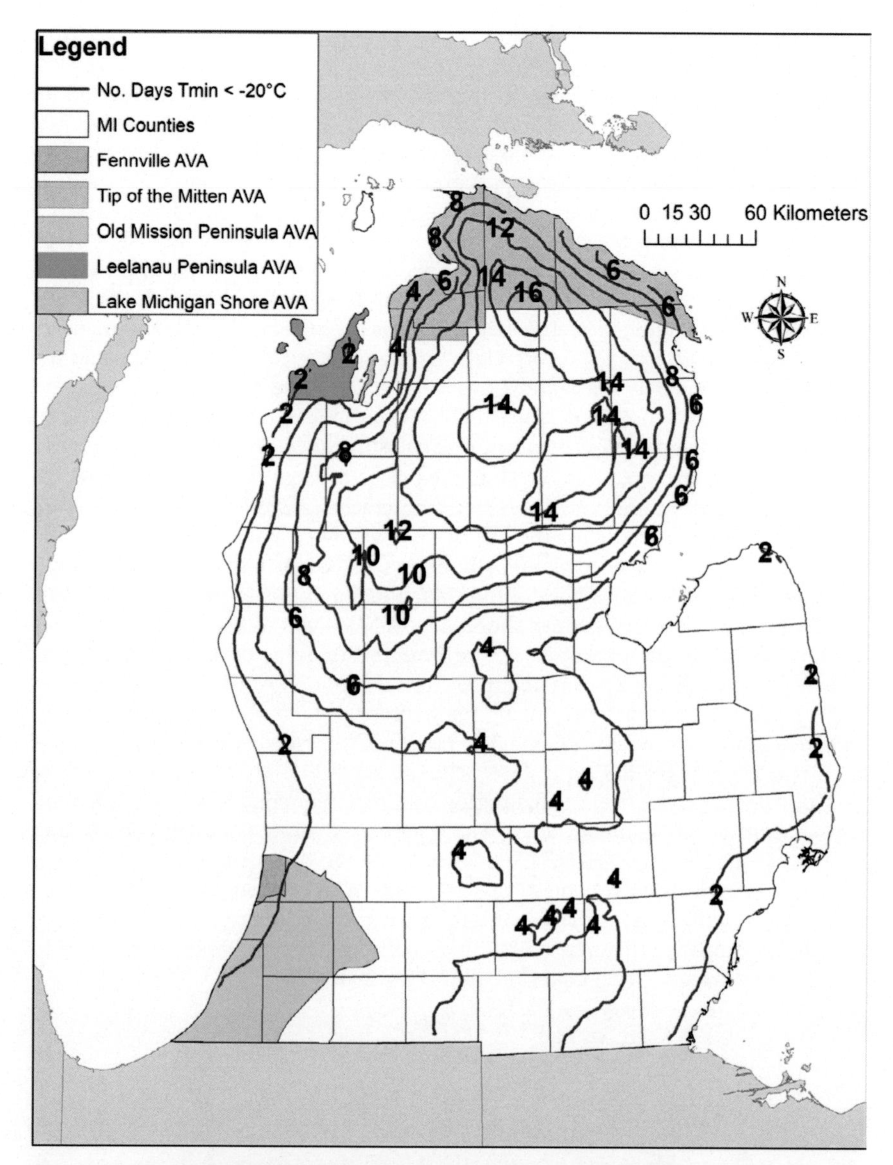

Fig. 4. Designated American Viticultural Areas in the Lower Peninsula of Michigan and the mean annual number of days with minimum temperatures of -20 °C or lower, 1981 to 2010. Temperature data courtesy of PRISM Climate Group, Oregon State University.

While the assessment of an individual site for overall suitability for an orchard or vineyard is complex, with a near infinite set of possible combinations of topography, soil, vegetation, etc., researchers in some production areas have developed quantitative ratings or suitability scores to help identify best potential sites (e.g., SCS USDA, 1972; Batholic and Martsolf, 1979). Recent related efforts have included areal mapping utilizing high resolution digital elevation models and simulation of cold air drainage patterns (Chung et al., 2006). For greatest utility to the decision-maker, such information must integrate climatic factors on macro-, meso-, and microscale levels (Yun, 2010).

At the same time, growers must also select cultivars that balance cool season chilling requirements with the climatological risk of freezing temperatures during spring development. This requires some knowledge of a region's climate. For example, planting a low-chilling cultivar in a high-chill region increases the risk of cold damage when an early bloom is hit by a spring frost. This decision is further complicated by the fact that areas with relatively cold climates in which the air temperature falls below 0 °C for long periods of time actually accumulate fewer chilling hours than regions where winter temperatures persist in the 0 to 10 °C range. In general, the longer and later the initial phenological development can be delayed, the better the chance of the crop escaping cold injury (Rodrigo and Herrero, 2000).

In addition to the passive frost and freeze control strategy of site selection, there are a number of active strategies employed in orchard and vineyard operations around the world to reduce the risk of cold injury. These include under-tree and over-tree-sprinkling irrigation (Heinemann et al., 1992), forced convection wind machines (Battany, 2012), flooding (George, 1979), heaters (Angus, 1955), and fogging (Mee and Bartholic, 1979). Given economic and environmental constraints, the most commonly used strategies worldwide are water sprinkling and wind or forced convection, both of which can potentially provide several degrees Centigrade of temperature modification in a orchard or vineyard environment (Snyder and de Melo-Abreu, 2005). Water-based systems are attractive because of their potential effectiveness, low costs, and the dual purpose of irrigation (Perry, 1986). During operation, sprinkler irrigation is turned on before the freezing point is reached, and left on until the air warms again. As long as additional water is applied and releases latent heat while freezing, the temperature of the vegetation being protected remains close to freezing. Drawbacks of this strategy include the potential demand for large amounts of water, the potential for excessive ice buildup and resulting damage to vegetation, less than desired effectiveness during windy conditions, saturated soils and leaching of nutrients, and increased risk of plant disease (Perry et al., 1980).

One related indirect strategy is the application of water to reduce vegetative tissue temperature through evaporative cooling which in turn delays phenological development and increases relative resilience to cold injury. In a recent study with apples and cherries in Michigan, Rijal (2017) applied water mist through a solid set canopy delivery system after endo-dormancy to king bloom and full bloom of non-misted buds. The misting delayed the bloom by 4 to 9 d in apple and 7 to 11 d in cherry using 84 to 260 mm ha^{-1} in apple and 55 to 108 mm ha^{-1} in sweet cherry, which is substantially less water than reported in previous related studies with conventional sprinklers. A deterministic heat transfer model of a tree fruit bud was then developed, calibrated, and validated using growth chamber data and validated using potted plant and field data collected earlier in the study. The model was run with 10 years of hourly observed climate data at three locations in major fruit-producing regions of Michigan for the period 2006 to 2015. Overall, the model estimated a delay in bloom of misted buds by more than a week compared to non-misted buds, which translated into a potential reduction in the frequency of damaging freeze events of 50 to 75%, and a decrease in freeze injury severity by 10 to 60% in misted apple buds and 45 to 100% in misted cherry buds.

Forced convection wind machines are used extensively for frost protection due to their effectiveness, simplicity of operation, and the suitability of their use with large canopy crops such as orchards and vineyards (Battany, 2012). The effectiveness of wind machines for frost protection is in large part a function of the strength of low-level temperature inversions, with observed crop canopy air temperature increases of approximately one-third to one-half the inversion strength (Ribeiro et al., 2006). However, knowledge of the presence and strength of inversions is critical in frost protection operations, as their usage without the presence of an inversion may actually accentuate the potential cold injury (Reese and Gerber, 1969).

Solar Radiation

Solar radiation is the primary source of energy for the production of plant dry matter through photosynthesis. Yields of temperate perennial fruit crops are largely dependent on the light use efficiency of canopies (Robinson and Lakso, 1991), while fruit quality is related more to the distribution and amount of light in the plant canopy (Buler and Mika, 2009). The morphological architecture of fruit tree canopies and vineyards plays a key role, with favorable light penetration and wind ventilation conditions in plant canopies achieved through various pruning and training systems (Willaume et al., 2004).

Radiation-use efficiency (RUE) of perennial fruit crops, defined as the ratio of dry matter produced per unit of radiant energy used in its production during a growing season are typically on the order of 1.0 g MJ^{-1} (Monteith, 1977). Radiation-use efficiency is directly dependent on the rate of insolation, temperature, vapor pressure deficit, and to other nonclimatic factors such as time of season, nutrient status, and age (Rosati et al., 2004). As C-3 species, RUE values for tree fruit and grape species are generally smaller than those for herbaceous species because of the high energy cost of woody biomass and the respiration of supporting organs (Kiniry, 1998). As such, solar radiation is typically not a limiting climate-related factor for production, although persistent high or low levels of solar insolation at certain times during the growing season can lead to changes in fruit quality such as sun scald (Glenn et al., 2002). In terms of total net carbon flux exchange, tree fruit crops have been found to be comparable in magnitude to deciduous forests growing in similar climate conditions (Zanotelli et al., 2015).

Precipitation and Water Usage

While water is an essential element in metabolic processes of perennial fruit crops, the primary function of water uptake by the plant is the physical moderation of leaf temperatures through evaporative cooling during transpiration. Water use and demand of these fruit crops depends greatly on plant age, canopy size, phenological stage, time of year, and pruning system (Fernández et al., 2008; Wang and Wang, 2017; Girona et al., 2011; Fereres et al., 1982; Dragoni et al., 2005). The impacts of water stress also vary greatly depending on phenological stage, and range from cessation of growth and leaf abscission during vegetative growth to fruit drop and early leaf senescence, and development of poor fruit quality during later reproductive stages (Ebel et al., 1995).

In many of the temperate climates that perennial tree fruit is produced, there is sufficient water between growing season precipitation and stored soil moisture to produce a crop each year. Many production areas around the world are preferentially located on coarse-textured soils which have relatively greater hydraulic conductivity, relatively rapid drainage rates, and high trafficability characteristics, but lower water holding capacities (Kistner et al., 2018). To improve crop yield and quality, reduce the risk of water stress situations (especially during orchard establishment) and to expand the range of production into more arid areas, irrigation is a common strategy as both a primary and secondary water source in many commercial production systems. Water management may be complicated however, as variation in canopy structure may result in high variability of water use from tree to tree in the same orchard and from field to field (Li et al., 2002). Rooting depths for water uptake may vary depending on soil type, tree age and variety, and rootstock. Previous research has suggested typical rooting depths of mature orchards on the order of 1.5 to 3 m (Westwood, 1993). However, the bulk of tree roots extracting water are in the top meter. Dwarf rootstocks generally have shallower root systems than standard rootstocks and may require additional water through irrigation as a result (Fereres et al., 1990).

Excessive amounts of precipitation are also a potential problem for growers, especially in species where high soil moisture levels can actually decrease fruit quality (Glenn et al., 2006). One notable excess water-related problem for sweet cherries just prior to harvest is cracking or splitting of the fruit due to untimely precipitation, rendering it unsaleable. Early studies suggested osmotically-driven water penetration through the wetted fruit surface was a primary factor, although more recent research concluded that a complex combination of tensile forces acting on the fruit surface from inside the fruit and the loss of bearing structure of the fruit skin was the primary cause (Balbontín et al., 2013; Sekse, 1995).

In contrast to many crops in which evapotranspiration is driven primarily by net radiation load, water use for tree fruits including apple is controlled by a more balanced combination of stomatal conductance, net radiation, and vapor pressure deficit (Osroosh et al., 2015; Jarvis, 1985; Lakso, 2003). While the actual rates of water use vary greatly depending on the factors described above, it is interesting to consider estimated rates of potential evapotranspiration. In their comprehensive work detailing crop evapotranspiration rates and water requirements Allen et al. (1998) derived Kc crop coefficients which are multiplied by standard daily reference evapotranspiration rates to obtain representative daily potential evapotranspiration rates for mature tree fruit crops. Their values range from 0.45 to 0.80 during developmental phenological stages, to 0.85 to 1.20 during mid-season, to 0.65 to 0.95 during late season stages, which places their potential daily water usage at rates near or above many other crop types. For comparison, the same values for maize grown as an annual crop for grain are 0.30 for the initial stage, 1.20 for mid-season, and 0.35 to 0.60 for late-season (Allen et al., 1998).

With total annual precipitation in many production areas less than 750 mm, supplemental irrigation is necessary to ensure consistent fruit yields and quality. An overarching objective goal of most irrigation strategies for perennial fruit crops is to supply just enough water at the right times to meet plant needs if water supplied though precipitation is insufficient. The application of water is based on crop water demand, the amount of fruit produced, and its quality (Perry, 1986). Chenafi et al. (2016) identified two major extreme strategies when irrigating orchards. The first, when water is non-limiting, is to irrigate with the aim of maximizing production. A second strategy is needed when water is limited by drought or limited availability of irrigation water. In this case, the limited water may be used to keep the crop alive. In between these two extremes is the intentional limited use of water or Regulated Deficit Irrigation strategy which has been observed to effectively reduce water usage and control tree growth with few if any negative impacts on yield or quality (Chalmers et al., 1984; Caspari et al., 1994).

Water management strategies for grape production (especially wine production) are quite different from those of tree fruit, as water usage by grapes is generally less. The derived Kc crop coefficients from Allen et al. (1998) range from 0.30 during developmental phenological stages to 0.70 to 0.85 during mid-season stages to 0.45 during late season stages. Second, grapes tend to have deep rooting systems which allows then to avoid serious water deficits during periods of drought (Alsina et al., 2011). Producers have long known from experience that higher grape quality tended to be associated with relatively low rainfall rates, with subsequent positive impacts on wine quality and additional benefits through a reduction in plant disease pathogen risk (Jackson, 2014).

Protective Canopy Covers

Given the need for consistent high quality and yields for most specialty crops, there has been increasing use of protective covers or high tunnels to help reduce the risk and severity of many types of weather-related limitations (Carey et al., 2009). High tunnels include a range of designs from single-span to multi-span structures typically consisting of steel-tubed frames covered with a single layer of polyethylene greenhouse film. They are typically unheated and passively ventilated and must have water for irrigation (Lamont et al., 2003). Due to their cost and the physical size of many perennial fruit crops, the tunnels are not feasible options for some fruit crops. However, for high value crops such as sweet cherries, they have been found to be economically feasible with increased yield and quality, longer and more predictable fruit shelf life, advance and prolong the crop harvest season, and reduce insect and disease pressure (Lang, 2009; Jirgena et al., 2013). In a study of cherry production in the Great Lakes region of the United States, Ho et al. (2018) simulated the economic implications for production systems responding with and without high tunnels, crop insurance, and weather insurance. They concluded that all three strategies provided better economic returns than a control system, and that high tunnels generated the largest net returns, especially if price premiums existed for earlier and larger fruit.

Hail nets are another type of cover direct mounted above orchards and vineyards utilized to protect vegetation and fruit from damage resulting from hail. While sometimes more costly than the original establishment costs of the orchards they protect (Treder et al., 2016), the nets have been shown to increase potential profitability in areas where hail is a frequent hazard (Whitaker and Middleton, 1999). In addition to their protective function, hail nets may also modify crop microclimate, with potential reductions in incident solar radiation (Funk and Blanke, 2003). Depending on the crop type and cultivar, this reduction may be beneficial, such as in a reduction of fruit sunburn (McCaskill et al., 2016) or adverse, such as hindrance to normal growth and color development (Girona et al., 2011).

Recent Observed and Projected Future Trends

While potential impacts of a changing climate on many annual grain and oilseed food crops have been studied extensively, relatively less is known about the effects on perennial fruit crops. Climate in most global production areas has become warmer in recent decades, with general warming of mean temperatures from 0.1 to 0.4 °C per decade (Hartmann et al., 2013). There have also been related changes in seasonality in many areas, especially with an earlier onset of the spring season (Andresen et al., 2013). In Japan, for example, Primack et al. (2009) reconstructed a climate series based on cherry blossom festivals back to the ninth century and concluded that cherries are currently flowering earlier than they have at any time during the previous 1200 yr. In the Great Lakes region of North America, the seasonal spring warm-up is also occurring earlier on average than in the past, trending as much as 10 d earlier in just the past 30 yr, which has in turn has advanced the dates of budbreak and other phenological stages of most perennial crops, potentially leaving them more vulnerable to subsequent cold injury (Fig. 5). Rill (2016) found that the frequency of potentially damaging spring freeze events following initial vegetative crop development of sour cherry is increasing in some areas of the region. This is in agreement with the finding by Augspurger (2013) that frost damage to various woody species in Illinois increased since 1980 due to advancing phenology. In a study of temporal patterns and trends across southwestern Lower Michigan during the period 1950 to 2011, Schultze et al. (2014) found an average increase over the region of more than 3.7 growing degree days (base 10 °C) per year as well as an increase in the length of the frost-free growing season by 28 d in length since 1971.

With overall warmer temperatures, many production areas around the world have experienced declines in chilling hours during the cool season. Locations include portions of the United States (Baldocchi and Wong, 2008), Europe (Atkinson et al., 2013), and Australia (Darbyshire et al., 2011). In some low latitude locations

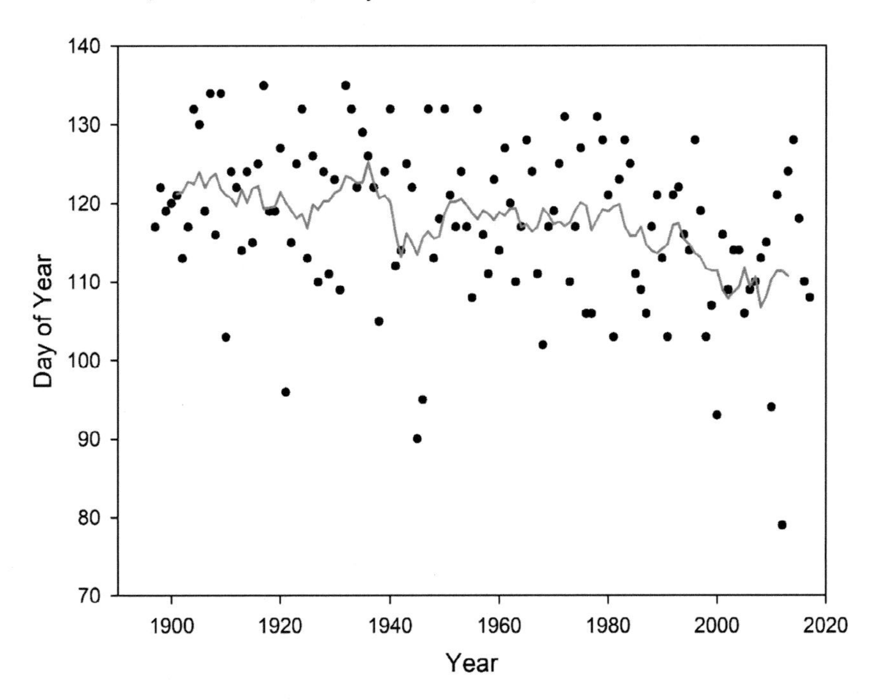

Fig. 5. Simulated dates of Stage 2 (side green) for sour cherry based on daily maximum and minimum temperatures, 1898 to 2015, at Traverse City, MI. Dates in individual years are plotted as solid circles. The solid line depicts a nine-year moving average of the individual years (Rill, 2016).

with already climatologically low chilling hour accumulations, recent milder winters have led to negative crop impacts such as in the southeastern United States (Couvillon, 1995) and in Tunisia, where Ghrab et al. (2014) observed delayed flowering, extended flowering duration, increased bud abscission and affected fruit set, and substantial decreases of fruit yield and quality in peach during abnormally mild winters. In contrast, in a climatologically cooler production area at a higher latitude in central Europe, the number of total accumulated chilling hours was observed to have not changed significantly over time during the recent decades, and this is likely to also be the case in many other temperate climates (Luedeling et al., 2011a).

For viticulture, the observed warming of the past few decades has been largely beneficial with longer and warmer growing seasons and less risk of frost. However, the impacts have been mixed by region. Warmer temperatures have provided more consistent ripening conditions for existing cultivars and have allowed production of warm season cultivars in poleward latitude areas including where such cultivation was previously impossible (e.g., Great Britain). In contrast, in already warm climates, recent trends toward warmer and drier climates have led to challenges in ripening balanced fruit. (Jones et al., 2012).

Trends in precipitation during recent decades vary greatly by region (Hartmann et al., 2013), although overall impacts are generally relatively less than those associated with temperature trends given the growing availability of irrigation in many specialty crop production areas (Walthall et al., 2012). For high-value specialty crops, there is also risk of too much water. Kistner et al. (2018) found the largest number of weather-related economic losses in specialty crops across the Midwestern United States during recent decades was associated with excessive moisture. Similarly, Rill (2016) found increases in the number of poor pollination days during the last two to three decades, which she attributed to an increase in the frequency of wet days. Even in areas with observed increases in precipitation, supplemental irrigation is required in many years to fully meet crop water requirements for maximum yield and quality of high-value crops (Dragoni and Lakso, 2011).

Projected long term climatic trends through the end of the century obtained with a series of model-derived projections vary somewhat by region and greenhouse gas emissions scenario, but many are consistent with recent historical trends. From projections associated with the most recent assessment of the Intergovernmental Panel on Climate Change, warming is projected across almost all current international tree fruit production areas with general mean annual temperature increases generally ranging from 0.5° to 3.0° for the 2046 to 2065 mid-century time frame and 1.0 to 5.0 °C for the late century 2081 to 2100 period (Collins et al., 2013). Annual precipitation is generally expected to increase in mid latitude production areas, with decreases in most subtropical areas. Given projected warmer temperatures, potential and actual evapotranspiration rates are expected to increase in all production areas, with corresponding decreases in mean soil moisture, especially in Mediterranean sections of southern Europe and northern Africa, and during the warm season (Collins et al., 2013).

Direct impacts associated with warming temperatures include accelerated crop growth rates and phenology, reductions in cold chilling hours, and possible changes in crop quality. Specific Responses may vary between growing regions and cultivars. In a model simulation-based study of projected impacts of a warmer climate on apples in New South Wales, Australia, Parkes et al. (2017) found advances of bloom date at some location relative to recent historical dates but delays in others. An additional related problem is the potential for shifts in phenology when the flowering period of pollinating trees no longer overlaps with that of the cultivar being grown for fruit.

Some earlier studies of the impact of a warming climate on perennial fruit production have also suggested an increase in frost damage risk associated with advanced phenology (Rochette et al., 2004). In a more recent study examining potential future impacts on sour cherry production in the Great Lakes region of North America, Rill (2016) found a consistent and uniform phenological advance of the growing season in mid-21st century and late 21st century timeframes, which is in agreement with similar studies in Europe (Hoffmann and Rath, 2013; Molitor et al., 2014). However, she found projected changes in the frequency and severity of damaging freezes relative to historical observations to be both complex and inconsistent, with variations in impacts by both location and climate projection. These findings are consistent with those of Winkler et al. (2013) who concluded, based on a large ensemble of downscaled CMIP3 simulations, that future changes in frost risk for sour cherry production in Michigan are uncertain.

As previously noted, trends toward warmer temperatures will also lead to increasing lack of winter chilling hours in some current production areas and potential reductions in fruit yield and quality. Reduced winter chill is likely to affect the timing and quality of bud dormancy release and flowering, which may increase the risk of subsequent cold damage and yield (Luedeling et al., 2011a). However, the potential magnitude of impacts associated with this problem is still not certain, as substantial knowledge gaps exist in the field of fruit tree dormancy (Luedeling et al., 2011b). Better understanding of the release of endodormancy is thought to be critical (Beauvieux et al., 2018). Possible adaptive strategies for lack of chilling include plant breeding to develop new low chill-cultivars as well as the introduction of cultural practices for crops to better tolerate low chill environments (Atkinson et al., 2013).

Given projections for decreasing frequencies of extreme low temperatures events and increases in high temperature extremes, it is also important to consider potential impacts, especially in those climates where high extremes are already more frequent. High temperature extremes are capable of leading to ovule blossom damage during flower bud initiation (Tomita et al., 2016) and to reductions in crop yield and quality through sunburn damage, poor blush development, watercore, rapid fruit ripening, and reduced fruit growth (Parkes et al., 2017).

In terms of moisture, some current production areas of the world are projected to become wetter while others may become drier with more frequent droughts. Under future climate change projection scenarios, rainfall patterns in current production areas are generally expected to shift and become more erratic. Overall, the number of days with heavy precipitation during the future is expected to increase in most areas, coupled with an increased number of dry days between rainfall events (Collins et al., 2013). Precipitation projections also suggest potential changes in the seasonality of precipitation. For example, in the Midwestern USA, winter and spring seasonal totals are projected to increase while warm season totals are projected to remain constant or decrease (Pryor et al., 2014). In general, warm season moisture deficits could be exacerbated by increased temperatures and longer growing seasons which increase the overall crop water use (Kistner et al., 2018).

One potential positive impact on perennial fruit crops will be increasing carbon dioxide levels in the atmosphere. As C-3 plant species, these crops are expected to benefit from increasing radiation use and water use efficiencies

(Hatfield et al., 2008). The enrichment effect may be temporary, however, with warmer nights expected to increase plant respiration rates and consume fixed carbon, potentially negating any yield increases for most commodities by the mid-twenty-first century (Eigenbrode et al., 2013). Additional offsetting factors include expected increases in potential evapotranspiration, pest pressure, heat stress, and water shortages (Houston et al., 2018).

Pests

Insect pests, plant pathogens and weeds are a major constraint to crop production and driven to a large extent by weather and climatic conditions (Oerke 2006). They have also been identified by specialty crop growers as a primary production issue concern associated with a changing climate, particularly the spread of exotic species (Johnson and Morton, 2015). With projected increases in temperature and the frequency of extreme weather events, pest outbreaks, and subsequent damage to cropping systems are expected to become more frequent and severe in the future (Deutsch et al., 2018; Hatfield et al., 2014). Most research suggests an overall increase in the number of outbreaks and poleward migration of a wide variety of weeds, insects, and pathogens in future decades (Gregory et al., 2009) associated at least in part to warmer winters and increased survival of overwintering of insects and other pathogens (Tobin et al., 2008).

In theory, projected warmer temperatures in the future should result in an increase in growth and development rates of poikilothermic organisms such as insects and many weeds. In a study of the potential impacts of projected warming temperatures on apple codling moth (*Cydia Pomonella* [Linneaus]) pest pressure on apple production in the Great Lakes Region of the United States, Winkler et al. (2002) estimated an average of one extra generation of the pest per growing season due to during the late 21st century timeframe at their study locations. Just as importantly, they identified potential new management challenges for growers given the need for up to two additional control sprays resulting from the increased number of generations, including some that would fall within current pre-harvest intervals.

Not all projected changes in future pest pressure are straightforward. Baule et al. (2017) examined the potential changes in plant disease risk associated with a changing climate in Michigan. They considered cherry leaf spot, (*Blumeriella jaapii*), a major fungal pathogen affecting sour cherry production, with an operational disease simulation in historical and projected future time frames (Fig. 6). Relative to the historical period, they found both a shift of greatest disease pressure in the future simulations to earlier periods during the growing season and a general decrease in disease infection rates which was associated with lower relative humidity values, despite projected future increases in specific humidity.

While these types of studies help identify some of the potential links between changing climate and pest pressure, many aspects remain unclear. Recently published studies suggest that the interactions among environmental and biotic factors linked with pest pressure are highly complex, and hinder attempts to generalize pest response (Juroszek and von Tiedemann, 2013; Thomson et al., 2010). In addition, within field variability of environmental conditions related to disease and insect pressure is very high, making direct attribution of causal mechanisms and identification of potential management strategies difficult (Penrose and Nicol,

1996). Because of these challenges, careful, regular monitoring of pest populations will be imperative as the climate changes (Wolfe et al., 2017).

Adaptive Strategies

Given changing environmental limitations and constraints on their production systems, growers and the agricultural industry will need to make strategic and tactical adaptive changes in their operations to better manage the changing types and levels of weather and climate-related risks. For example, adopting new technologies to monitor orchard climate and tree performance will improve growers' capacity to understand and manage the impacts of a changing climate. (Parkes et al., 2017). Adaptive capacity is also influenced by the relative availability of effective adaptation options and the capacity of individuals to implement adaptation options and be supported in their application (Jones and Webb, 2010).

One critical overarching issue in the adaptation process is the assessment of a given area or region to successfully allow fruit production. Collectively, recent research suggests a potential future migration of tree fruit production both out of some existing areas and into new, cooler regions (Jones et al., 2012; Wolfe et al., 2017). However, such expansion would require substantial economic investment and may be limited by other constraints such as the lack of infrastructure and conditions in other production areas (Winkler et al., 2013). In addition, unlike traditional row crops, there is often little to no tolerance for weather-induced reductions in commodity quality, which makes effective climate resiliency and adaptation

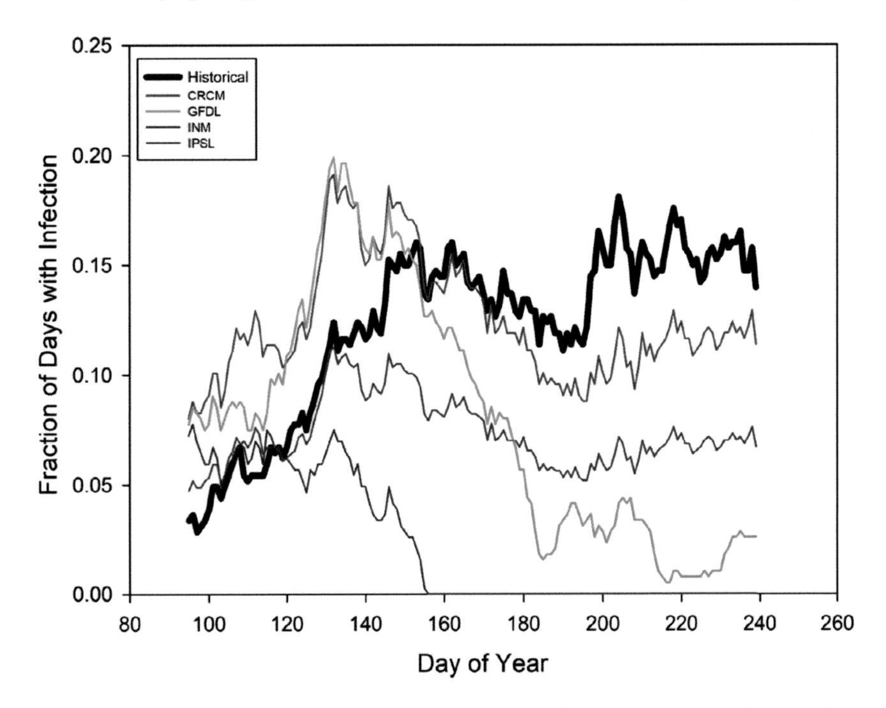

Fig. 6. Fraction of days each year at Grand Rapids, MI with infections of cherry leaf spot for historical (1976–2015, in black) & projected future (2071–2100, colors from four separate global climate models) time frames (from Baule et al., 2017).

practices all the more critical for sustainable production (Kistner et al., 2018). Given the relatively long time scales for growth and production of tree fruit crops, variety selection and choice of production site will be important strategies. However, it may be difficult to introduce new varieties of fruit crops because consumers recognize and value specific varieties that they are accustomed to (Wolfe et al., 2017).

As noted earlier, many adaption strategies that may be useful under a changing climate are already being used globally to improve production and quality in some production areas (Houston et al., 2018). While they are in general expensive, they have been found to be cost effective, especially with high value crops (McFerson, 2015). Examples of heat-related strategies include application of water from overhead sprinkler irrigation or reflective materials to reduce crop and fruit temperature and heat damage risk (Evans, 2004). Similarly, cooling technologies for harvest and storage operations will also become useful given warmer ambient environmental temperatures (Moretti et al., 2010).

One primary adaptive strategy for perennial fruit crops is the development of cultivars better suited for particular climatic conditions through plant breeding and/or genetic manipulation (Parkes et al., 2017). For example, Andersen et al. (2018) recently examined the susceptibility of sour cherry to the cherry leaf spot fungal disease and developed a two-gene model for crop resistance and tolerance, potentially allowing the future breeding of sour cherry cultivars with durable resistance to the disease. One major challenge with plant breeding for adapting to climate change is that breeding programs for perennial fruit crops typically require long periods of time, potentially failing to keep up with changing environmental conditions (Iwata et al., 2016). A related challenge is the relative lack of genetic diversity of some current crop species such as wine grapes, which have been maintained over long periods of time to ensure true breeding cultivars. As a result, the crop remains especially vulnerable to many types of pathogen pressures, and long-term sustainable production in the future will depend on exploitation of its natural genetic diversity (Myles et al., 2011).

Finally, fruit growers have identified the need for crop-specific weather, production, and financial risk management tools including crop insurance to help manage the increased production risk under a changing climate (Kistner et al., 2018). Insurance can play an important role in the process of adaptation to climate change by spreading risk and buffering the financial implications of unexpected crop failure following extreme weather events (Di Falco et al., 2014), and demand is expected to increase in the future given projected climatic changes (Garrido et al., 2011). Diversification has been also been identified as a related approach to help reduce weather-related risk (Walthall et al., 2012), as a substitute strategy for insurance (Di Falco et al., 2014) and as potential means of buffing climatic impacts on crop quality (Ahmed and Stepp, 2016).

References

Aggelopoulou, K.D., D. Wulfsohn, S. Fountas, T.A. Gemtos, G.D. Nanos, and S. Blackmore. 2010. Spatial variation in yield and quality in a small apple orchard. Precis. Agric. 11(5):538–556. doi:10.1007/s11119-009-9146-9

Ahmed, S., and J.R. Stepp. 2016. Beyond yields: Climate change effects on specialty crop quality and agroecological management. Elementa 4(0):000092. doi:10.12952/journal.elementa.000092

Ahmed, S., J.R. Stepp, C. Orians, T. Griffin, C. Matyas, A. Robbat, S. Cash, D. Xue, C. Long, U. Unachukwu, S. Buckley, D. Small, and E. Kennelly. 2014. Effects of extreme climate events on tea (Camellia sinensis) functional quality validate indigenous farmer knowledge and sensory preferences in tropical China. PLoS One 9(10): E109126. doi:10.1371/journal.pone.0109126.

Allen, R.G., L.S. Pereira, D. Raes, and M. Smith. 1998. Crop evapotranspiration-Guidelines from computing crop water requirements. FAO Irrigation and Drainage Paper 56. United Nations Food Agric. Organ., Rome..

Alsina, M.M., D.R. Smart, T. Bauerle, F. de Herralde, C. Biel, C. Stockert, C. Negron, and R. Save. 2011. Seasonal changes of whole root system conductance by a drought-tolerant grape root system. J. Exp. Bot. 62(1):99–109. doi:10.1093/jxb/erq247

Andersen, K.L., A.M. Sebolt, G.W. Sundin, and A.F. Iezzoni. 2018. Assessment of the inheritance of resistance and tolerance in cherry (Prunus sp.) to Blumeriella jaapii, the causal agent of cherry leaf spot. Plant Pathol. 67(3):682–691. doi:10.1111/ppa.12765

Andresen, J.A., D.G. McCullough, B.E. Potter, C.N. Koller, L.S. Bauer, and C.W. Ramm. 2001. Effects of winter temperatures on gypsy moth egg masses in the Great Lakes region of the United States. Agric. For. Meteorol. 110:85–100. doi:10.1016/S0168-1923(01)00282-9

Andresen, J., and J.A. Winkler. 2009. Weather and climate. In: R.J. Schaetzl, J.T. Darden, and D.S. Brandt, editors, Michigan geography and geology. Pearson Publishing, London. p. 672.

Angus, D.E. 1955. The use of heaters for frost prevention in a pineapple plantation. Aust. J. Agric. Res. 6:186–195. doi:10.1071/AR9550186

Atkinson, C.J., R.M. Brennan, and H.G. Jones. 2013. Declining chilling and its impact on temperate perennial crops. Environ. Exp. Bot. 91:48–62. doi:10.1016/j.envexpbot.2013.02.004

Augspurger, C.K. 2013. Reconstructing patterns of temperature, phenology, and frost damage over 124 years: Spring damage risk is increasing. Ecology 94(1):41–50. doi:10.1890/12-0200.1

Balbontín, C., H. Ayala, R.M. Bastías, G. Tapia, M. Ellena, C. Torres, J.A. Yuri, J. Quero-García, J.C. Ríos, and H. Silva. 2013. Cracking in sweet cherries: A comprehensive review from a physiological, molecular, and genomic perspective. Chil. J. Agric. Res. 73(1):66–72. doi:10.4067/S0718-58392013000100010

Baldocchi, D., and S. Wong. 2008. Accumulated winter chill is decreasing in the fruit growing regions of California. Clim. Change 87(S1):153–166. doi:10.1007/s10584-007-9367-8

Bates, T., and J. Morris. 2009. Mechanical cane pruning and crop adjustment decreases labor costs and maintains fruit quality in New York 'Concord'grape production. Horttechnology 19(2):247–253. doi:10.21273/HORTSCI.19.2.247

Batholic, J.F., and J.D. Martsolf. 1979. Site selection. In: B.J. Barfield and J.F. Gerber, editors, Modification of the areal environment of plants. ASAE, St. Joseph, MI. p. 281–290.

Battany, M.C. 2012. Vineyard frost protection with upward-blowing wind machines. Agric. For. Meteorol. 157:39–48. doi:10.1016/j.agrformet.2012.01.009

Baule, W.J., J. Andresen, A. Pollyea, and L. Briley. 2017. Influence of the Great Lakes on hydro-climatic variables in CMIP5 models and implications for weather-driven applications. In: AMS 23rd Conference on Applied Climatology, Asheville, NC. 25–27 June. American Meteorological Society, Boston, MA.

Beauvieux, R., B. Wenden, and E. Dirlewanger. 2018. Bud dormancy in perennial fruit tree species: A pivotal role for oxidative cues. Front. Plant Sci. 9:657. doi:10.3389/fpls.2018.00657

Buler, Z., and A. Mika. 2009. The influence of canopy architecture on light interception and distribution in 'Sampion' apple trees. J. Fruit Ornamental Plant Res. 17(2):45–52.

Carey, E.E., L. Jett, W.J. Lamont, T.T. Nennich, M.D. Orzolek, and K.A. Williams. 2009. Horticultural crop production in high tunnels in the United States: A snapshot. Horttechnology 19(1):37–43. doi:10.21273/HORTSCI.19.1.37

Caspari, H.W., M.H. Behboudian, and D.J. Chalmers. 1994. Water use, growth, and fruit yield of Hosui'Asian pears under deficit irrigation. J. Am. Soc. Hortic. Sci. 119(3):383–388. doi:10.21273/JASHS.119.3.383

Chalmers, D.J., P.D. Mitchell, and P.H. Jerie. 1984. The physiology of growth control of peach and pear trees using reduced irrigation. Acta Hortic. 146:143–150. doi:10.17660/ActaHortic.1984.146.15

Che, D. 2010. Value-added agricultural products and entertainment in Michigan's fruit belt.. In: G. Halseth, S. Markey, and D. Bruce, editors, The next rural economies: Constructing rural place in global economies. p. 102–114.

Chenafi, A., P. Monney, E. Arrigoni, A. Boudoukha, and C. Carlen. 2016. Influence of irrigation strategies on productivity, fruit quality and soil-plant water status of subsurface drip-irrigated apple trees. Fruits 71(2):69–78. doi:10.1051/fruits/2015048

Chung, U., H.H. Seo, K.H. Hwang, B.S. Hwang, J. Choi, J.T. Lee, and J.I. Yun. 2006. Minimum temperature mapping over complex terrain by estimating cold air accumulation potential. Agric. For. Meteorol. 137(1–2):15–24. doi:10.1016/j.agrformet.2005.12.011

Collins, M., R. Knutti, J. Arblaster, J.-L. Dufresne, T. Fichefet, P. Friedlingstein, X. Gao, W.J. Gutowski, T. Johns, G. Krinner, M. Shongwe, C. Tebaldi, A.J. Weaver, and M. Wehner. 2013. Long-term climate change: projections, commitments and irreversibility In: T.F. Stocker, D. Qin, G.K. Plattner, M. Tignor, S.K. Allen, J. Boschung, A. Nauels, Y. Xia, V. Bex, and P.M. Midgley, editors, Climate Change 2013: The physical science basis. Contribution of Working Group I to the Fifth Assessment Report of the Intergovernmental Panel on Climate Change. Cambridge Univ. Press, Cambridge, U.K.

Considine, M.J., and J.A. Considine. 2016. On the language and physiology of dormancy and quiescence in plants. J. Exp. Bot. 67(11):3189–3203. doi:10.1093/jxb/erw138

Couvillon, G.A. 1995. Temperature and stress effects on rest in fruit trees: A review. Acta Hortic. 395:11–20. doi:10.17660/ActaHortic.1995.395.1

Darbyshire, R., L. Webb, I. Goodwin, and S. Barlow. 2011. Winter chilling trends for deciduous fruit trees in Australia. Agric. For. Meteorol. 151(8):1074–1085. doi:10.1016/j.agrformet.2011.03.010

Dennis, F.G., and G.S. Howell. 1974. Cold hardiness of tart cherry bark and flower buds. Michigan State University Agricultural Experiment Station, East Lansing, MI.

Deutsch, C.A., J.J. Tewksbury, M. Tigchelaar, D.S. Battisti, S.C. Merrill, R.B. Huey, and R.L. Naylor. 2018. Increase in crop losses to insect pests in a warming climate. Science 361(6405):916–919. doi:10.1126/science.aat3466

Dragoni, D., and A.N. Lakso. 2011. An apple-specifc ET model. Acta Hortic. 903:1175–1180. doi:10.17660/ActaHortic.2011.903.164

Dragoni, D., A.N. Lakso, and R.M. Piccioni. 2005. Transpiration of apple trees in a humid climate using heat pulse sap flow gauges calibrated with whole-canopy gas exchange chambers. Agric. For. Meteorol. 130(1–2):85–94. doi:10.1016/j.agrformet.2005.02.003

Dzhangaliev, A.D. 2003. The wild apple tree of Kazakhstan. Hortic. Rev. (Am. Soc. Hortic. Sci.) 29:63–304.

Ebel, R.C., J.P. Mattheis, and D.A. Buchanan. 1995. Drought stress of apple trees alters leaf emissions of volatile compounds. Physiol. Plant. 93(4):709–712. doi:10.1111/j.1399-3054.1995.tb05120.x

Eigenbrode, S., S. Capalbo, L. Houston, J. Johnson-Maynard, C. Kruger, and B. Olen. 2013. Agriculture. In: M.M. Dalton, P. Mote, and A. Snover, editors, Climate change in the Northwest: Implications for our landscapes, waters, and communities. Island Press, Washington. p. 149–180. doi:10.5822/978-1-61091-512-0_6

Epstein, D.J., J. Andresen, G. Bird, J. Flore, L. Gut, P. McManus, J. Nugent, R. Isaacs, A. Schilder, M. Whalon, and J. Sanchez. 2007. Tart cherry systems. Chapter 6 in Building a sustainable future: Ecologically based farming systems, S. Deming, L. Johnson, D. Lehnert, D. Mutch, L. Probyn, K. Renner, J. Smeenk, S. Thalmann, and L. Worthington, editors. Extension Bull. E-2983. East Lansing, Mich.: Michigan State University.

Evans, R.G. 2004. Energy balance of apples under evaporative cooling. Trans. ASAE 47(4): 1029–1037.

Di Falco, S., F. Adinolfi, M. Bozzola, and F. Capitanio. 2014. Crop insurance as a strategy for adapting to climate change. J. Agric. Econ. 65(2):485–504. doi:10.1111/1477-9552.12053

FAO. 2018. Food and agriculture data. Food Agric. Organ. United Nations, Rome, Italy. http://faostat.fao.org/faostat (Accessed 2 May 2019).

Fereres, E., D.A. Goldhamer, B.A. Stewart, and D.R. Nielsen. 1990. Deciduous fruit and nut trees. In: R.J. Lascano and R.E. Sojka, Irrigation of agricultural crops, Agronomy 30, pg. 987-1017. ASA, CSSA, SSSA, Madison, WI.

Fereres, E., D.A. Martinich, T.M. Aldrich, J.R. Castel, E. Holzapfel, and H. Schulbach. 1982. Drip irrigation saves money in young almond orchards. Calif. Agric. 36(9): 12–13. http://www.ucanr.edu/sites/calagjournal/archive/?article=ca.v036n09p12.

Fernández, J.E., S.R. Green, H.W. Caspari, A. Diaz-Espejo, and M.V. Cuevas. 2008. The use of sap flow measurements for scheduling irrigation in olive, apple and Asian pear trees and in grapevines. Plant Soil 305:91–104. doi:10.2307/42951841

Funke, K., and M. Blanke. 2003. Can reflective ground cover compensate for light losses under hail nets. Erwerbs-Obstbau 45:137–144.

Garrido, A., M. Bielza, D. Rey, M.I. Minguez, and M. Ruiz-Ramos. 2011. Insurance as an adaptation to climate variability in agriculture. In: P. Hazell, R. Mendelsohn, and A. Dinar, editors, Handbook on climate change and agriculture. Edward Elgar Publishing, Cheltenham, UK. p. 420–445. doi:10.4337/9780857939869.00029

George, J.G. 1979. Frost protection by flood irrigation. In: B.J. Barfield and J.F. Gerber, editors, Modification of the areal environment of plants. ASAE, St. Joseph, MI. p. 368–372.

Ghrab, M., M. Ben Mimoun, M.M. Masmoudi, and N. Ben Mechlia. 2014. Chilling trends in a warm production area and their impact on flowering and fruiting of peach trees. Sci. Hortic. (Amsterdam) 178:87–94. doi:10.1016/j.scienta.2014.08.008

Girona, J., J. del Campo, M. Mata, G. Lopez, and J. Marsal. 2011. A comparative study of apple and pear tree water consumption measured with two weighing lysimeters. Irrig. Sci. 29(1):55–63. doi:10.1007/s00271-010-0217-5

Glenn, D.M., E. Prado, A. Erez, J. McFerson, and G.J. Puterka. 2002. A reflective, processed-kaolin particle film affects fruit temperature, radiation reflection, and solar injury in apple. J. Am. Soc. Hortic. Sci. 127(2):188–193. doi:10.21273/JASHS.127.2.188

Glenn, D.M., R. Scorza, and W.R. Okie. 2006. Genetic and environmental effects on water use efficiency in peach. J. Am. Soc. Hortic. Sci. 131(2):290–294. doi:10.21273/JASHS.131.2.290

Gregory, P.J., S.N. Johnson, A.C. Newton, and J.S.I. Ingram. 2009. Integrating pests and pathogens into the climate change/food security debate. J. Exp. Bot. 60(10):2827–2838. doi:10.1093/jxb/erp080

Hampson, C.R., H.A. Quamme, and R.T. Brownlee. 2002. Canopy growth, yield, and fruit quality of 'Royal Gala' apple trees grown for eight years in five tree training systems. HortScience 37(4):627–631. doi:10.21273/HORTSCI.37.4.627

Hartmann, D.L., A.M.G. Klein Tank, M. Rusticucci, L.V. Alexander, S. Brönnimann, Y. Charabi, F.J. Dentener, E.J. Dlugokencky, D.R. Easterling, A. Kaplan, B.J. Soden, P.W. Thorne, M. Wild, and P.M. Zhai. 2013. Observations: Atmosphere and surface In: T.F. Stocker, D. Qin, G.-K. Plattner, M. Tignor, S.K. Allen, J. Boschung, A. Nauels, Y. Xia, V. Bex, and P.M. Midgley, editors, Climate Change 2013: The physical science basis. Contribution of Working Group I to the Fifth Assessment Report of the Intergovernmental Panel on Climate Change. Cambridge Univ. Press, Cambridge, U.K.

Hatfield, J.L., K.J. Boote, P. Fay, L. Hahn, C. Izaurralde, B.A. Kimball, T. Mader, J. Morgan, D. Ort, and W. Polley. 2008. Agriculture. The effects of climate change on agriculture, land resources, water resources, and biodiversity in the United States. U.S. Climate Change Science Program Synthesis and Assessment Product 4.3. Global Change Research Information Office, Washington, D.C.

Hatfield, J., G. Takle, R. Grotjahn, P. Holden, R.C. Izaurralde, T. Mader, E. Marshall, and D. Liverman. 2014. Ch. 6: Agriculture. In: J.M. Melillo, T.C. Richmond, G.W. Yohe, editor, Climate change impacts in the United States: The Third National Climate Assessment, In: U.S. Global Change Research Program, Washington, D.C. p. 150–174.

Heinemann, P.H., C.T. Morrow, T.S. Stombaugh, B.L. Goulart, and J. Schlegel. 1992. Evaluation of an automated irrigation system for frost protection. Appl. Eng. Agric. 8(6):779–785. doi:10.13031/2013.26113

Hewett, E.W. 1976. Seasonal variation of cold hardiness in apricots. N. Z. J. Agric. Res. 19(3):353–358. doi:10.1080/00288233.1976.10429078

Hewett, E.W. and J.E. Hawkins. 1968. Sprinkler irrigation to protect apricots from frost. New Zealand Journal of Agricultural Research 11:4, 927-938, doi:10.1080/00288233.1968.10422426.

Ho, S.-T., J.E. Ifft, B.J. Rickard, and C.G. Turvey. 2018. Alternative strategies to manage weather risk in perennial fruit crop production. Agric. Resour. Econ. Rev.: 1–25. doi:10.1017/age.2017.29.

Hoffmann, H., and T. Rath. 2013. Future bloom and blossom frost risk for Malus domestica considering climate model and impact model uncertainties. PLoS One 8(10): E75033. doi:10.1371/journal.pone.0075033.

Houston, L., S. Capalbo, C. Seavert, M. Dalton, D. Bryla, and R. Sagili. 2018. Specialty fruit production in the Pacific Northwest: Adaptation strategies for a changing climate. Clim. Change 146(1–2):159–171. doi:10.1007/s10584-017-1951-y

Iwata, H., M.F. Minamikawa, H. Kajiya-Kanegae, M. Ishimori, and T. Hayashi. 2016. Genomics-assisted breeding in fruit trees. Breed. Sci. 66(1):100–115. doi:10.1270/jsbbs.66.100

Jackson, R.S. 2014. Wine science: Principles and applications. Elsevier Academic Press, Amsterdam, The Netherlands.

Jarvis, P.G. 1985. Coupling of transpiration to the atmosphere in horticultural crops: The omega factor. Acta Hortic. 171:187–206. doi:10.17660/ActaHortic.1985.171.17

Jirgena, H., J. Hazners, E. Kaufmane, S. Strautina, D. Feldmane, and M. Skrivele. 2013. Risks and returns in strawberry, raspberry and cherry production with various methods. Economics and Rural Development 9(2):16–26.

Johnson, A., and L.W. Morton. 2015. Midwest climate and specialty crops: Specialty crop leader views and priorities for Midwest specialty crops. Sociology Technical Report 1039. Department of Sociology, Iowa State University, Ames, IA.

Jones, G.V., and E. Hellman. 2003. Site assessment. In: E. Hellman, editor, Oregon viticulture. 5th ed. Oregon State Univ. Press, Corvallis, OR. p. 44–50.

Jones, G.V., R. Reid, and A. Vilks. 2012. Climate, grapes, and wine: Structure and suitability in a variable and changing climate In: P.H. Dougherty, The geography of wine. Springer Netherlands, Dordrecht. p. 109–133. doi:10.1007/978-94-007-0464-0_7

Jones, G.V., and L.B. Webb. 2010. Climate change, viticulture, and wine: Challenges and opportunities. J. Wine Res. 21(2–3):103–106. doi:10.1080/09571264.2010.530091

Juroszek, P., and A. von Tiedemann. 2013. Plant pathogens, insect pests and weeds in a changing global climate: A review of approaches, challenges, research gaps, key studies and concepts. J. Agric. Sci. 151(2):163–188. doi:10.1017/S0021859612000500

Kiniry, J.R. 1998. Biomass accumulation and radiation use efficiency of honey mesquite and eastern red cedar. Biomass Bioenergy 15(6):467–473. doi:10.1016/S0961-9534(98)00057-9

Kistner, E., O. Kellner, J. Andresen, D. Todey, and L.W. Morton. 2018. Vulnerability of specialty crops to short-term climatic variability and adaptation strategies in the Midwestern USA. Clim. Change 146(1–2):145–158. doi:10.1007/s10584-017-2066-1

Knudson, W.A. 2012. The economic impact of this spring's weather on the fruit and vegetable sectors. Working Paper 01-052012. Michigan State University, East Lansing, MI. http://legislature.mi.gov/documents/2011-2012/CommitteeDocuments/House/Agriculture/Testimony/Committee1-5-30-2012.pdf (Accessed 3 May 2019).

Labe, Z., T. Ault, and R. Zurita-Milla. 2017. Clim. Dyn. 48:3949 doi:10.1007/s00382-016-3313-2

Lakso, A.N. 2003. Water relations of apples. In: D.C. Ferree and I.A. Warrington, editors, Apples: Botany, production and uses. CAB International, Wallingford, Oxon, UK. p. 167–194. doi:10.1079/9780851995922.0167

Lamont, W.J., M.D. Orzolek, E.J. Holcomb, K. Demchak, E. Burkhart, L. White, and B. Dye. 2003. Production system for horticultural crops grown in the Penn State high tunnel. Horttechnology 13(2):358–362. doi:10.21273/HORTTECH.13.2.0358

Lang, G.A. 2009. High tunnel tree fruit production: The final frontier? Horttechnology 19(1):50–55. doi:10.21273/HORTSCI.19.1.50

Lang, G., J. Early, G. Martin, and R. Darnell. 1987. Endodormancy, paradormancy, and ecodormancy—physiological terminology and classification for dormancy research. Hortscience 22: 371–377. https://www.scienceopen.com/document?vid=72cb57c5-ade2-4529-b2ff-f7987d4b215e.

Luedeling, E., E.H. Girvetz, M.A. Semenov, and P.H. Brown. 2011a. Climate change affects winter chill for temperate fruit and nut trees (A Traveset, Ed.). PLoS One 6(5): E20155. doi:10.1371/journal.pone.0020155.

Luedeling, E., A. Kunz, and M. Blanke. 2011b. Mehr chilling für obstbäume in wärmeren wintern? Erwerbs-Obstbau 53(4):145–155. doi:10.1007/s10341-011-0148-1

Li, F., S. Cohen, A. Naor, K. Shaozong, and A. Erez. 2002. Studies of canopy structure and water use of apple trees on three rootstocks. Agric. Water Manage. 55(1):1–14. doi:10.1016/S0378-3774(01)00184-6

McCaskill, M.R., L. McClymont, I. Goodwin, S. Green, and D.L. Partington. 2016. How hail netting reduces apple fruit surface temperature: A microclimate and modelling study. Agricultural and Forest Meteorology 226–227: 148-160. doi:10.1016/j.agrformet.2016.05.017.

McFerson, J.R. 2015. Why we're breeding apples for climate change. Growing Produce, 26 June. https://www.growingproduce.com/fruits/apples-pears/why-were-breeding-apples-for-climate-change/ (Accessed 1 Oct. 2018).

Mee, T.R., and J.F. Bartholic. 1979. Frost protection by flood irrigation. In: B.J. Barfield and J.F. Gerber, editors, Modification of the areal environment of plants. ASAE, St. Joseph, MI. p. 368–372.

Molitor, D., A. Caffarra, P. Sinigoj, I. Pertot, L. Hoffmann, and J. Junk. 2014. Late frost damage risk for viticulture under future climate conditions: A case study for the Luxembourgish winegrowing region. Aust. J. Grape Wine Res. 20(1):160–168. doi:10.1111/ajgw.12059

Monfreda, C., N. Ramankutty, and J.A. Foley. 2008. Farming the planet: 2. Geographic distribution of crop areas, yields, physiological types, and net primary production in the year 2000. Global Biogeochem. Cycles 22:GB1022. doi:10.1029/2007GB002947

Monteith, J.L. 1977. Climate and efficiency of crop production in Britain. Philos. Trans. R. Soc. Lond. B Biol. Sci. 281:277–294. doi:10.1098/rstb.1977.0140

Moretti, C.L., L.M. Mattos, A.G. Calbo, and S.A. Sargent. 2010. Climate changes and potential impacts on postharvest quality of fruit and vegetable crops: A review. Food Res. Int. 43(7):1824–1832. doi:10.1016/j.foodres.2009.10.013

Myles, S., A.R. Boyko, C.L. Owens, P.J. Brown, F. Grassi, M.K. Aradhya, B. Prins, A. Reynolds, J.-M. Chia, D. Ware, C.D. Bustamante, and E.S. Buckler. 2011. Genetic structure and domestication history of the grape. Proc. Natl. Acad. Sci. USA 108(9):3530–3535. doi:10.1073/pnas.1009363108

NASS USDA. 2017. Quick Stats. USDA National Agriculural Statistics Service, Washington, D.C. https://quickstats.nass.usda.gov/ (Accessed 3 May 2019).

Oerke, E. 2006. Crop losses to pests. J. Agric. Sci. 144(01):31. doi:10.1017/S0021859605005708

Osroosh, Y., T.R. Peters, and C.S. Campbell. 2015. Estimating potential transpiration of apple trees using theoretical non-water-stressed baselines. J. Irrig. Drain. Eng. 141(9):04015009. doi:10.1061/(ASCE)IR.1943-4774.0000877

Parkes, H., N. White, L. Goodwin, J. Treeby, and S. MurphyWhite. 2017. Understanding apple and pear production systems in a changing climate. http://era.daf.qld.gov.au/id/eprint/6037/ (Accessed 21 Oct. 2018).

Pearce, R. 2001. Plant freezing and damage. Ann. Bot. (Lond.) 87(4):417–424. doi:10.1006/anbo.2000.1352

Penrose, L.J., and H.I. Nicol. 1996. Aspects of microclimate variation within apple tree canopies and between sites in relation to potential Venturia inaequalis infection. N. Z. J. Crop Hortic. Sci. 24(3):259–266. doi:10.1080/01140671.1996.9513960

Perry, K.B. 1986. FROSTPRO, a model of overhead irrigation rates for frost/freeze protection of apple orchards. HortScience 21(4): 1060–1061.

Perry, K.B., J.D. Martsolf, and C.T. Morrow. 1980. Conserving water in sprinkling for frost protection by intermittent application. J. Am. Soc. Hortic. Sci. 105(5):657–660.

Primack, R.B., H. Higuchi, and A.J. Miller-Rushing. 2009. The impact of climate change on cherry trees and other species in Japan. Biol. Conserv. 142(9):1943–1949. doi:10.1016/j.biocon.2009.03.016

Proebsting, E.L., Jr., and H.H. Mills. 1978. Low temperature resistance [frost hardiness] of developing flower buds of six deciduous fruit species. J. Am. Soc. Hortic. Sci. 103:192–198.

Pryor, S.C., D. Scavia, C. Downer, M. Gaden, L. Iverson, R. Nordstrom, J. Patz, and G.P. Robertson. 2014. Midwest. Climate change impacts in the United States: The third national climate assessment. U.S. Global Change Research Program, Washington, D.C. doi:10.7930/J0J1012N.

Reese, R.L., and J.F. Gerber. 1969. Empirical description of cold protection provided by a wind machine. J. Am. Soc. Hortic. Sci. 91: 697–700.

Ribeiro, A.C., J.P. De Melo-Abreu, R.L. Snyder. 2006. Apple orchard frost protection with wind machine operation. Agricultural and Forest Meteorology 141:71–81. doi:10.1016/j.agrformet.2006.08.019

Rill, L. 2016. Climatology of springtime freeze events in the Great Lakes region and their impact on sour cherry yields in historical and projected future time frames. M.S. ths. Michigan State University, East Lansing, MI.

Rijal, I. 2017. Use of water mist to protect tree fruits from spring frost damage. Ph.D. dissertation, Department of Geography, Environment, and Spatial Sciences, Michigan State University, East Lansing, MI.

Robinson, T.L., A.M. DeMarree, and S.A. Hoying. 2007. An economic comparison of five high density apple planting systems. Acta Hortic. 732:481–489. doi:10.17660/ActaHortic.2007.732.73

Robinson, T.L., and A.N. Lakso. 1991. Bases of yield and production efficiency in apple orchard systems. J. Am. Soc. Hortic. Sci. 116(2):188–194. doi:10.21273/JASHS.116.2.188

Rochette, P., G. Bélanger, Y. Castonguay, A. Bootsma, and D. Mongrain. 2004. Climate change and winter damage to fruit trees in eastern Canada. Can. J. Plant Sci. 84(4):1113–1125. doi:10.4141/P03-177

Rodrigo, J., and M. Herrero. 2002. Effects of pre-blossom temperatures on flower development and fruit set in apricot. Sci. Hortic. (Amsterdam) 92(2):125–135. doi:10.1016/S0304-4238(01)00289-8

Rodrigo, J. 2000. Spring frosts in deciduous fruit trees— morphological damage and flower hardiness. Sci. Hortic. (Amsterdam) 85(3):155–173. doi:10.1016/S0304-4238(99)00150-8

Rosati, A., S.G. Metcalf, and B.D. Lampinen. 2004. A simple method to estimate photosynthetic radiation use efficiency of canopies. Ann. Bot. (Lond.) 93(5):567–574. doi:10.1093/aob/mch081

Ruml, M., A. Vuković, and D. Milatović. 2010. Evaluation of different methods for determining growing degree-day thresholds in apricot cultivars. Int. J. Biometeorol. 54(4):411–422. doi:10.1007/s00484-009-0292-6

Santibanes, F. 1994. Crop requirements: Temperate crops. In: J.F. Griffiths, editor, Handbook of agricultural meteorology. Oxford university press, Oxford., New York. p. 380.

Schultze, S.R., P. Sabbatini, J.A. Andresen, and J.W. Myers. 2014. Spatial and temporal study of climatic variability on grape production in Southwestern Michigan. Am. J. Enol. Vitic. 65(2):179–188. doi:10.5344/ajev.2013.13063

SCS USDA. 1972. Red tart cherry site inventory for Grand Traverse County, Michigan. United States Department of Agriculture, Washington, D.C.

Sekse, L. 1995. Fruit cracking in sweet cherries (Prunus avium L.). Some physiological aspects–a mini review. Sci. Hortic. (Amsterdam) 63:135–141. doi:10.1016/0304-4238(95)00806-5

Silva, G.J., T.M. Souza, R.L. Barbieri, and A. Costa de Oliveira. 2014. Origin, domestication, and dispersing of pear (Pyrus spp.). Adv. Agric. 2014:1–8. doi:10.1155/2014/541097

Snyder, R.L., and J.P. de Melo-Abreu. 2005. Frost protection: Fundamentals, practice and economics. FAO, Rome, Italy.

Thomson, L.J., S. Macfadyen, and A.A. Hoffmann. 2010. Predicting the effects of climate change on natural enemies of agricultural pests. Biol. Control 52(3):296–306. doi:10.1016/j.biocontrol.2009.01.022

Tobin, P.C., S. Nagarkatti, G. Loeb, and M.C. Saunders. 2008. Historical and projected interactions between climate change and insect voltinism in a multivoltine species. Glob. Change Biol. 14(5):951–957. doi:10.1111/j.1365-2486.2008.01561.x

Tomita, A., E. Hagihara, M. Dobashi-Yamashita, and K. Shinya. 2016. Effects of high temperature during flowering on ovule degeneration of sweet cherry (Prunus avium L.). Hortic. Res. 15(3):291–295. doi:10.2503/hrj.15.291

Treder, W., A. Mika, Z. Buler, and K. Klamkowski. 2016. Effects of hail nets on orchard light microclimate, apple tree growth, fruiting, and fruit quality. Acta Scientiarum Polonorum - Hortorum Cultus 15(3):17–27.

VanderWeide, J., I.G. Medina-Meza, T. Frioni, P. Sivilotti, R. Falchi, and P. Sabbatini. 2018. Enhancement of fruit technological maturity and alteration of the flavonoid metabolomic profile in Merlot (Vitis vinifera L.) by early mechanical leaf removal. J. Agric. Food Chem. 66(37):9839–9849. doi:10.1021/acs.jafc.8b02709

Walthall, C.L., J. Hatfield, P. Backlund, L. Lengnick, E. Marshall, M. Walsh, S. Adkins, M. Aillery, E.A. Ainsworth, C. Ammann, C.J. Anderson, I. Bartomeus, L.H. Baumgard, F. Booker, B. Bradley, D.M. Blumenthal, J. Bunce, K. Burkey, S.M. Dabney, J.A. Delgado, J. Dukes, A. Funk, K. Garrett, M. Glenn, D.A. Grantz, D. Goodrich, S. Hu, R.C. Izaurralde, R.A.C. Jones, S.-H. Kim, A.D.B. Leaky, K. Lewers, T.L. Mader, A. McClung, J. Morgan, D.J. Muth, M. Nearing, D.M. Oosterhuis, D. Ort, C. Parmesan, W.T. Pettigrew, W. Polley, R. Rader, C. Rice, M. Rivington, E. Rosskopf, W.A. Salas, L.E. Sollenberger, R. Srygley, C. Stöckle, E.S. Takle, D. Timlin, J.W. White, R. Winfree, L.W. Morton, and L.H. Ziska. 2012. Climate change and agriculture in the United States: Effects and adaptation. USDA Technical Bulletin 1935, United States Dep. of Agriculture, Washington, D.C.

Wang, D., and L. Wang. 2017. Dynamics of evapotranspiration partitioning for apple trees of different ages in a semiarid region of northwest China. Agric. Water Manage. 191:1–15. doi:10.1016/j.agwat.2017.05.010

Watkins, R. 1995. Cherry, plum, peach, apricot and almond: Prunus spp. (Rosaceae). In: J. Smartt and N.W. Simmonds, editors, Evolution of crop plants. 2nd ed. Longman Scientific & Technical, Essec, England. p. 423–429.

Webster, A.D. 2005. The origin, distribution, and genetic diversity of temperate treed fruits. In: J. Tromp, A.D. Webster, and S.J. Wertheim, editors, Fundamentals of temperate zone tree fruit production. Backhuys Publishers, Leiden. p. 400.

Westwood, M.N. 1993. Temperate-zone pomology physiology and culture. Timber Press, Portland, OR.

Whitaker, K., and S. Middleton. 1999. The profitability of hail netting in apple orchards. 43rd Annual Conference of the Australian Agricultural and Resource Economics Society, Christchurch. 20–22 Jan. 1999. p. 20–22.

Willaume, M., P.É. Lauri, and H. Sinoquet. 2004. Light interception in apple trees influenced by canopy architecture manipulation. Trees (Berl.) 18:705 doi:10.1007/s00468-004-0357-4

Winkler, J.A., J. Andresen, J. Bisanz, R. Black, G. Guentchev, J. Nugent, K. Piromsopa, N. Rothwell, C. Zavalloni, J. Clark, H. Min, A. Pollyea, and H. Prawrinata. 2013. Michigan's tart cherry industry: Vulnerability to climate variability and change. p. 104–116. In: S.C. Pryor, editor, Climate change in the Midwest: Impacts, risks, vulnerability, and adaptation. Indiana Univ. Press, Bloomington.

Winkler, J.A., J.A. Andresen, G. Guentchev, and R.D. Kriegel. 2002. Possible impacts of projected temperature change on commercial fruit production in the Great Lakes Region. J. Great Lakes Res. 28(4):608–625. doi:10.1016/S0380-1330(02)70609-6

Wolfe, D.W., A.T. DeGaetano, G.M. Peck, M. Carey, L.H. Ziska, J. Lea-Cox, A.R. Kemanian, M.P. Hoffmann, and D.Y. Hollinger. 2017. Unique challenges and opportunities for northeastern US crop production in a changing climate. Clim. Change 146:1–15. doi:10.1007/s10584-017-2109-7

Yun, J.-I. 2010. Agroclimatic maps augmented by a GIS technology. Korean Journal of Agricultural and Forest Meterology 12(1):63–73. doi:10.5532/KJAFM.2010.12.1.063

Zabadal, T.J., and J.A. Andresen. 1997. Vineyard establishment: Preplant decisions. Michigan State University Extension, Ann Arbor, MI.

Zabadal, T., P. Sabbatini, and D. Elsner. 2009. Wine grape varieties for Michigan and other cold climate viticultural regions. Ext. Bull. CD-007. Michigan State Univ. East Lansing, MI.

Zanotelli, D., L. Montagnani, G. Manca, F. Scandellari, and M. Tagliavini. 2015. Net ecosystem carbon balance of an apple orchard. Eur. J. Agron. 63:97–104. doi:10.1016/j.eja.2014.12.002

Zavalloni, C., J.A. Andresen, and J.A. Flore. 2006. Phenological Models of Flower Bud Stages and Fruit Growth of Montmorency' Sour Cherry Based on Growing Degree-day Accumulation. J. Am. Soc. Hortic. Sci. 131(5):601–607. doi:10.21273/JASHS.131.5.601

Methods of Agroclimatology: Modeling Approaches for Pests and Diseases

Simone Orlandini,* Roger D. Magarey, Eun Woo Park, Marc Sporleder, and Jürgen Kroschel

Abstract

Models represent very powerful tools to describe biophysical responses to environmental conditions. Particularly, the application of modeling for the description of pests and diseases in agriculture represents an important step toward the better understanding of pest epidemiology as well as an opportunity to support farmers in optimizing crop protection management. This chapter aims to give a full overview of agroclimatological models for pests and diseases affecting agricultural production. For pests and diseases, we describe analytical and simulation approaches to modeling including phenological, population and atmospheric transport models. The steps to develop reliable pest models are also discussed including IT infrastructure, model design, environmental inputs, parametrization, and validation. We also describe model products, delivery methods, and the expected benefits for the different categories of stakeholders. Finally, we give an overview of future developments in the field of pest modeling. In particular, this includes the role of precision agriculture including big-pest data and the coupling of pest and disease models with crop models to estimate pest impacts as a part of climate change studies.

Models represent very powerful tools to improve forecasting, decision-making, and planning of agricultural activities. They play an increasing role in the context of climate change adaptation supporting the identification and evaluation of short- and long-term strategies (Garrett et al., 2013). Based on meteorological variables, they simulate the dynamics of agricultural systems (crop growth and productivity, water balance, weed, pest, disease development and dynamics, etc.), producing a wide range of information for stakeholders (farmers, extension

Abbreviations: CDM, complex disease model; DD, degree-day; DSS, decision support system; GIS, geographic information systems; ILCYM, insect life cycle modeling; IT, information technology; SDM, species distribution model; WMO CAgM, World Meteorological Organization Commission of Agricultural Meteorology.

S. Orlandini, Dep. of Agrifood Production and Environmental Sciences, Univ. of Florence, Piazzale delle Cascine 18. 50144, Florence, Italy; R.D. Magarey, Center for Integrated Pest Management, North Carolina State Univ., Raleigh, NC, 27606; E.W. Park, Interdisciplinary Program in Agricultural and Forest Meteorology and Dep. of Agricultural Biotechnology, Seoul National Univ., Seoul, 08826, Republic of Korea; M. Sporleder, Independent scholar, Lange Str. 59, 37139 Adelebsen, Germany; J. Kroschel, International Potato Center (CIP), DCE Crop Systems Intensification and Climate Change (CSI-CC), Apartado 1558, Lima 12, Peru. Received 22 Jan. 2016. Accepted 24 July 2016. *Corresponding author (simone.orlandini@unifi.it)

doi:10.2134/agronmonogr60.2016.0027

© ASA, CSSA, and SSSA, 5585 Guilford Road, Madison, WI 53711, USA.
Agroclimatology: Linking Agriculture to Climate, Agronomy Monograph 60.
Jerry L. Hatfield, Mannava V.K. Sivakumar, John H. Prueger, editors.

services, policymakers, etc.) to improve decision-making by integrating their experience and knowledge with objective data obtained by the application of models. The management of insect pests and diseases represents one of the most important fields of model applications. Most of the biological processes of insect and pathogen development and reproduction are strongly related to the meteorological conditions, which can be considered as one of the main driving factors of pest's specific distribution in agroecosystems and their infestation and severity during growing seasons. The use of model algorithms that take into account the relationships among crop, pathogen or insect, and weather variables allow the simulation of risks to crops and how this may change over time. Model outputs can be used to understand potential range expansions of pests, assist surveillance programs to detect new pests, and to alert and guide growers to determine when to apply crop protection measures. The incorporation of simulation models into decision support systems (DSS) can further increase the potential benefit to farmers, allowing available scientific knowledge to be integrated with the practical, regional, or site-specific experience of technicians and extension workers. At present, technological trends are favorable for the development of new simulation models and particularly for their operational application to support farmers in crop and pest and disease management activities. An important application of disease modeling is in the simulation of crop losses. A good example is RICEPEST, designed to simulate impacts of six pests and diseases as well as weeds on rice (Oryza sativa L.) yield (Willocquet et al., 2002). Crop loss modeling has important applications for global food security and climate change research (Esker et al., 2012).

Recent advances in information technology (IT) provide tremendous opportunities for modeling complicated biological systems at different spatiotemporal scales and incorporating these models into the management of insect pests and diseases for crop production also considering the use of different climate change scenarios in the modeling framework. Two salient features of IT are huge data processing and mobile computing, which are used for the management, storage, analysis, presentation, and delivery of large amount of data in pest forecast systems. Powerful computing resources and technologies, such as cloud computing, geographic information systems (GIS), and global positioning systems, internet, wireless communication, smart phone, etc., make it easier than ever to implement various forecast models in support of protecting crops from insect pests and diseases. Automated weather stations and numerical weather forecast for short- and long-term periods also improve practical use of models by enhancing site specificity of pest forecasts and providing crop growers with increased time windows to practice appropriate control techniques. Pest forecasts at high spatial resolution and location-based information services are pivotal for the preparedness and adaptation planning to climate change as well as to precision agriculture, a new technology that can incorporate the spatial characteristics of pest information generated by pest and disease models.

On this basis, the aim of the chapter is to introduce approaches to modeling for two groups—pests (including arthropods and to a lesser extent weeds) and diseases—using agrometeorological data for crop protection. For each group, we introduce modeling concepts and analytical approaches to modeling before discussing more complex simulation models. Atmospheric transport models are also briefly discussed. For pest and disease models, the requirements in terms of data, IT, model design, requirement, and validation are addressed. We also describe the main products, delivery methods, and the expected benefits for the different categories of stakeholders. Finally, we give an overview of future developments in the field of pest modeling

Types of pest and disease models

There is a vast difference in approaches to modeling pest and diseases, and their specific use depends on the current research stage and intended application. Principally, each model should be used only within the scope for which it was developed. Pest models can be broadly classified into analytical (often referred to mathematical or theoretical models; e.g., May and McLean [2007]) or simulation (complex) models. The aim of analytical models is to explain population or disease intensity change in a general descriptive way through mathematical analysis by a single or a few most important mechanisms, typically following the principle of parsimony, that is, the modeling process starts with the simplest possible structure and includes complexity only when absolutely necessary (Berryman et al., 1990; Royama, 1992). In contrast, the goal of simulation models is to provide a detailed representation of the pest or pathosystem in which population change is determined by different interacting ecological processes. Simulation model components (i.e., ecological processes) are generally based on mechanistic models that explain the biophysical or biochemical pattern of an underlying biological process within the system. While analytical models are beneficial for operational forecasting because of the limited number of input parameters, complex simulation models may improve simulations because of their broader inclusion of factors affecting an insect life system and biological significance of model components (Logan, 1994). Usually, mechanistic models are more likely to perform better than empirical models that are based on statistical relationships with, for example, weather variables, when extrapolating beyond the observed conditions.

Overview of pest model concepts

In pest management, the purpose of pest models can be further subclassified based on two tasks to predict pest population performance (i.e., size or characteristics like age–class distribution) in time and to predict pest distribution potential in space. Interests in the latter have grown rapidly in the last two decades and refer to species distribution models (SDMs, also referred to as species distribution mapping, climate envelope-modeling, habitat modeling, and environmental niche-modeling) (Venette et al., 2010). The methods to generate SDMs may be either inductive (a common application of this method is to predict species ranges using climate data as predictor variables, generally the analysis based on species presence and/or absence data) or deductive (e.g., based on pest-specific phenological models).

Analytical Pest Models

Principally, analytical (simple mathematical or theoretical) models are applied for two purposes in insect pest management. One purpose is to predict the emergence time of specific pest insect life stages, like overwintering egg hatch or the first emergence of adults from overwintering pupae, and the information is used to optimize timing of control measures that target specific life stages of the pest in the field (Finch et al., 1996; Trnka et al., 2007). The second purpose is to predict pest populations.

Analytical Pest Phenology Models

Since insects are poikilotherms, that is, organisms that cannot maintain a constant body temperature, they require a certain amount of heat to develop from one point in their life cycle to another. In other words, the measure of accumulated heat provides a reference for the physiological age and development of insects. Insect phenological models based on temperature, particularly heat accumulation, have been widely used since the 1970s (Welch et al., 1978). These models are particularly applicable in temperate climates where most insect pest species exhibit univoltine life cycles with discrete generations and an overwintering phase. Because of seasonal regulation, most emergences take place over a relatively short period of time and are easily monitored. However, in tropical regions, where most pest species are multivoltine with little seasonality, such an approach is limited to aid in pest management decisions because pest species produce several overlapping generations with a relative short life cycle feeding on constantly growing vegetation. In the tropics, the warning is generally for the first occurrence of the pest in the crop (Krishnaiah et al., 1997; Rahman and Khalequzzaman, 2004).

The degree-day (DD) approach is frequently used for this purpose because of its simplicity and practical advantages. Hereby, the insect pest development is modeled by only two parameters describing the linear relation between development and temperature, that is, the theoretical low temperature threshold for development, T_{min}, and the temperature DD accumulation, required to finalize a development stage. For some species, an upper temperature threshold for development, T_{max}, at which development begins to decrease or stop has been determined; different cutoff methods are used for limiting DD accumulation above this threshold temperature depending on the physiological interpretation of the upper threshold. These parameters are available for >500 insect species (pests and beneficial insects) (Nietschke et al., 2007) in databases of North Carolina State University (NAPPFAST; see Nietschke et al., 2007) and the University of California (http://www.ipm.ucdavis.edu/MODELS). These databases provide a useful resource for seeking DD models for agricultural pests and can be applied by decision makers to schedule management operations such as scouting or the application of pesticide treatments. As an example, the Japanese beetle (*Popillia japonica* Newman) has a base (T_{min}) and upper (T_{max}) developmental threshold of 10 and 34°C, respectively, and on average requires 524 DD from 1 January before adults will begin to emerge (Ludwig, 1928; Régnière et al., 1981; Magarey et al., 2015). After 2 d, the beetle would accumulate 20 DD when daily temperatures average 20°C and 48 DD when daily temperatures average 34°C. At average temperatures above 34°C, by using the horizontal development cutoff method, the beetle accumulates only 24 DD d^{-1} because development is limited at the upper thermal threshold ($DD_{max} = T_{max} - T_{min}$). In total, three cutoff methods exist—horizontal, vertical, and intermediate—for determining heat accumulation at temperatures above the upper temperature threshold. The horizontal cutoff assumes that development continues at a constant maximal rate if temperature exceeds T_{max}. The intermediate cutoff assumes that development slows, but does not stop, at temperatures above the threshold [mathematically, the area above T_{max} is subtracted twice from the area above T_{min}, e.g. if $T > T_{max}$, then $DD = T - 2(T - T_{max}) - T_{min}$; in the example at 36°C, DD = 36 − 2(36 − 34) − 10 = 22], while the vertical cutoff assumes that no development takes place when temperature is above T_{max}. All three methods are to be used in conjunction with the DD calculation method, that is, the method to estimate DD based on daily minimum and maximum temperature (single triangle, double triangle, sine, or double sine method). These methods only approximate the real accumulated DD for a given set of daily temperatures and development thresholds through calculating the area below the temperature curve within the temperature thresholds. The DD accumulation based on simple daily averages of the minimum and maximum temperature [i.e., $(T_{min} + T_{max})/2$] is used for pragmatic reason only; more complex calculation methods, like the triangle or sine method, provide quite useful estimates of the daily temperature curve considering the error in temperature data for a given area and the precision required for pest management decisions. However, the preferred DD calculation method depends on the location and season (Allen, 1976). For example, the University of California Integrated Pest Management Program (http://ipm.ucanr.edu) recommends using the single sine method with a horizontal cutoff at the upper threshold in California. For other regions and pests, other recommendations might exist. For example, Rodrigues Caicedo et al. (2012) suggested a logistic function for estimating daily temperature profiles within Colombia, while Huld et al. (2006) presented a new method for estimating such profiles within Europe. Since temperature data are becoming more and more available through increasing numbers of meteorological stations, simulations can also be based on real temperature records (e.g., measurements at 1-h intervals). In the future, remote sensing technology offers good potential to improve the current temperature monitoring and increase the spatial and temporal resolutions of temperature data (Marques Da Silva et al., 2015).

In addition, DD models are commonly used to estimate germination for weed species (Steinmaus et al., 2000). In some instances, DD models are used as building blocks in more complex pest models. An example is a three-phase model to predict the emergence of adults of orange wheat blossom midge [*Sitodiplosis mosellana* (Géhin)] (Jacquemin et al., 2014). The first phase comprised a temperature accumulation of 250 DD above 3°C, starting from 1 January. Once this initial condition is satisfied, the second phase starts, and it lasts until the occurrence of a double signal consisting of a rise in the mean daily temperature up to 13°C, followed by rainfall. This rainfall event triggers an accumulation phase of 160 DD above 7°C. Once this last condition is met, the adults emerge.

However, insects' heat requirements are not linear with temperature, particularly at high and low temperatures. Therefore, DD models might be poor predictors when daily temperature fluctuates to extremes (Stinner et al., 1974). Worner (1992) pointed out that the interaction of cyclical temperatures with nonlinear development could introduce significant deviations in linear development rate models especially at extreme low and high temperature values of the development rate function. An alternative method is the rate summation approach, which enables the use of nonlinear models (e.g., Logan et al. [1976] and others), including mechanistic models of biophysical significance (Sharpe and DeMichele, 1977), which has perhaps proved to be the most viable approach (Stinner et al., 1974). Most of the earlier models lack a stochastic function for variability in

Table 1. Analytical models to forecast pest population growth using difference or differential equations.

Model	Difference equations†	Differential equations†	Source
(1) Exponential model; populations growing at a constant rate	$N_t = N_{t-1}\ \exp(r)$	$\frac{\Delta N}{\Delta t} = rN$	Malthusian model
(2) Logistic population model; population growing to maximum size	$N_t = N_{t-1}\ \exp\{r_{max}\left[1-\left(\frac{N_{t-1}}{K}\right)^{\omega}\right]\}$	$\frac{\Delta N}{\Delta t} = rN\left[1-\left(\frac{N}{K}\right)^{\omega}\right]$	Ricker (1954)
(3) Model for the sterile male technique	$N_t = N_{t-1}\ \exp(r)\left(\frac{N_{t-1}}{N_{t-1}+S}\right)$		Knipling (1955)
(4) Predator-prey model (Lotka-Volterra model)	$N_t = N_{t-1} + rN_{t-1} - aN_{t-1}P_{t-1}$ $P_t = P_{t-1} - sP_{t-1} + bN_{t-1}P_{t-1}$	$\frac{\Delta N_P}{\Delta t} = r_P N_P - aN_P N_A$ $\frac{\Delta N_A}{\Delta t} = -r_A N_A + bN_P N_A$	Volterra (1926)
(5) Modified Lotka-Volterra model		$\frac{\Delta N_P}{\Delta t} = N_P\left(r_P - a_{11}N_P - a_{12}N_A\right)$ $\frac{\Delta N_A}{\Delta t} = N_A\left(-r_A - a_{21}N_P\right)$	
6) Sterile male technique combined with carrying capacity	$N_t = N_{t-1}\left(\frac{N_{t-1}}{N_{t-1}+S}\right)\left[\frac{\lambda K}{K+(\lambda-1)N_{t-1}\left(\frac{N_{t-1}}{N_{t-1}+S}\right)}\right]$		Prout (1978)

† *t*, time (independent variable); *N*, population size (dependent variable) at time *t*; *r*, proportionality constant (parameter) between the rate of population growth and the size of the population; in difference equations the parameter *r* corresponds with the intrinsic rate of increase, and the expression exp(*r*) with the finite rate of increase (λ) while in differential equations; in contrast, the parameter *r* corresponds with the finite rate λ minus 1, ($\lambda - 1$); *K*, maximum population size (carrying capacity); ω, shape parameter; *S*, sterile number released (in Eq. [3] the threshold release level S^* for a given initial pest population P_0 is given by $S^* = P_0(r - 1)$]; *a* and *b* are positive constants.

development times among individuals within a population (Sharpe et al., 1981; Wagner et al., 1984b), which is responsible for the variability in pest emergence and, hence, important for the pest spread (Régnière, 1984; Phelps et al., 1993). Progressive models handle mean rate vs. temperature relationships (Wagner et al., 1984a) and include a distribution function for development (Wagner et al., 1984b, 1985). Rather than treating rate summation deterministically, development rate is considered as random variables (Stinner et al., 1975). Stochastic insect development models vary in the choice of the random variable (i.e., development rate or time) and in the form of the frequency distribution applied to the random variable (Sharpe et al., 1977; Curry et al., 1978). The coefficient of variation of rate distributions is relatively independent of temperature (Sharpe et al., 1977), and a single temperature-independent distribution of the normalized development time suffices to adequately describe the distribution at all temperature (see one-shape theory of Curry et al. [1978]). The one-shape theory has been validated for ~80% of 194 sets of published data on 113 species of insects and mites (Shaffer, 1983). Insect species that exhibit seasonality generally have resting phases (diapause or aestivation) in their life cycles, which can be accommodated in Monte Carlo simulation modeling (Phelps et al., 1993).

Analytical Population Models

The second purpose of theoretical models is to forecast pest population development. These models have been developed for more than a century starting from exponential and logistic models (Malthus, 1798; Verhulst, 1838, 1845; Lotka 1939, 1945; Volterra, 1926) and have been extended by adding other features like delayed density dependence, asymmetric population growth, group effects, and competitive equilibrium (Royama, 1992). Elaborated theories exist for single population dynamics (Coulson and Godfray, 2007), for predator–prey interaction (Hassell, 1978), and for pest management measures like the sterile insect technique (Knipling, 1955). In general, these models consist of coupled differential or difference equations with only a few state variables (examples are provided in Table 1). For further information on merits of these models see the reviews by Ruesink (1976), Shoemaker (1981), and Getz and Gutierrez (1982). Huffaker (1980) describes, in detail, many applications of mathematical ideas in pest management. May (1974, 1981) and Hassell (1978) provide a detailed discussion of and access to the literature on analytical models in population dynamics and pest management.

As an example, a population that grows at a constant rate can be mathematically expressed as follows:

$$N_t = N_{t-1}\exp(r) \quad \text{Eq. [1]}$$

where N is the population size at the time t, and r is the intrinsic rate of population increase (Malthusian growth model). If the population size is restricted in a given environment the following logistic mathematical expression by Ricker (1954) is commonly used:

$$N_t = N_{t-1}\exp\left\{r_{\max}\left[1-\left(\frac{N_{t-1}}{K}\right)^{\omega}\right]\right\} \quad \text{Eq. [2]}$$

where r_{max} is the maximum rate of increase for the species, K is the carrying capacity of the environment, and ω is a shape parameter (if $w = 1$ it is Ricker's model, Ricker [1954]). Knipling (1955) published a simple model for the sterile male technique, that is, it is helpful to optimize the strategy of releasing sterile insects in a given situation. Assuming that a constant number, S, of sterile males are released in each time interval, Eq. [1] is replaced by the following:

$$N_t = N_{t-1}\exp(r)\left(\frac{N_{t-1}}{N_{t-1}+S}\right) \quad \text{Eq. [3]}$$

The added term in brackets denotes the dilution of fertile males that lowers the growth rate. The threshold release level S^* for a given initial population P_0 can be calculated by $S^* = P_0\ (e^r - 1)$. Such threshold was observed by Baumhover et al. (1955) in applying the sterile insect technique to eradicate the screw worm [*Callitroga hominivorax* (Coquerel)], which preys on warm-blooded animals, from the United States. When sterile insects are released over a period of several generations at constant numbers above the threshold, the target pest population declines. The behavior of the above functions is illustrated in Fig. 1. Different features can be included in a single model by combining two or more models. As an example, Prout (1978) suggested a model that combines Knipling's model for the sterile male technique with carrying capacity:

$$N_t = N_{t-1}\left(\frac{N_{t-1}}{N_{t-1}+S}\right)\left[\frac{\lambda K}{K+(\lambda-1)N_{t-1}\left(\frac{N_{t-1}}{N_{t-1}+S}\right)}\right] \quad \text{Eq. [4]}$$

in which λ is the finite rate of population increase and other parameters are as in Eq. [2] and [3].

The effects of biological control agents have been generally described as a two-species interaction model using the Lotka–Volterra function, which is defined as two combined differential equations:

$$\frac{dN_P}{dt} = N_P\left(r_P - a_{11}N_P - a_{12}N_A\right)$$
$$\frac{dN_A}{dt} = N_A\left(-r_A + a_{21}N_P\right) \quad \text{Eq. [5]}$$

where N_P and N_A are the population densities of the pest and the antagonist species, respectively, and a_{11}, a_{12}, and a_{21} are positive parameters describing the interaction between the two species.

When using such discrete-time population models for predicting future population sizes, possible observation errors, or process noises are ignored. State–space models that account for both process noise and observation error exist but have been little used. For merits on including stochasticity in these types of models see Humbert et al. (2009), Dennis et al. (2006), Geritz and Kisdi (2004), and de Valpine and Hastings (2002).

Pest Simulation Models

The mathematical models described above can be combined with the assumption that population change is driven by interaction of different ecological processes (biotic, abiotic) and pest management practices that significantly affect pest populations. Sharov (1992) refers to this concept as a systems approach in population ecology. These models, in contrast to theoretical models, aspire to explicitly handle all ecological processes that have a significant effect on the pest population. Such life-system models vary in their level of complexity. Sharov (1992) differentiates between medium and huge models; huge models are supposed to be universal and should integrate all available knowledge and satisfy any need for information, while medium models are designed to integrate the most important information and to satisfy a number of important purposes. As insect development is a self-ruling subsystem, it can be simulated independently from population dynamics using temperature as input factor. Therefore, life-system models frequently include phenological submodels for the target pest species. Thus, evolutionarily life-system models have their roots in both theoretical and phenological models (Sharov, 1992).

Complex phenological models form a thorough basis for further life-system modeling or deductive SDMs. Modeling the regulatory effects of temperature on an arthropod pest's life history is fundamental to better understanding pest population dynamics, managing pest populations, and studying the effects of climate change on the pest (Bale et al., 2002). Development, survival, and reproduction, as well as mechanisms of sex determination and resting phases in some species, are dominated by temperature (Brown et al., 2004). If process–temperature functions

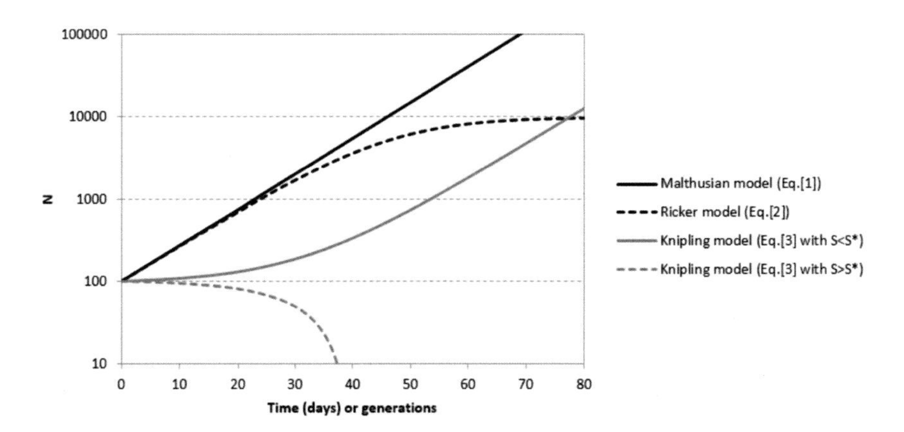

Fig. 1. Behavior of Eq. [1–3]. Parameters used were r = 0.1 for the Malthusian model. Ricker's model was calculated using r_{max} = 0.1, K = 10000, and ω = 1, and for the Knipling model (r = 0.1), S was set to 9.9 (solid line) and 10.9 (scattered line), 0.6 above and 0.4 below the threshold (=10.5), respectively.

for these parameters are established, it permits assembling full life-cycle pest models with temperature as the main input variable. Literature on temperature functions for specific pests is constantly rising, although most research assesses the effects on development only. Examples for experimentally establishing all required functions describing an insect's life history are the models established for the potato tuber moth [*Phthorimaea operculella* (Zeller)] (Sporleder et al., 2004); the white fly [*Bemisia tabaci* (Gennadius)] (Muñiz and Nombela, 2001); the psocid (*Liposcelis tricolor* Badonnel) (Dong et al., 2007); the vegetable leafminer (*Liriomyza sativae* Blanchard) (Haghani et al., 2007); the cotton bollworm (*Helicoverpa armigera* Hübner) (Mironidis and Savopoulou-Soultani, 2008); the maize stem borers [*Chilo partellus* (Swinhoe) (Khadioli et al., 2014a), *Busseola fusca* (Fuller), and *Sesamia calamistis* Hampson (Khadioli et al., 2014b)]; and important parasitoids used as biological control agents, for example, *Lysiphlebia mirzai* Shuja-Uddin, a parasitoid of *Toxoptera citricida* (Kirkaldy) (Liu and Tsai, 2002), *Anagyrus pseudococci* Girault attacking the vine mealybug [*Planococcus ficus* (Signoret)] (Daane, 2004), and *Diadegma anurum* (Thomson) attacking *Plutella xylostella* L. (Golizadeh et al., 2008). The feasibility that such models can be established is described by Régnière et al. (2012). A similar approach is used in the model builder of the Insect Life Cycle Modeling (ILCYM) software developed by the International Potato Center (www.cipotato.org/ilcym; Sporleder et al., 2013; Tonnang et al., 2013). The ILCYM software is a generic open-source computer-aided tool that facilitates the development of phenology models and predicts and maps pest activity on global, regional, and national scales (Sporleder et al., 2013). The ILCYM software compiles all temperature functions (for development, survival, and reproduction established) with the parameters for variability in these processes and makes it possible to generate life table parameters for a given range of fluctuating temperatures as input data. The software applies a rate summation approach and cohort-updating algorithm that simulates pest population development with their specific age–stage structure that allows temporal and special simulations of population parameters according to real or interpolated temperature data. The model was used to simulate the geographic distribution potential of potato tuber moth under current (Sporleder et al., 2008) and future climates (Kroschel et al., 2013) and the effect of different baculovirus application intervals and rates for controlling the moth (Sporleder and Kroschel, 2008); however, for the latter simulation, factors like virus transmission in the field were ignored because information on such processes was missing.

Overview of Disease Model Concepts

The central theorem for disease models is the disease triangle concept in plant pathology, which states that if pathogen inoculum is present at the susceptible stage of host plant growth, then infection may occur under favorable environmental conditions. In each disease development process, such as overwintering, dispersal, primary infection, incubation, secondary inoculum production, and secondary infection, the interaction among host plant, pathogen, and environment determines success and rate of the process. Many disease models predict risk or intensity of disease development by estimating inoculum potential and pathogen infection periods based on information about the weather, crop, and pathogen. The inoculum potential is the inoculum density being modified by capacity (or environmental) factors (Baker, 1978). The infection period is the period during which environmental conditions favor infection (Campbell and

Table 2. Goidanich table: Duration of the incubation period of *Plasmopara viticola* (in days) depending on the temperature and relative humidity.

Air temperature	Low relative humidity (≤65%)	High relative humidity (>65%)
°C		
14	15	11
15	13	9.5
16	11.5	8.5
17	10	7.5
18	9	6.5
19	8	6
20	7	5
21	6.5	4.5
22	6	4.5
23	5.5	4
24	5.5	4
25	6	4.5
26	6	4.5

Madden, 1990). Since both inoculum potential and infection period are significantly affected by environment, disease models are commonly driven by one or more environmental factors.

The most important environmental factors are weather conditions (temperature, humidity, dew, rain, etc.), soil characteristics (moisture, porosity, pH, texture, fertility, etc.), and the presence of predators or biological vectors of the pathogen (aphids, plant hoppers, thrips, etc.). Temperature exerts its main effect by varying the rate of disease development (van Heemst, 1986). The availability of water, as measured by relative humidity, temperature of dew point, leaf wetness duration, water vapor pressure deficit, and amount of precipitation (Dalla Marta et al., 2005), is a decisive factor for outbreak of diseases since bacterial and fungal pathogens require the presence of free water for infection (Magarey et al., 2005a). Evaluation of coefficients of dispersion and diffusion of pathogens (spores, sporangia, insects, etc.) represents a fundamental point of the analysis of an epidemic and is generally performed on the basis of the movements of air masses, described by speed and wind direction (Friesland and Orlandini, 2006). Finally, the photoperiod and the quality of solar radiation also influence development and pathogenesis (Canessa et al., 2013) and biotic and abiotic stress responses of host plant (Karpinski et al., 2003).

Many weather-driven disease forecast models use observed weather data as input and identify past infection or conditions for such infection. The disease forecast models do not forecast but hindcast in the literal sense. However, the model output is interpreted as forecasting on future appearance of disease symptoms after an incubation period of pathogen in the infected host plant. In the practical applications of forecast models for disease management, it is important to take actions for disease control prior to infection or during the incubation period. The disease control actions should have curative effects in this case. In this regard,

Table 3. Goidanich table: daily percentage of progress of the incubation period of *Plasmopara viticola* depending on temperature and relative humidity.

Air temperature	Low relative humidity (≤65%)	High relative humidity (>65%)
°C		
14	6.6	9.0
15	7.6	10.5
16	8.6	11.7
17	10.0	13.3
18	11.1	15.3
19	12.5	16.6
20	14.2	20.0
21	15.3	22.2
22	16.6	22.2
23	18.1	25.0
24	18.1	25.0
25	16.6	22.2
26	16.6	22.2

weather forecast data along with observed data would facilitate applications of weather-driven disease models in better disease management and integrated pest management. The scheduling of protectant fungicides would be improved by predicting infection periods using forecast rather than near-real-time observed weather data. The accuracy of forecast data has been an issue, since unnecessary treatment applications may be made for infection periods that do not occur. However, weather forecasts are increasing in accuracy. Every 10 yr, forecasts are expected to increase in accuracy by 1 d (Bauer et al., 2015). For example, a 4-d forecast is expected to be as accurate as a 3-d forecast in 10-yr time.

Analytical and Empirical Disease Models

Several methods of model development in plant disease epidemiology have been outlined by Teng (1985), Zadoks and Rabbinge (1985), Madden and Ellis (1988), Magarey and Sutton (2007), Madden et al. (2007) and De Wolf and Isard (2007). The simplest models are usually based on tables or a single function relating one or more weather variables with a key disease process. They are simple to implement and users do not need relevant computer science knowledge for efficient practical use. Simple disease models may attempt to predict infection periods and the length of the incubation and the rate of disease development through a simplistic mathematical description of the impact of weather on plant disease progress. Simple models have some advantages over more complex models in that they are often easier to parameterize and implement for operational forecasting. One example is the 10:10:24 rule for grapevine downy mildew [*Plasmopara viticola* (Berk. & M.A. Curtis) Berl. & De Toni]. According to the rule, downy mildew primary infection is likely to have occurred if temperatures remain above 10°C, there has been more than 10 mm of rain, and the soil was wet for 24 h (Magarey et al., 2002). This rule provides farmers with a practical way to tell if infection has taken place using simple data collection methods such as a thermometer and a rain gauge.

The next step up in complexity is a lookup table. One of the most popular and well known is the Mills apple scab table. This table allows growers to determine if primary apple scab [*Venturia inaequalis* (Cooke) G. Winter] infection has taken place based on the duration of a wetness period and the average temperature during this period (MacHardy and Gadoury, 1989). This table was extremely useful in combination with simple data collection devices such as a hydrothermograph. Another example is the Goidanich tables based on the 1923 work of Müller describing a calendar of incubation for grapevine downy mildew (Goidanich, 1959). Named Table of Goidanich, it provides the duration of the infection as a function of temperature at two different levels of relative humidity. Practically, at each temperature, the table determines the duration of the incubation period (Table 2) in days or its rate of advancement (obtained by multiplying by 100 the reciprocal of the duration in days) (Table 3).

Another well-known example that also uses temperature and relative humidity is the Wallin Table for prediction of potato late blight [*Phytophthora infestans* (Mont.) de Bary] (Krause and Massie, 1975). In this table, rows are the average temperature during the wetness period and columns are hours above 90% relative humidity. For potato late blight, lower temperatures and longer moisture periods yield higher disease severity combinations. This approach is also commonly known as the risk-index approach and has been used for a number of plant diseases including TOM-CAST for forecasting tomato early blight (*Alternaria* sp.) , Septoria leaf spot (*Septoria lycopersici* Speg.), and anthracnose fruit rot (*Colletotrichum coccodes*) (Pitblado, 1992). Another example of a simple infection models is the temperature–moisture response function (Magarey et al., 2005b; Magarey and Sutton, 2007). This function uses the cardinal temperatures and a minimum wetness duration requirement to predict the rate of infection at any temperature when leaves are wet. A similar concept was developed by Pfender (2003) who developed degree wet hours, which is analogous to the concept of a DD used for insect phenology. There have also been more sophisticated functions developed for describing the leaf wetness and temperature interactions (Duthie, 1997). With the development of computers, these types of tables or rules could be converted into an algorithm that greatly improves their ease of use. Simple disease models are also commonly constructed from statistical relationships between disease variables and weather variables such as temperature, relative humidity, and precipitation. Examples are prediction models for Fusarium head blight [*Gibberella zeae* (Schwein.) Petch] (De Wolf et al., 2003) and Sclerotinia blight (*Sclerotinia minor* Jagger) on peanut (Smith et al., 2007). There are fewer examples of disease forecasting for bacterial pathogens. One pathogen that is extensively modeled is fire blight on apple caused by *Erwinia amylovora* (Burrill) Winslow et al.), which requires a certain number of DD for pathogen multiplication and then splash to infect a susceptible host (Dewdney et al., 2007). Another good example is *Psuedomonas syringae* Van Hall forecasting, where Jardine and Stephens (1987) identified dry periods as being factors limiting epidemics of bacterial speck in tomato.

Another approach is empirical disease forecasting, which is the statistical correlation of weather variables with disease intensity observations from field or other studies (De Wolf and Isard, 2007). Some researchers see statistical models as the future of disease modeling (Madden et al., 2007). Statistical models can use a variety of inputs such as past-disease prevalence and cropping history to create forecasts. One example is the Fusarium head blight model, developed from 50 location–year data using temperature, relative humidity, and rainfall as predictor variables (De Wolf et al., 2003). Other examples are the tomato spotted wilt virus (*Tospovirus*) and thrip (Thysanoptera: *Thripidae*) vector model (Chappell et al., 2013) and a system created for Sclerotina stem rot [*Sclerotinia sclerotiorum* (Lib.) de Bary] (Twengström et al., 1998). An advantage of the statistical approach is that there is little or no need for controlled laboratory experimental data sets to construct and parameterize the model. The disadvantage of the statistical correlation approach is that usually historical data sets of disease prevalence are required to construct the model. Extra care should be taken with statistical models when transferring to a new climate or region as this may involve extrapolation outside of the range of weather inputs for which the model was constructed.

Disease Simulation Models

For the sake of distinction from simple disease model, complex disease models (CDMs) are considered as process-based models consisting of more than one process of disease development such as overwintering, sporulation, dispersal, primary infection, incubation, and secondary infections. One can develop CDMs by combining mechanistic and empirical approaches to pathosystems. Development of CDMs for plant disease warning requires analyses of empirical and theoretical information on individual processes of disease development and integration of the processes into a system simulating partial- or whole-disease cycle.

Submodels of CDMs represent processes of disease development and are driven by certain measurable factors whose variation explains variations in the model output. The measured factors are commonly weather variables. Submodels generally consist of one or a few mathematical equations such as regression equations based on empirical data and differential equations derived from theoretical principles. The mathematical equations describe quantitative relationships between pathogen, host plant, and environment. Since mathematical equations in CDMs are simplified abstractions of biological processes, one or more assumptions are made about predictor (or independent) variables, response (or dependent) variables, and their relationships. Response variables of the mathematical equations represent quantity of biological phenomena related to disease development. Empirical or theoretical equations to estimate additional predictor variables from observed weather data are often included as submodels for CDM. For example, wetness period, which is not commonly measured by the national weather service stations, can be estimated by statistical and physical models (Magarey et al., 2005a; Bregaglio et al., 2011). Final output variables of CDM are mainly possible infection risk (De Wolf et al., 2003; Fry et al., 1983; Do et al., 2012; Xu, 1999), buildup of inoculum or disease (Clarkson et al., 2014; Gilles et al., 2004; Magnien et al., 1991; Orlandini et al., 1993; Rossi et al., 1997; Tixier et al., 2006; Xu, 1999), and recommendation for fungicide treatment (Fry et al., 1983; Pellegrini et al., 2010). One can find lists of CDMs from websites of integrated pest management programs (e.g., www.ipm.ucdavis.edu/DISEASE/DATABASE/diseasemodeldatabase.html).

Infection processes of airborne foliar diseases have a distinction between the primary and the secondary infection cycles in which a series of infection processes can be identified and measured quantitatively. Grape downy mildew caused by P. viticola is a well-known example for which many process-based CDMs were developed and employed as warning systems for local grape growers (Gessler et al., 2011). The grape downy mildew models are driven by air temperature, relative humidity, leaf wetness duration, and rainfall (Rosa et al., 1995). Some of the models estimate primary infection risk (Rossi et al., 2008), whereas others predict infection risk on the secondary cycles (Magnien et al., 1991). There are also models predicting the first date of primary infection followed by risk on the secondary infection cycles (Magarey et al., 1991; Park et al., 1997; Orlandini et al., 1993).

Mathematical modeling for soilborne diseases has not been addressed as much as for airborne diseases. Gilligan (1983) noted in his review that simple transposition of models for airborne pathogens is generally unlikely to be satisfactory and that inoculum density, survival of inoculum, and the infection court are all of greater importance in the soil than in the air. In addition, soil data, like soil temperature and soil water content, are not as easily available as above -ground weather data in the public domain and have to be estimated by empirical or theoretical models if observed data are not available.

Modeling of soilborne diseases has concentrated on primary infection because of limited knowledge of secondary infection and latent and infectious periods (Gilligan, 1983). Primary infection by active inoculum accumulated within the rhizosphere of host plants is the main target for model development. For example, a CDM for Phytophthora blight of chili pepper developed by Do et al. (2012), referred to by Ahn et al. (2013) as PBcast, predicts the date of primary infection after overwintering. The model takes into account daily accumulation processes of active inoculum of *Phytophthora capcisi* Leonian in soil by simulating production of zoosporangia, zoospore release and germtube formation from zoosporangia, and survival of zoospores and germinated zoosporangia. It is driven by daily air temperature and relative humidity, rainfall, and soil texture, which are used to estimate daily soil temperature and soil water content (Do et al., 2012).

Since the creation of complex models is time consuming, there have been efforts to develop generic structures for CDMs. One notable effort is BioMA, which has four software components that implement models to simulate the dynamics of generic polycyclic fungal epidemics and interactions with crop physiological processes (Bregaglio and Donatelli, 2015). Each of these components describes a unique simulation component and can be used separately or together, making the model both powerful and highly flexible. The InoclumPressure component describes the production of primary inoculum and the occurrence of primary infections, DiseaseProgress describes the development of secondary infection cycles during the cropping season, ImpactsOnPlants describes the interactions between epidemic development and crop physiological processes, and AgromanagementDisease describes the impact of agricultural management practices on disease development.

Atmospheric Transport Models

The use of aerobiological models can support the application of pest and disease models by considering micro-, meso-, and macroscale processes. The most common biota that are atmospherically transported are spores and pollen; however, the study of organism flight is also important to agricultural production systems including numerous species of insects, birds, and bats. Aerobiological techniques have been used successfully in several areas such as tracking the spread of foot-and-mouth disease (Moutou and Durand, 1994), locusts, bush flies, the seasonal migration of pests or the entry of nonindigenous species (Parry et al., 2015). One well-documented example is cucurbit downy mildew [caused by Pseudoperonospora cubensis (Berkeley & Curtis) Rostovtsev], which requires a live host to survive and hence cannot survive winter in the northern United States (Ojiambo et al., 2011). The interdisciplinary approach to aerobiology incorporates the sampling routines and instrumental observations of entomologists, plant pathologists, and other biologists, together with real-time weather or climatic data of meteorologists, for use in models specifically designed to simulate certain disease infections or insect infestations (Isard and Gage 2001). In addition, the aerobiological techniques may include monitoring and modeling of airborne movement of beneficial biota and their impact on pest populations.

Isard et al. (2005) describes an aerobiological model that provides a unifying approach to synthesize knowledge of processes associated with the aerial movement of biota. It includes five stages: preconditioning, takeoff and ascent, horizontal transport, descent, and impact. The time scales associated with these processes typically range from days to years for preconditioning and impact, minutes to weeks for horizontal transport, and seconds to hours for takeoff and ascent as well as descent and landing. Isard et al. (2005) used soybean rust (Phakopsora pachyrhizi Syd. & P. Syd.) as an example to describe processes that impact the movement of fungal spores at each of five stages. One important variable for soybean rust is that ultraviolet radiation, which kills spores and as a consequence, cloud cover is required for long distance transport. A key variable in aerobiological models for all biota is wind speed and direction data, which determines the horizontal and vertical movement of biota. Wind shear and gustiness at the surface of plants can determine spore release, uptake, dispersal, and deposition. Long-range transport (possible for some biota such as rust pathogens, e.g., Tromp [1980]) is reported for dispersal of yellow rust spores over 1000 km.

A number of models can be used for simulation of atmospheric transport (Parry et al., 2015). HYSPLIT, an atmospheric transport model developed by the National Air Resources Laboratory (Draxler and Rolph, 2015), continues to increase in functionality and is a good first choice because it can be freely downloaded and does not require programming skills. For more advanced applications, where spatial and temporal quantification of viable propagules or individuals is desired, a purpose built model may be required.

How to build pest and disease models

Information Technology Infrastructure

Common use of personal computers with powerful computing resources, web-based Internet systems, and wireless communication infrastructures facilitated instant data processes for collecting weather data from automated weather stations installed at remote places, analyzing huge data sets to generate disease forecast information, and distributing the information to crop growers at near-real-time basis (Kang et al., 2010; Magarey et al., 1997, 2007). Application of software for GIS and simulation models for weather data analysis has also made significant contributions to improving temporal and spatial resolution of weather data, consequently making plant disease forecasts applicable to disease management for local crop growers (Kang et al., 2010; Magarey et al., 2001; Russo and Zack, 1997; Seem et al., 2000).

Another example is the IT infrastructure used to support the Integrated Pest Information Platform for extension and education (Russo 2009, Isard et al., 2015) (Fig. 2). The flow of information begins at the left hand side with ingestion of data (including weather data and pest observations) and ends with delivery of products to users. Although models for research applications are easily implemented on a desktop computer, the operational delivery of products to a large number of users in different geographies requires a sophisticated infrastructure to ensure the integrity and reliability of product delivery. One of the most important steps is converting diverse data sets into a common format to enable their seamless integration into analytical processes. Quality control is also a critical element, since weather station data contains not only missing records but also spurious records as a consequence of sensor failure. A third facet to highlight is a controlled system for delivery of products such that sensitive information is restricted to authorized users.

Model Inputs and Monitoring

Simulation models can be structured in different ways, based on empirical or mechanistic approaches, from simple manual calculation to integration with GIS. As complexity increases, agrometeorological data have to be collected and transmitted in real time to allow the available system to be correctly applied. Sometimes, crop characteristics (phenology, leaf area index), agronomic management, topographical, and geographical information may be required.

Pest and disease model outputs can be applied directly by farmers with evident benefits in the evaluation of real epidemiological conditions and microclimate evaluation. On the other hand, management of simulations and updating of decision support systems represent large obstacles. The use of agrometeorological station is one option for supplying the weather inputs for pest models. However, an agrometerological station can be time consuming to maintain to

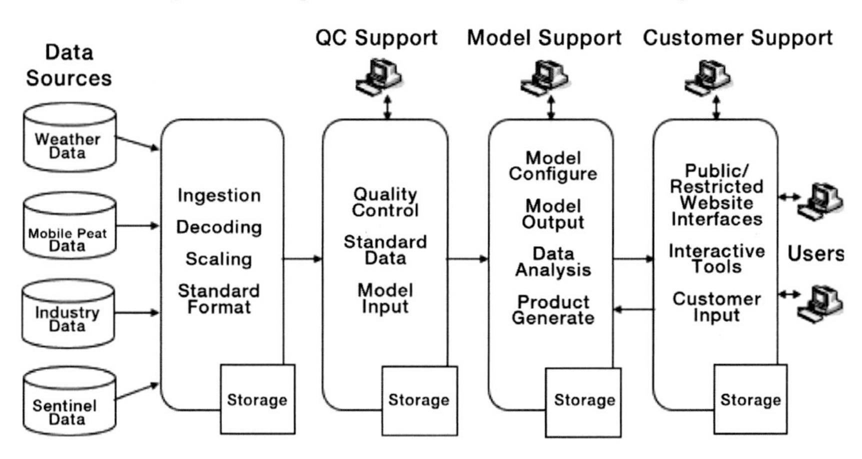

Fig. 2. An IT infrastructure for modeling (Russo, 2009, used with permission).

ensure good quality data. Another solution is to use a territorial application where weather inputs and model outputs are managed by extension services, local or national administrations, cooperatives of farmers, etc. (Jones et al., 2010). This solution is also suitable for strategic applications considering current and future agroclimatic conditions. In the territorial application, weather data are provided by a network of weather stations or numerical grid models, remotely sensed data including radar, or a combination of different sources (Cicogna et al., 2005; Shafer et al., 2000). Examples of gridded data sets that can be used for plant disease forecasting include the Real Time Mesoscale Analysis system (RTMA) in the United States (De Pondeca et al., 2011) and AGRI4CAST in Europe (Joint Research Centre, 2015), and Climate Forecast System Reanalysis (CFSR) globally (Saha et al., 2014). Additionally, numerical weather models can provide gridded data for forecasting future conditions (3–5 d) and to obtain current data with suitable spatial resolution. For plant disease forecasting, leaf wetness has been a limitation since the data has historically been collected by leaf wetness sensors (Gleason et al., 2008). This requires the use of an agricultural weather station and, for this reason, some researchers have suggested using relative humidity as the standard for wetness duration (Rowlandson et al., 2015). However, the use of leaf wetness simulation models in combination with grid weather data sets is now proving to be a practical alternative to station-based sensors (Magarey et al., 2005a). One advantage of using the simulation approach over using station-based relative humidity data is that it can take into account changing crop height and leaf area index, which impacts the duration of wetness inside canopies.

Another important input for pest modeling is pest observations. In some cases, observations may be used to initialize a model, for example, a biofix based on first adult trap catch or a phenological stage. As will be discussed below, pest observations are also critical for model validation. Pest observations can also be combined with model output to make subfield maps of pest intensity (Merrill et al., 2015). Although farmers frequently do their own pest monitoring, agricultural professionals, including extension agents, crop consultants, and agronomists, are also valuable sources of data. There has also been the development of smart trapping technology that can report observations over a wireless network (Loughlin, 2013). Unfortunately, unlike weather data, pest observations are rarely shared to allow farmers to see a more comprehensive picture of pest activity. A new project, the Integrated Pest Information Platform for Extension and Education, is attempting to change this culture by building a platform for sharing pest data along with model outputs (Isard et al., 2005).

Model Parameterization: Advice for Model Parameterization

Many pest models are parameterized from experiments conducted under controlled environmental conditions. For example, many experiments measure an insect's development time (total and stage specific), mortality, fecundity, and longevity under different temperatures. These data can be used to parameterize a variety of models including phenology models based on DD and population models that predict the proportion of individuals in each life stage and the total population. Likewise, experiments where plants are inoculated under different temperature and wetness regimes can be used to parameterize infection models (Magarey et al., 2005b; Madden and Ellis, 1988). Sometimes caution is needed in interpretation of experiments; for example, mortality rates estimated in hot water bath treatments may not equate to mortality in hot air. Whenever possible, it is recommended to find the original experimental data rather than solely relying on functions fitted by third parties. There may be considerable differences in experimental design and the observed variable, so it is recommended that parameter definitions be precisely defined to ensure consistency in interpretation between studies (Kim et al., 2010). For example, the minimum wetness requirement for infection was defined for a specific level of disease severity and incidence (Magarey et al., 2005b). Software is now commonly available (such as Matlab's Grabit) that allow a modeler to digitize graphs and extract the values to reduce transcription errors. For DD models, the base threshold can be estimated from a linear regression of temperature (x-axis) vs. the inverse of stage duration (in days) (y-axis). The base is the x-intercept of the linear equation when $1/d = 0$ and DD requirements are the inverse of the slope (Nietschke et al., 2007). For infection response in plant pathogens, it is possible to define a unique response equation (Madden and Ellis, 1988) or use an existing generic function (Magarey et al., 2005b; Launay et al., 2014).

There have been a few efforts to compile parameter libraries; for example, developmental data including thresholds and DD requirements for insects (Jarosik et al., 2011, Nietschke et al., 2007) and infection requirements for pathogens (Magarey et al., 2005b). A common approach when data for a particular species is lacking is to identify parameters from closely related species. In this case, field studies may also be helpful when controlled data is absent, particularly by allowing a modeler to see if estimated parameters fit observed data. In the case of pest models developed from statistical relationships between weather variables and observed pest incidence in the field, extra caution should be used in transferring these models to other sites.

Input data for developing full temperature-based phenology models using ILCYM software are based on life table studies under a wide range of constant temperatures (from 5, 10, 15, 20, 25, 30 to 35°C) in which a species could develop to understand each species minimum and maximum thresholds for development. Likewise, life-table studies under fluctuation temperature conditions are used for model validations. Life-table studies are based on insect cohort (i.e., a group of individual insects of the same age and development stage, e.g., egg, larva) or complete (i.e., a group of individuals observed from the beginning of their egg stage until the death of all adults) and phenological events (i.e., development time of immature life stages, mortality, longevity and fecundity) life-table experiments. Details on how different data types were collected and used as inputs in ILCYM are described in Tonnang et al. (2013).

Model Validation: Advice for Validating Models

Strategy for model validation involves a number of sequential steps. In entomology and plant pathology, pest model validation is usually done by a single party, but Paez (2009) recommends that the model validation include at least three parties: evaluators, modelers, and stakeholders. In addition to this list we would also include those with field expertise such as agronomists, plant pathologists, or entomologists. In the process of model validation, all parties should share the design of the model validation process, but these parties will also have unique and separate responsibilities that contribute to a more robust validation. The first step is to identify the model question, which might be predicting the first appearance of a pest, its relative abundance, the timing of infection periods or a particular pest stage, or the timing of a pesticide spray. Care is needed in relating observations to predictions. A specific pest stage is often observed, so pest abundance and crop impacts are influenced by both pest and host phenology. Likewise, disease appearance may be delayed by a latent period that occurs after infection. Abundance is also influenced by many other factors including migration, alternative hosts, pesticide treatments, etc. Once the model question is defined, criteria for success or failure of the prediction should be established. For example, for first disease appearance for Cercospora leaf spot on sugar beet (caused by Cercospora beticola Sacc.), a window of ±7 d between observations and predictions was defined as a success (Racca et al., 2010). The third step is to define the experimental context under which the validation is conducted and the type of field observations that are available to perform the model validation. Sometimes, observation data are available for individual farms on specific dates, other times data might be regional and annual summaries. In many cases, existing observations are used to evaluate a model, other times, especially for an operational test of a model vs. a calendar program, new experiments are conducted (Grünwald et al., 2002). The fourth step is to define model evaluation statistics. The biological and meteorological literature provides useful statistics for categorical forecast validation. These techniques include those based on two-by-two contingency tables, those based on the difference between the forecast and observation (e.g., root mean square, mean error, mean absolute error), and regression techniques (Wilks, 1995; Welch et al., 1981; Racca et al., 2010). The fifth step is to collect the weather data that corresponds to the field observations. Site-specific weather data is preferred, but it is important that these data sets are cleaned and checked for missing values and errors. When such data sets are not available, gridded data sets can be used. The sixth step is modeling. When possible, multiple models should be compared, as this will likely be more informative than a single model. Modelers are provided with training data sets that include both weather data and field observations to allow for their model to be calibrated. Once model calibration has been done, modelers should be provided with test data sets by the evaluator that does include the field observations that are used for model validation. This blind comparison prevents overfitting of models, which may artificially inflate prediction scores. The model outputs are returned to the evaluator who conducts the appropriate statistical tests. The final step is to share the results with stakeholders and field experts and seek their interpretation of the results

Products from pest and disease models

Risk Products

Risk products, typically maps, have two general applications. The first application is to predict the potential distribution and impact of nonindigenous pests. Pest risk maps are used by phytosanitary agencies or other decision makers to support pest risk analysis, pest surveillance, and emergency programs. As part of this effort, risk maps created from climate data are often combined with maps predicting pest entry and density of susceptible hosts (Magarey et al., 2011) or visualizing the potential establishment and abundance of pests under current and future climates (Kroschel et al., 2013). A second application of risk maps is to aid growers or other decision makers in preseason planning for pest and disease management. In particular, risk products might show the expected frequencies of years that a pest might be expected to cause concern in a given location (Magarey et al., 2015). This kind of information can be invaluable for companies or other organizations seeking to stockpile a right amount of chemicals or other resources. Risk products might also be used to determine the most likely dates when particular activities, such as survey or trap placement, might take place. An example is the *Bacillus thuringiensis* Berliner (Bt) evaluation tool that provides growers with the expected benefits of using Bt-corn (Hellmich et al., 2005). Pest risk maps are typically created from 30-yr climate data sets or from weather data sets averaged over 10 yr or more.

Risk maps for nonindigenous species are generally created under two main constraints, that is, limited time to produce the risk map product and sometimes an absence of biological data to fit and validate the model. There are a variety of techniques to develop risk maps including inductive techniques where a pest's potential distribution is inferred from a mathematical relationship between climate variables and the known distribution (Venette et al., 2010). One popular example is maximum entropy modeling, which uses distribution data in combination with derived background observations. However, maximum entropy modeling and other species-distribution models may not extrapolate reliably especially into novel climates (Elith and Leathwick, 2009). In contrast to the inductive approach, a deductive model predicts pest behavior from mathematical modeling of biological processes, for example, infection, population increase, and spread, etc., using parameters derived from controlled experiments or field observations. By running a process-based model from 10 or more years of weather data, a user can create a risk product that describes the frequency of conducive years that a pest population may exceed a threshold where it can cause damage to crops or environmental hosts. There are also approaches such as CLIMEX, using the Compare Locations model (Sutherst and Maywald, 1985, Kriticos et al., 2015), which is a simple process-based model that uses literature and distribution data to fit the model parameters. CLIMEX is very widely used, in part, because the model parameters can be inferred from the distribution data itself. While the inductive models usually define the potentially suitable area, deductive or process-based models can provide outputs of potential pest abundance or damage. Risk maps may also provide other information, such as the frequency of suitable years, which can be used as an input into a probabilistic pathway model.

Although risk maps have been used for many years, they have become increasingly sophisticated as more advanced pest models with outputs tailored to a specific crop–pest system have been developed and used for generating risk maps. For example, ILCYM is linked to uDig platform (https://github.com/locationtech/udig-platform), which is a GIS-based application that contains basic tools for mapping and managing geographic information (Sporleder et al., 2013). ILCYM's potential population distribution and risk mapping module implements phenology models (pests or beneficial insects of a wide range of insect orders and families) and, based on simulated life-table parameters, ILCYM calculates and maps three risk indices (establishment risk index, generation index, and activity index) for assessing the potential establishment and population distribution, abundance, and spread of insect species on global, regional, and national scales. The risk indices can be simulated under current and future climate change scenarios. For spatial simulations under present climates, ILCYM uses temperature data obtained from WorldClim (http://www.worldclim.org/), which is well documented in Hijmans et al. (2005). For studying the effects of climate change, ILCYM provides downscaled future climate scenarios from different climate change projections (IPCC, 2007) and predictions based on the WorldClim database described by Govindasamy et al. (2003). Further, ILCYM allows the use of several GIS functionalities for vector and raster analysis as in the use of shape files to map pest information only in those regions where target crops are grown. This has been applied to understand pest problems for specific crops, for example, the potato tuber moth under current and future climates (Kroschel et al., 2013), and to compile a Pest Distribution and Risk Atlas for Africa to support pest risk analysis and the preparedness of policymakers and farmers to implement timely adaptation strategies (Kroschel et al., 2016; http://cipotato.org/riskatlasforafrica/).

Pest Forecasts

In contrast to risk maps, pest forecasts are generated throughout the growing season for day-to-day decision-making. Model products can be distributed using a variety of distribution networks and formats such as word of mouth, newsletters, maps, recorded phone messages, newspapers and magazines, radio and television, videotel, televideo, fax, e-mail, mobile phone text, web sites, and smart phone applications. The last two have wider benefits (fast dissemination and utilization of information, interaction and feedback with the users, immediate visualization of information, easy comprehension of information and advice, increased computer use by farmers, cost reduction, fast updating and improving of the system, control of system performance, and application of multimedia tools [texts, graphics, maps, figures, audio, video, etc.]). Delivery systems have been classified as either push, pull, or plug (Russo, 2000). Push and pull systems provide model outputs from a remote source. Push systems deliver information to the user, while pull systems require the user to request the information. Push systems work best for supplying daily information, updates, and other types of dynamic information. Pull systems work best for allowing users to access specific databases or resources. Plug systems provide model products to be built into an on-site device, for example, a disease predictor. As an example of a push system, the SkyBit Apple IPM and apple disease products provide daily summaries of weather variables and pest alerts (Russo, 2000). The products can be delivered by e-mail and are initiated using a grower-provided biofix. The disease and pest risk is shown either as quantitative score (e.g., for apple scab percentage ascospore maturity, wetness hours, temperature during wetness hours, and for insect pests percentage hatch) and also indicated at four risk levels—no risk, wait, watch, and warning—for easy interpretation by farmers. The product also provides a 5-d forecast so growers can apply protectant pesticides before an infection event occurs. As an example of a plug system, pest or disease forecast products are also commonly provided as part of a weather station package, where a user (or group of users) purchases a weather station and a software package that provides forecasts created from input weather data monitored by the station. As an example of a pull and push system, the Fusarium head blight warnings developed by a multiuniversity and industry partnership can be viewed online as a map, but growers can also sign up to receive a text alert that will be sent to their mobile phone (McMullen et al., 2012).

Model products can also be incorporated into a variety of DSS from very simple (such as production guides) to complex (such as computerized expert system). The advantages of a simple DSS include low cost, low technology, generic application, ease of delivery in multiple ways, and limited time requirements for learning and use of the DSS. In contrast, a more sophisticated DSS can provide greater integration of knowledge and allow growers to choose management tools and their associated risks. The adoption of DSS depends on user's exposure and tolerance to risk and how much a DSS can assist in the risk management compared with a grower or consultant's intuition (Shtienberg, 2013; Gent et al., 2013). The need for a more sophisticated DSS is driven by the increasing complexity of agricultural decision making such as multicriteria decisions that impact maximum residue limits and specific market requirements, minimizing spray drift, rotation of pesticide classes to avoid resistance, treatment costs and impact on beneficial insects, and the environment. Balanced against this is the problem that more complex systems require greater time to use. If a DSS is too slow, users will interact with it less frequently. As such, the user may ultimately be unaware of the model output itself but only interact with the management options such as pesticide selection and timing (Rosenberger, 2003). A good example of DSS is the Washington State University Decision Aid System for tree fruit, which provides management recommendations based on model output for 10 insect pests and four diseases (Jones et al., 2010). The DSS provides growers with the pest predictions based on current and 10-d weather forecast. Based on the predictions, the DSS makes management recommendations, including when to scout and apply preferred treatment options, and contains links to more detailed information. It is very important to include specific training activities, involving model developers, services (meteorological, extension, etc.), and final users, to improve model understanding, to establish the important feed-back interactions, and generally to create a modeling culture (Eitzinger et al., 2009).

Research Products

The expected impacts of climate change in many regions can increase the need for early warning systems to support agricultural decision makers. Large variations in weather conditions between years, as well as during the growing season, can make decision-making more difficult and can reduce the value of previous experience.

When models are applied in future-scenario analyses, another important source of uncertainty is how effectively managers will adapt to changes in disease and pest risks. Higher pest and disease risk imposes greater demands on all people involved in agricultural production, from plant breeders, entomologists, and plant pathologists to extension agents and farmers. The demands will be higher on farmers in systems where support from research and extension is not readily available as in developing countries. The question remains whether effective management will be widely implemented and whether management can be formulated so that it does not substantially reduce profitability or reduce other ecosystem services. In the context of fragile environments, decision support systems at the farm level have to be developed and disseminated to increase a sustainable level of productivity and income at country and farm levels. Moreover, the spatial and temporal correlation resulting from potential pest and disease spread means that decisions made by some farmers will influence pest and disease problems experienced in other cultivated areas or regions. Forecasting future distributions of diseases and pests from current known species climate relationships is highly problematic. This is because the observed distribution of diseases and pests alone provides no clear information about how the species might respond under completely novel environmental conditions. Thus, model outputs based on extrapolations may lead to substantial errors in managing disease and pest invasions and climate change impacts (Garrett et al., 2013).

Economic and Environmental Benefits of Models

The main direct benefit of using models is the reduction of plant protection costs. In the absence of a model or other DSS, the system is generally managed by the classic turn-based drives, also called a calendar (World Meteorological Organization Commission of Agricultural Meteorology [WMO CAgM], 1988), which provides prespecified dates for the application of pesticides based on the characteristics of persistence of the active ingredient used regardless of the actual risk from pests and diseases. An objective economic assessment must take into account the savings in manpower, machinery, energy, and chemicals as well as other expenses incurred to implement the program of pest management based on model output and climatic conditions. In this sense, the costing has different attributes if the application is made at the farm level, which requires farmers to buy weather stations and incur data processing costs, or the district level, in which case the system may be managed by a dedicated body and users pay a subscription fee (WMO CAgM, 1988). Numerous studies have shown, experimentally, many benefits. The reduction in the number of pesticide treatments is, on average, between 15 and 50%, and a corresponding reduction of the costs could be expected (Fabre et al., 2007). There are also numerous indirect benefits whose economic quantification is quite complex but which further confirm the effectiveness of these methodologies. Globally, they allow agriculture to reduce its environmental impact in terms of greenhouse gas emissions, water consumption, and loss of biodiversity. First, the entry of chemicals into the environment is reduced, resulting in less pollution of soils, groundwater, and the atmosphere. This results in benefits for the whole terrestrial ecosystem including the maintenance of fertility of soils and in a lower content of pesticide residues in harvested commodities. A more controlled use of pesticides results in a lower number of applications and the alternation of products with different chemistries, which also reduces the risk of emergence of resistant strains to the active ingredients used. The reduced disruption of natural ecosystems favors the presence of natural biological control and reduces the appearance of new pests. Finally, less frequent applications in fields cause lower soil compaction and degradation (WMO CAgM, 1988). In practice, there are many barriers that can prevent the adoption of pest and disease models by users including complexity and the time required to use the model (or decision support system) (Magarey et al., 2002). McRoberts et al. (2011) outlines a research framework to improve the delivery of pest and disease models such that the skill associated with the forecasted probability of disease occurrence is weighted against the expected regret, that is, the expected impacts of an incorrect forecast. The proposed interdisciplinary research framework could allow epidemiologists, social scientists, economists, and risk analysts to identify barriers that prevent many users from adopting pest and disease models.

Future directions and trends

Models represent powerful tools to support end users in the field of crop protection management. A combination of favorable conditions (biological knowledge, technological development, etc.) has stimulated the formulation and application of these decision support systems. A large number of models have been proposed in the last decades to simulate pest attacks for the most important crops at a global level. Models can be applied with tactical and strategic aims by using simple tools (table) as well as more complicated software or integrated systems providing information at territorial levels (GIS and maps output). The application at farm level as well as the activity of extension services allows a reduction of pesticide use (~10–25%) with economic and environmental benefits. The increasing use and potentiality of the Internet and mobile phones represent fundamental tipping points, allowing a fast and widespread diffusion of the information produced by model simulations. Other trends we foresee are an increase in application of models as a climate change adaptation and mitigation tool. Earlier, we noted that crop loss modeling has important applications for global food security and climate change research (Esker et al., 2012). One important future development for predicting these losses is the coupling of pest and disease models with crop models, enabling powerful simulations of crop loss under changing environmental conditions (Kropff et al., 1995). Progress on developing this coupling has been inhibited for several reasons. One is that many pest models do not have outputs that are easy to translate into pest impacts and another is that the high diversity in modeling approaches impedes the ease of coupling. However, there are now more comprehensive efforts to alleviate these problems through generic pest modeling (Bregaglio and Donatelli, 2015; Savary et al., 2015) and the inclusion of pest and disease modelers into existing crop modeling research projects. For example, in 2015, the Agricultural Modeling Inter-Comparison (AgMip) Project organized a pest- and disease-modeling workshop to advance this process.

Probably the most important future development is represented by precision agriculture based on direct linkage between monitoring, models, and field equipment. This development is facilitated by increasing resolution of grid weather data sets, remote sensing, and positioning systems. Mobile phones and tablets allow the user to constantly receive high-quality information and make decisions about crop protection management. It is likely that there will also be an increasing quantity of pest observations, moving toward the concept of big data, which refers to datasets whose size is beyond the ability of typical database software tools to capture, store, manage, and analyze. Large quantities of pest observations and other agronomic data will enable researchers to quickly make connections and discoveries that would have taken far more time and resources to complete otherwise. This revolution will, in part, be driven by the increasing use of automated sensors attached to farm machinery (Mahlein et al., 2012), smart trapping technology (Loughlin, 2013), drones (Rieder et al., 2014), simulated pest observations (Merrill et al., 2015), and improved sharing of pest data (Isard et al., 2015). In this case, pest observations as part of a big-data analysis could be used to verify and calibrate pest models. An information engine would extract predicted pest values according to the location and date of each observation. These outputs, mined over many seasons and geographies, could be used in a big-data analysis to (i) provide error checking that pest observations fall into an acceptable range in terms of spatial and temporal occurrence, (ii) provide a way to fill in missing pest observations, and (iii) provide a technique to predict and forecast future pest observations including those for long-range pest forecasting. Even more radical is the on-coming reality where farm machinery becomes fully automated. In such a system, high-resolution pest models could be linked directly to variable-rate application based on environmental conditions that vary within a field (Russo, 2015; Merrill et al., 2015).

Although we believe big pest data will increase the ability for model validation, there are still many obstacles to overcome including incompatibilities between models results and disease assessments. Uncertainty of models and their output can be minimized only within the limits of biological understanding while extending model products into finer spatial resolutions and longer forecast periods. We see a trend toward improved model performance as constraints (e.g., missing microclimate data input, deficits in information transfer and lack of acceptance from users) are overcome. Finally, as new agricultural stakeholders with validated models will grow in practical importance, an increasing number of them will be used in routine procedures for the benefit of agricultural users.

References

Ahn, M.I., K.S. Do, K.H. Lee, W.S. Kang, and E.W. Park. 2013. Application of the PBcast model for timing fungicide sprays to control Phytophthora blight of pepper. In: Abstract and Programme Book, the 19th Australasian Plant Pathology Conference, Auckland, New Zealand.

Allen, J.C. 1976. A modified sine wave method for calculating degree days. Environ. Entomol. 5:388–396. doi:10.1093/ee/5.3.388

Baker, R. 1978. Inoculum potential. In: J.G. Horsfall and E.B. Cowling, editors, Plant disease. An advanced treatise. Vol. 2. Academic Press, New York. p. 137–156.

Bale, J., G.J. Masters, I.D. Hodkins, C. Awmack, T.M. Bezemer, V.K. Brown, J. Buterfield, A. Buse, J.C. Coulson, J. Farrar, J.E.G. Good, R. Harrigton, S. Hartley, T.H. Jones, R.L. Lindroth, M.C. Press, I. Symrnioudis, A.D. Watt, and J.B. Whittaker. 2002. Herbivory in global climate change research: Direct effects of rising temperature on insect herbivores. Glob. Change Biol. 8:1–16. doi:10.1046/j.1365-2486.2002.00451.x

Bauer, P., A. Thorpe, and G. Brunet. 2015. The quiet revolution of numerical weather prediction. Nature 525:47–55. doi:10.1038/nature14956

Baumhover, A.H., A.J. Graham, B.A. Bitter, D.E. Hopkins, W.D. New, F.H. Dudley, and R.C. Bushland. 1955. Screw-worm control through release of sterilized flies. J. Econ. Entomol. 48:462–466. doi:10.1093/jee/48.4.462

Berryman, A.A., J.A. Millstein, and R.R. Mason. 1990. Modelling Douglas-fir tussock moth population dynamics: The case for simple theoretical models. In: A.D. Watt, editor, Population dynamics of forest insects. Intercept, Andover, UK. p. 369–380.

Bregaglio, S., and M. Donatelli. 2015. A set of software components for the simulation of plant airborne diseases. Environ. Model. Softw. 72:426–444. doi:10.1016/j.envsoft.2015.05.011

Bregaglio, S., M. Donatelli, R. Confalonieri, M. Acutis, and S. Orlandini. 2011. Multi metric evaluation of leaf wetness models for large-area application of plant disease models. Agric. For. Meteorol. 151:1163–1172. doi:10.1016/j.agrformet.2011.04.003

Brown, J.H., J.F. Gillooly, A.P. Allen, V.M. Savage, and G.B. West. 2004. Toward a metabolic theory of ecology. Ecology 85:1771–1789. doi:10.1890/03-9000

Campbell, C.L., and L.V. Madden. 1990. Introduction to plant disease epidemiology. John Wiley & Sons, New York.

Canessa, P., J. Schumacher, M.A. Hevia, P. Tudzynski, and L.F. Larrondo. 2013. Assessing the effects of light on differentiation and virulence of the plant pathogen *Botrytis cinerea*: Characterization of the white collar complex. PLoS One 8:e84223. doi:10.1371/journal.pone.0084223

Chappell, T.M., A.L. Beaudoin, and G.G. Kennedy. 2013. Interacting virus abundance and transmission intensity underlie tomato spotted wilt virus incidence: An example weather-based model for cultivated tobacco. PLoS One 8:e73321. doi:10.1371/journal.pone.0073321

Cicogna, A., S. Dietrich, M. Gani, R. Giovanardi, and M. Sandra. 2005. Use of meteorological radar to estimate leaf wetness as data input for application of territorial epidemiological model (downy mildew—*Plasmopara viticola*). Phys. Chem. Earth Parts ABC 30:201–207. doi:10.1016/j.pce.2004.08.015

Clarkson, J.P., L. Fawcett, S.G. Anthony, and C. Young. 2014. A model for *Sclerotinia sclerotiorum* infection and disease development in lettuce, based on the effects of temperature, relative humidity and ascospore density. PLoS One 9:e94049. doi:10.1371/journal.pone.0094049

Coulson, T., and H.C.J. Godfray. 2007. Single-species dynamics. In: R.M. May and R.R. McLean, editors, Theoretical ecology. Principles and applications. 3rd ed. Oxford Univ. Press, Oxford, NY. p. 17–34.

Curry, G.L., R.M. Feldman, and P.J. Sharpe. 1978. Foundations of stochastic development. J. Theor. Biol. 74:397–410. doi:10.1016/0022-5193(78)90222-9

Daane, K. 2004. Temperature-dependent development of *Anagyrus pseudococci* (Hymenoptera: Encyrtidae) as a parasitoid of the vine mealybug, *Planococcus ficus* (Homoptera: Pseudococcidae). Biol. Control 31:123–132. doi:10.1016/j.biocontrol.2004.04.010

Dalla Marta, A., R.D. Magarey, and S. Orlandini. 2005. Modelling leaf wetness duration and downy mildew simulation on grapevine in Italy. Agric. For. Meteorol. 132:84–95. doi:10.1016/j.agrformet.2005.07.003

Dennis, B., J.M. Ponciano, S.R. Lele, M.L. Taper, and D.F. Staples. 2006. Estimating density dependence, process noise, and observation error. Ecol. Monogr. 76:323–341. doi:10.1890/0012-9615(2006)76[323:EDDPNA]2.0.CO;2

De Pondeca, M.S., G.S. Manikin, G. DiMego, S.G. Benjamin, D.F. Parrish, R.J. Purser, W.S. Wu, J.D. Horel, D.T. Myrick, and Y. Lin. 2011. The real-time mesoscale analysis at NOAA's National Centers for Environmental Prediction: Current status and development. Weather Forecast. 26:593–612. doi:10.1175/WAF-D-10-05037.1

de Valpine, P., and A. Hastings. 2002. Fitting population models incorporating process noise and observation error. Ecol. Monogr. 72:57–76. doi:10.1890/0012-9615(2002)072[0057:FPMIPN]2.0.CO;2

Dewdney, M.M., A.R. Biggs, and W.W. Turechek. 2007. A statistical comparison of the blossom blight forecasts of MARYBLYT and Cougarblight with receiver operating characteristic curve analysis. Phytopathology 97:1164–1176. doi:10.1094/PHYTO-97-9-1164

De Wolf, E.D., and S.A. Isard. 2007. Disease cycle approach to plant disease prediction. Annu. Rev. Phytopathol. 45:203–220. doi:10.1146/annurev.phyto.44.070505.143329

De Wolf, E.D., L.V. Madden, and P.E. Lipps. 2003. Risk assessment models for wheat Fusarium head blight epidemics based on within season weather data. Phytopathology 93:428–435. doi:10.1094/PHYTO.2003.93.4.428

Do, K.S., W.S. Kang, and E.W. Park. 2012. A forecast model for the first occurrence of Phytophthora blight on chili pepper after overwintering. Plant Pathol. J. 28:172–184. doi:10.5423/PPJ.2012.28.2.172

Dong, P., J.J. Wang, F.X. Jia, and F. Hu. 2007. Development and reproduction of the Psocid *Liposcelis tricolor* (Psocoptera: Liposcelididae) as a function of temperature. Ann. Entomol. Soc. Am. 100:228–235. doi:10.1603/0013-8746(2007)100[228:DAROTP]2.0.CO;2

Draxler, R.R., and G.D. Rolph. 2015. HYSPLIT (HYbrid Single-Particle Lagrangian Integrated Trajectory) Model. NOAA Air Resources Laboratory, Silver Spring, MD. http://ready.arl.noaa.gov/HYSPLIT.php (accessed 17 Jan. 2017).

Duthie, J.A. 1997. Models of the response of foliar parasites to the combined effects of temperature and duration of wetness. Phytopathology 87:1088–1095. doi:10.1094/PHYTO.1997.87.11.1088

Eitzinger, J., S. Thaler, S. Orlandini, P. Nejedlik, V. Kazandjiev, T. Sivertsen, and D. Mihailovic. 2009. Applications of agroclimatic indices and process oriented crop simulation models in European agriculture. Idojaras 113:1–12.

Elith, J., and J.R. Leathwick. 2009. Species distribution models: Ecological explanation and prediction across space and time. Annu. Rev. Ecol. Evol. Syst. 40:677–697. doi:10.1146/annurev.ecolsys.110308.120159

Esker, P. D., S. Savary, and N. McRoberts. 2012. Crop loss analysis and global food supply: Focusing now on required harvests. CAB Rev. 7:1–14. doi:10.1079/PAVSNNR20127052

Fabre, F., M. Plantegenest, and J. Yuen. 2007. Financial benefit of using crop protection decision rules over systematic spraying strategies. Phytopathology 97:1484–1490. doi:10.1094/PHYTO-97-11-1484

Finch, S., R.H. Collier, and K. Phelps. 1996. A review of work done to forecast pest insect attacks in UK horticultural crops. Crop Prot. 15:353–357. doi:10.1016/0261-2194(95)00135-2

Friesland, H., and S. Orlandini. 2006. Simulation models of plant pests and diseases. In: G. Maracchi, A. Mestre, L. Toulios, L., Kajfez-Bogataj, and A.A. Hocevar, editors, COST Action 718: Meteorology applications for agriculture. European Cooperation in the field of Scientific and Technical Research, Belgium. p. 81–98.

Fry, W.E., A.E. Apple, and J.A. Bruhn. 1983. Evaluation of potato late blight forecasts modified to incorporate host resistance and fungicide weathering. Phytopathology 73:1054–1059. doi:10.1094/Phyto-73-1054

Garrett, K.A., A.D.M. Dobson, J. Kroschel, B. Natarajan, S. Orlandini, H.E.Z. Tonnang, and C. Valdivia. 2013. The effects of climate variability and the color of weather time series on agricultural diseases and pests, and on decisions for their management. Agric. For. Meteorol. 170:216–227. doi:10.1016/j.agrformet.2012.04.018

Gent, D.H., W.F. Mahaffee, N. McRoberts, and W.F. Pfender. 2013. The use and role of predictive systems in disease management. Annu. Rev. Phytopathol. 51:267–289. doi:10.1146/annurev-phyto-082712-102356

Geritz, S.A.H., and É. Kisdi. 2004. On the mechanistic underpinning of discrete-time population models with complex dynamics. J. Theor. Biol. 228:261–269. doi:10.1016/j.jtbi.2004.01.003

Gessler, C., I. Pertot, and M. Perazzolli. 2011. *Plasmopara viticola*: A review of knowledge on downy mildew of grapevine and effective disease management. Phytopathol. Mediterr. 50:3–44.

Getz, W.M., and A.P. Gutierrez. 1982. A perspective on systems analysis in crop production and insect pest management. Annu. Rev. Entomol. 27:447–466. doi:10.1146/annurev.en.27.010182.002311

Gilles, T., K. Phelps, J.P. Clarkson, and R. Kennedy. 2004. Development of MILIONCAST, an improved model for predicting downy mildew sporulation on onions. Plant Dis. 88:695–702. doi:10.1094/PDIS.2004.88.7.695

Gilligan, C.A. 1983. Modeling of soilborne pathogens. Annu. Rev. Phytopathol. 21:45–64. doi:10.1146/annurev.py.21.090183.000401

Gleason, M.L., K.B. Duttweiler, J.C. Batzer, S.E. Taylor, P.C. Sentelhas, J.E.B.A. Monteiro, and T.J. Gillespie. 2008. Obtaining weather data for input to crop disease-warning systems: Leaf wetness duration as a case study. Sci. Agric. 65:76–87. doi:10.1590/S0103-90162008000700013

Goidanich, G. 1959. Manuale di Patologia Vegetale. Edagricole, Bologna, 713 pp.

Golizadeh, A., K. Kamali, Y. Fathipour, and H. Abbasipour. 2008. Life table and temperature-dependent development of *Diadegma anurum* (Hymenoptera: Ichneumonidae) on its host *Plutella xylostella* (Lepidoptera: Plutellidae). Environ. Entomol. 37:38–44. doi:10.1603/0046-225X(2008)37[38:LTATDO]2.0.CO;2

Govindasamy, B., P.B. Buffy, and J. Coquard. 2003. High-resolution simulations of global climate, part 2: Effects of increased greenhouse gases. Clim. Dyn. 21:391–404. doi:10.1007/s00382-003-0340-6

Grünwald, N.J., G.R. Montes, H.L. Saldaña, O.A.R. Covarrubias, and W.E. Fry. 2002. Potato late blight management in the Toluca Valley: Field validation of SimCast modified for cultivars with high field resistance. Plant Dis. 86:1163–1168. doi:10.1094/PDIS.2002.86.10.1163

Haghani, M., Y. Fathipour, A.A. Talebi, and V. Baniameri. 2007. Thermal requirement and development of *Liriomyza sativae* (Diptera: Agromyzidae) on cucumber. J. Econ. Entomol. 100:350–356. doi:10.1603/0022-0493(2007)100[350:TRADOL]2.0.CO;2

Hassell, M.P. 1978. The dynamics of arthropod predator-prey systems. Princeton Univ. Press, Princeton, NJ.

Hellmich, R.L., D.D. Calvin, J.M. Russo, and L.C. Lewis. 2005. Integration of BT maize in IPM systems: A U.S. perspective. In: M.S. Hoddle, compiler, Second International Symposium on Biological Control of Arthropods. Davos, Switzerland, 12–16 Sept. 2005. The Bugwood Network, Tifton, GA. p. 356–361.

Hijmans, R.J., S.E. Cameron, J.L. Parra, P.G. Jones, and A. Jarvis. 2005. Very high resolution interpolated climate surfaces for global land areas. Int. J. Climatol. 25:1965–1978. doi:10.1002/joc.1276

Huffaker, C.B. 1980. New technology of pest control. John Wiley & Sons, New York.

Huld, T.A., M. Šúri, E.D. Dunlop, and F. Micale. 2006. Estimating average daytime and daily temperature profiles within Europe. Environ. Model. Softw. 21:1650–1661. doi:10.1016/j.envsoft.2005.07.010

Humbert, J.Y., L. Scott Mills, J.S. Horne, and B. Dennis. 2009. A better way to estimate population trends. Oikos 118:1940–1946. doi:10.1111/j.1600-0706.2009.17839.x

IPCC. 2007. Climate Change 2007: Impacts, adaptation and vulnerability. Contribution of Working Group II to the Fourth Assessment Report of the Intergovernmental Panel on Climate Change. Cambridge Univ. Press, Cambridge, UK.

Isard, S.A., and S.H. Gage. 2001. Flow of life in the atmosphere. Michigan State Univ. Press, East Lansing.

Isard, S.A., S.H. Gage, P. Comtois, and J.M. Russo. 2005. Principles of the atmospheric pathway for invasive species applied to soybean rust. Bioscience 55:851–861. doi:10.1641/0006-3568(2005)055[0851:POTAPF]2.0.CO;2

Isard, S.A., J.M. Russo, R.D. Magarey, J. Golod, and J.R. VanKirk. 2015. Integrated pest information platform for extension and education (iPiPE): Progress through sharing. J. Integ. Pest Manage. 6. doi:10.1093/jipm/pmv013

Jacquemin, G., S. Chavalle, and M. De Proft. 2014. Forecasting the emergence of the adult orange wheat blossom midge, *Sitodiplosis mosellana* (Géhin) (Diptera: Cecidomyiidae) in Belgium. Crop Prot. 58:6–13. doi:10.1016/j.cropro.2013.12.021

Jarosik, V., A. Honek, R.D. Magarey, and J. Skuhrovec. 2011. Developmental database for phenology models: related insect and mite species have similar thermal requirements. J. Econ. Entomology 104:1870–1876. doi:10.1603/EC11247

Jardine, D.J., and C.T. Stephens. 1987. A predictive system for timing chemical applications to control *Pseudomonas syringae* pv. *tomato,* causal agent of bacterial speck. Phytopathology 77:823–827. doi:10.1094/Phyto-77-823

Jones, V.P., J.F. Brunner, G.G. Grove, B. Petit, G.V. Tangren, and W.E. Jones. 2010. A web-based decision support system to enhance IPM programs in Washington tree fruit. Pest Manag. Sci. 66:587–595.

Joint Research Centre. 2015. AGRI4CAST interpolated meteorological data. Inst. for Environ. and Sustain. Monitoring Agric. Resources. http://mars.jrc.ec.europa.eu/mars/Aboutus/AGRI4CAST/Data-distribution/AGRI4CAST-Interpolated-Meteorological-Data

Kang, W.S., S.S. Hong, Y.K. Han, K.R. Kim, S.G. Kim, and E.W. Park. 2010. A web-based information system for plant disease forecast based on weather data at high spatial resolution. Plant Pathol. J. 26:37–48. doi:10.5423/PPJ.2010.26.1.037

Karpinski, S., H. Gabrys, A. Mateo, B. Karpinska, and P.M. Mullineaux. 2003. Light perception in plant disease defense signaling. Curr. Opin. Plant Biol. 6:390–396. doi:10.1016/S1369-5266(03)00061-X

Khadioli, N., Z.E.H. Tonnang, E. Muchugu, G. Ong'amo, T. Achia, I. Kipchirchir, J. Kroschel, and B. Le Ru. 2014a. Effect of temperature on the phenology of *Chilo partellus* (Swinhoe) (Lepidoptera, Crambidae): Simulation and visualization of the potential future distribution of *C. partellus* in Africa under warmer temperatures through the development of life-table parameters. Bull. Entomol. Res. 104:809–822. doi:10.1017/S0007485314000601

Khadioli, N., Z.E.H. Tonnang, G. Ong'amo, T. Achia, I. Kipchirchir, J. Kroschel, and B. Le Ru. 2014b. Effect of temperature on the life history parameters of noctuid lepidopteran stem borers, *Busseola fusca* and *Sesamia calamistis*. Ann. Appl. Biol. 165:373–386. doi:10.1111/aab.12157

Kim, K.S., S.E. Taylor, M.L. Gleason, F.W. Nutter Jr., L.B. Coop, W.F. Pfender, R.C. Seem, P.C. Sentelhas, T.J. Gillespie, A. Dalla Marta, and S. Orlandini. 2010. Spatial portability of numerical models of leaf wetness duration based on empirical approaches. Agric. For. Meteorol. 150:871–880. doi:10.1016/j.agrformet.2010.02.006

Knipling, E.F. 1955. Possibilities of insect control or eradication through the use sexually sterile male. J. Econ. Entomol. 48:459–462. doi:10.1093/jee/48.4.459

Krause, R.A., and L.B. Massie. 1975. Predictive systems: Modern approaches to disease control. Annu. Rev. Phytopathol. 13:31–47. doi:10.1146/annurev.py.13.090175.000335

Krishnaiah, N.V., I.C. Pasalu, L. Padmavathi, K. Krishnaiah, and A. Ram Prasad. 1997. Day degree requirement of rice yellow stem borer, *Scirpophaga incertulas* (Walker). Oryza (Cuttack, India) 34:185–186.

Kriticos, D.J., G.F. Maywald, T. Yonow, E.J. Zurcher, N.I. Herrmann, and R.W. Sutherst. 2015. CLIMEX Version 4: Exploring the effects of climate on plants, animals and diseases. CSIRO, Canberra.

Kropff, M.J., P.S. Teng, and R. Rabbinge. 1995. The challenge of linking pest and crop models. Agric. Syst. 49:413–434. doi:10.1016/0308-521X(95)00034-3

Kroschel, J., M. Sporleder, N. Mujica, and P. Carhuapoma. 2016. Pest distribution and risk atlas for Africa: Potential global and regional distribution and abundance of agricultural and horticultural pests and associated biocontrol agents under current and future climates. International Potato Center, Lima, Peru. doi:10.4160/9789290604761-4

Kroschel, J., M. Sporleder, H. Tonnang, H. Juarez, P. Carhuapoma, J.C. Gonzales, and R. Simon. 2013. Predicting climate-change-caused changes in global temperature on potato tuber moth *Phthorimaea operculella* (Zeller) distribution and abundance using phenology modeling and GIS mapping. Agric. For. Meteorol. 170:228–241. doi:10.1016/j.agrformet.2012.06.017

Launay, M., J. Caubel, G. Bourgeois, F. Huard, I.G. de Cortazar-Atauri, M.O. Bancal, and N. Brisson. 2014. Climatic indicators for crop infection risk: Application to climate change impacts on five major foliar fungal diseases in Northern France. Agric. Ecosyst. Environ. 197:147–158. doi:10.1016/j.agee.2014.07.020

Liu, Y.H., and J.H. Tsai. 2002. Effect of temperature on development, survivorship, and fecundity of *Lysiphlebia mirzai* (Hymenoptera: Aphidiidae), a parasitoid of *Toxoptera citricida* (Homoptera: Aphididae). Environ. Entomol. 31:418–424. doi:10.1603/0046-225X-31.2.418

Logan, J.A. 1994. In defense of big ugly models. Am. Entomologist. 40:202-207.

Logan, J.A., D.J. Wollkind, S.C. Hoyt, and L.K. Tanigoshi. 1976. An analytic model for description of temperature dependent rate phenomena in arthropods. Environ. Entomol. 5:1133–1140. doi:10.1093/ee/5.6.1133

Lotka, A.J. 1939. On an integral equation in population analysis. Ann. Math. Stat. 10:144–161. doi:10.1214/aoms/1177732213

Lotka, A.J. 1945. Population analysis as a chapter in the mathematical theory of evolution. In: W.E. LeGros Clark and P.B. Medawar, editors, Essays on growth and form. Oxford Univ. Press, Oxford. p. 355–395.

Loughlin, D. 2013. Developments in the world of insect detection. Int. Pest Control 55:88–90.

Ludwig, D. 1928. The effects of temperature on the development of an insect (*Popillia japonica Newman*). Physiol. Zool. 1:358–389. doi:10.1086/physzool.1.3.30151052

MacHardy, W.E., and D.M. Gadoury. 1989. A revision of Mills' criteria for predicting apple scab infection periods. Phytopathology 79:304–310. doi:10.1094/Phyto-79-304

Madden, L.V., and M.A. Ellis. 1988. How to develop plant disease forecasters, In: J. Kranz, and J. Rotem, editors, Experimental techniques in plant disease epidemiology. Springer–Verlag, Heidelberg. p. 191–208. doi:10.1007/978-3-642-95534-1_14

Madden, L.V., G. Hughes, and F.V.D. Bosch. 2007. The Study of Plant Disease Epidemics. The American Phytopathological Society, St. Paul, MN. 421 pp.

Magarey, R.D., D.M. Borchert, J.S. Engle, M. Colunga-Garcia, F.H. Koch, and D. Yemshanov. 2011. Risk maps for targeting exotic plant pest detection programs in the United States. Bull. OEPP 41:46–56. doi:10.1111/j.1365-2338.2011.02437.x

Magarey, R.D., D.M. Borchert, G.A. Fowler, and S.C. Hong. 2015. The NCSU/APHIS plant pest forecasting system (NAPPFAST). In: R. Venette, editor, Pest risk modeling and mapping for invasive alien species. CABI, Wallingford, UK. p. 82–96. doi:10.1079/9781780643946.0082

Magarey, P.A., R.W. Emmett, N.I. Herrmann, M.F. Wachtel, and J.W. Travis. 1997. Development of AusVit, a computerized decision support system for integrated management of diseases, pests and other production factors in Australian viticulture. Viticulture Enology Sci. 52:175–179.

Magarey, R.D., G.A. Fowler, D.M. Borchert, T.B. Sutton, M. Colunga-Garcia, and J.A. Simpson. 2007. NAPPFAST: An internet system for the weather-based mapping of plant pathogens. Plant Dis. 91:336–345. doi:10.1094/PDIS-91-4-0336

Magarey, R.D., R.C. Seem, J.M. Russo, J.W. Zack, K.T. Waight, J.W. Travis, and P.V. Oudemans. 2001. Site-specific weather information without on-site sensors. Plant Dis. 85:1216–1226. doi:10.1094/PDIS.2001.85.12.1216

Magarey, R.D., R.C. Seem, A. Weiss, T.J. Gillespie, and L. Huber. 2005a. Estimating surface wetness on plants. In: J.L. Hatfield, J.M. Baker, and M.K. Viney, editors, Agronomy Monograph, Micrometeorological measurements in agricultural systems 47:199-226. Agronomy Monograph. ASA, Madison, WI.

Magarey, R.D., and T.B. Sutton. 2007. How to create and deploy infection models for plant pathogens. In: A. Ciancio and K.G. Mukerji, editors, Integrated management of plant pests and diseases. Vol. I. General concepts in integrated pest and disease management. Springer, Dordrecht. p. 3–25. doi:10.1007/978-1-4020-6061-8_1

Magarey, R.D., T.B. Sutton, and C.L. Thayer. 2005b. A simple generic infection model for foliar fungal plant pathogens. Phytopathology 95:92–100. doi:10.1094/PHYTO-95-0092

Magarey, R.D., J.W. Travis, J.M. Russo, R.C. Seem, and P.A. Magarey. 2002. Decision support systems: Quenching the thirst. Plant Dis. 86:4–14. doi:10.1094/PDIS.2002.86.1.4

Magarey, P.A., M.F. Wachtel, P.C. Weir, and R.C. Seem. 1991. A computer-based simulator for rational management of grapevine downy mildew (*Plasmopara viticola*). Plant Prot. Q. 6:29–33.

Magnien, C., D. Jacquin, N. Muckensturm, and P. Guillemard. 1991. MILVIT: A descriptive quantitative model for the asexual phase of grapevine downy mildew. Bull. OEPP 21:451–459. doi:10.1111/j.1365-2338.1991.tb01275.x

Mahlein, J., E.C. Oerke, U. Steiner, and H.W. Dehne. 2012. Recent advances in sensing plant diseases for precision crop protection. Eur. J. Plant Pathol. 133:197–209. doi:10.1007/s10658-011-9878-z

Malthus, T. 1798. An essay on the principles of population, as it affects the future improvement of society with remarks on the speculation of Mr. Godwin, M. Condorcet, and other writers, printed for J. Johnson. St. Paul's Church-Yard, London.

Marques Da Silva, J.R., C.V. Damásio, A.M. Sousa, L. Bugalho, L. Pessanha, and P. Quaresma. 2015. Agriculture pest and disease risk maps considering MSG satellite data and land surface temperature. Int. J. Appl. Earth Obs. Geoinf. 38:40–50. doi:10.1016/j.jag.2014.12.016

May, R.M. 1974. Biological populations with nonoverlapping generations: Stable points, stable cycles, and chaos. Science 186:645–647. doi:10.1126/science.186.4164.645

May, R.M. 1981. Models for single populations. In: R.M. May, editor, Theoretical ecology: Principles and applications. 2nd ed. Blackwell Science, Oxford, UK. p. 5–29.

May, R.M., and A.R. McLean. 2007. Theoretical ecology: Principles and applications. 3rd ed. Oxford Univ. Press, Oxford, UK.

McMullen, M., G. Bergstrom, E. De Wolf, R. Dill-Macky, D. Hershman, G. Shaner, and D. Van Sanford. 2012. A unified effort to fight an enemy of wheat and barley: Fusarium head blight. Plant Dis. 96:1712–1728. doi:10.1094/PDIS-03-12-0291-FE

McRoberts, N., C. Hall, L.V. Madden, and G. Hughes. 2011. Perceptions of disease risk: From social construction of subjective judgments to rational decision making. Phytopathology 101:654–665. doi:10.1094/PHYTO-04-10-0126

Merrill, S.C., T.O. Holtzer, F.B. Peairs, and P.J. Lester. 2015. Validating spatiotemporal predictions of an important pest of small grains. Pest Manag. Sci. 71:131–138. doi:10.1002/ps.3778

Mironidis, G.K., and M. Savopoulou-Soultani. 2008. Development, survivorship, and reproduction of *Helicoverpa armigera* (Lepidoptera: Noctuidae) under constant and alternating temperatures. Environ. Entomol. 37:16–28. doi:10.1603/0046-225X(2008)37[16:DSAROH]2.0.CO;2

Moutou, F., and B. Durand. 1994. Modelling the spread of foot-and-mouth disease virus. Vet. Res. 25:279–284.

Muñiz, M., and G. Nombela. 2001. Differential variation in development of the B- and Q-biotypes of *Bemisia tabaci* (Homoptera: Aleyrodidae) on sweet pepper at constant temperatures. Environ. Entomol. 30:720–727. doi:10.1603/0046-225X-30.4.720

Nietschke, B., R.D. Magarey, D.M. Borchert, D.D. Calvin, and E.M. Jones. 2007. A developmental database to support insect phenology models. Crop Prot. 26:1444–1448. doi:10.1016/j.cropro.2006.12.006

Ojiambo, P.S., G.J. Holmes, W. Britton, T. Keever, M.L. Adams, M. Babadoost, S.C. Bost, R. Boyles, M. Brooks, J. Damicone, M.A. Draper, D.S. Egel, K.L. Everts, D.M. Ferrin, A.J. Gevens, B.K. Gugino, M.K. Hausbeck, D.M. Ingram, T. Isakeit, A.P. Keinath, S.T. Koike, D. Langston, M.T. McGrath, S.A. Miller, R. Mulrooney, S. Rideout, E. Roddy, K.W. Seebold, E.J. Sikora, A. Thornton, R.L. Wick, C.A. Wyenandt, and S. Zhang. 2011. Cucurbit downy mildew ipmPIPE: A next generation web-based interactive tool for disease management and extension outreach. Plant Health Prog. doi:10.1094/PHP-2011-0411-01-RV

Orlandini, S., B. Gozzini, M. Rosa, E. Egger, P. Storchi, G. Maracchi, and F. Miglietta. 1993. PLASMO: A simulation model for control of *Plasmopara viticola* on grapevine. Bull. OEPP 23:619–626. doi:10.1111/j.1365-2338.1993.tb00559.x

Paez, T.L. 2009. Introduction to model validation. 2009 IMAC-XXVII: A Conference & Exposition on Structural Dynamics, Orlando, FL. 9–12 Feb. Society for Experimental Mechanics, Orlando, FL.

Park, E.W., R.C. Seem, D.M. Gadoury, and R.C. Pearson. 1997. DMCAST: A prediction model for grape downy mildew development. Viticulture Enology Sci. 52:182–189.

Parry, H.R., D. Eagles, and D.J. Kriticos. 2015. Simulation modelling of long-distance windborne dispersal for invasion ecology. In: R.C. Venette, editor, Pest risk modelling and mapping for invasive alien species. CABI, Wallingford, UK. p. 49–64. doi:10.1079/9781780643946.0049

Pellegrini, A., D. Prodorutti, A. Frizzi, C. Gessler, and I. Pertot. 2010. Development and evaluation of a warning model for the optimal use of copper in organic viticulture. J. Plant Pathol. 92:43–55.

Pfender, W.F. 2003. Prediction of stem rust infection favorability, by means of degree-hour wetness duration, for perennial ryegrass seed crops. Phytopathology 93:467–477. doi:10.1094/PHYTO.2003.93.4.467

Phelps, K., R.H. Collier, R.J. Reader, and S. Finch. 1993. Monte Carlo simulation method for forecasting the timing of pest insect attacks. Crop Prot. 12:335–342. doi:10.1016/0261-2194(93)90075-T

Pitblado, R.E. 1992. The development and implementation of TOM-CAST a weather timed fungicide spray program for field tomatoes. Ministry of Agriculture and Food, Ontario, Canada.

Prout, T. 1978. The joint effects of the release of sterile males and immigration of fertilized females on a density regulated population. Theor. Popul. Biol. 13:40–71. doi:10.1016/0040-5809(78)90035-7

Racca, P., T. Zeuner, J. Jung, and B. Kleinhenz. 2010. Model validation and use of geographic information systems in crop protection warning service. In: E.C. Oerke, R. Gerhards,

G. Menz, and R. Sikora, editors, Precision crop protection: The challenge and use of heterogeneity. Springer, Heidelberg. p. 259–276. doi:10.1007/978-90-481-9277-9_16

Rahman, M.T., and M. Khalequzzaman. 2004. Temperature requirements for the development and survival of rice stemborers in laboratory conditions. Insect Sci. 11:47–60. doi:10.1111/j.1744-7917.2004.tb00179.x

Régnière, J. 1984. A method of describing and using variability in development rates for the simulation of insect phenology. Can. Entomol. 116:1367–1376. doi:10.4039/Ent1161367-10

Régnière, J., J. Powell, B. Bentz, and V. Nealis. 2012. Effects of temperature on development, survival and reproduction of insects: Experimental design, data analysis and modeling. J. Insect Physiol. 58:634–647. doi:10.1016/j.jinsphys.2012.01.010

Régnière, J., R.L. Rabb, and R. Stinner. 1981. *Popillia japonica*: Simulation of temperature-dependent development of the immatures. Environ. Entomol. 10:290–296. doi:10.1093/ee/10.3.290

Ricker, W.E. 1954. Stock and recruitment. J. Fish. Res. Board Can. 11:559–623. doi:10.1139/f54-039

Rieder, R., W. Pavan, J.M.C. Maciel, J.M.C. Fernandes, and M.S. Pinho. 2014. A virtual reality system to monitor and control diseases in strawberry with drones: A project. In: D.P. Ames, N.W.T. Quinn, and A.E. Rizzoli, editors, Bold visions for environmental modelling. Proc. of the 7th Int. Congress on Environ. Modelling and Software, San Diego, CA. 15–19 June 2014. The International Environmental Modelling and Software Society, Manno, Switzerland. p. 919–926.

Rodríguez Caicedo, D., J.M. Cotes Torres, and J.R. Cure. 2012. Comparison of eight degree-days estimation methods in four agroecological regions in Colombia. (In Portuguese with English abstract.) Bragantia 71:299–307. doi:10.1590/S0006-87052012005000011

Rosa, M., B. Gozzini, S. Orlandini, and L. Seghi. 1995. A computer program to improve the control of grapevine downy mildew. Comput. Electron. Agric. 12:311–322. doi:10.1016/0168-1699(95)00007-Q

Rosenberger, D.A. 2003. Factors limiting IPM-compatibility of new disease control tactics for apples in eastern United States. Plant Health Prog. doi:10.1094/PHP-2003-0826-01-RV

Rossi, V., T. Caffi, S. Giosue, and R. Bugiani. 2008. A mechanistic model simulating primary infections of downy mildew in grapevine. Ecol. Modell. 212:480–491. doi:10.1016/j.ecolmodel.2007.10.046

Rossi, V., P. Racca, S. Giosue, D. Pancaldi, and I. Alberti. 1997. A simulation model for the development of brown rust epidemics in winter wheat. Eur. J. Plant Pathol. 103:453–465. doi:10.1023/A:1008677407661

Rowlandson, T., M. Gleason, P. Sentelhas, T. Gillespie, C. Thomas, and B. Hornbuckle. 2015. Reconsidering leaf wetness duration determination for plant disease management. Plant Dis. 99:310–319. doi:10.1094/PDIS-05-14-0529-FE

Royama, T. 1992. Analytical population dynamics. 1st ed., Chapman & Hall, London, New York. doi:10.1007/978-94-011-2916-9

Ruesink, W.G. 1976. Status of the systems approach to pest management. Annu. Rev. Entomol. 21:27–44. doi:10.1146/annurev.en.21.010176.000331

Russo, J.M. 2000. Weather forecasting for IPM. In: G.G. Kennedy and T.B. Sutton, editors, Emerging technologies for integrated pest management: Concepts, research, and implementation. American Phytopathological Society, St. Paul, MN. p. 453–473.

Russo, J.M. 2009. Epidemiology in a cyberinfrastructure world. In: D.M. Gadoury, editor, Proceedings of the 10th International Epidemiology Workshop, New York State Agricultural Experiment Station, Geneva, NY. p. 140–142.

Russo, J.M. 2015. A Framework for managing information in precision agriculture. Precision Agriculture, Meister Media, 24 Jan. 2015. http://www.precisionag.com/opinion/a-framework-for-managing-information-in-precision-ag / (accessed 17 Jan. 2017).

Russo, J.M., and J.W. Zack. 1997. Downscaling GCM output with a mesoscale model. J. Environ. Manage. 49:19–29. doi:10.1006/jema.1996.0113

Saha, S., S. Moorthi, X. Wu, J. Wang, S. Nadiga, P. Tripp, D. Behringer, Y.T. Chuang, M. Iredell, M. Ek, J. Meng, R. Yang, M.P. Mendez, H. van den Dool, Q. Zhang, W. Wang,

M. Chen, and E. Becker. 2014. The NCEP climate forecast system version 2. J. Clim. 27:2185–2208. doi:10.1175/JCLI-D-12-00823.1

Savary, S., S. Stetkiewicz, F. Brun, and L. Willocquet. 2015. Modelling and mapping potential epidemics of wheat diseases—examples on leaf rust and *Septoria tritici* blotch using EPIWHEAT. Eur. J. Plant Pathol. 142:771–790. doi:10.1007/s10658-015-0650-7

Seem, R.C., R.D. Magarey, J.W. Zack, and J.M. Russo. 2000. Estimating disease risk at the whole plant level with general circulation models. Environ. Pollut. 108:389–395. doi:10.1016/S0269-7491(99)00218-3

Shafer, M.A., C.A. Fiebrich, D.S. Arndt, S.E. Fredrickson, and T.W. Hughes. 2000. Quality assurance procedures in the Oklahoma Mesonetwork. J. Atmos. Ocean. Technol. 17:474–494. doi:10.1175/1520-0426(2000)017<0474:QAPITO>2.0.CO;2

Shaffer, P.L. 1983. Prediction of variation in development period of insects and mites reared at constant temperatures. Environ. Entomol. 12:1012–1019. doi:10.1093/ee/12.4.1012

Sharov, A.A. 1992. The life-system approach: A system paradigm in population ecology. Oikos 63:485–494. doi:10.2307/3544976

Sharpe, P.J., G.L. Curry, D.W. DeMichele, and C.L. Cole. 1977. Distribution model of organism development times. J. Theor. Biol. 66:21–38. doi:10.1016/0022-5193(77)90309-5

Sharpe, P.J., and D.W. DeMichele. 1977. Reaction kinetics of poikilotherm development. J. Theor. Biol. 64:649–670. doi:10.1016/0022-5193(77)90265-X

Sharpe, P.J., R.M. Schoolfield, and G.D. Butler, Jr. 1981. Distribution model of *Heliothis zea* (Lepidoptera: Noctuidae) development times. Can. Entomol. 113:845–856. doi:10.4039/Ent113845-9

Shoemaker, C. 1981. Applications of dynamic programming and other optimization methods in pest management. IEEE Trans. Automat. Contr. 26:1125–1132. doi:10.1109/TAC.1981.1102782

Shtienberg, D. 2013. Will decision-support systems be widely used for the management of plant diseases? Annu. Rev. Phytopathol. 51:1–16. doi:10.1146/annurev-phyto-082712-102244

Smith, D.L., J.E. Hollowell, T.G. Isleib, and B.B. Shew. 2007. A site-specific, weather-based disease regression model for Sclerotinia blight of peanut. Plant Dis. 91:1436–1444. doi:10.1094/PDIS-91-11-1436

Sporleder, M., and J. Kroschel. 2008. The potato tuber moth granulovirus (*Po*GV): Use limitations and possibilities for field applications. In: J. Kroschel and L.A. Lawrence, editors, Integrated pest management for the potato tuber moth, *Phthorimaea operculella* (Zeller). Margraf, Weikersheim. p. 49–72.

Sporleder, M., J. Kroschel, M.R. Gutierrez Quispe, and A. Lagnaoui. 2004. A temperature-based simulation model for the potato tuberworm, *Phthorimaea operculella Zeller* (Lepidoptera; Gelechiidae). Environ. Entomol. 33:477–486. doi:10.1603/0046-225X-33.3.477

Sporleder, M., R. Simon, H. Juarez, and J. Kroschel. 2008. Regional and seasonal forecasting of the potato tuber moth using a temperature-driven phenology model linked with geographic information systems. In: J. Kroschel and L.A. Lawrence, editors, Integrated pest management for the potato tuber moth, *Phthorimaea operculella* (Zeller). Margraf, Weikersheim. p. 15–30.

Sporleder, M., H.E.Z. Tonnang, P. Carhuapoma, J.C. Gonzales, H. Juarez, and J. Kroschel. 2013. Insect Life Cycle Modelling (ILCYM) software: A new tool for regional and global insect pest risk assessments under current and future climate change scenarios. In: J.E. Peña, editor, Potential invasive pests of agricultural crops. CABI, Wallingford, UK. p. 412–427. doi:10.1079/9781845938291.0412

Steinmaus, S.J., T.S. Prather, and J.S. Holt. 2000. Estimation of base temperatures for nine weed species. J. Exp. Bot. 51:275–286. doi:10.1093/jexbot/51.343.275

Stinner, R.E., J.S. Bacheler, C. Tuttle, and G.D. Butler, Jr. 1975. Simulation of temperature-dependent development in population dynamics models. Can. Entomol. 107:1167–1174. doi:10.4039/Ent1071167-11

Stinner, R.E., A.P. Gutierrez, and G.D. Butler, Jr. 1974. An algorithm for temperature-dependent growth rate simulation. Can. Entomol. 106:519–524. doi:10.4039/Ent106519-5

Sutherst, R.W., and G.F. Maywald. 1985. A computerised system for matching climates in ecology. Agric. Ecosyst. Environ. 13:281–299. doi:10.1016/0167-8809(85)90016-7

Teng, P.S. 1985. A comparison of simulation approaches to epidemic modeling. Annu. Rev. Phytopathol. 23:351–379. doi:10.1146/annurev.py.23.090185.002031

Tixier, P., J.M. Risede, M. Dorei, and E. Malezieux. 2006. Modelling population dynamics of banana plant-parasitic nematodes: A contribution to the design of sustainable cropping systems. Ecol. Modell. 198:321–331. doi:10.1016/j.ecolmodel.2006.05.003

Tonnang, E.Z.H., H. Juarez H, P. Carhuapoma, J.C. Gonzales, D. Medoza, M. Sporleder, R. Simon, and J. Kroschel. 2013. ILCYM: Insect Life Cycle Modeling. A software package for developing temperature-based insect phenology models with applications for regional and global analysis of insect population and mapping (user guide). Int. Potato Center, Lima, Peru.

Trnka, M., F. Muška, D. Semerádová, M. Dubrovský, E. Kocmánková, and Z. Žalud. 2007. European corn borer life stage model: Regional estimates of pest development and spatial distribution under present and future climate. Ecol. Modell. 207:61–84. doi:10.1016/j.ecolmodel.2007.04.014

Tromp, S.W. 1980. Biometeorology: The impact of the weather and climate on humans and their environment (animals and plants). Heyden & Son Ltd., London.

Twengström, E., R. Sigvald, C. Svensson, and J. Yuen. 1998. Forecasting Sclerotinia stem rot in spring sown oilseed rape. Crop Prot. 17:405–411. doi:10.1016/S0261-2194(98)00035-0

van Heemst, H.D.J. 1986. Physiological principles. In: H. van Keulen and J. Wolf, editors, Modelling of agricultural production: Weather, soils and crops. Pudoc, Wageningen, Netherlands. p. 13–26.

Venette, R.C., D.J. Kriticos, R.D. Magarey, F.H. Koch, R.H.A. Baker, S.P. Worner, G. Raboteaux, N. Nadilia, D.W. McKenney, E.J. Dobesberger, D. Yemshanov, P.J. De Barro, W.D. Hutchison, G. Fowler, T.M. Kalaris, and J. Pedlar. 2010. Pest risk maps for invasive alien species: A roadmap for improvement. Bioscience 60:349–362. doi:10.1525/bio.2010.60.5.5

Verhulst, P.F. 1838. Instructions on the law that the population follows in its growth. (In French.) Correspondances Mathématique et Physique 10:113–121.

Verhulst, P.F. 1845. Mathematical researches into the law of population growth increase. (In French.) Nouveaux Mémoires de l'Académie Royale des Sciences et Belles-Lettres de Bruxelles 18:1–42.

Volterra, V. 1926. Variations and fluctuations in the number of individuals in cohabiting animal species. (In Italian) Mem. Acad. Lincei Roma 2:31–113.

Wagner, T.L., H.I. Wu, R.M. Feldman, P.J.H. Sharpe, and R.N. Coulson. 1985. Multiple-cohort approach for simulating development of insect populations under variable temperatures. Ann. Entomol. Soc. Am. 78:691–704. doi:10.1093/aesa/78.6.691

Wagner, T.L., H.I. Wu, P.J.H. Sharpe, R.M. Schoolfield, and R.N. Coulson. 1984a. Modeling insect development rates: A literature review and application of a biophysical model. Ann. Entomol. Soc. Am. 77:208–220. doi:10.1093/aesa/77.2.208

Wagner, T.L., H.I. Wu, P.J.H. Sharpe, and R.N. Coulson. 1984b. Modeling distributions of insect development time: A literature review and application of the Weibull function. Ann. Entomol. Soc. Am. 77:475–483. doi:10.1093/aesa/77.5.475

Welch, S.M., B.A. Croft, J.F. Brunner, and M.F. Michels. 1978. PETE: An extension phenology modeling system for management of multi-species pest complex. Environ. Entomol. 7:487–494. doi:10.1093/ee/7.4.487

Welch, S., B. Croft, and M. Michels. 1981. Validation of pest management models. Environ. Entomol. 10:425–432. doi:10.1093/ee/10.4.425

Wilks, D.S. 1995. Statistical methods in the atmospheric sciences. Academic Press, New York.

Willocquet, L., S. Savary, L. Fernandez, F.A. Elazegui, N. Castilla, D. Zhu, Q. Tang, S. Huang, X. Lin, H.M. Singh, and R.K. Srivastsava. 2002. Structure and validation of RICEPEST, a production situation-driven, crop growth model simulating rice yield response to multiple pest injuries for tropical Asia. Ecol. Modell. 153:247–268. doi:10.1016/S0304-3800(02)00014-5

World Meteorological Organization Commission of Agricultural Meteorology. 1988. Agrometeorological aspects of operational crop protection. Tech. note 192, WMO. Geneva, Switzerland.

Worner, S.P. 1992. Performance of phenological models under variable temperature regimes: Consequences of the Kaufmann or rate summation effect. Environ. Entomol. 21:689–699. doi:10.1093/ee/21.4.689

Xu, X.M. 1999. Modelling and forecasting epidemics of apple powdery mildew (*Podosphaera leucotricha*). Plant Pathol. 48:462–471. doi:10.1046/j.1365-3059.1999.00371.x

Zadoks, J.C., and R. Rabbinge. 1985. Modelling to a purpose. In: C.A. Gilligan, editor, Advances in plant pathology. Vol. 3. Mathematical modelling of crop diseases. Academic Press, London. p. 231–244.

Measurement Techniques

Allan Howard,* Aston Chipanshi, Andrew Davidson, Raymond Desjardins, Andrii Kolotii, Nataliia Kussul, Heather McNairn, Sergii Skakun, and Andrii Shelestov

Abstract

Measurement techniques as they apply to agroclimate look beyond conventional instrumentation and methodologies, which are used to derive new data from direct observations of specific agrometeorological variables. They address the integration of the meteorological, hydrologic and biophysical variables critical for understanding the processes governing agricultural production and the agricultural interaction with the environment. Agroclimate measurement techniques also consider the temporal and spatial scales relevant to agriculture.

Soil moisture is a key variable for crop productivity, crop management practices, flood and excess moisture risk, and can control greenhouse gas emissions from farming operations. Crop condition and drought monitoring practices have been used as early warning for production and food security issues. Greenhouse gas flux is a critical indicator of the degree to which agriculture is either a source or a sink for greenhouse gases. Within each of these areas are multiple state of the art operational or near-operational techniques for measurement of indices and elements that pertain to spatial and temporal scales that are important for agriculture.

In an era of rapidly increasing availability of data, there are opportunities to better describe and measure the complexities of interactions influencing agricultural productivity and

Abbreviations: AVHRR, Advanced Very High Resolution Radiometer; CAR, Crop Census Agricultural Regions; E, Soil dialectic permittivity; EO, Earth Observation; ESI, Evaporative Stress Index; ESU, Elementary Sampling Unit; FAPAR, fraction of absorbed photosynthetically active radiation; FCOVER, fraction of vegetation cover; FDR, Frequency Domain Reflectometry; GHG, greenhouse gases; GRACE, Gravity Recovery and Climate Experiment; ICCYF, Integrated Canadian Crop Yield Forecaster; IEM, integral equation model; LAI, Leaf Area Index; LUT, Look Up Table; MODIS, Moderate resolution Imaging Spectroradiometer; NDVI, Normalized Difference Vegetation Index; RCI, Rapid Change Index; SAR, synthetic aperture radar; SMAP, L-band Soil Moisture Active Passive; SPI , Standardized Precipitation Index; TDR, Time Domain Reflectometry; T_B, Brightness Temperature; VegDRI, Vegetation Drought Response Index; VI, Vegetation Indices; VWC, Vegetation Water Content; s_o, backscatter; τ, optical depth.

A. Howard and A. Chipanshi, Agriculture and Agri-Food Canada, Science and Technology Branch, Regina, SK, Canada; A. Davison, R. Desjardins, and Heather McNairn, Agriculture and Agri-Food Canada, Science and Technology Branch, Ottawa, ON, Canada; A. Kolotii and A. Shelestov, National University of Life and Environmental Sciences of Ukraine, Kyiv, Ukraine; A. Kolotii, N. Kussul, S. Skakun, and A. Shelestov, Space Research Institute National Academy of Sciences and State Space Agency of Ukraine, Kyiv, Ukraine; A. Kolotii, N. Kussul, and A. Shelestov, National Technical University of Ukraine, Kyiv Polytechnic Institute, Kyiv, Ukraine. Received 7 Jan. 2016. Accepted 7 Jan. 2016. *Corresponding author (allan.howard@agr.gc.ca)

doi:10.2134/agronmonogr60.2014.0056.5

© ASA, CSSA, and SSSA, 5585 Guilford Road, Madison, WI 53711, USA.
Agroclimatology: Linking Agriculture to Climate, Agronomy Monograph 60.
Jerry L. Hatfield, Mannava V.K. Sivakumar, John H. Prueger, editors.

the agrienvironmental footprint. Measurement techniques are increasingly relying on sophisticated modeling and analysis techniques that integrate data from several sources to derive new information at the temporal and spatial scales required to support the agriculture sector's needs for science-based early warning and decision support.

Measurement of agrometeorological variables has been comprehensively covered (Hatfield and Baker, 2005; World Meteorological Organization, 2008; Petropoulos, 2014). These books cover standard operational procedures and equipment, siting and data recording. Agroclimatology focuses on the interaction of atmospheric, soil and biological factors that influence production of agricultural commodities and the environmental footprint associated with agricultural production. Measurement techniques for agroclimate must consider the integration of those meteorological, hydrologic and biophysical variables that are critical for agricultural production and the agricultural interaction with the environment, and do so at temporal and spatial scales relevant to agriculture.

The purpose of this chapter is to identify key state of the art operational or near-operational techniques and discuss them at scales that are relevant to agricultural production. The chapter focuses on three main types of techniques that are critical for determination of agricultural productivity and the agri-environmental footprint. Soil moisture is a key variable for crop productivity, crop management practices, flood and excess moisture risk, and is a controlling factor in greenhouse gas emissions from farming operations. Crop condition and drought monitoring techniques have been used as early warning for production and food security issues. Greenhouse gas flux is a critical indicator of the degree to which agriculture is either a source or a sink for greenhouse gases. This has implications for the sustainability of agriculture. Discussion is supported with case studies that apply the concepts and techniques toward measurement of each of these elements.

Soil Moisture

Soil moisture is a critical variable for several agrienvironmental and agricultural productivity factors. Stored moisture in the soil is a direct supply of water to the crop, and supplements growing season precipitation. It impacts latent heat fluxes that control the crop microclimate. It controls several biochemical processes related to pests, disease, rootzone oxygen exchange and fertility and is a controlling factor in runoff and groundwater recharge. In drier areas, where potential evapotranspiration exceeds precipitation, knowledge of spring soil moisture is a key element in determining several management practices, such as rates of fertilizer application, crop seeding rate and in some cases, crop selection. It is a key factor in assessment of drought risk and for establishing boundary conditions in weather forecast modeling.

Soil moisture is highly dependent on soil properties, landscape conditions, precipitation variability, land cover and freeze-thaw conditions. Consequently, soil moisture is highly spatially variable in most agricultural fields and cost effective means for determining soil moisture at field scales and depths relevant to agriculture remains a challenge.

This section covers soil moisture measurement techniques that are available over a range of scales. Typically in situ measurements offer the greatest spatial precision as they can be taken from direct contact with the soil and sampling or instrumentation can be placed at specified depths. However the main limitation of in situ measurement is that the pixel size for a measurement is typically a volume of a few cubic centimeters and therefore they are difficult to extrapolate to field scale without extensive replication and cost. A limited number of intermediate scale surface observation techniques are available that offer the potential to assist in scaling up to field scale. Data from remote sensing techniques are relatively inexpensive and available at time intervals suitable for monitoring, but are available at scales coarse enough that a pixel represents several square kilometers.

In Situ Measurements

A wide variety of nondestructive or minimally destructive techniques have been developed over the past forty years that can be considered for use. Most are well-adapted to multiple measurements and/or monitoring of soil moisture at one location over a period of several seasons. All come with a variety of strengths and limitations. Recent comprehensive reviews of soil moisture measurement techniques (Robinson et al., 2008; Dobriyal et al., 2012; Romano, 2014) show that there is no preferred method for determination of in situ soil moisture and that sensor technology is continually developing. A summary of some common techniques is presented in Table 1.

Electromagnetic techniques offer the most promising means of measurement of soil moisture because this category contains a range of techniques that measure the same soil water content proxy, the bulk soil dielectric permittivity (E), at scales ranging from localized in situ sensors, to intermediate (field) scale and remote sensing techniques (Huisman et al., 2003). In situ sensors in this category are typically well-suited to wireless observation networks where automated observations can be taken at temporal resolutions as fine as is required (e.g., from one measurement every several hours to measurements at subminute frequencies) for almost any application.

As the dielectric permittivity of water in the liquid phase (E ~ 80) is considerably larger than that of the soil matrix (E ~ 4 to 5) and air (E ~ 1), soil moisture has a dominant influence on soil dielectric permittivity. The soil permittivity is strongly determined by the content of water in the liquid phase in the soil. Because the dielectric permittivity of water in the frozen state drops to values comparable to the soil matrix, freezing and thawing of soil water can be determined by the appearance of a sudden "drying" of the soil when frozen and a sudden spike in the apparent water content during thawing.

The relationship between dielectric permittivity and soil is complex. When subjected to an alternating electrical field, preferred molecular alignment of the water molecules in the soil to that field requires the application of sufficient energy to overcome the random movement from thermal motion. The alignment process stores electrical energy, which becomes evident as dielectric permittivity. Dielectric permittivity is composed of two parts: (i) the real permittivity, or that stored energy that overcomes the random movement of the molecules, and (ii) the imaginary permittivity, or the influence of the ionic makeup of the soil solution that acts to acts to dissipate the stored energy. The imaginary permittivity is referred to as dielectric loss (Robinson et al., 2003).

Dielectric loss can be attributed to two main processes: (i) electrical conduction and (ii) molecular relaxation (Robinson et al., 2003). Soil properties that enhance electrical conduction include salinity and exchangeable cations, whereas properties that influence molecular relaxation of the soil water are often associated with strong interactions between the soil surface and the solution (Seyfried et al., 2005). Dielectric loss can greatly complicate the calibration of sensors that use electromagnetic principles. A more detailed explanation of the relationship

Table 1. A summary of some common in-situ soil moisture measurement techniques.

Method	Mechanism	Strengths	Limitations
Acoustic Wave Oscillation (Meisami-asl et al., 2012)	Some properties of sound waves, including sweep frequencies (10 to 300MHz) and multiple tone sound have been correlated to soil moisture.	Less influence from soil properties and the potential for sampling larger volumes of soil that most sensors.	Still in an experimental stage.
Capacitance Sensors (Bogena et al., 2007; Dean et al., 1987)	Bulk permittivity of the soil is measured by an oscillating current. The magnitude of the resonant frequency is a function of soil moisture content.	Several types available. Geometry of sensors is adaptable for boreholes. Some are low cost; most require low maintenance.	Specific calibration of the soil is required. Susceptible to interference by bulk electrical conductivity and temperature.
Electrical Conductivity (EC) Sensors (Stenitzer, 1993)	EC of the soil solution is calibrated to soil moisture. Porous median such as gypsum blocks are placed in contact with the soil.	Suitable for continuous monitoring. Soil matric potential measured.	Does not measure volumetric soil moisture. Effective only in the wetter ranges of soil moisture.
Frequency Domain Reflectrometry (Robock et al., 2000)	Bulk permittivity measured by reflected electromagnetic pulse reaching a set voltage (operate at 0.10 to 0.25 GHz)	Similar to Time Domain Reflectometry but with lower frequency and faster response time	Influence of clay dispersion on permittivity therefore unique soil calibration required. Well suited to remote automated data collection.
Gravimetric measurement (Gardner, 1965)	A soil sample is removed from the ground weighed in the moist state and then dried at 105°C to a constant weight.	Long history of use. It enables the determination of soil properties within the landscape when sampling for moisture.	The soil is disturbed. It is labor intensive and not well suited to monitoring. The soil bulk density is required to establish volumetric moisture content.
Ground Penetrating Radar (Huisman et al., 2002, 2003; Tran et al., 2015)	Bulk permittivity of the soil is measured using high frequency electromagnetic waves.	Suitable for use at field scales by moving the equipment across the soil using a sled or ATV taking multiple measurements.	Signals are complex to interpret, and complicated by surface roughness, salinity and variations in soil stratigraphy.
Neutron Moisture Meter (Chanasyk and Naeth, 1996)	Release a pulse of fast neutrons and count slow (thermalized) returned neutrons that become thermalized when in contact with hydrogen atoms	Reliable technology with a long history of use. Suitable for deep boreholes for monitoring deep rooted crops (e.g. alfalfa)	Requires use of radioactive materials that are restricted and require licensing, Not well-suited to automated data collection.
Tensiometers (Schmugge et al., 1980)	Measure soil water tension through negative pressure on water filled tube.	Direct measurement of soil matric potential	Volumetric soil moisture not measured, useful only in the wetter ranges of soil moisture.
Thermal Dispersion (Matile et al., 2013)	Release a pulse of heat and measure change in temperature in surrounding soil, with heat transmission closely related moisture	Alternative to electromagnetic methods	Sensitive to significant fraction of organic matter or coarse gravel
Time Domain Reflectrometry (Topp et al., 1980)	Bulk permittivity of the soil is measured by time for EM pulse (frequency > 0.5 GHz) to travel along buried waveguide	Nondestructive and accurate. Less susceptible to interference by bulk electrical conductivity	Soils with high cation exchange capacity and organic soils require specific calibration.

between soil properties, frequency and dielectric permittivity can be found in Robinson et al. (2003), Seyfried et al. (2005) and Chen and Or (2006).

Sensors using dielectric properties to estimate soil moisture content in situ are typically in the form of probes that can be installed directly into the soil. They have been generally referred to as impedance probes (Cosh, 2005; Ojo et al., 2015). These probes measure the response of electromagnetic waves propagated along a coaxial cable to the bulk soil dielectric permittivity. Each probe has a set of parallel rods that are inserted in an undisturbed soil. The soil water acts as a resistance (impedance) that reflects the wave or a portion of it back to the source. These systems of measurement are nondestructive, suited to automated measurement and data collection, and accurate to within ± 1% when the soil water is in the liquid phase over the range of field moisture conditions; although, measurements during the frozen state are not accurate without a specific calibration for that purpose.

There are two main methods of measuring the wave response. Time domain reflectometry (TDR) is based on the relationship between the travel time of the wave and the length of the rods in the sensor (Topp et al., 1980). Frequency domain reflectometry (FDR) probes use variations in the frequency of the signal resulting from the soil permittivity to estimate soil moisture content (Dobriyal et al., 2012). The higher frequencies of the TDR reduce the sensitivity of the response to soils properties such as salinity, texture or temperature (Robinson et al., 2008). FDR probes operate at lower frequencies (e.g., 50 to 150 MHz compared to over 1000 MHz for TDR probes) and are therefore more susceptible to influence from these soil properties. Some FDR probe models come with hardware and software to independently measure temperature and electrical conductivity, thus enabling them to more accurately determine the influence from soil properties and salinity, which comprise the dielectric loss or imaginary permittivity component.

Frequency domain reflectometry–type probes are popular because of their lower cost, in terms of both capital investment and time. Their lower power consumption requirements coupled with the ability of the probes to be multiplexed with dataloggers make them attractive for remote monitoring.

Impedance probes are the standard soil moisture sensor for the Canadian Real-time In situ Soil Monitoring for Agriculture (RISMA) network (Adams et al., 2015), and the United States national cooperative network, the United States Department of Agriculture Natural Resources Conservation Service (USDA-NRCS), and Soil Climate Analysis Network (SCAN) (Schafer et al., 2007). The publicly available data from these networks contain surface meteorological and soil moisture data at hourly resolution. Data from SCAN were correlated with (i) satellite-based active and passive microwave signatures in an agricultural landscape (Nghiem et al., 2012), (ii) in validation of the Variable Infiltration Capacity (VIC), Decision Support System for Agrotechnology Transfer (DSSAT), and Climatic Water Budget (CWB) models (Meng and Quiring, 2008) and (iii) in validation of drought indicators derived from water storage data from the Gravity Recovery and Climate Experiment (GRACE) satellites (Houborg et al., 2012). As explained below in "Drought Monitoring", the GRACE-derived drought indicators were particularly useful as proxy for the sparse availability of ground-based observations of soil moisture and groundwater for drought monitoring (Houborg et al., 2012).

Intermediate Scale Measurements

Cosmic Ray Techniques

Determination of field soil moisture by the use of sensors to detect the intensity of passive neutrons generated by the interaction of cosmic rays with terrestrial atoms has been discussed in Zreda et al. (2008). The ratio of high energy neutrons to low energy neutrons above the landscape surface is inversely correlated with the number of hydrogen (H) atoms in the soil and therefore can be related to area averaged soil moisture content. The method is relatively insensitive to variations in soil chemical properties, although sensitivity to variations in organic matter within the soil or to vegetative growth on the landscape merits consideration. The cosmic ray probe is mounted above the soil surface and measures the flux of high energy (fast) neutrons.

The Cosmic ray Soil Moisture Observing System (COSMOS) is a continental-scale network consisting of instruments designed to improve the availability of continental-scale soil moisture measurements by ultimately deploying a network of 500 cosmic ray probes across the United States (Zreda et al., 2012).

One potential source of uncertainty is the influence of the H content of biomass. The largest and most variable pool of H is from soil moisture. The H content of biomass is considered to be relatively constant however the contribution from variation in atmospheric humidity is significant enough that a correction factor is recommended (Zreda et al., 2012). Franz et al. (2013) was able to separate the contribution from soil moisture to estimate biomass water equivalent in a pine forest and in a maize field; however, they acknowledge that uncertainties arose from several factors including humidity and assumptions used in determining the forest component of the landscape. A dry bias in cosmic ray derived near-surface soil moisture data in a mixed forest following snowmelt has also been observed (Lv et al., 2014). Snowcover over 6 cm deep or the presence of surface water can make determination of soil moisture impossible (Zreda et al., 2012).

One significant feature of the cosmic ray method is that the sampling footprint at sea level can be 300 m in radius (Baatz et al., 2014). The sensitivity to thermalized neutrons attenuates with moisture content and depth, and the depth of measurement varies from 12 cm in wet soils to 70 cm in dry soils (Franz et al., 2013).

The cosmic ray techniques offer an opportunity to measure a broader footprint of soil moisture that is more representative of field scale, and therefore offers an intermediary scale between the point measurements of the previously mentioned techniques and the broader-scale remote sensing techniques.

Remote Sensing Measurements

Remote sensing measures the amount of radiation emitted, reflected and transmitted by a target. Sensors record this energy in one or more electromagnetic frequencies and through modeling the power of the detected energy can be related to a target parameter, such as soil moisture. The bulk soil permittivity (E) can be detected by sensors when the land surface is subjected to applied electromagnetic fields at microwave frequencies (wavelengths of 1 to100 cm). Soils with higher moisture have greater reflectivity, as the power of the energy reflected (and by reciprocity emitted) is related to the dielectric permittivity through the Fresnel equations (Ulaby et al., 1986).

For time-sensitive applications, microwave sensors have a distinct advantage over optical sensors operating at shorter visible-infrared wavelengths. At longer microwave wavelengths atmospheric contributions to emission and scattering are minimal, enabling the collection of data even in the presence of clouds and haze.

Microwave remote sensing can be applied either passively or actively, both with advantages and disadvantages. Tables 2 and 3 present passive and active satellites suitable for estimating soil moisture. Whether passive or active approaches are used, these sensors measure moisture in only the near surface volume (top few centimeters). The depth of sensing (penetration depth) is not set, but is dependent primarily on the frequency and incident angle of the sensor, and on the soil wetness. This

Table 2. Specifications of Selected Space-borne Passive Radiometers.

	Special Senor Microwave/ Imager (SSM/I)	Soil Moisture and Ocean Salinity (SMOS)	Advanced Microwave Scanning Radiometer 2 (AMSR-2)	Soil Moisture Active Passive (SMAP)
Country	United States	European Space Agency	Japan	United States
Launch Date	1987	2009	2012	2015
Frequencies (GHz)	19.3, 22.2, 37.0, and 85.5	1.4	6.9, 7.3, 10.6, 18.7, 23.8, 36.5, 89.0	1.4
Approximate Ground Resolution (km)	37 by 28 (37 GHz) 15 by 13 (85.5 GHz)	35 to 50	62 by 35 (6.9 GHz) 5 by 3 (89.0 GHz)	40
Swath (km)	1400	1000	1450	1000

Table 3. Specifications of Selected Space-borne Synthetic Aperture Radars.

	TerraSAR-X	RADARSAT-2	Radar Imaging Satellite (RiSAT)	Advanced Land Observing Satellite 2 (ALOS-2) PALSAR	Sentinel-1
Country	Germany	Canada	India	Japan	European Space Agency
Launch Date	2007 2010 (TanDEM-X)	2007	2012	2014	2014
Band (wavelength-cm)	X (3.1)	C (5.6)	C (5.6)	L (22.9)	C (5.6)
Frequency (GHz)	9.7	5.4	5.35	1.2	5.4
Approximate Ground Resolution (m)	1 to 16	3 to 100	2 to 50	1 to 100	5 to 100
Nominal Swath Width (km)	1 to 100	10 to 500	10 to 240	25 to 490	20 to 400
Exact Repeat Cycle (days)	11	24	25	14	12

depth is a fraction of the incident wavelength and is estimated between one-tenth (modeled) to one-quarter (measured in the field) of a wavelength (Jackson, 2002).

Measurement by Passive Microwave Techniques

Passive radiometers detect microwave energy naturally emitted by the Earth. The magnitude of emitted energy at microwave frequencies is quite small and thus, radiometers must integrate over large footprints to record a strong enough signal relative to background and system noise (Jensen, 2007). Hence space-based radiometers have very coarse resolutions, on the order of tens of kilometers. Passive microwave satellites image very large swaths and thus provide soil moisture products at regional and national scales at relatively frequent temporal intervals (1–2 d at high latitudes and 3 d at the Equator) (Pacheco et al., 2015).

Passive radiometers record responses as brightness temperature (T_B). T_B is a function of the emissivity (e) and physical temperature (T) of the soil ($T_B = eT$). Soils with higher moisture content have lower emissivity and accordingly, lower T_B. If present, vegetation attenuates soil emissions and contributes to its own microwave emissions, complicating soil moisture retrieval (Jackson, 2002). Attenuation is characterized by the optical depth (τ) which is empirically related to the vegetation water content (VWC); τ is vegetation-type specific (Elachi and van Zyl, 2006). Soil roughness also affects the T_B. Roughness increases surface area and emissivity. In almost all cases, approaches to retrieve soil moisture from T_B use an approximation of the radiative transfer equation known as the tau-omega (τ- ω) model, where ω is the single scattering albedo (Mladenova et al., 2014). In the absence of vegetation, estimating emissivity is easily accomplished using radiometer-measured T_B and a measure of temperature. When a canopy is present, τ must be estimated from a measure of VWC to adjust for attenuation effects. While T_B is provided by one radiometer polarization, measures of temperature, VWC and roughness are determined from ancillary sources (i.e., single channel approach where for example, VWC is estimated from the optically-derived Normalized Difference Vegetation Index) or from a second polarization on the same radiometer (dual channel approach) (Mladenova et al., 2014). Radiometers measure a large dynamic range in brightness temperature. At L-band (1.4 GHz) T_B decreases by ~70 K from dry to saturated soils (Elachi and van Zyl, 2006). Considering this sensitivity, T_B can be inverted to estimate soil moisture at accuracies of about 0.04 g cm^{-3} when vegetation present has a VWC less than 5 kg m^{-2} (Elachi and van Zyl, 2006).

Measurement by Active Microwave Techniques

In contrast, active microwave sensors (Synthetic Aperture Radars or SARs) generate their own energy, propagating pulses of microwaves and detecting the power of the energy scattered backto the sensor. Spatial resolutions of SARs are much finer, on the order of meters, relative to passive sensors. However, the width of swaths imaged by SARs is much smaller and therefore more overpasses are required to provide the same spatial coverage as passive sensors. With a smaller swath, a SAR satellite re-images a specific area less frequently. Constellations (such as the proposed RADARSAT Constellation) are needed to achieve an equivalent temporal frequency and spatial coverage as that of passive systems.

Active sensors measure the power of energy scattered back to the sensor (backscatter (s^o)) proportionate to the power propagated by the radar. This two-way transmission results in complex scattering and a more challenging soil moisture retrieval. Backscatter is highly sensitive to the incident angle of the transmitted wave and polarization of the transmitted and received wave. The geometry (soil and vegetation) affects scattering behavior (single, double or multiple scattering), while dielectric properties affect the scattering power (Dobson and Ulaby, 1998). Both higher soil moisture content and rougher soils lead to greater scattering. As such, retrieval approaches must model both dielectric and roughness contributions to s^o. The Integral Equation Model (IEM) is physically based and integrates the small perturbation, geometric and physical optics models (Fung and Chen, 1992). The IEM is appropriate for a wide range of moisture and roughness conditions. Inversion of the model is complex and thus Look Up Table (LUT) approaches have been used, yielding soil moisture errors of about 0.04 g cm^{-3} when two incident angles and polarizations are exploited (Merzouki and McNairn, 2015). Semi-empirical models such as the Oh model and the Dubois model simplify the scattering problem, primarily by reducing roughness parameters (Oh et al., 1992; Dubois et al., 1995). Accuracies with these models, for non-vegetated soils, are reported in the range of 0.04 g cm^{-3} (Dubois model) and 0.08 g cm^{-3} (Oh model) (Merzouki et al., 2011). Vegetation creates multiple two-way scattering, greatly complicating soil moisture retrieval. The semi-empirical Water Cloud Model (WCM) represents the backscatter power as the incoherent sum of contributions from vegetation (s^o_{veg}) and soil (s^o_{soil}) (Attema and Ulaby, 1978). However using C-band data and the WCM, Jiao et al. (2011) found limited sensitivity to soil moisture under established canopies. L-band microwaves penetrate deeper into the canopy. The L-band Soil Moisture Active Passive (SMAP) satellite will estimate soil moisture under vegetation by inverting 3-dimensional crop specific LUTs of complex forward radar models. Prelaunch validation using L-band airborne data yielded retrieval accuracies from 0.037 to 0.086 g m^{-3} depending on crop type (McNairn et al., 2015).

Crop Condition and Drought Monitoring

Crop condition monitoring refers to repeated measurement and reporting of the changing growth and development aspects of crops and pastures during the growing season. Regional and national scale crop monitoring and reporting are increasingly based on satellite based optical sensors such as the Moderate resolution Imaging Spectroradiometer (MODIS). Optical sensors have an advantage over microwave sensors for crop condition applications by sensing those wavelengths reflected as a result of plant biophysical processes. However these wavelengths are obscured by cloud cover, limiting the time available for measurement. MODIS data at the 250-meter resolution is supported by weather data obtained from land-based climate stations or satellite platforms. A reporting time frame such as weekly or biweekly is chosen to assess growth and development elements.

Apart from providing vital scientific data on plant growth and development, crop condition monitoring is driven by (i) the increasing societal awareness and the need to know the adverse impacts of the environment on the food production systems, (ii) information demand from producers, grain traders, and government policymakers as well the agricultural industry as a whole to assist their decision making, and (iii) concerns about the future global food insecurity and the attendant social problems. The proliferation of satellite based sensors with global coverage has made data suitable for crop condition assessment data widely

Table 4. Examples of crop condition monitoring activities and products.

Country	Agency	Product	More information
Australia	Australian Bureau of Agricultural and Resource Economics and Science (ABARES)	National commodity forecasts	
	Queensland Alliance for Agriculture and Food Innovation (QAAFI), and Department of Agriculture and Food of Western Australia (DAFWA)	State and shire commodity forecasts	Nikolova et al. (2012)
Canada	Statistics Canada	Crop Condition Assessment Program (CCAP)	Reichert and Caissy (2002)
	Agriculture and Agri-Food Canada	Canadian Crop Yield Forecaster	Chipanshi et al. (2012)
China	Institute of Remote Sensing and Digital Earth (RADI) at the Chinese Academy of Sciences	China Crop Watch	Wu et al. (2014)
Europe	Joint Research Centre (JRC) of European Commission	Monitoring Agriculture with Remote Sensing (MARS) Crop Yield Forecasting System	Joint Research Centre (2012)
United States	US Department of Agriculture National Agricultural Statistics Service (NASS)	Cropscape	Han et al. (2012)
	NASS in collaboration with the Joint Agricultural Weather Facility (JAWF) of USDA and NOAA	World Agricultural Supply and Demand Estimates Report (WASDE)	USDA (2012)
Global	U.S. Agency for International Development (USAID)	Famine Early Warning Systems Network (FEWS)	http://www.fews.net
	United Nations Food and Agriculture Organization (FAO)	Global Information and Early Warning System (GIEWS)	http://www.fao.org/giews
	Group on Earth Observation (GEO)	GEO Global Agricultural Monitoring (GEOGLAM)	http://geoglam-crop-monitor.org/
	GEO Joint Experiment on Crop Assessment and Monitoring (JECAM)	JECAM Annual Progress Report	http://www.jecam.org

available at frequencies and resolutions suitable for crop monitoring, inexpensive and reliable. As a result, many countries have developed crop monitoring systems from satellite-based data supported by ground-based data, or collectively Earth Observation (EO) data. Examples of such systems are presented in Table 4.

Vegetation Indices

Remotely-sensed data collection has the potential to provide quantitative information on the amount, condition, and type of vegetation, provided that the effects of physical and physiological processes on the spectral characteristics of canopies are fully understood.

One of the greatest challenges in the remote sensing of agricultural systems has been the reliable estimation of biophysical variables (such as aboveground biomass, net primary productivity and yield) from satellite platforms. This is largely a consequence of the "mixed pixel" problem, where factors other than the presence and amount of green vegetation (e.g., senescent vegetation, soil, shadow) combine to form composite spectra (Asner, 1998; Asner et al., 1998; Fourty et al., 1996; Goel, 1988; Myeni et al., 1989; Ross, 1981). Spectral mixing often makes the discrimination of green vegetation difficult and has prompted the development of numerous spectral vegetation indices (VIs). VIs are dimensionless radiometric measures that combine two or more spectral bands to enhance the vegetative signal, while simultaneously minimizing background effects. Vegetation indices are one of the most widely used remote sensing measurements, and thus, many exist. The most common VIs utilize red (*R*) green (*G*) blue (*B*), near-infrared (NIR) and/or shortwave infrared (SWIR) canopy reflectance. The indices are described in Table 5. Although many indices are well correlated with various plant biophysical parameters some, such as the Normalized Difference Vegetation Index, have received more attention than others.

The Normalized Difference Vegetation Index (NDVI) can be calculated from the red and near infrared data acquired by several satellite systems (Table 5). The principle behind the NDVI is based on the relationship between the physiological properties of healthy vegetation and the type and amount of radiation it can absorb and reflect (Gitelson and Kaufman 1998). More specifically, plant chlorophyll strongly absorbs solar radiation in the red portion of the electromagnetic spectrum, while plant spongy mesophyll strongly reflects solar radiation in the near-infrared region of the spectrum (Jackson and Ezra 1985; Tucker 1979; Tucker et al., 1991). As a result, vigorously growing healthy vegetation has low red-light reflectance and high near-infrared reflectance, and hence, high NDVI values.

The NDVI produces output values in the range of -1.0 to 1.0. Increasing positive NDVI values indicate increasing amounts of green vegetation, while NDVI values near zero and decreasing negative values are characteristic of non-vegetated surfaces such as barren surfaces (rock and soil) and water, snow, ice, and clouds (Jensen 2007). It is important to note, however, that because the NDVI becomes less sensitive to plant chlorophyll at high chlorophyll contents, the NDVI approaches saturation asymptotically under moderate-to-high biomass conditions (Baret and Guyot 1991; Gitelson and Kaufman 1998; Huete et al., 2002; Myneni et al., 2002; Sellers 1985). As a result, although the NDVI has been shown to correlate well with many canopy biophysical properties, including vegetation abundance (Hurcom and Harrison 1998; Purevdorj et al., 1998), aboveground biomass (Boutton et al., 1980; Davidson and Csillag 2001; Weiser et al., 1986), green leaf area (Asrar et al., 1986; Baret and Guyot 1991; Weiser et al., 1986), photosynthetically active radiation (PAR) (Asrar et al., 1986; Baret and Guyot 1991; Hatfield et al., 1984; Tucker et al., 1986; Weiser et al., 1986), and productivity (Box et al., 1989; Prince 1991; Running et al., 1989), it generally does so in a nonlinear fashion across low-to-high productivity gradients.

The NDVI has emerged as one of the most robust tools for monitoring natural vegetation and crop conditions. The most commonly-used products are n-day (e.g., 7 or 10 d) maximum-value NDVI (Max-NDVI) composites and their associated anomalies (the associated NDVI differences from "normal conditions") (Cracknell 2001). While the detailed methodologies for creating these datasets vary, maximum-value compositing usually involves (i) examining each NDVI value pixel by pixel for each observation date during the n-day compositing period, (ii) determining the maximum-value NDVI for each pixel during the n-day period, and (iii) creating a single-output image that contains only the maximum NDVI value for each pixel for the n-day period. Maximum-value NDVI compositing has become a popular resource management tool because it captures the dynamics of green vegetation and minimizes problems common to single-date NDVI data, such as those associated with interference from cloud cover, atmospheric attenuation, surface directional reflectance, and view and illumination geometry (Holben, 1986).

The Advanced Very High Resolution Radiometer (AVHRR) instruments that have been flown onboard 14 of NOAA's Polar Orbiting Satellites since 1978 have been considered the longest-lived and most influential series of Earth-observing satellites ever launched (Hastings and Emery, 1992). Since some VIs that have been applied, including NDVI, are based on anomalies during the period of record for observations, the long data record of AVHRR data has been particularly useful. The AVHRR, originally designed for meteorological applications, senses in the visible, near-infrared, and thermal infrared portions of the electromagnetic spectrum at a spatial resolution of 1.1 km. However, because the AVHRR sensor was not originally designed for monitoring vegetation, it suffers from limitations regarding the design of its red and near infrared channels when formulating NDVI (Fensholt and Sandholt, 2005). Two particularly important limitations of the AVHRR are (i) the overlap of the near infrared channel (0.725 to 1.100μm) with a region of considerable atmospheric water vapor absorption (0.9 to 0.98 μm) that can introduce noise to the remotely sensed signal (Huete et al., 2002; Justice et al., 1991); and (ii) the relatively "quick" saturation of the red channel, and hence VIs derived from it, over medium-to-dense vegetation (Gitelson and Kaufman, 1998; Huete, 1988; Jensen, 2007; Myneni et al., 1997).

These limitations were directly addressed with the development of a new generation of EO platforms including the moderate resolution imaging spectroradiometer (MODIS) launched onboard NASA's Terra satellite in Dec. 1999. MODIS, which has been acquiring data in 36 narrow spectral bands since Feb. 2000, was designed to provide data for vegetation and land cover mapping applications. The MODIS sensor offers a number of improvements over the AVHRR for NDVI calculation (Fensholt and Sandholt, 2005; Huete et al., 2002; Trishchenko et al., 2002). These include improved (i) spectral resolution, (ii) radiometric resolution, (iii) spatial resolution, (iv) geolocation accuracy, and (v) on-board radiometric calibration for producing scaled reflectances (Jensen, 2007). The MODIS red and near-infrared channels were selected to avoid the spectral regions of water absorption that constitute a major limitation of the AVHRR (Justice et al., 1991; Vermote and Saleous, 2006). Furthermore, the unprecedented radiometric resolution of MODIS-Terra makes its red and near-infrared channels more sensitive to small variations in chlorophyll content, thereby lessening how quickly its NDVI saturates over denser vegetation. As a result of these improvements, MODIS-Terra holds promise for environmental monitoring in general and the estimation of vegetation indices in particular (Fensholt and Sandholt, 2005).

Yield Estimation and Forecasting

Traditionally, regional or national crop yield estimates were made by field or farmer surveys conducted during or after the crop growing season (e.g., USDA, 2012; Statistics Canada, 2012). The survey method was resource intensive and significant time lags in data processing meant that reliable estimates were not normally available until long after the growing season. For example, Statistics Canada conducted national crop yield surveys in July, September and November and the last (or the most reliable) yield estimates were released in early December while most crops were harvested 2 to 3 mo earlier. To reduce the costs associated with surveys and to increase the lead time of the crop yield estimates, tremendous efforts have been made by several countries to incorporate EO-based VI methods for in season crop monitoring which provide the capability of producing crop yield forecasts (e.g., Potgieter et al., 2006; de Wit and van Diepen, 2007; Semenov and Doblas-Reyes, 2007; Qian et al., 2009; Mkhabela et al., 2011; Bornn and Zidek, 2012; Chipanshi et al., 2012; Nikolova et al., 2012).

In Canada the Integrated Canadian Crop Yield Forecaster (ICCYF), a statistical yield forecasting tool integrates remote sensing and agroclimate data in near real time to forecast grain and oil seed crops with a lead time of 2 to 3 mo starting in July (Newlands et al., 2014; Chipanshi et al., 2015). The ICCYF was first calibrated using NDVI as derived from the AVHRR sensor at 1 km resolution, combined with the corresponding agroclimatic indices from ground-based climate stations (e.g., water stress, accumulated precipitation and growing degree days) against regional crop yield as reported by Statistics Canada from 1987 to 2012. The basic unit for comparison was the Crop Census Agricultural Regions (CARs), units of approximately 1000 km^2. The calibrated model from each CAR uses the near real time NDVI and climatic indices from seeding to the prediction date as model inputs. Beyond the prediction date to the end of the season, the unobserved variables that are required to make a forecast are estimated from a statistical procedure called random forest (Liaw and Wiener, 2002).The use of the random forest scheme takes advantage of the posterior statistical distribution of the predictor variables which is generated from the Markov-chain Monte Carlo algorithm (Dowd, 2006). Since outlook projections of crop yields from EO data started in 2013, the ICCYF has consistently generated yield results that are not statistically different at the CAR level from the final observed yield numbers that are released in November by Statistics Canada. Due to the similarity between official numbers and ICCYF simulations and the gain in lead time by the ICCYF over the survey methods, there have been discussions to replace some of the survey results with simulated values in the near future.

The skill in yield predictions with the ICCYF is expected to improve further when crop specific masks are used to generate NDVI values in near real time. Currently, a generalized agriculture crop mask is used for all crops. Changes to the ICCYF algorithms under extreme weather (when results are less reliable) are being tested.

Drought Monitoring

Measurement and monitoring of drought have challenges because the physical factors such as precipitation variability do not always predict the impacts of drought; the onset and recession of drought is imprecise and the spatial and temporal variability of drought can be large. The common factor in drought is that droughts develop from a deficiency of precipitation that results in water shortages for some activity or for some group (Wilhite and Glanz, 1985).

Quantification of drought severity was originally based on meteorological and/or hydrological data. Several indices have been developed and have been reviewed for use in the United States (Heim 2002, Keyantash and Dracup 2002) and globally (Vicente-Serrano et al., 2012). Some of the most commonly used indices include the (i) Palmer Drought Severity Index (Palmer, 1965) based on soil conditions and current and prior climatological conditions, (ii) the Surface Water Supply Index (Shafer and Dezman, 1982; Garen, 1993), based on non-exceedance probabilities of normalized data for snowpack, precipitation, streamflow and reservoir storage, and (iii) the Standardized Precipitation Index (SPI) (McKee et al., 1993; World Meteorological Organization, 2012) based on the non-exceedance probabilities based on the normalized variance of regional precipitation over the period of record for a specified time period, such as monthly or yearly. The SPI does not include a temperature component and therefore ignores the contribution of temperature variability in drought severity, which could be a limitation in areas where potential evapotranspiration (PE) exceeds precipitation (P). The Standardized Precipitation and Evaporative Index (Vicente-Serrano et al., 2010) was developed with the incorporation of a P–PE component where PE can be calculated from temperature using established PE models such as the Penman–Monteith or Thornthwaite models, depending on availability of input data.

Recent increases in data handling and modeling techniques and the increasing availability of remote sensing data have resulted in development of indices that integrate meteorological and biophysical expressions in the landscape and allow satellite-based monitoring of drought conditions. The United States Drought Monitor (USDM) (Svoboda et al., 2002) uses a hybrid of meteorological and hydrological indices combined with remotely-sensed vegetation indices to provide a weekly national assessment of drought for the United States. It has been the model for other drought monitoring efforts such as the North American Drought Monitor (NADM) https://www.ncdc.noaa.gov/temp-and-precip/drought/nadm/.

The Vegetation Drought Response Index (VegDRI) (Brown et al., 2008) integrates traditional climate-based drought index information, SPI and PDSI, with satellite-based Normalized Difference Vegetation Index (NDVI). The VegDRI therefore assesses the effects of drought on vegetation by observing the vegetation conditions from satellite and the level of dryness from climate data. Brown et al. (2008) reported that more spatially-detailed drought information could be obtained using VegDRI than what was available with the USDM.

Anderson et al. (2007) used remotely sensed land surface temperature from thermal infrared imagery from the GOES satellite to derive temporal anomalies in the ratio of actual evapotranspiration (ET) to PE, which were expressed as the Evaporative Stress Index (ESI). The ESI provides anomalies in the values of the ratio of ET to PE compared to historic data. Patterns of water stress can be determined at a spatial resolution of 5to 10 km over continental scales. The ESI compared favorably with drought conditions as determined by the USDM (Anderson et al., 2007) and it is currently used operationally. Its application to United States drought conditions can be observed at the National Oceanic and Atmospheric Administration's National Integrated Drought Information System website https://www.drought.gov/drought/content/products-current-drought-and-monitoring-remote-sensing/evaporative-stress-index.

Rapid onset of drought or "flash drought" conditions were observed in the Central Plains of the United States during the summer of 2012 (NOAA Drought Task Force, 2013). The combination of below normal rainfall, above normal temperatures, sunshine and wind greatly increase PE rates resulting in rapid drying conditions (Otkin et al., 2013). The Rapid Change Index (RCI) (Otkin et al., 2014) was developed to assess the cumulative magnitude of weekly ESI changes, which were correlated to the onset of drought conditions as determined by the USDM. During the 2012 drought they provided sensitivity to drought onset a month before the USDM drought assessment rapidly deteriorated from no drought to extreme drought. Otkin et al. (2015) applied the RCI to changes in evapotranspiration, precipitation and soil moisture to provide early warning of drought across the continental United States, they noted that while proving the concept that rapid decline in the three indices can be used to identify areas susceptible to drought onset and intensification, further validation is required before it can be used with confidence for operational purposes.

Soil moisture and drought indicators have been generated from the NASA and GRACE satellites. These twin satellites are sensitive enough to detect small changes in the earth's gravitational field caused by water redistribution at or beneath the earth's surface. Total terrestrial water storage data from the GRACE system have a monthly temporal resolution, and a coarse (150, 000 km^2) spatial resolution. To make the information useful for drought monitoring, the GRACE Data Assimilation System (Zaitchik et al., 2008) was used to integrate GRACE TWS data with meteorological observations from both ground observation stations and satellite based sensors. This enabled the downscaling and stratification of the TWS data into basin scale surface soil moisture, rootzone soil moisture and groundwater storage for the continental United States as near real time drought indices and compared with the United States and North American drought monitors. The data substituted for the sparse availability of ground-based observations of soil moisture and groundwater. The groundwater data in particular contributed a longer-term component to the anomalies associated with drought (Houborg et al., 2012). Operational weekly water storage data map products based on the hybridized GRACE data are available at the National Drought Mitigation Centre website http://drought.unl.edu/MonitoringTools/NASAGRACEDataAssimilation.aspx.

Case Study: Crop Condition Assessment in the Ukraine

Crop condition assessment is an important component of agriculture resource monitoring. Globally available products on crop condition assessment provide an extremely important input to food security within, for example, the Global Agriculture Monitoring (GLAM) initiative. In the Ukraine, such information allows for the identification of crop phenological indicators and could be used for prediction of both crop yield (Kogan et al., 2013) and crop production (Kussul et al., 2013; Kussul et al., 2014).

Leaf area index (LAI), fraction of absorbed photosynthetically active radiation (FAPAR) and fraction of vegetation cover (FCOVER) are indicators that characterize the crop state (Camacho et al., 2013). Coarse resolution images acquired by SPOT-VEGETATION, MODIS, and PROBA-V are used to provide regular and timely products on biophysical parameters at global scale. To provide consistent and reliable information these products should be validated using ground measurements. Hence, the particular objectives of this case study are: (i) to validate global biophysical products with use of in situ data, and (ii) to assess the efficiency (in terms of prediction error minimization) of satellite based data when assimilated into winter wheat crop yield forecasting models.

A study area in Onufriivka county of Kirovohrad region was selected for winter wheat forecasting. For the validation of the required biophysical parameters, a JECAM (Joint Experiment of Crop Assessment and Monitoring) test site was chosen. All surveys on the JECAM test site were conducted on two scales: local subsite (Pshenichne test site) of 3 by 3 km and a medium scale (county-level) site of approximately 1000 km^2 (Camacho et al., 2013).

Three field campaigns in 2013 (14 to 17 May, 12 to 15 June and 14 to17 July) and two field campaigns in 2014 (12 June and 31 July) were conducted to characterize the

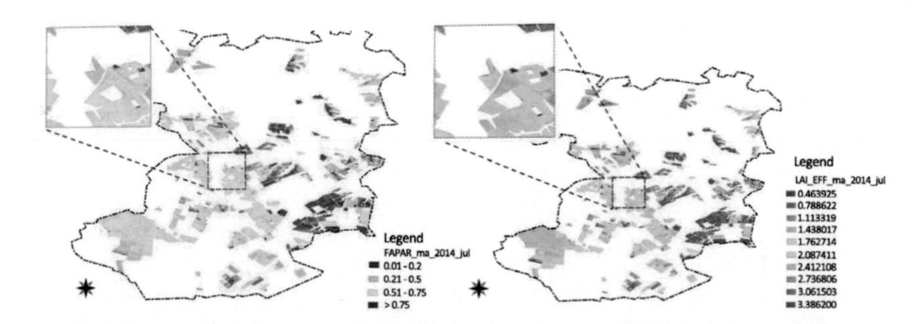

Fig. 1. Biophysical maps for maize, derived from Landsat-8, fraction of absorbed photosynthetically active radiation (FAPAR) as for 31 July, 2014 (left); Leaf Area Index (LAI) effective as for 31 July, 2014 (right).

vegetation biophysical parameters at the Pshenichne test site. Digital Hemispheric Photographic (DHP) images were acquired with a NIKON D70 camera by staff standing above the crop. Hemispherical photos allow the calculation of LAI and FCOVER by measuring gap fraction through an extreme wide-angle camera lens (i.e., 180°). The hemispherical images acquired during the field campaign are processed with the CAN-EYE software (http://www.avignon.inra.fr/can_eye) to derive LAI, FAPAR and FCOVER estimations. The in situ biophysical values are used for producing LAI, FCOVER and FAPAR maps from optical satellite images, and provide cross-validation, and validation of global remote sensing products (Morissette et al., 2006).

Satellite imagery acquired from Landsat-8 (at 30 m spatial resolution) was used to support ground observations and provide high-resolution biophysical maps. SPOT Vegetation products LAI and FAPAR (at 1 km resolution) used in this study for yield forecast model calibration were obtained from Copernicus Global Land Service (http://land.copernicus.eu).

Ground observations follow the Validation of Land European Remote sensing Instruments (VALERI) protocol in which the measurements are made for a set of elementary sampling units (ESUs) (Baret et al., 2005). The center of each ESU is georeferenced using a Global Positioning System device. A pseudoregular sampling grid is used within each ESU of approximately 20 by 20 m. The number of hemispherical photos per ESU ranges between 12 and 15. The number of ESUs varied from year to year depending on available resources. During the three 2013 campaigns 30, 34, and 37 ESUs were sampled whereas during the two campaigns in 2014, 28 and 25 ESUs were sampled.

The NDVI specific for winter wheat specific, with use of a dynamic crop mask (a mask developed every year) is used as the main variable to derive three biophysical values (LAI, FAPAR, FCOVER) from satellite images. Two types of models are considered to relate NDVI to each of the three biophysical parameters estimated from ground measurements: linear ($Y = b_0 + b_1$*NDVI) and exponential ($Y = b_0$ * exp(b_1*NDVI)), where Y = either LAI, FAPAR or FCOVER and b_0 and b_1 are adjustable parameters of the regression model. The following metrics are used to assess efficiency of the models: (i) root mean square error (RMSE); (ii) cross-validation RMSE with a leave-one-out method (RC); (iii) model's adjusted coefficient of determination r^2. Yield is estimated as a sum of the trend component and deviation from trend, caused by the current situation with vegetation development.

Deviation is estimated with a linear single-factor regression model (Kogan et al., 2013; Kussul et al., 2014; Camacho et al., 2013).

Relationships between satellite-derived NDVI values and ground measurements of biophysical parameters were developed using both linear and exponential models. The best results for winter wheat have been achieved with a single factor exponential model for LAI (up to $r^2 = 0.84$) and linear single factor models for FAPAR (up to $r^2 = 0.84$). Samples of created biophysical maps are shown in Fig. 1. These maps are used for validation of global LAI and FAPAR products derived from Copernicus Global Land Service.

Biophysical products (FAPAR and LAI) are more preferable to be used as predictors in crop yield forecasting regression models. Corresponding models possess much better statistical properties and are more reliable than the NDVI based model. The most accurate result in the study to date has been obtained for LAI values derived from SPOT-VGT (at 1 km resolution) on county scale averaged using the crop mask (with $r^2 = 0.86$).

Therefore, we have concluded that for the Ukraine, LAI and FAPAR are the best variables for developing accurate, reliable regression based models for winter wheat yield forecasting at the county level. Models calibrated with biophysical parameters are much more accurate than models calibrated with classical vegetation indices (NDVI) and global biophysical products agree sufficiently with in situ data to allow them to be used confidently for yield forecasting in an operational mode.

Greenhouse Gas Flux

The three main greenhouse gases (GHGs) emitted from agricultural sources are carbon dioxide (CO_2), methane (CH_4) and nitrous oxide (N_2O). The source of emissions (e.g., soil-based vs. animal-based) usually dictates the most appropriate measurement technique to quantify GHG emissions. Agricultural soils, which are diffuse (nonpoint) sources, can either emit or absorb CO_2 and CH_4 depending on the management practices and the environmental conditions. They are also an important source of N_2O because of the increased application of nitrogen (N) fertilizer and manure. Livestock and animal waste treatment systems are significant sources of both CH_4 and N_2O. We will briefly present examples of GHG flux measurements from agricultural sources. We will show how the combination of measurements and models is being used to improve our understanding of the interactions between management practices and GHG emissions. We will also show how the amount of GHG emissions associated with an agricultural product can help quantify the impact of a production system on the environment.

Chamber Measurements

Closed and open chambers are frequently used to quantify the impact of a change in management practices on GHG emissions from agricultural soils (Rochette et al., 1992). They are inexpensive and easy to use. Their main limitation is that they provide information on a very small area. Lessard et al. (1996; 1997) used dynamic closed chamber systems to quantify the influence of manure applications on N_2O and CH_4 emissions over a growing season. Zou et al. (2005) used chambers over rice paddies to show that there are trade-offs between CH_4 and N_2O emissions for certain management practices. For example, in contrast with continuous flooding, mid-season drainage caused a drop in CH_4 emissions, while concurrently increasing N_2O emissions. Rochette et al. (1992) showed that respiration from soil under barley was 25%

lower than from fallow (uncropped) soil and that the afternoon soil respiration averaged 22 and 17% more than morning on fallow and barley fields, respectively.

Measurements from Point Sources

Livestock operations emit a substantial amount of CH_4 from enteric fermentation and manure management. An inverse dispersion technique in conjunction with open-path instruments is ideal for measuring GHG emissions from multiple sources with large spatial and temporal variability, as is common on typical livestock operations (Flesch et al., 2005; Laubach et al., 2013). It uses a backward in time particle dispersion model which requires line-averaged concentration measurements of the gas of interest downwind and upwind of the source as well as the wind statistics provided by a sonic anemometer. Gao et al. (2011) used such a system to measure the average hourly methane emissions from a dairy feedlot for the fall and winter seasons. They showed that it is important to consider the diurnal pattern to assess the mitigation potential of a mitigation strategy. In a study on dairy farms, VanderZaag et al. (2014) showed that in the fall when the manure storage tank was full, 60% of the whole farm emissions came from the manure storage. They also reported that whole farm CH_4 emissions were 40% higher in the fall than in the spring.

Tower and Aircraft-based Flux Measurements

Tower and aircraft-based flux measurements provide useful information on GHG emissions at a wide range of scales (Pattey et al., 2006a). Ma et al. (2007) estimated the net carbon exchange, at a grassland site in California, from 2000 to 2006 using the eddy covariance technique. The net annual exchange, which varied from −88 to 141 g C m^{-2} y^{-1}, depended primarily on the amount of rain during the growing period. Wagner-Riddle et al. (2007) used the flux-gradient technique to quantify the N_2O emissions associated with the spring thaw period. The flux-gradient technique is sometimes used rather than the eddy covariance technique for cases when there are no fast response sensors for the gas of interest. Over a 5 yr period in a corn–soybean–wheat rotation, they showed that N_2O emissions during November to April comprised between 30% and 90% of the annual emissions, mostly due to the large N_2O emissions during soil thawing. Pattey et al. (2006b) demonstrated the management and weather impact on N_2O fluxes during the growing season. Periods after N fertilizer applications coincided with increased N_2O fluxes. They also showed the considerable daytime variability of N_2O emissions that peaked during the midmorning and decreased in the afternoon as the soil dried. Tower-based flux measurements of N_2O have also been an excellent source of information for testing and improving biogeochemical models (Smith et al., 2002). Using the DNDC model, the climatology of N_2O emissions in Canada during spring thaw was examined by Smith et al. (2004). They estimated on average, over the seven soil groups, that the N_2O emissions during spring thaw were about 30% of the annual emissions. This information is used in Canada's national agricultural GHG inventory (Rochette et al., 2008) to scale annual emissions to account for the spring thaw period. Desjardins et al. (2010) used aircraft-based flux measurements to verify N_2O emission estimates at the regional scale. By comparing these flux measurements to process-based model estimates they were able to verify the magnitude of indirect N_2O emissions.

Process-based Models

Regardless of the temporal and spatial scales covered using various flux measurement techniques, it is clear that it is impossible to measure continuously and compare all management options. Models are then essential to fill the gap left by direct flux measurements. They help improve our understanding of the interactions between management practices and environmental conditions for a wide range of GHG sources. They also allow us to separate the environmental impacts of human activities and year to year variability due to climate. Better information on the impact of management practices is essential to improve GHG emission inventory (VanderZaag et al., 2013).

The importance of models, especially those at the farm scale are gaining prominence in tools such as HOLOS (Little et al., 2008), the "Cool Farm Tool" (Hillier et al., 2011) and ULICEES (Vergé et al., 2012). These tools are being used to provide information on the carbon footprint of agricultural products. Because of consumer demands for environmentally sustainable food production producers of agricultural products with a lower carbon footprint will likely have access to either a wider range of markets or may get a preferential treatment such as less tariffs or taxes.

Case Study: The Carbon Footprint of Beef in Canada

An on-farm model has been developed to estimate the carbon footprint of agricultural products (Vergé et al., 2012). This model was developed using GHG flux measurements obtained using the techniques described above. The cradle to farm gate carbon footprint associated with beef production in Canada was quantified using this model for each Census year from 1981 to 2006 (Desjardins et al., 2012). It was obtained by first calculating the carbon footprint of all crops in Canada (Dyer et al., 2010). Then based on the diets of cattle, we estimated the GHG emission per kg of live weight at the exit gate of the farm. We note a substantial decrease in the carbon footprint (CFb) of beef from 1981 to 2006 (Table 4.1). This is due to better breeds, better diets bigger carcass weight and improved soil management practices.

Carbon footprint estimates are very dependent on what GHG emissions are included. If we include the impact of soil carbon change in the carbon footprint (CFbc) calculation, we obtain a slightly larger carbon footprint for the period between 1981 and 1991 and a reduction in the carbon footprint for the period between 1996 and 2006. This is because of improved soil conservation practices, such as reduced tillage and reduced summer fallowing. The carbon footprint of beef also changes if the GHG emissions from cattle from the dairy sector are also included. In this case, an extra 14.4% of the GHG emissions from the dairy sector need to be allocated to meat production (IDF-FIL, 2010). The carbon footprint of all beef production in Canada including the beef from the dairy sector (CFbcd) is given in Table 6. This value is larger than the other estimates because originally all the GHG emissions from the dairy sector were associated with milk production. This is one example of the range of carbon footprint values that can be obtained depending what is included in the calculation.

Looking Ahead

Agroclimatology is an integration of meteorological, biophysical, and hydrological parameters over spatial and temporal scales that are relevant for understanding agricultural production. This chapter discussed measurement techniques which

Table 6. Cradle to farm gate carbon footprint associated with beef production in Canada (Updated from Desjardins et al., 2012).

Year	Carbon footprint	Carbon footprint including soil carbon change	Carbon footprint including dairy sector
	kg CO^2e per kg LW		
1981	16.6	16.9	18.2
1986	15.3	15.4	16.6
1991	13.7	13.9	14.9
1996	12.4	12.1	12.8
2001	10.3	9.7	10.2
2006	10.0	9.0	9.5

utilize combinations of in situ instrumentation, remote sensing instrumentation, and modeling to solve the complex interactions associated with measurements of soil moisture, crop conditions, expressions of drought, and the flux associated with greenhouse gases in agricultural landscapes.

The rapid expansion of satellite-based remote sensing technology is vastly increasing the availability of timely, inexpensive data with increasingly temporal and spatial finer resolutions. With the development and adoption of drone technology, low-level aerial observations will add even higher resolution to monitoring efforts. Scaling of data will become even more important as independent monitoring activities conducted within a field merge with more conventional coarser scale monitoring. Competitive advantages will be gained from the ability to process and interpret large data volumes with increasing timeliness for decision support.

Crowd sourced data will add complexity as to how data quality is assessed, managed, processed, and interpreted. Measurement techniques in this era of "big data" have extended beyond sensors and methodologies which derive new data from observations of the natural system to techniques that can capture the value of existing data. The measurement techniques of the future will have increasing reliance on the analytical powers of modeling, land data assimilation systems, neural network systems, and other means of optimizing the information contained in data to develop indices that better describe and measure the complex interactions of agroclimatologic processes.

References

Adams, J.R., H. McNairn, A.A. Berg, and C. Champagne. 2015. Evaluation of near-surface soil moisture data from an AAFC monitoring network in Manitoba, Canada: Implications for L-band satellite validation. J. Hydrol. 521:582–592. doi:10.1016/j.jhydrol.2014.10.024

Hatfield, J.L., and J.M. Baker, editors. 2005. Micrometeorology in agricultural systems. Agron. Monogr. 47. ASA, CSSA, and SSSA, Madison, WI.

Anderson, M.C., C.R. Hain, B.D. Wardlow, A. Pimstein, J.R. Mecikalski, and W.P. Kust. 2007. Thermal-based evaporative stress index for monitoring surface moisture depletion. In: B.D. Wardlow, M.C. Anderson, and J.P. Verdin, editors, Remote sensing of drought: Innovative monitoring approaches. CRC Press, Boca Raton, FL. p. 145–167.

Asner, G.P. 1998. Biophysical and biochemical sources of variability in canopy reflectance. Remote Sens. Environ. 64:234–253. doi:10.1016/S0034-4257(98)00014-5

Asner, G.P., C.A.Wessman, D.S. Schimel and S. Archer. 1998. Variability in leaf and litter optical properties:implications for BRDF model inversions using AVHRR, MODIS and MISR. Remote Sens. Environ. 63:243–257. doi:10.1016/S0034-4257(97)00138-7

Asrar, G., R.L. Weiser, D.E. Johnson, E.T. Kanemasu, and J.M. Milleen. 1986. Distinguishing among tallgrass prairie cover types from measurements of multispectral reflectance. Remote Sens. Environ. 19:159–169. doi:10.1016/0034-4257(86)90069-6

Attema, E.P.W., and F.T. Ulaby. 1978. Vegetation modelled as a water cloud. Radio Sci. 13:357–364. doi:10.1029/RS013i002p00357

Baatz, R., H.R. Bogena, H. J. Hendricks Franssen, J.A. Huisman, W. Qu, C. Montzka, and H. Vereecken. 2014. Calibration of a catchment scale cosmic-ray probe network: A comparison of three parameterization methods. J. Hydrol. 516:231–244. doi:10.1016/j.jhydrol.2014.02.026

Baret, F., and G. Guyot. 1991. Potentials and limits of vegetation indices for LAI and APAR assessment. Remote Sens. Environ. 35:161–173. doi:10.1016/0034-4257(91)90009-U

Baret, F., M. Weiss, D. Allard, S. Garrigues, M. Leroy, H. Jeanjean, et al. 2005. VALERI: A network of sites and a methodology for the validation of medium spatial resolution land satellite products. Remote Sens. Environ. 96 (1):69-89.

Bogena, H.R., J.A. Huisman, C. Oberdorster, and H. Vereecken. 2007. Evaluation of a low-cost soil water content sensor for wireless network applications. J. Hydrol. 344:32–42. doi:10.1016/j.jhydrol.2007.06.032

Borrn, L. and J.V. Zidek. 2012. Efficient stabilization of crop yield prediction in the Canadian Prairies. Agricultural and Forest Meteorology 152:223–232.

Boutton, T.W., A.T. Harrison, and B.N. Smith. 1980. Distribution of biomass of species differing in photosynthetic pathway along an altitudinal gradient in southeastern Wyoming grassland. Oecologia 45:287–298. doi:10.1007/BF00540195

Box, E.O., B.N. Holben, and H. Kalb. 1989. Accuracy of the AVHRR vegetation index as a predictor of biomass, primary productivity and net CO2 flux. Vegetatio 80:71–89. doi:10.1007/BF00048034

Brown, J.F., B.D. Wardlow, T. Tadesse, M.J. Hayes, and B.C. Reed. 2008. The Vegetation Drought Response Index(VegDRI): A new integrated approach for monitoring drought stress in vegetation. GIsci. Remote Sens. 45(1):16–46. doi:10.2747/1548-1603.45.1.16

Camacho, F., J. Cernicharo, R. Lacaze, F. Baret, and M. Weiss. 2013. GEOV1: LAI, FAPAR Essential Climate Variables and FCOVER global time series capitalizing over existing products. Part 2: Validation and intercomparison with reference products. Remote Sens. Environ. 137:310–329. doi:10.1016/j.rse.2013.02.030

Ceccato, P., S. Flasse, and J.-M. Grégoire. 2002a. Designing a spectral index to estimate vegetation water content from remote sensing data: Part 2. Validation and applications. Remote Sens. Environ. 82:198–207. doi:10.1016/S0034-4257(02)00036-6

Ceccato, P., N. Gobron, S. Flasse, B. Pinty, and S. Tarantola. 2002b. Designing a spectral index to estimate vegetation water content from remote sensing data: Part 1. Theoretical approach. Remote Sens. Environ. 82:188–197. doi:10.1016/S0034-4257(02)00037-8

Chanasyk, D.S., and M.A. Naeth. 1996. Field measurement of soil moisture using neutron probe. Can. J. Soil Sci. 76:317–323. doi:10.4141/cjss96-038

Chen, Y., and D. Or. 2006. Effects of Maxwell-Wagner polarization on soil complex dielectric permittivity under variable temperature and electrical conductivity. Water Resour. Res. 42:W06424. doi:10.1029/2005WR004590

Chipanshi, A., Y. Zhang, L. Kouadio, N. Newlands, A. Davidson, H. Hill, et al. 2015. Evaluation of the Integrated Canadian Crop Yield Forecast model for in-season prediction of crop yield across the agricultural landscape. Agric. For. Meteorol. 206:137–150. doi:10.1016/j.agrformet.2015.03.007

Chipanshi, A.C., Y. Zhang, N.K. Newlands, H. Hill, and D.S. Zamar. 2012. Canadian Crop Yield Forecaster (CCYF)–a GIS and statistical integration of agro-climates and remote sensing information. In: Proc. Workshop on the application of remote sensing and GIS technology on crops productivity among APEC economies, 30-31 July 2012, Asia – Pacific Economic Corporation, Beijing, People's Republic of China.

Cosh, M.H., T.J. Jackson, R. Bindlish, J.S. Famiglietti, and D. Ryu. 2005. Calibration of an impedance probe for estimation of surface soil water content over large regions. J. Hydrol. 311:49–58. doi:10.1016/j.jhydrol.2005.01.003

<conf>Cosh, M.H., T. Ochsner, J. Basara, and T.J. Jackson. 2010. The SMAP in-situ soil moisture sensor testbed:comparing in-situ sensors for satellite validation. In: Proc. IEEE International Geoscience and Remote Sensing Symposium, July 25-30, 2010. Honolulu, HI. p. 699-701. doi: 10.1109/IGARSS.2010.5652389.</conf>

Cracknell, A.P. 2001. The exciting and totally unanticipated success of the AVHRR in applications for which it was never intended. Adv. Space Res. 28:233–240. doi:10.1016/S0273-1177(01)00349-0

Davidson, A., and F. Csillag. 2001. The influence of vegetation index and spatial resolution on a two-date remote sensing derived relation to C4 species coverage. Remote Sens. Environ. 75:138–151. doi:10.1016/S0034-4257(00)00162-0

de Wit, A.J.W., and C.A. van Diepen. 2007. Crop model data assimilation with the Ensemble Kalman filter for improving regional crop yield forecasts. Agric. For. Meteorol. 146(1-2):38–56. doi:10.1016/j.agrformet.2007.05.004

Dean, T.J., J.P. Bell, and A.J.B. Baty. 1987. Soil moisture measurement by an improved capacitance technique, Part I. Sensor design and performance. J. Hydrol. 93:67–78. doi:10.1016/0022-1694(87)90194-6

Desjardins, R.L., E. Pattey, W.N. Smith, D. Worth, B. Grant, R. Srinivasan, J.I. MacPherson, and M. Mauder. 2010. Multiscale estimates of N2O emissions from agricultural lands. Special Issue of Agr. Forest Meteorol. 150(6):817–824. doi:10.1016/j.agrformet.2009.09.001

Desjardins, R.L., D.E. Worth, X.P.C. Verge, D. Maxime, J.A. Dyer, and D. Cerkowniak. 2012. Carbon footprint of beef cattle. Sustainability 4:3279–3301. doi:10.3390/su4123279

Dobriyal, P., A. Qureshi, R. Badola, and S. Ainul Hussain. 2012. A review of the methods available for estimating soil moisture and its implications for water resource management. J. Hydrol. 458–459:110–117. doi:10.1016/j.jhydrol.2012.06.021

Dobson, M.C., and F.T. Ulaby. 1998. Mapping soil moisture distribution with imaging radar. In: F.M. Henderson and A.J. Lewis, editors, Principles & applications of imaging radar. John Wiley & Sons, New York.

Dowd, M. 2006. A sequential Monte Carlo approach for marine ecological prediction. Environmetrics 17(5):435–455. doi:10.1002/env.780

Dubois, P.C., J. Van Zyl, and T. Engman. 1995. Measuring soil moisture with imaging radars. IEEE Trans. Geosci. Rem. Sens. 33:4915–4926.

Dyer, J.A., X.P.C. Vergé, R.L. Desjardins, D.E. Worth, and B.G. McConkey. 2010. The impact of increased biodiesel production on the greenhouse gas emissions from field crops in Canada. Energy Sustain. Dev. 14(2):73–82. doi:10.1016/j.esd.2010.03.001

Elachi, C., and J.J. van Zyl. 2006. Introduction to the physics and techniques of remote sensing. Wiley, Hoboken. doi:10.1002/0471783390

Fensholt, R., and I. Sandholt. 2005. Evaluation of MODIS and NOAA AVHRR vegetation indices with in situ measurements in a semi-arid environment. Int. J. Remote Sens. 26:2561–2594. doi:10.1080/01431160500033724

Flesch, T.K., J.D. Wilson, L.A. Harper, and B.P. Crenna. 2005. Estimating gas emissions from a farm with an inverse-dispersion technique. Atmos. Environ. 39:4863–4874. doi:10.1016/j.atmosenv.2005.04.032

Fourty, T., F. Baret, S. Jacquemoud, G. Schmuck, and J. Verdebout. 1996. Leaf optical properties with explicit description of its biochemical composition:direct and inverse problems. Remote Sens. Environ. 56:104–117. doi:10.1016/0034-4257(95)00234-0

Franz, T.E., M. Zreda, R. Rosolem, B.K. Hornbuckle, S.L. Irvin, H. Adams, et al. 2013. Ecosystem-scale measurements of biomass water using cosmic ray neutrons. Geophys. Res. Lett. 40:1–5. 10.1002/grl.50791

Fung, A.K., and K.S. Chen. 1992. Dependence of the surface backscattering coefficients on roughness, frequency and polarization states. Int. J. Remote Sens. 13:1663–1680. doi:10.1080/01431169208904219

Gardner, W.H. 1965. Water Content. In: C.A. Black, editor, Methods of soil analysis. Part 1. Physical and mineralogical properties, including statistics of measurement and sampling. Amer. Soc. Agron., Soil Science Soc. of America, Madison, Wi. p. 82–127, 10.2134/agronmonogr9.1.c7.

Garen, D.C. 1993. Revised surface water supply index for western United States. J. Water Resour. Plan. Manage. 119(4):437–454. doi:10.1061/(ASCE)0733-9496(1993)119:4(437)

Gao, Z., Y. Huijun, W. Ma, J. Li, X. Liu, and R.L. Desjardins. 2011. Diurnal and seasonal patterns of methane emissions from a dairy operation in North China plain Adv. Meteorology. doi:10.1155/2011/190234

Gitelson, A.A., and Y.J. Kaufman. 1998. MODIS NDVI optimization to fit the AVHRR data series: Spectral considerations. Remote Sens. Environ. 66:343–350. doi:10.1016/S0034-4257(98)00065-0

Goel, N.S. 1988. Models of vegetation canopy reflectance and their use in estimation of biophysical parameters from reflectance data. Remote Sens. Environ. 4:1–212.

Han, W., Z. Yang, L. Di, and R. Mueller. 2012. CropScape: A web service based application for exploring and disseminating US conterminous geospatial cropland data products for decision support. Comput. Electron. Agric. 84:111–123. doi:10.1016/j.compag.2012.03.005

Hardisky, M.A., V. Klemas, and R.M. Smart. 1983. The influence of soil salinity, growth form, and leaf moisture on the spectral reflectance of spartina alterniflora species. Photogramm. Eng. Remote Sensing 49:77–83.

Hastings, D.A., and W.J. Emery. 1992. The Advanced Very High Resolution Radiometer (AVHRR): A brief reference guide. Photogramm. Eng. Remote Sensing 58:1183–1188.

Hatfield, J.L., G. Asrar, and E.T. Kanemasu. 1984. Intercepted photosynthetically active radiation estimated by spectral reflectance. Remote Sens. Environ. 14:65–75. doi:10.1016/0034-4257(84)90008-7

Heim, R.R. 2002. A review of twentieth-century drought indices used in the United States. Bull. Am. Meteorol. Soc. (August):1149–1165.

Hillier, J., C. Walter, D. Malin, T. Garcia-Suarez, L. Mila-i-Canals, and P. Smith. 2011. A farm-focused calculator for emissions from crop and livestock production. Environ. Model. Softw. 26(9):1070–1078. doi:10.1016/j.envsoft.2011.03.014

Holben, B. 1986. Characteristics of maximum value composite images from temporal AVHRR data. Int. J. Remote Sens. 7:1417–1434. doi:10.1080/01431168608948945

Houborg, R., M. Rodell, B. Li, R. Reichle, and B.F. Zaitchik. 2012. Drought indicators based on model assimilated GRACE terrestrial water storage observations. Water Resour. Res. 48:W07525. doi:10.1029/2011WR011291

Huete, A., K. Didan, T. Miura, E.P. Rodriguez, X. Gao, and L.G. Ferreira. 2002. Overview of the radiometric and biophysical performance of the MODIS vegetation indices. Remote Sens. Environ. 83:195–213. doi:10.1016/S0034-4257(02)00096-2

Huete, A.R., H.Q. Liu, K. Batchily, and W.J. van Leeuwen. 1997. A comparison of vegetation indices over a global set of TM images for EOS-MODIS. Remote Sens. Environ. 59:440–451. doi:10.1016/S0034-4257(96)00112-5

Huete, A.R., and C. Justice. 1999. MODIS vegetation index (MOD13) algorithm theoretical basis document. NASA Goddard Flight Center, Greenbelt, MD. p. 129.

Huete, A.R. 1988. A soil-adjusted vegetation index (SAVI). Remote Sens. Environ. 25:295–309. doi:10.1016/0034-4257(88)90106-X

Huisman, J.A., S.S. Hubbard, J.D. Redman, and A.P. Annan. 2003. Measuring soil water content with ground penetrating radar: A review. Vadose Zone J. 2:476–491. doi:10.2136/vzj2003.4760

Huisman, J.A., J.J.J.C. Snepvangers, W. Bouten, and G.B.M. Heuvelinket. 2002. Mapping spatial variation in surface soil water content:comparison of ground-penetrating radar and time domain reflectometry. J. Hydrol. 269:194–207. doi:10.1016/S0022-1694(02)00239-1

Hunt, E.R., Jr., and B.N. Rock. 1989. Detection of changes in leaf water content using Near- and Middle-Infrared reflectances. Remote Sens. Environ. 30:43–54. doi:10.1016/0034-4257(89)90046-1

Hurcom, S.J., and A.R. Harrison. 1998. The NDVI and spectral decomposition for semi-arid vegetation abundance estimation. Int. J. Remote Sens. 19:3109–3125. doi:10.1080/014311698214217

IDF-FIL. 2010. A common carbon footprint approach for dairy. The IDF guide to standard life cycle assessment methodology for the dairy sector. Bulletin of the International Dairy Federation. Brussels, Belgium. 45(1):46. http://www.fil-idf.org/Public/PublicationsPage.php?ID = 27121#list (20 July 2018).

Jackson, T.J. 2002. Passive microwave remote sensing methods. In: J.H. Dane and G.C. Topp, editors, Methods of soil analysis: Part 4- Physical methods. 3rd ed. SSSA. Madison, Wisconsin.

Jackson, T.J., and C.E. Ezra. 1985. Spectral response of cotton to suddenly induced water stress. Int. J. Biometeorol. 6:177–185.

Jensen, J.R. 2007. Remote sensing of the environment: An Earth resource perspective. Prentice Hall Upper Saddle River, NJ.

Jiao, X., H. McNairn, J. Shang, E. Pattey, J. Liu, and C. Champagne. 2014. The sensitivity of RADARSAT-2 polarimetric SAR data to corn and soybean Leaf Area Index. Can. J. Rem. Sens. 37:69–81. doi:10.5589/m11-023

Jordan, C.F. 1969. Derivation of Leaf Area Index from quality of light on the forest floor. Ecology 50:663–666. doi:10.2307/1936256

Joint Research Centre. 2012. MARS crop yield forecasting system (MCYFS). European Commission. http://www.marsop.info/marsop3/.

Justice, C.O., T.F. Eck, D. Tanré, and B.N. Holben. 1991. The effect of water vapour on the normalized difference vegetation index derived for the Sahelian region from NOAA AVHRR data. Int. J. Remote Sens. 12:1165–1187. doi:10.1080/01431169108929720

Keyantash, J., and J.A. Dracup. 2002. The quantification of drought: An evaluation of drought indices. Bull. Am. Meteorol. Soc. (August):1167–1180.

Kogan, F., N. Kussul, T. Adamenko, S. Skakun, O. Kravchenko, O. Kryvobok, A. Shelestov, A. Kolotii, O. Kussul, and A. Lavrenyuk. 2013. Winter wheat yield forecasting in Ukraine based on Earth observation, meteorological data and biophysical models. Int. J. Appl. Earth Obs. Geoinf. 23:192–203. doi:10.1016/j.jag.2013.01.002

Kussul, O., N. Kussul, S. Skakun, O. Kravchenko, A. Shelestov, and A. Kolotii. 2013. Assessment of relative efficiency of using MODIS data to winter wheat yield forecasting in Ukraine. In: 2013 IEEE International Geoscience and Remote Sensing Symposium, 21-26 July 2013, Melbourne, Australia. IEEE, New York. p. 3235-3238. doi:10.1109/IGARSS.2013.6723516

Kussul, N., A. Kolotii, S. Skakun, A. Shelestov, O. Kussul, and T. Oliynuk. 2014. Efficiency estimation of different satellite data usage for winter wheat yield forecasting in Ukraine. In: 2014 IEEE Geoscience and Remote Sensing Symposium, 13-18 July 2014, Quebec City, QC. IEEE, New York. p. 5080-5082. doi:10.1109/IGARSS.2014.6947639

Laubach, J., M. Bai, C.S. Pinares-Patino, F.A. Phillips, T.A. Naylor, G. Molano, et al. 2013. Accuracy of micrometeorological techniques for detecting a change in methane emissions from a herd of cattle. Agric. For. Meteorol. 176:50–63. doi:10.1016/j.agrformet.2013.03.006

Lessard, R., P. Rochette, E.G. Gregorich, E. Pattey, and R.L. Desjardins. 1996. N2O fluxes from manure-amended soil under maize. J. Environ. Qual. 25:1371–1377. doi:10.2134/jeq1996.00472425002500060029x

Lessard, R.L., P. Rochette, E.G. Gregorich, R.L. Desjardins, and E. Pattey. 1997. CH4 fluxes from a soil amended with dairy cattle manure and ammonium nitrate. Can. J. Soil Sci. 77:179–186. doi:10.4141/S96-108

Liaw, A., and M. Wiener. 2002. Classification and regression by random forest. R. News. 2(3):18–22.

Little, S., J. Lindeman, K. Maclean, and H. Janzen. 2008. HOLOS- a tool to estimate and reduce greenhouse gases from farms. Methodology and algorithms for version 1.1.x. Agriculture and Agri-Food Canada, Ottawa, ON. Cat. No. A52-136/2008-PDF, p. 158.

Lv, L., T.E. Franz, D.A. Robinson, and S.B. Jones. 2014. Measured and modeled soil moisture compared with cosmic-ray neutron probe estimates in a mixed forest. Vadose Zone J. 13(1) doi:10.2136/vzj2014.06.0077.

Ma, S., D. Baldocchi, L. Xu, and T. Hehn. 2007. Inter-annual variability in carbon dioxide exchange of an oak/grass savanna and open grassland in California. Agric. For. Meteorol. 147:157–171. doi:10.1016/j.agrformet.2007.07.008

Matile, L., R. Berger, D. Wächter, and R. Krebs. 2013. Characterization of a new heat dissipation matric potential sensor. Sensors (Basel Switzerland). 13:1137–1145. doi:10.3390/s130101137

McKee, T.B., N.J. Doesken, and J. Kleist. 1993. The relationship of drought frequency and duration to time scales. In: Proc. Eighth Conf. on Applied Climatology, Anaheim, CA. 17–22 Jan. 1993. Amer. Meteorol. Soc., Washington, D.C. p. 179–184.

McNairn, H., T.J. Jackson, G. Wiseman, S. Bélair, A. Berg, P.R. Bullock, et al. 2015. The soil moisture active passive validation experiment 2012 (SMAPVEX12): Pre-launch calibration and validation of the SMAP satellite. IEEE Trans. Geosci. Rem. Sens. 53:2784–2801. doi:10.1109/TGRS.2014.2364913

Meisami-asl, E., A. Sharifi, H. Mobli, A. Eyvani, and R. Alimardani. 2012. On-site measurement of soil moisture content using an acoustic system. Agric. Eng. Int. CIGR J. 15(4):1–8 http://www.cigrjournal.org.

Meng, L., and S.M. Quiring. 2008. A comparison of soil moisture models using soil climate analysis network observations. J. Hydrometeorol. 9:641–659. doi:10.1175/2008JHM916.1

Merzouki, A., H. McNairn, and A. Pacheco. 2011. Mapping soil moisture using RADARSAT-2 data and local autocorrelation statistics. IEEE J. Sel. Top. Appl. Earth Obs. Remote Sens. 4:128–137. doi:10.1109/JSTARS.2011.2116769

Merzouki, A., and H. McNairn. 2015. Toward operational use of Synthetic Aperture Radar (SAR) for surface soil moisture monitoring for agriculture: Preparing for the RADARSAT Constellation Mission. Can. J. Rem. Sens.

Mkhabela, M.S., P. Bullock, S. Raj, S. Wang, and Y. Yang. 2011. Crop yield forecasting on the Canadian Prairies using MODIS NDVI data. Agric. For. Meteorol. 151:385–393. doi:10.1016/j.agrformet.2010.11.012

Mladenova, I.E., T.J. Jackson, E. Njoku, R. Bindlish, S. Chan, M.H. Cosh, et al. 2014. Remote monitoring of soil moisture using passive microwave-based techniques–Theoretical basis and overview of selected algorithms for AMSR-E. Remote Sens. Environ. 144:197–213. doi:10.1016/j.rse.2014.01.013

Morisette, J.T., F. Baret, J.L. Privette, R.B. Myneni, J.E. Nickeson, S. Garrigues, et al. 2006. Validation of global moderate-resolution LAI products: A framework proposed within the CEOS land product validation subgroup. IEEE Trans. Geosci. Rem. Sens. 44(7):1804–1817. doi:10.1109/TGRS.2006.872529

Myneni, R.B., J. Ross, and G. Asrar. 1989. A review on the theory of photon transport in leaf canopies. Agric. For. Meteorol. 45:1–153. doi:10.1016/0168-1923(89)90002-6

Myneni, R.B., S. Hoffman, Y. Knyazikhin, J.L. Privette, J. Glassy, Y. Tian, Y. Wang, X. Song, Y. Zhang, and G.R. Smith. 2002. Global products of vegetation leaf area and fraction absorbed PAR from year one of MODIS data. Remote Sens. Environ. 83:214–231. doi:10.1016/S0034-4257(02)00074-3

Myneni, R.B., R. Ramakrishna, R.R. Nemani, and S.W. Running. 1997. Estimation of global leaf area index and absrobed PAR using radiative transfer models. IEEE Trans. Geosci. Rem. Sens. 35:1380–1393. doi:10.1109/36.649788

Newlands, N., D. Zamar, A.L. Kouadio, Y. Zhang, A. Chipanshi, A. Potgieter, S. Toure, and H. Hill. 2014. An integrated, probabilistic model for improved seasonal forecasting of agricultural crop yield under environmental uncertainty. Front. Environ. Sci 2:17. doi:10.3389/fenvs.2014.00017

Nghiem, S.V., B.D. Wardlow, D. Allured, M.D. Svoboda, D. LeComte, M. Rosencrans, S.K. Chan and G. Neumann. 2012. Microwave remote sensing of soil moisture: Science and applications. In: B.D. Wardlow, M.C. Anderson, and J.P. Verdin, editors, Remote sensing of drought: Innovative monitoring approaches. CRC Press, Boca Raton, FL. p. 197-226.

Nikolova, S., S. Bruce, L. Randall, G. Barrett, K. Ritman, and M. Nicholson. 2012. Using remote sensing data and crop modeling to improve production forecasting: A scoping study. Technical Report 12.3. Department of Agriculture, Fisheries and Forestry,

Australian Bureau of Agricultural and Resource Economics and Sciences (ABARES), Australian Government, Canberra, Australia.

NOAA Drought Task Force. M. Hoerling, Lead; S. Schubert and K. Mo, Co-Leads. 2013. An interpretation of the origins of the 2012 Central Great Plains drought. NOAA National Integrated Drought Information System, Washington, D.C. www.drought.gov/drought/content/resources/reports (Accessed 16 July 2019).

Oh, Y., K. Sarabandi, and F.T. Ulaby. 1992. An empirical model and an inversion technique for radar scattering from bare soil surfaces. IEEE Trans. Geosci. Rem. Sens. 30:370–381. doi:10.1109/36.134086

Ojo, E.R., P.R. Bullock, J. L'Heureux, J. Powers, H. McNairn and A. Pacheco. 2015. Calibration and evaluation of a frequency domain reflectometry sensor for the real-time in-situ monitoring. Vadose Zone J. 14 (3) doi:10.2136/vzj2014.08.0114.

Otkin, J.A., M.C. Anderson, C. Hain, and M. Svoboda. 2014. Examining the relationship between drought development and rapid changes in the evaporative stress index. J. Hydrometeorol. 15:938–956. doi:10.1175/JHM-D-13-0110.1

Otkin, J.A., M.C. Anderson, C. Hain, and M. Svoboda. 2015. Using temporal changes in drought indices to generate probabilistic drought intensification forecasts. J. Hydrometeorol. 16:88–105. doi:10.1175/JHM-D-14-0064.1

Otkin, J.A., M.C. Anderson, C. Hain, I. Mladenova, J. Basara, and M. Svoboda. 2013. Examining rapid onset drought development using the thermal infrared–based evaporative stress index. J. Hydrometeorol. 14:1057–1074. doi:10.1175/JHM-D-12-0144.1

Pacheco, A., H. McNairn, A. Mahmoodi, C. Champagne, and Y. Kerr. 2015. The impact of national land cover and soils data on soil moisture and ocean salinity (SMOS) soil moisture retrieval over Canadian agricultural landscapes. IEEE J. Sel. Top. Appl. Earth Obs. Remote Sens. doi:10.1109/JSTARS.2015.2417832

Palmer, W.C. 1965. Meteorological drought. U.S. Weather Bureau Research Paper 45, U.S. Department of Commerce, Washington, D.C.

Paltridge, G.W., and J. Barber. 1988. Monitoring grassland dryness and fire potential in Australia with NOAA/AVHRR data. Remote Sens. Environ. 25:381–394. doi:10.1016/0034-4257(88)90110-1

Pattey, E., G. Edwards, I.B. Strachan, R.L. Desjardins, S. Kaharabata, and C. Wagner Riddle. 2006a. Towards standards for measuring greenhouse gas flux from agricultural fields using instrumented towers. Can. J. Soil Sci. 86:373–400. doi:10.4141/S05-100

Pattey, E., I.B. Strachan, R.L. Desjardins, G.C. Edwards, D. Dow, and I.J. MacPherson. 2006b. Application of a tunable diode laser to the measurement of CH4 and N2O fluxes from field to landscape scale using several micrometeorological techniques. Agric. For. Meteorol. 136:222–236. doi:10.1016/j.agrformet.2004.12.009

Peñuelas, J., and Y. Inoue. 1999. Reflectance indices indicative of changes in water and pigment contents of peanut and wheat leaves. Photosynthetica 36:355–360. doi:10.1023/A:1007033503276

Petropoulos, G.P. 2014. Remote sensing of energy fluxes and soil moisture Content. CRC Press, Taylor and Francis Group, Boca Raton, FL.

Potgieter, A.B., G.L. Hammer and A. Doherty.2006. Oz-Wheat: A regional-scale crop yield simulation model for Australian wheat. Information Series. Queensland Department of Primary Industries & Fisheries, Brisbane, Australia. p. 20.

Prince, S.D. 1991. A model of regional primary production for use with coarse resolution satellite data. Int. J. Remote Sens. 6:1313–1330. doi:10.1080/01431169108929728

Purevdorj, T., R. Tateishi, T. Ishiyama, and Y. Honda. 1998. Relationships between percent vegetation cover and vegetation indices. Int. J. Remote Sens. 19:3519–3535. doi:10.1080/014311698213795

Qian, B., R. De Jong, R. Warren, A. Chipanshi, and H. Hill. 2009. Statistical spring wheat yield forecasting for the Canadian Prairie provinces. Agric. For. Meteorol. 149:1022–1031. doi:10.1016/j.agrformet.2008.12.006

Reichert, G.C., and D. Caissy. 2002. A reliable crop condition assessment program (CCAP) incorporating NOAA AVHRR data, a geographical information system, and the internet. Statistics Canada. Ottawa, ON. http://www26.statcan.ca/ccap-peec/esri-2002conf-eng.jsp (verified 20 July 2018).

Robinson, D.A., S.B. Jones, J.M. Wraith, D. Or, and S.P. Friedman. 2003. A review of advances in dielectric and electrical conductivity measurement in soils using Time Domain Reflectometry. Vadose Zone J. 2:444–475. doi:10.2136/vzj2003.4440

Robock, A., K.Y. Vinnikov, G. Srinivasan, J.K. Entin, S.E. Hollinger, N.A. Speranskaya, and A. Namkhai. 2000. The global soil moisture data bank. Bull. Am. Meteorol. Soc. 81:1281–1299. doi:10.1175/1520-0477(2000)081<1281:TGSMDB>2.3.CO;2

Rochette, P., E.G. Gregorich, and R.L. Desjardins. 1992. Comparison of static and dynamic closed chambers for measurement of soil respiration under field conditions. Can. J. Soil Sci. 72:605–609. doi:10.4141/cjss92-050

Rochette, P., D. Worth, E. Huffman, J.A. Brierley, B.G. McConkey, J.Y. Yang, J.J. Hutchinson, R.L. Desjardins, R. Lemke, and S. Gameda. 2008. Inventory of agricultural N2O emissions in Canada using a country-specific methodology. Can. J. Soil Sci. 88(5):655–669. doi:10.4141/CJSS07026

Ross, J.K. 1981. The radiation regime and architecture of plant stands. : Kluwer Boston. Hingham, MA doi:10.1007/978-94-009-8647-3

Romano, N. 2014. Soil moisture at local scale: Measurements and simulations. J. Hydrol. 516:6–20. doi:10.1016/j.jhydrol.2014.01.026

Rouse, J.W., R.H. Haas, J.A. Schell, and D.W. Deering. 1973. Monitoring vegetation systems in the Great Plains with ERTS-1. Proceedings of the Third Earth Resources Technology Satellite Symposium, 10-14 Dec. 1973. Goddard, MD. NASA, Washington, D.C. p. 309-317.

Running, S.W., R.R. Nemani, D.L. Peterson, L.E. Band, D.F. Potts, L.L. Pierce, and M.A. Spanner. 1989. Mapping regional evapotranspiration and photosynthesis by coupling satellite data with ecosystem simulation. Ecology 70:1090–1101. doi:10.2307/1941378

Schaefer, G.L., M.H. Cosh, and T.J. Jackson. 2007. The USDA natural resources conservation service Soil Climate Analysis Network (SCAN). J. Atmos. Ocean. Technol. 24:2073–2077. doi:10.1175/2007JTECHA930.1

Schmugge, T., T.J. Jackson, and H.L. McKim. 1980. Survey of methods for soil moisture determination. Water Resour. Res. 16(6):961–979. doi:10.1029/WR016i006p00961

Sellers, P.J. 1985. Canopy reflectance, photosynthesis and transpiration. Int. J. Remote Sens. 6:1335–1372. doi:10.1080/01431168508948283

Semenov, M., and F.J. Doblas-Reyes. 2007. Utility of dynamical seasonal forecasts in predicting crop yield. Clim. Res. 34:71–81. doi:10.3354/cr034071

Seyfried, M.S., L.E. Grant, E. Du, and K. Humes. 2005. Dielectric loss and calibration of the hydra probe soil water sensor. Vadose Zone J. 4:1070–1079. doi:10.2136/vzj2004.0148

Shafer, B.A., and L.E. Dezman. 1982. Development of a surface water supply index (SWSI) to assess the severity of drought conditions in snowpack runoff areas. In: Proceedings of the 50th Annual Western Snow Conference, Reno, NV. Western Snow Conference, Brush Prairie, WA. p. 164-175.

Smith, W.N., R.L. Desjardins, B. Grant, C. Li, M. Corre, P. Rochette, and R. Lemke. 2002. Testing the DNDC model using N2O emissions at two experimental sites. Can. J. Soil Sci. 82:365–374. doi:10.4141/S01-048

Smith, W.N., B. Grant, R.L. Desjardins, R. Lemke, and C. Li. 2004. Estimates of the interannual variations of N2O emissions from agricultural soils in Canada. Nutr. Cycl. Agroecosyst. 68:37–45. doi:10.1023/B:FRES.0000012230.40684.c2

Statistics Canada. 2012. Definitions, data sources and methods of Field Crop Reporting Series. Record number: 3401. Agriculture Division, Statistics Canada, Ottawa, ON. http://www.statcan.gc.ca/imdb-bmdi/3401-eng.htm (verified 20 July 2018).

Stenitzer, E. 1993. Monitoring soil moisture regime of field crops with gypsum blocks. Theor. Appl. Climatol. 48:159–165. doi:10.1007/BF00864922

Svoboda, M., D. LeComte, M. Hayes, R. Heim, K. Gleason, J. Angel, et al. 2002. The drought monitor. Bull. Am. Meteorol. Soc. 83:1181–1190. doi:10.1175/1520-0477(2002)083<1181:TDM>2.3.CO;2

Topp, G.C., J.L. Davis, A.P. Annan. 1980. Electromagnetic determination of soil water: Measurements in coaxial transmission lines. Water Resour. Res. 16:574–582. doi:10.1029/WR016i003p00574

Tran, A.P., P. Bogaert, F. Wiaux, M. Vanclooster, and S. Lambot. 2015. High-resolution space–time quantification of soil moisture along a hillslope using joint analysis of ground penetrating radar and frequency domain reflectometry data. J. Hydrol. 523:252–261. doi:10.1016/j.jhydrol.2015.01.065

Trishchenko, A.P., J. Cihlar, and Z. Li. 2002. Effects of spectral response function on surface reflectance and NDVI measured with moderate resolution satellite sensors. Remote Sens. Environ. 81:1–18. doi:10.1016/S0034-4257(01)00328-5

Tucker, C.J. 1979. Red and photographic infrared linear combinations for monitoring vegetation. Remote Sens. Environ. 8:127–150. doi:10.1016/0034-4257(79)90013-0

Tucker, C.J., I.Y. Fung, C.D. Keeling, and R.H. Gammon. 1986. Relationship between atmospheric CO2 variations and satellite-derived vegetation index. Nature 319:195–199. doi:10.1038/319195a0

Tucker, C.J., W.W. Newcomb, S.O. Los, and S.D. Prince. 1991. Mean and inter-year variation of growing-season normalized difference vegetation index for the Sahel 1981-1989. Int. J. Remote Sens. 12:1133–1135. doi:10.1080/01431169108929717

Ulaby, F.T., R.K. Moore, and A.K. Fung. 1986. Microwave remote sensing: Active and passive. Vol. III. From theory to applications. Artech House, Dedham, MA.

VanderZaag, A.C., J.D. MacDonald, L. Evans, X.P.C. Verge, and R.L. Desjardins. 2013. Towards an inventory of methane emissions from manure management that is responsive to changes on Canadian farms. Environ. Res. Lett. 8(3):035008. doi:10.1088/1748-9326/8/3/035008

VanderZaag, A.C., T.K. Flesch, R.L. Desjardins, H. Baldé, and T. Wright. 2014. Methane emissions from two dairy farms in eastern Ontario: Seasonal and manure-management effects. Agric. For. Meteorol. 194:259–267. doi:10.1016/j.agrformet.2014.02.003

Vergé, X.P.C., J.A. Dyer, D.E. Worth, W.N. Smith, R.L. Desjardins, and B.G. McConkey. 2012. A greenhouse gas and soil carbon model for estimating the carbon footprint of livestock production in Canada. Animals (Basel) 2:437–454. doi:10.3390/ani2030437

Vermote, E.F., and N.Z. Saleous. 2006. Calibration of NOAA16 AVHRR over a desert site using MODIS data. Remote Sens. Environ. 105:214–220. doi:10.1016/j.rse.2006.06.015

Vicente-Serrano, S.M., B. Santiago, J. Lorenzo-Lacruz, J.J. Camarero, J.I. López-Moreno, C. Azorin-Molina, et al. 2012. Performance of drought indices for ecological, agricultural, and hydrological applications. Earth Interactions. 16(10):1-27. http://EarthInteractions.org

Wagner-Riddle, C., A. Furon, N.L. McLaughlin, I. Lee, J. Barbeau, S. Jayasaundara, et al. 2007. Intensive measurement of nitrous oxide emissions from a corn-soybean-wheat rotation under two contrasting management systems over 5 years. Glob. Change Biol. 13:1722–1736. doi:10.1111/j.1365-2486.2007.01388.x

Weiser, R.L., G. Asrar, G.P. Miller, and E.T. Kanemasu. 1986. Assessing grassland biophysical characteristics from spectral measurements. Remote Sens. Environ. 20:141–152. doi:10.1016/0034-4257(86)90019-2

Wilhite, D.A., and M.H. Glanz. 1985. Understanding the drought phenomenon: The role of definitions. Water Int. 10:111–120. doi:10.1080/02508068508686328

World Meteorological Organization. 2012. Standardized precipitation index user guide. WMO-No. 1090. World Meteorological Organization (WMO), Geneva, Switzerland. http://www.wamis.org/agm/pubs/SPI/WMO_1090_EN.pdf (verified 20 July 2018).

World Meteorological Organization. 2008. Guide to meteorological instruments and methods of observation WMO-No. 8. World Meteorological Organization (WMO), Geneva, Switzerland. https://www.wmo.int/pages/prog/www/IMOP/CIMO-Guide.html (verified 20 July 2018).

Wu, B., J. Meng, Q. Li, N. Yan, X. Du, and M. Zhang. 2014. Remote sensing-based global crop monitoring: Experiences with China's CropWatch system. Int. J. Digit. Earth 7(2):113–137. doi:10.1080/17538947.2013.821185

Zaitchik, B.F., M. Rodell, and R.H. Reichle. 2008. Assimilation of GRACE terrestrial water storage into a land surface model: Results for the Mississippi River basin. J. Hydrometeorol. 9:535–548. doi:10.1175/2007JHM951.1

Zou, J., Y. Huang, J. Jiang, X. Zhang and R.L. Sass. 2005. A 3-year field measurement of methane and nitrous oxide emissions from rice paddies in China: Effects of water regime, crop residue, and fertilizer application. Global Biogeochemical Cycles. 19(2):GB2021. doi:1029/2004GB002401.

Zreda, M., D. Desilets, T.P.A. Ferre′, and R.L. Scott. 2008. Measuring soil moisture content non-invasively at intermediate spatial scale using cosmic-ray neutrons. Geophys. Res. Lett. 35:L21402. doi:10.1029/2008GL035655

Zreda, M., W.J. Shuttleworth, X. Zeng, C. Zweck, D. Desilets, T. Franz, and R. Rosolem. 2012. COSMOS: The cosmic-ray soil moisture observing system. Hydrol. Earth Syst. Sci. 16:4079–4099. doi:10.5194/hess-16-4079-2012

Crop Models as Tools for Agroclimatology

Heidi Webber*, Munir Hoffmann, Ehsan Eyshi Rezaei

Abstract

Climate changes will bring average warmer temperatures, elevated atmospheric CO_2 and more frequent extreme weather events. These are expected to result in impacts on crop growth and cropping systems. However, traditional field and controlled environment experiments are limited in studying climate change impacts on crops due to the very large number of combinations of precipitation, CO_2 and temperature changes possible. As such, crop models are the main tool to study climate change impacts on crops, overcoming the limitations of field experimentation. However, given the complexity and scientific uncertainty around crop response to multiple stressors expected to prevail under elevated CO_2 and climate change, crop model projections still contain many uncertainties limiting their utility in risk assessments or informing adaptations. Model improvement is needed to represent processes driving responses to multiple stressors. Methodological advances are also required to assess the adequacy of using field-based models with aggregate climate, soil, and crop management data over large areas in impact assessments. Despite these limitations in assessing risks of climate change to cropping systems, process-based crop models do offer capacity to aid in understanding crop responses observed in field conditions as they can be used to test hypotheses about which processes drive observed responses and can aid in helping inform experimental design.

The study of potential climate change impacts on crops and cropping systems is challenging. While controlled environment studies provide important insights about mechanisms of crop response to environmental variables that drive crop growth and development (Sadras and Calderni, 2014), knowledge about the extent to which sequences of weather events, exposure to multiple stressors, previous stressor exposure, and acclimation remains fragmented (Hatfield et al., 2011). Indeed, field experiments considering canopy-level dynamics are required to capture interactions and feedbacks with resource capture and use. Nevertheless, results derived from field experiments are sometimes contradictory (Suzuki et al., 2014) likely reflecting differences in soils, background climate, management, genotypes and methods used to impose stressors. Indeed, future climate change is highly uncertain and comprehensive experimentation for possible scenarios is prohibitively expensive. Crop models which simulate crop growth and

---

Abbreviations: GCM, global climate model; RCM, regional climate model; RAP, representative agricultural pathway; RUE, radiation use efficiency; SOC, soil organic carbon; SSP, shared socioeconomic pathway; VPD, vapor pressure deficit.

H.Weber and M. Hoffman, Leibniz-Centre for Agricultural Landscape Research (ZALF), 15374 Müncheberg, Germany; E.E. Rezaei, Department of Crop Sciences, University of Göttingen, Göttingen 37075, Germany. *Corresponding author: webber@zalf.de

doi:10.2134/agronmonogr60.2016.0025

© ASA, CSSA, and SSSA, 5585 Guilford Road, Madison, WI 53711, USA.
Agroclimatology: Linking Agriculture to Climate, Agronomy Monograph 60.
Jerry L. Hatfield, Mannava V.K. Sivakumar, John H. Prueger, editors.

development as a function of weather, soil, genotype and management variables are a widely used alternative to field experiments to investigate climate change impacts on crops (White et al., 2011a) particularly at large scale assessments. A main challenge in their use relates to interpretation of their results, as studies differ widely in model skill, processes included, climate change scenarios considered and underlying assumptions about crop management and adaptations (Webber et al., 2014). While process-based crop models have been developed over decades (Boote et al., 2013; van Ittersum et al., 2003), their original purpose was not to address extreme events, though these are now understood as posing large risks to crop production and food security (Ewert et al., 2015).

In this chapter, we present an overview of the main modeling approaches and associated methodologies to assess climate change impacts, including quantification of uncertainties in impact assessments. We briefly discuss challenges around: accounting extreme events and multiple stressors; pests, weeds and disease; and use of field scale models at larger scales. Crop models also have the potential to aid in understanding the role of cropping systems in climate change mitigation (Ehrhardt et al., 2018), though this is beyond the scope of this chapter.

Approaches to Modeling

Mathematical models used to relate climate and other variables to crop growth and yield are broadly termed *crop models*. They range from very detailed, mechanistic representations of hourly processes to highly empirical models regressing a single growing season variable (e.g., precipitation) to grain yields. While most models lie somewhere along this spectrum, we can distinguish two main groups of models: process-based crop growth models and statistical models (Lobell and Asseng, 2017). For most of this chapter we focus on the former as they have the potential to identify underlying mechanisms of crop responses to climate (Boote et al., 2013; White et al., 2011a), though the later are widely applied in climate change impact studies (e.g., Schlenker and Lobell, 2010; Schlenker and Roberts, 2009; Tack et al., 2017b). Statistical crop models can vary in complexity, but are generally regression-based relationships that associate variation in observed crop yields with variation in weather or other variables (Lobell and Asseng, 2017). In associating observed variation in yield to climate or other factors, statistical crop models can make use of variation in time at a single location, across locations at a single time in cross-sectional studies, or using a combination of these in panel studies (Auffhammer et al., 2013). The various strengths of the different types of statistical models is beyond the scope of this chapter (Auffhammer and Schlenker, 2013). A main advantage of statistical over process-based models is that they can relate climate variables to yield observations that reflect farmer's management conditions and factors like pest and disease pressure that most process-based model do or cannot adequately represent (Roberts et al., 2017). On the other hand, process-based models can better identify the causal mechanisms explaining observed responses. The ultimate choice of model type depends on the research question, though the benefits of comparing and combining approaches is increasingly recognized (Roberts et al., 2017; Siebert et al., 2017).

Process-based Crop Models: Overview and Data Requirements

Process-based crop models have a long history of use in aiding understanding of crop response to environmental conditions, dating to the models of de Wit (1965) and (Monteith, 1965) relating photosynthesis response to light interception. Models have since evolved to consider crop growth and yield formation processes in response to a wider range of variables, including temperatures, soil texture and nutrient status, and crop management such as fertilization and irrigation (Boote et al., 2013). Many of the most widely used models can be considered modeling frameworks (Holzworth et al., 2015) as they allow selecting which limitations to consider (e.g., nitrogen, water or phosphorus) or selection of sub-modules with differing degrees of complexity, for example, soil water balance in APSIM (Keating et al., 2003) or assimilation model and soil carbon dynamics in DSSAT (Jones et al., 2003). These models are now widely applied to assess the implications of alternate crop management and increasingly in climate change impact assessments (Ewert et al., 2015). Finally, it should be emphasized that process-based models are characterized by varying levels as empiricism, particularly for complex underlying mechanisms (Wallach et al., 2013).

Common to most process-based crop models is that weather, soil texture, crop genotypes and basic crop management are exogenous inputs to the model that drive cropping system dynamics (Fig. 1). The processes considered by any particular model vary widely as does the level of detail captured. However, most models will simulate crop growth and possibly related soil water and nutrient processes on a daily or sub-daily basis which in turn may affect the subsequent day's radiation capture or water use, for example. Their dynamic nature which can capture feedbacks is one of the most important features distinguishing process-based from statistical crop models. Final grain yield or biomass production is a common end season output variable of interest, though many other output variables are also available. Depending on the model, leaf area, water use, grain protein content or soil carbon content can be output variables as daily or end season values. In Fig. 1, output variables are called impact indicators as they are analyzed to assess climate change impacts.

Temperature, precipitation and radiation are the primary inputs of all crop models (Sinclair and Seligman, 1996), generally as daily values, but in some cases hourly. Wind speed and a measure of air humidity are required inputs for many models (Hoogenboom, 2000). As availability and quality issues with humidity data are common, the FAO has produced a set of recommendations to aid with cases of missing data for estimating evapotranspiration (Allen et al., 1998). Other required input information include site latitude and elevation, crop management (e.g., sowing date) and genotype (required temperature sum to reach flowering and maturity or related life cycle indicators). Some models may require additional genotype information, but these generally have default values and can be considered as model parameters for a specified genotype. Models considering water or nutrient limitation additionally require soil (physical and chemical characteristics) and associated management data (fertilizer and irrigation application amounts and dates). Cropping system models that consider the dynamics of soil carbon and nitrogen also require information on crop rotations and residue management and possibly tillage method. Finally, process-based crop models require specification of initial values for any cumulative state variables (e.g., seed mass, initial soil water or mineral nitrogen content).

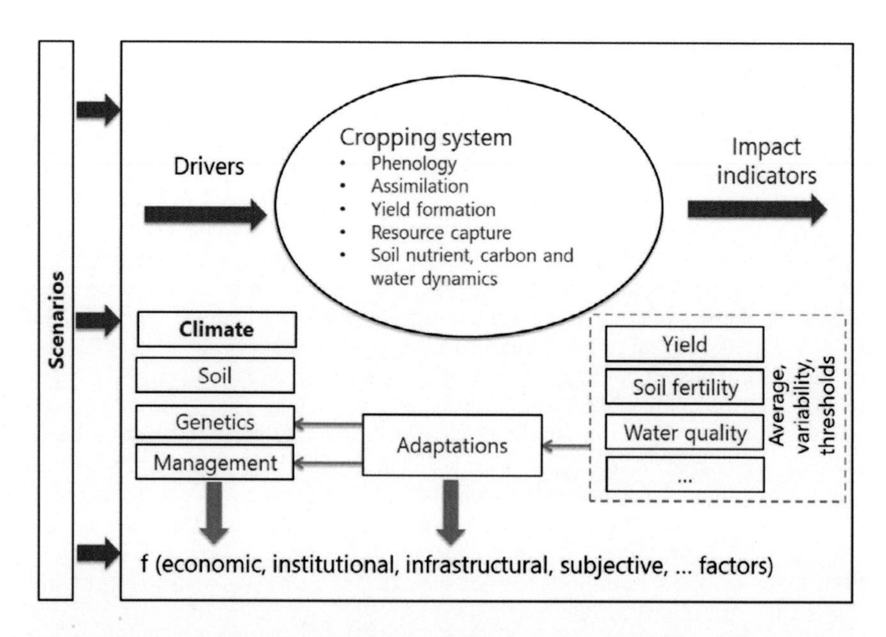

Fig. 1. Process-based crop models use mathematical formulations to represent processes of crop growth and in some cases crop systems (e.g., soil water and nutrient dynamics). Exogenous input variables (e.g., daily weather, soil) drive crop growth and system dynamics while a range of output variables indicate the state of the crop or soil. In climate change impact assessments, scenarios are used to vary the drivers, primarily climate, and the impact assessed for changes in average, variability or frequency of surpassing specified thresholds. Adaptations may be identified to minimize negative impact of climate change, though the economic, institutional and infrastructure requirements or suitability need to evaluation with other tools

This section outlines the main processes of crop growth and development common to many process-based models. It is certainly not comprehensive, but intended to provide a basic understanding of their functioning. In a following section, we detail the responses to these processes to climate variables.

Almost all process-based models simulate crop phenological development and lifecycle. This is a critical step as it sets the duration over which crops can intercept radiation as well as determines the timing of when they are most sensitive to various stressors (Boote et al., 2013). Most models employ the concept of thermal time to drive leaf initiation and crop development from sowing, to emergence, to anthesis and maturity, which may be further moderated by a vernalization response, and effects of daylength (McMaster et al., 2008; White et al., 2008).

To simulate biomass assimilation and growth, one of two approaches is usually implemented. The first involves the explicit simulation of photosynthesis and respiration. Within this approach, most models use fairly empirical equations of leaf level photosynthesis which can be limited by light or internal CO_2. Most do not include explicit linking of stomatal conductance and photosynthesis, but instead use empirical relationships to reduce photosynthesis as a function of soil water content or the ratio of actual to potential transpiration. An exception to this is the GEOCROS model (Xinyou and Van Laar, 2005) with explicit coupling of photosynthesis, leaf nitrogen, transpiration, and stomatal conductance. The mechanistic basis of respiration is less well understood and therefore captured in crop growth models (Xinyou and Van Laar, 2005). It is generally simulated as two distinct processes: growth and maintenance respiration (Boote et al., 2013). However, respiration is now understood as being associated with growth, nitrogen, and other ion uptake and phloem translocation, as well as harder to estimate protein turnover, cell ion contents among others (Cannell and Thornley, 2000). The second and more widely applied approach for biomass assimilation makes use of the concept of crop radiation use efficiency (RUE) in which incident radiation intercepted by the crop canopy to converted to biomass (Monteith, 1977). Radiation use efficiency is an empirical conversion factor that can vary with crop species, age, nitrogen and water status, and ambient CO_2 and temperature (Sinclair and Horie, 1989). A smaller number of models, most notably AQUACROP (Raes et al., 2009), considers water productivity (WP) as a conversion factor, which varies with species and CO_2 to convert water use to biomass. Whichever method is used to simulate biomass, most models are source limited in that they partition biomass between roots, stems, tubers, leaves, and seeds as a function of crop development stage (Goudriaan and Van Laar, 2012). Biomass allocated to leaves contributes to leaf area and radiation capture. A smaller number of models, termed sink limited, explicitly model leaf growth or seed formation potentially limiting assimilation (Boote et al., 2013; Kumar et al., 2016).

There are many approaches to determine water and nitrogen demand, or requirements, for potential growth of a particular crop, in a particular location under ambient CO_2 (Van Ittersum and Rabbinge, 1997). Growth can be limited when there is inadequate supply of water, nitrogen, or other factors. Models employ a number of methods of varying complexity to simulate soil water and nitrogen availability and uptake by the crop. Soil water balances can range from single layer tipping bucket approaches which assume one soil horizon and no water movement within the profile (Wolf, 2012), to multilayer tipping buckets accounting for heterogeneous horizons (Ritchie, 1998) and time constrained nature of water infiltration (Addiscott and Whitmore, 1991), to analytical solutions of Richard's equation as in APSIM-SWIM (Huth et al., 2012). The degree of limitation by water is commonly estimated as the ratio of actual to potential water supply or soil water content falling below a lower threshold. The degree of nitrogen limitation is based on low values of crop nitrogen content which can be quantified in a variety of ways. From these limitations, empirical coefficients are estimated and applied reduce radiation use efficiency or photosynthesis and leaf expansion, and also to alter partitioning of assimilates between roots, shoots and seeds (Ewert et al., 2015).

Model Parameterization and Calibration

Parametrization is critical to ensure the accuracy of simulated responses to climate in process-based crop models (Wang et al., 2017). In reality parameters do not represent purely genotypic traits, but rather genotype by environment interactions, as models are necessarily simplifications of real systems which are based on experimental evidence generated in a small number of environments. Calibration is the adjustment of genotype parameters such that model prediction errors are minimized under specified management and environmental conditions. However, the process of model calibration is suggested to constitute a large source of uncertainty crop modeling (Seidel et al., 2018). Sources of uncertainty in calibration stem from measurement errors in observation, limited number of observations relative to parameters, and the method used for calibration (Kersebaum et al., 2015). Standard statistical methods often do not lead to the best parameter set owing to the complex nonlinear structure of many processes and the existence of cross correlation between them and there is no generally accepted best calibration procedure (Seidel et al., 2018; Wallach et al., 2013). The relative advantages and shortcomings of a number of parameter estimation methods, such as Bayesian, generalized likelihood, Markov Chain Monte Carlo and manual trial and error, have been extensively discussed (Dumont et al., 2014; Stedinger et al., 2008).

Adding model complexity by including more mechanistic process representation is another approach to improving prediction accuracy across environments. However, adding model complexity does not always improve accuracy and can even produce increased uncertainty associated with increased parameterization (Ramirez-Villegas et al., 2017). On the other hand, not including appropriate mechanistic processes can lead to models simulating accurate outcomes for the wrong reason through calibration, missing important feedbacks of intermediate processes and potentially leading to errors under climate change or under other production conditions (Siebert et al., 2017; White et al., 2011a). Therefore, modellers should critically assess required model complexity considering both their research questions as well as data availability for model parametrization (Adam et al., 2011; Passioura, 1996).

Whatever the approach to calibration or degree of model complexity, a number of measures can be used to quantify model performance in simulating observed crop growth against independent datasets (Jame and Cutforth, 1996; Wallach et al., 2013). Measures include simple comparison of means and correlation coefficients between simulated and observed values to more complex indices such as root mean square error (RMSE) (Maiorano et al., 2017), each with specific advantages and disadvantages (Bellocchi et al., 2011). Root mean square error is a widely used index to evaluate crop models, with particular sensitivity to large errors (Chai and Draxler, 2014; Willmott et al., 2009). Combinations of RMSE and other indices such as the mean absolute error, coefficient of determination, or relative errors, are suggested to provide a comprehensive overview of model performance (Wallach et al., 2013).

Responsiveness to Climate Variables Relevant for Climate Change Studies

Temperatures

Warmer temperatures projected with climate change will have a number of consequences for crop production, with the significance differing between crops, tropical versus temperate locations, and the capacity to adapt cropping systems (Hatfield et al., 2011). Before discussing crop model's ability to assess temperature changes, a distinction is made here between mean temperature effects and those related to more frequent temperature extremes, recognizing that warmer average temperatures imply more days with very hot temperatures. We discuss both in this section below.

In simulating crop phenology, thermal time accumulation is commonly treated as a linear function of temperatures (daily or hourly) above a base temperature, below which development ceases. Many models consider optimum temperatures above which development is not accelerated, and some include functions to decrease development at temperatures higher than optimum, as documented for a number of wheat models (Wang et al., 2017). Parameterization of phenology models is difficult due to large number of temperature and photoperiod related parameters (White et al., 2011b). Furthermore, there is limited theoretical basis on which plants development and other physiological rates are understood (Parent et al., 2010). The diversity of approaches and parameter values demonstrated by Wang et al. (2017) results in very large uncertainty in wheat phenology development. Some models additionally consider vernalization or cold temperature requirements in early vegetative phases using a similar thermal time concept. Leaf initiation and elongation rate, which are also highly sensitive to temperature, were determined by Parent et al. (2010) to follow a common temperature response with development rates after normalization. Their thermodynamically-based phenology model is proposed as an alternative to linear thermal time equations used in most models.

Assimilation is a second main process in crop models that contains average temperature responses. For more detailed models considering leaf-level photosynthesis, algorithms have been developed to account for temperature sensitivities of Rubisco (Bernacchi et al., 2001) and RuBP regeneration - (Bernacchi et al., 2003) limited photosynthesis (Medlyn et al., 2002). However, currently very few models capture the effects of stomatal conductance rate controlling leaf temperature feeding back onto photosynthesis. In principle this could be implemented for GECROS (Xinyou and Van Laar, 2005), in a manner similar to other applications (e.g., Kimball et al., 2015). Respiration associated with growth, nutrient uptake, uptake of other substances, and translocation is understood as temperature insensitive and treated as such in crop models (Boote et al., 2013). However, for the commonly termed 'maintenance' respiration, use of a constant Q10 factor to increase respiration with higher temperature is common (Wang et al., 2017) despite evidence that Q10 decreases as temperatures rise (Tjoelker et al., 2001). The mechanistic GECROS model does not include a temperature dependence on maintenance respiration (Xinyou and Van Laar, 2005) based on evidence that short-term responses of respiration to temperature are generally not equivalent to longer-term responses (Gifford, 1995). For models implementing RUE approaches, many include temperature responses that act to increase RUE from a base temperature to a peak (or plateau) before decreasing again at higher temperature (Wang et al., 2017).

While models have long been used to address effects of average temperatures on crop growth or development, they were not originally designed to assess the effects of extreme hot or freezing temperatures (Ewert et al., 2015). Until recently, few models explicitly considered heat stress effects (White et al., 2011a), though recent years have seen concerted efforts to improve models' high temperature responses (Maiorano et al., 2017). Heat stress refers to the large yield losses associated with failure of flowering or fertilization, grain abortion, or accelerated leaf senescence when even very short periods (hours to days) of temperatures above a critical, crop-specific threshold are experienced (Barlow et al., 2015; Eyshi Rezaei et al., 2015b). Effects considered may include those on flowering (Lizaso et al., 2017), harvest index (Moriondo et al., 2011), yield formation (Challinor et al., 2005; Gabaldón-Leal et al., 2016; Lobell et al., 2015) and grain filling rate (in some versions of APSIM) and duration via accelerated leaf senescence (Asseng et al., 2011). While most models use air temperature to simulate heat stress, various models use simulated crop canopy temperature (Webber et al., 2018b).

Risk of crop damage from freezing temperatures may increase if temperatures become more variable (Rigby and Porporato, 2008) and flowering events advance with warmer temperatures (Eyshi Rezaei et al., 2017). Frost is currently a major constraint to production in Northern Europe (Olesen et al., 2011), though is little considered in crop modeling studies. A few models can simulate frost damage, though there are few examples of published studies evaluating their response. Responses in current models include crop failure as a function of the length of time of freezing temperatures. Many models reduce leaf expansion rates, again depending on the duration of exposure to freezing temperatures. Finally, crop models are also able to assess effects of longer average frost free periods but rely on rules or scenarios of altered crop management to specify earlier sowing dates.

Indeed, the power of crop models is that they enable assessment of how the various processes affected by temperature combine over the course of the growing season to impact crop yields. For example, a global wheat model ensemble that projected a 6% decrease in wheat production for each 1 °C increase in temperature reproduced observed responses over a mean temperature range of 15 to 32 °C (Asseng et al., 2015), though model accuracy decreased at higher temperatures. Beyond projecting yield changes, process-based modeling studies are able decompose the response and attribute it to underlying processes. Studies by Webber et al. (2018a) and Faye et al. (2018) associated yield losses under climate change to earlier maturity for cereal crops in Europe and the Sudan Savanna of West Africa, respectively. For both studies, sensitivity to photoperiod emerged as an important factor appearing to moderate the temperature effects on crop development. Webber et al. (2018a) attributed additional losses in maize due to warming to increased drought, even after controlling for earlier maturity and detecting little change in precipitation. Drought stress did not increase for winter wheat. Atmospheric water demand increases with the vapor pressure deficit (VPD) of the atmosphere, and for a given amount of humidity in the air, VPD will increase relatively more for higher summer temperatures than in winter (Monteith and Unsworth, 2007). These results highlight the power of process-based modeling to attribute impacts of warmer temperatures to underlying process, though we deal more with responsiveness to drought in the section on precipitation.

Precipitation

While too little precipitation may lead to drought stress and water deficit in crops, excessive rains and rainfall intensity lead to waterlogging, increased disease loads and harvest losses due to lodging (Kristensen et al., 2011). Here we discuss first how crop models consider the main effects of drought and then excess rain and waterlogging. While intense rainfall can also cause lodging, this is rarely included in crop models and we do not treat it here. Similarly, high humidity can affect disease loads, though it is beyond the scope of this chapter.

Dry soils leading to drought stress in crops result not only from lack of precipitation but also from high temperatures which drive crop water use via high vapor pressure deficit, as discussed in the previous section. As such dry soils can be thought to result from the demand for water exceeding the supply. Many crop models use a variant of this concept, supply not meeting demand- but applied on a daily basis- to simulate drought stress effects (Boote et al., 2013). Others consider the fraction of available water (between a threshold for extraction without stress and wilting pointing) which remains in the soil (Sinclair et al., 2005). The latter probably better represents how crops perceive and respond to drought, which is understood as resulting from roots sensing drying soil which lead to a combination of hormonal signaling (Davies and Zhang, 1991) and altered water potential in the plant (related to the cumulative force enabling water to flow through plants, with elements related to gravitation and osmotic and pressure gradients). This signaling and changed water potential is a short-term response to be distinguished from the longer-term emergent consequences of these responses (Tardieu et al., 2018).

One of the earliest observed crop responses to dry soil is the reduction in leaf elongation rate that is understood to be controlled by changes in leaf water potential (Caldeira et al., 2014; Tardieu et al., 2018). This is reflected in some models (e.g., DSSAT-CSM, CERES-Wheat and AFRCWHEAT2), which reduce leaf expansion rates or directly reducing leaf area index even when daily supply exceeds demand and before other processes are effected, though with no explicit modeling of crop water potentials. Stomatal closure acting to reduce transpiration and photosynthesis and/or assimilation is another short-term response to dry soil. However, most models reduce photosynthesis or RUE with drought stress (Ewert et al., 2015), essentially imposing the response on photosynthesis, though it is known to be more tolerant than stomatal functioning or tissue elongation to drought stress (Boyer, 1970), with growth becoming sink limited as drought conditions persist (Muller et al., 2011). As a consequence of not directly simulating stomatal conductance, models miss feedbacks between stomatal closure, transpiration and canopy temperature (for an exception, an earlier variant of DSSAT-CSM Pickering et al., 1995). Instead, approaches to model transpiration are usually physically based, driven by crop water extraction as a function of the atmospheric evaporative demand (Allen et al., 1998) and actual soil water content. Finally, many models increase partitioning of new assimilates to the roots or remobilize a greater portion of water soluble carbohydrates from stems to grains under drought.

These responses act over short time periods with various feedbacks and interactions (Tardieu et al., 2018), and it is their collective time-integrated response that are observed under production conditions in terms of rooting patterns, transpiration rates, leaf area and associated dynamics, water use efficiency, and seed abortion, among others. A complication with observation of drought response is dependence on drought type (duration, intensity, timing), as a successful adaptive trait in a very dry environment may come with a yield penalty under non-water limiting conditions (Richards et al., 2002; Tardieu, 2012). For example, Lopes and Reynolds (2010) found wheat genotypes with increased partitioning to roots that maintain relatively higher transpiration rates (evidenced by cooler canopy temperature) yield higher than varieties with lower root biomass under drought conditions. However, this strategy can only be successful when soil water at depth is available. In environments with no residual water, partitioning of assimilates to stem reserves for later remobilization to grain is a more successful strategy in terms of grain yield (Dreccer et al., 2009).

Similarly, early vigor can limit evaporation losses and save water for later in the season in environments with frequent rainfall (Rebetzke and Richards, 1999), but may lead to yield penalties in environments where stored soil water is relied on and evaporative demand is high (Chenu et al., 2013; Tardieu et al., 2018). Process-based crop models are well suited to model the short-term responses and emergent longer-term crop state. However, accurate simulation of drought stress response is sensitive to the methods used to simulate evaporative demand of the atmosphere, soil water balance, partitioning between evaporation and transpiration, soil water evaporation, root growth, root water uptake, water stress effects on transpiration rates, and the various feedbacks between these responses and above ground growth (Jamieson et al., 1998; Webber et al., 2016a). Realistic simulation of crop water supply and demand is further complicated by the need to specify soil properties which can be highly variable even in a single field and the fact that models are rarely evaluated on soil water use or root growth. The reliance on above ground biomass and yield for model calibration under drought stress has inevitably resulted in models getting the right yield for the wrong reason, as suggested in a maize multi-model comparison (Bassu et al., 2014).

Waterlogging or flooding, occurs when the root and/or shoot is in saturated soil or submerged, respectively, with the magnitude of damage largely dependent on duration of saturated conditions (Ahmed et al., 2013). Damage from waterlogging occurs due to oxygen deprivation (Armstrong et al., 1983), which plants require for respiration and oxidation reactions (Armstrong et al., 2009), with root growth and function being most severely affected (Malik et al., 2001). It is fairly common for crop models to account for effects of oxygen deprivation on root and shoot growth and reducing transpiration using a range of empirical approaches (Shaw and Meyer, 2015; Shaw et al., 2013). APSIM (Asseng et al., 1998) and SWAGMAN Destiny (Yang et al., 2016) include effects of waterlogging on root and canopy growth. AquaCrop (Vanuytrecht and Thorburn, 2017) decreases transpiration when soil water content exceeds saturation. Finally models such as DRAINMOD (Skaggs et al., 2012) directly reduce daily grain yield with waterlogged conditions.

Elevated CO_2

Elevated atmospheric CO_2 will increase biomass production and crop yield by stimulating photosynthesis under well-watered and fertilized conditions for C3 crops (Ainsworth and Long, 2005; Kimball, 2016). On the other hand, photosynthesis is not stimulated for C4 crops under well-watered conditions with little or no impact on biomass production (Kimball, 2016; Leakey et al., 2009). Process-based crop models largely capture this response for C3 crops by using factors to increase RUE with more mechanistic photosynthesis models increasing assimilation as a function of CO_2, though some models contain similar responses for C4 crops, which is not supported by empirical evidence or theory (Vanuytrecht and Thorburn, 2017). For both C3 and C4s, elevated CO_2 reduces stomatal conductance and therefore leaf level transpiration, which results in reduced water use, particularly for C4 crops (Kimball, 2016; Leakey et al., 2009). For C3s, some evidence suggests that water use at a whole plant level may increase if enhanced assimilation leads to higher leaf area, though the effect will likely depend on within season dynamics of water availability and plant growth (Gray et al., 2016). How reduced transpiration rates translates to reduced negative impacts on crop yield will depend on the duration and severity of water limitation as explained by Kimball (2016). Models differ in their consideration and treatment of effects of elevated CO_2 on transpiration (Vanuytrecht and Thorburn, 2017). Many maize (C4) models were not able to reproduce the water savings and reductions in yield losses under drought and elevated CO_2 (Durand et al., 2018), with this study leading to model improvements as reported in model descriptions in the impact study of Webber et al. (2018a). The same study reported that two of eight winter wheat models applied in Europe did not consider effects of elevated CO_2 on transpiration. Indeed, when properly parameterized, crop models are well suited to capture the feedbacks between increased growth, reduced transpiration rates, and patterns of crop soil water use as they combine to determine final grain yield (Jin et al., 2017). Leaf, photosynthetic nitrogen use efficiency will increase with elevated CO_2 (Leakey et al., 2009), though few models adjust leaf nitrogen content with elevated CO_2 (Vanuytrecht and Thorburn, 2017), for an exception see Olesen et al. (2004).

Ozone

Atmospheric pollutants such as ozone also pose a significant risk to crop production (Ainsworth, 2016). Surface ozone has increased four to five times over the last two centuries (Brauer et al., 2016), and is estimated to put downward pressure on global wheat and maize yields of 4 to 15% and 2 to 5%, respectively (Avnery et al., 2011). Despite its importance in constraining yields, it is not treated by most crop models (Emberson et al., 2018). Accelerated senescence resulting in the shortening of the grain filling duration (Gelang et al., 2008) is one of the largest negative effects of ozone on crop growth, though the mechanisms are not well understood (Emberson et al., 2018). This is captured in both AFRCWHEAT2-O3 and MCWLA-Wheat. Earlier senescence leading to lower grain weight (Feng and Kobayashi, 2009) largely explains the yield reductions associated with ozone, though some crops experience relatively larger reductions in grain number (Feng and Kobayashi, 2009). A slightly larger group of models reduce photosynthesis with elevated ozone (Emberson et al., 2018).

Crop Model Applications in Climate Change Impact Assessments

Climate Scenarios

The future climate is unknown due to both uncertainties in greenhouse gas emissions which will depend on economic and political developments, as well as knowledge gaps in understanding the behavior of the global climate system. Climate change impact assessments use scenarios to account for uncertainty in economic developments, while inputs from multiple climate models are also sometimes used to account for scientific uncertainties about the climate system (Knutti and Sedláček, 2013). Scenarios are not meant as future predictions, but rather represent a range of plausible future conditions (Moss et al., 2010).

Climate scenarios can be produced in a number of ways: altering one or more key climatic input variables from a reference or baseline value (e.g., decrease precipitation by 10% or increase temperature by 1 °C); spatial analogs by considering climate from region in another; or by using the outputs of physically based climate models forced to approximate different levels of greenhouse gas inputs (Moss et al., 2010). Various climate change impact studies using process-based (e.g., Asseng et al., 2015; Bassu et al., 2014) and statistical (e.g., Lobell et al., 2011; Tack et al., 2017a) models have considered uniform temperature increases or precipitation changes to baseline climate records. The use of outputs of physically-based climate models (e.g., various types of global climate models, GCMs, or regional climate models, RCMs) has the advantage of representing possible future climates in which precipitation and temperature changes are physically consistent and associated with specific CO_2. In this case, the scenarios correspond to different greenhouse gas forcing which serve as inputs to the climate models. Presently, representative concentration pathways (RCPs), which produce radiative forcing levels of 8.5, 6, 4.5 and 2.6 W m^{-2} by 2100, are the community standard input to climate models (van Vuuren et al., 2011). Unlike previous climate scenarios, RCPs provide greenhouse gas concentrations without the need to specify the emissions sources. Indeed, a separate process is used to prescribe the economic developments that could produce the concentrations in each RCP, referred to as the Shared Socioeconomic Pathways (SSP) (O'Neill et al., 2014). The RCPs serve as input for large international coordinated experiments, such as CMIP5, which are conducted to understand the global climate system and model differences (Taylor et al., 2012) as well as provide climate scenario projections (Knutti and Sedláček, 2013) for use in crop and other impact models. However, before use in crop models, GCM data must be downscaled either with weather generators, regional climate models, or using a "delta" method to remove any bias between the GCM and historical climate observations (Hempel et al., 2013).

Impact Versus Adaptation Studies

The distinction between studying climate change impacts and adaptations is not straightforward and both types of studies are fraught with difficulties. Climate change impacts denote a change in yield (or other variable of interest) that occur under a scenario climate relative to yield levels in a reference climate. Adaptations refer to the ability of a measure to reduce the impact of climate change– doing so requires to already consider how much the measure would have changed yields levels already in the reference climate, which is rarely considered in crop modeling studies (Lobell, 2014). Zimmermann et al. (2017) demonstrated that optimizing sowing dates and varieties already had the potential to increase yield levels in current climates, and they assumed conservatively that factors not considered in the crop model (field workability or labor and machinery constraints) represent unachievable yield gain under the adapted sowing dates and varieties. Indeed, many studies presented as climate change impact or adaptation studies confound true climate impacts with other factors related to altered crop management as discussed by Lobell (2014).

Indeed even estimating climates change impacts (for use in later economic analysis) is challenging as the expected impact of the climate scenario may interact with cropping intensity and varieties (Lobell, 2014). While current GCM climate data can be generated without specification of the economic and emissions scenario using RCPs, future crop management (intensity of nitrogen and irrigation among others) and possible adaptation options will depend on socio–economic and technological developments at both global and local levels (Webber et al., 2014). However, large area climate change impact studies using crop models rarely consider specification of future irrigation or nitrogen management (for recent exceptions see Webber et al., 2015; Zhao et al., 2015b). Similarly, adjusting crop sowing dates or varieties is rare in crop impact studies, though Zimmermann et al. (2017) demonstrated considerable difference in projected climate change impacts on yields for six major crops in Europe introduced by not considering technology progress and adaptations in varieties or sowing dates– even after correcting for their potential to improve yields in the reference climate. While the Shared Socioeconomic pathways (SSPs) provide a very broad picture of global population and economic growth, they are not specific to agriculture and cannot inform crop management (O'Neill et al., 2014). For regional and local studies, development of representative agricultural pathways (RAPs) have been pursued in AgMIP (Valdivia et al., 2015) and the CGIAR Program on Climate Change, Agriculture and Food Security (CCAFS) (Palazzo et al., 2017). Depending on the scale of analysis, region and research question, RAPs will specify developments related to markets, incomes, labor availability, technology, policy and landuse, among others, that are translated into parameters or input variables for models (Rosenzweig et al., 2013). For crop models this might include intensity of fertilizers and irrigation, crop varieties as well as crop rotations and residue management practices. Given the high level of uncertainty in future socio–economic conditions, defining RAPs is a stakeholder process which not only generates RAPs, but also serves as a space in which various stakeholders can come together in defining the most critical aspects of their farming systems, assess risks, and inform decision making processes (Rosenzweig et al., 2013). The stakeholder engagement process is challenging requiring significant time and resources, but ultimately judged as the most important part creating regional scenarios (Valdivia et al., 2015).

Despite difficultly quantifying uncertainty associated with projecting future climate change impacts, crop models can help to clarify underlying mechanisms of crop response to climate (Hochman et al., 2017; Lobell et al., 2015). As discussed by Webber et al. (2018a), this can inform design of appropriate climate risk management or adaptation strategies, such as insurance against specific weather variables (Dalhaus et al., 2018), investments in infrastructure (drainage or irrigation) or breeding efforts (Chenu et al., 2009) supporting dialogue across stakeholders groups together with scientists (Webber et al., 2014).

Challenges and Future Research Needs

Uncertainties

Process-based crop models have been the main tools to explore the impact of climate change on crop production since the initial IPCC reports (Rosenzweig and Parry, 1994). Quantifying the uncertainty stemming from model input data for climate, soils, and crop management, model structure, and model parameterization has been the focus of numerous research efforts in the past decade, with relatively little attention given to assessing the uncertainties prior to this (for an exception see Wolf et al., 1996). In the following, we discuss model structure and parameterization uncertainty, as well as efforts and options to limit them.

Pioneering crop model intercomparison studies in terms of their extent of tested wheat and barley models (Palosuo et al., 2011; Rötter et al., 2012) and their spatial scope (Asseng et al., 2013) showed the extent of model error in reproducing field trial results. These studies indicated that crop models need substantial improvement to incorporate state of the art biophysical knowledge on plant growth and yield determining processes (model structure) as well as model parameterization (Rötter et al., 2011). Second, the ensemble median best reproduced the observed data (Martre et al., 2015).

These results set the scene for a range of similar studies within the AgMIP (Rosenzweig et al., 2013) and Modeling European Agriculture with Climate Change for food Security (MACSUR; https://macsur.eu/) projects. In particular, the use of multimodel ensemble means proved to be a robust tool to provide predictions across different crops and cropping systems (Table 1).

Based on the perturbation of climate input into individual models, the structural dimension of prediction uncertainty was determined and reduced by the multimodel ensemble (Asseng et al., 2015). While in the first studies all participating crop models were used equally in the design of multimodel ensembles, Wallach et al. (2016)

Table 1. Key studies in which the multimodel ensemble median was evaluated as the best predictor of observed data.

Crop or crop rotation	Models	Scope	Reference
Wheat	8	Lednice (CZ), Verovany (CZ), Bratislava (SK), Foulum (DK), Flakkebjerg (DK), Jyndevad (DK), Jokioinen (FI), Müncheberg (GE)	Palosuo et al., 2011
Barley	9	Lednice (CZ), Verovany (CZ), Bratislava (SK), Foulum (DK), Flakkebjerg (DK), Jyndevad (DK), Jokioinen (FI)	Rötter et al., 2012
Wheat	27	Four sites in IN, NL, AG, AU	Martre et al., 2015
Maize	23	Lusignan (FR), Ames (US), Rio Verde (BZ), Morogoro (TZ)	Bassu et al., 2014
Rice	13	Los Banos (PH), Ludhiana (IN), Nanjing (CH), Shizukuishi (JP)	Li et al., 2015
Potato	9	Chinoli, (BO), Gisozi, (BU), Jyndevad (DK), Washington (US)	Fleisher et al., 2017
Sugar cane	3	16 sites in Brazil	Dias and Sentelhas, 2017
Crop rotations	15	Foloum (DK), Müncheberg (GER, Braunschweig (GER), Hirschstetten (AT), Thibie (FR)	Kollas et al., 2015
Grain nitrogen uptake	12	Foloum (DK), Müncheberg (GER, Braunschweig (GER), Thibie (FR)	Yin et al., 2017

outlined an agenda for designing improved ensemble studies investigating specific criteria for including models in an ensemble and degree of independence between individual models. Further, they addressed the need to include parameter uncertainty in such ensembles. Tao et al. (2018) explored the combined uncertainty from climate, crop model and parameterization in a probabilistic assessment, in which model structure emerged as the main key driver of overall prediction uncertainty.

To summarize, substantial improvements have been made in the last decade to quantify uncertainty in model prediction and the design of multi-model ensembles. This acts to reduce uncertainty arising from model structure and to a lesser extent from model parameterization as well as improving prediction capabilities. This approach has been widely adopted to provide input for assessments needed at short notice, such as the 1.5 °C IPCC report (Liu et al., 2018). Significantly, projections from process-based multimodel ensembles for the effect of increasing temperature on major crops are largely agree with findings from statistical approaches (Liu et al., 2016; Zhao et al., 2017).

Less progress is found in the literature on efforts to reduce uncertainty resulting from individual model structure and parameterization. Multimodel ensembles serve, to a certain extent, as black boxes, providing limited insight in terms of which model processes drive responses. However, more recent efforts focus on learning from previous model comparisons to inform model improvement model structure and parameterization (Brown et al., 2018; Wang et al., 2017). It is significant that the two studies mentioned address wheat, illustrating the narrow research focus on major cereals and soybean. In particular, important smallholder crops such as pearl millet, cowpea, mung bean, and many others have not been included in such assessments. (Rötter et al., 2018) outlined an iterative cycle of model structure and parameter improvement based on targeted field experiments and extensive statistical analysis of larger data sets. Such combined approach should result in a better mechanistic understanding of climate change and the impact of extreme conditions on crop production. A key limitation will remain the available resources and commitment of funding agencies for process-based model improvement.

Soil data and module-related uncertainties have received less attention in uncertainty research efforts. Folberth et al. (2016) showed that soil parameterization may introduce more uncertainty in yield simulations than climate change signals. While there is urgent need to identify most appropriate model structures, quantifying uncertainty due to parameterization remains important. Considering the effort required to derive suitable descriptions of soil water holding capacity (Dalgliesh and Foale, 1998) or soil organic matter fractioning for the whole profile, fast and automated approaches for soil mapping (Gamma-ray spectrometry for soil texture (Heggemann et al., 2017), Vis-NIR for soil organic carbon (Rodionov et al., 2016), and Temperature-Moisture-Sensor (TMS) for continuous soil moisture and temperature monitoring (Wild et al., 2019) provide promising opportunities to reduce model uncertainty. These technologies are particularly important in the context of the high spatial heterogeneity of soil conditions that exist even within a single field.

Finally, another critical source of uncertainty in climate change impact studies relates to methodologies to initialize model simulations. Most large-area impact studies to date reinitialize soil water and soil nitrogen values each year, though Basso et al. (2015) demonstrated that this can lead to large uncertainties in simulated crop yield changes. Indeed climate change impacts on crop growth are projected to lead to changes in soil organic carbon as the organic residue input amount changes, with feedback on crop yield through the role of SOC (soil organic carbon) affecting soil water holding capacity water and nitrogen availability (Basso et al., 2018). Despite the importance of considering SOC, considerable challenges to do so include the scarcity of data available to parameterize process based SOC models.

Multiple Stressors

There is mounting evidence of the risk that extreme weather events pose to crop production and expected to increase with climate change (Field, 2012). Crop models' capacity to assess the effects of drought, heat stress, flooding, and heavy intensity rainfalls and frost was discussed for each individually in a previous section. However, under real production conditions, they can be expected to occur at the same time (e.g., drought and heat stress), in the same crop cycle (e.g., flooding followed by frost followed by drought), or from year to year, where they will likely co-occur also with soil-related and atmospheric limitations, as well as elevated CO_2. Scientific understanding of stress combinations is quite limited (Suzuki et al., 2014) and relies heavily on controlled environment studies which are difficult to abstract to field conditions (Mittler, 2006).

Much of the challenge of studying effects of multiple stressors is to account for changing sensitivity of crops to stressors over their development cycle and the dynamic nature of how stressors, in which effects on leaf area or water extraction modify subsequent stressor exposure and responses (Smith and Dukes, 2013). Used together with experimentation, process-based crop models can potentially aid in understanding mechanisms of multiple stressors (Rötter et al., 2018) by controlling for many of the emergent responses that occur as feedbacks between previous stress exposure and resource use and inform experimental design.

However, in their current state models may be relatively limited in their capacity to assess impacts of multiple stressors. There are few reported studies that evaluate model responses to combinations of stressors (Rötter et al., 2018), though this is probably due to the fairly limited set of experiments that evaluate multiple stressors in field experiments. For example, we expect that combinations of drought and heat stress under elevated CO_2 will become increasingly common, though few (no) models currently account for feedbacks expected between them. However, experimental results are sometimes contradictory at canopy and crop levels, with questions remaining regarding response mechanisms are not fully understood (Ainsworth et al., 2008; Carmo-Silva et al., 2012; Fitzgerald et al., 2016).

A related question is: which temperature (air or crop) should be considered in crop models when assessing temperature and drought responses? In a study of temperature responses across 30 wheat models, Wang et al. (2017) reported that many models contained outdated parameterization of temperature responses for many processes. However, another possible explanation of the varied temperature responses in models may relate to the fact that many of the responses have been derived under controlled conditions in which air and crop temperature were the same, whereas in field conditions they can differ by many degrees (Kimball et al., 2015; Siebert et al., 2014). Indeed some observational evidence suggests that crop temperature is a better predictor of development than air temperature (Craufurd et al., 2013), at least under irrigated conditions (Dingkuhn et al., 1995; Kimball et al., 2012). While almost all models use air temperature to drive crop development, other models use crop canopy or apex temperatures (e.g., STICS, Brisson et al., 2003). Indeed, the shoot apex is where leaf growth initiates and is understood to be where plants sense temperature to drive development (McMaster et al., 2003). This has led many to experiment and conclude that soil temperature is a better predictor of leaf growth in wheat, as until jointing the shoot apex is located near the soil surface (Peacock, 1975). Other studies suggest that crop development is likely driven by an integrated temperature signal considering temperatures at different parts of the plant (McMaster et al., 2003). Crop temperature follows a gradient within the canopy depending on canopy architecture, crop water status, and ambient atmospheric conditions (Rattalino Edreira and Otegui, 2012).

Indeed crop temperature and its simulation are a means to integrate possible heat and drought stress effects– as under drought stress, when plants reduce transpiration, evaporative cooling is also reduced, leading to relatively warmer canopies (Clawson et al., 1989). Accurate simulation of canopy temperature requires iteration as canopy temperature itself determines heat transfer in the canopy and affects plant boundary layer stability conditions (Webber et al., 2016b; Yoshimoto et al., 2011), though simple empirical equations can perform almost as well as more analytical solutions when evaluated in the conditions they were developed (Webber et al., 2018b). Beyond integrating heat and drought, methods to simulate canopy temperature can also capture the effects of CO_2 leading to warmer crop temperatures (Webber et al., 2018b; Yoshimoto et al., 2005) observed in field experiments (Gray et al., 2016; Kimball et al., 1999). Nevertheless, even the most mechanistic methods of determining canopy temperature do not currently allow for feedbacks of crop temperature on transpiration (Kimball et al., 2015). Use of simulated crop canopy temperatures is currently limited to crop development stages where the canopy is closed unless more complex 2- and 3-source energy balance methods are used, such as in an earlier version of DSSAT-CSM

(Jagtap and Jones, 1989). This means that most simulated crop temperature currently cannot easily be used to simulate processes before the crop has a closed canopy (e.g., early development and photosynthesis, frost damage).

Pest and Diseases

Crop management practices from fertilization, cultivar selection, and pest and disease control can also change the crop response to climatic variables (Chen et al., 2015; Shi et al., 2018). Modeling of pest and disease effects on crop growth has received very little attention in crop modeling impact assessments, which limits the validity of crop model–based assessments to cases where good pest control can be assumed (Donatelli et al., 2017). This is mainly because of complex nature and species-specific damaging mechanism of pests on crop growth and yield (Whish et al., 2015). There is also a strong correlation between increase in temperature and precipitation quantities and severity and potential risk of pest and disease breakout in cropping systems (Kropff et al., 1995) which can be a potential for simulation of pest and disease appearance at the canopy level (Dillehay et al., 2005). This is a critical research area though is not dealt with in more detail here.

Scale of Model Application

Crop models are widely applied at large scales in climate change impact assessments, though most crop models used in these assessments have originally been developed considering crop growth at the field scale (Van Bussel et al., 2011). Using field-scale models at larger scales may lead to two main sources of uncertainties: inaccuracy of physiological processes represented and lost information when spatially heterogeneous input data (climate, soils, crop varieties and management) are aggregated (Hansen and Jones, 2000). Model structure generally is not changed when field scale models are applied at larger scales, with usually only parameters being adjusted (Adam et al., 2011). A few studies have attempted to simplify field-scale models to apply at larger scales by considering most important growth processes as summary model (Anderson et al., 2003; Enquist et al., 2003).

The uncertainty introduced when using aggregate data will depend on the underlying heterogeneity over the study area and if nonlinear processes dominate. Aggregation of input data acts to level out local extremes with unknown bias on model outputs (Van Bussel et al., 2011). However, quantifying of aggregation error is challenging due to the nonlinear and complex nature of crop models as well as the lack of information on heterogeneity of target area (Kuhnert et al., 2017). To date, most studies attempting to quantify aggregation errors have used grid base climate, soil and management inputs at a number of different resolutions, comparing the error at coarser resolutions to the finest resolution (Zhao et al., 2015a), but these results are hard to generalize to new regions. High resolution climate data are required for those regions that are characterized by heterogeneous topography to simulate realistic spatial patterns of crop yield (Zhao et al., 2015a). Studies in Northwest of Germany revealed that aggregation of soil data had a larger impact on model yield simulations as compared with climate data aggregation (Hoffmann et al., 2016), which may be explained of different aggregation concepts for soil data and a larger contribution of spatial heterogeneity of soil on yield variability in temperate climates (Wassenaar et al., 1999). Soil information is aggregated based on dominant soil characteristics because a simple averaging may result in nonsensical soil profiles, which do not exist at higher resolution (Eyshi Rezaei et al., 2015a).

Results of a crop model ensemble showed that soil aggregation can introduce 15% in bias for yield simulations (Hoffmann et al., 2016) in simulations in a Northwest state of Germany. Using of the new technologies such as high-resolution remotely sensed information could be an alternative to typical field experiments for parametrization of crop models at large scales.

Conclusions

Climate change affects both average growing conditions of crops as well as their exposure to extreme weather events like heat stress, intense rainfall, flooding, and prolonged droughts. Understanding climate change impacts on crops is difficult to assess in traditional field or controlled environment settings as crop growth is varyingly sensitive to weather dynamics over the entire crop growth cycle with a very large number of possible combinations of precipitation and temperature changes with complex interactions and feedbacks. Crop models overcome limitations of field experimentation and are now widely used in climate change impact assessments. However, given the complexity and scientific uncertainty around crop response to multiple stressors expected to prevail under elevated CO_2 and climate change, crop model projections still contain many uncertainties limiting their utility in risk assessments or informing adaptations. Model improvement is needed to represent processes driving responses to multiple stressors. Methodological advances are also required to assess the adequacy of using field-based models with aggregate climate, soil, and crop management data over large areas in impact assessments. It is expected that improved computing power with new data sources from remote sensing and other big data sources will enable finer resolution simulations overcoming some issues of input data aggregation. If crop models are currently somewhat limited in assessing risks of climate change to cropping systems, process-based crop models do offer capacity to aid in understanding crop responses observed in field and production conditions as they can be used to test hypotheses about which processes drive observed responses and can aid in helping inform experimental design.

References

Adam, M., L. Van Bussel, P. Leffelaar, H. Van Keulen, and F. Ewert. 2011. Effects of modelling detail on simulated potential crop yields under a wide range of climatic conditions. Ecol. Modell. 222:131–143.

Addiscott, T., and A. Whitmore. 1991. Simulation of solute leaching in soils of differing permeabilities. Soil Use Manage. 7:94–102.

Ahmed, F., M.Y. Rafii, M.R. Ismail, A.S. Juraimi, H.A. Rahim, R. Asfaliza, and M.A. Latif. 2013. Waterlogging tolerance of crops: Breeding, mechanism of tolerance, molecular approaches, and future prospects. BioMed Res. Int. 2013:963525.

Ainsworth, E.A. 2016. Understanding and improving global crop response to ozone pollution. Plant J. 90:886–897.

Ainsworth, E.A., and S.P. Long. 2005. What have we learned from 15 years of free-air CO_2 enrichment (FACE)? A meta-analytic review of the responses of photosynthesis, canopy properties and plant production to rising CO_2. New Phytol. 165:351–372.

Ainsworth, E.A. 2008. Rice production in a changing climate: A meta-analysis of responses to elevated carbon dioxide and elevated ozone concentration. GCB Bioenergy 14(7):1642–1650. doi:10.1111/j.1365-2486.2008.01594.x

Allen, R.G., L.S. Pereira, D. Raes, and M. Smith. 1998. Crop evapotranspiration-Guidelines for computing crop water requirements. FAO Irrigation and drainage paper 56. FAO, Rome.

Anderson, M.C., W.P. Kustas, and J.M. Norman. 2003. Upscaling and downscaling—A regional view of the soil–plant–atmosphere continuum. Agron. J. 95:1408–1423.

Armstrong, W., M.T. Healy, S. Lythe, N. Phytologist, and N. Aug. 1983. Oxygen Diffusion in Pea. II. Oxygen Concentrations in the Primary Pea Root Apex as Affected by Growth, the Production of Laterals and Radial Oxygen Loss 11. Production 94:549–559.

Armstrong, W., T. Webb, M. Darwent, and P.M. Beckett. 2009. Measuring and interpreting respiratory critical oxygen pressures in roots. Ann. Bot. (Lond.) 103(2):281–293.

Asseng S., F. Ewert, P. Martre, R.P. Rotter, D.B. Lobell, D. Cammarano, B.A. Kimball, M.J. Ottman, G.W. Wall, J.W. White, M.P. Reynolds, P.D. Alderman, P.V.V. Prasad, P.K. Aggarwal, J. Anothai, B. Basso, C. Biernath, A.J. Challinor, G. De Sanctis, J. Doltra, E. Fereres, M. Garcia-Vila, S. Gayler, G. Hoogenboom, L.A. Hunt, R.C. Izaurralde, M. Jabloun, C.D. Jones, K.C. Kersebaum, A.K. Koehler, C. Muller, S. Naresh Kumar, C. Nendel, G. O'Leary, J.E. Olesen, T. Palosuo, E. Priesack, E. Eyshi Rezaei, A.C. Ruane, M.A. Semenov, I. Shcherbak, C. Stockle, P. Stratonovitch, T. Streck, I. Supit, F. Tao, P.J. Thorburn, K. Waha, E. Wang, D. Wallach, J. Wolf, Z. Zhao, and Y. Zhu. 2015. Rising temperatures reduce global wheat production. Nat. Clim. Chang. 5:143–147.

Asseng, S., F. Ewert, C. Rosenzweig, J.W. Jones, J.L. Hatfield, A.C. Ruane, K.J. Boote, P.J. Thorburn, R.P. Rotter, D. Cammarano, N. Brisson, B. Basso, P. Martre, P.K. Aggarwal, C. Angulo, P. Bertuzzi, C. Biernath, A.J. Challinor, J. Doltra, S. Gayler, R. Goldberg, R. Grant, L. Heng, J. Hooker, L.A. Hunt, J. Ingwersen, R.C. Izaurralde, K.C. Kersebaum, C. Muller, S.N. Kumar, C. Nendel, G. O'Leary, J.E. Olesen, T.M. Osborne, T. Palosuo, E. Priesack, D. Ripoche, M.A. Semenov, I. Shcherbak, P. Steduto, C. Stockle, P. Stratonovitch, T. Streck, I. Supit, F. Tao, M. Travasso, K. Waha, D. Wallach, J.W. White, J.R. Williams, and J. Wolf. 2013. Uncertainty in simulating wheat yields under climate change. Nat. Clim. Chang. 3:827–832.

Asseng, S., I. Foster, and N.C. Turner. 2011. The impact of temperature variability on wheat yields. Glob. Change Biol. 17(2):997–1102.

Asseng, S., B.A. Keating, I.R.P. Fillery, P.J. Gregory, J.W. Bowden, N.C. Turner, J.A. Palta, and D.G. Abrecht. 1998. Performance of the APSIM-wheat model in Western Australia. Field Crops Res. 57:163–179.

Auffhammer, M., S.M. Hsiang, W. Schlenker, and A. Sobel. 2013. Using weather data and climate model output in economic analyses of climate change. Rev. Environ. Econ. Policy 7:181–198.

Auffhammer, M., and W. Schlenker. 2013. It's not just the statistical model. A comment on Seo (2013). Clim. Change 121:125–128.

Avnery, S., D.L. Mauzerall, J. Liu, and L.W. Horowitz. 2011. Global crop yield reductions due to surface ozone exposure: 1. Year 2000 crop production losses and economic damage. Atmos. Environ. 45:2284–2296.

Barlow, K.M., B.P. Christy, G.J. O'Leary, P.A. Riffkin, and J.G. Nuttall. 2015. Simulating the impact of extreme heat and frost events on wheat crop production: A review. Field Crops Res. 171(1):109–119.

Basso, B., B. Dumont, B. Maestrini, I. Shcherbak, G.P. Robertson, J.R. Porter, P. Smith, K. Paustian, P.R. Grace, S. Asseng, et al. 2018. Soil organic carbon and nitrogen feedbacks on crop yields under climate change. Agric. Environ. Lett. 3(1):180026.

Basso, B., D.W. Hyndman, A.D. Kendall, P.R. Grace, and G.P. Robertson. 2015. Can impacts of climate change and agricultural adaptation strategies be accurately quantified if crop models are annually re-initialized? PLoS One 10:e0127333.

Bassu, S., N. Brisson, J.-L. Durand, K. Boote, J. Lizaso, J.W. Jones, C. Rosenzweig, A.C. Ruane, M. Adam, C. Baron, B. Basso, C. Biernath, H. Boogaard, S. Conijn, M. Corbeels, D. Deryng, G. De Sanctis, S. Gayler, P. Grassini, J. Hatfield, S. Hoek, C. Izaurralde, R. Jongschaap, A.R. Kemanian, K.C. Kersebaum, S.-H. Kim, N.S. Kumar, D. Makowski, C. Mueller, C. Nendel, E. Priesack, M.V. Pravia, F. Sau, I. Shcherbak, F. Tao, E. Teixeira, D. Timlin, and K. Waha. 2014. How do various maize crop models vary in their responses to climate change factors? Glob. Change Biol. 20:2301–2320.

Bellocchi, G., M. Rivington, M. Donatelli, and K. Matthews. 2011. Validation of biophysical models: Issues and methodologies. Sustainable Agriculture Volume 2: Springer, Amsterdam, The Netherlands. p. 577-603.

Bernacchi, C., C. Pimentel, and S. Long. 2003. In: vivo temperature response functions of parameters required to model RuBP-limited photosynthesis. Plant Cell Environ. 26:1419–1430.

Bernacchi, C., E. Singsaas, C. Pimentel, A. Portis, Jr., and S. Long. 2001. Improved temperature response functions for models of Rubisco-limited photosynthesis. Plant Cell Environ. 24:253–259.

Boote, K.J., J.W. Jones, J.W. White, S. Asseng, and J.I. Lizaso. 2013. Putting mechanisms into crop production models. Plant Cell Environ. 36:1658–1672.

Boyer, J. 1970. Differing sensitivity of photosynthesis to low leaf water potentials in corn and soybean. Plant Physiol. 46:236–239.

Brauer, M., G. Freedman, J. Frostad, A. van Donkelaar, R.V. Martin, F. Dentener, R.V. Dingenen , K. Estep, H. Amini, J.S. Apte, K. Balakrishnan, L. Barregard, D. Broday, V. Feigin, S. Ghosh, P.K. Hopke, L.D. Knibbs, Y. Kokubo, Y. Liu, S. Ma, L. Morawska, J.L.T. Sangrador, G. Shaddick, H.R. Anderson, T. Vos, M.H. Forouzanfar, R.T. Burnett, and A. Cohen. 2016. Ambient air pollution exposure estimation for the global burden of disease 2013. Environ. Sci. Technol. 50(1):79–88.

Brisson, N., C. Gary, E. Justes, R. Roche, B. Mary, D. Ripoche, D. Zimmer, J. Sierra, P. Bertuzzi, P. Burger, F. Bussière, Y.M. Cabidoche, P. Cellier, P. Debaeke, J.P. Gaudillère, C. Hénault, F. Maraux, B. Seguin, and H. Sinoquet. 2003. An overview of the crop model STICS. Eur. J. Agron. 18:309–332.

Brown, H., N. Huth, and D. Holzworth. 2018. Crop model improvement in APSIM: Using wheat as a case study. Eur. J. Agron. 100:141–150.

Caldeira, C.F., M. Bosio, B. Parent, L. Jeanguenin, F. Chaumont, and F. Tardieu. 2014. A hydraulic model is compatible with rapid changes in leaf elongation rate under fluctuating evaporative demand and soil water status. Plant Physiol. 164(4):1718–1730.

Carmo-Silva, A.E., M.A. Gore, P. Andrade-Sanchez, A.N. French, D.J. Hunsaker, and M.E. Salvucci. 2012. Decreased CO_2 availability and inactivation of Rubisco limit photosynthesis in cotton plants under heat and drought stress in the field. Environmental and Experimental Botany. 83:1–11. doi:10.1016/j.envexpbot.2012.04.001

Cannell, M., and J. Thornley. 2000. Modelling the components of plant respiration: Some guiding principles. Ann. Bot. (Lond.) 85:45–54.

Chai, T., and R.R. Draxler. 2014. Root mean square error (RMSE) or mean absolute error (MAE)?–Arguments against avoiding RMSE in the literature. Geosci. Model Dev. 7:1247–1250.

Challinor, A., T. Wheeler, P. Craufurd, and J. Slingo. 2005. Simulation of the impact of high temperature stress on annual crop yields. Agric. For. Meteorol. 135:180–189.

Chen, D., S. Wang, B. Xiong, B. Cao, and X. Deng. 2015. Carbon/nitrogen imbalance associated with drought-induced leaf senescence in *sorghum bicolor*. PLoS One 10(8):e0137026.

Chenu, K., S.C. Chapman, F. Tardieu, G. McLean, C. Welcker, and G.L. Hammer. 2009. Simulating the yield impacts of organ-level quantitative trait loci associated with drought response in maize-A 'gene-to-phenotype' modeling approach. Genetics 183(4):1507–1523.

Chenu, K., R. Deihimfard, and S.C. Chapman. 2013. Large-scale characterization of drought pattern: A continent-wide modelling approach applied to the Australian wheatbelt–spatial and temporal trends. New Phytol. 198:801–820.

Clawson, K., R. Jackson, and P. Pinter. 1989. Evaluating plant water stress with canopy temperature differences. Agron. J. 81:858–863.

Craufurd, P.Q., V. Vadez, S.K. Jagadish, P.V. Prasad, and M. Zaman-Allah. 2013. Crop science experiments designed to inform crop modeling. Agric. For. Meteorol. 170:8–18.

Dalgliesh N., and M.A. Foale. 1998. Soil matters: Monitoring soil water and nutrients in dryland farming. CSIRO, Canberra, Australia.

Dalhaus, T., O. Musshoff, and R. Finger. 2018. Phenology information contributes to reduce temporal basis risk in agricultural weather index insurance. Sci. Rep. 8:46.

Davies, W.J., and J. Zhang. 1991. Root signals and the regulation of growth and development of plants in drying soil. Annu. Rev. Plant Biol. 42:55–76.

de Wit, C.T. 1965. Photosynthesis of leaf canopies. Pudoc, Wageningen, The Netherlands.

Dias, H.B., and P.C. Sentelhas. 2017. Evaluation of three sugarcane simulation models and their ensemble for yield estimation in commercially managed fields. Field Crops Res. 213:174–185.

Dillehay, B.L., Calvin, D.D., Voight, D.G., Kuldau, G.A., Roth, G.W., Hyde, J.A., Russo, J.M., Kratochvil, R.J., 2005. Verification of a European corn borer (Lepidoptera: Crambidae) loss equation in the major corn production region of the Northeastern United States. J. Econ. Entomol. 98:103–112. doi:10.1093/jee/98.1.103

Dingkuhn, M., A. Sow, A. Samb, S. Diack, and F. Asch. 1995. Climatic determinants of irrigated rice performance in the Sahel—I. Photothermal and micro-climatic responses of flowering. Agric. Syst. 48:385–410.

Donatelli, M., R.D. Magarey, S. Bregaglio, L. Willocquet, J.P. Whish, and S. Savary. 2017. Modelling the impacts of pests and diseases on agricultural systems. Agricultural Systems 155: 213-224.

Dreccer, M.F., A.F. van Herwaarden, and S.C. Chapman. 2009. Grain number and grain weight in wheat lines contrasting for stem water soluble carbohydrate concentration. Field Crops Res. 112:43–54.

Dumont, B., V. Leemans, M. Mansouri, B. Bodson, J.-P. Destain, and M.-F. Destain. 2014. Parameter identification of the STICS crop model, using an accelerated formal MCMC approach. Environ. Model. Softw. 52:121–135.

Durand, J.-L., K. Delusca, K. Boote, J. Lizaso, R. Manderscheid, H.J. Weigel, A.C. Ruane, C. Rosenzweig, J. Jones, L. Ahuja, S. Anapalli, B. Basso, C. Baron, P. Bertuzzi, C. Biernath, D. Deryng, F. Ewert, T. Gaiser, S. Gayler, F. Heinlein, K.C. Kersebaum, S.-H. Kim, C. Müller, C. Nendel, A. Olioso, E. Priesack, J.R. Villegas, D. Ripoche, R.P. Rötter, S.I. Seidel, A. Srivastava, F. Tao, D. Timlin, T. Twine, E. Wang, H. Webber, and Z. Zhao. 2018. How accurately do maize crop models simulate the interactions of atmospheric CO_2 concentration levels with limited water supply on water use and yield? Eur. J. Agron. 100:67–75.

Ehrhardt, F., J.F. Soussana, G. Bellocchi, P. Grace, R. McAuliffe, S. Recous, R. Sándor, P. Smith, V. Snow, and M. de Antoni Migliorati. 2018. Assessing uncertainties in crop and pasture ensemble model simulations of productivity and N_2O emissions. Glob. Change Biol. 24:e603–e616.

Emberson, L.D., H. Pleijel, E.A. Ainsworth, M. van den Berg, W. Ren, S. Osborne, G. Mills, D. Pandey, F. Dentener, P. Büker, F. Ewert, R. Koeble, and R. Van Dingenen. 2018. Ozone effects on crops and consideration in crop models. Eur. J. Agron. 100:19–34.

Enquist, B.J., E.P. Economo, T.E. Huxman, A.P. Allen, D.D. Ignace, and J.F. Gillooly. 2003. Scaling metabolism from organisms to ecosystems. Nature 423:639.

Ewert, F., R. Rötter, M. Bindi, H. Webber, M. Trnka, K. Kersebaum, J.E. Olesen, M. van Ittersum, S. Janssen, M. Rivington, M. Semenov, D. Wallach, J. Porter, D. Stewart, J. Verhagen, T. Gaiser, T. Palosuo, F. Tao, C. Nendel, P. Roggero, L. Bartošová, and S. Asseng. 2015. Crop modelling for integrated assessment of climate change risk to food production. Environ. Model. Softw. 72:287–303.

Eyshi Rezaei, E., S. Siebert, and F. Ewert. 2017. Climate and management interaction cause diverse crop phenology trends. Agric. For. Meteorol. 233:55–70.

Eyshi Rezaei, E., S. Stefan, and E. Frank. 2015a. Intensity of heat stress in winter wheat—phenology compensates for the adverse effect of global warming. Environ. Res. Lett. 10:024012.

Eyshi Rezaei, E., H. Webber, T. Gaiser, J. Naab, and F. Ewert. 2015b. Heat stress in cereals: Mechanisms and modelling. Eur. J. Agron. 64:98–113.

Faye B., H. Webber, J.B. Naab, D.S. MacCarthy, M. Adam, F. Ewert, J.P.A. Lamers, C.F. Schleussner, A.C. Ruane, U. Gessner, G. Hoogenboom, K. Boote, V. Sheilia, F. Saeed, D. Wisser, S. Hadir, P. Laux, and T. Gaiser. 2018. Impacts of 1.5 versus 2.0°C on cereal yields in the West African Sudan Savanna. Environ. Res. Lett. 13:034014.

Feng, Z., and K. Kobayashi. 2009. Assessing the impacts of current and future concentrations of surface ozone on crop yield with meta-analysis. Atmos. Environ. 43:1510–1519.

Field, C.B. 2012. Managing the risks of extreme events and disasters to advance climate change adaptation: Special report of the intergovernmental panel on climate change. Cambridge Univ. Press, Cambridge, U.K.

Fitzgerald, G.J., M. Tausz, G. O'Leary, M.R. Mollah, S. Tausz-Posch, S. Seneweera, and D. McNeil. 2016. Elevated atmospheric [CO_2] can dramatically increase wheat yields in semi-arid environments and buffer against heat waves. GCB Bioenergy 22(6):2269-2284.

Fleisher, D.H., B. Condori, R. Quiroz, A. Alva, S. Asseng, C. Barreda, M. Bindi, K.J. Boote, R. Ferrise, and A.C. Franke. 2017. A potato model intercomparison across varying climates and productivity levels. Glob. Change Biol. 23:1258–1281.

Folberth, C., R. Skalský, E. Moltchanova, J. Balkovič, L.B. Azevedo, M. Obersteiner, and M. Van Der Velde. 2016. Uncertainty in soil data can outweigh climate impact signals in global crop yield simulations. Nat. Commun. 7:11872.

Gabaldón-Leal, C., H. Webber, M.E. Otegui, G.A. Slafer, R.A. Ordóñez, T. Gaiser, I.J. Lorite, M. Ruiz-Ramos, and F. Ewert. 2016. Modelling the impact of heat stress on maize yield formation. Field Crops Res. 198:226–237.

Gelang, J., H. Pleijel, E. Sild, H. Danielsson, S. Younis, and G. Selldén. 2008. Rate and duration of grain filling in relation to flag leaf senescence and grain yield in spring wheat (*Triticum aestivum*) exposed to different concentrations of ozone. Physiol. Plant. 110:366–375.

Gifford, R.M. 1995. Whole plant respiration and photosynthesis of wheat under increased CO_2 concentration and temperature: Long-term vs. short-term distinctions for modelling. Glob. Change Biol. 1:385–396.

Goudriaan, J., and H. Van Laar. 2012. Modelling potential crop growth processes: Textbook with exercises: Springer Science & Business Media, Amsterdam, The Netherlands.

Gray, S.B., O. Dermody, S.P. Klein, A.M. Locke, J.M. McGrath, R.E. Paul, D.M. Rosenthal, U.M. Ruiz-Vera, M.H. Siebers, R. Strellner, E. Ainsworth, C. Bernacchi, S. Long, D. Ort, and A. Leakey. 2016. Intensifying drought eliminates the expected benefits of elevated carbon dioxide for soybean. Nat. Plants (London, U. K.)2:16132.

Hansen, J., and J. Jones. 2000. Scaling-up crop models for climate variability applications. Agric. Syst. 65:43–72.

Hatfield, J.L., K.J. Boote, B. Kimball, L. Ziska, R.C. Izaurralde, D. Ort, A.M. Thomson, and D. Wolfe. 2011. Climate impacts on agriculture: Implications for crop production. Agron. J. 103:351–370.

Heggemann, T., G. Welp, W. Amelung, G. Angst, S.O. Franz, S. Koszinski, K. Schmidt, and S. Pätzold. 2017. Proximal gamma-ray spectrometry for site-independent in situ prediction of soil texture on ten heterogeneous fields in Germany using support vector machines. Soil Tillage Res. 168:99–109.

Hempel, S., K. Frieler, L. Warszawski, J. Schewe, and F. Piontek. 2013. A trend-preserving bias correction: The ISI-MIP approach. Earth System Dynamics 4:219–236.

Hochman, Z., D.L. Gobbett, and H. Horan. 2017. Climate trends account for stalled wheat yields in Australia since 1990. Glob. Change Biol. 23:2071–2081.

Hoffmann, H., G. Zhao, S. Asseng, M. Bindi, C. Biernath, J. Constantin, E. Coucheney, R. Dechow, L. Doro, and H. Eckersten. 2016. Impact of spatial soil and climate input data aggregation on regional yield simulations. PLoS One 11:e0151782.

Holzworth, D.P., V. Snow, S. Janssen, I.N. Athanasiadis, M. Donatelli, G. Hoogenboom, J.W. White, and P. Thorburn. 2015. Agricultural production systems modelling and software: Current status and future prospects. Environ. Model. Softw. 72:276–286.

Hoogenboom, G. 2000. Contribution of agrometeorology to the simulation of crop production and its applications. Agric. For. Meteorol. 103:137–157.

Huth, N., K. Bristow, and K. Verburg. 2012. SWIM3: Model use, calibration, and validation. Trans. ASABE 55:1303–1313.

Jagtap, S.S., and J. Jones. 1989. Evapotranspiration model for developing crops. Trans. ASAE 32:1342–1350.

Jame, Y., and H. Cutforth. 1996. Crop growth models for decision support systems. Can. J. Plant Sci. 76:9–19.

Jamieson, P.D., J.R. Porter, J. Goudriaan, J.T. Ritchie, H. van Keulen, and W. Stol. 1998. A comparison of the models AFRCWHEAT2, CERES-wheat, Sirius, SUCROS2 and SWHEAT with measurements from wheat grown under drought. Field Crops Res. 55:23–44.

Jin Z, Q. Zhuang, J. Wang, S.V. Archontoulis, Z. Zobel, and V.R. Kotamarthi. 2017. The combined and separate impacts of climate extremes on the current and future US rainfed maize and soybean production under elevated CO_2. GCB Bioenergy 23(7):2687–2704.

Jones, J.W., G. Hoogenboom, C.H. Porter, K.J. Boote, W.D. Batchelor, L. Hunt, P.W. Wilkens, U. Singh, A.J. Gijsman, and J.T. Ritchie. 2003. The DSSAT cropping system model. Eur. J. Agron. 18:235–265.

Keating, B.A., P.S. Carberry, G.L. Hammer, M.E. Probert, M.J. Robertson, D. Holzworth, N.I. Huth, J.N.G. Hargreaves, H. Meinke, Z. Hochman, G. McLean, K. Verburg, V. Snow, J.P. Dimes, M. Silburn, E. Wang, S. Brown, K.L. Bristow, S. Asseng, S. Chapman, R.L. McCown, D.M. Freebairn, and C.J. Smith. 2003. An overview of APSIM, a model designed for farming systems simulation. Eur. J. Agron. 18(3–4):267–288.

Kersebaum, K., K. Boote, J. Jorgenson, C. Nendel, M. Bindi, C. Frühauf, T. Gaiser, G. Hoogenboom, C. Kollas, and J.E. Olesen. 2015. Analysis and classification of data sets for calibration and validation of agro-ecosystem models. Environ. Model. Softw. 72:402–417.

Kimball, B., J. White, M.J. Ottman, G. Wall, C. Bernacchi, J. Morgan, and D. Smith. 2015. Predicting canopy temperatures and infrared heater energy requirements for warming field plots. Agron. J. 107:129–141.

Kimball, B., J. White, G. Wall, and M. Ottman. 2012. Infrared-warmed and unwarmed wheat vegetation indices coalesce using canopy-temperature–based growing degree days. Agron. J. 104:114–118.

Kimball, B.A. 2016. Crop responses to elevated CO_2 and interactions with H_2O, N, and temperature. Curr. Opin. Plant Biol. 31:36–43.

Kimball, B., R.L. LaMorte, P.J. Pinter, Jr., G.W. Wall, D.J. Hunsaker, F.J. Adamsen, S.W. Leavitt, T.L. Thompson, A.D. Matthias, and T.J. Brooks. 1999. Free-air CO_2 enrichment and soil nitrogen effects on energy balance and evapotranspiration of wheat. *Water Resour. Res.*, 35(4): 1179-1190.

Knutti, R., and J. Sedláček. 2013. Robustness and uncertainties in the new CMIP5 climate model projections. Nat. Clim. Chang. 3:369.

Kollas, C., K.C. Kersebaum, C. Nendel, K. Manevski, C. Müller, T. Palosuo, C.M. Armas-Herrera, N. Beaudoin, M. Bindi, and M. Charfeddine. 2015. Crop rotation modelling—A European model intercomparison. Eur. J. Agron. 70:98–111.

Kristensen, K., K. Schelde, and J.E. Olesen. 2011. Winter wheat yield response to climate variability in Denmark. J. Agric. Sci. 149:33–47.

Kropff, M., P. Teng, and R. Rabbinge. 1995. The challenge of linking pest and crop models. Agricultural Systems 49(4):413-434.

Kuhnert, M., J. Yeluripati, P. Smith, H. Hoffmann, M. Van Oijen, J. Constantin, E. Coucheney, R. Dechow, H. Eckersten, and T. Gaiser. 2017. Impact analysis of climate data aggregation at different spatial scales on simulated net primary productivity for croplands. Eur. J. Agron. 88:41–52.

Kumar, U., M.R. Laza, J.-C. Soulié, R. Pasco, K.V. Mendez, and M. Dingkuhn. 2016. Compensatory phenotypic plasticity in irrigated rice: Sequential formation of yield components and simulation with SAMARA model. Field Crops Res. 193:164–177.

Leakey, A.D., E.A. Ainsworth, C.J. Bernacchi, A. Rogers, S.P. Long, and D.R. Ort. 2009. Elevated CO_2 effects on plant carbon, nitrogen, and water relations: Six important lessons from FACE. J. Exp. Bot. 60:2859–2876.

Li, T., T. Hasegawa, X. Yin, Y. Zhu, K. Boote, M. Adam, S. Bregaglio, S. Buis, R. Confalonieri, and T. Fumoto. 2015. Uncertainties in predicting rice yield by current crop models under a wide range of climatic conditions. Glob. Change Biol. 21:1328–1341.

Liu, B., S. Asseng, C. Müller, F. Ewert, J. Elliott, D.B. Lobell, P. Martre, A.C. Ruane, D. Wallach, and J.W. Jones. 2016. Similar estimates of temperature impacts on global wheat yield by three independent methods. Nat. Clim. Chang. 6:1130.

Liu, B., P. Martre, F. Ewert, J.R. Porter, A.J. Challinor, C. Müller, A.C. Ruane, K. Waha, P.J. Thorburn, P.K. Aggarwal, M. Ahmed, J. Balkovič, B. Basso, C. Biernath, M. Bindi, D. Cammarano, G. De Sanctis, B. Dumont, M. Espadafor, E. Eyshi Rezaei, R. Ferrise, M. Garcia-Vila, S. Gayler, Y. Gao, H. Horan, G. Hoogenboom, R.C. Izaurralde, C.D. Jones, B.T. Kassie, K.C. Kersebaum, C. Klein, A.-K. Koehler, A. Maiorano, S. Minoli, M. Montesino San Martin, S.N. Kumar, C. Nendel, G.J. O'Leary, T. Palosuo, E. Priesack, D. Ripoche, R.P. Rötter, M.A. Semenov, C. Stöckle, T. Streck, I. Supit, F. Tao, M. Van der Velde, D. Wallach, E. Wang, H. Webber, J. Wolf, L. Xiao, Z. Zhang, Z. Zhao, Y. Zhu, and S. Asseng. 2018. Global wheat production with 1.5 and 2.0°C above pre-industrial warming. Glob. Change Biol. 24:1291–1307.

Lizaso, J., M. Ruiz-Ramos, L. Rodríguez, C. Gabaldon-Leal, J. Oliveira, I. Lorite, A. Rodríguez, G. Maddonni, and M. Otegui. 2017. Modeling the response of maize phenology, kernel set, and yield components to heat stress and heat shock with CSM-IXIM. Field Crops Res. 214:239–252.

Lobell, D.B. 2014. Climate change adaptation in crop production: Beware of illusions. Glob. Food Secur. 3:72–76.

Lobell, D.B., and S. Asseng. 2017. Comparing estimates of climate change impacts from process-based and statistical crop models. Environ. Res. Lett. 12:015001.

Lobell, D.B., M. Bänziger, C. Magorokosho, and B. Vivek. 2011. Nonlinear heat effects on African maize as evidenced by historical yield trials. Nat. Clim. Chang. 1:42–45.

Lobell, D.B., G.L. Hammer, K. Chenu, B. Zheng, G. McLean, and S.C. Chapman. 2015. The shifting influence of drought and heat stress for crops in northeast Australia. Glob. Change Biol. 21:4115–4127.

Lopes, M.S., and M.P. Reynolds. 2010. Partitioning of assimilates to deeper roots is associated with cooler canopies and increased yield under drought in wheat. Funct. Plant Biol. 37:147–156.

Maiorano, A., P. Martre, S. Asseng, F. Ewert, C. Müller, R.P. Rötter, A.C. Ruane, M.A. Semenov, D. Wallach, and E. Wang. 2017. Crop model improvement reduces the uncertainty of the response to temperature of multi-model ensembles. Field Crops Res. 202:5–20.

Malik, A., T.D. Colmer, H. Lambers, and M. Schortemeyer. 2001. Changes in physiological and morphological traits of roots and shoots of wheat in response to different depths of waterlogging. Aust. J. Plant Physiol. 28(11):1121–1131. doi:10.1071/Pp01089

Martre, P., D. Wallach, S. Asseng, F. Ewert, J.W. Jones, R.P. Rötter, K.J. Boote, A.C. Ruane, P.J. Thorburn, and D. Cammarano. 2015. Multimodel ensembles of wheat growth: Many models are better than one. Glob. Change Biol. 21:911–925.

McMaster, G.S., J.W. White, L. Hunt, P. Jamieson, S. Dhillon, and J. Ortiz-Monasterio. 2008. Simulating the influence of vernalization, photoperiod and optimum temperature on wheat developmental rates. Ann. Bot. (Lond.) 102:561–569.

McMaster, G.S., W. Wilhelm, D. Palic, J.R. Porter, and P. Jamieson. 2003. Spring wheat leaf appearance and temperature: Extending the paradigm? Ann. Bot. (Lond.) 91:697–705.

Medlyn, B., E. Dreyer, D. Ellsworth, M. Forstreuter, P. Harley, M. Kirschbaum, X. Le Roux, P. Montpied, J. Strassemeyer, and A. Walcroft. 2002. Temperature response of parameters of a biochemically based model of photosynthesis. II. A review of experimental data. Plant Cell Environ. 25:1167–1179.

Mittler, R. 2006. Abiotic stress, the field environment and stress combination. Trends Plant Sci. 11:15–19.

Monteith, J. 1965. Light distribution and photosynthesis in field crops. Ann. Bot. (Lond.) 29:17–37.

Monteith, J., and M. Unsworth. 2007. Principles of environmental physics. Academic Press, Burlington, VT.

Monteith, J.L. 1977. Climate and the efficiency of crop production in Britain. Philos. Trans. R. Soc. Lond. B Biol. Sci. 281:277–294.

Moriondo, M., C. Giannakopoulos, and M. Bindi. 2011. Climate change impact assessment: The role of climate extremes in crop yield simulation. Clim. Change 104:679–701.

Moss, R.H., J.A. Edmonds, K.A. Hibbard, M.R. Manning, S.K. Rose, D.P. Van Vuuren, T.R. Carter, S. Emori, M. Kainuma, and T. Kram. 2010. The next generation of scenarios for climate change research and assessment. Nature 463:747.

Muller, B., F. Pantin, M. Génard, O. Turc, S. Freixes, M. Piques, and Y. Gibon. 2011. Water deficits uncouple growth from photosynthesis, increase C content, and modify the relationships between C and growth in sink organs. J. Exp. Bot. 62:1715–1729.

O'Neill, B.C., E. Kriegler, K. Riahi, K.L. Ebi, S. Hallegatte, T.R. Carter, R. Mathur, and D.P. van Vuuren. 2014. A new scenario framework for climate change research: The concept of shared socioeconomic pathways. Clim. Change 122:387–400.

Olesen, J.E., G.H. Rubæk, T. Heidmann, S. Hansen, and C.D. Børgensen. 2004. Effect of climate change on greenhouse gas emissions from arable crop rotations. Nutr. Cycl. Agroecosyst. 70:147–160.

Olesen, J.E., M. Trnka, K. Kersebaum, A. Skjelvåg, B. Seguin, P. Peltonen-Sainio, F. Rossi, J. Kozyra, and F. Micale. 2011. Impacts and adaptation of European crop production systems to climate change. Eur. J. Agron. 34:96–112.

Palazzo, A., J.M. Vervoort, D. Mason-D'Croz, L. Rutting, P. Havlík, S. Islam, J. Bayala, H. Valin, H.A.K. Kadi, and P. Thornton. 2017. Linking regional stakeholder scenarios and shared socioeconomic pathways: Quantified west African food and climate futures in a global context. Glob. Environ. Change 45:227–242.

Palosuo, T., K.C. Kersebaum, C. Angulo, P. Hlavinka, M. Moriondo, J.E. Olesen, R.H. Patil, F. Ruget, C. Rumbaur, and J. Takáč. 2011. Simulation of winter wheat yield and its variability in different climates of Europe: A comparison of eight crop growth models. Eur. J. Agron. 35:103–114.

Parent, B., O. Turc, Y. Gibon, M. Stitt, and F. Tardieu. 2010. Modelling temperature-compensated physiological rates, based on the co-ordination of responses to temperature of developmental processes. J. Exp. Bot. 61:2057–2069.

Passioura, J.B. 1996. Simulation models: Science, snake oil, education, or engineering? Agron. J. 88:690–694.

Peacock, J. 1975. Temperature and leaf growth in Lolium perenne. II. The site of temperature perception. J. Appl. Ecol. 115(1):1–5.

Pickering N.B., J.W. Jones, and K.J. Boote. 1995. Adapting SOYGRO V5. 42 for prediction under climate change conditions. In: C. Rosenweig, editor, Climate change and agriculture: Analysis of potential international impacts. ASA, Madison, WI. p. 77–98.

Raes, D., P. Steduto, T.C. Hsiao, and E. Fereres. 2009. AquaCrop—the FAO crop model to simulate yield response to water: II. Main algorithms and software description. Agron. J. 101:438–447.

Ramirez-Villegas, J., A.-K. Koehler, and A.J. Challinor. 2017. Assessing uncertainty and complexity in regional-scale crop model simulations. Eur. J. Agron. 88:84–95.

Rattalino Edreira, J.I., and M.E. Otegui. 2012. Heat stress in temperate and tropical maize hybrids: Differences in crop growth, biomass partitioning and reserves use. Field Crops Res. 130:87–98.

Rebetzke, G., and R. Richards. 1999. Genetic improvement of early vigour in wheat. Aust. J. Agric. Res. 50:291–302.

Richards, R., G. Rebetzke, A. Condon, and A. Van Herwaarden. 2002. Breeding opportunities for increasing the efficiency of water use and crop yield in temperate cereals. Crop Sci. 42:111–121.

Rigby, J., and A. Porporato. 2008. Spring frost risk in a changing climate. Geophys. Res. Lett. 35(12): L12703.

Ritchie, J. 1998. Soil water balance and plant water stress. In: G.Y. Tsuji, G. Hoogenboom, and P.K. Thornton, editors, Understanding options for agricultural production. Springer, Amsterdam, The Netherlands. p. 41-54.

Roberts, M.J., N.O. Braun, T.R. Sinclair, D.B. Lobell, and W. Schlenker. 2017. Comparing and combining process-based crop models and statistical models with some implications for climate change. Environ. Res. Lett. 12:095010.

Rodionov, A., S. Pätzold, G. Welp, R. Pude, and W. Amelung. 2016. Proximal field Vis-NIR spectroscopy of soil organic carbon: A solution to clear obstacles related to vegetation and straw cover. Soil Tillage Res. 163:89–98.

Rosenzweig, C., J.W. Jones, J.L. Hatfield, A.C. Ruane, K.J. Boote, P. Thorburn, J.M. Antle, G.C. Nelson, C. Porter, S. Janssen, S. Asseng, B. Basso, F. Ewert, D. Wallach, G. Baigorria, and J.M. Winter. 2013. The agricultural model intercomparison and improvement project (AgMIP): Protocols and pilot studies. Agric. For. Meteorol. 170:166–182.

Rosenzweig, C., and M.L. Parry. 1994. Potential impact of climate change on world food supply. Nature 367:133.

Rötter, R.P., M. Appiah, E. Fichtler, K.C. Kersebaum, M. Trnka, and M.P. Hoffmann. 2018. Linking modelling and experimentation to better capture crop impacts of agroclimatic extremes—A review. Field Crops Res. 221:142–156.

Rötter, R.P., T.R. Carter, J.E. Olesen, and J.R. Porter. 2011. Crop-climate models need an overhaul. Nat. Clim. Chang. 1:175–177.

Rötter, R.P., T. Palosuo, K.C. Kersebaum, C. Angulo, M. Bindi, F. Ewert, R. Ferrise, P. Hlavinka, M. Moriondo, and C. Nendel. 2012. Simulation of spring barley yield in different climatic zones of Northern and Central Europe: A comparison of nine crop models. Field Crops Res. 133:23–36.

Sadras, V.O. and D.F. Calderini. 2014. Crop physiology: applications for breeding and agronomy. In: V.O. Sadras and D.F. Calderini. Crop physiology: Applications for genetic improvement and agronomy: Second Edition. pp. 1–16. doi:10.1016/C2012-0-07386-3

Schlenker, W., and D.B. Lobell. 2010. Robust negative impacts of climate change on African agriculture. Environ. Res. Lett. 5:014010.

Schlenker, W., and M.J. Roberts. 2009. Nonlinear temperature effects indicate severe damages to US crop yields under climate change. Proc. Natl. Acad. Sci. USA 106:15594–15598.

Seidel, S.J., T. Palosuo, P. Thorburn, and D. Wallach. 2018. Towards improved calibration of crop models–Where are we now and where should we go? Eur. J. Agron. 94:25–35.

Shaw, R.E., and W.S. Meyer. 2015. Improved empirical representation of plant responses to waterlogging for simulating crop yield. Agron. J. 107(5):1711–1723.

Shaw, R.E., W.S. Meyer, A. McNeill, and S.D. Tyerman. 2013. Waterlogging in Australian agricultural landscapes: A review of plant responses and crop models. Crop Pasture Sci. 64(6):549–562.

Shi, P., L. Tang, C. Lin, L. Liu, H. Wang, W. Cao, and Y. Zhu. 2015. Modeling the effects of post-anthesis heat stress on rice phenology. Field Crops Research 177:26-36.

Siebert, S., F. Ewert, E.E. Rezaei, H. Kage, and R. Graß. 2014. Impact of heat stress on crop yield—on the importance of considering canopy temperature. Environ. Res. Lett. 9:044012.

Siebert, S., H. Webber, and E.E. Rezaei. 2017. Weather impacts on crop yields-searching for simple answers to a complex problem. Environ. Res. Lett. 12:081001.

Sinclair, T., and T. Horie. 1989. Leaf nitrogen, photosynthesis, and crop radiation use efficiency: A review. Crop Sci. 29:90–98.

Sinclair, T.R., G.L. Hammer, and E.J. Van Oosterom. 2005. Potential yield and water-use efficiency benefits in sorghum from limited maximum transpiration rate. Funct. Plant Biol. 32:945–952.

Sinclair, T.R., and G. Seligman. 1996. Crop modeling: From infancy to maturity. Agron. J. 88:698–704.

Skaggs, R.W., M. Youssef, and G.M. Chescheir. 2012. Drainmod: Model use, calibration, and validation. Trans. ASABE 55(4):1509–1522.

Smith, N.G., and J.S. Dukes. 2013. Plant respiration and photosynthesis in global-scale models: Incorporating acclimation to temperature and CO_2. Glob. Change Biol. 19:45–63.

Stedinger, J.R., R.M. Vogel, S.U. Lee, and R. Batchelder. 2008. Appraisal of the generalized likelihood uncertainty estimation (GLUE) method. Water Resour. Res. 44(12): W00B06.

Suzuki, N., R.M. Rivero, V. Shulaev, E. Blumwald, and R. Mittler. 2014. Abiotic and biotic stress combinations. New Phytol. 203:32–43.

Tack, J., A. Barkley, and N. Hendricks. 2017a. Irrigation offsets wheat yield reductions from warming temperatures. Environ. Res. Lett. 12:11407.

Tack, J., J. Lingenfelser, and S.K. Jagadish. 2017b. Disaggregating sorghum yield reductions under warming scenarios exposes narrow genetic diversity in US breeding programs. Proc. Natl. Acad. Sci. USA 114:9296–9301.

Tao, F., R.P. Rötter, T. Palosuo, C. Gregorio Hernández Díaz-Ambrona, M.I. Mínguez, M.A. Semenov, K.C. Kersebaum, C. Nendel, X. Specka, H. Hoffmann, F. Ewert, A. Dambreville, P. Martre, L. Rodríguez, M. Ruiz-Ramos, T. Gaiser, J.G. Höhn, T. Salo, R. Ferrise, M. Bindi, D. Cammarano, and A.H. Schulman. 2018. Contribution of crop model structure, parameters and climate projections to uncertainty in climate change impact assessments. Glob. Change Biol. 24:1291–1307.

Tardieu, F. 2012. Any trait or trait-related allele can confer drought tolerance: Just design the right drought scenario. J. Exp. Bot. 63:25–31.

Tardieu, F., T. Simonneau, and B. Muller. 2018. The physiological basis of drought tolerance in crop plants: A scenario-dependent probabilistic approach. Annu. Rev. Plant Biol. 69:733–759.

Taylor, K.E., R.J. Stouffer, and G.A. Meehl. 2012. An overview of CMIP5 and the experiment design. Bull. Am. Meteorol. Soc. 93:485–498.

Tjoelker, M.G., J. Oleksyn, and P.B. Reich. 2001. Modelling respiration of vegetation: Evidence for a general temperature-dependent Q10. Glob. Change Biol. 7:223–230.

Valdivia, R.O., J.M. Antle, C. Rosenzweig, A.C. Ruane, J. Vervoort, M. Ashfaq, I. Hathie, S.H.-K. Tui, R. Mulwa, and C. Nhemachena. 2015. Representative agricultural pathways and scenarios for regional integrated assessment of climate change impacts, vulnerability, and adaptation. In: C. Rosenweig and D. Hillel, editors, Handbook of climate change and agroecosystems: The agricultural model intercomparison and improvement project integrated crop and economic assessments. ICP Series on climate change impacts, adaptation, and mitigation, Part 1. Vol. 3: Imperial College Press, London. p. 101-156.

Van Bussel, L., F. Ewert, and P. Leffelaar. 2011. Effects of data aggregation on simulations of crop phenology. Agric. Ecosyst. Environ. 142:75–84.

van Ittersum, M., P. Leffelaar, H. van Keulen, M. Kropff, L. Bastiaans, and J. Goudriaan. 2003. On approaches and applications of the Wageningen crop models. Eur. J. Agron. 18:201–234.

van Ittersum, M., and R. Rabbinge. 1997. Concepts in production ecology for analysis and quantification of agricultural input-output combinations. Field Crops Res. 52:197–208.

van Vuuren, D.P., J. Edmonds, M. Kainuma, K. Riahi, A. Thomson, K. Hibbard, G.C. Hurtt, T. Kram, V. Krey, J.-F. Lamarque, T. Masui, M. Meinshausen, N. Nakicenovic, S.J. Smith, and S.K. Rose. 2011. The representative concentration pathways: An overview. Clim. Change 109:5–31.

Vanuytrecht, E., and P.J. Thorburn. 2017. Responses to atmospheric CO2 concentrations in crop simulation models: A review of current simple and semicomplex representations and options for model development. Glob. Change Biol. 23:1806–1820.

Wallach, D., D. Makowski, J.W. Jones, and F. Brun. 2013. Working with dynamic crop models: Methods, tools and examples for agriculture and environment. Academic Press, Waltham, MA.

Wallach, D., L.O. Mearns, A.C. Ruane, R.P. Rötter, and S. Asseng. 2016. Lessons from climate modeling on the design and use of ensembles for crop modeling. Clim. Change 139:551–564.

Wang, E., P. Martre, Z. Zhao, F. Ewert, A. Maiorano, R.P. Rötter, B.A. Kimball, M.J. Ottman, G.W. Wall, and J.W. White. 2017. The uncertainty of crop yield projections is reduced by improved temperature response functions. Nat. Plants (London, U. K.) 3:17102.

Wassenaar, T., P. Lagacherie, J.-P. Legros, and M. Rounsevell. 1999. Modelling wheat yield responses to soil and climate variability at the regional scale. Clim. Res. 11:209–220.

Webber, H., F. Ewert, J.E. Olesen, C. Müller, S. Fronzek, A.C. Ruane, M. Bourgault, P. Martre, B. Ababaei, M. Bindi, R. Ferrise, R. Finger, N. Fodor, C. Gabaldón-Leal, T. Gaiser, M. Jabloun, K.-C. Kersebaum, J.I. Lizaso, I.J. Lorite, L. Manceau, M. Moriondo, C. Nendel, A. Rodríguez, M. Ruiz-Ramos, M.A. Semenov, S. Siebert, T. Stella, P. Stratonovitch, G. Trombi, and D. Wallach. 2018a. Diverging importance of drought stress for maize and winter wheat in Europe. Nat. Commun. 9:4249.

Webber, H., T. Gaiser, and F. Ewert. 2014. What role can crop models play in supporting climate change adaptation decisions to enhance food security in Sub-Saharan Africa? Agric. Syst. 127:161–177.

Webber, H., T. Gaiser, R. Oomen, E. Teixeira, G. Zhao, D. Wallach, A. Zimmermann, and F. Ewert. 2016a. Uncertainty in future irrigation water demand and risk of crop failure for maize in Europe. Environ. Res. Lett. 11:074007.

Webber, H., J. White, B. Kimball, F. Ewert, S. Asseng, E.E. Rezaei, P. Pinter, J.L. Hatfield, M.P. Reynolds, B. Ababaei, M. Bindi, J. Doltra, R. Ferrise, H. Kage, B. Kassie, K.C. Kersebaum, A. Luig, J.E. Olesen, M.A. Semenov, P. Stratonovitch, A. Ratjen, R. LaMorte, S.W. Leavitt, D. Hunsaker, G.W. Wall, and P. Martre. 2018b. Physical robustness of canopy temperature models for crop heat stress simulation across environments and production conditions. Field Crops Res. 216:75–88.

Webber, H., G. Zhao, W. Britz, W. de Vries, J. Wolf, T. Gaiser, H. Hoffmann, and F. Ewert. 2015. Specification of nitrogen use in regional climate impact assessment studies. 5th International Symposium for Farming Systems Design, Montpellier, France. 7–10 Sept. 2015. Southern African Program on Ecosystem Change and Society, Cape Town, South Africa.

Webber, H.A., F. Ewert, B. Kimball, S. Siebert, J.W. White, D. Trawally, G.W. Wall, M.J. Ottman, and T. Gaiser. 2016b. Simulating canopy temperature for modelling heat stress in cereals. Environ. Model. Softw. 77:143–155.

Whish, J.P., N.I. Herrmann, N.A. White, A.D. Moore, and D.J. Kriticos. 2015. Integrating pest population models with biophysical crop models to better represent the farming system. Environmental Modelling & Software 72: 418-425.

White, J.W., M. Herndl, L. Hunt, T.S. Payne, and G. Hoogenboom. 2008. Simulation-based analysis of effects of and loci on flowering in wheat. Crop Sci. 48:678–687.

White, J.W., G. Hoogenboom, B.A. Kimball, and G.W. Wall. 2011a. Methodologies for simulating impacts of climate change on crop production. Field Crops Res. 124:357–368.

White, J.W., B.A. Kimball, G.W. Wall, M.J. Ottman, and L. Hunt. 2011b. Responses of time of anthesis and maturity to sowing dates and infrared warming in spring wheat. Field Crops Res. 124:213–222.

Wild, J., M. Kopecký, M. Macek, M. Šanda, J. Jankovec, and T. Haase. 2019. Climate at ecologically relevant scales: A new temperature and soil moisture logger for long-term microclimate measurement. Agric. For. Meteorol. 268:40–47.

Willmott, C.J., K. Matsuura, and S.M. Robeson. 2009. Ambiguities inherent in sums-of-squares-based error statistics. Atmos. Environ. 43:749–752.

Wolf, J. 2012. LINTUL5: Simple generic model for simulation of crop growth under potential, water limited and nitrogen, phosphorus and potassium limited conditions. Plant Production Systems Group, Wageningen University, Wageningen, The Netherlands.

Wolf, J., L. Evans, M. Semenov, H. Eckersten, and A. Iglesias. 1996. Comparison of wheat simulation models under climate change. I. Model calibration and sensitivity analyses. Clim. Res. 7:253–270.

Xinyou, Y., and H. Van Laar. 2005. Crop systems dynamics: An ecophysiological simulation model for genotype-by-environment interactions. Wageningen Academic Publishers, Wageningen, The Netherlands.

Yang, H., Y. Chen, F. Zhang, T. Xu, and X. Cai. 2016. Prediction of salt transport in different soil textures under drip irrigation in an arid zone using the SWAGMAN Destiny model. Soil Res. 54(7)869–879.

Yin, X., K.C. Kersebaum, C. Kollas, S. Baby, N. Beaudoin, K. Manevski, T. Palosuo, C. Nendel, L. Wu, and M. Hoffmann. 2017. Multi-model uncertainty analysis in predicting grain N for crop rotations in Europe. Eur. J. Agron. 84:152–165.

Yoshimoto, M., M. Fukuoka, T. Hasegawa, M. Utsumi, Y. Ishigooka, and T. Kuwagata. 2011. Integrated micrometeorology model for panicle and canopy temperature (IM2PACT) for rice heat stress studies under climate change. Agricultural Meteorology 67:233–247.

Yoshimoto, M., H. Oue, N. Takahashi, and K. Kobayashi. 2005. The effects of FACE (Free-Air CO_2 Enrichment) on temperatures and transpiration of rice panicles at flowering stage. Agricultural Meteorology 60(5):597-600.

Zhao, C., B. Liu, S. Piao, X. Wang, D.B. Lobell, Y. Huang, M. Huang, Y. Yao, S. Bassu, P. Ciais, J.-L. Durand, J. Elliott, F. Ewert, I.A. Janssens, T. Li, E. Lin, Q. Liu, P. Martre, C. Müller, S. Peng, J. Peñuelas, A.C. Ruane, D. Wallach, T. Wang, D. Wu, Z. Liu, Y. Zhu, Z. Zhu, and S. Asseng. 2017. Temperature increase reduces global yields of major crops in four independent estimates. Proc. Natl. Acad. Sci. USA 114:9326–9331.

Zhao, G., S. Siebert, A. Enders, E.E. Rezaei, C. Yan, and F. Ewert. 2015a. Demand for multi-scale weather data for regional crop modelling. Agric. For. Meteorol. 200:156–171.

Zhao, G., H. Webber, H. Hoffmann, J. Wolf, S. Siebert, and F. Ewert. 2015b. The implication of irrigation in climate change impact assessment: A European wide study. Glob. Change Biol. 21:4031–4048.

Zimmermann, A., H. Webber, G. Zhao, F. Ewert, J. Kros, J. Wolf, W. Britz, and W. de Vries. 2017. Climate change impacts on crop yields, land use and environment in response to crop sowing dates and thermal time requirements. Agric. Syst. 157:81–92.

Modeling Soil Dynamic Processes

Bruno Basso*, Davide Cammarano, Massimiliano De Antononi Migliorati, Bernardo Maestrini, Benjamin Dumont, and Peter R. Grace

Process-based crop models are computerized representations of a real plant growing in soil and influenced by its interactions with weather and management. They are only an approximation of the real world, and regardless of their level of detail, they cannot take into account every single factor (Jones et al., 2017). However, they are useful tools because users can extrapolate temporal patterns of such interactions beyond a single location or year. They can complement field or laboratory experiments and help gain new knowledge. Most of the models that simulate the biophysical interactions of soil, plant, and atmosphere simulate how each relationship evolves over time. The predictive ability of a given model depends on how well it is designed, its degree of complexity, and the quality of model input used in the simulation.

The goal of this chapter is to discuss different approaches used to model soil dynamic processes: physical (temperature and water), chemical, and biological (nitrogen and carbon). We also present insights on the spatial variability of soils and their impact on yield variations as well as a case study of a multi-model ensemble to predict changes of soil organic carbon (SOC) and its feedback to yield in sites across the world.

Approaches to Modeling Soil Dynamic Processes

Soil Temperature

Soil temperature is an important determinant of many below-ground processes, such as seed germination, plant nutrient uptake, microbial processes, evaporation

---

Abbreviations: DUL, drained upper limit; ES, soil water evaporation; FC, field capacity; LL, lower limit; SAT, saturated water content; WFPS, water-filled pore space;

B. Basso and B. Maestrini, Department of Earth and Environmental Sciences, Michigan State University, East Lansing, MI; B. Basso, W.K. Kellogg Biological Station, Michigan State University, Hickory Corners, MI; B. Basso, M.M. D'Antoni, and P.R. Grace, Institute for Future Environments, Queensland University of Technology, Brisbane, Australia; D. Cammrano, James Hutton Institute, Scotland; B. Dumont, University of Liege, Gambloux, Belgium.*Corresponding author (Basso@msu.edu)

doi:10.2134/agronmonogr60.2018.0018

© ASA, CSSA, and SSSA, 5585 Guilford Road, Madison, WI 53711, USA.
Agroclimatology: Linking Agriculture to Climate, Agronomy Monograph 60.
Jerry L. Hatfield, Mannava V.K. Sivakumar, John H. Prueger, editors.

of water from soil, and soil chemical processes (Coelho and Dale, 1980; Sharpley and Williams, 1990; Kaspar and Bland, 1992; Boone et al., 1998). The main weather variables that influence soil temperature are solar radiation and air temperature. However, other factors can also affect soil temperature, such as rainfall, soil water content, texture, surface residues, and tillage. Novak (2005) classified the factors influencing soil temperature into two groups: i) the ones that influence the amount of heat that enters the soil by conduction at its surface and ii) the ones influencing the volumetric heat capacity and the soil thermal conductivity. There is also a small upward flux of 50 n W m^{-2} due to the radioactivity and cooling of the earth's core that influences soil temperature. Respiration from plant roots and soil microorganisms adds further sources of heat (Loomis and Connor, 1998). Measurements of soil temperature are not always available, especially if soil temperature is needed spatially across a field, but methods for estimating such parameters have been developed. Luo et al. (1992) divided the soil temperature models into three categories: i) empirical models, ii) mixed empirical and mechanistic models, and iii) mechanistic models based on physical processes. The models belonging to the first group are simple empirical models based on relationships between soil temperature, climate, and soil variables. The models belonging to the second group utilize the concepts for heat flow as described in detail in Wierenga and De Wit (1970). Luo et al. (1992) pointed out that for these kinds of models, the upper boundary temperature has to be given or at least empirically derived. The soil temperature models in the third category are built on the concepts of radiative energy balance and heat fluxes, such as sensible, latent, and ground conductive. However, most of these models were developed for bare soil only, and when used in systems where plants are growing, additional assumptions and interactions have to be considered.

An early soil temperature model developed by Gupta et al. (1983) was modified by Dwyer et al. (1990) to estimate soil surface temperature (5 cm) from air temperature to predict corn emergence dates. Results of that study showed that the model was able to estimate emergence dates within one to two days of the field observations. In contrast to such a simple approach, Luo et al. (1992) developed a model (SLTMP) based on energy balance and heat conduction in the soil. The model calculates net radiation and heat fluxes as modified by canopy leaves and cumulative evaporation to obtain the energy balance of the soil surface under vegetation cover. The soil surface temperature is estimated at a one-hour time-step. The approach developed by Potter and Williams (1994) has been used in several crop simulation models. The authors modified the EPIC soil surface temperature model by replacing this submodule with the one developed by Parton (1984), which utilizes daily maximum and minimum air temperatures, plant biomass, crop residues, and total daily solar radiation. Soil temperature at depth is then calculated, taking into account the long-term average annual air temperature, the soil surface temperature on a given day, and a coefficient that allows weighing the effects of yesterday's temperature with the current day's soil temperature estimate. Also included is a depth factor influenced by soil bulk density and water content.

Another family of soil temperature models utilizes heat conduction as discussed in Hillel (1982):

$$\frac{\partial T}{\partial t}=\frac{\partial}{\partial z}\left(\alpha\frac{\partial T}{\partial z}\right) \quad [1]$$

Where T is the soil temperature at a given depth z (cm) and at a time t (h), is the soil thermal diffusivity (cm^2 h^{-1}). Two boundaries are needed: one deals with the assumption that surface soil temperature is equal to air temperature and varies according to a sinusoidal function (Hillel, 1982), while the other condition states that the soil temperature is constant and at deeper layers is equal to mean air temperature (Elias et al., 2004).

The soil temperature calculation of mechanistic models like the HYDRUS-1D (Šimůnek et al., 2008) numerically solves the convection–dispersion equation where the one-dimensional heat transfer is calculated as follows:

$$\frac{\partial C_p(\theta)T}{\partial T}=\frac{\partial}{\partial x}\left(\lambda(\theta)\frac{\partial T}{\partial x}\right)-C_w\frac{\partial T}{\partial x}-C_w\times S\times T \quad [2]$$

In this equation q is the volumetric soil water content, $C_p(\theta)$ is the volumetric heat capacity of the porous medium, and Cw is the volumetric heat capacity of the liquid phase. $\lambda(\theta)$ is the soil apparent thermal conductivity. It is calculated as follows:

$$\lambda(\theta)=\lambda_0(\theta)+\beta_i C_w|q| \quad [3]$$

b_i is the thermal dispersivity and $\lambda_0(\theta)$ is the thermal conductivity of the soil (Chung and Horton, 1987) calculated as follows:

$$\lambda_0(\theta)=b_1+b_2\theta+b_3\theta^{0.5} \quad [4]$$

Crop simulation models can include any of the generic approaches described above. For example, the DSSAT crop simulation model (Jones et al., 2003) utilizes a modified version of the EPIC soil temperature routine. The SALUS model (Basso and Ritchie, 2015) uses an approach based on the work of Vinocur and Ritchie (2001). The APSIM model requires information on soil water content, air temperature, soil water evaporation, and incident net radiation for the upper boundary conditions. Their model is conceptualized as a series of resistances and nodes where the former represents the resistance to heat transfer, and the latter represents the heat storage. In the STICS crop model, soil temperature is a function of the surface conditions that drives the daily thermal wave and the heat inertia of the soil into deeper layers. Daily crop temperature and air temperature amplitude are used as upper limits for the calculations of soil temperatures. At a given depth, the amplitude and the soil temperatures are calculated using the model of McCann et al. (1991). However, a recent study comparing multiple grassland models (Sándor et al., 2017) found that no matter what approach is utilized in those models, soil temperature is satisfactorily estimated. Therefore, the main issue to consider when choosing a soil temperature model is the availability of data for the specific calculation and computational time.

Soil Water

When considering the main approaches used in the modeling of soil water, it is important to make a distinction between the "statics" of soil water and the "dynamics" of it. These two components have different definitions and are associated with different aspects of the water in the soil. In this chapter we consider the movements of water within the soil and how it is simulated. Water can flow upward through a crop (through the transpiration process), evaporate from the bare soil, and move downward in the soil (e.g., drainage). Soil water dynamics can be modeled using either empirical or mechanistic models; however, within each of these categories lie varying degrees of complexity. As a matter of fact, White et al. (1993) concluded that there is not really a given separation between the empirical and the mechanistic approach because most models, even the simplest, contain a given degree of physically-based logic.

Models of soil water dynamics can be classified as either *simple models* or *complex models.* Simple models have a fixed number of soil layers and use a "tipping bucket," or capacity, approach for water that flows in and out of each soil layer. Within this family of models, some approaches consider the soil as a single layer and others as multiple layers. Complex models consider the soil profile as a continuum, so they can deal with a one- or two-dimensional flow of water.

The water flow is also tightly bounded to the amount of water that infiltrates the soil, runs off the soil, evaporates from the soil surface, and is transpired by the crop. All of these aspects can be modeled in many ways, and Table S2 in Asseng et al. (2013) provides a listing of existing wheat models and the different approaches employed to simulate the water balance.

In the capacity, or "tipping bucket," approach, the simplest representation is to consider one uniform soil layer where the profile is filled by water from precipitation and/or irrigation and cleared by water lost from evaporation and/ or transpiration, which is not distinguished because vegetation is not obviously modeled. If the profile is full, the water is lost by surface runoff. This approach was employed by early versions of models of soil water dynamics, such as WATBAL, which was developed by Fitzpatrick and Nix (1969). Multiple soil layers were later incorporated where soil and water inputs were required for each layer. Most of these multi-layer capacity models originated from an early version of the Ritchie model (Ritchie 1972; Ritchie 1981a; Ritchie 1981b; Ritchie, 1998). In this approach water from an upper layer moves to the lower layer, similar to moving from a series of buckets. Water infiltrates the soil as a function of rainfall/irrigation minus the runoff. Runoff is estimated with the SCS methodology (Soil Conservations Service, 1972). For a given soil layer, the available soil water content is determined by the drained upper limit (DUL), lower limit (LL) and saturated water content (SAT). Downward saturated flow happens when, for a given soil layer, water content is above the DUL. Upward flow is caused by plant transpiration and soil evaporation. The tipping-bucket approach, written as a simple differential equation as stated by Emerman (1995) is

$$\frac{d\partial}{dt} = -\alpha(\theta - FC) \quad [5]$$

where is the soil water content, t is the time step, FC is the Field Capacity, and is assumed to be the drainage coefficient.

In complex models, soil water dynamics is a continuous movement within the soil as opposed to the cascade movement in the tipping-bucket approach. These models can have a one- or two-dimensional water flow within the soil. One of the most popular implementations of this approach is the Richards' equation (Richards, 1931). However, where this approach is utilized, there are no analytical solutions, and as a consequence the modelers have implemented numerical solutions. The other hydraulic principle that could potentially be applied is Darcy's Law. Although it is not applicable for unsaturated flow, it has been coupled with the Richards' equation to achieve these outcomes (Broadbridge and White, 1988). The main assumption is that hydraulic conductivity and diffusivity are dependent on volumetric water content rather than soil depth (Philip, 1966). Lee and Abriola (1999) proposed a modification of the Richards' approach to simplify the usage of the equation, but their model only proposed downward fluxes of water and did not consider the effects of evaporation and transpiration.

The Richards' equation for a given soil profile for describing one-dimensional vertical water movement is

$$\frac{\partial \theta}{\partial t}=\frac{\partial}{\partial z}\left[K(\theta)\left(\frac{\partial h}{\partial z}+1\right)\right] \quad [6]$$

where θ is the volumetric soil water content, $K(\theta)$ the hydraulic conductivity of the soil, z is the depth, and is the soil water pressure, relative to the atmosphere, expressed in cm of water.

Following the approach of van Genuchten (1980) and Mualem (1976), the hydraulic functions are

$$\Theta=\frac{\theta-\theta_r}{\theta_s-\theta_r}=\left[\frac{1}{1+|\alpha h|^n}\right]^m \quad [7]$$

$$K(\theta)=K_s\Theta^{0.5}\left[1-\left(1-\Theta^{1/m}\right)^m\right]^2 \quad [8]$$

Where is the relative saturation, θ_s is the saturated water content, θ_r the residual water content, α and n the parameters of the retention and conductivity functions, respectively, $m = 1-1/n$, and is the saturated hydraulic conductivity. To solve Eq. [5] requires a complex numerical solution similar to that provided by Šimůnek et al. (1992). The process involves solving a series of linear equations instantaneously in both spatial and temporal domains.

While the majority of crop models use the empirical approach (Table 1), the APSIM and CropSyst models are capable of simulating soil water dynamics with either the tipping-bucket or the Richards approach.

A comparison of the two approaches of soil water dynamics (capacity, or tipping-bucket, approach and Richards' approach) was illustrated in the work of Buttler and Riha (1992) who concluded that both methods were suitable for estimating water content of the soil profile under consideration. The authors stated that while the Richards equation was preferred for the simulation of drainage fluxes and soil water distribution, it proved difficult to obtain the necessary parameters under field conditions for estimating the relationship between soil water content and water potential as required by the Richards equation. On the other hand, the tipping-bucket approach was very sensitive to changes in FC values, and if this information is not available, it should be carefully estimated. However, as reported in Table 1, most crop simulation models utilize the capacity approach because of its simplicity in deriving the key parameters required. Emerman (1995) combined the tipping-bucket approach and the Richards equation, and while they found an improvement in the simulated soil water content, their results are obviously limited to the dataset under consideration and to several assumptions made for that particular study.

Table 1. **Twenty-seven wheat crop models utilized in the AgMIP Wheat Study (Asseng et al., 2013) and the water dynamics modeling approach utilized by each model.**

Models	Water model
Apsim-NWheat	Capacity Approach
Apsim-Wheat	Capacity Approach/Richards Approach
AquaCrop	Capacity Approach
CropSyst	Capacity Approach/Richards Approach
DSSAT-CERES	Capacity Approach
DSSAT-CROPSIM	Capacity Approach
Ecosys	Richards Approach
EPIC Wheat	Capacity Approach
ExpertN-CERES	Richards Approach
ExpertN-GECROS	Richards Approach
ExpertN-SPASS	Richards Approach
ExpertN-SUCROS	Richards Approach
FASSET	Capacity Approach
GLAM-Wheat	Capacity Approach
HERMES	Capacity Approach
InfoCrop	Capacity Approach
LINTUL-4	Capacity Approach
LINTUL-FAST	Capacity Approach
LPJmL	Capacity Approach
MCWLA-Wheat	Richards Approach
MONICA	Capacity Approach
O'Leary Model	Capacity Approach
SALUS	Capacity Approach
Sirius	Capacity Approach
Sirius Quality	Capacity Approach
STICS	Capacity Approach
WOFOST	Capacity Approach

In modeling the interactions between soil-crop-atmosphere, these two approaches to simulating soil water dynamics can be included as sub-routines in more sophisticated crop simulation models. Once a sub-routine is part of such an integrated system, it is difficult to identify the causes of any bias in simulation results. For example, when comparing 16 wheat growth models, Cammarano et al. (2016) demonstrated that the relatively simple FAO-56 method for estimating potential evapotranspiration did not show significant differences when compared to other equations in the context of a crop model. In another study where crop simulation models were compared for the energy-balance approach, the empirical methods performed as well or better than the more mechanistic approaches (Webber et al., 2017). These two studies highlight the fact that the results of model intercomparisons should be viewed and analyzed carefully. However, Asseng et al. (2013) demonstrated that when these intercomparisons are carefully designed, convergence of model outputs can provide more robust and conclusive outcomes that in turn aid the overall improvement of model accuracy.

The soil water balance, in its simplest form can be written as follows:

$$\Delta SW = P + Irr + UF - Dr - Rf - Tp - Es \quad [9]$$

where ΔSW is the soil water change, P is the precipitation, Irr is the amount (if any) of irrigation, UF is the upward fluxes (contrinution from water table or capillary fringe), Dr is the drainage, Rf is the runoff, Tp is the plant transpiration, and Es is the soil water evaporation. Cammarano et al. (2016) found that the simulated crop transpiration accounted for more than 50% of the total variability among models. However, crop transpiration is affected by many other processes and subroutines. Sándor et al. (2017) compared grassland models across many soils and environments of Europe and found a large uncertainty in simulating soil water content and yield. Durand et al. (2017) compared maize models and found that under well-watered conditions, the models reproduced the absence of yield response to increased CO_2. However, when soil water deficit was coupled with increased CO_2, the uncertainty in simulated yield was higher, and the models could not accurately simulate the soil water content at anthesis. In both the Cammarano et al. (2016) and Durand et al. (2017) studies, there was a mixture of models using either the tipping-bucket or the Richards approach.

In Cammarano et al. (2016), regardless of the underlying approach, the majority of models are capable of correctly simulating the trend of ET, with only one crop model underestimating ET 60 d after sowing (Fig. 1). The variability in simulated soil water evaporation (ES) was not dependent on whether the capacity or Richards' approach was used. The simulation of ES depends only in part on soil water dynamics, as vegetation dynamics (e.g., canopy cover and leaf area index) also play an important part in its simulation.

In conclusion, the choice between the capacity and the Richards approaches for soil water dynamics is dependent on the aim and specific circumstances of the user, the ease of deriving the necessary parameters needed, and their fit within a higher hierarchical model.

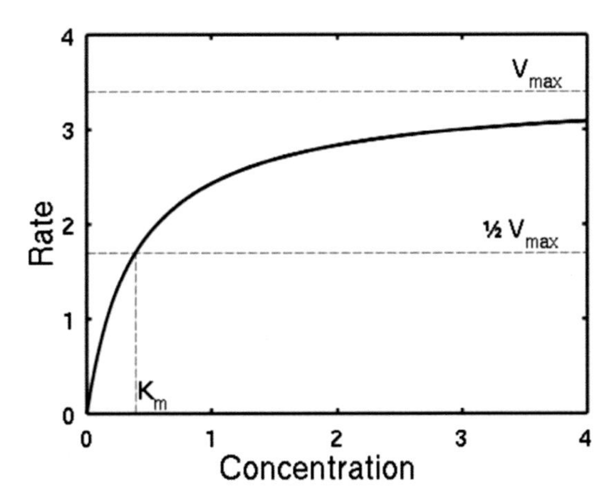

Fig. 1. General representation of how Michaelis-Menten kinetics determine nitrification rates at varying soil NH_4^+ concentrations. (Modified from Meier et al., 2006).

Factors Controlling N_2O Production in Soils

Soils play a fundamental role in the release of N_2O to the atmosphere. It is estimated that soils contribute approximately 50% of all global N_2O emissions (Ehhalt et al., 2001; US-EPA, 2010). The concentration of N_2O in the atmosphere has increased by 20% since the preindustrial era (Park et al., 2012). This is mainly due to the application of inorganic and organic nitrogen fertilizers to soils (Shcherbak et al., 2014). The Global Warming Potential (GWP) of N_2O is 298 times that of CO_2 and almost 14 times that of CH_4 over a 100-yr time horizon, and it is estimated to contribute 8.1% of the total global warming effect caused by GHGs (Myhre et al., 2013). Nitrous oxide also depletes stratospheric ozone, which has an essential role in reducing the mutagenic and carcinogenic effects of solar ultraviolet radiation (Crutzen, 1981; Ravishankara et al., 2009). Therefore, it is critical to identify farming practices that support agricultural production while minimizing N_2O emissions.

Field experiments have been crucial to advancing the knowledge of the processes and management practices that regulate N_2O emissions. However, field trials are time-consuming, expensive, and designed to investigate a limited number of treatments (Jones et al., 2003) at a time. Moreover, the N cycle is complex and influenced by climate, land-use management, and soil properties, which by their nature are variable. Plant and soil interactions based on soil water and N availability also influence the N cycle, resulting in elevated variability of N_2O emissions in both time and space (Folorunso and Rolston, 1984; Schelde et al., 2012). The results of field trials conducted at a particular point in time and space are, therefore, difficult to generalize beyond their original application.

In this respect, models have become indispensable tools to the study of how crop production and N_2O emissions are affected by various agricultural practices and changes in environmental conditions.

Nitrous oxide is produced in the soil as an obligate intermediate of two microbial-mediated mechanisms: nitrification and denitrification (Fig. 2). Nitrification occurs in aerobic conditions and consists of the oxidation of ammonium (NH_4^+) to nitrate (NO_3^-) by chemoautotrophic bacteria. Nitrification is a two-step process, where NH_4^+ is oxidized to nitrite (NO_2^-), which is then converted into NO_3^- (Conrad, 2001). Although studies have shown that heterotrophic bacteria can be a source of N_2O under certain conditions, chemoautotrophic microorganisms are largely, if not entirely, responsible for N_2O losses due to nitrification in most soils (Hutchinson and Davidson, 1993; Bremner, 1997).

Denitrification is the reduction of NO_3^- to nitrogen gas (N_2) under anaerobic conditions by heterotrophic bacteria. These bacteria use NO_3^- as a substitute for oxygen as a terminal electron acceptor, thereby reducing NO_3^- to N_2 (Bollmann and Conrad, 1998). Nitrous oxide production related to denitrification can be significantly larger than that from nitrification, because denitrification produces N_2O as an obligate intermediate. This means that if conditions are not favorable for the completion of the entire denitrification process, large amounts of denitrified NO_3^- can be lost as N_2O, not as N_2.

Soil conditions regulate the extent to which nitrification and denitrification occur, and therefore, influence the amount of N_2O emitted by soil (Firestone and Davidson, 1989; Bouwman, 1998). Primary factors that govern the size of the microbial pool are N and C availability and soil temperature. Secondary factors include soil moisture and pH, which control the partitioning of soil N into NO, N_2O, or

N_2. In addition, the soil properties that restrict or enhance diffusion of the gases produced by microbial activity, such as soil structure or texture, also affect N_2O emissions. The availability of N in the soil is usually the main factor that limits N_2O production, specifically the inorganic form of NH_4^+. Nitrification rates in uncultivated environments are limited by the slow mineralization of N produced by the decomposition of plant and animal residues (Dalal et al., 2003). In agricultural soils, however, this process is substantially accelerated. Nitrification is enhanced by the addition of rapidly nitrifiable forms of N (as in NH_4^+-based fertilizers) and the aeration caused by tillage, both of which increase the amount of available N and provide the conditions required by nitrifying bacteria (Robertson and Groffman, 2007). In contrast, N_2O emissions that are related to denitrification are mainly controlled by the amount of NO_3^- available. Low NO_3^- concentrations prolong the reduction of N_2O to N_2 by denitrifying microorganisms, thereby decreasing the N_2O/N_2 ratio. Conversely, elevated concentrations of NO_3^- almost completely inhibit the reduction of N_2O to N_2 and result in larger amounts of N lost as N_2O.

Denitrification rates are also affected by the availability of labile organic C because denitrifying bacteria are heterotrophic and require C as an energy source. In anaerobic conditions, the presence of high amounts of labile C or readily decomposable organic matter can, therefore, significantly increase denitrification rates (Dalal et al., 2003; Li et al., 2005a). The effect of organic materials on denitrification varies with their resistance to decomposition. Readily decomposable substrates such as glucose promote denitrification, while materials such as lignin are difficult to decompose because of their aromatic structure (Bremner, 1997). As with other biological processes, nitrification and denitrification rates are directly influenced by temperature. Increased N_2O emissions have been observed at temperatures exceeding 25 °C (Dalal et al., 2003).

Soil aeration is the second most important factor that regulates the activity of nitrifying and denitrifying bacteria as it relates to soil water content. Nitrification is typically the main source of N_2O emissions when water-filled pore space (WFPS) is below 40% (Fig. 3). Denitrification rates rise with increasing water content, and become the predominant process over 70% WFPS (Bouwman, 1998;

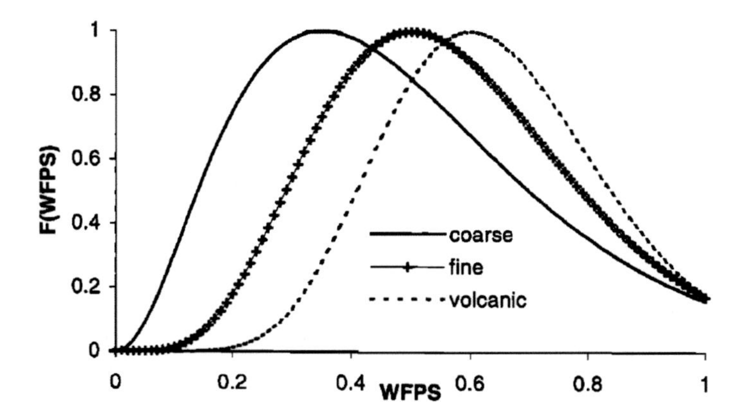

Fig. 2. Function curves illustrating the effects of water-filled pore space (WFPS) on nitrification rates (F(WFPS)) using first-order kinetics (Parton et al., 2001). Different function curves can be used to adjust the effect of each factor in different types of soils.

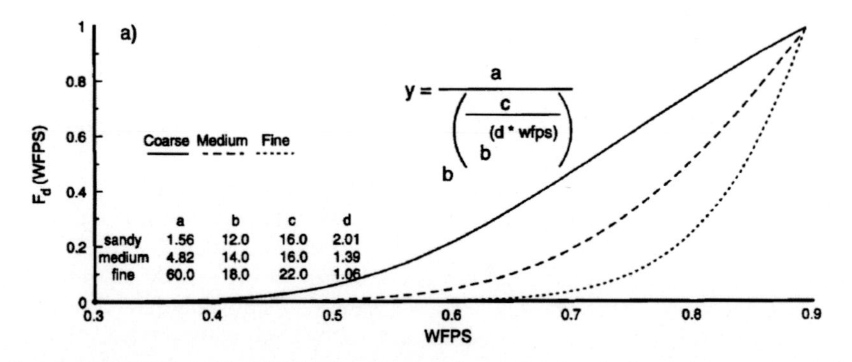

Fig. 3. Example of function used to calculate the effect of water-filled pore space on denitrification (Parton et al., 1996). Different function curves can be used to describe the effect of the variable in different types of soils.

Kiese and Butterbach-Bahl, 2002; Werner et al., 2006). Anaerobic soil conditions occur above 65% to 75% WFPS, which promote denitrification and the production of both N_2O and N_2 (Panek et al., 2000). When soil water content exceeds 80–90% WFPS, denitrifying bacteria are able to complete the NO_3^- reduction process and emit mainly N_2 as the end product. Consequently, the N_2O/N_2 ratio decreases as the soil water content exceeds 75% WFPS (Dalal et al., 2003).

Microbial activity related to both nitrification and denitrification is negatively affected by acid soil conditions, with the highest rates of N_2O production observed at soil pH between 7 and 8.5. This inhibitory effect of low pH has been associated with the decreased availability of organic C and mineral N under acid conditions (Šimek and Cooper, 2002). Although N_2O production tends to diminish at low soil pH, several studies have reported increased N_2O/N_2 production ratios when soil pH was below 7 (Dalal et al., 2003; Liu et al., 2010). Low pH in fact inhibits the reduction of N_2O to N_2 more than it inhibits the reduction of NO_3^- to N_2O, thus favoring emissions of N_2O over N_2. Consequently, as pH increases, denitrification products tend to be more completely reduced toward N_2 production (Chapuis-Lardy et al., 2007).

Additional influences on the release of N_2O are those factors that determine the volume of soil pores, such as texture, bulk density, aggregate stability, and organic matter content. These parameters are pivotal in the regulation of soil water content and consequently, soil aeration, gas production, and diffusion. For example, the larger pore space in coarse, sandy soils tends to favor nitrification, while the greater quantity of water that can be stored in fine-textured, clay soils promotes denitrification (Bollmann and Conrad, 1998). The smaller pore size of fine-textured soils also reduces hydraulic conductivity and creates more persistent waterlogging conditions (Granli, 1995). Moreover, smaller pores rapidly fill with water and can create anaerobic microsites that enable denitrification to occur at lower soil water contents than coarse-textured soils (Parton et al., 2001). Well-developed soil structure can also create microenvironments where bacteria are physically isolated from nutrients, water, or oxygen.

Simulation of N_2O Emissions

Process-based models are among the most accurate of the numerous approaches that have been used to develop models that predict N dynamics and N_2O emissions in agroecosystems. These models use a mechanistic approach to represent the complex biophysical processes that influence greenhouse gas emissions and plant growth, such as soil organic matter, soil water content, fertilizer N management, and plant N availability (Del Grosso et al., 2009). Only a few process-based models simulate crop yields and N_2O emissions at field-scale on daily time steps, and among these even fewer have been tested with the environmental conditions that represent a variety of locations around the globe. A detailed review of plant–soils models that simulate N_2O can be found in Chen et al. (2008).

This section examines the main approaches currently adopted to simulate N_2O emissions related to nitrification and denitrification. The concepts outlined here are based on the three most extensively tested process-based models currently used in the Agricultural Model Intercomparison and Improvement Project (AgMIP.org; Agricultural Model Intercomparison and Improvement Project, 2019): DNDC, APSIM, and DAYCENT.

N_2O Emissions from Nitrification

Process-based models simulate N_2O emissions derived from nitrification by assuming that a fixed proportion of the amount of soil N that is nitrified daily is lost as N_2O as calculated in the following generic equation:

$$N_2O_{nitr} = R_{nitr} \times K_1 \quad [10]$$

where N_2O_{nitr} is the amount of N_2O emitted daily via nitrification, R_{nitr} is the daily nitrification rate, and K_1 is the fraction of nitrified N lost as N_2O. Different models have adopted values of K_1 that range from 0.2% to 2% for agricultural soils (Li et al., 2000; Parton et al., 2001; Thorburn et al., 2010). The large range that is used for K_1 reveals that the chemo-physical processes leading to N_2O losses from nitrification are still not completely understood.

Most process-based models simulate nitrification for each soil layer defined in the model. Nitrification is calculated in each layer as a function of soil NH_4^+ concentration, water content, temperature, and pH using Michaelis-Menten or first-order kinetics. Michaelis-Menten kinetics are described by the equation:

$$R_{nitr} = \frac{d[NO_3^-]}{dt} = \frac{V_{max}[NH_4^+]}{K_m + [NH_4^+]} \times F(\text{WFPS}) \times F(t) \times F(\text{pH}) \quad [11]$$

where R_{nitr} is the daily nitrification rate, $[NO_3^-]$ is the concentration of the product of the reaction, V_{max} is the maximum nitrification rate achievable by the system, $[NH_4^+]$ is the concentration of the initial substrate, and K_m represents the substrate concentration at which the reaction rate is half of V_{max}. *F(t), F(*WFPS*),* and *F*(pH*)* are functions that define the effect of soil water-filled pore space percentage (WFPS), temperature (*t*), and pH on nitrification.

The amount of NO_3^- formed in models that adopt Michaelis-Menten kinetics is determined as a function of time, with nitrification rates approaching V_{max} asymptotically (Fig. 1). As a result, NO_3^- levels in the soil increase with time until the reaction equilibrium is attained. After equilibrium there is no net increase of soil NO_3^-, as NH_4^+ is converted into NO_3^- only when NO_3^- levels decrease.

In models where nitrification is regulated by first-order kinetics, the reaction proceeds at rates that are linearly correlated with soil NH_4^+ concentration until a maximum nitrification rate is reached. First-order kinetics functions also account for the effects of different environmental conditions on nitrification rates, as illustrated by the function described by Parton et al. (2001):

$$R_{nitr} = \text{Net}_{min} \times K_2 + [\text{NH}_4] \times K_{max} \times F(t) \times F(\text{WFPS}) \times F(\text{pH}) \quad [12]$$

where R_{nitr} is the daily nitrification rate, $\text{Net}_{min} \times K_2$ is the amount of N mineralized daily, $[NH_4^+]$ is the soil ammonium concentration, K_{max} is the maximum fraction of NH_4^+ that can be nitrified daily, and *F(t)*, *F(*WFPS*)*, and *F(*pH*)* represent the effect of soil temperature, water-filled pore space, and pH on nitrification, respectively.

For example, nitrification can be inhibited by water stress when soil moisture (e.g., WFPS) is too low or by oxygen availability when soil moisture is too high (Fig. 4). Soil temperature functions can be parameterized to have a positive or negative exponential effect on nitrification rates when temperatures are below or above optimum soil temperatures and pH functions can exponentially limit nitrification when soils become acidic.

N_2O Emissions from Denitrification

Compared with nitrification, denitrification is more complex in terms of its biochemistry (Fig. 1), and it produces N_2O as an obligate intermediate. This means that the amount of N_2O emitted depends largely on how soil conditions affect the completion of the process (Section 2). Models adopt two radically different mechanistic approaches to reflect this level of complexity.

Traditionally, models simulate denitrification as a first-order reaction that is dependent on the following factors: availability of the electron donor (active C) and acceptor (NO_3^-), as well as appropriate temperature, WFPS conditions, and soil physical properties that influence gas diffusivity (Parton et al., 1996; Del Grosso et al., 2000; Thorburn et al., 2010). These models assume that denitrification does not occur below field capacity.

Models that use this approach simulate total N gas losses (N_2O + N_2) as a function of the above-mentioned variables and regulate denitrification rates based on the molecular species (NO_3^- or active C) or environmental condition (oxygen availability) that is most limiting. N_2O fluxes are then determined with a N_2/N_2O ratio function as reported by Del Grosso et al. (2000):

$$\frac{\text{N}_2}{\text{N}_2\text{O}} = K_1 \times F\left(-\frac{\text{NO}_3^-}{\text{CO}_2}\right) \times F(\text{WFPS}) \quad [13]$$

where K_1 describes gas diffusivity in the soil at field capacity, NO_3^- is the soil nitrate pool, and CO_2 is the heterotrophic CO_2 respiration, which is used as a proxy for active C concentration in the soil. In these models the N_2/N_2O ratio increases as the ratio of NO_3^- to active C decreases and as conditions approach saturation (gas diffusivity and O_2 availability decrease).

Another approach for simulating denitrification was proposed by Li et al. (2004), who developed the concept of an "anaerobic balloon" to describe the range of redox environments present in soil. This approach divides the soil matrix into aerobic and anaerobic compartments based on the redox potential of each layer as calculated with the Nernst equation (Eq. [14]):

$$Eh = E_0 + RT/nF \times \ln([ox]/[red]) \qquad [14]$$

where Eh is the redox potential, E_0 is the standard electromotive force, R and F are constants, T is temperature, n is the transferred electron number, and $[ox]$ and $[red]$ are the oxidant and reductant concentrations in the soil, respectively.

This model allocates the substrates (e.g., active C, NO_3^-, and NH_4^+) into aerobic and anaerobic compartments with denitrification taking place only in the anaerobic fraction. When soil water content increases as a result of rainfall or irrigation, the anaerobic balloon swells, and more substrate is allocated to the anaerobic compartment for denitrification, leaving less substrate in the aerobic fraction for nitrification. This dynamic is reversed when the soil dries.

In this approach N_2O is simulated by direct conversion of NO_3^- into N_2O at pre-defined rates. Additionally, N_2O is produced by reducing a portion of NO_2^- (Li et al., 2005b) according to the reactions described in Section 2 ($NO_3^- \rightarrow NO_2^- \rightarrow NO \rightarrow N_2O \rightarrow N_2$). N_2O produced by nitrification or denitrification is subject to further transformation during its diffusion within the same soil layer. As the gas diffusion rates in the soil are regulated by soil porosity, soil temperature, and air-filled porosity, the swelling of the anaerobic balloon delays the emissions of the N_2O from the soil matrix, increasing the probability of N_2O to be further reduced to N_2 through denitrification.

Looking Ahead: Future Trends to Improve Simulation of N_2O

Current process-based models display a range of complexity in the representation of nitrification and denitrification and have been tested for a variety of crops and environments (Stehfest and Müller, 2004; Abdalla et al., 2010; Thorburn et al., 2010). In the majority of cases, the algorithms that regulate nitrification and denitrification were developed with datasets from laboratory incubations under static environmental conditions or from field measurements that adopted manual gas sampling methods (Li et al., 1992; Del Grosso et al., 2000; Meier et al., 2006). Laboratory incubations rarely replicate the complex biochemical dynamics that occur under field conditions, and N_2O emissions measured in the laboratory can differ substantially from those obtained in the field (Williams et al., 1998). Moreover, single-point manual sampling can neglect diurnal patterns of daily N_2O fluxes that are related to temperature, causing it to considerably underestimate overall N_2O emissions (Scheer et al., 2013).

New datasets for different agroecosystems, crop rotations, and agronomic practices that include high-frequency soil water, soil mineral N, and N_2O measurements are currently being developed (Brümmer et al., 2008; Barton et al., 2011; De Antoni Migliorati et al., 2015). These field-based datasets form ideal validation data to rigorously test algorithms that have been developed using laboratory studies under a defined range of environmental conditions. Significant improvements in simulating daily N_2O emissions can also be expected from the inclusion of algorithms that simulate gaseous diffusion and the exchange of nitrogenous gases between soil layers (Clough et al., 2005; Shcherbak et al., 2014; Friedl et al., 2016). A more realistic prediction of daily N_2O and N_2 losses would be obtained in these models if the N_2O that is produced in the lower soil layers was subject to further transformation as it diffused from those lower layers to the soil surface.

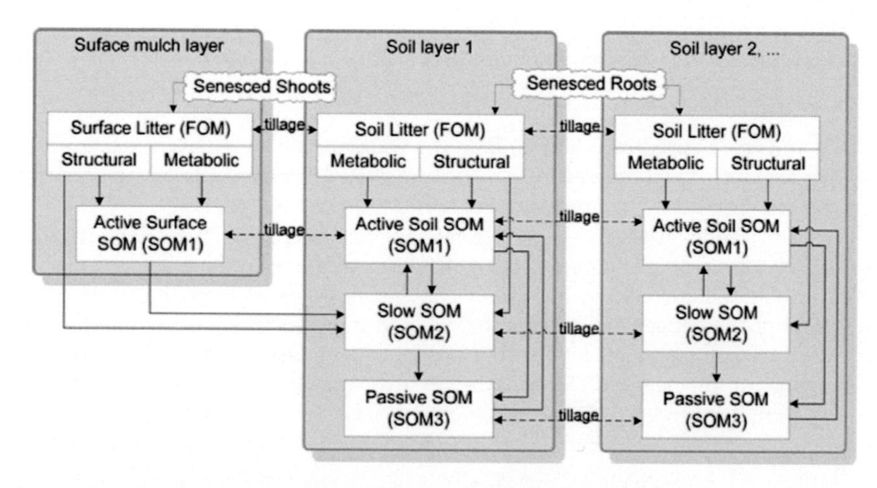

Fig. 4. Conceptual diagram of the subdivision and flows of SOM pools as simulated in the DSSAT-CENTURY organic matter sub-module. (Porter et al., 2010)

However, improving the capability of models to replicate the effects of soil conditions on nutrient dynamics and microbial processes should not be the only objective of future model development efforts. To date, most models have focused on the simulation of site-specific characteristics and have neglected the importance of water and nutrient exchange between different landscape positions (Haas et al., 2013). For example, N can be lost from a fertilized soil via ammonia (NH_3) volatilization or NO_3^- leaching and then relocated to another part of the landscape (a river, a lake, or another soil) where it can be further transformed and emitted as N_2O (Mosier et al., 2004). To address this issue, process-based models would benefit from changing to grid-based systems where topographical differences can be taken into account. Geographical information systems (GIS) and hydrological nutrient transport models should be coupled and synchronized to account for lateral and vertical matter exchange with adjacent grids. Such models would be better able to predict the spatial and temporal distribution as well as the magnitude and variability in N_2O emissions at a landscape scale, thereby increasing our capacity to correctly estimate N balances in agroecosystems.

Soil Carbon

Another important aspect of the soil dynamics is the simulation of the Soil Organic Matter (SOM), because it has a particular importance due to the changes that occur as a result of organic and inorganic fertilizer or residues application. Most models simulate SOM using multiple pools of carbon and nitrogen based on their decomposability. For example, with the CENTURY approach (Parton et al., 1994), SOM is partitioned into active, slow, and passive pools of decomposability, with the latter being the most resistant to change (Fig. 4).

With this approach organic matter in the form of plant residues or manures enters the soil system in the form of metabolic or structural litter, defined as easily decomposable organic matter (e.g., proteins, sugars) or recalcitrant material (e.g., lignin), respectively. The products of metabolic or structural litter decomposition that have a low C to N ratio are allocated to the *active* SOM pool. This pool has a turnover time that ranges from a few days to one year and includes the microbial biomass and other highly labile by-products of decomposition. Decomposition products with a higher C to N ratio are allocated to the *slow* SOM pool and have turnover times of 10 to 15 yr. Decomposed material assigned to the slow pool includes cell walls and decomposition products from the active SOM pool as well as stabilized microbial biomass that is physically isolated from decomposition by textural features of soil structure. Humus, inert material, and stabilized microbial biomass become part of the *passive* SOM pool, which has turnover times in the order of hundreds of years.

The allocation of SOM into these three pools plays a pivotal role in determining soil nutrient dynamics (Basso et al., 2011, Basso et al., 2018). A large *active* SOM pool can provide plants with significant amounts of inorganic nutrients in a single year, while soils with slowly decomposing SOM may provide little nutritional support for growth during that time period. Decomposition of organic matter is associated with soil mineralization processes which, in turn, affect the release of N and P into soils. When SOM decomposes, a proportion of carbon by the microbial biomass is respired as CO_2, and the remainder flows between pools. These organic C flows are accompanied by organic N and P in proportion to the C/N and C/P ratios of the decomposing material. If the N or P content of the decomposing SOM pool exceeds that of the recipient pool, inorganic N or P is released to the soil through mineralization. Conversely, when the N or P contents of the decomposing organic matter are less than that required by the recipient pool, inorganic N or P is immobilized as the system attempts to reach equilibrium with the C/N and C/P ratios of the recipient pool. Immobilization results in such nutrients being temporarily unavailable for plant growth and eventually entering the SOM pools. As a result, decomposition rates and nutrient turnover are reduced when the amounts of soil inorganic N or P are not sufficient to support the potential C flow between the pools.

The most accurate representation of allocation of soil nutrients into the various pools is achieved when long-term nutrient flows between pools are close to equilibrium (Del Grosso et al., 2011). Management practices and climate variability are additional factors that significantly influence long-term soil nutrient dynamics. For example, SOM levels can vary substantially over a period of decades in response to changes in crop rotation or fertilization strategies. Tillage practices can also affect SOM mineralization rates and soil compaction, which directly impact water retention, infiltration, and drainage rates.

The long-term effects of soil management practices on SOM pools are vital to understanding the influences of agriculture on nutrient availability and, more specifically, the emissions of greenhouse gases such as CO_2 and nitrous oxide (N_2O). Long-term (or residual) effects are most accurately modeled by using a "continuous mode" of simulation where soil parameters are not reinitialized to the same condition at the beginning of each cropping season. Continuous mode modeling incorporates changes to initial conditions based on values obtained from the simulation of the previous season. A comparative study by Basso et al. (2018) demonstrated that when climate change mitigation strategies are evaluated, substantial bias can occur if simulations fail to include long-term changes in soil water and nutrient conditions due to interactions that occur from one year to the next. A continuous simulation provides a more realistic assessment of management factors that can attenuate or exacerbate greenhouse gas emissions over the longer periods of time, and it can be used to develop appropriate long-term mitigation strategies.

Modeling nitrate leaching is not discussed in this chapter. We recommend reading the paper by van der Laan et al. (2014) to learn how different model deal with the simulation of this important source of nitrogen losses to the environment.

Case Study: Accounting for Soil Organic Carbon Feedbacks to Crop Yield under Climate Change

Crop models are often used to simulate the impacts of management, weather, and soils on the yields and the environment under current and future climate conditions (Asseng et al., 2013; Bassu et al., 2014; Asseng et al., 2015; Basso et al., 2015). However, most crop model simulations have been computed by reinitializing soil water and nutrient conditions to the same state each year (Asseng et al., 2013; Bassu et al., 2014; Asseng et al., 2015). *Reinitialization* means that after each growing season, the model reinitializes the values of soil water and N to the ones defined by the user and treats every growing season as a "single season." In this way any changes to the soil C will be impacted by the boundary conditions specified (e.g., initial values of soil N, water, and residue) and that growing season's weather. Such a practice is justifiable if the goal is to isolate the effects of climate on crop yield from other variables. Yield reductions or increases in response to different weather conditions will, however, be reflected in a dynamic feedback to the soil through the growth of roots and the eventual restitution of residues. Furthermore, the way soils and crops are managed, along with the amount of soil water, N, and C left available after harvest, will deeply impact the initial conditions characterizing the next growing season.

For these reasons the annual reinitialization of soil conditions to the same state each year hinders the potential for soil and crop management, including soil C sequestration, to mitigate the effects of climate change (Basso et al., 2018). Therefore, such an approach should not be used to assess impacts on long-term food security with a specific focus on mitigation.

For example, Basso et al. (2018) quantified the deviations in yield predictions associated with the pre-season reinitialization of soil conditions compared with the sequential run of the model. Within the AgMIP several specific studies were set up; some dealt with specific crops (e.g., wheat, maize, potato), others on specific processes (e.g., wheat evapotranspiration comparison), and one in particular focused on the impact of continuous simulations on soil organic C (Basso et al., 2018).

Among the models that were part of the AgMIP wheat- and maize-pilot studies (Asseng et al., 2013; Bassu et al., 2014), a subgroup of five wheat and four maize models were used to simulate SOC dynamics in a study led by Basso et al. (2018) (Table 2). Yields were simulated for maize–fallow and wheat–fallow cropping systems at eight sites around the globe (four sites for each crop) used in previous AgMIP crop-specific studies (Asseng et al., 2013; Bassu et al., 2014)(Table 3). Modelers were asked to initialize models with the same SOC (Table 3), NO_3- N- and water content and to run the models in sequential mode for a 30-year period (1980–2010). The models were run using daily weather data as input as well as a factorial temperature change (Table 2). As suggested by Martre et al. (2014), median outputs, considered to be the best estimator of the model ensembles, were used to analyze the outputs.

Model ensembles of the simulated yields obtained under the two different running modes (reinitialized every year vs. continuous), showed similarities in their trends and patterns over time, over sites, and between different temperature increase scenarios. Figures 8 and 9 report simulation results for the baseline scenario.

However, there was significant divergence in absolute simulation results, even though the models were initialized with the same soil conditions. In some cases simulations started diverging even over the first few growing seasons (as early as the second or the third season). For wheat in baseline conditions (Fig. 8), differences ≥ 1 tonne ha^{-1} between two simulation modes were reported for AR (Argentina), IN (India), and NL (the Netherlands) as early as 1982. After this initial split, the differences between the two modes tended to remain stable over time (1983–2010). Maize simulations resulted in two yield curves that were reasonably close for FR (France) and US (the United States), regardless of the temperature treatment. TZ (Tanzania) showed constant yield differences between the two running modes under baseline conditions. In particular, BR (Brazil) showed an important yield difference that furthermore increased with time. These differences were associated with the soil variable status at the start of each growing season in the sequential mode, which differed from the a priori values in the reinitialized mode.

Accordingly, with the pilot studies (Asseng et al., 2013; Bassu et al., 2014), the coefficients of variation for grain yields from the model ensembles ranged between 24% and 39% for wheat simulations and between 20% and 46% for maize, under baseline conditions with no climate change. Similar to the pilot studies, these coefficients were found to increase in the temperature scenarios diverging from the baseline conditions.

To quantify these divergences, the cumulative difference in the median yields between model simulations run in reinitialized and sequential mode was calculated following Eq. [1], where i is the year index for which the cumulative difference is computed (between 1981 and 2010).

$$\text{Cumul. } \Delta \text{ Yield}_i = \sum_{j=1981}^{i} \left(\text{Yield}_{j,\text{Seq.}} - \text{Yield}_{j,\text{ Reinit.}} \right) \quad [15]$$

The cumulated differences for each site under the baseline scenario are graphically presented for wheat (Fig. 5A) and maize (Fig. 5B). Figure 5A and 5B also present the mean over the respective four sites. Figures 5C and 5D present the sole mean results but for the different temperature treatments.

Under baseline conditions (Fig. 5A) the sequential wheat yield simulations were higher relative to reinitialized simulations for Argentina, where cumulative yield differences rose up to +50 tonnes ha^{-1} after thirty years. In contrast, the cumulative yield difference was negative in India (−45 tonnes ha^{-1}) and the Netherlands (-36 tonnes ha^{-1}), indicating lower simulated yields under sequential mode compared with the reinitialized mode. In Australia, where wheat had the lowest N fertilization, low precipitation, and poor yields, the differences were negligible (-1 tonne ha^{-1} over the 30 yr).

Over 30 yr, the cumulative maize yield differences between reinitialized and sequential mode rose between -15 tonnes ha^{-1} for Brazil and −86 tonnes ha^{-1} for

Table 1. Crop models used in the AgMIP Soils and Crop Rotation Initiative (Basso et al. 2018).

Model (Version) Crop† Documentation (*Reference*)
APSIM (V7.3) M http://www.apsim.info *(6)*
APSIM-NWheat (V1.55) W http://www.apsim.info/Wiki/(7)
MONICA (V1.0) M, W http://monica.agrosystem-models.com (8)
SALUS M, W http://salusmodel.glg.msu.edu (9)
STICS (V8.1) M, W http://www6.paca.inra.fr/stics_eng/*(10)*
Expert-N (V3.0.10)– SPASS (2.0) W http://www.helmholtz-muenchen.de/en/iboe/expertn/*(11)*
† M: maize; W: wheat

Table 3. Factors studied in simulations (adapted from Basso et al., 2018).

Factors Factor levels
Site Maize: Brazil, France, Tanzania, USA Wheat: Argentina, Australia, India, Netherlands Temperature [°C] Baseline, -3, +3, +6, +9 Simulation mode Reinitialized, sequential

Crop Site SOC [%]
Argentina 2.7 Wheat Australia 0.57 India 0.42 Netherlands 2.4
Brazil 1.1 Maize France 0.9 Tanzania 1.4 USA 2.4

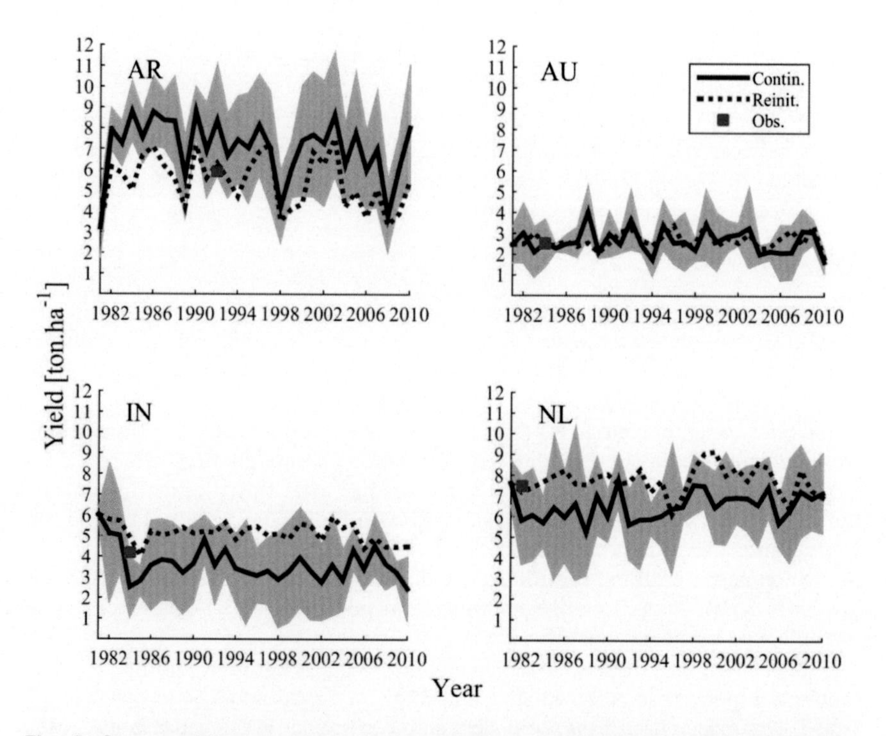

Fig. 5. Crop model ensembles of wheat yield at different sites. Ensemble model simulation of wheat yield for the baseline temperature treatment under reinitialized (dashed black line) and sequential (solid black line) modes. Observations are reported in red. Gray area shows the model dispersion (percentile 15% and 85%) when run in sequential mode (adapted from Basso et al., 2018).

the United States under baseline conditions (Fig. 5B). Yield differences between the two simulation modes were characterized by higher yield in the sequential mode by +8 tonnes ha^{-1} for France and +52 tonnes ha^{-1} for Tanzania, which are the two sites with irrigation.

In an effort to summarize the information, results were averaged over sites for the different temperature treatments. For wheat, yield differences between the two simulation modes ranged from −5 tonnes ha^{-1} (at -3 °C) to -13 tonnes ha^{-1} (at +9 °C) globally (Fig. 10C). Simulated yields tended to decrease with a temperature increase under both running modes. However, these results highlighted a yield reduction that is exacerbated by a temperature increase under the sequential mode compared with the reinitialized one.

Contrarily, no systematic pattern was found in the values averaged over sites for the maize cropping system, unless the results were averaged per cropping system, dissociating irrigated (France and Tanzania) and rainfed systems (Brazil and United States; Fig. 10D). The continuous mode was found to predict higher yield for the irrigated sites and lower yields for the rainfed sites in comparison to the reinitialized simulations. At -3 °C the cumulated difference between the two simulation modes was the highest for the irrigated systems and the lowest for the rainfed system. Under rainfed conditions at +6 °C and above, there was little difference between results from the two modes, and it was mainly associated with very low yield levels predicted under both running modes. Under the irrigated treatment at

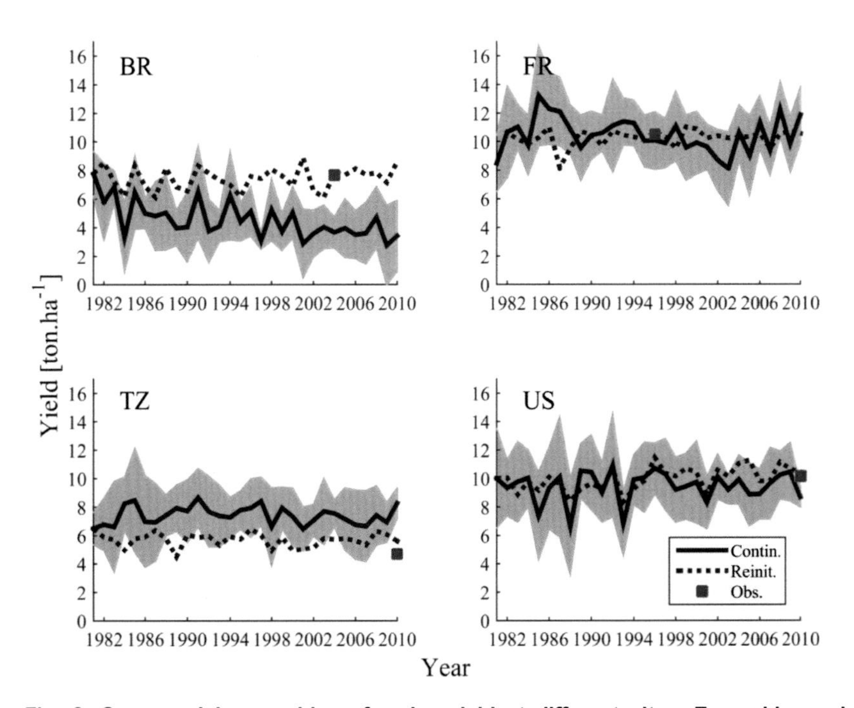

Fig. 6. Crop model ensembles of maize yield at different sites. Ensemble model simulation of maize yield for baseline temperature scenario under reinitialized (dashed black line) and sequential (solid black line) simulation modes. Observations are reported in red. Gray area shows the model dispersion (percentile 15% and 85%) when run in sequential mode (adapted from Basso et al., 2018).

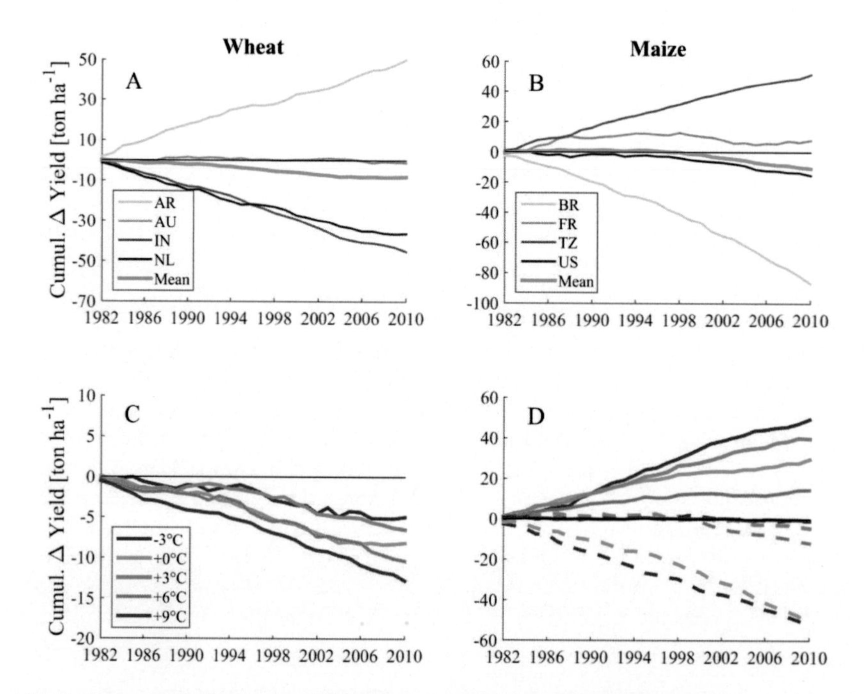

Fig. 7. Grain yield divergence between sequential and reinitialized mode.

+0 °C, +3 °C and +6 °C, the differences between the two simulation modes were all positive, reaching respectively ~ +30, ~ +40, and ~ +15 tonnes ha^{-1} after 30 yr.

These results illustrate the need to account for the carry-over effects in the off-season of soil dynamic processes related to soil water, C, and N. Soil water content can dramatically differ in the fallow period when rainfall does not refill the soil profile, which is often the case in dryland environments. This aspect has compounding effects on SOC content, which is driven by the amount of residues returned to the soil. Yields were higher in the irrigated sites and where rainfall was not a limiting factor, as more root biomass and crop residues were returned to the soil. This resulted in more nutrients being mineralized, allowing for more N to be available to the next cropping season.

This study demonstrated the importance of soil-crop models run in sequential mode to account for the carry-over effects of soil and crop management and their impact on soil carbon storage as a prerequisite to design climate change mitigation strategy (13).

Simulated cumulative grain yield differences (Δ Yield [ton ha^{-1}] = Yield$_{\text{Cont. mode}}$ - Yield$_{\text{Reinit. mode}}$) at different sites and temperature treatments between pre-season reinitialized soils state and sequential mode, for wheat and maize simulations. Results are reported for each site for baseline conditions (A, B– gray lines). Mean over the four sites for each crop are also presented (A, B– cyan line). Cumulative averaged yield differences are reported for each temperature treatment and for each crop (C, D– colored lines). In (D) cumulative yield differences are reported separately for irrigated (France and Tanzania– solid line) and rainfed maize sites (United States and Brazil– dashed line) (adapted from Basso et al., 2018).

Soil Spatial Variability

In a popular editorial for the European Journal of Soil Science, Webster (2001, p. 322) argues that *"`Variety is the spice of life', and soil science would be pretty uninteresting if there were no variation. How would field pedologists entertain us if all soil were uniformly drab grey loam to 3 m?"* Understanding the causes of the variations of soil properties in space, and ultimately being able to predict them, is a research topic that is the research subject of several disciplines, among them pedology, soil mapping and classification, agronomy, and more recently, precision agriculture. Spatial variability of soils is an inherent property of soils and can occur at a continuum of scales, from the microscale of the difference in oxygen concentration between the interior and the exterior of a soil aggregate to the differences between soil orders at continental scale. In this section we briefly discuss the differences in soil that can be observed at field scale.

Measurement of Soil Properties Variability at Field Scale

The main limitation to capturing the soil spatial variability of a soil property is often the availability of a sufficiently dense network of measurements. Nonetheless, the determination of a sufficient number of measuring points cannot be established by a single golden rule. From a brief review we found that this number varies significantly from 260 points for a 0.2 ha plot for the measurement of soil mineral nitrogen (O'Halloran et al., 2004) to 110 points for a 50-ha field (Gaston et al., 2001). Because agricultural systems are characterized by the a number of interactions between system state variables (e.g., plant biomass, soil mineral N content, soil organic matter stock), the measurement of each variable over dense networks of field measurement is often impractical, and there is a concrete risk of investing a considerable amount of resources to investigate a variable that has little influence on the behavior of the system. Therefore, measurements are more often performed over relatively few samples, and proxy measurements are used to predict the variable of interest over the whole field.

Mapping Soil at Field Level: Extending a Few Soil Samples to a Soil Map

Creating soil maps at field scale is a process that relies on two main pillars: i) the association between soil state variables (e.g., soil organic matter, soil water availability) that would require punctual resource intensive measurements and proxy measurements that can be carried over a large area, and ii) Tobler's first law of geography (near things are more related than distant things). For example, electrical conductivity, which can be measured extensively over a field and at different depths, is used as a proxy for a number of soil properties such as soil water content, soil texture, and organic matter, which in turn are obviously correlated among themselves.

Other proxies to predict soil properties are infrared spectra, either by traditionally measuring them in the laboratory or via a mobile apparatus capable of acquiring spectra on-the-go directly in the field (Nawar and Mouazen, 2017) using aerial and remote sensing images.

Using Existing Soil Maps: The Case of Soil Survey Geographic Database

In some cases, existing maps are already available to farmers; however, the scale and accuracy at which they were developed is often the main limit to their use.

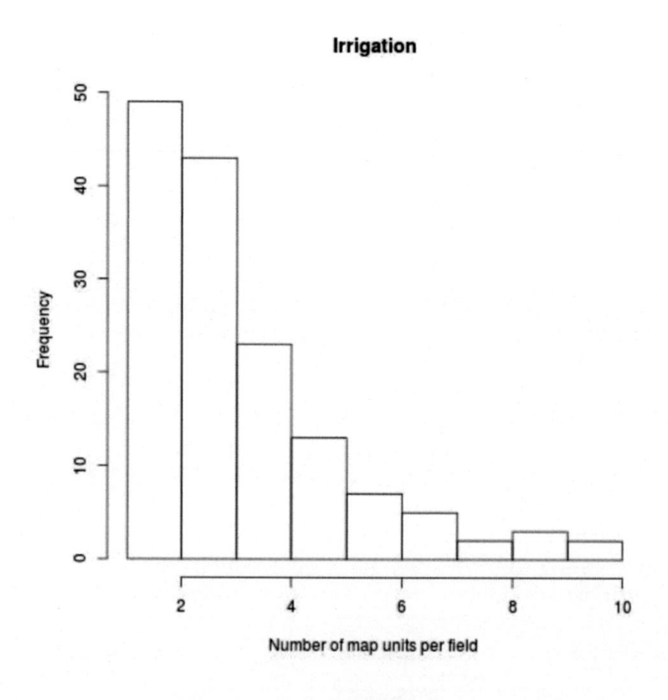

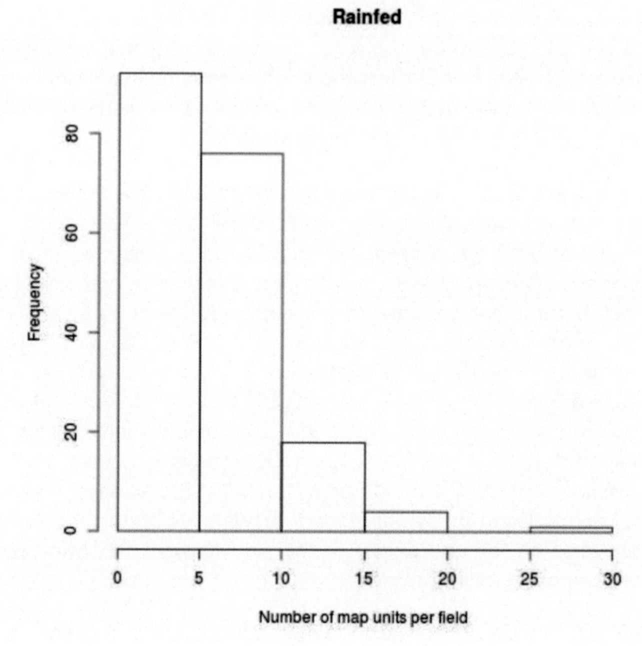

Fig. 8. Distribution of the number of map units for each field identified by the Soil Survey Geographic Database (Maestrini and Basso, 2018).

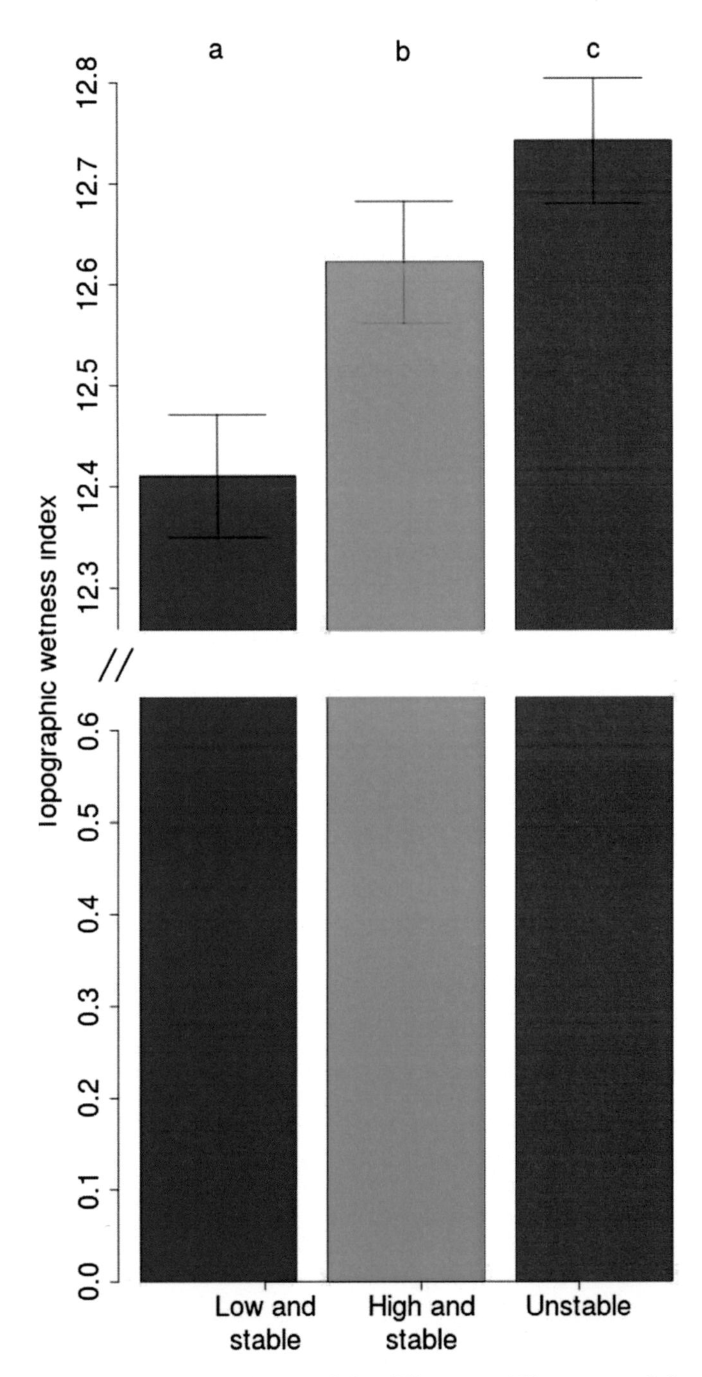

Fig. 9. Topographic wetness index of the different stability zones of the field. The zones marked as unstable are characterized by larger fluctuations in yield between the years (Maestrini and Basso, 2018).

Here we examine the case of SSURGO (Soil Survey Geographic Database) and describe how it can be used to manage the crops at within-field level.

Soil Survey Geographic Database is a soil survey that covers the entire United States. The map consists of a polygons map keys, at a scale such that often a field is contains more than one polygon. Nonetheless, each polygon (map key unit) is described as a relatively uniform area that contains numerous components; however, though each component is described in terms of typical physical features (e.g., slope and aspect), it is not spatially explicit (i.e., it is not mapped). From the perspective of an end user, SSURGO can be seen as a set of polygons, each containing a set of possible soils (i.e., the components) that are not explicitely located. However, even if the position is not determined, an estimate is given of the area occupied by each component within a polygon.

For practical purposes, once the polygons that are included in the area are established, there are several available approaches to determine the soil properties for each polygon:

i) Simply pick the most representative component for each map unit (polygon).

ii) Make a weighted average of all the components comprised in a map unit (polygon).

iii) Try to identify the component that more closely matches the physiography of the terrain. (This approach may be difficult for agricultural fields that are usually characterized by modest slopes.)

In a study on yield temporal stability at field scale, Maestrini and Basso (2018) analyzed more 400 fields located in the U.S. Midwest and found that approximately half of the fields contained more than five map units (Fig. 1).

The usefulness of SSURGO for precision agriculture is debatable, both because there is a substantial uncertainty in the parameters and because the delimitation between contiguous areas usually follows the underlying landforms; therefore, it carries information that is probably already available to the farmer.

Among the five factors that influence soil formation: Parent material, Climate, Biota, Topography, and Time (Jenny, 1941), topography is the one that shows the largest variability at within-field scale; therefore, it is often regarded as the most important factor shaping soil variability at within-field scale. As a result, topography often represents a valuable resource for managing soil spatial variability. In fact, topography is an important driver of yield spatial variability because it controls water routing and is an important proxy for soil properties. Kravchenko and Bullock (2000) showed that areas characterized by a lower topographic index—the ratio between contributing area and the tangent of the slope—are also characterized by a lower yield. Several mechanisms may contribute to this result: first of all the areas that have a lower topographic wetness index (TWI) usually sit on top of local reliefs, are drier, and are a source of erosion; therefore, they are usually characterized by soils that are shallower and poorer in organic matter, whereas the areas that are characterized by a higher topographic index are located in concave areas and characterized by deeper soils and wetter soils.

Maestrini and Basso (2018) further expanded this theory on the importance of topography and proposed that topography does not control only the yield level (i.e., if the yield is relatively high or low) but also the temporal stability of the yield. They hypothesized that areas characterized by the highest topographic index are also generally more wet and unstable because they can be waterlogged during rainy years (therefore exposed to losses, especially during germination) and relatively wetter in dry years.

References

Abdalla, M., M. Jones, J. Yeluripati, P. Smith, J. Burke, and M. Williams. 2010. Testing DayCent and DNDC model simulations of N_2O fluxes and assessing the impacts of climate change on the gas flux and biomass production from a humid pasture. Atmos. Environ. 44:2961–2970. doi:10.1016/j.atmosenv.2010.05.018

Agricultural Model Intercomparison and Improvement Project. 2019. Agricultural Model Intercomparison and Improvement Project. Columbia University, New York, NY. http://www.agmip.org/ (Accessed 2 Apr. 2019). *[2019 is year accessed].*

Asseng, S., F. Ewert, P. Martre, R.P. Rötter, D.B. Lobell, D. Cammarano, B. Kimball, M. Ottman, G. Wall, J.W. White, M.P. Reynolds, P.D. Alderman, P.V. Prasad, P.K. Aggarwal, J. Anothai, B. Basso, C. Biernath, A.J. Challinor, G. De Sanctis, J. Doltra, E. Fereres, M. Garcia-Vila, S. Gayler, G. Hoogenboom, L.A. Hunt, R.C. Izaurralde, M. Jabloun, C.D. Jones, K.C. Kersebaum, A.-K. Koehler, C. Müller, S. Naresh Kumar, C. Nendel, G. O'Leary, J.E. Olesen, T. Palosuo, E. Priesack, E. Eyshi Rezaei, A.C. Ruane, M.A. Semenov, I. Shcherbak, C. Stöckle, P. Stratonovitch, T. Streck, I. Supit, F. Tao, P.J. Thorburn, K. Waha, E. Wang, D. Wallach, J. Wolf, Z. Zhao, and Y. Zhu. 2015. Rising temperatures reduce global wheat production. Nat. Clim. Chang. 5:143–147. doi:10.1038/nclimate2470

Asseng, S., F. Ewert, C. Rosenzweig, J.W. Jones, J.L. Hatfield, A.C. Ruane, K.J. Boote, P.J. Thorburn, R.P. Rotter, D. Cammarano, N. Brisson, B. Basso, P. Martre, P.K. Aggarwal, C. Angulo, P. Bertuzzi, C. Biernath, A.J. Challinor, J. Doltra, S. Gayler, R. Goldberg, R. Grant, L. Heng, J. Hooker, L.A. Hunt, et al. 2013. Uncertainty in simulating wheat yields under climate change. Nat. Clim. Chang. 3:827–832. doi:10.1038/nclimate1916

Barton, L., K. Butterbach-Bahl, R. Kiese, and D.V. Murphy. 2011. Nitrous oxide fluxes from a grain–legume crop (narrow-leafed lupin) grown in a semiarid climate. Glob. Change Biol. 17:1153–1166. doi:10.1111/j.1365-2486.2010.02260.x

Basso, B., B. Dumont, B. Maestrini, I. Shcherbak, G.P. Robertson, J.R. Porter, P. Smith, K. Paustian, P.R. Grace, S. Asseng, S. Bassu, C. Biernath, K.J. Boote, D. Cammarano, G. De Sanctis, J.-L. Durand, F. Ewert, S. Gayler, D.W. Hyndman, J. Kent, P. Martre, C. Nendel, E. Priesack, D. Ripoche, A.C. Ruane, J. Sharp, P.J. Thorburn, J.L. Hatfield, J.W. Jones, and C. Rosenzweig. 2018. Soil organic carbon and nitrogen feedbacks on crop yields under climate change. Agric. Environ. Lett. 3. doi:10.2134/ael2018.05.0026

Basso, B., and J.T. Ritchie. 2015. Simulating crop growth and biogeochemical fluxes in response to land management using the SALUS model. In: S.K. Hamilton, J.E. Doll, and G.P. Robertson, editors, The Ecology of agricultural landscapes: Long-term research on the path to sustainability. Oxford Univ. Press, New York. p. 252–274.

Basso, B., D.W. Hyndman, A.D. Kendall, P.R. Grace, and G.P. Robertson. 2015. Can impacts of climate change and agricultural adaptation strategies be accurately quantified if crop models are annually re-initialized? PLoS One 10:e0127333. doi:10.1371/journal.pone.0127333

Basso, B., O. Gargiulo, K. Paustian, G.P. Robertson, C. Porter, P.R. Grace, and J.W. Jones. 2011. Procedures for initializing soil organic carbon pools in the DSSAT-CENTURY model for agricultural systems. Soil Sci. Soc. Am. J. 75. doi:10.2136/sssaj2010.0115

Bassu, S., N. Brisson, J.-L. Durand, K. Boote, J. Lizaso, J.W. Jones, C. Rosenzweig, A.C. Ruane, M. Adam, C. Baron, B. Basso, C. Biernath, H. Boogaard, S. Conijn, M. Corbeels, D. Deryng, G.D. Sanctis, S. Gayler, P. Grassini, J. Hatfield, S. Hoek, C. Izaurralde, R. Jongschaap, A.R. Kemanian, K.C. Kersebaum, S.-H. Kim, N.S. Kumar, D. Makowski, C. Muller, C. Nendel, E. Priesack, M.V. Pravia, F. Sau, I. Shcherbak, F. Tao, E. Teixeira, D. Timlin, and K. Waha. 2014. How do various maize crop models vary in their response to climate change factors? Glob. Change Biol. 20:2301–2320.

Bollmann, A., and R. Conrad. 1998. Influence of O_2 availability on NO and N2O release by nitrification and denitrification in soils. Glob. Change Biol. 4:387–396. doi:10.1046/j.1365-2486.1998.00161.x

Boone, R.D., K.J. Nadelhoffer, J.D. Canary, and J.P. Kaye. 1998. Roots exert a strong influence on the temperature sensitivity of soil respiration. Nature 396:570–572. doi:10.1038/25119

Bouwman, A.F. 1998. Environmental science: Nitrogen oxides and tropical agriculture. Nature 392:866–867. doi:10.1038/31809

Bremner, J. 1997. Sources of nitrous oxide in soils. Nutr. Cycl. Agroecosyst. 49:7–16. doi:10.1023/A:1009798022569

Broadbridge, P., and I. White. 1988. Constant rate rainfall infiltration: A versatile non-linear model. Analytical solution. Water Resour. Res. 24:145–154. doi:10.1029/WR024i001p00145

Brümmer, C., N. Brüggemann, K. Butterbach-Bahl, U. Falk, J. Szarzynski, K. Vielhauer, R. Wassmann, and H. Papen. 2008. Soil-Atmosphere Exchange of N_2O and NO in Near-Natural Savanna and Agricultural Land in Burkina Faso (W. Africa). Ecosystems 11:582–600. doi:10.1007/s10021-008-9144-1

Buttler, I.W., and S.J. Riha. 1992. Water fluxes in oxisols: A comparison of approaches. Water Resour. Res. 28:221–229. doi:10.1029/91WR02197

Cammarano, D., R.P. Rotter, S. Asseng, F. Ewert, D. Wallach, P. Martre, J.L. Hatfiled, J.W. Jones, C. Rosenzweig, A.C. Ruane, K.J. Boote, P.J. Thorburn, K.C. Kersebaum, P.K. Aggarwal, C. Angulo, B. Basso, P. Bertuzzi, C. Biernath, N. Brisson, A.J. Challinor, J. Doltra, S. Gayler, R. Goldberg, L. Heng, J. Hooker, L.A. Hunt, J. Ingwersen, R.C. Izaurralde, C. Muller, S.N. Kumar, C. Nendel, G.J. O'Leary, J.E. Olesen, T.M. Osborne, T. Palosuo, E. Priesack, D. Ripoche, M.A. Semenov, I. Schecherbak, P. Steduto, C.O. Stockle, P. Stratonovitch, T. Streck, I. Supit, F. Tao, M. Travasso, K. Waha, J.W. White, and J. Wolf. 2016. Uncertainty of wheat water use: Simulated patterns and sensitivity to temperature and CO_2. Field Crops Res. 198:80–92. doi:10.1016/j.fcr.2016.08.015

Chapuis-Lardy, L., N. Wrage, A. Metay, J.-L. Chotte, and M. Bernoux. 2007. Soils, a sink for N2O? A review. Glob. Change Biol. 13:1–17. doi:10.1111/j.1365-2486.2006.01280.x

Chen, D.C., Y. Li, P.R. Grace, and A. Mosier. 2008. N_2O emissions from agricultural lands: A synthesis of simulation approaches. Plant Soil 309:169–189.

Chung, S.O., and R. Horton. 1987. Soil heat and water flow with a partial surface mulch. Water Resour. Res. 23:2175–2186.

Clough, T.J., R.R. Sherlock, and D.E. Rolston. 2005. A review of the movement and fate of N2O in the subsoil. Nutr. Cycl. Agroecosyst. 72:3–11. doi:10.1007/s10705-004-7349-z

Coelho, D.T., and R.F. Dale. 1980. An energy-crop growth variable and temperature function for predicting maize growth and development: Planting to silking. Agron. J. 72:503–510. doi:10.2134/agronj1980.00021962007200030023x

Conrad, R. 2001. Evaluation of data on the turnover of NO and N_2O by oxidative versus reductive microbial processes in different soils. Phyton (Horn) 41:61–72.

Crutzen, P.J. 1981. Atmospheric chemical processes of theoxides of nitrogen including nitrous oxide. In: C.C. Delwiche, editor, Denitrification, nitrification and atmospheric nitrous oxide. John Wiley & Sons, New York. p. 17–44.

Dalal, R.C., W. Wang, G.P. Robertson, and W.J. Parton. 2003. Nitrous oxide emission from Australian agricultural lands and mitigation options: A review. Soil Res. 41:165–195. doi:10.1071/SR02064

De Antoni Migliorati, M., M. Bell, P.R. Grace, C. Scheer, D.W. Rowlings, and S. Liu. 2015. Legume pastures can reduce N_2O emissions intensity in subtropical cereal cropping systems. Agric. Ecosyst. Environ. 204:27–39. doi:10.1016/j.agee.2015.02.007

Del Grosso, S.J., D.S. Ojima, W.J. Parton, E. Stehfest, M. Heistemann, B. DeAngelo, and S. Rose. 2009. Global scale DAYCENT model analysis of greenhouse gas emissions and mitigation strategies for cropped soils. Global Planet. Change 67:44–50. doi:10.1016/j.gloplacha.2008.12.006

Del Grosso, S.J., W.J. Parton, C.A. Keough, and M. Reyes-Fox. 2011. Special features of the DayCent modeling package and additional procedures for parameterization, calibration, validation, and applications. In: L.R. Ahuja and L. Ma, editor, Methods of introducing system models into agricultural research. American Society of Agronomy, Crop Science Society of America, Soil Science Society of America, p. 155–176. doi:10.2134/advagricsystmodel2.c5

Del Grosso, S.J., W.J. Parton, A.R. Mosier, D.S. Ojima, A.E. Kulmala, and S. Phongpan. 2000. General model for N_2O and N_2 gas emissions from soils due to dentrification. Global Biogeochem. Cycles 14:1045–1060.

Durand, J.L., K. Delusca, K. Boote, J. Lizaso, R. Manderssceid, H.J. Weigel, A.C. Ruane, C. Rosenzweig, J. Jones, L. Ahuja, S. Anapalli, B. Basso, C. Baron, P. Bertuzzi, C. Biernath, D. Deryng, F. Ewert, T. Gaiser, S. Gayler, F. Heinlein, K.C. Kersebaum, S.H. Kim, C. Muller, C. Nendel, A. Olioso, E. Priesack, K.R. Villegas, D. Ripoche, R.P. Rotter, S.I. Seidel, A. Srivastava, F. Tao, D. Timlin, T. Twine, E. Wang, H. Webber, and Z. Zhao. 2017. How accurately do maize crop models simulate the interactions of atmospheric CO_2 concentration levels with limited water supply on water use and yield? Eur. J. Agron. doi:10.1016/j.eja.2017.01.002

Dwyer, L.M., H.N. Hayhoe, and J.L.B. Culley. 1990. Prediction of soil temperature from air temperature for estimating corn emergence. Can. J. Plant Sci. 70:619–628. doi:10.4141/cjps90-078

Ehhalt, D., M. Prather, F. Dentener, R. Derwent, E. Dlugokencky, E. Holland, I. Isaksen, J. Katima, V. Kirchhoff, P. Matson, P. Midgley, and M. Wang. 2001. Atmospheric chemistry and greenhouse gases. Climate change 2001: The scientific basis. In: J.T. Houghton, Y. Ding, D.J. Griggs, M. Noguer, P.J. van der Linden, X. Dai, K. Maskell, and C.A. Johnson, editors, Contribution of working group I to the Third Assessment Report of the Intergovernmental Panel on Climate Change. Cambridge Univ. Press, Cambridge. p. 239–287.

Elias, E.A., R. Cichota, H.H. Torriani, Q. Lier, and J. van Lier. 2004. Analytical soil–temperature model correction for temporal variation of daily amplitude. Soil Sci. Soc. Am. J. 68:784–788. doi:10.2136/sssaj2004.7840

Emerman, S.H. 1995. The tipping bucket equations as a model for macropore flow. J. Hydrol. 171:23–47. doi:10.1016/0022-1694(95)02735-8

Firestone, M.K., and E.A. Davidson. 1989. Microbiological basis of NO and N_2O production and consumption in soils. In: Andreae, M.O., and D.S. Schimel, editors, Exchanges of trace gases between terrestrial ecosystems and the atmosphere. John Wiley & Sons, New York.

Fitzpatrick, E.A., and H.A. Nix. 1969. A model for simulating soil water regime in alternating fallow-crop systems. Agric. Meteorol. 6:303–319. doi:10.1016/0002-1571(69)90023-5

Folorunso, O.A., and D.E. Rolston. 1984. Spatial variability of field-measured denitrification gas fluxes. Soil Sci. Soc. Am. J. 48:1214–1219. doi:10.2136/sssaj1984.03615995004800060002x

Friedl, J., C. Scheer, D.W. Rowlings, H.V. McIntosh, A. Strazzabosco, D.I. Warner, and P.R. Grace. 2016. Denitrification losses from an intensively managed sub-tropical pasture–Impact of soil moisture on the partitioning of N_2 and N_2O emissions. Soil Biol. Biochem. 92:58–66. doi:10.1016/j.soilbio.2015.09.016

Gaston, L. A., Locke, M. A., Zablotowicz, R. M., & Reddy, K. N. 2001. Spatial variability of soil properties and weed populations in the Mississippi Delta. Soil Science Society of America Journal, 65: 449–459. https://doi.org/10.2136/sssaj2001.652449x

Granli, T. and O.C. Bøckman. 1995. Nitrous oxide emissions from soils in warm climates. Fert. Res. 42:159–163. doi:10.1007/BF00750510

Gupta, S.C., W.E. Larson, and D.R. Linden. 1983. Tillage and surface residue effects on soil upper boundary temperatures. Soil Sci. Soc. Am. J. 47:1212–1218. doi:10.2136/sssaj1983.03615995004700060030x

Haas, E., S. Klatt, A. Fröhlich, P. Kraft, C. Werner, R. Kiese, R. Grote, L. Breuer, and K. Butterbach-Bahl. 2013. Landscape DNDC: A process model for simulation of biosphere–atmosphere–hydrosphere exchange processes at site and regional scale. Landsc. Ecol. 28:615–636.

Hillel, D. 1982. Introduction to soil physics. Academic Press, San Diego.

Hutchinson, G.L., and E.A. Davidson. 1993. Processes for production and consumption of gaseous nitrogen oxides in soil. In: D.E. Rolston, J.M. Duxbury, L.A. Harper, and A.R. Mosier, editors, Agricultural ecosystem effects on trace gases and global climate change. ASA, Madison, WI.

Jenny, H. 1941. Factors of soil formation: A system of quantitative pedology, Soil Science. Dover Publications, New York.

Jones, J.W., J.M. Antle, B. Basso, K.J. Boote, R.T. Conant, I. Foster, H.C.J. Godfray, M. Herrero, R.E. Howitt, S. Janssen, B.A. Keating, R. Munoz-Carpena, C.H. Porter, C. Rosenzweig, and T.R. Wheeler. 2017. Toward a new generation of agricultural system data, models, and knowledge products: State of agricultural system science. Agric. Syst. 155:269–298. doi:10.1016/j.agsy.2016.09.021

Jones, J.W., G. Hoogenboom, C.H. Porter, K.J. Boote, W.D. Batchelor, L. Hunt, P.W. Wilkens, U. Singh, A.J. Gijsman, and J.T. Ritchie. 2003. The DSSAT cropping system model. Eur. J. Agron. 18:235–265. doi:10.1016/S1161-0301(02)00107-7

Kaspar, T.C., and W.L. Bland. 1992. Soil temperature and root growth. Soil Sci. 154:290–299. doi:10.1097/00010694-199210000-00005

Kiese, R., and K. Butterbach-Bahl. 2002. N_2O and CO_2 emissions from three different tropical forest sites in the wet tropics of Queensland, Australia. Soil Biol. Biochem. 34:975–987. doi:10.1016/S0038-0717(02)00031-7

Lee, D.H., and L.M. Abriola. 1999. Use of the Richards equation in land surface parameterizations. J. Geophys. Res. 104:27519–27526. doi:10.1029/1999JD900951

Li, C., S. Frolking, and K. Butterbach-Bahl. 2005a. Carbon sequestration in arable soils is likely to increase nitrous oxide emissions, offsetting reductions in climate radiative forcing. Clim. Change 72:321–338. doi:10.1007/s10584-005-6791-5

Li, Y., D. Chen, Y. Zhang, R. Edis, and H. Ding. 2005b. Comparison of three modeling approaches for simulating denitrification and nitrous oxide emissions from loam-textured arable soils. Global Biogeochem. Cycles 19:GB3002. doi:10.1029/2004GB002392

Li, C., J. Cui, G. Sun, and C. Trettin. 2004. Modeling impacts of management on carbon sequestration and trace gas emissions in forested wetland ecosystems. Environ. Manage. 33:S176–S186. doi:10.1007/s00267-003-9128-z

Li, C., J. Aber, F. Stange, K. Butterbach-Bahl, and H. Papen. 2000. A process-oriented model of N_2O and NO emissions from forest soils: 1. Model development. J. Geophys. Res., D, Atmospheres 105:4369–4384. doi:10.1029/1999JD900949

Li, C., S. Frolking, and T.A. Frolking. 1992. A model of nitrous oxide evolution from soil driven by rainfall events: 1. Model structure and sensitivity. J. Geophys. Res., D, Atmospheres 97:9759–9776. doi:10.1029/92JD00509

Liu, B., P.T. Mørkved, Å. Frostegård, and L.R. Bakken. 2010. Denitrification gene pools, transcription and kinetics of NO, N_2O and N_2 production as affected by soil pH. FEMS Microbiol. Ecol. 72:407–417. doi:10.1111/j.1574-6941.2010.00856.x

Loomis, R.S., and D.J. Connor. 1998. Crop ecology: Productivity and management in agricultural systems. Cambridge Univ. Press., Cambridge.

Luo, V., R.S. Loomis, and T.C. Hsiao. 1992. Simulation of soil temperature in crops. Agric. For. Meteorol. 61:23–38. doi:10.1016/0168-1923(92)90023-W

Maestrini, B., & Basso, B. 2018. Drivers of within-field spatial and temporal variability of crop yield across the US Midwest. Scientific Reports, 8(1), 14833. https://doi.org/10.1038/s41598-018-32779-3

Martre, P., D. Wallach, S. Asseng, F. Ewert, J.W. Jones, R.P. Rötter, K.J. Boote, A.C. Ruane, P.J. Thorburn, D. Cammarano, J.L. Hatfield, C. Rosenzweig, P.K. Aggarwal, C. Angulo, B. Basso, P. Bertuzzi, C. Biernath, N. Brisson, A.J. Challinor, J. Doltra, S. Gayler, R. Goldberg, R.F. Grant, L. Heng, J. Hooker, L.A. Hunt, J. Ingwersen, R.C. Izaurralde, K.C. Kersebaum, C. Müller, S.N. Kumar, C. Nendel, G. Oleary, J.E. Olesen, T.M. Osborne, T. Palosuo, E. Priesack, D. Ripoche, M.A. Semenov, I. Shcherbak, P. Steduto, C.O. Stöckle, P. Stratonovitch, T. Streck, I. Supit, F. Tao, M. Travasso, K. Waha, J.W. White, and J. Wolf. 2014. Multimodel ensembles of wheat growth: Many models are better than one. Glob. Change Biol. 21:911–925. doi:10.1111/gcb.12768

McCann, I.R., M.J. Mcfarland, and J.A. Witz. 1991. Near-surface bare soil temperature model for biophysical models. Trans. ASAE 34:748–755. doi:10.13031/2013.31726

Meier, E.A., P.J. Thorburn, and M.E. Probert. 2006. Occurrence and simulation of nitrification in two contrasting sugarcane soils from the Australian wet tropics. Soil Res. 44:1–9. doi:10.1071/SR05004

Mosier, A.R., R. Wassmann, L. Verchot, J. King, and C. Palm. 2004. Methane and nitrogen oxide fluxes in tropical agricultural soils: Sources, sinks and mechanisms. Environ. Dev. Sustain. 6:11–49. doi:10.1023/B:ENVI.0000003627.43162.ae

Mualem, Y. 1976. A new model for predicting the hydraulic conductivity of unsaturated porous media. Water Resour. Res. 12:513–522. doi:10.1029/WR012i003p00513

Myhre, G., D. Shindell, F.-M. Bréon, W. Collins, J. Fuglestvedt, J. Huang, D. Koch, J.-F. Lamarque, D. Lee, B. Mendoza, T. Nakajima, A. Robock, G. Stephens, T. Takemura, and H. Zhang. 2013. Anthropogenic and natural radiative forcing. Climate Change 2013: The physical science basis. In: T.F. Stocker, D. Qin, G.K. Plattner, M. Tignor, S.K. Allen, J. Boschung, A. Nauels, Y. Xia, V. Bex, and P.M. Midgley, editors, Contribution of Working Group I to the fifth assessment report of the Intergovernmental Panel on Climate Change. Cambridge Univ. Press, Cambridge.

Nawar, S., & Mouazen, A. M. (2017). Predictive performance of mobile vis-near infrared spectroscopy for key soil properties at different geographical scales by using spiking and data mining techniques. Catena, 151, 118–129. https://doi.org/10.1016/j.catena.2016.12.014

Novak, M. 2005. Soil temperature. In: J.L. Hatfield, J.M. Baker, and M.K. Viney, editors, Micrometeorology in agricultural systems. Agronomy Monograph 47. American Society of Agronomy, Madison, WI.

O'Halloran, I. P., von Bertoldi, A. P., & Peterson, S. 2004. Spatial variability of barley (*Hordeum vulgare*) and corn (*Zea mays* L .) yields , yield response to fertilizer N and soil N test levels. Canadian Journal of Soil Science, 84, 307–316.

Panek, J.A., P.A. Matson, I. Ortiz-Monasterio, and P. Brooks. 2000. Distinguishing nitrification and denitrification sources of N_2O in a Mexican wheat system using 15N. Ecol. Appl. 10:506–514.

Park, S., P. Croteau, K.A. Boering, D.M. Etheridge, D. Ferretti, P.J. Fraser, K.R. Kim, P.B. Krummel, R.L. Langenfelds, T.D. Van Ommen, and L.P. Steele. 2012. Trends and seasonal cycles in the isotopic composition of nitrous oxide since 1940. Nat. Geosci. 5:261–265. doi:10.1038/ngeo1421

Parton, W.J., E.A. Holland, S.J.D. Grosso, M.D. Hartman, R.E. Martin, A.R. Mosier, D.S. Ojima, and D.S. Schimel. 2001. Generalized model for NOx and N_2O emissions from soils. J. Geophys. Res. 106:17403–17419.

Parton, W.J., A.R. Mosier, D.S. Ojima, D.W. Valentine, D.S. Schimel, K. Weier, and A.E. Kulmala. 1996. Generalized model for N_2 and N_2O production from nitrification and denitrification. Global Biogeochem. Cycles 10:401–412. doi:10.1029/96GB01455

Parton, W.J., D.S. Ojima, C.V. Cole, and D.S. Schimel. 1994. A general model for soil organic matter dynamics: Sensitivity to litter chemistry, texture and management. In: R.B. Bryant and R.W. Arnold, editors, Quantitative modeling of soil forming processes. Soil Science Society of America, Madison, WI. p. 147–167.

Parton, W.J. 1984. Predicting soil temperature in a shortgrass steppe. Soil Sci. 138:93–101. doi:10.1097/00010694-198408000-00001

Philip, J.R. 1966. Plant water relations: Some physical aspects. Annu. Rev. Plant Physiol. 17:245–268. doi:10.1146/annurev.pp.17.060166.001333

Porter, C.H., J.W. Jones, S. Adiku, A.J. Gijsman, O. Gargiulo, and J.B. Naab. 2010. Modeling organic carbon and carbon-mediated soil processes in DSSAT v4.5. Operational Research 10:247–278. doi:10.1007/s12351-009-0059-1

Potter, K.N., and J.R. Williams. 1994. Predicting daily mean soil temperatures in the EPIC simulation model. Agron. J. 86:1006–1011. doi:10.2134/agronj1994.00021962008600060014x

Ravishankara, A., J.S. Daniel, and R.W. Portmann. 2009. Nitrous oxide (N_2O): The dominant ozone-depleting substance emitted in the 21st century. Science 326:123–125. doi:10.1126/science.1176985

Richards, L.A. 1931. Capillary conductivity of liquid through porous media. Physics 1:318–333. doi:10.1063/1.1745010

Ritchie, J.T. 1972. A model for predicting evaporation from a row crop with incomplete cover. Water Resour. Res. 8:1204–1213. doi:10.1029/WR008i005p01204

Ritchie, J.T. 1981a. Water dynamics in the soil-plant-atmosphere. Plant Soil 58:81–96. doi:10.1007/BF02180050

Ritchie, J.T. 1981b. Soil water availability. Plant Soil 58:327–338. doi:10.1007/BF02180061

Ritchie, J.T. 1998. Soil water balance and plant water stress. In: G.Y. Tsuji, G. Hoogenboom, and P.K. Thornton, editors, Understanding options of agricultural production. Kluwer Academic Publishers and International Consortium for Agricultural Systems Applications, Dordrecht, The Netherlands. p. 41–54. doi:10.1007/978-94-017-3624-4_3

Robertson, G.P., and P.M. Groffman. 2007. Nitrogen transformations. Soil microbiology, ecology, and biochemistry. p. 341–364. doi:10.1016/B978-0-08-047514-1.50017-2

Sándor, R., Z. Barcza, M. Acutis, L. Doro, D. Hidy, M. Kochy, J. Minet, E. Lellei-Kovacs, S. Ma, A. Perego, S. Rolinski, F. Ruget, M. Sanna, G. Seddaiu, L. Wu, and G. Bellocchi. 2017. Multi-model simulation of soil temperature, soil water content and biomass in Euro-Mediterranean grasslands: Uncertainties and ensemble performance. Eur. J. Agron. 88:22–40. doi:10.1016/j.eja.2016.06.006

Scheer, C., P.R. Grace, D.W. Rowlings, and J. Payero. 2013. Soil N_2O and CO_2 emissions from cotton in Australia under varying irrigation management. Nutr. Cycl. Agroecosyst. 95:43–56. doi:10.1007/s10705-012-9547-4

Schelde, K., P. Cellier, T. Bertolini, T. Dalgaard, T. Weidinger, M. Theobald, and J.E. Olesen. 2012. Spatial and temporal variability of nitrous oxide emissions in a mixed farming landscape of Denmark. Biogeosciences 9:2989–3002. doi:10.5194/bg-9-2989-2012

Sharpley, A.N., and J.R. Williams. 1990. EPIC erosion/productivity impact calculator: 1. Model documentation. United States Department of Agriculture Technical Bulletin No. 1768, U.S. Government Printing Office, Washington, D.C.

Shcherbak, I., N. Millar, and G.P. Robertson. 2014. Global meta-analysis of the nonlinear response of soil nitrous oxide (N2O) emissions to fertilizer nitrogen. Proc. Natl. Acad. Sci. USA 111:9199–9204. doi:10.1073/pnas.1322434111

Šimek, M., and J.E. Cooper. 2002. The influence of soil pH on denitrification: Progress towards the understanding of this interaction over the last 50 years. Eur. J. Soil Sci. 53:345–354. doi:10.1046/j.1365-2389.2002.00461.x

Šimůnek, J., M.T. Van Genuchten, and M. Sejna. 2008. Development and applications of the HYDRUS and STANMOD software packages and related codes. Vadose Zone J. 7:587–600. doi:10.2136/vzj2007.0077

Šimůnek, J., T. Vogel, and M.Th. Van Genuchten. 1992. The SWMS_2D code for simulating water flow and solute transport in two-dimensional variably saturated media, version 1.1. Research Report No. 126. U. S. Salinity Lab, ARS USDA, Riverside, CA.

Soil Conservations Service (SCS). 1972. National engineering handbook, Hydrology Section 4. United States Department of Agriculture, Washington, D.C.

Stehfest, E., and C. Müller. 2004. Simulation of N_2O emissions from a urine-affected pasture in New Zealand with the ecosystem model DayCent. J. Geophys. Res.: Atmos. 109(D3): D03109.

Thorburn, P.J., J.S. Biggs, K. Collins, and M.E. Probert. 2010. Using the APSIM model to estimate nitrous oxide emissions from diverse Australian sugarcane production systems. Agric. Ecosyst. Environ. 136:343–350. doi:10.1016/j.agee.2009.12.014

US-EPA. 2010. Methane and nitrous oxide emissions from natural sources. Office of Atmospheric Programs, Washington, D.C.

Van Genuchten, M.T. 1980. A closed-form equation for predicting the hydraulic conductivity of unsaturated soils. Soil Sci. Soc. Am. J. 44:892–898. doi:10.2136/sssaj1980.03615995004400050002x

van der Laan, M., J.G. Annadale, K.L. Bristow, R.J. Stirzaker, C.C. du Preez, and P.J. Thorburn. 2014. Modeling nitrogen leaching: Are we getting the right answer for the right reason? Agric. Water Manage. 133:74–80. doi:10.1016/j.agwat.2013.10.017

Vinocur, M.G., and J.T. Ritchie. 2001. Maize leaf development biases caused by air-apex temperature differences. Agron. J. 93:767–772. doi:10.2134/agronj2001.934767x

Webber, H., P. Martre, S. Asseng, B. Kimball, J. White, M. Ottman, G.W. Wall, G. De Sanctis, J. Doltra, R. Grant, B. Kassie, A. Maiorano, J.E. Olesen, D. Ripoche, E.E. Rezaei, M.A. Semenov, P. Stratonovitch, and F. Ewert. 2017. Canopy temperature for simulation of heat stress in irrigated wheat in a semi-arid environment: A multi-model comparison. Field Crops Res. 202:21–35. doi:10.1016/j.fcr.2015.10.009

Webster, R. 2001. Statistics to support soil research and their presentation. Eur. J. Soil Sci. 52:331–340. doi:10.1046/j.1365-2389.2001.00383.x

Werner, C., X. Zheng, J. Tang, B. Xie, C. Liu, R. Kiese, and K. Butterbach-Bahl. 2006. N_2O, CH_4 and CO_2 emissions from seasonal tropical rainforests and a rubber plantation in Southwest China. Plant Soil 289:335–353. doi:10.1007/s11104-006-9143-y

White, D.H., S.M. Howden, and H.A. Nix. 1993. Modelling agricultural and pastoral systems. In: A.J. Jakeman, M.B. Beck, and M.J. McAleer, editors, Modelling change in environmental systems. John Wiley & Sons Ltd, Australia.

Wierenga, P.J., and C.T. de Wit. 1970. Simulation of heat flow in soils. Soil Sci. Soc. Am. Proc. 34:845–848. doi:10.2136/sssaj1970.03615995003400060012x

Williams, P.H., S.C. Jarvis, and E. Dixon. 1998. Emission of nitric oxide and nitrous oxide from soil under field and laboratory conditions. Soil Biol. Biochem. 30:1885–1893. doi:10.1016/S0038-0717(98)00052-2

The Role of Fish in a Globally Changing Food System

A.J. Lynch* and J.R. MacMillan

Abstract

Though humans have been fishing for food since they first created tools to hunt, modern food systems are predominately terrestrial focused and fish are frequently overlooked. Yet, within the global food system, fish play an important role in meeting current and future food needs. Capture fisheries are the last large-scale "wild" food, and aquaculture is the fastest growing food production sector in the world. Currently, capture fisheries and aquaculture provide 4.3 billion people with at least 15% of their animal protein. In addition to providing protein and calories, fish are important sources of critical vitamins and vital nutrients that are difficult to acquire through other food sources. As the climate changes, human populations will continue to grow, cultural tastes will evolve, and fish populations will respond. Sustainable fisheries and aquaculture are poised to fill demand for food not met by terrestrial food systems. Climate change and other global changes will increase, decrease, or modify many wild fish populations and aquaculture systems. Understanding the knowledge gaps around these implications for global change on fish production is critical. Applied research and adaptive management techniques can assist with the necessary evolution of sustainable food systems to include a stronger emphasis on fish and other aquatic organisms.

As humans have evolved, populations have remained close to water. Even now, 60% of the world's population lives within 100 km of a coastline (Rick and Erlandson, 2008), and over 50% lives closer than 3 km to a source of surface freshwater (Kummu et al., 2011). By shear proximity, many have looked to the water for food. Records of fishing date back more than 40,000 yr (Jackson et al., 2001). In more recent years, food production systems have increasingly focused on terrestrial output. But, in the face of global change, capture fisheries and aquaculture will likely be increasingly important to maintaining sustainable sources of food. Wild capture and aquacultured fish[1] play a significant role in promoting human health, eliminating hunger, reducing poverty, and creating wealth in the world.

[1] Following Food and Agriculture Organization (2014), "fish" includes finfish, crustaceans, mollusks, miscellaneous aquatic animals, but excludes aquatic plants and algae.

A.J. Lynch, U.S. Geological Survey, National Climate Change and Wildlife Science Center, 12201 Sunrise Valley Drive, MS-516, Reston, VA 20192; and J.R. MacMillan, Clear Springs Foods, Inc., P.O. Box 712, Buhl, ID 83316 (randy.macmillan@clearsprings.com). *Corresponding author (ajlynch@usgs.gov).

doi:10.2134/agronmonogr60.2014.0059

© ASA, CSSA, and SSSA, 5585 Guilford Road, Madison, WI 53711, USA.
Agroclimatology: Linking Agriculture to Climate, Agronomy Monograph 60.
Jerry L. Hatfield, Mannava V.K. Sivakumar, John H. Prueger, editors.

Capture fisheries are defined by the Food and Agriculture Organization of the United Nations (FAO) as the sum of activities related to harvest of a given fish resource. Fishery resources are not exclusive to finfish but, rather, encompass all aquatic organisms that are harvested from a wild population. Capture fisheries are the last, industrial-scale wild-sourced food harvest. In 2012, total global capture production was 91.3 million tonnes (FAO, 2014). This is 91.3 million tonnes of animal protein produced without any investment in cultivation, care, or husbandry. This is no trivial contribution to the global food supply.

Aquaculture is defined by the FAO as the farming of aquatic organisms (e.g., fish, mollusks, crustaceans) and harvest by an owner who is responsible for the maintenance, protection, and enhancement of the cultivated stock. As the fastest growing food production sector, global aquaculture production of food fish expanded at an average annual rate of 6.2% in the period 2000 to 2012 from 32.4 million to 66.6 million tonnes (FAO, 2014). Farmed fish now comprise about 50% of the global seafood supply (FAO, 2014).

As a sector, employment in capture fisheries and aquaculture has grown faster than the traditional terrestrial agricultural sector (FAO, 2014). In 2012, 58.3 million people participated in capture fisheries and aquaculture (FAO, 2014). While this is a small fraction (4.4%) of the 1.3 billion people active in the agricultural sector, the contributions to food are significant: capture fisheries and aquaculture provide 2.9 billion people with almost 20% of their animal protein and a total of 4.3 billion people with at least 15% of their animal protein (FAO, 2014).

Fish population dynamics, as well as fisheries and aquaculture production, are closely tied to climate. Climate can drive spatial and temporal fluctuations of fish populations directly (e.g., metabolic and reproductive rates) or indirectly (e.g., predators, competitors, and prey) and can determine the distribution, abundance, and migration patterns of fish (Stenseth et al., 2002). Natural climate variability can occur on a seasonal, interannual, interdecadal, multidecadal, or even centennial scale (Lehodey et al., 2006). The complex interactions between climate and fish manifest through direct effects on survival, growth, fitness, behavior, population abundance, and species range; indirect effects manifest through habitat, predators, prey, competitors, and pathogens (Brander, 2010). The biological regime shift between sardine (*Sardinops sagax*) and anchoveta (*Engraulis ringens*) populations in the Pacific based on air and ocean temperatures, atmospheric carbon dioxide, nutrient supply, and ocean productivity is one well-documented example of climate variability impacts on fish populations (e.g., Chavez et al., 2003).

Fish as Food

Of the 158 million tonnes of fish produced in 2012, 136 million went directly to human consumption (FAO, 2014). World per capita fish consumption has increased from 9.9 kg in the 1960s to 19.2 kg in 2012 (FAO, 2014), and fish are increasingly recognized as an important component of a modern healthy diet, a critical food source for developing countries, and an important source of employment and income (High Level Panel of Experts, 2014).

Nutrition

Fish provide key macro- and micronutrients and protein, are low in saturated fat, and have been linked to a wide array of health benefits for the developing fetus, infants, and adults (Millen et al., 2015). Diet quality is not only important for reducing nutrient deficiency, but is now regarded as essential for preventing chronic diseases such as obesity, type 2 diabetes, hypertension, coronary artery disease, and cancers (Kennedy, 2006). Fish can supply 50% or more of high-quality protein, niacin, zinc, and vitamin B6 and at least 10% of vitamins E and B12, thiamin, riboflavin, phosphorus, magnesium, iron, copper, potassium, and linoleic acid (Nesheim and Yaktine, 2007). They are the primary source for omega-3 fatty acids in human diets and certain species, such as salmons, trouts, and albacore (*Thunnus alalunga*), having a higher concentration of omega-3 fatty acids per serving than any other food sources (Nesheim and Yaktine, 2007). Consumption of omega-3 fatty acids is, among other things, associated with both reduced cardiovascular disease (Kris-Etherton, 2002) and chronic obstructive pulmonary disease (Varraso et al., 2015). Consumption of as few as one to two servings of fish a week reduces risk of coronary death by 36% and total mortality by 17% (Mozaffarian and Rimm, 2006). Fish are also generally lower in saturated fats compared with other animal proteins. Reduced consumption of saturated fats has been associated with reduction in cardiovascular disease and diabetes (Schwab et al., 2014). By providing a good source of animal protein containing all essential amino acids, valuable essential fatty acids, as well as important vitamins and minerals, fish contribute to a healthy diet and reduce the risks of malnutrition and disease.

Food Security

Though capture fisheries are often absent from agricultural food production discussions because they are wild sourced, capture fisheries contribute significantly to food supply, particularly in food insecure situations. In 2010, fish accounted for roughly 19.6% of animal protein intake in developing countries and 24.7% in low income food deficit countries (FAO, 2014). In 2012, global capture fishery production was reported to be 91.3 million tonnes, 79.7 million tonnes from marine waters and 11.6 million tonnes from inland waters (FAO, 2014). More than 86% of capture fisheries production is used for direct human consumption and the remainder is repurposed into fish meal and fish oil (FAO, 2014). Fish meal and fish oil are also an integral component of food systems: they are predominately utilized as a specialized feed additive for terrestrial livestock and aquaculture and as a health supplement for humans (e.g., fish oil capsules).

Small-scale aquaculture also contributes to poverty alleviation and promotes food and nutrition security in less developed parts of the world (Bondad-Reantaso and Prein, 2009; Hishamunda et al., 2009; Belton et al., 2014; FAO, 2014). More than 80% of global aquaculture production comes from small- to medium-sized fish farms (which may be affiliated with larger businesses), with nearly 90% in Asia (High Level Panel of Experts, 2014). With the global human population anticipated to increase from the current 7 billion to 9–10 billion by 2050 and most wild fish capture fisheries at or above maximum sustainable yield (FAO, 2014), aquaculture production will need to significantly increase if fish supplies are to adequately meet demand by midcentury.

Livelihood

Capture fisheries and aquaculture are also important as a source of individual and national wealth. Between 660 and 820 million people are estimated to depend totally or partially on fisheries and aquaculture for income and employment (High Level Panel of Experts, 2014). Fish are one of the most traded food commodities in the world (US$129.8 billion in exports), sometimes worth half the total value of traded commodities in certain developing countries (FAO, 2014). The global aquaculture industry alone produces products valued at over $144 billion (FAO, 2014). About 80% of aquaculture targets low trophic (planktivorous) species (e.g., grass carp *Ctenopharyngodon idella,* silver carp *Hypophthalmichthys molitrix,* bighead carp *H. nobilis,* and tilapia *Tilapia* spp.), with an impressive annual economic value of $44 billion. But the unit value is relatively low compared with carnivorous species (e.g., sea breams, salmonids, and shrimp) which account for $54 billion in economic value from only 13% of aquaculture production (FAO, 2014).

Small-scale fisheries and aquaculture are significant, but often undervalued, sources of livelihood. Nearly 90% of the estimated 34 million full-time or part-time fishers and between 70 and 80% of aquaculture operations are small scale, meaning artisanal with low levels of technology and capital investment per operation (High Level Panel of Experts, 2014). However, the economic value of these ventures is difficult to estimate because the outputs are often very local in scope: fish are traded locally or consumed directly by fishing families. As one small example, a study of six river basins in West and Central Africa found that local fisheries supported 227,000 full-time fishers and had a first-sale value of $295 million (Neiland and Bene, 2006). Perhaps more importantly, the value of small-scale fisheries transcends economic statistics because these fisheries also serve a critical nonmonetary role in the case of subsistence where no financial transactions occur.

Environmental Impacts

As in all forms of food production, capture fisheries and aquaculture can have environmental consequences. Depending on the fishery method, capture fisheries can be done in a sustainable and environmentally conscious manner or, on the other hand, can be responsible for overfishing, habitat destruction, food web restructuring, and bycatch. Overfishing is one of the most obvious environmental impacts where targeted efforts deplete stocks to the point where they can no longer support harvest. Rebuilding stocks can be quite difficult; for example, Atlantic cod (*Gadus morhua*) stocks struggle to reestablish long after the controversial Grand Banks closures of the 1990s (Froese, 2004). Trawl fisheries are particularly disruptive: in the North Sea, as an example, the trawl fleet removes an estimated 56% of benthic biomass and 21% of benthic production (Hiddink et al., 2006). Simplification of food webs, by harvest of important species, can also cause cascading effects to ecosystems (Jackson et al., 2001). Bycatch, in which unwanted fishes are captured and discarded, is also a significant problem. Applying a conservative estimate, bycatch may represent 40.4% of global marine catches (Davies et al., 2009).

Aquaculture operations can discharge nutrients (phosphorus and nitrogen), biosolids (waste feed and manure), fish pathogens, and drugs and chemicals; they can compete with shoreline recreation, shipping lanes, and marine protected areas (Ferreira et al., 2014); and they can alter the viewshed, introduce nonnative species, and spread disease (Kapuscinski and Brister, 2000; Youngson et al., 2001). The type of aquaculture, however, fundamentally determines potential impacts. Extensive aquaculture (e.g., bivalve mollusks) in which animal density is relatively low and relies on natural feed resources will have less impact than intensive farms where animals are confined at high stocking densities and manufactured feeds are used (Pernet et al., 2014). For example, oyster farms often remove more nutrients from the system (i.e., filter feeding) than they provide (i.e., excretion), while net pen operations (e.g., Atlantic salmon—*Salmo salar*) pollute excess feed and waste into surrounding waters. Environmental impacts of intensive farming can be mitigated by judicious farm siting, various operational controls, and elevated levels of management.

Carnivorous fish, marine shrimp, and even early life stages of planktivorous or omnivorous species require nutrients that are found in wild fish. Aquaculture operations for these species often rely on fish meal and fish oil from reduction fisheries of small pelagic fish (e.g., anchovy, mackerels, sardines, and herring). The supply–demand dynamic can have implications for these wild stocks unless the capture fisheries are managed sustainably (Tacon and Metian, 2008), and, increasingly, aquaculture operations are examining more cost-effective feed sources that do not rely on wild fish stocks (e.g., vegetarian or insect feed).

Climate Considerations

Climate change and climate variability influence aquatic ecosystems and fish populations. Fish species have developed the capacity to adapt to changing conditions or have gone extinct (Brander, 2010). There are well-documented relationships between natural fish production and climate. Climate variability has been linked with fish production variability with high levels of confidence in both marine and freshwater systems (e.g., Chavez et al., 2003; Lynch et al., 2015a). Climate may strongly influence latitudinal and depth distribution and abundance of marine and freshwater fish through survival, growth, reproduction, and interactions with habitat and other species (Perry et al., 2005).

However, it is important to note that with current and projected climate change, the rate of change and addition of stressors (e.g., fishing, pollution, and invasive species) is unprecedented. The resilience or adaptive ability of species to new climate scenarios and new externalities can only be examined and projected. In some cases, the changes may have a positive effect on fish production (e.g., increased survival) and in others a negative one (e.g., reduction in habitat). Less mobile species and species with slower life histories are likely to be more vulnerable to changing conditions (Perry et al., 2005). How these effects will translate to fisheries production and human consumption are difficult to estimate, in part, because of the impacts of climate change on terrestrial food systems and the potential for additional demands on fisheries systems (Table 1).

Capture Fisheries

For marine systems, increasing sea temperatures, thermal expansion, glacial melting, acidification, and dissolved oxygen are changing the habitat for marine fish. Between 1971 and 2010, ocean temperatures have warmed on a global scale by 0.11°C per decade; since the beginning of the industrial era, oceans have increased in acidity by 26%, and between 1901 and 2010 global mean sea level has risen by 0.19 m (IPCC, 2014).

Increasing temperatures are already shifting distributions of species poleward. Forty-five out of 61 species (74%) examined in southeastern Australia, for example, exhibited major distributional shifts from climate (Last et al., 2011).

Table 1. Selected examples of climate change impacts on capture fisheries and aquaculture.†

Climate change	Impact to capture fisheries	Impact to aquaculture
Temperature increase	shifting species distribution with latitude and depth (e.g., Shuter and Post, 1990; Chu et al., 2005; Perry et al., 2005; Last et al., 2011)	increased or decreased production potential (and profit potential) depending on operation location and farmed species (e.g., Morgan et al., 2001; Handisyde et al., 2006)
	emerging new foodborne diseases and redistribution of existing diseases and zoonoses may increase risks associated with food safety (Jaykus et al., 2008)	increased susceptibility to pathogens and diseases (and related costs) for some farmed species (e.g., Pörtner et al., 2010; Burge et al., 2014) with implications for food safety (e.g., Jaykus et al., 2008; Tirado et al., 2010)
Sea level rise	anthropogenic modifications to combat sea level rise may inhibit wild populations from appropriate habitat (e.g., Harley et al., 2006)	damage to coastal aquaculture facilities and increased costs for relocation (e.g., Cochrane et al., 2009)
Changes in precipitation and water availability	reduced freshwater habitat availability and shifting distributions (e.g., Brander, 2007; Magalhães et al., 2007)	increased conflict over water resources and reduced fish production capacity (e.g., Dai, 2011)
	variable responses of fish to nutrients and contaminants (e.g., Morgan et al., 2001)	escapement of farmed fish with flooding events (e.g., Walsh and Hanafusa, 2013)
Changes in water pH and salinity	reduced strength of carbonate structures in invertebrates (e.g., Kroeker et al., 2013)	increased production costs for shellfish to account for defective shell formation (e.g., Barton et al., 2012)
	impairment of larval fish (e.g., Munday et al., 2010)	
	reduced fecundity of certain fish species (e.g., Nissling and Westin, 1997)	
Decreased oxygen solubility	habitat restrictions for marine and freshwater fish (e.g., Pörtner and Knust, 2007)	potential for decreased fish production capacity (e.g., Handisyde et al., 2006)
Increased storm intensity and frequency		destruction of aquaculture infrastructure (e.g., Cochrane et al., 2009)
		increased bacterial loads from flooded septic and sewage systems (e.g., Iwamoto et al., 2010)

† For additional example impacts, see Table 1 from WorldFish Center (2007).

Likewise, 21 out of 36 demersal species (58%) in the North Sea have shown boundary shifts with latitude and depth (Perry et al., 2005). Temperature-driven shifts in diseases and parasites, particularly if the pathogens have zoonotic potential, add an additional layer of complexity to capture fisheries with regard to food safety (Jaykus et al., 2008).

Thermal expansion and glacial melt are increasing sea level and, in some cases, decreasing salinity of marine environments. While many fish species can keep pace with sea level rise, anthropogenic modifications, such as seawalls, limit the migration in certain habitat types, like intertidal zones (Harley et al., 2006).

Freshening of the Arctic, as an example, reduces fecundity for Atlantic cod (*Gadus morhua*) because cod spermatozoa are not active in salinity <11 psu and eggs sink to anoxic conditions because they are denser than the fresher water (Nissling and Westin, 1997).

Ocean acidification is most commonly cited as being detrimental to marine invertebrates and algae that build carbonate structures. But reduced seawater pH levels have also been shown to alter behavior of larval fish, making them more susceptible to predation and decreasing their survival to recruitment in adult populations. In experimental settings with reduced pH, the ability of larvae to sense predators was highly impaired and, in some instances, they were even attracted to the smell of predators (Munday et al., 2010).

Oxygen solubility in water has an inverse relationship with temperature. In both salt and freshwater, as temperature increases, oxygen solubility (i.e., the amount of biologically available oxygen) decreases. The common eelpout (*Zoarces viviparus*) is one example species which has been restricted by thermally limited oxygen supply in the North and Baltic Seas (Pörtner and Knust, 2007).

As with marine systems, temperature is an extremely important factor governing growth, survival, fecundity, and distribution of many freshwater species (Shuter and Post, 1990). While there is generally greater flexibility for climate change–induced migration in marine systems, many freshwater species, such as those in headwater streams, are limited in their ability to expand. Chu et al. (2005), for example, project that brook trout (*Salvelinus fontinalis*) distribution will decrease by 49% by 2050.

Reduced precipitation and greater evaporation will directly change water levels and cause agricultural, industrial, and municipal users to acquire a larger proportion of available water, further altering freshwater habitats (Brander, 2007). With changed habitats, species abundances and distributions will shift, as will parasites and pathogens, with implications for food safety (Jaykus et al., 2008). In the Mediterranean, stream fish assemblages are known to shift with drought conditions (Magalhães et al., 2007).

Because of their proximity to anthropogenic terrestrial inputs, many freshwater and coastal fishes are also exposed to chronic contamination with toxic pollutants. Reduced precipitation and increased evapotranspiration is expected to increase the concentration of these contaminants. Pollutants can have differential effects on fishes; Morgan et al. (2001) found a positive growth effect of low-level ammonia on rainbow trout (*Oncorhynchus mykiss*), but no acclimatization to sublethal acidification in an experimental 2°C warming scenario.

For both marine and freshwater systems, climate change–related shifts in thermal, chemical, and physical habitat will alter fish populations and capture fishery productivity. In some cases, the fisheries will have the potential to expand; in other cases, the fisheries will likely diminish; and in yet others, the fisheries will remain stable but may shift location.

Aquaculture

Many of the potential climate stressors on wild fisheries can also impact aquaculture. Climate change is altering sea surface and inland water temperatures, sea levels, precipitation and water availability, intensity of storms, and pH (e.g., Handisyde et al., 2006; Cochrane et al., 2009; FAO, 2014). Known and predicted aquatic resource changes associated with climate change are likely to differentially impact current aquaculture operations and potential for aquaculture growth. Such impacts will depend on the severity of specific changes which are likely site specific. Some changes may actually be advantageous to aquaculture if growing or farming seasons are extended and water temperature increases foster optimal physiology with subsequent improved feed conversion and growth rates. The impacts are nevertheless likely to vary locally and regionally, as will the ability of producers to invest in adaptive measures (Handisyde et al., 2006).

Impact will vary with the aquatic animal species farmed and their physiologic plasticity (Crozier and Hutchings, 2014), the method of farming (Barton et al., 2012), production intensity (whether extensive or intensive), environmental carrying capacity (e.g., for setting net pens [Ross et al., 2013]), type of feed and its availability (Handisyde et al., 2006), production costs, and success of efforts to mitigate and adapt to the impacts of climate change. Climate change has the potential to disrupt the stability of feed ingredient supplies whether obtained from marine waters or from terrestrial plant-based proteins (Cochrane et al., 2009). Other factors such as salinity, carbon dioxide, pH, trace metals, turbidity, pollutants, and underwater irradiance directly impact the physiology of the farmed animals (Pörtner and Peck, 2010; Boyd and Hutchins, 2012). Consequently, the impacts of climate change will likely be compounded by increasing levels of pollution in near-shore, open-ocean, and freshwater environments (Boyd and Hutchins, 2012).

With rare exception (e.g., tuna), farmed aquatic animals are obligatory ectotherms (poikilothermic), their body temperatures match the temperature of their environment. Temperature directly affects the rates of virtually all physiological and biochemical processes, and environmental temperature is one of the most important factors defining the fundamental growing conditions of the animal (Brett, 1979) and production success. Elevated water temperature may be beneficial in some locations and with some species, and because growth rates increase with temperature, growing seasons may be extended and primary productivity may be enhanced to increase food supply for low trophic level species (Morgan et al., 2001; Handisyde et al., 2006).

However, elevated water temperatures can also challenge the physiological limits (Pörtner et al., 2010) of aquatic animals, increasing aquatic animal susceptibility to disease (e.g., Burge et al., 2014), changing animal behavior, shifting pathogen prevalence, changing pathogen virulence, and otherwise causing significant changes in thermal tolerance of aquatic animals (Pörtner, 2012). This is particularly important in the context of food safety; the emergence of new diseases or redistribution of existing diseases could impact the occurrence of foodborne disease, zoonoses (Jaykus et al., 2008), and toxicities (Tirado et al., 2010). Recent mortality events in some New Zealand salmon farms, for example, has been attributed to elevated water temperature (Powell, 2015), and mortality in some bivalve shellfish farms has been attributed to ocean warming and acidification, compounded by the introduction of *Vibrio tubiashii* into the Pacific Northwest (Elston et al., 2008; Vezzulli et al., 2013). Elevated water temperature will decrease the oxygen content of rearing water and increase the prevalence of harmful algal blooms (Handisyde et al., 2006). While the decreased dissolved oxygen concentration may decrease fish production capacity. Farmers may be able to selectively breed more thermally tolerant fish (e.g., Muñoz et al., 2014), and there is suggestion that diets could be modified to enhance thermal tolerance

(Tejpal et al., 2014; Kumar et al., 2014). In some cases, farmers may be able to farm more elevated temperature-tolerant species or other species not previously suitable for that location. But how elevated water temperatures will impact net pen fouling is still currently unknown.

Sea level rise has potential to eliminate space currently used for aquaculture (Cochrane et al., 2009). This is particularly important in areas in southeast Asia and the Polynesian islands where freshwater aquaculture is valued for its alleviation of poverty and food security (FAO, 2014). Rising sea levels will likely diminish existing mangrove wetlands which are important nurseries for some farmed marine species. Mangrove wetlands also physically protect inland areas during storm surges (Cochrane et al., 2009). The rise in sea level combined with reduced supplies of freshwater may also alter salinity characteristics of estuaries, adversely impacting bivalve mollusk farming and aquaculture seed, in addition to the wild ecosystems (Handisyde et al., 2006).

Changes in the availability of freshwater through diminished precipitation, drought, or over-appropriation of water resources (Dai, 2011) will limit freshwater aquatic animal production capacity or fish farming altogether. Production unit stocking density is, among other factors, significantly influenced by the amount of freshwater and dissolved oxygen available (Person-Le Ruyet et al., 2008; Ronald et al., 2014). Various regions of the world (most of Africa, southern Europe, the Middle East, the Americas, Australia, and southeast Asia) are projected to have increased aridity during the 21st century (Dai, 2011), and megadroughts are projected in the American southwest and central plains (Cook et al., 2015). Drought can significantly impact water availability for aquaculture operations. Rainbow trout farms in Idaho, for example, are dependent on spring flows originating from the Eastern Snake Plain Aquifer. Spring flows are dependent on ground water levels in the aquifer which are dependent on recharge from snow melt and incidental recharge from surface water irrigation. Spring flows and ground water levels are declining (as is the case in many U.S. aquifers) as water is pumped out for other purposes. The impact on aquaculture of these short- or long-term diminished freshwater supplies may depend on efforts to sustain the local or regional water supply. In some areas, recharge of groundwater supplies is feasible when supply is plentiful to reserve supply for when water is limited. Groundwater pumping, however, increases the cost of aquaculture production. As freshwater supplies diminish, competition for the resource will likely intensify (FAO, 2014).

Increased intensity of storms associated with climate change will also impact aquaculture (Cochrane et al., 2009). Extreme weather events, such as storms and floods, have potential to harm or eliminate aquaculture infrastructure. Net pens may break free from moorings and nets may be damaged, releasing farmed fish (e.g., see Walsh and Hanafusa [2013] for impacts from the 2011 Japanese tsunami). Floods have potential to overfill ponds, causing loss of fish with, in some cases, the potential for aquaculture introductions into the surrounding ecosystems. Floods may also overwhelm on-land septic and sewage systems, leading to elevated bacterial loads at coastal bivalve mollusk farming sites (Iwamoto et al., 2010).

Ocean acidification will be particularly damaging to commercially important aquaculture organisms, such as bivalve mollusks, that build carbonate structures (Barton et al., 2012; Kroeker et al., 2013). Collectively, ocean acidification and water temperature increases can negatively impact animal respiration, motility, fertility, embryonic development, hatching, growth, an animal's adaptability, and food quality and quantity (Pörtner, 2008; Pörtner et al., 2010; Elliott and Elliott, 2010; Boyd and Hutchins, 2012). The impact varies with species and life stage, but the early life stages are most susceptible. Oyster health is being adversely impacted by changes in the natural food quality due to warming sea surface temperature and changes in phytoplankton food quality (Pernet et al., 2014). Similarly, decreased water pH has been shown to have a negative effect on farmed Pacific oyster (*Crassostrea gigas*) causing mortality and defective shell formation (Barton et al., 2012). To alleviate this problem, oyster hatcheries have improved their carbonate chemistry management by water quality manipulation.

Climate change will change the potential productivity for fish farms based on the species cultured, method of operation, and location. In some instances, the changing conditions will provide the opportunity for business expansion, increased productivity, and higher profitability; other operations will need to adapt to conditions to maintain profitability.

Future Opportunities

The global food sector is rapidly changing with economic growth, human population growth, urbanization, poverty, and changing cultures. Globalization is increasing availability of fishery products to previously restricted markets. Climate change is poised to alter food systems in additional, potentially synergistic, ways. Fisheries and aquaculture provide an important nutritional and economic resource to people across the globe. The sector will need to evolve and adapt to changing conditions to meet current and future food needs while maintaining sustainable enterprises.

Tools for Sustainability

To ensure success of capture fisheries and aquaculture operations, now and in the future, it is important to recognize knowledge gaps in the relationships between fish and climate change, anticipate changes, and be willing to continually adapt to changing conditions. This is not an easy task but, in some cases, decision support and monitoring tools exist to support these objectives. Particularly useful in considering the unknowns surrounding climate change, decision support tools help decision makers assess options, weigh uncertainties, and evaluate priorities in a structured context (Lynch et al., 2015b). Monitoring tools can track status and changes in an important resource over time. For example, increased traceability technologies for fishery resources, ranging from basic documentation to genetic testing and geospatial referencing, will help identify fish production and harvest in relation to climate change impacts.

Decision making, even with the assistance of tools, is a difficult process, and there is generally a stigma to revisiting decisions even if conditions change or new information is available. Applying adaptive management practices from the beginning of a decision-making process can remove that stigma. Adaptive management is an iterative process (i.e., learn-by-doing) where a resource problem is identified and management action is applied to address it; the response to management is monitored and adjustments are made based on new knowledge of the situation (Fröcklin et al., 2013). This concept can work particularly well with climate change. For example, Burge et al. (2014) found adaptive management practices (e.g., temporarily reducing harvest when disease events are discovered) to be effective means of addressing climate change–induced Dermo disease in Eastern oysters (*Crassostrea virginica*). The advantage of using an adaptive approach to management is that no decision is "wrong" and the process is continually revisable with changing conditions.

Growth Potential for Aquaculture

While capture fisheries are largely maximized, aquaculture continues to grow in prominence as a food sector. Its growth rate eclipses that of poultry, pork, dairy, beef, and grains and "it is arguably the most vibrant sector of the global food system" (Troell et al., 2014). Over 600 different kinds of aquatic animals and plants are farmed globally but only a fraction of that total are cultured on a large scale (FAO, 2014). Research continues to identify requirements and develop the techniques to farm new species. Potential for increased farmed fish production remains strong, and the sector currently provides more opportunities for efficient protein production than terrestrial sources (Troell et al., 2014), although other costs associated with production (e.g., fuel prices) may increase.

Aquaculture lends itself to controls not available to wild capture fisheries. Depending on the production technique, control of the aquatic environment for farming aquatic animal species can be significant. Recirculating aquaculture exemplifies the potential (Lawson, 1995). In this type of aquaculture, minimal amounts of water are used and technically sophisticated tools are employed to recondition the water and recirculate it. Production intensity for recirculating aquaculture is high, but costs and managerial intensity of production compared with wild capture and other types of aquaculture (e.g., pond, net pen, serial reuse flow through raceways, and most forms of extensive aquaculture) are also high. Though all aquaculture will be affected by climate change, extensive aquaculture, in which aquatic animals are placed in natural environments and food comes from natural supplies, lacks environmental control and will be most impacted. Better predictive tools for changes in the environment will be helpful.

Considerable research is ongoing to improve the sustainability of aquaculture and such efforts will likely increase aquaculture operations' resilience and adaptability to climate change. For example, efforts for many species are directed at selective breeding for a variety of phenotypes, including thermal tolerance (Elliott and Elliott, 2010; Lind et al., 2012), alternative proteins (to fish meal) for carnivorous fish (Ayadi et al., 2012), thermal adaptability of fish through dietary manipulation (Kumar et al., 2014), and vaccinology (Evensen, 2014). FAO (2014) suggests aquaculture will need to increase 33% by 2021 to meet anticipated demand. The increase will need to be even greater to supply demand from 9 to 10 billion people by midcentury.

Conclusions

Fish are a vitally important component of the global food system. Fish are high in protein and macro- and micronutrients, and low in fat. They are particularly important in providing food security to many impoverished communities and providing livelihoods around the world. Climate and climate variability influence wild and farmed fish productivity. Recognizing the anticipated changes and being ready to adapt to changing conditions will be important to maintaining sustainable fisheries and aquaculture to meet increasing food demands from growing global populations, evolving cultural tastes, and a changing climate.

Acknowledgments

We thank editors J. Hatfield, J. Prueger, and M. Sivakumar for the opportunity to contribute a fish chapter to this project. We thank Robert Jones (Program Coordinator, NOAA Fisheries, Office of Aquaculture) for conducting an internal U.S. Geological Survey review and an anonymous peer reviewer; both provided insightful comments for improving this chapter. We also thank our family for supporting our professional endeavors. Any use of trade, firm, or product names is for descriptive purposes only and does not imply endorsement by the U.S. Government.

References

Ayadi, F.Y., K.A. Rosentrater, and K. Muthukumarappan. 2012. Alternative protein sources for aquaculture feeds. J. Aquac. Feed Sci. Nutr. 4:1–26.

Barton, A., B. Hales, G.G. Waldbusser, C. Langdon, and R.A. Feely. 2012. The Pacific oyster, *Crassostrea gigas*, shows negative correlation to naturally elevated carbon dioxide levels: Implications for near-term ocean acidification effects. Limnol. Oceanogr. 57(3):698–710.

Belton, B., N. Ahmed, and K. Murshed-e-Jahan. 2014. Aquaculture, employment, poverty, food security and well-being in Bangladesh: A comparative study. CGIAR Research Program on Aquatic Agricultural Systems. Penang, Malaysia. Program Report: AAS-2014-39. http://www.aas.cgiar.org/publications/aquaculture-employment-poverty-food-security-and-well-being-bangladesh-comparative (accessed 10 Nov. 2015).

Bondad-Reantaso, M.G., and M. Prein, editors. 2009. Measuring the contribution of small-scale aquaculture. FAO Fisheries and Aquaculture Technical Paper 534. FAO, Rome.

Boyd, P., and D. Hutchins. 2012. Understanding the responses of ocean biota to a complex matrix of cumulative anthropogenic change. Mar. Ecol. Prog. Ser. 470:125–135.

Brander, K.M. 2007. Global fish production and climate change. Proc. Natl. Acad. Sci. U. S. A. 104(50):19709–19714.

Brander, K. 2010. Impacts of climate change on fisheries. J. Mar. Syst. 79(3–4):389–402.

Brett, J.R. 1979. 10 environmental factors and growth. Fish Physiol. 8:599–675.

Burge, C.A., C. Mark Eakin, C.S. Friedman, B. Froelich, P.K. Hershberger, E.E. Hofmann, L.E. Petes, K.C. Prager, E. Weil, B.L. Willis, S.E. Ford, and C.D. Harvell. 2014. Climate change influences on marine infectious diseases: Implications for management and society. Ann. Rev. Mar. Sci. 6:249–77.

Chavez, F.P., J. Ryan, S.E. Lluch-Cota, and M. Niquen C. 2003. From anchovies to sardines and back: Multidecadal change in the Pacific Ocean. Science 299(5604):217–21.

Chu, C., N.E. Mandrak, and C.K. Minns. 2005. Potential impacts of climate change on the distributions of several common and rare freshwater fishes in Canada. Divers. Distrib. 11(4):299–310.

Cochrane, K.L., C. De Young, D. Soto, and T. Bahri. 2009. Climate change implications for fisheries and aquaculture: Overview of current scientific knowledge. FAO, Rome.

Cook, B.I., T.R. Ault, and J.E. Smerdon. 2015. Unprecedented 21st century drought risk in the American Southwest and Central Plains. Sci. Adv. 1(1):E1400082–E1400082.

Crozier, L.G., and J.A. Hutchings. 2014. Plastic and evolutionary responses to climate change in fish. Evol. Appl. 7(1):68–87.

Dai, A. 2011. Drought under global warming: A review. Wiley Interdiscip. Rev. Clim. Chang. 2(1):45–65.

Davies, R.W.D., S.J. Cripps, A. Nickson, and G. Porter. 2009. Defining and estimating global marine fisheries bycatch. Mar. Policy 33(4):661–672.

Elliott, J.M., and J.A. Elliott. 2010. Temperature requirements of Atlantic salmon *Salmo salar*, brown trout *Salmo trutta* and Arctic charr *Salvelinus alpinus*: Predicting the effects of climate change. J. Fish Biol. 77(8):1793–817.

Elston, R.A., H. Hasegawa, K.L. Humphrey, I.K. Polyak, and C.C. Häse. 2008. Re-emergence of Vibrio tubiashii in bivalve shellfish aquaculture: Severity, environmental drivers, geographic extent and management. Dis. Aquat. Organ. 82(2):119–34.

Evensen, Ø. 2014. Future fish vaccinology. In: R. Gudding, A. Lillehaug, and Ø. Evensen, editors, Fish vaccination. John Wiley & Sons, Chichester, UK. p. 162–171.

Ferreira, J.G., C. Saurel, J.D. Lencart e Silva, J.P. Nunes, and F. Vazquez. 2014. Modelling of interactions between inshore and offshore aquaculture. Aquaculture 426–427:154–16.

Food and Agriculture Organization. 2014. The state of world fisheries and aquaculture. FAO, Rome.

Fröcklin, S., M. de la Torre-Castro, L. Lindström, and N.S. Jiddawi. 2013. Fish traders as key actors in fisheries: Gender and adaptive management. Ambio 42(8):951–62.

Froese, R. 2004. Keep it simple: Three indicators to deal with overfishing. Fish Fish. 5(1):86–91.

Handisyde, N.T., L.G. Ross, M.-C. Badjeck, and E.H. Allison. 2006. The effects of climate change on world aquaculture: A global perspective. Stirling, UK. http://www.fao.org/fishery/gisfish/cds_upload/1176298188002_Climate_Exec_full.pdf (accessed 4 Nov. 2015).

Harley, C.D.G., A. Randall Hughes, K.M. Hultgren, B.G. Miner, C.J.B. Sorte, C.S. Thornber, L.F. Rodriguez, L. Tomanek, and S.L. Williams. 2006. The impacts of climate change in coastal marine systems. Ecol. Lett. 9(2):228–241.

Hiddink, J.G., S. Jennings, M.J. Kaiser, A.M. Queirós, D.E. Duplisea, and G.J. Piet. 2006. Cumulative impacts of seabed trawl disturbance on benthic biomass, production, and species richness in different habitats. Can. J. Fish. Aquat. Sci. 63(4):721–736.

High Level Panel of Experts. 2014. Sustainable fisheries and aquaculture for food security and nutrition. A report by the High Level Panel of Experts on Food Security and Nutrition. Committee on World Food Security, Rome.

Hishamunda, N., J. Cai, and C.S. Leung. 2009. Commercial aquaculture and economic growth, poverty alleviation and food security. Assessment framework. FAO, Rome.

IPCC. 2014. Climate Change 2014: Synthesis Report. Contribution of Working Groups I, II and III to the Fifth Assessment Report of the Intergovernmental Panel on Climate Change. Core Writing Team, R.K. Pachauri, and L.A. Meyer, editors. IPCC, Geneva, Switzerland.

Iwamoto, M., T. Ayers, B.E. Mahon, and D.L. Swerdlow. 2010. Epidemiology of seafood-associated infections in the United States. Clin. Microbiol. Rev. 23(2):399–411.

Jackson, J.B., M.X. Kirby, W.H. Berger, K.A. Bjorndal, L.W. Botsford, B.J. Bourque, R.H. Bradbury, R. Cooke, J. Erlandson, J.A. Estes, T.P. Hughes, S. Kidwell, C.B. Lange, H.S. Lenihan, J.M. Pandolfi, C.H. Peterson, R.S. Steneck, M.J. Tegner, and R.R. Warner. 2001. Historical overfishing and the recent collapse of coastal ecosystems. Science 293(5530):629–37.

Jaykus, L.-A., M. Woolridge, J.M. Frank, M. Miraglia, A. McQuatters-Gollop, C. Tirado, R. Clarke, and M. Friel. 2008. Climate change: Implications for food safety. FAO, Rome.

Kapuscinski, A.R., and D.J. Brister. 2000. In: Black, K.D., editor, Environmental impacts of aquaculture. Sheffield Academic Press, Sheffield, UK.

Kennedy, E.T. 2006. Evidence for nutritional benefits in prolonging wellness. Am. J. Clin. Nutr. 83(suppl):410S–414S.

Kris-Etherton, P.M. 2002. Fish consumption, fish oil, omega-3 fatty acids, and cardiovascular disease. Circulation 106(21):2747–2757.

Kroeker, K.J., R.L. Kordas, R. Crim, I.E. Hendriks, L. Ramajo, G.S. Singh, C.M. Duarte, and J.-P. Gattuso. 2013. Impacts of ocean acidification on marine organisms: Quantifying sensitivities and interaction with warming. Glob. Chang. Biol. 19(6):1884–96.

Kumar, N., P.S. Minhas, K. Ambasankar, K.K. Krishnani, and R.S. Rana. 2014. Dietary lecithin potentiates thermal tolerance and cellular stress protection of milk fish (Chanos Chanos) reared under low dose endosulfan-induced stress. J. Therm. Biol. 46:40–6.

Kummu, M., H. de Moel, P.J. Ward, and O. Varis. 2011. How close do we live to water? A global analysis of population distance to freshwater bodies. PLoS One 6(6):e20578.

Last, P.R., W.T. White, D.C. Gledhill, A.J. Hobday, R. Brown, G.J. Edgar, and G. Pecl. 2011. Long-term shifts in abundance and distribution of a temperate fish fauna: A response to climate change and fishing practices. Glob. Ecol. Biogeogr. 20(1):58–72.

Lawson, T.B. 1995. Fundamentals of aquacultural engineering. Springer Science & Business Media, New York.

Lehodey, P., J. Alheit, M. Barange, T. Baumgartner, G. Beaugrand, K. Drinkwater, J.-M. Fromentin, S.R. Hare, G. Ottersen, R.I. Perry, C. Roy, C.D. van der Lingen, and F. Werner. 2006. Climate variability, fish, and fisheries. J. Clim. 19(20):5009–5030.

Lind, C.E., R.W. Ponzoni, N.H. Nguyen, and H.L. Khaw. 2012. Selective breeding in fish and conservation of genetic resources for aquaculture. Reprod. Domest. Anim. 47 Suppl 4:255–63.

Lynch, A.J., W.W. Taylor, T.D. Beard, and B.M. Lofgren. 2015a. Climate change projections for lake whitefish (Coregonus clupeaformis) recruitment in the 1836 Treaty Waters of the Upper Great Lakes. J. Great Lakes Res. 41(2):415–422.

Lynch, A.J., E. Varela-Acevedo, and W.W. Taylor. 2015b. The need for decision-support tools for a changing climate: Application to inland fisheries management. Fish. Manag. Ecol. 22:14–24.

Magalhães, M.F., P. Beja, I.J. Schlosser, and M.J. Collares-Pereira. 2007. Effects of multi-year droughts on fish assemblages of seasonally drying Mediterranean streams. Freshw. Biol. 52(8):1494–1510. doi:10.1111/j.1365-2427.2007.01781.x

Millen, B., A.H. Lichtenstein, S. Abrams, L. Adams-Campbell, C. Anderson, J.T. Brenna, W. Campbell, S. Clinton, G. Foster, F. Hu, M. Nelson, M. Neuhouser, R. Pérez-Escamilla, A.M. Siega-Riz, and M. Story. 2015. Scientific Report of the 2015 Dietary Guidelines Advisory Committee. Off. of Disease Prev. and Health Promotion Washington, DC.

Morgan, I.J., D.G. McDonald, and C.M. Wood. 2001. The cost of living for freshwater fish in a warmer, more polluted world. Glob. Chang. Biol. 7(4):345–355.

Mozaffarian, D., and E.B. Rimm. 2006. Fish intake, contaminants, and human health: Evaluating the risks and the benefits. J. Am. Med. Assoc. 296(15):1885–99.

Munday, P.L., D.L. Dixson, M.I. McCormick, M. Meekan, M.C.O. Ferrari, and D.P. Chivers. 2010. Replenishment of fish populations is threatened by ocean acidification. Proc. Natl. Acad. Sci. U. S. A. 107(29):12930–12934.

Muñoz, N.J., A.P. Farrell, J.W. Heath, and B.D. Neff. 2014. Adaptive potential of a Pacific salmon challenged by climate change. Nat. Clim. Chang. 5(2):163–166.

Neiland, A., and C. Bene. 2006. Review of river fisheries valuation in West and Central Africa. Int. Water Manage. Inst., Colombo, Sri Lanka.

Nesheim, M.C., and A.L. Yaktine, editors. 2007. Seafood choices: Balancing benefits and risks. The Natl. Acad. Press, Washington, DC.

Nissling, A., and L. Westin. 1997. Salinity requirements for successful spawning of Baltic and Belt Sea cod and the potential for cod stock interactions in the Baltic Sea. Mar. Ecol. Prog. Ser. 152:261–271.

Pernet, F., F. Lagarde, N. Jeannée, G. Daigle, J. Barret, P. Le Gall, C. Quere, and E.R. D'Orbcastel. 2014. Spatial and temporal dynamics of mass mortalities in oysters is influenced by energetic reserves and food quality. PLoS One 9(2):e88469.

Perry, A.L., P.J. Low, J.R. Ellis, and J.D. Reynolds. 2005. Climate change and distribution shifts in marine fishes. Science 308(5730):1912–1915.

Person-Le Ruyet, J., L. Labbé, N. Le Bayon, A. Sévère, A. Le Roux, H. Le Delliou, and L. Quéméner. 2008. Combined effects of water quality and stocking density on welfare and growth of rainbow trout (Oncorhynchus mykiss). Aquat. Living Resour. 21(2):185–195.

Pörtner, H.-O. 2008. Ecosystem effects of ocean acidification in times of ocean warming: A physiologists view. Mar. Ecol. Prog. Ser.:203–217.

Pörtner, H.-O. 2012. Integrating climate-related stressor effect on marine organisms: Unifying principles linking molecule to ecosystem-level changes. Mar. Ecol. Prog. Ser. 470:273–290.

Pörtner, H.O., and R. Knust. 2007. Climate change affects marine fishes through the oxygen limitation of thermal tolerance. Science 315(5808):95–97.

Pörtner, H.O., and M.A. Peck. 2010. Climate change effects on fishes and fisheries: Towards a cause-and-effect understanding. J. Fish Biol. 77(8):1745–1779.

Pörtner, H.O., P.M. Schulte, C.M. Wood, and F. Schiemer. 2010. Niche dimensions in fishes: An integrative view. Physiol. Biochem. Zool. 83(5):808–826.

Powell, S. 2015. Millions lost after warm seas kill salmon. http://www.stuff.co.nz/marlborough-express/news/67314620/Millions-lost-after-warm-seas-kill-salmon (accessed 13 Apr. 2015).

Rick, T.C., and J. Erlandson, editors. 2008. Human impacts on ancient marine ecosystems: A global perspective. Univ. of Calif. Press, Berkeley, CA.

Ronald, N., B. Gladys, and E. Gasper. 2014. The effects of stocking density on the growth and survival of Nile tilapia (Oreochromis niloticus) fry at Son fish farm, Uganda. J. Aquac. Res. Dev. 5(2). doi:10.4172/2155-9546.1000221

Ross, L.G., T.C. Telfer, L. Falconer, D. Soto, and J. Aguilar-Majarrez, editors. 2013. Site selection and carrying capacities for inland and coastal aquaculture. In: Site selection and carrying capacities for inland and coastal aquaculture, FAO Fisheries and Aquaculture Proceedings No. 21. FAO, Rome. p. 46.

Schwab, U., L. Lauritzen, T. Tholstrup, T. Haldorssoni, U. Riserus, M. Uusitupa, and W. Becker. 2014. Effect of the amount and type of dietary fat on cardiometabolic risk factors and risk of developing type 2 diabetes, cardiovascular diseases, and cancer: A systematic review. Food Nutr. Res. 58:25145. http://dx.doi.org/10.3402/fnr.v58.25145.

Shuter, B.J., and J.R. Post. 1990. Climate, population viability, and the zoogeography of temperate fishes. Trans. Am. Fish. Soc. 119(2):314–336.

Stenseth, N.C., A. Mysterud, G. Ottersen, J.W. Hurrell, K.-S. Chan, and M. Lima. 2002. Ecological effects of climate fluctuations. Science 297(5585):1292–6129.

Tacon, A.G.J., and M. Metian. 2008. Global overview on the use of fish meal and fish oil in industrially compounded aquafeeds: Trends and future prospects. Aquaculture 285(1-4):146–158.

Tejpal, C.S., E.B. Sumitha, A.K. Pal, H. Shivananda Murthy, N.P. Sahu, and G.M. Siddaiah. 2014. Effect of dietary supplementation of l-tryptophan on thermal tolerance and oxygen consumption rate in Cirrhinus mrigala fingerlings under varied stocking density. J. Therm. Biol. 41:59–64.

Tirado, M.C., R. Clarke, L.A. Jaykus, A. McQuatters-Gollop, and J.M. Frank. 2010. Climate change and food safety: A review. Food Res. Int. 43(7):1745–1765. doi:10.1016/j.foodres.2010.07.003

Troell, M., R.L. Naylor, M. Metian, M. Beveridge, P.H. Tyedmers, C. Folke, K.J. Arrow, S. Barrett, A.-S. Crepin, P.R. Ehrlich, A. Gren, N. Kautsky, S.A. Levin, K. Nyborg, H. Osterblom, S. Polasky, M. Scheffer, B.H. Walker, T. Xepapadeas, and A. de Zeeuw. 2014. Does aquaculture add resilience to the global food system? Proc. Natl. Acad. Sci. 111(37):13257–13263.

Varraso, R., S.E. Chiuve, T.T. Fung, R.G. Barr, F.B. Hu, W.C. Willett, and C.A. Camargo. 2015. Alternate Healthy Eating Index 2010 and risk of chronic obstructive pulmonary disease among US women and men: Prospective study. BMJ 350:H286.

Vezzulli, L., R.R. Colwell, and C. Pruzzo. 2013. Ocean warming and spread of pathogenic vibrios in the aquatic environment. Microb. Ecol. 65(4):817–825.

Walsh, M.L., and K. Hanafusa. 2013. Two years after: Post-tsunami recovery of fisheries and aquaculture in northeastern Japan. World Aquac. Mag. 44:20–25.

WorldFish Center. 2007. The threat to fisheries and aquaculture from climate change. Penang, Malaysia. http://pubs.iclarm.net/resource_centre/ClimateChange2.pdf (accessed 5 Nov. 2015).

Youngson, A.F., A. Dosdat, M. Saroglia, and W.C. Jordon. 2001. Genetic interactions between marine finfish species in European aquaculture and wild conspecifics. J. Appl. Ichthyol. 17(4):153–162.

Climate Change Effects on Arthropod Diversity and its Implications for Pest Management and Sustainable Crop Production

Hari C. Sharma and Mukesh K. Dhillon*

Abstract

Current estimates of climate change indicate an increase in global mean annual temperature, CO_2 concentration, and erratic and non-uniform distribution in annual precipitation and fluctuation in crop productivity, which might result in unusual effects on agro-ecosystems and lead to changes in cropping systems, crop diversity, and their interactions with biotic stress factors. Climate change and global warming will also influence arthropod diversity, geographical distribution, population dynamics, herbivore-plant interactions, activity and abundance of natural enemies, emergence of new biotypes of insect pests, and crop losses associated with insect pests. Changes in geographical distribution, diversity and abundance of insect pests will also be influenced by changes in the cropping patterns triggered by climate change. Major insect pests such as cereal stem borers (*Chilo, Sesamia,* and *Scirpophaga*), the pod borers (*Helicoverpa, Maruca,* and *Spodoptera*), aphids, and white flies may move to temperate regions, leading to greater damage in cereals, grain legumes, vegetables, and fruit crops. Global warming will also reduce the effectiveness of pest-resistant cultivars, transgenic plants, natural enemies, biopesticides, and synthetic chemicals for pest management. As a result, economic relationships between the costs and benefits of pest control will undergo a major change. Therefore, there is a need to generate information on the likely effects of climate change on insect pests to develop robust technologies that will be effective and economic in future under global warming and climate change.

Farming communities in the semiarid tropics (SAT) are poor and cannot afford increased crop losses due to insect pests as a result of global warming and climate change. Insect pests have been estimated to cause over $243.4 billion (USD) losses annually in eight major field crops (Oerke and Dehne, 2004). The extent of losses due to insect pests are likely to increase as a result of climate-induced changes in pest distribution, incidence, increased rate of population growth, and pest outbreaks (Sharma, 2014). Host-plant resistance, natural plant products, biopesticides, natural enemies, agronomic practices, and rational application of insecticides offer

---

Abbreviations: NPV, nuclear polyhedrosis virus; SAT, semiarid tropics; TNC, total nonstructural carbohydrates.

H.C. Sharma, YSP University of Horticulture & Forestry, Nauni, 173,230, Solan, Himachal Pradesh, India; M.K. Dhillon, Division of Entomology, ICAR-Indian Agricultural Research Institute (IARI), New Delhi 110012, India. *Corresponding author: mukeshdhillon@rediffmail.com.

doi:10.2134/agronmonogr60.2016.0019

© ASA, CSSA, and SSSA, 5585 Guilford Road, Madison, WI 53711, USA.
Agroclimatology: Linking Agriculture to Climate, Agronomy Monograph 60.
Jerry L. Hatfield, Mannava V.K. Sivakumar, John H. Prueger, editors.

a potentially viable option for integrated pest management. However, the relative efficacy of many of these control measures is likely to change as a result of global warming (Parmesan et al., 2013; Singh and Reddy, 2013). Changes in extent and pattern of precipitation are possibly of greater importance for agriculture than changes in temperature alone, especially in regions where rainfall may be a limiting factor for crop production (Parry, 1990). Amount of precipitation may increase in certain regions as a result of intensification of hydrological cycle (Rowntree, 1990), which may have a major bearing on cultivars grown and the cropping patterns, and influence the composition of arthropods in different agro-ecosystems (Porter et al., 1991; Sutherst, 1991). Greater mobility, increased population growth rate, and changes in geographical distribution of pests will be some of the first visible effects of climate change. Geographical distribution of many tropical and subtropical insect pests will extend, along with shifts in production areas of their host plants (Sharma, 2014; Gonzalez and Bell, 2013). Increase in length of crop growing season in subtropical and temperate regions will increase abundance and severity of insect pests, and reduce the developmental period resulting in more generations in a cropping season or year, and hence, increased losses. Distribution and relative abundance of some insect pests in the temperate regions, which are vulnerable to high temperatures, may decrease (Gomulkiewicz and Shaw, 2013). These species may find suitable alternative habitats at greater latitudes. Many insect species may also have their diapause strategies disrupted as the linkages between temperatures and moisture regimes and the daylengths are altered (Dhillon et al., 2017). Genetic variation and multi-factor inheritance of innate recognition of environmental signals by insect pests may mean that many species can adapt readily to disruption caused by climate change.

The Intergovernmental Panel for Climate Change (IPCC) has provided evidence of accelerated global warming as an average temperature in the last 100 to 150 yr, which has increased by 0.76 °C (IPCC, 2007). Doubling of CO_2 will take place between 2025 and 2070, depending on the level of greenhouse gas emissions. Mean annual temperature changes between 3 and 6 °C are estimated to occur across Europe, with the greatest increase occurring at high latitudes (Patz, 2013). The effects of change in abiotic factors such as CO_2, O_3 and temperature on agricultural systems in mid- to high-latitude regions are predicted to be less severe than in low-latitude regions, and may even be beneficial (Cannon, 1998). Climate change also has implications for distribution and abundance of arthropod species and communities. Temperature increases associated with climatic change could result in: (i) extension of geographical range of insect pests; (ii) increased over-wintering; (iii) changes in population growth rates; (iv) increased number of generations; (v) extension of development season; (vi) changes in crop–pest synchrony; (vii) changes in interspecific interactions; (vii) increased risk of invasion by migrant pests; and (ix) introduction of alternative hosts and 'green bridges' or over-wintering hosts (Porter et al., 1991). These changes will have a major bearing on food security, particularly in developing countries in the SAT where the need to increase and sustain food production is most urgent. As a result, as and when the damage to crops does increase, it will result in a significant increase in economic losses. Long-term monitoring of population levels and insect behavior, particularly in identifiably sensitive regions, may provide some of the first indications of a biological response to climate change. In addition, it will be important to keep ahead of undesirable pest adaptations. For this reason, it is critical that possible climatic changes should be considered in research to develop pest management systems that are effective under global warming and climate change.

Climate Change Effects on Geographic Distribution of Arthropods

The effects of climate change on living organisms are recognized from the level of individual species to communities (Hickling et al., 2006). Climate change may have positive, negative, or no impact on individual insect species. Very low and very high temperatures are critical in determining geographical distribution of insect pests (Hill, 1987). For species whose geographical distribution is limited by low temperatures, increase in temperature may result in a greater ability to overwinter at higher latitudes and may increase the pest's chances of extending its geographical range (EPA, 1989; Hill and Dymock, 1989; Singer and McBride, 2012). Spatial shifts in the distribution will also occur under changing climatic conditions (Parry and Carter, 1989). However, insect species with wide geographical range will tend to be less affected. Some of the ecological issues facing the world today are the climate change and invasive species, which have largely been viewed independently. To prevent large-scale economic and environmental damage due to pest invasions, the first step is to understand the factors affecting climate induced pest invasions. Life history, development rate, and climate change could be useful indicators of pest pressure, and the communities into which the pest invaders will arrive. Current predictions of range and abundances of insect pests, following global warming, tend to be made solely on the basis of physiology and ignore the dispersal and interactions between species. Global warming effects on insect abundance and dispersal cannot be easily derived from the physiology of individual insect species, which might increase in abundance at physiologically non-optimum temperatures, raising the concern that rare species currently not regarded as pests may become economically important with global warming (Jenkinson et al., 1996; Jeltsch et al., 2013). Univoltine temperate species with long maturation periods and low vagility could face regional extinction (Rooney and Smith, 1996). Higher mean temperatures will increase pest developmental rate, fecundity, frequency of pest outbreaks, and distribution (Lawton, 1991). Pest outbreaks attributed to climate change are already occurring, for example bean leaf beetles are increasing in abundance in the United States (UNEP, 2006), and expansion of soybean cyst nematode and corn gray leaf blight has been observed (Rosenzweig and Solecki, 2000). More recently, outbreaks of the rice brown planthopper have occurred in Yangtze River Delta of China (Hu et al., 2014).

Simulation Modeling for Predicting Effects of Climate Change

Meta-analysis has shown that climate-induced community changes are likely to increase niche-availability in future to exacerbate the problem of invasive species (Ward and Masters, 2007). Therefore, it is timely and important to link further research with these important ecological threats. Field and laboratory observation based prediction model can be used to forecast the effect of global warming on the regional distribution of typical temperate insect pests. Paleoecologically- based models predicted that with global warming, animals in temperate regions would migrate poleward to remain within their temperature tolerance ranges (Rooney and Smith, 1996; La Sorte and Jetz, 2010; Buckley et al., 2013). However, the likely effects of global warming on invertebrates are of greater concern because of their critical role in ecosystem structure and function, as migration poses a problem for many species due to limited dispersal abilities. Most pest population prediction models and climatic variables operate at a different spatial and temporal scale than do the global climate models. Improvements in methodology are necessary to realistically assess the impact of climate change on a global scale. In cool subtropics such as those of Japan and northern China, elevation of temperature above normal will result in severe pest outbreaks (Luo et al., 1998), while conditions in the humid tropics will result in significantly more severe disease and pest epidemics.

CLIMEX-modeling software has been developed to predict the future distribution ranges of nun moth, *Lymantria monacha* L. and the gypsy moth, *L. dispar* L. (Vanhanen et al., 2007). The software calculates an ecoclimatic index based on the life cycle requirements of a species, and representing the probability of a viable population to exist at a certain location. Simulations with temperature increases of 1.4, 3.6, and 5.8 °C corresponded well with current distributions of these species. The climate warming scenarios shifted the northern boundary of the distribution for both of these species to the north by 500 to 700 km. The southern edge of the ranges retracted northwards by 100 to 900 km. These being very serious pest species, the shift in their distribution pose a potential threat to silviculture, and therefore, should be considered in planning of forest pest management practices. Studies on the European maize stem borer, *Ostrinia nubilalis* (Hübner) in Europe indicated that with the estimated climatic change, the maize stem borer would shift northward up to 1220 km, with an additional generation expected in nearly all regions where it was known to occur in 1991 (Porter et al., 1991). Future insect population dynamics in response to climate change might be predicted by understanding the mechanisms involved, based on long-term data. Yamamura et al. (2006) used such a system for rice stem borer, *Chilo suppressalis* (Walker), green rice leafhopper, *Nephotettix cincticeps* (Uhler), and small brown planthopper, *Laodelphax striatellus* (Fallén) using 50-yr dynamics of annual light-trap catches of these insects from rice fields. They analyzed long-term population dynamics using state-space model to understand dynamics of populations and the influence of the past changes in the environment, and suggested that the number of light-trap catches of *C. suppressalis* and *N. cincticeps* in summer increased with an increase in temperature in the previous winter, and there was carryover effect of temperature to next year. Future research should consider insect herbivore phenotypic and genotypic flexibility, their responses to global change parameters operating in concert, and the awareness that some patterns may only become apparent in the longer term (Bale et al., 2002). Every year in Asia during rainy season in June and July, a strong southwesterly air stream develops 200 to 300 km South of the Bai-u front. The air stream known as the low-level jet stream (LLJ), has maximum wind speeds at 1000 to 3000 m above sea level (Watanabe and Seino, 1991). The LLJ wind stream transports insects, fungal spores, and weed seeds from tropical to temperate regions. If the front shifts northward, more rice insect pests could reach the temperate rice-growing areas, resulting in serious outbreaks over wider areas. Heavy pesticide use, altered wind patterns, and higher temperatures in 2005 caused the brown planthopper, *Nilaparvata lugens* (Stål) to outbreak in China, and resulted loss of about 1.88 million t (Hu et al., 2014). At the moment, summer typhoons are relatively predictable, but little is known about how climate change may alter wind patterns, which may increase the abundance of insect pests carried by summer winds. The disturbance caused by insects may substantially change Canadian forests due to global warming, especially in case of insects whose distribution depend largely on climate (Fleming and Candau, 1998). For example, spruce budworm, *Choristoneura fumiferana* (Clem.) on *Picea glauca* (Moench) Voss and *Abies balsamea* (L.) Mill., might react to global warming (Fleming and Candau, 1998). Likelihood of wildfires increases after insect attack, which could be unpredictable because of uncertain insect disturbance patterns. Insect disturbance may also influence biodiversity indirectly.

Migratory Insect Pests and Climate Change

Migrant pests are expected to respond more quickly to climate change than plants, and the distribution of insect pests will be greatly influenced by changes in the range of host crops because the distribution of a pest is also dependent on the availability of a host to colonize (Chapman et al., 2011; Jeltsch et al., 2013). However, whether or not a pest would move with a crop into a new area would depend on other environmental conditions such as presence of overwintering sites, soil type, and moisture, for example, populations of the corn earworm, *Helicoverpa zea* (Boddie) in the North America could reach higher levels, leading to greater damage in maize and other crops (EPA, 1989). Range expansions (including biotic as well as abiotic factors), and the removal of edge effects, could result in increased abundance of species near the northern limits of their ranges in the UK (Cannon, 1998). Distribution of native European butterfly populations has been observed to extend into northern ranges and recede in southern ranges, and similar changes may occur in case of forest insect pests (Evans et al., 2002; Jeltsch et al., 2013). Under a warmer climate, exotic pests such as southern pine beetle, *Dendroctonus frontalis* Zimmermann in the United States, could establish populations in Europe and may expose U.K. forests to damage by this pest and other bark beetles such as *Ips typographus* (L.), which is present in some parts of Europe, but not in the U.K., could become a serious problem (Evans et al., 2002). Therefore, combined effects of increased global trafficking of timber and wood products and climate change are likely to result in exotic pests such as Asian longhorn beetle, *Anoplophora glabripennis* (Motschulsky) to become more prevalent. Global warming will lead to early infestation of soybean by *H. zea*, with potentially significant levels of economic loss (EPA, 1989). More importantly, it is difficult to predict pest outbreaks related to environmental and climate changes. Rising temperatures are likely to result in the availability of new niche to insect pests. There will be increased dispersal of airborne species in response to atmospheric disturbances. Many insects are migratory, and therefore, may be well adapted to exploit new opportunities by moving rapidly into areas favorable to such species.

Climate Change and Insect Multiplication Rates

Climate change might alter population dynamics of insects, especially for insects whose occurrence in time and space is severely limited by climatic factors, which may increase the uncertainties associated with planning control measures, hazard rating models, and projections for the sustainability of the agro-ecosystems (Karuppaiah and Sujayanad, 2012). Potential changes in damage patterns can affect ecosystem resilience due to changes in rates of various processes in nutrient and biogeochemical cycling as a result of insect damage. Fleming and Volney (1995) emphasized that natural selection due to change in *C. fumiferana* life cycle,

Table 1. Likely effects of climate change on insect pest management practices and possible modifications to reduce vulnerability to climate change.

Climatic effect	Effect on pests	Pest management practices	Possible modification or new methods (alternative)
Elevated temperature	Increase in insect populations	Insecticides Resistant varieties	Crop diversification. Develop varieties with stable resistance across environments.
	Disruption in biological control	Biological control	Developing natural enemies. Develop refugia. Introduce entomopathogenic natural enemies tolerant to high temperatures.
Increased frequency of drought	Increase in crop susceptibility to insect pests	Insecticides	Enhance biological control by increasing habitat biodiversity.
		Resistant varieties	Develop varieties that are resistant across environments.
Change in wind patterns	Increase in migratory pests	Insecticides	Increase system diversity.
	Increase in alien invasive species.	Insecticides	Introduce new natural enemies.
Increased rainfall	Increase in insect incidences	Resistant varieties Insecticides	Increase genetic biodiversity.

threshold effects, and equilibrium for forest–pest systems might respond to climate change. Temperature is the dominant abiotic factor, which directly affects development, survival, and abundance of insects (Matter et al., 2011). However, other abiotic factors such as CO_2 or ultraviolet light, and precipitation have been largely neglected in current research on climate change, which might change the pest population dynamics (Bale et al., 2002; Sun et al., 2011a). Insect pests will generally become more abundant with an increase in temperature through range extensions, phenological changes, increased development rate, migration, and overwintering (Cannon, 1998). A critical role of temperature has also been observed in induction and termination of diapause in *Chilo partellus* (Swinhoe), wherein a particular range of temperature decides the type of diapause, survival, and its duration (Dhillon and Hasan, 2017). Further, these studies also showed that the development in diapausing larvae of *C. partellus* is a linear function of temperature, and the relationship between temperature and development rate of diapausing larvae of *C. partellus* was nonlinear in the temperature range 10 to 38 °C (Dhillon and Hasan, 2017). Continuous rise in atmospheric CO_2 will affect pest species with varying degrees of response, and the consequent effects on these species would be strongly mediated via the host, for example, under elevated CO_2, aphids may become more serious pests (Cannon, 1998). Elevated CO_2 prolong the development of *Helicoverpa armigera* (Hübner), by decreasing the foliar nitrogen of host plants. Elevated CO_2 negatively affect *H. armigera* larval survival, larval weight, larval period, pupation, and adult emergence, but positive effect on pupal weight, pupal period, and fecundity under laboratory conditions (Akbar et al., 2015). Elevated CO_2 and temperature increased food consumption and metabolism of larvae by enhancing the activity of midgut proteases, carbohydrases (amylase and cellulase), and mitochondrial enzymes. Elevated CO_2 and global warming will affect insect growth and development, which will change the interactions between the insect pests and their crop hosts. However, in contrast, the phloem-sucking insects such as aphids and whitefly had species-specific responses to elevated CO_2 because of complex interactions that occur in the phloem sieve elements of the plants (Sun et al., 2011a). The main effect of temperature in temperate regions will be winter survival. Higher temperatures would extend the summer season resulting an increase in thermal budget for growth and reproduction of insect pests (Bale et al., 2002; Matter et al., 2011). The predicted changes due to climate change have been illustrated in Table 1. Warmer temperatures may increase the number of generations during the season (CCC, 2007). Climate change can lead to emergence of preexisting pests as major pests or provide climatic conditions required for introduced pests to develop and survive.

Warmer geographic regions and warmer periods in paleontological history have been associated with insect herbivory (Bale et al., 2002). Climate influences insect life span, growth rate, inter-specific interaction, and availability of host plants. For example, higher winter temperature increases the abundance of stripped stem borer, *C. suppressalis* and green leafhopper, *Nephotettix virescens* (Distant) in rice (Coley and Barone, 1996), while high summer temperatures between 1999 and 2001 have been found associated with rice bug, *Leptocorisa oratorius* (F.) outbreaks (Yamamura et al., 2006). *Spodoptera exempta* (Walker) outbreaks have been reported to be negatively associated with rainfall of the proceeding six to eight months in Africa (Haggis, 1996). However, positive spatial and temporal association has been observed between rainstorms and *S. exempta* outbreaks (Tucker and Pedgley, 1983; Rose et al., 2000). Heavy rainfall followed by drought during July to October often leads to Oriental armyworm, *Mythimna separata* (Walker) outbreaks in south central India (Sharma et al., 2002). Rainfall promotes the growth of grasses, including cultivated crops, which may lead to an increase in local populations. Weather conditions may also result in aggregation of *M. separata* moths, and then to heavy infestations (Sharma et al., 2002).

Effect of Global Warming on Pest Emergence, Development, and Generation Turnover

Global warming will lead to earlier infestation by *H. zea* (Boddie) in North America (EPA, 1989), and *H. armigera* (Hübner) in North India (Sharma, 2014), resulting in increased crop loss. Rising temperatures are likely to result in availability of new niches for insect pests. Temperature has a strong influence on the viability and incubation period of *H. armigera* eggs, which can be predicted based on day degrees required for egg hatching (Dhillon and Sharma, 2007). Decreased egg hatching and reduced larval survival because of lower humidity and lower or higher temperatures also result in low *M. separata* population densities during the postrainy seasons (Sharma et al., 2002). Increase in winter temperature will result in increased winter survival of *Elatobium abietinum* (Walker), leading to intense tree defoliation, resulting in decline of Sitka spruce, *Picea sitchensis* (Rafinesque) productivity (Evans et al., 2002). An increase of 3 °C in the mean daily temperature would cause the carrot fly, *Delia radicum* (L.), to become active a month earlier than at present (Collier et al., 1991). Temperature increases of 5 to 10 °C would result in the completion of four generations each year, which would necessitate adoption of new control strategies. An increase of 2 °C will reduce the time from birth to reproductive maturity of the bird cherry aphid, *Rhopalosiphum padi* (Linn.), by varying amounts depending on the mean temperature (Morgan, 1996). An increase of 1 and 2 °C in daily maxima and minima will make codling moth, *Cydia pomonella* (Linn.) active about 10 to 20 d earlier than expected. Daily increase in temperatures by 1 °C would result 10-d advancement in adult moth activity and 25 d advancement in larval emergence of the summer fruit tortrix, *Adoxophyes orana* (Fischer von Rosslertamm), while temperature increases of 2 °C would bring both adult activity and larval emergence forward by 10 d and the pest would complete three generations. Overwintering of pests will increase as a result of climate change, producing larger spring populations as a base for a build-up in numbers in the following season. These may be more vulnerable to parasitoids and predators if the latter also overwinter more readily, for example, diamond back moth, *Plutella xylostella* (L.) overwintered in 1991 to 1992 in Alberta (Dosdall, 1994). If overwintering becomes common in western Canada, the pest status of this insect could increase dramatically. Insect populations during outbreaks can cause tree mortality over vast areas in Canada's forest (Fleming and Volney, 1995).

Rapid changes in climate may impose a strong selection pressure on organisms in response to climate change, as has been observed in case of leaf beetle, *Chrysomela aeneicollis* Schaeffer (Rank and Dahlhoff, 2002). There was 11% increase in the *Pgi*-1 allele frequency in the beetles collected before 1988, which might be because of induced expression of a 70 kDa heat shock protein (HSP) at lower temperatures. Females expressed higher levels of HSP70 than males after exposure to heat, and recovery by female *Pgi* 1–1 homozygotes after exposure to cold (–5 °C) was significantly better than the beetle populations collected during 1996, suggesting that the cooler climate of the mid-1990s might have caused an increase in frequency of the *Pgi*-1 allele, due to a more robust physiological response to cold by those genotypes (Rank and Dahlhoff, 2002).

Climate Change and Insect-Plant Interactions

Problems with new agricultural pests will increase if climatic changes favor the introduction of crop cultivars susceptible to insect pests. Introduction of new crops and cultivars to take advantage of new environmental conditions is one of the adaptive methods suggested as a possible response to climatic changes (Parry and Carter, 1989). The most desirable form of insect resistance is the one that is stable across locations and seasons. However, climatic and edaphic factors influence the level and nature of resistance to insect pests (Kogan, 1982). Inherited characters, especially those involving physiological characteristics, are influenced by the environmental factors (Castagneyrol et al., 2014). Such an interaction in insect–plant relationship can be studied by testing a set of diverse sources of resistance across locations and seasons. Resistance to sorghum midge breaks down under high humidity and moderate temperatures in Kenya (Sharma et al., 1999). There are indications that stem rot (*Sclerotium rolfsii* Sacc.) resistance in groundnut is temperature dependent (Pande et al., 1994). Plant resistance to pests is one of the most environment friendly components of pest management. However, climate change may alter the interactions between the pests and their host plants. Hence, development of cultivars with stable resistance to insect pests would certainly provide an effective approach in Integrated Pest Management. The mechanisms governing escalation in herbivory are elusive and represent a complex interplay of temperature and CO_2 effects on both insects and the host plants (DeLucia et al., 2008). The insect density in an area would increase as we move from cold (northern latitudes) to the equatorial warm regions, and from high mountain peaks to mountain bases (Gaston and Williams, 1996; Wilf and Labandeira, 1999). Increase in global temperature, atmospheric CO_2, and the length of the dry season are likely to have ramifications for plant–herbivore interactions in the tropics (Coley and Markham, 1998; Guo et al., 2013). There will be an increase in the impact of insect pests, which benefit from reduced host defenses as a result of stress caused by a lack of adaptation to suboptimal climatic conditions.

Effect of Temperature on Expression of Resistance to Insect Pests

Surface temperatures are expected to increase by several degrees in the next century, affecting both growth and flowering of plants (McKone et al., 1998). Temperature is one of the most important physical factors of the environment, affecting the behavioral and physiological interactions of insects and plants (Benedict and Hatfield, 1988). Temperature induced stress can cause changes in plant physiology, resulting in changes in the levels of biochemical and morphological defenses or nutritional quality of the host–plant, resulting in a dramatic increase in rate of development of pest populations (White, 1984). Temperature not only affects plant growth, but also the biology, behavior, and population dynamics of insect pests (Tingey and Singh, 1980). In general, low temperatures have a negative effect on plant resistance to insects (Kogan, 1982). Differences between resistant and susceptible genotypes of sorghum to greenbug increase with an increase in temperature (Schweissing and Wilde, 1978). In alfalfa, the levels of resistance to pea aphid, *Acyrthosiphon pisum* (Harris) and alfalfa aphid *Therioaphis maculata* (Buckton) are enhanced at higher temperatures (Kogan, 1982). Sorghum midge-resistant genotypes suffer greater damage at lower temperatures (Sharma et al., 1999). Increase in temperature may disrupt the synchrony in phenology between the plants and the insects within a food chain, for example, date of winter moth, *Operophtera brumata* (L.) egg hatch has advanced over bud burst in pedunculate oak, *Quercus robur* L. over the past two decades (van Asch et al., 2007). Disruption in synchrony will lead to genetic selection in the species in question, but the prerequisite for such a change is to have sufficient genetic variation. van Asch et al. (2007) predicted a rapid response to selection, leading to a restoration of synchrony of egg hatch with *Q. robur* bud opening, which shows a clear potential of winter moth to adapt rapidly to environmental change. As a result of air pollution and climate change, a new type of forest decline has appeared in Europe and North America due to diseases related to pollution, resulting in growth anomalies and general weakening, which leads to greater vulnerability to insect pests and plant pathogens (Nuorteva, 1997). Twenty-two–year data for *Chionochloa* spp. in New Zealand indicated that high temperature in summer before flowering resulted in heavy flowering and seed infestation (McKone et al., 1998). Higher temperatures below the species upper lethal limit may lead to faster development rates, resulting in a rapid increase in pest populations.

Effect of Photoperiod on Expression of Resistance to Insect Pests

Photoperiod is although dominant cue for the seasonal synchrony of temperate insects, their thermal requirements may differ at different times of the year. However, it is not necessary that these factors operate in tandem as insect herbivores show a number of distinct life history strategies to exploit plants with different growth forms, which will be differentially affected by climate warming (Bale et al., 2002). Photoperiod eventually affects the temperature, which also affects the development of both the insects and the crop plants. Photoperiod also alters the physicochemical characteristics of crop plants, and thus, influences the interaction between insects and crop plants. Failure or inability to grow certain crop plants during the off-season may be largely because of increased susceptibility to insects and diseases. Intensity and quality of light have been reported to influence biosynthesis of phenylpropanoids and anthocyanins (Carew and Krueger, 1976). Prolonged exposure to high intensity light induces susceptibility in soybean plants (otherwise resistant) to cabbage looper, *Trichoplusia ni* (Walker), by influencing the flavonoid composition of soybean leaves (Khan et al., 1986). Susceptibility in sorghum to midge, *Stenodiplosis sorghicola* (Coquillett) increases under longer daylength (Sharma et al., 1999, 2001).

Effect of Elevated CO_2 on Expression of Resistance to Insect Pests

Pest–plant interactions will change in response to the effects of increased concentration of CO_2 on plants. With enriched CO_2 in the atmosphere, as expected in the next century, many species of herbivorous insects will confront with less nutritious host plants, thus resulting in prolonged larval development and greater mortality (Coviella and Trumble, 1999). Increased atmospheric CO_2 will not only be highly species-specific, but also specific to each insect–plant system. The extent and diversity of plant damage by insects increases with an abrupt rise in atmospheric CO_2 and global temperature that occurred > 55 million years ago (Currano et al., 2008; Casteel et al., 2012; Guo et al., 2012). Many insects respond directly to CO_2 as a cue for identifying favorable oviposition sites or desirable food sources (O'Neill et al., 2011). Population growth of *Aphis glycines* Matsumura under elevated CO_2 was significantly greater than those reared under ambient CO_2 (O'Neill et al., 2011). In addition, there are indirect effects of elevated CO_2 on leaf chemistry, which affect plant growth and palatability by the insects (Bezemer and Jones, 1998; Casteel et al., 2012). A rise in CO_2 generally increases the C/N ratio of plant tissues (Lincoln et al., 1984; Coley and Markham, 1998; Ainsworth et al., 2002), reducing nutritional quality for protein-limited insects (Coley and Markham, 1998; Coviella and Trumble, 1999). Increased CO_2 may also cause a slight decrease in nitrogen-based defenses (e.g., alkaloids) and a slight increase in carbon-based defenses (e.g., tannins). As a result, insects may accelerate their food intake to compensate for reduced leaf nitrogen content (Coviella and Trumble, 1999; Kopper et al., 2001; Holton et al., 2003), although this may not always be the case (Kopper and Lindroth, 2003; Knepp et al., 2005). Some plants also change chemical composition in direct response to pest damage to make less suitable for pest growth (Rhoades, 1985). There may also be an increase in damage by some phytophagous insects, but CO_2 may also increase plant growth. For example, the Japanese beetles, *Popillia japonica* Newman, and the Mexican bean beetles, *Epilachna varivestis* Mulsant prefer to feed on soybean leaves grown under high CO_2 (Hamilton et al., 2005; Casteel et al., 2012). Elevation of CO_2 could force mean annual temperature to rise from 10.5 °C to 20.1 °C and the severity of insect damage to increase from 38% to 57%, reflecting fundamental change in insect–plant interactions.

Yield increases predicted by most CO_2 enrichment studies in many cases will be offset by yield losses caused by phytophagous insects, pathogens, and weeds (Antle et al., 2001). Fifteen studies on crop plants have shown 30% decrease in tissue nitrogen under high CO_2, reduction in tissue quality, and 80% increased feeding by insects (Lincoln et al., 1984, 1986; Osbrink et al., 1987; Coviella and Trumble, 1999). Conversely, seeds and their associated herbivores appear to be unaffected by CO_2 (Akey et al., 1988). In general, foliage feeders (e.g., Lepidoptera) tend to perform poor under high CO_2 (Osbrink et al., 1987; Akey and Kimball, 1989; Tripp et al., 1992; Boutaleb Joutei et al., 2000), whereas population of sap suckers (e.g., aphids) increases under high CO_2 (Heagle et al., 1994; Awmack et al., 1997; Bezemer and Jones, 1998), indicating feeding habit-specific insect pest outbreaks. Most of the studies on climate change performed earlier have excluded insect pests (Coakley et al., 1999). For example, soybeans grown under elevated CO_2 suffered 57% more damage from insects, attributing to increases in the levels of simple sugars in soybean leaves (Hamilton et al., 2005). Following infestation of chickpea plants with pod borer, *H. armigera*, the activities of defensive enzymes [peroxidase (POD), polyphenol oxidase (PPO), phenylalanine ammonia lyase (PAL) and tyrosine ammonia lyase (TAL)], and amounts of total phenols and condensed tannins increased with an increase in CO_2 concentration (Sharma et al., 2016). The *H. armigera*-infested plants had higher H_2O_2 content; and the amounts of oxalic and malic acids were greater at 750 than at 350 ppm CO_2. Plant damage was greater at 350 ppm than at 550 and 750 ppm CO_2.

Effect of Relative Humidity on Expression of Resistance to Insect Pests

Atmospheric humidity also influences insect-host plant interactions (Sharma et al., 1999). Temperature and photoperiod influence the growth (Sharma et al., 1999) and chemical composition of sorghum grain (Price et al., 1978, 1979), which in turn may affect the expression of resistance to sorghum midge (Sharma et al., 1999). Karaman et al. (1998) observed that reduced water availability and lower relative humidity affect *Chilo agamemnon* Blesz. activity in sugarcane. High humidity increases the ease of detection of odors, and thus, may influence host finding and non-preference mechanism of resistance to insects. Environmental conditions not only affect the survival and development of insects, but also affect physico-chemical characteristics of the plants, which in turn may affect the expression of genotypic resistance and/or susceptibility in sorghum to *Calocoris angustatus* Lethiery (Sharma and Lopez, 1994). Rhoades (1985) suggested that pest outbreaks are more likely to occur with stressed plants, because under such circumstances, the plant defense system is compromised and the resistance to pest infestation is

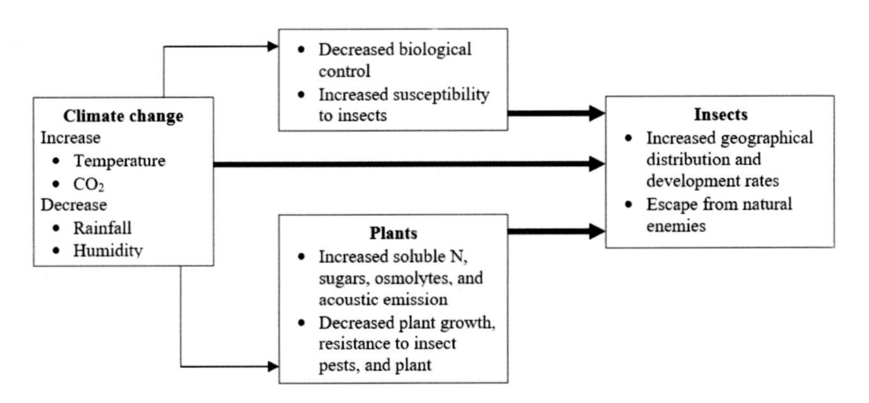

Fig. 1. Influence of drought on plants, insects, and natural enemies.

lowered. Summer rainfall resulted in an increase in vegetation cover, leading to an increase in abundance of the Auchenorrhyncha (Masters et al., 1998). However, summer drought caused a decrease in vegetation cover, but did not lead to a corresponding decrease in abundance of the Auchenorrhyncha. Egg hatch and the termination of nymphal hibernation occurred earlier in warm winters, but rate of nymphal development was unaffected. Drought-stressed plants may lead to pest and disease outbreaks because water-stressed plants tend to have increased soluble nitrogen, increased sugars, decreased growth, and decreased resistance to insect pests (Fig. 1) (Mattson and Haack, 1987). Phloem-feeding insects, such as brown planthopper and green leafhopper, particularly benefit from the higher sugars and free amino acids found in drought-stressed rice plants. Analysis of long-term records of grasshopper outbreaks in Kansas has shown that the greatest damage occurs during dry years (Smith, 1954). The 1997 to 1998 El Niño, which caused a severe drought, resulted in the loss of 140,000 ha of rice in Northern Vietnam, and the crop was further ravaged by rats and insect pests in India (NDMC, 1998). Studies on damage by pine needle scale, *Matsucoccus acalyptus* (Herbert), stem-boring moth, *Dioryctria albovittella* (Hulst.) and *Pinus edulis* Engelm. indicated that scale insect attack on needles of juvenile trees resulted in open crown needles, while stem-boring moth killed the terminal shoots of mature trees resulting in a dense crown (Classen et al., 2005). Herbivory by the scale insects reduced leaf area index of susceptible trees by 39%, while increased soil moisture and temperature beneath these trees by 35 and 26%, respectively.

Effect of Climate Change on Transgene Expression in Insect-Resistant Transgenic Plants

There is a big question mark on the effect of global warming and climate change on the stability of *Bacillus thuringiensis* (*Bt*) toxin genes in transgenic plants, and the possibilities for the breakdown of resistance are real. The *Bt* gene expression is variable, and is influenced by genetic and environmental factors. Epistatic and environmental effects on foreign gene expression could influence the stability, efficacy, and durability (Sachs et al., 1996). It is therefore important to understand the effects of climate change on the efficacy of transgenic plants in pest management. There have been some failures in insect control through the transgenic crops under extreme environmental conditions (Wang et al., 2014). The *Bt*-transgenic cotton cultivars had lower nitrogen, and higher total nonstructural carbohydrates (TNC): nitrogen ratio (C to N ratio), carbon defensive compounds, condensed tannins, and gossypol under elevated CO_2, and resulted in a significant decrease in *Bt* toxin, suggesting that expected elevated CO_2 concentrations will influence plant–herbivore interactions (Coviella et al., 2002; Chen et al., 2005). Changes in chemical components in the host plants due to increased CO_2 affected the growth, development, survival, and reproduction of *H. armigera* and *Spodoptera exigua* (Hübner). Digestibility of *H. armigera* larvae reared on *Bt*-transgenic cotton under elevated CO_2 was greater than those fed on nontransgenic cotton grown under ambient CO_2 conditions. The damage inflicted by cotton bollworm, regardless of the presence or absence of insecticidal genes, is predicted to be more serious under elevated CO_2 (Chen et al., 2005). Higher temperatures and prolonged drought lead to increased susceptibility of *Bt*-cotton to bollworms (Hilder and Boulter, 1999; Sharma and Ortiz, 2000). Survival of *H. armigera* larvae on *Bt*-transgenic plants was influenced when exposed to higher or lower temperatures, either for short periods or throughout growth, although δ-endotoxin concentration did not change (Olsen et al., 2005a,b). Cotton crop flooded with 3 to 4 cm deep water for 12 d lost resistance to insects as compared to the control plants irrigated normally (Wu et al., 1997). Under flooded conditions, the activity of superoxide dismutase in *Bt* cotton first greatly increased and then continuously dropped.

Effect of Edaphic Factors on Expression of Resistance to Insect Pests

Change in the fertility status of soils as a result of climate change may also affect expression of resistance to some insect species. For example, in some cases high levels of nutrients increase the level of insect resistance, while in others, they increase the susceptibility (Yu et al., 2013). Availability of high levels of N in the soil decreases the damage by shoot fly, *Atherigona soccata* (Rondani) and spotted stem borer, *Chilo partellus* (Swinhoe) in sorghum (Reddy and Narasimha Rao, 1975; Chand et al., 1979). Shoot fly damage also decreases in soils with high amounts of phosphorus pentoxide (P_2O_5) (Channabasavanna et al., 1969). Similarly, availability of high levels of K decreases incidence of sugarcane top borer, *Scirpophaga excerptalis* Walker. However, high levels of nitrogen lead to greater damage by cotton jassids, *Amrasca biguttula biguttula* Ishida (Purohit and Deshpande, 1992), and brown plant hopper, *N. lugens* (Stål.) (Lu et al., 2004; Yu et al., 2013). Changes in nutrient supply also affect the expression of resistance to greenbug, *Schizaphis graminum* (Rondani) in sorghum (Schweissing and Wilde, 1979).

Climate change will affect the availability of water in the soil, which indirectly affect plant growth and insect-host plant interactions. The *Aphis fabae* Scopoli has a lower reproduction rate on water stressed plants (McMurtry, 1962). High levels of water stress reduce the damage by shoot fly, *A. soccata* in sorghum (Soman et al., 1994) and lower the damage by *C. agamemnon* in sugarcane (Karaman et al., 1998). However, water stressed plants of sorghum suffer greater damage by spotted stem borer, *C. partellus* and sugarcane aphid, *Melanaphis sacchari* (Zehntner) (Faris et al., 1979; Sharma et al., 1988; Sharma and Lopez, 1991). Although, the leaf feeding by *C. partellus* larvae in sorghum was found to be increased under irrigated conditions, there was less deadheart formation, suggesting that moisture availability in the soil increases plant growth, and pushes the growing point upward at a relatively faster rate. Because of this, the larvae are not able to cause deadheart formation (Sharma et al., 2005). Moisture content of sorghum seedlings, plant biomass, and humidity have positive association with leaf feeding and survival of *C. partellus* larvae in sorghum (Sharma et al., 1997). Increase in head bug, *C. angustatus* population, and grain damage are greater during the rainy season as compared to dry season (Sharma and Lopez, 1994).

Effect of Climate Change on Insect Pest–Natural Enemy Interactions

Tritrophic interactions between host plants, insects, and the natural enemies have resulted from a long co-evolutionary process, and are specific to particular environmental conditions. Relationships between insect pests and their natural enemies will change as a result of global warming, resulting in both increases and decreases in the statuses of individual pest species. The higher trophic levels are more likely to be affected by climate change because they depend on the capacity of lower trophic levels to adapt to these changes. Changes in temperature will also alter the timing of diurnal activity patterns of different groups of insects (Young, 1982), and changes in inter-specific interactions could also alter the effectiveness of natural enemies for pest management (Hill and Dymock, 1989). Climate change might also disturb the predator–prey and parasitoid–host interactions, and even alter the balance between insect pests, natural enemies, and the host plants because of alteration in their synchrony. Rises in temperature will generally favor insect development and survival in winter (Evans et al., 2002). Therefore, quantifying the effect of climate change on the activity and effectiveness of natural enemies will be a major concern in future pest management programs. A majority of insects are benign to agroecosystems, and there is considerable evidence to suggest that this is due to population control through interspecific interactions among insect pests and their natural enemies–pathogens, parasites, and predators (Price, 1987). Oriental armyworm (*M. separata*) populations increase during extended periods of drought (detrimental to natural enemies) followed by heavy rainfall because of the adverse effect of drought on activity and abundance of its natural enemies (Sharma et al., 2002). Aphid abundance increases with an increase in CO_2 and temperature; however, parasitism rates remain unchanged at elevated CO_2 level. Climate change can also disturb predator–prey relationships, as higher trophic levels appear to be more sensitive to climatic variability (Voigt et al., 2003). Temperatures up to 25 °C will enhance the control of aphids by coccinellids (Freier and Triltsch, 1996). Parasitoids depend on the developmental rate of their insect hosts, and the exposure of these insects to either of the stressful temperatures will induce lethal and sublethal effects on the parasitoids. Temperature not only affects the rate of insect development, but also has a profound effect on fecundity and sex ratio of the parasitoids (Dhillon and Sharma, 2008, 2009). The endosymbiont bacteria associated with either parasitoids or their host insects are also suppressed by a short exposure to high temperature. Since climate change affects hosts and parasitoids differently, we expect the distribution range of each species to change with the environment, leading to rearrangements of current communities, including the adaptation of parasitoids to new host species (Hance et al., 2007). Abiotic stresses such as heat and drought may or may not increase plant susceptibility to pathogens (Garrett et al., 2006), but droughts reduce the effectiveness of biological control and weaken the host plants, causing crops to become more susceptible to insect pests (Wilson, 2008). Climate change may influence the complex system functions and change the ecosystem resilience to pest outbreaks or a systems resistance to perturbation (MEA, 2005). For example, rice ecosystems are extremely complex and rich in species, which creates the potential to provide a high level of biological control (Settle et al., 1996). Elevated CO_2 decreased the developmental time of *Lysiphlebia japonica* Ashmead, a hymenoptera parasitoid of *Aphis gossypii* Glover, but did not affect parasitism or rate of emergence (Sun et al., 2011b). However, the parasitism of caterpillars decreases as climatic variability increases (Stireman et al., 2005). Little is known regarding how climate change may affect competitive interactions between pathogens and insect pests. Therefore, the interactions between insect pests and their natural enemies need to be studied carefully to devise appropriate methods for using natural enemies in pest management.

Effect of Climate Change on the Effectiveness of Insecticides, Biopesticides, and Natural Plant Products for Pest Management

A hot and humid climate may increase the population of insect pests as well as crop protection costs as more frequent application of pesticides are required to control the pests (Chen and McCarl, 2001). There is a need for better understanding of the effect of climate change on efficacy and persistence of pesticides in the environment, and development of insect resistance to insecticides. Temperature has shown a positive effect on the efficacy of organochlorines, organophosphates, and carbamates, but a negative effect on the synthetic pyrethroids (Wang and Shen, 2007). Moreover, high temperatures and low humidity reduce the fumigation period of phosphine against *Liposcelis bostrychophila* (Badonnel) (Nayak and Collins, 2008). Early emergence of light brown apple moth, *Epiphyas postvittana* (Walker) in New Zealand may lead to problems with the timing of insecticide applications (Hill and Dymock, 1989). Increased rainfall will decrease pesticide efficacy, and thus, require more frequent applications. Leaf area consumed by *S. exigua* in cotton increased due to elevated CO_2 as a result of increased C/N ratio, which improved the efficacy of foliar applications of *Bacillus thuringiensis* (Coviella and Trumble, 2000). However, elevated CO_2 and N fertilization significantly influenced food consumption resulting in differential *Bt* toxin intake, which affected larval mortality.

Although, biopesticides and natural plant products are among the widely used eco-friendly methods of pest control, their efficiency is highly variable across environments (Abbaszadeh et al., 2011). Relative humidity is associated with high mortality in insects infected with fungal and viral pathogens; however, such effects on hosts infected with bacteria are variable, as environmental influence on host-pathogen interactions operate through the biology of the pathogen, immune response of the host, and the rate of pathogen entry into the insect host. Temperature,

Table 2. Effect of environmental factors on the efficacy of biopesticides and plant products in insect pest management.

Climatic components	Efficacy of biopesticides and plant products†				
	EPF	EPB	EPN	EPV	Botanicals
Temperature	Optimum between 25-30 °C	Optimum up to 30 °C	Optimum is 25°C; 10-15°C host searching problem; and >30 °C result in mortality	> 90 °C decrease the efficacy	Temperature has negative effect on botanicals
Relative humidity	Positively correlated with RH	Effect of RH is variable	75% = Optimum soil moisture; <75% = decrease searching ability and survival; >75% = causes oxygen depletion and mortality	Positively correlated with RH	Positively correlated with RH
Elevated atmospheric CO_2	–	Increase in efficacy	–	Increase in efficacy	Increase in efficacy
UV light	Cause degradation and mortality	Cause degradation and mortality	Cause degradation and mortality	Cause degradation and mortality	Cause degradation and mortality
Soil, water and plant pH	Variable	Variable	Variable	Variable	Variable

†EPF, Entomopathogenic fungi; EPB, Entomopathogenic bacteria; EPN, Entomopathogenic nematodes; and EPV, Entomopathogenic viruses. Reproduced from Abbaszadeh et al. (2011).

relative humidity, elevated atmospheric CO_2, UV radiation, pH, rainfall, and soil moisture, etc., are the most influential climatic factors determining the efficiency of biopesticides and plant products (Table 2). Temperature also plays a significant role in the effectiveness of biocontrol agents (Sharma, 2014). Studies on biological control of insects with *Schistocerca gregaria* (Forskål) and *Metarrhizium anisopliae* subsp. *acridium* (Metchinkoff) Sorokin, have shown that pathogenicity, latent period of infection, and host recovery rate vary dramatically across seasons due to changes in the environment (Blanford and Thomas, 1999). Temperature and host behavior strongly influenced the effectiveness of the entomopathogens, *M. anisopliae* var. *acridum* and *Beauveria bassiana* (Bals.) Vuill. used for the control of locusts and grasshoppers (Gardner and Thomas, 2002). Under hotter conditions, *B. bassiana* is substantially less effective than the other fungal strains. The threshold germination temperature for *B. bassiana* has been reported to be 16 to 30 °C (Martin et al., 1999). There was a successful control of Colorado potato beetle, *Leptinotarsa decemlineata* (Say) larvae with the microbial bioagent, *B. bassiana* in potato at 5 to 23 °C. However, it was not successful at temperatures > 45 °C at the canopy level (Martin et al., 1999). As a consequence of inability of pathogens to sporulate at high temperatures, *B. bassiana* would not be effective to control pest insects in hot summers.

Climate Change and Pest Management

Host plant resistance, natural enemies, entomopathogens and their products and synthetic chemicals are potential tools for insect pest management. Economic relationships between the costs and benefits of these control measures will change as a result of change in climate. There will be increased variability in insect damage as a result of greater variability in climate. Higher temperatures will make dry seasons drier, and conversely may increase the amount and intensity of rainfall, making wet seasons wetter than at present. Spinosads and avermectins produced by fungi, nuclear polyhedrosis virus (NPV), and *Bt* toxins are now widely-used environment-friendly products. However, many of these methods of pest control are highly sensitive to temperature, relative humidity and UV radiation. Increase in temperatures and UV radiation, and a decrease in relative humidity may render many of these control tactics to be ineffective. Therefore, there is a need to develop appropriate formulations of biopesticides that will be effective under situations of global warming in future.

Farming communities need to be aware of climate change and its impacts. Farmers and local communities will need a set of pest control strategies that can produce sustainable yields under future climatic variability. In this regard, there is a great deal of traditional wisdom that can be shared. Many traditional farming communities in marginal areas have long coped with farming under unpredictable climatic conditions. The use of diverse cropping systems in such areas reflects a strategy to reduce the risks to climatic variability (Stigter et al., 2005). Some crops yield more under irrigated conditions, while others are more tolerant to drought. Thus, a paradigm shift is needed from optimizing yields using a single cultivar under ideal conditions to diversify cropping systems against climatic variability, to multiple cultivars and crops with different levels of resistance to insect pests. Other strategies may include enhancing natural biological control by adding soil organic matter, increasing habitat complexity to increase predator abundance, and altered cropping practices to reduce weeds. Successful adaptation measures will involve a blending of local knowledge with modern pest management technology. For example, modern rice varieties with different genetic backgrounds can be successfully grown together to manage rice blast outbreaks (Zhu et al., 2000).

Conclusions

There is a need to predict and map potential changes in geographical distribution of insect pests, and study how climatic changes will affect incidence and population dynamics of insect pests. An understanding of how climate change and shifts in agroecosystems influence insect pests and their natural enemies will be very important. We need to understand effect of climate change on expression of resistance to insect pests, efficacy of transgenic crops in pest management, and assess the efficacy of various pest management technologies under diverse environmental conditions. Development of crop cultivars with resistance to insect pests is of utmost importance to reduce the impact of climate change on crop protection and food security. Thus, there is an urgent need to develop strategies to mitigate the effects of climate change for sustainable crop production and food security.

References

Abbaszadeh, G., M.K. Dhillon, C. Srivastava, and R.D. Gautam. 2011. Effect of climatic factors on bioefficacy of biopesticides in insect pest management. Biopestic. Int. 7(1):1–14.

Ainsworth, A.E., P.A. Davey, C.J. Bernacchi, O.C. Dermody, E.A. Heaton, D.J. Moore, P.B. Morgan, S.L. Naidu, H. Yoo Ra, X. Zhu, P.S. Curtis, and S.P. Long. 2002. A meta-analysis of elevated [CO_2] effects on soybean (*Glycine max*) physiology, growth and yield. Glob. Change Biol. 8(8):695–709. doi:10.1046/j.1365-2486.2002.00498.x

Akbar, S.M.D., T. Pavani, T. Nagaraja, and H.C. Sharma. 2015. Influence of CO_2 and temperature on metabolism and development of *Helicoverpa armigera* (Noctuidae: Lepidoptera). Environ. Entomol. 45(1):229–236. doi:10.1093/ee/nvv144

Akey, D.H., and B.A. Kimball. 1989. Growth and development of the beet armyworm on cotton grown in enriched carbon dioxide atmospheres. Southwest. Entomol. 14:255–260.

Akey, D.H., B.A. Kimball, and J.R. Mauny. 1988. Growth and development of the pink bollworm, *Pectinophora gossypiella* (Lepidoptera: Gelechiidae), on bolls of cotton grown in enriched carbon dioxide atmospheres. Environ. Entomol. 17:452–455. doi:10.1093/ee/17.3.452

Antle, J., M. Apps, R. Beamish, T. Chapin, W. Cramer, J. Frangi, J. Laine, L. Erda, J. Magnuson, I. Noble, J. Price, T. Prowse, T. Root, E. Schulze, O. Sirotenko, B. Sohngen, and J. Soussana. 2001. Ecosystems and their goods and services. p. In: J.J. McCarthy, O.F. Canziani, and N.A. Leary, editors, Climate change 2001: Impacts, adaptation and vulnerability. Third Assessment Report (TAR) of the Intergovernmental Panel on Climate Change (IPCC). Cambridge Univ. Press, Cambridge, U.K.

Awmack, C.S., R. Harrington, and S.R. Leather. 1997. Host plant effects on the performance of the aphid *Aulacorthum solani* (Kalt.) (Homoptera: Aphidae) at ambient and elevated CO_2. Glob. Change Biol. 3:545–549. doi:10.1046/j.1365-2486.1997.t01-1-00087.x

Bale, J.S.B., G.J. Masters, I.D. Hodkinson, C. Awmack, T.M. Bezemer, V.K. Brown, J. Butterfield, A. Buse, J.C. Coulson, J. Farrar, J.E.G. Good, R. Harrington, S. Hartley, T.H. Jones, R.L. Lindroth, M.C. Press, I. Symrnioudis, A.D. Watt, and J.B. Whittaker. 2002. Herbivory in global climate change research: Direct effects of rising temperature on insect herbivores. Glob. Change Biol. 8:1–16. doi:10.1046/j.1365-2486.2002.00451.x

Benedict, J.H., and J.L. Hatfield. 1988. Influence of temperature induced stress on host plant suitability to insects. In: E.A. Heinrichs, editor, Plant stress-insect interactions. John Wiley & Sons, New York. p. 139–165.

Bezemer, T.M., and T.H. Jones. 1998. Plant-insect herbivore interactions in elevated atmospheric CO_2: Quantitative analyses and guild effects. Oikos 82:212–222. doi:10.2307/3546961

Blanford, S., and M.B. Thomas. 1999. Host thermal biology: The key to understanding host-pathogen interactions and microbial pest control? Agric. For. Entomol. 1(3):195–202.

Boutaleb Joutei, A., J. Roy, G. Van Impe, and P. Lebrun. 2000. Effect of elevated CO_2 on the demography of a leaf-sucking mite feeding on bean. Oecologia 123(1):75–81. doi:10.1007/s004420050991

Buckley, L.B., J.J. Tewksbury, and C.A. Deutsch. 2013. Can terrestrial ectotherms escape the heat of climate change by moving? Proc. R. Soc. Lond. 280:20131149. doi:10.1098/rspb.2013.1149

Cannon, R.J.C. 1998. The implications of predicted climate change for insect pests in the UK, with emphasis on non-indigenous species. Glob. Change Biol. 4:785–796. doi:10.1046/j.1365-2486.1998.00190.x

Carew, D.P., and J. Krueger. 1976. Anthocyanidins of *Catharanthus callus* cultures. Phytochemistry 15:442. doi:10.1016/S0031-9422(00)86854-5

Castagneyrol, B., H. Jactel, C. Vacher, E.G. Brockerhoff, and J. Koricheva. 2014. Effects of plant phylogenetic diversity on herbivory depend on herbivore specialization. J. Appl. Ecol. 51:134–141. doi:10.1111/1365-2664.12175

Casteel, C.L., O.K. Niziolek, A.D.B. Leakey, M.R. Berenbaum, and E.H. DeLucia. 2012. Effects of elevated CO_2 and soil water content on phytohormone transcript induction in *Glycine max* after Popillia japonica feeding. Arthropod–Plant Interact. 6:439–447. doi:10.1007/s11829-012-9195-2

Chand, P., M.P. Sinha, and A. Kumar. 1979. Nitrogen fertilizer reduces shoot fly incidence in sorghum. Sci. Cult. 45:61–62.

Channabasavanna, G.P., B.V. Venkat Rao, and G.K. Rajagopal. 1969. Preliminary studies on the effect of incremental levels of phosphatic fertilizer on the incidence of jowar shoot fly. Mysore J. Agric. Sci. 3:253–255.

Chapman, B.B., C. Bronmark, J.A. Nilsson, and L.A. Hansson. 2011. The ecology and evolution of partial migration. Oikos 120:1764–1775. doi:10.1111/j.1600-0706.2011.20131.x

Chen, C.C., and B.A. McCarl. 2001. An investigation of the relationship between pesticide usage and climate change. Clim. Change 50(4):475–487. doi:10.1023/A:1010655503471

Chen, F.J., G. Wu, M.N. Parajulee, and R.B. Shrestha. 2005. Effects of elevated CO_2 and transgenic *Bt* cotton on plant chemistry, performance, and feeding of an insect herbivore, the cotton bollworm. Entomol. Exp. Appl. 115(2):341–350. doi:10.1111/j.1570-7458.2005.00258.x

Classen, A.T., S.C. Hart, T.G. Whitman, N.S. Cobb, and G.W. Koch. 2005. Insect infestations linked to shifts in microclimate: Important climate change implications. Soil Sci. Soc. Am. J. 69:2049–2057. doi:10.2136/sssaj2004.0396

Climate Change Connection (CCC). 2007. A guide to creating climate friendly farms in Manitoba-Crop. 6th ed. Climate Change Connection, Winnipeg, MB. https://www.gov.mb.ca/agriculture/environment/guides-and-publications/pubs/crop-guide-web.pdf (verified 27 Mar. 2018).

Coakley, S.M., H. Scherm, and S. Charkraborty. 1999. Climate change and plant disease management. Annu. Rev. Phytopathol. 37:399–426. doi:10.1146/annurev.phyto.37.1.399

Coley, P.D., and A. Markham. 1998. Possible effects of climate change on plant/herbivore interactions in moist tropical forests. Clim. Change 39:455–472. doi:10.1023/A:1005307620024

Coley, P.D., and J.A. Barone. 1996. Herbivory and plant defenses in tropical forests. Annu. Rev. Ecol. Syst. 27:305–335. doi:10.1146/annurev.ecolsys.27.1.305

Collier, R.H., S. Finch, K. Phelps, and A.R. Thompson. 1991. Possible impact of global warming on cabbage root fly (*Delia radicum*) activity in the UK. Ann. Appl. Biol. 118:261–271. doi:10.1111/j.1744-7348.1991.tb05627.x

Coviella, C.E., and J.T. Trumble. 1999. Effects of elevated atmospheric carbon dioxide on insect-plant interactions. Conserv. Biol. 13:700–712.

Coviella, C.E., and J.T. Trumble. 2000. Effect of elevated atmospheric carbon dioxide on the use of foliar application of Bacillus thuringiensis. BioControl 45(3):325–336. doi:10.1023/A:1009947319662

Coviella, C.E., R.D. Stipanovic, and J.T. Trumble. 2002. Plant allocation to defensive compounds: Interactions between elevated CO_2 and nitrogen in transgenic cotton plants. J. Exp. Bot. 53:323–331. doi:10.1093/jexbot/53.367.323

Currano, E.D., P. Wilf, S.L. Wing, C.C. Labandeira, E.C. Lovelock, and D.L. Royer. 2008. Sharply increased insect herbivory during the Paleocene-Eocene thermal maximum. Proc. Natl. Acad. Sci. USA 105:1960–1964. doi:10.1073/pnas.0708646105

DeLucia, E.H., C.L. Casteel, P.D. Nabity, and B.F. O'Neill. 2008. Insects take a bigger bite out of plants in a warmer, higher carbon dioxide world. Proc. Natl. Acad. Sci. USA 105:1781–1782. doi:10.1073/pnas.0712056105

Dhillon, M.K., and H.C. Sharma. 2007. Effect of storage temperature and duration on viability of eggs of *Helicoverpa armigera* (Lepidoptera: Noctuidae). Bull. Entomol. Res. 97:55–59. doi:10.1017/S0007485307004725

Dhillon, M.K., and H.C. Sharma. 2008. Temperature and Helicoverpa armigera food influence survival and development of the ichneumonid parasitoid, *Campoletis chlorideae*. Indian J. Plant Prot. 36:240–244.

Dhillon, M.K., and H.C. Sharma. 2009. Temperature influences the performance and effectiveness of field and laboratory strains of the ichneumonid parasitoid, Campoletis chlorideae. BioControl 54:743–750. doi:10.1007/s10526-009-9225-x

Dhillon, M.K., and F. Hasan. 2017. Temperature-dependent development of diapausing larvae of *Chilo partellus* (Swinhoe) (Lepidoptera: Crambidae). J. Therm. Biol. 69:213–220. doi:10.1016/j.jtherbio.2017.07.016

Dhillon, M.K., F. Hasan, A.K. Tanwar, and A.P.S. Bhadauriya. 2017. Effects of thermo-photoperiod on induction and termination of hibernation in *Chilo partellus* (Swinhoe). Bull. Entomol. Res. 107(3):294–302. doi:10.1017/S0007485316000870

Dosdall, L.M. 1994. Evidence for successful overwintering of diamondback moth, *Plutella xylostella* (L.) (Lepidoptera: Plutellidae), in Alberta. Can. Entomol. 126:183–185. doi:10.4039/Ent126183-1

Environment Protection Agency (EPA). 1989. The potential effects of global climate change on the United States. Vol. 2: National studies. Review of the Report to Congress, US-EPA, Washington, D.C.

Evans, H., N. Straw, and A. Watt. 2002. Climate change: Implications for insect pests. Forestry Commission, Bristol, U.K. p. 99–118.

Faris, M.A., A.M. Lara, and A.F. de S. Leao Veiga. 1979. Stability of sorghum midge resistance. Crop Sci. 19:577–580. doi:10.2135/cropsci1979.0011183X001900050006x

Fleming, R.A., and J.N. Candau. 1998. Influences of climatic change on some ecological processes of an insect outbreak system in Canada's boreal forests and the implications for biodiversity. Environ. Monit. Assess. 49:235–249. doi:10.1023/A:1005818108382

Fleming, R.A., and W.J.A. Volney. 1995. Effects of climate change on insect defoliator population processes in Canada's boreal forest: Some plausible scenarios. Water Air Soil Pollut. 82:445–454. doi:10.1007/BF01182854

Freier, B., and H. Triltsch. 1996. Climate chamber experiments and computer simulations on the influence of increasing temperature on wheat-aphid-predator interactions. Aspects Appl. Biol. 45:293–298.

Gardner, S.N., and M.B. Thomas. 2002. Costs and benefits of fighting infection in locusts. Evol. Ecol. Res. 4(1):109–131.

Garrett, K.A., S.P. Dendy, E.E. Frank, M.N. Rouse, and S.E. Travers. 2006. Climate change effects on plant disease: Genomes to ecosystems. Annu. Rev. Phytopathol. 44:489–509. doi:10.1146/annurev.phyto.44.070505.143420

Gaston, K.J., and P.H. Williams. 1996. Spatial patterns in taxonomic diversity. In: K.J. Gaston, editor, Biodiversity, a biology of numbers and difference. Blackwell Science, London. p. 202–209.

Gomulkiewicz, R., and R.G. Shaw. 2013. Evolutionary rescue beyond the models. Philos. Trans. R. Soc., B. 368:20120093. doi:10.1098/rstb.2012.0093

Gonzalez, A., and G. Bell. 2013. Evolutionary rescue and adaptation to abrupt environmental change depends upon the history of stress. Philos. Trans. R. Soc., B. 368:20120079. doi:10.1098/rstb.2012.0079

Guo, H., Y. Sun, Y. Li, X. Liu, W. Zhang, and F. Ge. 2013. Elevated CO_2 decreases the response of the ethylene signaling pathway in Medicago truncatula and increases the abundance of the pea aphid. New Phytol. 201:279–291. doi:10.1111/nph.12484

Guo, H., Y. Sun, Q. Ren, K. Zhu-Salzman, C.Z. Wang, L. Kang, and F. Ge. 2012. Elevated CO_2 reduces the resistance and tolerance of tomato plants to *Helicoverpa armigera* by suppressing the JA signaling pathway. PLoS One 7:e41426.

Haggis, M.J. 1996. Forecasting the severity of seasonal outbreaks of African armyworm, Spodoptera exempta (Lepidoptera: Noctuidae) in Kenya from the previous years' rainfall. Bull. Entomol. Res. 86:129–136. doi:10.1017/S0007485300052366

Hamilton, J.G., O. Dermody, M. Aldea, A.R. Zangerl, M.R. Berenbaum, and E. Delucia. 2005. Anthropogenic changes in tropospheric composition increase susceptibility of soybean to insect herbivory. Environ. Entomol. 34:479–485. doi:10.1603/0046-225X-34.2.479

Hance, T., J. van Baaren, P. Vernon, and G. Boivin. 2007. Impact of extreme temperatures on parasitoids in a climate change perspective. Annu. Rev. Entomol. 52:107–126. doi:10.1146/annurev.ento.52.110405.091333

Heagle, A.S., R.L. Brandenburg, J.C. Burns, and J.E. Miller. 1994. Ozone and carbon dioxide effects on spider mites in white clover and peanut. J. Environ. Qual. 23:1168–1176. doi:10.2134/jeq1994.00472425002300060006x

Hickling, R., B.R. Roy, J.K. Hill, R. Fox, and C.T. Thomas. 2006. The distributions of a wide range of taxonomic groups are expanding polewards. Glob. Change Biol. 12:450–455. doi:10.1111/j.1365-2486.2006.01116.x

Hilder, V.A., and D. Boulter. 1999. Genetic engineering of crop plants for insect resistance- a critical review. Crop Prot. 18:177–191. doi:10.1016/S0261-2194(99)00028-9

Hill, D.S. 1987. Agricultural insects pests of temperate regions and their control. Cambridge Univ. Press, Cambridge, UK.

Hill, M.G., and J.J. Dymock. 1989. Impact of climate change: Agricultural/Horticultural Systems. DSIR Entomology Division Submission to the New Zealand Climate Change Programme. Department of Scientific and Industrial Research, Auckland, New Zealand.

Holton, M.K., R.L. Lindroth, and E.V. Nordheim. 2003. Foliar quality influences tree-herbivore-parasitoid interactions: Effects of elevated CO_2, O_3, and plant genotype. Oecologia 137:233–244. doi:10.1007/s00442-003-1351-z

Hu, G., F. Lu, B-P. Zhai, M-H. Lu, W-C. Liu, F. Zhu, X-W. Wu, G-H. Chen, and X-X. Zhang. 2014. Outbreaks of the Brown Planthopper Nilaparvata lugens (Stal) in the Yangtze River Delta: Immigration or Local Reproduction? PLoS ONE 9(2): e88973. doi:10.1371/journal.pone.0088973.

Intergovernmental Panel on Climate Change (IPCC). 2007. Summary for policymakers. In: S. Solomon, D. Qin, M. Manning, et al., editors, Climate change 2007: The physical science basis, contribution of working group I to the fourth assessment report of the IPCC. Cambridge Univ. Press, Cambridge, UK.

Jeltsch, F., D. Bonte, G. Pe'er, B. Reineking, P. Leimgruber, N. Balkenhol, B. Schröder, C.M. Buchmann, T. Mueller, N. Blaum, D. Zurell, K. Böhning-Gaese, T. Wiegand, J.A. Eccard, H. Hofer, J. Reeg, U. Egger, and S. Bauer. 2013. Integrating movement ecology with biodiversity research- exploring new avenues to address spatiotemporal biodiversity dynamics. Mov. Ecol. 1:6. doi:10.1186/2051-3933-1-6

Jenkinson, L.S., A.J. Davies, S. Wood, B. Shorrocks, and J.H. Lawton. 1996. Not that simple: Global warming and predictions of insect ranges and abundances- results from a model insect assemblage in replicated laboratory ecosystems. Aspects Appl. Biol. 45:343–348.

Karaman, G.A., A. Ghareb, A. Abdel-Naby, and M. Embaby. 1998. Effect of land leveling on Chilo agamemnon Blesz., infestation in sugarcane fields of middle Egypt. Arab Journal of Plant Prot. 16:60–65.

Karuppaiah, V., and G.K. Sujayanad. 2012. Impact of climate change on population dynamics of insect pests. World J. Agric. Sci. 8(3):240–246.

Khan, Z.R., D.M. Norris, H.S. Chaing, and A.S. Oosterwyk. 1986. Light induced susceptibility in soybean to cabbage looper, *Trichoplusia ni* (Lepidoptera: Noctuidae). Environ. Entomol. 15:803–808. doi:10.1093/ee/15.4.803

Knepp, R.G., J.G. Hamilton, J.E. Mohan, A.R. Zangerl, M.R. Barenbaum, and E.H. DeLucia. 2005. Elevated CO_2 reduces leaf damage by insect herbivores in a forest community. New Phytol. 167:207–218. doi:10.1111/j.1469-8137.2005.01399.x

Kogan, M. 1982. Plant resistance in pest management. In: R.L. Melcalf and W.H. Luckmann, editors, Introduction to insect pest management. 2nd ed. John Willey and Sons, New York. p. 93–134.

Kopper, B.J., and R.L. Lindroth. 2003. Effects of elevated carbon dioxide and ozone on the phytochemistry of aspen and performance of an herbivore. Oecologia 134:95–103. doi:10.1007/s00442-002-1090-6

Kopper, B.J., R.L. Lindroth, and E.V. Nordheim. 2001. CO_2 and O_3 effects on paper birch (Betulaceae: Betula papyrifera) phytochemistry and whitemarked tussock moth (Lymantriidae: Orgyia leucostigma) performance. Environ. Entomol. 30:1119–1126. doi:10.1603/0046-225X-30.6.1119

La Sorte, F.A., and W. Jetz. 2010. Projected range contractions of montane biodiversity under global warming. Proc. Biol. Sci. 277:3401–3410. doi:10.1098/rspb.2010.0612

Lawton, J.H. 1991. From physiology to population dynamics and communities. Funct. Ecol. 5:155–161. doi:10.2307/2389253

Lincoln, D.E., D. Couvet, and N. Sionit. 1986. Response of an insect herbivore to host plants grown in carbon dioxide enriched atmosphere. Oecologia 69:556–560. doi:10.1007/BF00410362

Lincoln, D.E., N. Sionit, and B.R. Strain. 1984. Growth and feeding response of *Psuedoplusia includens* (Lepidoptera: Noctuidae) to host plants grown in controlled carbon dioxide atmoshperes. Environ. Entomol. 13:1527–1530. doi:10.1093/ee/13.6.1527

Lu, Z.X., K.L. Heong, X.P. Yu, and C. Hu. 2004. Effects of plant nitrogen on fitness of the brown planthopper, *Nilaparvata lugens* Stal. In: rice. J. Asia Pac. Entomol. 7:97–104. doi:10.1016/S1226-8615(08)60204-6

Luo, Y., P.S. Teng, N.G. Fabellar, and D.O. TeBeest. 1998. The effects of global temperature change on rice leaf blast epidemics: A simulation study in three agroecological zones. Agric. Ecosyst. Environ. 68:187–196. doi:10.1016/S0167-8809(97)00082-0

Martin, P.A.W., R.F.W. Schroder, T.J. Poprawski, J.J. Lipa, D. Sosnowska, E. Hausvater, and V. Rasocha. 1999. The effect of high temperatures on the susceptibility of the Colorado potato beetle (Coleoptera: Chrysomelidae) to *Beauveria bassiana* (Balsamo) Vuillemin in Poland, the Czech Republic and the United States. Vedecke Prace Vyzkumny Ustav Bramborarsky Havlickuv Brod 13:69–77.

Masters, G.J., V.K. Brown, I.P. Clarke, J.B. Whittaker, and J.A. Hollier. 1998. Direct and indirect effects of climate change on insect herbivores: Auchenorrhyncha (Homoptera). Ecol. Entomol. 23:45–52. doi:10.1046/j.1365-2311.1998.00109.x

Matter, S.F., A. Doyle, K. Illerbrun, J. Wheeler, and J. Roland. 2011. An assessment of direct and indirect effects of climate change for populations of the Rocky Mountain Apollo butterfly (*Parnassius smintheus* Doubleday). Insect Sci. 18(4):385–392. doi:10.1111/j.1744-7917.2011.01407.x

Mattson, W.J., and R.A. Haack. 1987. The role of drought stress in provoking outbreaks of phytophagous insects. In: P. Barbosa and J.C. Schultz, editors, Insect outbreaks. Academic Press, NY. p. 365–407. doi:10.1016/B978-0-12-078148-5.50019-1

McKone, M.J., D. Kelly, and W.G. Lee. 1998. Effect of climate change on mast-seeding species: Frequency of mass flowering and escape from specialist insect seed predators. Glob. Change Biol. 4:591–596. doi:10.1046/j.1365-2486.1998.00172.x

McMurtry, J.A. 1962. Resistance of alfalfa to spotted alfalfa aphid in relation to environmental factors. Hilgardia 32:501–539. doi:10.3733/hilg.v32n12p501

Millennium Ecosystem Assessment (MEA). 2005. Ecosystems and human well-being: Biodiversity synthesis. World Resources Institute, Washington, D.C.

Morgan, D. 1996. Temperature changes and insect pests: A simulation study. Aspects Appl. Biol. 45:277–283.

National Drought Mitigation Center (NDMC). 1998. Reported drought-related effects of El Niño for March 1998. National Drought Mitigation Center, Lincoln, NE 68583-0749. http://enso.unl.edu/ndmc/enigma/elnino.html (verified 29 Mar. 2018).

Nayak, M.K., and P.J. Collins. 2008. Influence of concentration, temperature and humidity on the toxicity of phosphine to the strongly phosphine-resistant psocid *Liposcelis bostrychophila* Badonnel (Psocoptera: Liposcelididae). Pest Manag. Sci. 64(9):971–976.

Nuorteva, P. 1997. The role of air pollution and climate change in development of forest insect outbreaks. Acta Phytopathol. Entomol. Hung. 32:127–128.

O'Neill, B.F., A.R. Zangerl, E.H. DeLucia, C. Casteel, J.A. Zavala, and M.R. Berenbaum. 2011. Leaf temperature of soybean grown under elevated CO_2 increases *Aphis glycines* (Hemiptera: Aphididae) population growth. Insect Sci. 18(4):419–425. doi:10.1111/j.1744-7917.2011.01420.x

Oerke, E.C., and H.W. Dehne. 2004. Safeguarding production-losses in major crops and the role of crop protection. Crop Prot. 23:275–285. doi:10.1016/j.cropro.2003.10.001

Olsen, K.M., J.C. Daly, E.J. Fennegan, and R.J. Mathon. 2005b. Changes in Cry1Ac *Bt* transgenic cotton in response to two environmental factors: Temperature and insect damage. J. Econ. Entomol. 98:1382–1390. doi:10.1603/0022-0493-98.4.1382

Olsen, K.M., J.C. Daly, H.E. Holt, and E.J. Finnegan. 2005a. Season-long variation in expression of Cry1Ac gene and efficacy of Bacillus thuringiensis toxin in transgenic cotton against *Helicoverpa armigera* (Lepidoptera: Noctuidae). J. Econ. Entomol. 98:1007–1017. doi:10.1603/0022-0493-98.3.1007

Osbrink, W.L.A., J.T. Trumble, and R.E. Wagner. 1987. Host suitability of Phaseolus lunata for *Trichoplusia ni* (Lepidoptera: Noctuidae) in controlled carbon dioxide atmospheres. Environ. Entomol. 16:639–644. doi:10.1093/ee/16.3.639

Pande, S., J. Narayana Rao, M.V. Reddy, and D. McDonald. 1994. A technique to screen for resistance to stem rot caused by *Sclerotium rolfsii* in groundnut under greenhouse conditions. Indian J. Plant Prot. 22(2):151–158.

Parmesan, C., M.T. Burrows, C.M. Duarte, E.S. Poloczanska, A.J. Richardson, D.S. Schoeman, and M.C. Singer. 2013. Beyond climate change attribution in conservation and ecological research. Ecol. Lett. 16:58–71. doi:10.1111/ele.12098

Parry, M.L. 1990. Climate change and world agriculture. Earthscan, London.

Parry, M.L., and T.R. Carter. 1989. An assessment of the effects of climatic change on agriculture. Clim. Change 15:95–116. doi:10.1007/BF00138848

Patz, J. 2013. Climate variability and change: Food, water, and societal impacts. In: K.E. Pinkerton and W.N. Rom, editors, Global climate change and public health respiratory medicine. Springer Science + Business Media., New York. p. 211–235. doi:10.1007/978-1-4614-8417-2_12

Porter, J.H., M.L. Parry, and T.R. Carter. 1991. The potential effects of climatic change on agricultural insect pests. Agric. For. Meteorol. 57:221–240. doi:10.1016/0168-1923(91)90088-8

Price, M.L., A.M. Stremberg, and L.G. Butler. 1979. Tannin content as a function of grain maturity and drying conditions, in several varieties of *Sorghum bicolor* (L.) Moench. J. Agri. Food Chem. 27:1270–1274. doi:10.1021/jf60226a060

Price, M.L., L. van Scoyoc, and L.G. Butler. 1978. A critical evaluation of vanillin reaction as an assay in sorghum grain. J. Agric. Food Chem. 26:1214–1218. doi:10.1021/jf60219a031

Price, P.W. 1987. The role of natural enemies in insect populations. In: P. Barbosa and J.C. Schultz, editors, Insect outbreaks. Academic Press, San Diego, CA. p. 287–312. doi:10.1016/B978-0-12-078148-5.50016-6

Purohit, M., and A.D. Deshpande. 1992. Effect of nitrogenous fertilizer application on cotton leafhopper, Amrasca biguttula biguttula (Ishida). In: Proc. National Seminar on Changing Scenario in Pests and Pest Management in India, 31 Jan.–1 Feb. 1992. Plant Protection Association of India, Rajendranagar, Hyderabad, Andhra Pradesh, India.

Rank, N.E., and E.P. Dahlhoff. 2002. Allele frequency shifts in response to climate change and physiological consequences of allozyme variation in a montane insect. Evolution 56:2278–2289. doi:10.1111/j.0014-3820.2002.tb00151.x

Reddy, K.S., and D.V. Narasimha Rao. 1975. Effect of nitrogen application on shoot fly incidence and grain maturity in sorghum. Sorghum Newslett. 18:23–24.

Rhoades, D.F. 1985. Offensive-defensive interactions between herbivores and plants: Their relevance in herbivore population dynamics and ecological theory. Am. Nat. 125:205–238. doi:10.1086/284338

Rooney, T.P., and A.T. Smith. 1996. Global warming and the regional persistence of a temperate-zone insect (Tenodera sinensis). Am. Midl. Nat. 136:84–93. doi:10.2307/2426633

Rose, D.J.W., C.F. Dewhuse, and W.W. Page. 2000. The African armyworms: The status, biology, ecology, epidemiology, and management of Spodoptera exempta (Lepidoptera: Noctuidae). 2nd ed. Natural Resources Institute, Chatham, UK.

Rosenzweig, C., and W.D. Solecki, editors. 2000. Climate change and a global city: The metropolitan East Coast regional assessment. Columbia Earth Institute, New York, NY.

Rowntree, P.R. 1990. Estimate of future climatic change over Britain. Weather 45:79–89. doi:10.1002/j.1477-8696.1990.tb05059.x

Sachs, E.S., J.H. Benedict, J.F. Taylor, D.M. Stelly, S.K. Davis, and D.W. Altman. 1996. Pyramiding CryIA(b) insecticidal protein and terpenoids in cotton to resist tobacco budworm (Lepidoptera: Noctuidae). Environ. Entomol. 25:1257–1266. doi:10.1093/ee/25.6.1257

Schweissing, F.C., and G. Wilde. 1978. Temperature influence on greenbug resistance of crops in the seedling stage. Environ. Entomol. 7:831–834. doi:10.1093/ee/7.6.831

Schweissing, F.C., and G. Wilde. 1979. Temperature and plant nutrient effects on resistance of seedling sorghum to the greenbug. J. Econ. Entomol. 72:20–23. doi:10.1093/jee/72.1.20

Settle, W.H., H. Ariawan, E.T. Astuti, W. Cahyana, A.L. Hakim, D. Hindayana, and A.S. Lestari. 1996. Managing tropical rice pests through conservation of generalist natural enemies and alternative prey. Ecology 77(7):1975–1988. doi:10.2307/2265694

Sharma, H.C. 2014. Climate change effects on insects: Implications for crop protection and food security. J. Crop Improv. 28:229–259. doi:10.1080/15427528.2014.881205

Sharma, H.C., and R. Ortiz. 2000. Transgenics, pest management, and the environment. Curr. Sci. 79:421–437.

Sharma, H.C., and V.F. Lopez. 1991. Stability of resistance in sorghum to Calocoris angustatus (Hemiptera: Miridae). J. Econ. Entomol. 91:1088–1094. doi:10.1093/jee/84.3.1088

Sharma, H.C., and V.F. Lopez. 1994. Interactions between panicle size, insect density, and environment for genotypic resistance in sorghum to head bug, Calocoris angustatus. Entomol. Exp. Appl. 71:101–109. doi:10.1111/j.1570-7458.1994.tb01776.x

Sharma, H.C., B.U. Singh, and R. Ortiz. 2001. Host plant resistance to insects: Measurement, mechanisms, and insect-plant environment interactions. In: T.N. Ananthakrishna, editor, Insects and plant defence dynamics. Science Publishers Inc., Enfield, MH. p. 133–159.

Sharma, H.C., D.J. Sullivan, and V.S. Bhatnagar. 2002. Population dynamics of the Oriental armyworm, *Mythimna separata* (Walker) (Lepidoptera: Noctuidae) in South-Central India. Crop Prot. 21:721–732. doi:10.1016/S0261-2194(02)00029-7

Sharma, H.C., K.F. Nwanze, and V. Subramanian. 1997. Mechanisms of resistance to insects and their usefulness in sorghum improvement. In: H.C. Sharma, F. Singh, and K.F. Nwanze, editors, Plant resistance to insects in sorghum. International Crops Research Institute for the Semi-Arid Tropics, Patancheru, Andhra Pradesh, India. p. 81–100.

Sharma, H.C., M.K. Dhillon, J. Kibuka, and S.Z. Mukuru. 2005. Plant defense responses to sorghum spotted stem borer, *Chilo partellus* under irrigated and drought conditions. Int. Sorghum and Millets Newslett. 46:49–52.

Sharma, H.C., A.R. War, M. Pathania, S.M.D. Akbar, S.P. Sharma, and R.S. Munghate. 2016. Elevated CO_2 influences host plant defense response in chickpea against Helicoverpa armigera. Arthropod–Plant Interactions. 10:171–181. doi:10.1007/s11829-016-9422-3

Sharma, H.C., P. Vidyasagar, and K. Leuschner. 1988. No-choice cage technique to screen for resistance to sorghum midge (Diptera: Cecidomyiidae). J. Econ. Entomol. 81:415–422. doi:10.1093/jee/81.1.415

Sharma, H.C., S.Z. Mukuru, E. Manyasa, and J. Were. 1999. Breakdown of resistance to sorghum midge, *Stenodiplosis sorghicola*. Euphytica 109:131–140. doi:10.1023/A:1003724217514

Singer, M.C., and L. McBride. 2012. Geographic mosaics of species' association: A definition and an example driven by plant/insect phenological synchrony. Ecology 93(12):2658–2673. doi:10.1890/11-2078.1

Singh, R.P., and K.R. Reddy. 2013. Impact of climate change and farm management. Clim. Change and Environmental Sustain. 1(1):53–72. doi:10.5958/j.2320-6411.1.1.006

Smith, R.C. 1954. An analysis of 100 years of grasshopper population in Kansas (1854-1954). Trans. Kans. Acad. Sci. 57(4):397–433. doi:10.2307/3625918

Soman, P., K.F. Nwanze, K.B. Laryea, D.R. Butler, and Y.V.R. Reddy. 1994. Leaf surface wetness in sorghum and resistance to shoot fly, Atherigona soccata: Role of soil and plant water potentials. Ann. Appl. Biol. 124:97–108. doi:10.1111/j.1744-7348.1994.tb04119.x

Stigter, C.J., Z. Dawei, L.O.Z. Onyewotu, and M. Xurong. 2005. Using traditional methods and indigenous technologies for coping with climate variability. Clim. Change 70:255–271. doi:10.1007/s10584-005-5949-5

Stireman, J.O., III, L.A. Dyer, D.H. Janzen, M.S. Singer, J.T. Lill, R.J. Marquis, R.E. Ricklefs, G.L. Gentry, W. Hallwachs, P.D. Coley, J.A. Barone, H.F. Greeney, H. Connahs, P. Barbosa, H.C. Morais, and I.R. Diniz. 2005. Climatic unpredictability and parasitism of caterpillars: Implications of global warming. Proc. Natl. Acad. Sci. USA 102(48):17384–17387.

Sun, Y.C., J. Yin, F.J. Chen, G. Wu, and F. Ge. 2011a. How does atmospheric elevated CO_2 affect crop pests and their natural enemies? Case histories from China. Insect Sci. 18(4):393–400. doi:10.1111/j.1744-7917.2011.01434.x

Sun, Y.C., L. Feng, F. Gao, and F. Ge. 2011b. Effects of elevated CO_2 and plant genotype on interactions among cotton, aphids and parasitoids. Insect Sci. 18(4):451–461. doi:10.1111/j.1744-7917.2010.01328.x

Sutherst, R.W. 1991. Pest risk analysis and the greenhouse effect. Rev. Agric. Entomol. 79:1177–1187.

Tingey, W.M., and S.R. Singh. 1980. Environmental factors affecting the magnitude and expression of resistance. In: M.G. Maxwell and P.R. Jennings, editors, Breeding plants resistant to insects. John Wiley & Sons, New York. p. 87–113.

Tripp, K.E., W.K. Kroen, M.M. Peet, and D.H. Willits. 1992. Fewer whiteflies found on CO_2–enriched greenhouse tomatoes with high C:N ratios. HortScience 27:1079–1080.

Tucker, M.R., and D.E. Pedgley. 1983. Rainfall and outbreaks of the African armyworm, *Spodoptera exempta* (Walker) (Lepidoptera: Noctuidae). Bull. Entomol. Res. 73:195–199. doi:10.1017/S0007485300008804

UNEP. 2006. Global environmental outlook year book 2006. Earthprint, London.

van Asch, M., P.H. van Tienderen, L.J.M. Holleman, and M.E. Visser. 2007. Predicting adaptation of phenology in response to climate change, an insect herbivore example. Glob. Change Biol. 13:1596–1604. doi:10.1111/j.1365-2486.2007.01400.x

Vanhanen, H., T.O. Veteli, S. Päivinen, S. Kellomäki, and P. Niemelä. 2007. Climate change and range shifts in two insect defoliators: Gypsy moth and nun moth- a model study. Silva Fenn. 41:621–638. doi:10.14214/sf.469

Voigt, W., J. Perner, A.J. Davis, T. Eggers, J. Schumacher, R. Bährmann, B. Fabian, W. Heinrich, G. Köhler, D. Lichter, R. Marstaller, and F.W. Sander. 2003. Trophic levels are differentially sensitive to climate. Ecology 84(9):2444–2453. doi:10.1890/02-0266

Wang, F., S. Peng, K. Cui, L. Nie, and J. Huang. 2014. Field performance of *Bt* transgenic crops: A review. Aust. J. Crop Sci. 8(1):18–26.

Wang, X., and Z. Shen. 2007. Potency of some novel insecticides at various environmental temperatures on Myzus persicae. Phytoparasitica 35:414–422. doi:10.1007/BF02980705

Ward, N.L., and G.J. Masters. 2007. Linking climate change and species invasion: An illustration using insect herbivores. Glob. Change Biol. 13:1605–1615. doi:10.1111/j.1365-2486.2007.01399.x

Watanabe, T., and H. Seino. 1991. Correlation between the immigration area of rice plant hoppers and the low level jet stream in Japan. Appl. Entomol. Zool. (Jpn.) 26:457–462. doi:10.1303/aez.26.457

White, T.C.R. 1984. The abundance of invertebrate herbivores in relation to the availability of nitrogen in stressed food plants. Oecologia 63:90–105.

Wilf, P., and C.C. Labandeira. 1999. Response of plant-insect associations to Paleocene-Eocene warming. Science 284(5423):2153–2156. doi:10.1126/science.284.5423.2153

Wilson, R. 2008. Recognizing water stress in plants. Ext. Bull. No. 91-01. The Arboretum at Flagstaff, Flagstaff, AZ.

Wu, Y.R., D. Llewellyn, A. Mathews, and E.S. Dennis. 1997. Adaptation of *Helicoverpa armigera* (Lepidoptera: Noctuidae) to a proteinase inhibitor expressed in transgenic tobacco. Mol. Breed. 3:371–380. doi:10.1023/A:1009681323131

Yamamura, K., M. Yokozawa, M. Nishimori, Y. Ueda, and T. Yokosuka. 2006. How to analyze long-term insect population dynamics under climate change: 50-year data of three insect pests in paddy fields. Popul. Ecol. 48:31–48. doi:10.1007/s10144-005-0239-7

Young, A.M. 1982. Population biology of tropical insects. Plenum Press, New York. doi:10.1007/978-1-4684-1113-3

Yu, H., Y. Li, X. Li, J. Romeis, and K. Wu. 2013. Expression of Cry1Ac in transgenic *Bt* soybean lines and their efficiency in controlling lepidopteran pests. Pest Manag. Sci. 69:1326–1333. doi:10.1002/ps.3508

Zhu, Y., H. Chen, J. Fan, Y. Wang, Y. Li, J. Chen, J.X. Fan, S. Yang, L. Hu, H. Leung, T.W. Mew, P.S. Teng, Z. Wang, and C.C. Mundt. 2000. Genetic diversity and disease control in rice. Nature 406:718–722. doi:10.1038/35021046

Climate Extremes and Impacts on Agriculture

Mannava V.K. Sivakumar

Abstract

Earth's climate system is controlled by a complex set of interactions among the atmosphere, oceans, continents and living systems. The earth's climate fluctuates over seasons, decades, and centuries in response to both natural and human variables. The definition of climate extremes and a description of different atmospheric oscillations contributing to natural climate variability and climate extremes are presented. Evidence is pointing to the fact that human-driven changes in Earth's energy balance are driving a warmer and wetter atmosphere, with this trend superimposed on and magnifying natural variability. Extreme climate events include heavy precipitation and floods; droughts; high temperatures and heat waves; low temperatures and cold waves; tropical cyclones; hurricanes, thunderstorms and tornadoes; and winter storms. The impacts of each of these climate extremes on agriculture such as alteration of ecosystems; disruption of food production and water supply; damage to infrastructure and settlements; and human morbidity and mortality, are described with suitable examples. The growing concern with the impacts of climate extremes on agriculture has created new demands for information from, and assessment by agrometeorologists. Agrometeorological strategies for reducing the impacts of natural disasters in agriculture such as improved use of climate and weather information and forecasts, early warning systems, reorienting and recasting meteorological information, fine-tuning of climatic analysis and presentation in forms suitable for agricultural decision-making and helping marginal farmers cope with the adverse impact of climate extremes are described.

Climate in a narrow sense is usually defined as the average of weather data, or more rigorously, as the statistical description in terms of the mean and variability of relevant quantities over a period of time. Climate encapsulates the description of how the atmosphere behaves over a long period of time (typically defined as a 30-yr period) as well as the description of other aspects of weather patterns and parameters distribution, including anomalous, rare, and extreme weather events (WMO, 2016). However, earth's climate system is controlled by a complex set of interactions among the atmosphere, oceans, continents, and living systems. The Earth's climate fluctuates over seasons, decades, and centuries in response to both natural and human variables. Natural climate variability on different timescales is caused by cycles and trends in the Earth's orbit, incoming solar radiation, the atmosphere's chemical composition, ocean circulation, the biosphere, and much more.

In recent decades, people throughout the world have become increasingly concerned by extreme meteorological and hydrological events, which are becoming more frequent and more destructive. The number of weather-related natural catastrophes has risen on all continents since 1980, most notably in Asia and North America, which had a dramatic impact on the insurance industry (Munich, 2017).

Abbreviations: GIS, geographic information systems; ICTs, information and communication technologies.

Consultant, World Bank, Geneva, Switzerland. *Corresponding author (mannavas@gmail.com)

doi:10.2134/agronmonogr60.2016.0003

© ASA, CSSA, and SSSA, 5585 Guilford Road, Madison, WI 53711, USA.
Agroclimatology: Linking Agriculture to Climate, Agronomy Monograph 60.
Jerry L. Hatfield, Mannava V.K. Sivakumar, John H. Prueger, editors.

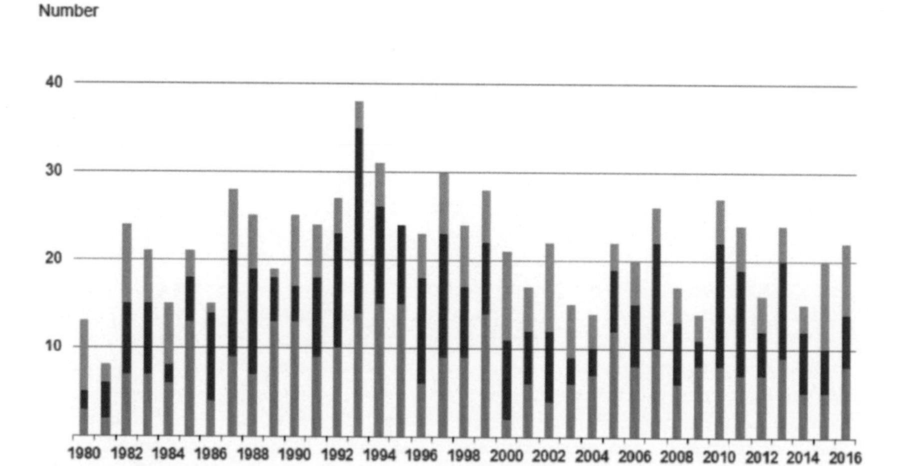

Fig. 1. Number of severe weather-related loss events worldwide during 1980 to 2016. Tropical storms, extratropical storms, convective storms, and local storms (Green); Floods and mass movement (Blue), and Extreme temperatures, droughts, and forest fires (Brown) (Source: Münchener Rückversicherungs-Gesellschaft, NatCatSERVICE).

The number of severe weather-related loss events worldwide during 1980 to 2016 included a number of tropical storms, extratropical storms, convective storms, and local storms, floods, extreme temperatures, droughts, and forest fires (Fig. 1). In the chapter dealing with observations, surface, and atmosphere in the Working Group I Fifth Assessment Report of the Intergovernmental Panel on Climate Change (IPCC), Hartmann et al. (2013) assessed that over land the number of warm days and nights had very likely increased, the number of cold days and nights had very likely decreased and that heavy precipitation events had likely increased in more regions than they had decreased. According to WMO (2013), the world experienced unprecedented high-impact climate extremes during the 2001 to 2010 decade, which was the warmest since the start of modern measurements in 1850. The decade ending in 2010 was an unprecedented era of climate extremes, as evidenced by heat waves in Europe and Russia, droughts in the Amazon Basin, Australia, and East Africa, and huge storms like Tropical Cyclone Nargis (2008) and Hurricane Katrina (2005). According to Perkins-Kirkpatrick et al. (2018), the year 2017 recorded more than 60 d of extreme daytime heat worldwide, nearly double the 1961 to 1990 average of 36.5, continuing the long-term trend toward more hot days each year across the globe.

Extreme events are responsible for many significant impacts, in terms of both casualties and economic effects. Developed countries today are better able to manage the impacts of a variable climate, but in developing countries the increasing frequency and magnitude of extreme weather events pose potentially disastrous consequences for agriculture and food security (Field, 2000; Wilhite, 2005; Sheffield and Wood, 2011; and Sivakumar et al., 2011). According to Zhu and Troy (2018), climate extremes can negatively impact crop production, and climate change is expected to affect the frequency and severity of extremes. Before 1980, temperature-related indices had few trends; after 1980, statistically significant warming trends exist for each crop in the majority of growing regions. In particular, crops have increasingly been exposed to extreme hot temperatures, above which yields have been shown to decline. According to Cabezon et al. (2015), the small developing states are disproportionately affected by extreme events, with the average annual cost being much greater than in larger countries.

Definition of Climate Extremes

According to the terminology of WMO and the Intergovernmental Panel on Climate Change (IPCC), an "Extreme Event" is the occurrence of a value of a weather or climate variable above (or below) a threshold value near the upper (or lower) ends of the range of observed values of the variable. According to Sillmann et al. (2017), weather and climate extreme events naturally cluster into two classes or categories related to the temporal scales on which they can occur, and which involve different models, processes, and research questions. Therefore, it is sensible to focus on short-duration (less than three days) and long-duration (weeks to months or even years) extreme events, their different mechanisms and differing approaches to evaluation and prediction. The short duration extreme events may include (i) Convective events leading to heavy precipitation, hail, lightning, tornadoes, violent downdrafts; (ii) Extra-tropical cyclones leading to wind storms, storm surges, extreme precipitation (rainfall or snowfall), freezing rain; (iii) Anticyclones leading to fog and air pollution, cold outbreaks, long-lived heat waves, and extended cold spells; and (iv) Tropical cyclones. The long duration extreme events include droughts, heat waves, cold spells and floods caused by persistent rainfall, but also extreme low Arctic sea ice extent, increased storminess and wildfire seasons.

A large amount of the available scientific literature on climate extremes is based on the use of so-called 'extreme indices,' which can either be based on the probability of occurrence of given quantities or on threshold exceedances (Seneviratne et al., 2012). Some advantages of using predefined extreme indices are that they allow some comparability across modeling and observational studies and across regions. Peterson and Manton (2008) discuss collaborative international efforts to monitor extremes by employing extreme indices.

Climate Variability and Extreme Events

Climate variability refers to variations in the mean state and other statistics (such as standard deviations, the occurrence of extremes, etc.) of the climate on all spatial and temporal scales beyond that of individual weather events (IPCC, 2014). Variability may be due to natural internal processes within the climate system (internal variability), or to variations in natural or anthropogenic external forcing (external variability). The United Nations Framework Convention on Climate Change (UNFCCC), in its Article 1, defines climate change as: 'a change of climate which is attributed directly or indirectly to human activity that alters the composition of the global atmosphere and which is in addition to natural climate variability observed over comparable time periods'.

The climate system responds to a range of influences that may be considered natural in origin, variability reflecting the essential nonlinearity of the underlying physics. Changes may arise on a variety of time scales, ranging from days or months to years or even hundreds of years in response to changes in the ocean. In addition to the quasi-steady patterns of atmospheric circulation such as the trade winds and the Hadley cell, there are transitory patterns that can persist for weeks, months, and years. These patterns are called modes or oscillations, but they are only quasi-periodic. The patterns or oscillations are driven by changes in sea surface temperature and atmospheric pressure that arise from natural variability and are shaped by the ocean's ability to store and transport heat, as well the effects of soil moisture, vegetation, and snow and ice. According to McElroy and Baker (2012), over the years, at least a dozen different oscillations were identified and followed with increasing insight. Important oscillations include (by time scale):

- Madden-Julian Oscillation (MJO): irregular tropical disturbance of winds and rainfall that travels eastward with a cycle of roughly 30 to 60 d.
- Monsoon: a major tropical wind system driven by differential heating of ocean and land that reverses direction twice yearly, leading to wet (summer) and dry (winter) seasons. Major monsoonal circulations occur in Asia and India, although smaller monsoons are also found in equatorial Africa, northern Australia, and the southwestern United States.
- El Niño/Southern Oscillation (ENSO): characterized by changes in tropical Pacific Ocean sea surface temperature and associated disruption of trade winds, rainfall, and droughts and is probably the best-known of the major drivers of interannual climate variability. The 2015/2016 El Niño event is one of the three strongest events since 1950 together with those of 1997/1998 and 1982/1983 (WMO, 2017). Both warm (El Niño) and cold (La Niña) events occur on an irregular cycle, with a warm event (El Niño) occurring about every four years.
- Indian Ocean Dipole (IOD): characterized by changing Indian Ocean sea surface temperatures that shift the normal convective winds to the west, leading to anomalous rainfall and droughts. IOD describes a mode of variability that affects the western and eastern parts of the ocean (WMO, 2017). It occurs on an irregular cycle of roughly 3 to 7 yr.
- Pacific Decadal Oscillation (PDO): characterized by sea surface temperature shifts over the entire Pacific Ocean area, with largest impact on the jet stream across North America and fishery ecosystems in the Northeastern Pacific. Changes appear on a time frame of 15 to 30 yr.
- North Atlantic Oscillation (NAO): the dominant mode of winter variability in the North Atlantic region characterized by an oscillation in atmospheric mass between the subtropical high and the polar low, reflecting the strength of westerly winds across the Atlantic into Europe. The index varies from year to year, but also can remain in one phase for several years. Related to the Arctic Oscillation, observed in the circumpolar Arctic.
- Atlantic Multi-decadal Oscillation (AMO): characterized by changes in the averaged North Atlantic sea surface temperature between the equator and 60 degrees north and associated with changes in the latitudinal overturning circulation. The cycle of warm and cool periods is multi decadal.

A review of the observational record demonstrates that the prevalence of recent climate extremes exceeds historical expectations and that their extent is worldwide, affecting people where they live, where they draw on fresh water resources, and where they grow food (McElroy and Baker, 2012):

- Global average land surface temperature has increased by about 0.9 °C since the 1950s. During the same period, the prevalence of extreme warm anomalies has increased and the prevalence of extreme cool anomalies has decreased.
- There is no obvious long-term trend in global annual precipitation over land. However, there is strong evidence that precipitation has occurred in more extreme events for most of the northern hemisphere.

- Higher temperatures combined with more extreme precipitation have combined to produce increasing prevalence of severe deficits of freshwater (which includes water in ice sheets, ice caps, glaciers, icebergs, bogs, ponds, lakes, rivers, streams, and groundwater) since about 1980 and a much smaller increase in the prevalence of freshwater surpluses since about 1990.
- Most of the permafrost observatories in the Northern Hemisphere show significant warming of permafrost since about 1980 to 1990.
- The minimum September arctic ice extent for each of the past five years (2007–2011) was lower than at any other year in the period of record and declining at an average rate of about 12% per decade.

Increasingly prevalent climate extremes such as droughts, floods, severe storms, and heat waves raise the specter of significant impacts due to changing climate in the near term. Evidence is pointing to the fact that human-driven changes in Earth's energy balance are driving a warmer and wetter atmosphere, with this trend superimposed on and magnifying natural variability. Small positive changes in the global mean annual temperature are causing an increased prevalence of local extreme weather conditions. The global temperature anomalies in 2016 show that in many parts of the world, temperatures in 2016 were 1 to 5 °C above the 1961 to 1990 reference period (Figs. 2 & 3).

Widespread changes in the instrumental record of extreme weather events such as droughts, heavy precipitation, heat waves, and the intensity of tropical cyclones were noted in the IPCC's Fourth Assessment Report, with these changes showing "discernable human influences" (IPCC, 2007). The IPCC's Fifth Assessment Report (IPCC, 2013) noted the substantial recent progress in the assessment of extreme weather and climate events, with the simulated global-mean trends in the frequency of extreme warm and cold days and nights over the second half of the 20th century now being generally

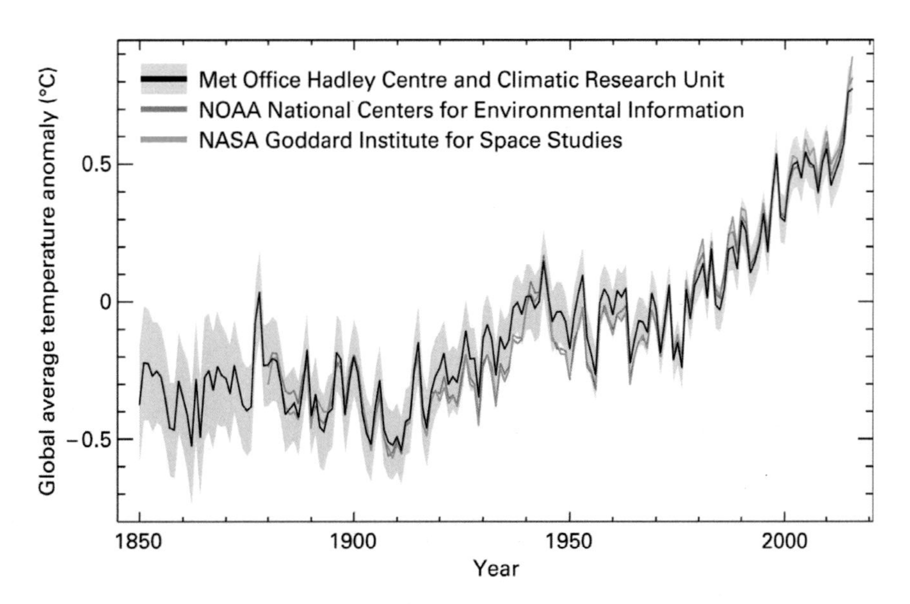

Fig. 2. Global average temperature anomalies in 2016 (1961–1990 reference period) for the three major datasets. The gray shading indicates the uncertainty in the HadCRU dataset. (Source: UK Met Office Hadley Centre).

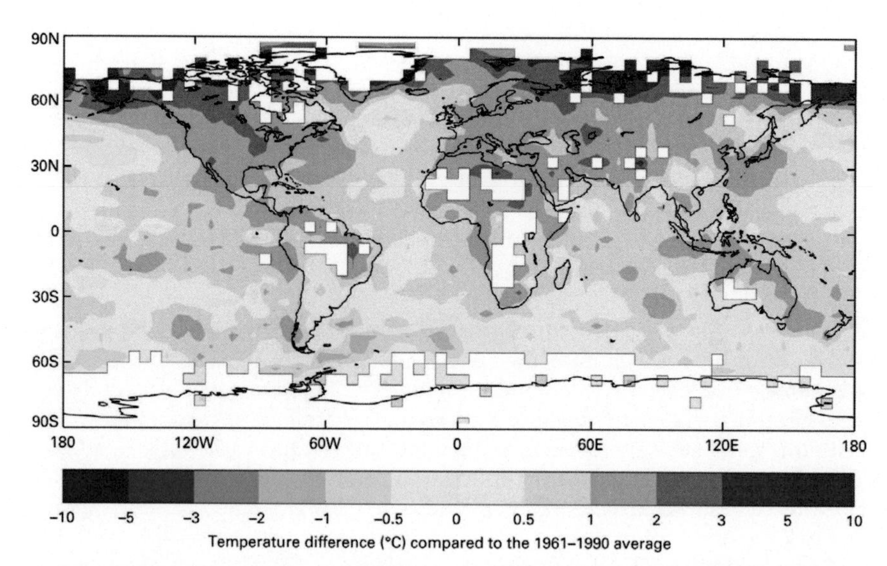

Fig. 3. Global temperature anomalies in 2016 (from 1961–1990 reference period) (Source: UK Met Office Hadley Centre).

consistent with observations. Human influence had been detected in changes in some climate extremes, with substantially stronger evidence compared to what was reviewed in the previous Assessment. It was considered likely that human influence had more than doubled the probability of occurrence of heat waves in some locations. More frequent hot and fewer cold temperature extremes over most land areas on daily and seasonal timescales were considered to be virtually certain, with it being very likely that heat waves will occur with a higher frequency and duration.

Impacts of Climate Extremes on Agriculture

Agriculture is a complex system, within which changes are driven by the joint effects of economic, environmental, political, and social forces (Olmstead, 1970; Bryant and Johnston, 1992). According to FAO (2017b), global food security could be in jeopardy, due to mounting pressures on natural resources and to climate change, both of which threaten the sustainability of food systems at large. The relationships between weather, climate, and production risk are well recognized (George et al., 2005).

According to Das (2003), the impact of extreme events on agriculture, rangeland, and forestry can be direct or indirect in their effect. Direct impacts arise from the direct physical damage on crops, animals and trees caused by the extreme hydrometeorological events, heat waves, cold waves, etc. The indirect damage refers to loss of potential production due to disturbed flow of goods and services, lost production capacities, reduction in the quality of yield, and increased costs of production. Statistical models have demonstrated that temperature trends have caused decreased maize and wheat production globally (Lobell et al., 2011). A recent meta-analysis has shown that decreases in global wheat, rice, and maize production will occur due to climate change without adaptation (Challinor et al., 2014).

According to IPCC (2015), impacts from recent climate-related extremes, such as heat waves, droughts, floods, cyclones and wildfires, reveal significant vulnerability and exposure of some ecosystems and many human systems to current climate variability (*very high confidence*). According to the statistics presented by Munich (2017), the total number of weather-related loss events during 1980 to 2016 were 15,000, the majority of which were storms and floods (Fig. 4). The overall losses due to these events were $ 3300 billions. The total fatalities during this period were 868,000 and the ensured losses reached $ 980 billions.

Localized extreme events tend to produce limited aggregate impacts, unlike countrywide natural events such as Hurricane Mitch (Charveriat, 2000). Sudden hazards such as thunderstorms usually have fewer long-lasting effects than droughts, which are often described as creeping in nature because of the slow rate at which they develop. Poor nations suffer the most from the natural disasters. As Devereux (2000) explained, poor people are more exposed because they tend to live in marginal areas and depend on high-risk, low-return livelihood systems such as rainfed agriculture and face many sources of economic vulnerability including little physical infrastructure.

Heavy Precipitation and Floods

Heavy precipitation and floods are a function of the climate (variability in rainfall pattern, occurrence of storms) as well as hydrology (shape of river beds, intensity of drainage, and debit flow of rivers) and soil characteristics (moisture absorption capacity). A flood is a temporary inundation of normally dry land with water, suspended matter and/or rubble caused by overflowing of rivers, precipitation, storm surge, tsunami, waves, mudflow, lahar, failure of water retaining structures, groundwater seepage, and water backup in sewer systems. Floods are among the greatest natural disasters known to mankind. According to historical record, 1092 flood disaster events occurred since 206 BC in China during a period of 2155 yr, averaging once in every two years (Heng, 2004). Heavy rains and floods result in certain constraints in agricultural activities but their impacts may vary greatly from one area to

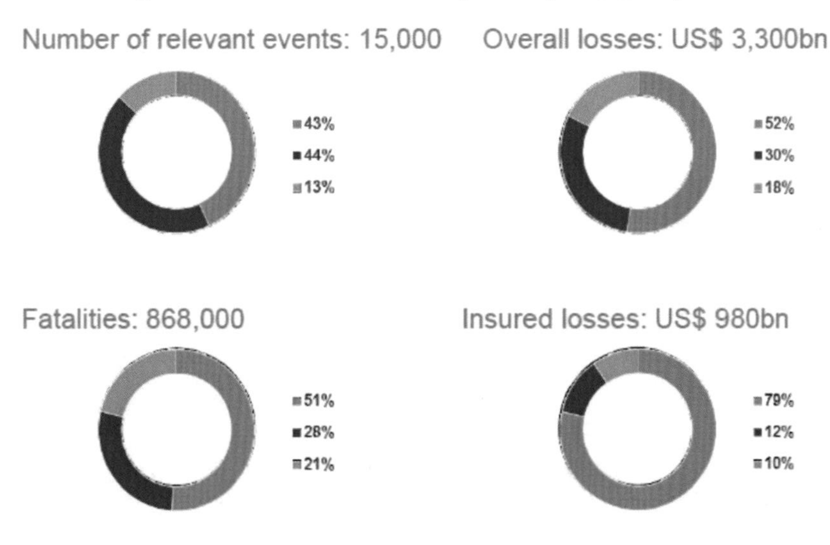

Fig. 4. Percentage distribution of weather-related loss events worldwide during 1980 to 2016. Tropical storms, extratropical storms, convective storms, and local storms (Green); Floods and mass movement (Blue), and Extreme temperatures, droughts and forest fires (Brown). (Source: Münchener Rückversicherungs-Gesellschaft, NatCatSERVICE).

the other depending on the rate of drainage seepage and drying. Indirect or direct negative effects, such as heavy rain and flood, can damage fragile plant organs like flowers and buds and cause soil erosion, water logging, and conditions favorable to crop and livestock pest and disease development as well as on pollution and pollinators (Gomez, 2005). More than three quarters of natural disasters in the Americas are high wind and floods. The countries most affected by floods are Brazil (15%), United States (12%) and Peru (11%). In monsoon areas such as southern Asia and coastal areas of West Africa, flashfloods are common, often caused by heavy rains associated with the monsoons. These heavy rains, apart from causing damage to physical structures such as roads, dams, and bridges, affect agricultural products in the areas.

Johnson (2003) categorized the direct and indirect effects of flooding into those that occur during the non-growing season or fallow period, and those that occur during the growing season. Important impacts during the non-growing season include loss of top soil, loss of soil nutrients, soil compaction, soil erosion, permanent damage to perennial crops, trees, livestock, buildings, and machinery, and permanent cessation of farming in floodplains. Impacts during the growing season include waterlogging of crops, lodging of standing crops, loss of soil nutrients, loss of pasture use, soil erosion, greater susceptibility to diseases and insects, interruptions to farm operations such as planting and harvesting, and permanent damage to perennial crops, trees, livestock, buildings, and machinery etc. Johnson (2003) also gave examples of the impacts of flooding on agriculture in the United States. In one example, flooding damage to Californian agriculture from winter rains in 1997 included loss of 24,000 ha of crop with an additional 38,500 ha damaged by rain and flooding, with the total losses estimated at $245 million.

A study of the impact of the 1998 Bangladesh floods revealed that 62% of all microfinance clients had lost their homes, nearly half had lost their everyday possessions, and over 75% had their ability to generate income at least suspended (Hassan and Hussain, 1998).

In 2016, the El Niño weakened and a negative Indian Ocean Dipole (IOD) phase became established, culminating in September. This phase gets established as westerly winds intensify along the equator, allowing warmer waters to concentrate near Australia. This sets up a temperature difference across the tropical Indian Ocean, with warmer than normal water in the east and cooler than normal water in the west. In September 2016, many parts of eastern Australia had record high monthly rainfall. The subsequent extensive flooding of inland rivers caused the main highway from Melbourne to Brisbane to be closed for more than a month. Damaging flooding occurred in early June on the east coast and in northern Tasmania (WMO, 2017).

In 2016, the Yangtze basin in China experienced its most significant flood season since 1999, with some tributaries experiencing record flood levels (Fig. 5). Rainfall was consistently high across the middle and lower Yangtze region from April to July, with total April through July rainfall over the region about 30% above average and similar to, or slightly above, that of 1998 and 1999. Over shorter timescales, very heavy rains from 18 to 20 July 2016 centered on the Beijing region also caused destructive flooding. In total, 310 deaths and damage amounting to US$14 billion were attributed to flooding in the Yangtze and Beijing regions (WMO, 2017).

Extreme flooding affected parts of the southern United States, especially Louisiana, from 9 to 15 August 2016. Seven-day rainfall totals in the worst-affected areas ranged from 500 mm to 800 mm, with 432 mm recorded in 15 h in Livingston on 12 August. Some rivers peaked at levels up to 1.5 m above previous records.

Thirteen deaths were reported and more than 50,000 homes and 20,000 businesses were damaged or destroyed. Total losses were estimated at US$10 billion.

Droughts

Droughts are the consequence of a natural reduction in the amount of precipitation over an extended period of time, usually a season or more in length, often associated with other climatic factors (such as high temperatures, high winds, and low relative humidity) that can aggravate the severity of the event. Drought is not a purely physical phenomenon, but instead is an interplay between natural water availability and human demands for water supply. The precise definition of drought is made complex due to political considerations, but there are generally three types of conditions that are referred to as drought.

- Meteorological drought is brought about when there is a prolonged period with below average precipitation.
- Agricultural drought is brought about when the moisture levels fall below critical levels due to low precipitation, soil conditions and higher evapotranspiration rates led by wind, humidity, solar radiation etc.,
- Hydrologic drought is brought about when the water reserves available in sources such as aquifers, lakes, and reservoirs falls below the critical levels. This condition can arise, even in times of average (or above average) precipitation, when increased usage of water diminishes the reserves.

Droughts affect the agriculture sector more than any other kind of climate extreme owing to their large scale and long-lasting nature. Drought or the threat of drought has become a constant problem in many parts of United States and droughts cost billions of dollars every year in United States (Ryu et al., 2010). From December 2011 to March 2017, the state of California experienced one of the worst droughts to occur in the region on record. The period between late 2011 and 2014 was the driest in California history since record-keeping began. According

Fig. 5. Flash floods in China in 2016.

to the U.S. Forest Service, a historic 129 million trees on 8.9 million acres have died due to drought and bark beetles in the state of California.

Droughts are generally distinguished from one another in three essential characteristics: intensity, duration, and spatial coverage (Wilhite, 2000). Drought severity is dependent not only on its duration, intensity, and geographical extent, but also on the demands made by human activities and vegetation on a region's water supplies. Droughts have particularly heavy impacts on rainfed smallholder farming systems in highland areas and in the tropics, which account for 80% of the world's cropland and produce about 60% of global agricultural output (FAO, 2011).

Numerous studies have been conducted on the impacts of droughts on crop growth and development at different levels including soil moisture uptake, root growth, shoot growth, various plant processes such as photosynthesis, respiration, plant water uptake and final yield and literature is replete with several good examples. But it is to be understood that the effects of droughts are seriously worsened by human factors such as population growth that forces people into drier and drier regions; overallocation of water resources; and inappropriate cropping and herding practices. The impacts of drought are likely to become ever more severe as a result of development processes and population increases (Squires, 2001). Droughts often stimulate sequences of actions and reactions leading to long-term land degradation. Droughts may also trigger local food shortages, speculation, hoarding, forced liquidation of livestock at depressed prices, personal stresses, social conflicts, and many other disasters associated with famines that may catastrophically affect numerous groups and strata of local populations.

Benson and Clay (1998) found that the aggregate impacts of droughts could be quite significant in terms of growth. A 50% fall in agricultural GDP would translate into a 10% decrease in GDP for an economy in which agriculture accounted for 20% of total activity in the predrought year. Gomes and Vergolino (1995) showed that during 1970, 1983, and 1993, the years of severe drought in Northeast Brazil, when the agricultural GDP decreased between 17.5 and 29.7%, the fluctuations in the region's GDP were explained almost entirely by the occurrence of droughts. For example, the estimated GDP per capita in the Northeast was $1494 in 1993, compared with $3010 in the rest of the country.

The year 2016 began with droughts associated with El Niño underway in several parts of the world (WMO, 2017). Much of southern Africa began the year 2016 in severe drought. For the second year in succession, rainfall was widely 20 to 60% below average for the summer rainy season (October through April) in 2015/2016. There were crop failures in many parts of the region. Drought emergencies were declared in all but one of South Africa's provinces, while, further north, poor agricultural production resulted in food shortages: WFP estimated that 18.2 million people would require emergency assistance by early 2017. Total cereal production in southern Africa in 2015/2016 was down by 13% from 2014/2015 and by 31% from 2013/2014.

In 2016, significant drought affected the Amazon basin in Brazil, as well as in the country's northeast (WMO, 2017). By the end of July, 24-mo rainfall was in the extremely dry category (Standardized Precipitation Index below -2) over almost the entire Amazon basin. Provisional figures show 2016 as the driest calendar year on record averaged over the Amazon basin. Crop production was reduced and rivers were at abnormally low levels. Central America also experienced ongoing drought in early 2016, with FAO estimating that 3.5 million people were experiencing food insecurity in El Salvador, Guatemala, Honduras, and Nicaragua. Very dry conditions also affected most of southern and central Chile with a prolonged period of below-average rainfall in central Chile, with Santiago's mean rainfall for the six years (2011–2016) being 40% below the long-term average. The dry conditions contributed to major forest fires, which broke out late in the year before worsening in January 2017.

According to the International Organization for Migration (IOM, 2016), El Niño-driven drought was the major factor contributing to the greatest number of people newly displaced in Ethiopia in the first quarter of 2016, compared with the same time frame in the three previous years (2013, 2014, and 2015).

High Temperatures and Heat Waves

Trends in maximum air temperatures exceeding a threshold value and persisting for a number of days or weeks and increasing minimum temperatures carry an impact on several sectors, including agriculture. It is clear from the observed records that there has been an increase in the global mean temperature of about 0.6°C since the start of the 20th century (Nicholls et al., 1996). A heat wave is a prolonged period of excessively hot weather, which may be accompanied by excessive humidity. According to IPCC (2015), the period from 1983 to 2012 was *very likely* the warmest 30-yr period of the last 800 yr in the Northern Hemisphere, where such assessment is possible (*high confidence*) and *likely* the warmest 30-yr period of the last 1400 yr (*medium confidence*). It is *likely* that the frequency of heat waves has increased in large parts of Europe, Asia, and Australia. For example, March 2017 was warmer than the 1981 to 2010 average over almost all of Europe, particularly so over the east of the continent (Fig. 6). Temperatures were substantially above average over northern Russia, where a peak value about 15 °C above normal was reached. Other relatively warm areas include the famine-threatened regions of Central and East Africa, and much of Australia.

The year 2016 started with an extreme heat wave in southern Africa, exacerbated by the ongoing drought. Many stations set all-time records in the first week of January; in some cases, these broke records which were only a few weeks old, following other heat waves in November and December 2015. On 7 January 2016, temperatures reached 42.7 °C in Pretoria and 38.9 °C in Johannesburg, both of which were 3 °C or more above the all-time records at those sites prior to November 2015. Extreme heat affected South and South-East Asia in April and May, prior to the start of the summer monsoon; South-East Asia was badly affected in April. Record or near-record temperatures occurred in parts of the Middle East and northern Africa on a number of occasions from late July to early September 2016 (WMO, 2017). The highest temperature observed was 54.0 °C at Mitribah (Kuwait) on 21 July which (subject to ratification) will be the highest temperature on record for Asia. Other extremely high temperatures included 53.9 °C at Basra (Iraq) and 53.0 °C at Delhoran (Islamic Republic of Iran–a national record), both on 22 July 2016.

A significant late-season heat wave affected many parts of western and central Europe in the first half of September (WMO, 2017). The highest temperatures occurred in southern Spain, where 45.4 °C was recorded at Cordoba on 6 September; September records were set at many other stations in Spain and in Portugal.

Growing evidence in the United States of America indicates that hot temperatures in excess of optimal thresholds for growth can be very harmful to major grain crops such as corn, soybeans, and wheat (Hatfield et al., 2014; Deryng et al., 2014).

Air temperatures in the Sahelian and Sudanian zones of West Africa (SSZ) are usually high because of the high radiation load. Mean maximum temperatures

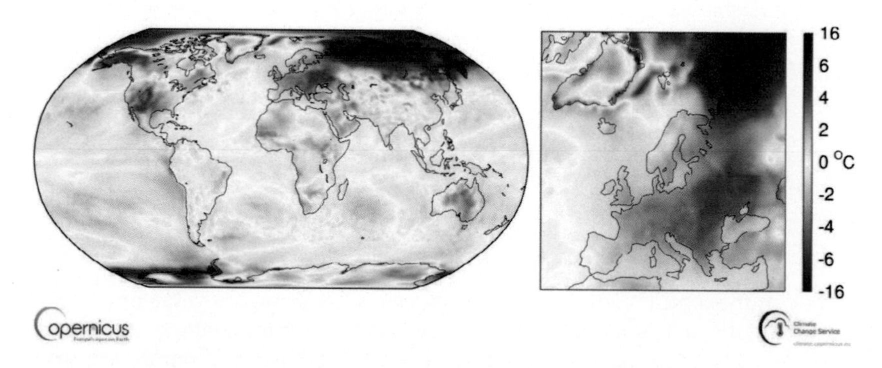

Fig. 6. Surface air temperature anomaly for March 2017 relative to the March average for the period 1981 to 2010. Source: ECMWF, Copernicus Climate Change Service.

could exceed 40 °C at the time of sowing and absolute temperatures could be much higher (Sivakumar, 1989). Diurnal variations in the air temperature and soil temperatures at the surface and 5-cm depth before the onset of rains show that the surface soil temperatures can increase rapidly from 27 °C at 0700 h to 56 °C at 1400 h. Agricultural productivity is sensitive to temperature increases. An increase in leaf-surface temperature would have significant effects on crop metabolism and yields, and it may make crops more sensitive to moisture stress (Riha et al., 1996). Experiments in India reported by Sinha (1994) found that higher temperatures and reduced radiation associated with increased cloudiness caused spikelet sterility and reduced yields. Similar studies conducted in Indonesia and the Philippines confirmed these results. Amien et al. (1996) found that rice yields in east Java could decline by 1% annually as a result of increases in temperature. Yaodong (2005) reported that if the temperature exceeds 32 °C, the rice crop in China has an adverse reaction to excessive heat. If the temperature exceeds 35 °C, the thousand-grain weight of rice can be decreased by as much as 0.42 to 3.2 g. As the temperature increases further, the thousand-grain weight also continues to decline. Full-blossom stage of rice is also a stage sensitive to high temperature. At full-blossom stage, seed-setting percentage of rice decreases with the intensity and time of high temperature. According to the study from the Plant Physiological Institute, Chinese Academy of Sciences, seed-setting percentage of rice is 80.9% at 28 °C. As temperature ascends to 38 °C, seed-setting percentage of rice declines to zero.

In tropical Asia, although wheat crops are likely to be sensitive to an increase in maximum temperature, rice crop would be vulnerable to an increase in minimum temperature. Simulations of the impact of climate change on crop yields using dynamic crop growth models indicated that productivity depended on the magnitude of temperature change.

Agriculture in the semiarid tropics of Africa, which is predominantly rainfed, is finely tuned to climate and even a slow, small change toward a worsening climate can increase climatic risks. Waggoner (1992) concluded that if the present climate is a productive one, warmer and drier will hurt. Growth is hindered by high temperatures, and plant metabolism for many cereal crops begins to break down above 40 °C. Burke et al. (1988) found that many crops manage heat stress (with ample water supply) through increased transpiration to maintain foliage temperatures at their optimal range. Because a large portion of African agriculture is rainfed, heat-related plant stress may reduce yields in several key crops such as wheat, rice, maize, and potatoes. Warmer night temperatures could impede vernalization in plants that require chilling, such as apples, peaches, and nectarines.

The survival rate of pathogens in winter or summer could vary with an increase in surface temperature (Patterson et al., 1999). Higher temperatures in winter will not only result in higher pathogen survival rates but also lead to extension of cropping area, which could provide more host plants for pathogens. Damage from diseases may be more serious because heat-stress conditions will weaken the disease-resistance of host plants and provide pathogenic bacteria with more favorable growth conditions.

Low Temperatures and Cold Waves

The low temperatures have a severe impact on crops. Extreme freezing temperatures (of both air and soil) and freezing durations under milder, subzero temperatures during winter exert different degrees of influence. According to Zheng et al. (2018), in the case of winter wheat, different responses of grain yield and yield components were observed under varying degrees of freezing injury. For plants, whose main stems and strong tillers both were killed by freezing, the spike number, grain number per spike, thousand-grain weight and grain yield decreased by 49.2%, 17.7%, 14.6%, and 64.3%, respectively (Zheng et al., 1989).

A cold wave is an influx of unusually cold air into middle or lower latitudes. Cold waves affect much larger areas than blizzards, ice storms, and other winter hazards. In the Northern Hemisphere, cold waves occur when very cold, dense air near the surface moves out of its source region in northern Canada or northern Asia. In 2016, the most significant cold wave occurred in late January in eastern Asia, with extreme low temperatures extending southwards from eastern China as far south as Thailand (WMO, 2017).

The recognized definition of frost in China is a phenomenon that temperature on the soil surface or near the plant canopy drops to below 0 °C in a short time, and crops suffer injuries (Binghai et al., 1985). Injury mechanism of frost is concluded as follows (Yangcai et al., 1991): frost does not result directly from low temperature, but from icing of plant tissue. The plant temperature below 0 °C is a necessary but not a sufficient condition for icing. Ice nucleation active bacteria (INA bacteria) is an inducement of icing. The density of INA bacteria is closely correlated with the icing temperature of plant tissues. Research results indicated that there are a lot of INA bacteria in many plants of China (Fuzai et al., 1989). These INA bacteria can induce plants to form ice under the condition of relatively high temperature (Yuxiang, 1990; Jianhua et al., 1990). There are three types of icing: cell-to-cell icing, intracellular icing, and plasmolysis. Only intracellular icing or cell-to-cell icing can cause cell to die thoroughly. If temperature ascends slowly in the course of defrosting, cell-to-cell icing or plasmolysis within the enduring ability does not cause cell to die. The distinction between frost and freezing injury, cool injury, and chilling injury (Duchang, 1999) is shown in Table 1.

Grain crops sown in the spring, such as maize, sorghum, millet, soybeans, buckwheat, naked oats, and potatoes, frequently suffer frost hazards in the northeast and northwest parts of China (Yaodong, 2005).

The impacts of low temperatures and cold waves differ on crops i.e., perennials vs. annuals. Annuals can be delayed in planting. Perennials can be impacted because dormancy break may not be controlled. In India, the 2002–2003 post-rainy

Table 1. Differences among frost, freezing, cool, and chilling injuries. †

Types	Temperature	Occurring season	Physiological reaction	Harmed crop	Crop state
	°C				
Frost	< 0	Comparative warm season (transition)	Dehydration, icing in short time	Winter crop, fruit tree and vegetable	Normal growth
Freezing	< 0	Winter, early spring or late autumn	Dehydration, icing	Winter crop, fruit tree	Wintering season
Cool	10–23	Warm season	Growth and development handicap	Thermophilic crop	Active growth
Chilling	0–10	Winter season	Physiological handicap	Tropical, subtropical crop	Slow growth

† Source: Cui Duzhang, 1999

season (*rabi season*) will be remembered for its severe cold wave conditions that prevailed for a prolonged period and had a significant impact on agriculture (Samra et al., 2003). The most distinguishable feature of the season was the sharp drop in maximum temperatures for a considerable period over an extended area in North India, which was a rare phenomenon. Reduced sunshine hours and limited availability of solar energy also compounded the impact of adverse weather conditions on agriculture. In some of the northern parts of the country day temperature fell lower than 15 °C while night temperature was below 0 °C. On many occasions the average temperature was less than 5 °C for consecutive days. Several crops and orchards in the north and northeastern regions of the country experienced the damaging effect of the cold wave and frost. The cold wave effect was also significant on livestock as well as shelterless human population.

Tropical Cyclones

Tropical cyclones, hurricanes, and typhoons are regional names for what is essentially the same phenomenon. The principal causes of the cyclones (Nuñez, 2005) are:

- strong winds,
- torrential rain and the associated flooding and
- high storm tides (the combined effect of storm surge and astronomical tides) leading to coastal saline inundation.

An average of 80 tropical cyclones form annually over the tropical oceans, of which, 30 occur in the typhoon region of the western North Pacific (Obasi, 1997). Once a tropical cyclone achieves surface wind strengths of at least 17 m s^{-1}, it is typically called a "tropical storm" or "tropical cyclone" and assigned a name (Xu et al., 2005). If the surface winds reach 33 m s^{-1}, the storm is called a "typhoon"(the northwest Pacific Ocean), a "hurricane" (the North Atlantic Ocean and the northeast Pacific Ocean), or a "severe tropical cyclone" (south-west Indian Ocean, the Bay of Bengal, the Arabian Sea, and the southwest Pacific Ocean). Tropical cyclones derive energy primarily from evaporation from the ocean and the associated condensation in convective clouds concentrated near their center, as compared with mid-latitude storms that primarily obtain energy from horizontal temperature gradients in the atmosphere.

The most destructive tropical cyclone of the year in 2016 (and also the most damaging meteorological disaster of any type) was Hurricane Matthew, which affected various parts of the North Atlantic in late September and early October (WMO, 2017). It reached Category 5 intensity south of Haiti, the first Atlantic hurricane to do so since 2007, and crossed south-western Haiti as a Category 4 system on 4 October. The heaviest casualties associated with Matthew occurred in Haiti, with at least 546 deaths reported. It also contributed to worsening existing issues of food insecurity and disease in the country, with cholera cases in the worst-affected provinces increasing by 50% from pre-hurricane levels. There were also major economic losses in the United States (mostly from flooding in North and South Carolina, Georgia, and Florida), Cuba, the Bahamas, and Haiti, amounting to more than US$15 billion.

Damage induced by cyclones is a phenomenon primarily of tropical and subtropical coastal regions. Tropical cyclones are among the most destructive of all natural hazards, causing considerable human suffering in about 70 countries around the world. The loss caused by a single storm may run into millions of dollars (Holland and Elsberry, 1993). This is particularly so in the case of developing countries, with coastal areas in developing countries suffering great loss of life.

The impact of tropical cyclones is greatest over coastal areas that bear the brunt of the strong winds and flooding from rainfall. The loss to an agriculture system from cyclones is due to direct and indirect effect (Nuñez, 2005). The direct effects are:

- destruction of vegetation, crops, orchards, and livestock
- damage to irrigation facilities such as canals, wells, and tanks
- long-term loss of soil fertility from saline deposits over land flooded by the sea
- disruption of the transportation system
- loss of a portion of the future harvests due to the destruction of standing crops
- loss of human life
- damage to property
- damage to fishing
- damage of off-shore and on-shore installations
- loss in productivity due to disruption of the work force and to other activities

Typhoons have been known to inflict severe damage on agriculture: for example, in southern Hainan on 2 October 1999, some 25 million timber and rubber trees were blown down (WMO, 1994). A typhoon that struck Thailand on 4 Nov. 1989 wiped out some 150,000 ha of rubber, coconut, and oil plantations and other crops (WMO, 1997).

In Orissa, India, cyclones on 17 to 18 Oct. 1999 and again on 29 to 30 Oct. 1999 caused devastating damage. The cyclones on 29 to 30 October with wind speeds of 270 to 300 km h^{-1} for 36 h were accompanied by torrential rain ranging from 400 to 867 mm over a period of 3 d. The two cyclones together severely affected around 19 million people in 12 districts (Roy et al., 2002). Sea waves reaching 7 m rushed 15 km inland. Two and a half million livestock perished and a total of 2.1 million ha of agricultural land was affected.

Located on the west coast of western North Pacific Ocean with a 18,000 km coast line, China is greatly affected by tropical cyclones. The number of the tropical cyclones landing in China reaches a maximum of 15 and a minimum 4 per year, with an average of 9.2 per year (Xu et al., 2005). Typhoons affecting China not only cause disasters through huge gales, rainstorms, storm-surge and rough sea tides brought by them, but also result in a group of disasters from their chain effects. For example, the typhoon rainstorms can create flood, water-logging as well as mud-rock flow, avalanche, and landslides and soil erosion. Tropical cyclone disasters seriously affect lowland agriculture and coastal fisheries in particular. In China, the total area of lowland, swamp and wetland is 25 million ha, about one quarter of area of arable land of the whole country. Coastal fisheries in China consist of three parts: off-shore fisheries, sea water aquaculture, and inland freshwater aquaculture. Tropical cyclones can cause sinking of fish boats, damage to aquafarms, and death of fishermen, thus leading to grave losses of fisheries (Xu et al., 2005).

Hurricanes, Thunderstorms, and Tornadoes

Hurricanes are one of the most physically destructive and economically disruptive extreme events that impact the United States. In addition to the torrential downpours and destructive winds associated with hurricanes, storm surges, and salt water intrusion into freshwater river systems are serious consequences (Motha, 2011). In 2005, Hurricane Katrina caused over $130 billion in property damage and over 1800 deaths. Hurricane Katrina killed or severely damaged 320 million large trees across over two million hectares of forest in the southern United States. Furthermore, coastal fisheries (e.g., oyster beds, shrimp) and harvesting infrastructure (e.g., boats, processing, and storage facilities) were also severely damaged.

According to the National Severe Storms Laboratory (NSSL) of NOAA, a thunderstorm is a rain shower during which thunder is heard. Worldwide, there are an estimated 16 million thunderstorms each year, and at any given moment, there are roughly 2000 thunderstorms in progress. It is well known that the United States has the greatest number of severe thunderstorms and tornadoes of any nation worldwide (Doswell, 2003). There are about 100,000 thunderstorms each year in the U.S. alone. About 10% of these reach severe levels. A thunderstorm is classified as “severe” when it contains one or more of the following: hail one inch or greater, winds gusting in excess of 50 knots (57.5 mph), or a tornado.

Hail up to the size of softballs damages cars and windows, and kills livestock caught out in the open. Strong (up to more than 120 mph) straight-line winds associated with thunderstorms knock down trees, power lines and mobile homes. Tornadoes (with winds up to about 300 mph) can destroy all but the best-built man-made structures.

Severe thunderstorms are one of the largest contributors to global losses in excess of USD $10 billion per year in terms of property and agriculture, as well as dozens of fatalities (Allen, 2018). Thunder storms can pull crops out from the ground or knock them over (Glanzer, 2018). They can also dry out wet plants, move soil, and cause erosion. Another big problem that comes from wind storms is the movement and dispersal of seeds. Tornadoes can injure or even kill livestock and can also cause the animals to get ill and pass away some time after a storm. For crops, the wind that comes with tornadoes can have damaging effects such as dispersal of seeds (Glanzer, 2018). Overall, with tornadoes come a lot of clean up and severe damage to farms.

Winter Storms

A winter storm is an event in which varieties of precipitation are formed that only occur at low temperatures, such as snow or sleet, or a rainstorm where ground temperatures are low enough to allow ice to form (i.e., freezing rain). A massive snowstorm with strong winds and other conditions meeting certain criteria is known as a blizzard. Large snowstorms could be quite dangerous as standing dead trees can also be brought down by the weight of the snow, especially if it is wet or very dense. The high damage potential of storms, ice and snow is illustrated by the worst ice storm catastrophe in Canada's history which occurred from 5 to 10 Jan. 1998 (Savage, 1998). Most of the precipitation fell as freezing rain and ice pellets with some snow. The representative temperature during the period was -10 °C. The reason that this ice storm turned into a catastrophe was its unusually long duration of around a week. The number of hours of freezing rain and drizzle was more than 80 h, nearly double the normal annual total. The Ice Storm of December 2002 in North Carolina resulted in massive power loss throughout much of the state, and property damage due to falling trees. The impact of winter storms on farms can involve a number of issues. Farm buildings can be damaged due to heavy snow or ice accumulation. Power failures or fuel shortages can impact animal production.

Agrometeorological Strategies for Reducing the Impacts of Climate Extremes on Agriculture

Agricultural research until the 1980s was preoccupied with issues of increased productivity to feed the growing populations and the success of green revolution in many developing countries had its foundations in increased use of external inputs such as improved seed, fertilizers, water, pesticides, fungicides, etc. As we moved into the 21st century, the agricultural research community is faced with the challenge of balancing the continuing need for increased productivity with the new concerns regarding the growing frequency of weather and climate extremes and the impacts of natural disasters on the agriculture sector. Looking ahead, the core question is whether today's agriculture and food systems are capable of meeting the needs of a global population that is projected to reach more than 9 billion by mid-century and may peak at more than 11 billion by the end of the century. Can we achieve the required production increases, even as the pressures on already scarce land and water resources and the negative impacts of climate extremes? This points to a new and important role for agrometeorologists around the world and some of the priority areas that need to be addressed are outlined below.

Improvement and Strengthening of Agrometeorological Networks

To assess the risks of climate extremes, the hazard characteristics, and their impacts on different agroecosystems, there is an urgent need for strengthening agrometeorological observation networks to support systematic data collection and analysis. Although agrometeorological networks have been in operation for many years now, in many developing countries and in some countries in transition, these networks are falling into disrepair due to lack of sustained funding. As more emphasis is now being placed on vulnerable regions including those in remote areas, a revaluation of the existing networks is needed to respond adequately to the needs of the priority regions and develop, periodically update and disseminate, as appropriate, location-based disaster risk information, including risk maps, to decision makers and the farming communities at risk of exposure to disasters in an appropriate format. It is important to integrate the deployment of modern technologies, such as automatic weather stations (AWSs), for data collection and transmission in the revaluation of existing networks. The use of AWSs is becoming more and more widespread, as they provide real-time meteorological data from places in agricultural areas with very scarce stations. Data from AWSs have become essential to provide information for the assessment of risk management and for decision making through the development of better decision tools and models and using social science to help encourage people to change decision-making.

Meteorologists from the United States are working on drone technology to better predict severe weather phenomena, which could save lives and limit property damage. The drones are able to monitor changes in the atmosphere from two to three kilometers above, an area both ground-based instruments and satellites have trouble tracking. This new system could improve storm-warning times by an hour or more, giving people time to stash their belongings and find a safe place to hide.

While most weather science has come from academia and research institutes, including those within National Meteorological and Hydrological Services (NMHSs), innovation in observing and computing has had major contributions from the private sector. Today, some of the world's largest companies at the heart of the global economy, for example in the space and computing industries, contribute to weather infrastructure (Thorpe, 2016). At the 68th Executive Council Meeting of WMO in June 2016, a special dialogue took place on cooperation between the public and private sectors in meteorology. Innovation in observing and computing technology and the existence of private capital is enabling the private sector component of the weather enterprise to grow rapidly. In September 2016, NOAA in the United States awarded Commercial Weather Data Pilot contracts to Pasadena, CA–based GeoOptics and San Francisco–based Spire Global. Under the contracts, the two companies provide GPS radio occultation data from commercial satellites. Such data provide profiles of atmospheric temperature and humidity that can be incorporated into forecasting models.

Establishment of Efficient Agrometeorological Databases including the Acquisition and Use of New Sources of Data for Natural Disaster Assessment and Mitigation

A basic requirement in natural disaster assessment and mitigation is an adequate agrometeorological database. This should include not only meteorological data, but also agricultural (phenological, crop management, pest and disease cycle), hydrological, land use, soil types, economic and other relevant information. Recent advances in operational satellite applications have improved our ability to address many issues of early drought warning and efficient monitoring. With help from satellites, drought, for instance, can be detected 4–6 wk earlier than before and delineated more accurately, and its impact on agriculture can be diagnosed far in advance of harvest (Zhao et al., 2005). The remote-sensing developments such as detection of soil moisture, estimation of evapotranspiration, rainfall etc., constitute new sources of data for many agrometeorological applications for disaster risk management in agriculture. These not only complement ground observations, but they also offer new types of data (like those of microwave satellites), provide global coverage, and can often be used to improve ground data (e.g., in area averaging). Usage of satellite-based precipitation data is necessary where in situ data are rare. In addition, atmospheric-model-based reanalysis data feature global data coverage and offer a full catalog of atmospheric variables including precipitation (Pfeifroth et al., 2013). It is important to take appropriate steps to promote the collection and use of these data in operational agrometeorology.

With efficient agrometeorological databases, including quality controlled historical data, it is feasible to estimate the risk of extreme events in quantitative terms, which is an important information in risk assessment (Guerreiro, 2005). With reference to the recent progress and trends related to preparedness and coping strategies for agricultural drought risk management, Wilhite (2007) described the useful information provided by the Drought Risk Atlas published by the National Drought Mitigation Center (NDMC) in the University of Nebraska.

Ensuring Timely Dissemination of Early Warnings of Natural Hazards to the Farming Community and Effective Monitoring

Contemporary early warning systems emerged in the 1970s and 1980s, as a response to drought-induced famines in the Sahel. The 2005 World Conference on Disaster Reduction in Kobe, Japan followed by the third early warning conference in Bonn, Germany in 2006, led to notable progress in linking early warning to early action and risk reduction. Currently systems exist to provide hazard forecasts and warnings against impending disasters induced by hydrometeorological hazards, but the hazard coverage at the country level is highly variable and reflect the countries' economic development level. The Sendai Framework for Disaster Risk Reduction (SFDRR), which succeeded the Hyogo Framework for Action 2005 to 2015: Building the Resilience of Nations and Communities to Disasters (HFA), and which was adopted at the Third World United Nations Conference on Disaster Risk Reduction (WCDRR-III) held in Sendai, Japan, in March 2015, emphasizes the priority "Enhancing disaster preparedness for effective response, and to "Build Back Better" in recovery, rehabilitation and reconstruction". The new consensus is that early warning is not only the production of technically accurate warnings, but also a system that requires an understanding of risk and a link between producers and consumers of warning information and an increase in social science work in agriculture in the United States and other locations, with the ultimate goal of triggering action to prevent or mitigate a disaster (IFRC, 2009). There is also a need for methodological work on the monitoring side, in particular regarding the identification of critical thresholds that should trigger early warnings. Close cooperation between agrometeorologists and their colleagues in the National Meteorological and Hydrological Services (NMHSs) can help ensure timely dissemination of early warnings of natural hazards to the farming community and effective monitoring. Agrometeorologists should promote wider utilization of existing warnings systems and disaster management information, for example by building links between climate and disaster databases and through better communication methods.

Promotion of Scientific Research on Climate Extreme Risk Patterns, Causes, and Effects in the Agriculture Sector

The analysis of data on current weather and historical climate, together with predictions and forward-looking analyses can help monitor climate extremes, quantify the risk and the causes and effects to set priorities for coping with the risk in the agricultural sector. One good example here is from Australia. There is a strong link between the incidence of severe drought in Australia and the El Niño-Southern Oscillation (ENSO) phenomenon (Nicholls, 1985). Conversely, La Niña events are usually accompanied by above average rainfall, an increased risk of flooding, and reduced frost incidence. Following a series of papers in the 1980s about the impacts of ENSO in Australia, and the potential for seasonal prediction based on indicators of ENSO (McBride and Nicholls, 1983; Nicholls, 1985), the Australian Bureau of Meteorology commenced issuing seasonal climate predictions in 1989. In addition to these statistical schemes, experimental seasonal predictions are also made using coupled atmosphere-ocean models. Such research efforts ensure the development of appropriate policies and measures to mitigate the risk of climate extremes in the agriculture sector.

In the southeastern United States, most climate variability is attributable to ENSO, with drier conditions associated with the La Niña phase and wetter conditions associated with El Niño (Schmidt et al., 2001). A Community Water Deficit Index (CWDI) was developed to forecast ENSO-based drought for the small to mid-sized communities in southeastern United States (Sharda et al., 2013). It was found that by using drought forecasts, and thus having a drought preparedness plan, communities can save both water and money (Sharda and Srivastava, 2016).

Facilitating Tactical Planning and Operational Decisions by the Farmers During the Crop Season Through the Provision of Improved Weather Forecasts (Both Short and Medium-range Forecasts) and Advisories

Catastrophic events like droughts, floods, and cyclones, spatial and temporal changes in important weather parameters like rainfall, temperature, wind, cloud cover, humidity, etc., effect crop yields by influencing farmers' decision about selection of cultivar, use of inputs, and crop management practices. Short-range forecasts are normally available one day in advance, but modern agricultural practices such as sowing of weather-sensitive high yielding varieties, need-based applications of fertilizers, pesticides, insecticides, efficient irrigation, and planning for harvest require weather forecast with higher lead time which enable the farmers to take ameliorative measures. Thus, for agricultural sector, location-specific weather forecasts in the medium range (3 to 10 d in advance) are very important. Agrometeorologists should ensure that these forecasts and advisories should be made available in a language that farmers can understand.

Promotion and Use of Seasonal to Interannual Climate Forecasts

One of the persistent demands of farmers is the provision of reliable forecasts of seasonal climate as it would help them take appropriate decisions as to which crops and/or cropping systems should be chosen well ahead of the sowing rains to avoid undue risks. International cooperation in studying the science behind detecting and forecasting natural and human-made hazards has led to advances in predictive accuracy and increased lead time. Improvements in the ability to forecast climate variability based on the advances in our understanding of ocean–atmosphere interactions over the past two decades offers opportunities to develop applications of seasonal-to-interannual climate predictions in the agricultural sector to deal more effectively with the effects of climate variability than ever before. The first International Workshop on Climate Prediction and Agriculture (CLIMAG), held in WMO in September 1999 (Sivakumar, 2000) considered a number of important issues relating to climate prediction applications in agriculture including capabilities in long-term weather forecasting for agricultural production, downscaling, scaling-up crop models for climate prediction applications, use of weather generators in crop modeling, economic impacts of shifts in ENSO event frequency and strengths, and economic value of climate forecasts for agricultural systems.

Agrometeorologists should make efforts to promote more active use of seasonal to inter-annual climate forecasts in agricultural planning and operations.

Promotion of Geographical Information Systems and Remote Sensing Applications and Agroecological Zoning for Sustainable Management of Farming Systems, Forestry, and Livestock

Agroecological zoning offers much scope for developing strategies for efficient natural resource management and in this context, recent advances in the geographical information systems (GIS) and remote sensing have made the task of integration and mapping of a wide range of databases much easier. Geographic information systems allow the collection, management, archival, analysis, and manipulation of large volumes of spatially referenced and associated attribute data. The advantages are manifold and highly important, especially for the fast cross-sector interactions and the production of synthetic and lucid information for decision-makers. Effective use of these techniques can promote the mainstreaming of the risk assessment of climate extremes, mapping, and management into rural development planning and management of mountains, rivers, coastal flood plain areas, drylands, wetlands, and all other areas prone to droughts and flooding. This can help strengthen the sustainable use and management of agroecosystems and implement integrated environmental and natural resource management approaches that incorporate disaster risk reduction.

Use of Improved Methods, Procedures, and Techniques for the Dissemination of Agrometeorological Information

Information and communication technologies (ICTs) broadly cover the set of activities that facilitates capturing, storage, processing, transmission, and display of information by electronic means. Current advances in the ICTs are changing the way farmers view information dissemination and exchange. The effectiveness of ICTs for agrometeorological information dissemination is being enhanced by linking them to other communication media, which are accessible to farmers. A good example of such an activity is the Radio and Internet (RANET) project implemented by the African Center of Meteorological Applications for Development (ACMAD) in Africa. Rapid advances made in the recent past in information technology, especially in audio–video media and mobile phones, need to be quickly operationalized to more effectively diffuse agrometeorological information to the user community. Here the development of a bottom-up approach of the full involvement of users is important to ensure that the methods and procedures so developed will adequately respond to the appropriate needs of the users.

Development of Agrometeorological Adaptation Strategies to Climate Variability and Climate Change

Food and fiber production is perhaps the sector most sensitive and vulnerable to climatic fluctuations. Agricultural growth and productivity depends on food production systems that are resilient against production failure due to shocks and climate variability (FAO, 2015). This requires a strong emphasis on sector-specific disaster risk reduction measures, technologies, and practices. There is a clear need identify the priority agrometeorological adaptation strategies for regions that are identified as being most vulnerable to the effects of climate variability and climate change and quickly diffuse this information to such regions. The prevailing philosophy regarding climate change mitigation, particularly in developing countries and at the subsistence farming level, has been one of "no-regrets", for example, only measures that make economic sense now should be adopted, because they reduce emissions from the agricultural sector or improve resilience of all sectors of agriculture against weather variability. All have a marked management component and could thus often be implemented at minimal cost. By systematically monitoring the impacts of natural disasters on different crops and/or cropping systems, agrometeorologists can help reduce the losses associated with the occurrence of natural hazards and provide input to the development of appropriate adaptation and mitigation strategies in response to changes in climate and climate variability.

Active Engagement in the Regional and Subregional Platforms for Disaster Risk Reduction and in the Thematic Platform for Agriculture Sector

The Sendai Framework for Disaster Risk Reduction (SFDRR) calls for forging effective partnerships for disaster risk reduction. Agrometeorologists should play an active role in the regional and subregional platforms for disaster risk reduction and the thematic platform for the agricultural sector to forge partnerships, periodically assess progress on implementation, and share the knowledge on disaster risk-informed policies and programs and to promote the integration of disaster risk management in other relevant sectors.

Promoting Weather Index Insurance for Coping with Climate Extremes in the Agricultural Sector

According to FAO (2015), humanitarian aid and official development assistance to the agriculture sector is small when compared with the economic impact and needs in the sector. More investment is needed in disaster risk reduction to build resilient livelihoods and food production systems. In this context, the potential for the use of index insurance products in agriculture is significant (Skees, 2003). Weather Index Insurance is currently proving to be a valuable instrument in many developing countries for transferring the financial impacts of low-frequency, high-consequence systemic risks out of rural areas. Agrometeorologists can facilitate the implementation of such index instruments to cope with risk of climate extremes in the agricultural sector by providing technical support and for monitoring and evaluation.

Conclusions

According to the available scientific evidence, climate extremes are on the rise and they continue to target the world's poorest and least-developed and there must be greater investment in the reduction of the risks from climate extremes. Despite a long history of climate extremes affecting agriculture, rangelands, and forestry, comprehensive documentation of these climate extremes at the national, regional, and international levels has been weak and a comprehensive assessment of their impacts on agriculture, forestry, and fisheries and strategies for their mitigation is critical for sustainable development, especially in the developing countries. To gain a better understanding of the climate variability and climate change and to cope with the impacts of climate extremes, improved agrometeorological databases are critical to develop more appropriate levels and forms of, mitigation and preparedness. Programs for improving prediction methods and dissemination of warnings should be expanded and intensified. Agrometeorologists should make more efficient use of the improvements in the information and communication technologies over the past two decades to provide timely and efficient agrometeorological information and products to the farming communities to help them cope with the impacts of climate extremes and improve the agricultural productivity.

References

Allen, J.T. 2018. Climate change and severe thunderstorms. Oxford Research Encyclopedia of Climate Science. Oxford University Press, Oxford, U.K. doi:10.1093/acrefore/9780190228620.013.62.

Amien, L., P. Rejekiningrum, A. Pramudia, and E. Susanti. 1996. Effects of interannual climate variability and climate change on rice yield in Java, Indonesia. In: L. Erda, W. Bolhofer, S. Huq, S. Lenhart, S.K. Mukherjee, J.B. Smith, and J. Wisniewski, editors, Climate change variability and adaptation in Asia and the Pacific. Kluwer Academic Publishers, Dordrecht, Netherlands. p. 29–39.

Benson, C., and E.J. Clay. 1998. The impact of drought on Sub-Saharan economies: A preliminary examination. World Bank Technical Paper No. 401. World Bank, Washington D.C.

Binghai, Z., W. Pengfei, and S. Jiaxin. 1985. Meteorological Dictionary. Shanghai Dictionary Press, Shanghai, China. p. 992–993.

Bryant, C.R., and T.R.R. Johnston. 1992. Agriculture in the city's countryside. University of Toronto Press, Toronto, Canada.

Burke, J.J., J.R. Mahan, and J.L. Hatfield. 1988. Crop-specific thermal kinetic windows in relation to wheat and cotton biomass production. Agron. J. 80:553–556. doi:10.2134/agronj1988.00021962008000040001x

Cabezon, E., L. Hunter, P. Tumbarello, K. Washimi, and Y. Wu. 2015. Enhancing macroeconomic resilience to natural disasters and climate change in the small states of the Pacific. IMF Working Paper WP/15/125, International Monetary Fund, Washington D.C.

Challinor, A.J., J. Watson, D.B. Lobell, S.M. Howden, D.R. Smith, and N. Chhetri. 2014. A meta-analysis of crop yield under climate change and adaptation. Nat. Clim. Chang. 4:287–291. doi:10.1038/nclimate2153

Charveriat, C. 2000. Natural disasters in Latin America and the Caribbean: An overview of risk. Research Department Working Paper No. 434. InterAmerican Development Bank, Washington D.C.

Das, H.P. 2003. Introduction. In: H.P. Das, T.I. Adamenko, K.A. Anaman, R.G. Gommes, and G. Johnson, editors, Agrometeorology related to extreme events. WMO No. 943. World Meteorological Organization, Geneva, Switzerland.

Deryng, D., D. Conway, N. Ramankutty, J. Price, and R. Warren. 2014. Global crop yield response to extreme heat stress under multiple climate change futures. Environ. Res. Lett. 9:034011. doi:10.1088/1748-9326/9/3/034011

Devereux, S. 2000. Famine in the Twentieth Century. IDS Working Paper 105. Institute of Development Studies, Brighton, U.K.

Doswell, C.A. 2003. Societal impacts of severe thunderstorms and tornadoes: Lessons learned and implications for Europe. Atmos. Res. 67-68:135–152. doi:10.1016/S0169-8095(03)00048-6

Duchang, C. 1999. Discuss on freezing injury, cool injury, chilling injury and frost (in Chinese). Chinese Agricultural Meteorology 20:56–57.

FAO. 2011. The state of the world's land and water resources for food and agriculture (SOLAW). FAO, Rome, Italy. www. fao.org/nr/solaw/solaw-home (accessed 24 Sept. 2018).

FAO. 2015. The impact of natural hazards and disasters on agriculture and food and nutrition security: a call for action to build resilient livelihoods. Food and Agriculture Organization, Rome, Italy.

FAO. 2017b. The future of food and agriculture: Trends and challenges. Food and Agriculture Organization, Rome, Italy.

Field, J.O. 2000. Drought, the famine process, and the phasing of interventions. In: D.A. Wilhite, editor, Drought: A global assessment. Vol. II. Routledge, London. p. 273–284.

Fuzai, S., Z. Hong, and W. He. 1989. A preliminary study on types and distribution of ice nucleation active bacteria in China. Chinese Agricultural Sciences 22:93–94.

George, D.A., C. Birch, D. Buckley, I.J. Partridge, and J.F. Clewett. 2005. Surveying and assessing climate risk to reduce uncertainty and improve farm business management. Ext. Farming Syst. J. 1:71–77.

Glanzer, N. 2018. How severe weather affects farmers. Crop Insurance Solutions, Milford, NE.

Gomes, G.M., and J.R. Vergolino. 1995. A microeconomia do desenvolvimento Nordestino: 1960–1994. IPEA Paper No. 372, IPEA, Rio de Janeiro, Brazil.

Gomez, B. 2005. Degradation of vegetation and agricultural productivity due to natural disasters and land use strategies to mitigate their impacts on agriculture, rangelands and forestry. In: M.V.K. Sivakumar, R.P. Motha, and H.P. Das, editors, Natural disasters and extreme events in agriculture. Springer, Berlin. p. 259–276. doi:10.1007/3-540-28307-2_15

Guerreiro, R.P.R. 2005. Accessibility of database information to facilitate early detection of extreme events to help mitigate their impacts on agriculture, forestry and fisheries. In: M.V.K. Sivakumar, R.P. Motha, and H.P. Das, editors, Natural disasters and extreme events in agriculture. Springer, Berlin. p. 51–70. doi:10.1007/3-540-28307-2_4

Hartmann, D.L., A.M.G. Klein Tank, M. Rusticucci, L.V. Alexander, S. Brönnimann, Y. Charabi, F.J. Dentener, E.J. Dlugokencky, D.R. Easterling, A. Kaplan, B.J. Soden, P.W. Thorne, M. Wild, and P.M. Zhai. 2013. Observations: Atmosphere and surface. In: T.F. Stocker, D. Qin, G.-K. Plattner, M. Tignor, S.K. Allen, J. Boschung, A. Nauels, Y. Xia, V. Bex, and P.M. Midgley, editors, Climate Change: The Physical Science Basis. Contribution of Working Group I to the Fifth Assessment Report of the Intergovernmental Panel on Climate Change. Cambridge Univ. Press, Cambridge, United Kingdom and New York, NY.</edb>

Hatfield, J., G. Takle, R. Grotjahn, P. Holden, R.C. Izaurralde, T. Mader, E. Marshall, and D. Liverman. 2014. Agriculture. In: J.M. Melillo, T.C. Richmond, and G.W. Yohe, editors, Climate change impacts in the United States: The Third National Climate Assessment. U.S. Global Change Research Program, Washington, D.C. p. 150–174.

Hassan, M.E., and F. Hussain. 1998. Effects and implications of high impact emergencies on microfinance: Experiences from the 1998 floods in Bangladesh. South Asian Network of Microfinance Initiatives, Dhaka, Bangladesh.

Heng, L. 2004. Flood management in China. Extended Summary. WMO/GWP Associated Programme on Flood Management. World Meteorological Organization, Geneva, Switzerland.

Holland, G.J., and R.L. Elsberry. 1993. Tropical cyclones a natural hazard: A challenge for the IDNDR. In: J. Lighthill, K. Emanuel, G.J. Holland, and Z. Zhang, editors, Tropical cyclone disasters. Peking Press, Peking, China.

IFRC. 2009. World Disasters Report 2009. International Federation of Red Cross and Red Crescent Societies. IFRC. Geneva, Switzerland. 204 pp.

IOM. 2016. Global Report on Internal Replacement (GRID 2016). Internal Displacement Monitoring Centre, International Organization for Migration, Geneva, Switzerland.

IPCC. 2007. Climate change 2007: The physical science basis. In: S. Solomon, D. Qin, M. Manning, Z. Chen, M. Marquis, K.B. Averyt, M. Tignor, and H.L. Miller, editors. Contribution of Working Group I to the Fourth Assessment Report of the Intergovernmental Panel on Climate Change. Cambridge Univ. Press, Cambridge, United Kingdom, and New York, NY.

IPCC. 2015. Climate Change 2014: Synthesis Report. Contribution of Working Groups I, II and III to the Fifth Assessment Report of the Intergovernmental Panel on Climate Change [Core Writing Team, R.K. Pachauri and L.A. Meyer (eds.)]. IPCC, Geneva, Switzerland, 151 pp.

IPCC. 2013. Climate Change 2013: The physical science basis. Contribution of Working Group I to the Fifth Assessment Report of the Intergovernmental Panel on Climate Change. Cambridge Univ. Press, Cambridge, U.K. and New York, NY.

IPCC. 2014. Summary for policymakers In: IPCC, editor, Climate Change 2014: Impacts, adaptation, and vulnerability. Contribution of Working Group II to the Fifth Assessment Report of the Intergovernmental Panel on Climate Change. Cambridge Univ. Press, Cambridge, UK and New York.

Jianhua, L., Y. Tao, W. He, Y. Fen, F. Sun, H. Zhu, and L. He. 1990. A study on relationship between ice nucleation active bacteria and frost of maize and soybean. Chinese Journal of Agrometeorology 11:1–6.

Johnson, G. 2003. Assessing the impact of extreme weather and climate events on agriculture, with particular reference to flooding and rainfall In: H.P. Das, T.I. Adamenko, K.A. Anaman, R.G. Gommes, and G. Johnson, editors, Agrometeorology related to extreme events. WMO No. 943. World Meteorological Organization, Geneva.

Lobell, D.B., W. Schlenker, and J. Costa-Roberts. 2011. Climate trends and global crop production since 1980. Science 333:616–620. doi:10.1126/science.1204531

McBride, J.L., and N. Nicholls. 1983. Seasonal relationships between Australian rainfall and the Southern Oscillation. Mon. Weather Rev. 111:1998–2004. doi:10.1175/1520-0493(1983)111<1998:SRBARA>2.0.CO;2

McElroy, M., and D.J. Baker. 2012. Climate extremes: Recent trends with implications for national security. Columbia University, New York.

Munich, R. 2017. NatCatSERVICE. Loss events worldwide 1980-2016. Munich Reinsurance, Munich, Germany.

Motha, R.P. 2011. The impact of extreme weather events on agriculture in the United States. Chapter 30, Publications from US Department of Agriculture and University of Nebraska Lincoln Faculty. Lincoln, NE. http://digitalcommons.unl.edu/usdaarsfacpub/1311 (accessed 24 Sept. 2018).

Nicholls, N. 1985. Towards the prediction of major Australian droughts. Aust. Meteorol. Mag. 33:161–166.

Nicholls, N., G.V. Gruza, J. Jouzel, T.R. Karl, L.A. Ogallo, and D.E. Parker. 1996. Observed climate variability and change In: Intergovernmental Panel on Climate Change, editor, Climate change 1995: The science of climate change. Intergovernmental Panel on Climate Change (IPCC). Cambridge Univ. Press, Cambridge, United Kingdom. p. 133–192.

Nuñez, L. 2005. Tools for forecasting or warning as well as hazard assessment to reduce impact of natural disasters on agriculture, forestry and fisheries. In: M.V.K. Sivakumar, R.P. Motha, and H.P. Das, editors, Natural disasters and extreme events in agriculture. Springer, Berlin. p. 71–92. doi:10.1007/3-540-28307-2_5

Obasi, G.O.P. 1997. Address at the opening of the Second Joint Session of the WMO/ESCAP Panel on Tropical Cyclones and the ESCAP/WMO Typhoon Committee, 20 Feb. 1997, Phuket, Thailand.

Olmstead, C.W. 1970. The phenomena, functioning units and systems of agriculture. Geogr. Pol. 19:31–41.

Patterson, D.T., J.K. Westbrook, R.J.V. Joyce, P.D. Lingren, and J. Rogasik. 1999. Weeds, insects and diseases. Clim. Change 43:711–727. doi:10.1023/A:1005549400875

Perkins-Kirkpatrick, S.E., M.G. Donat, and R.J.H. Dunn. 2018. Land surface temperature extremes. Bull. Am. Meteorol. Soc. 99:S15–S16.

Peterson, T.C., and M.J. Manton. 2008. Monitoring changes in climate extremes-A tale of international collaboration. Bull. Am. Meteorol. Soc. 89:1266–1271. doi:10.1175/2008BAMS2501.1

Pfeifroth, U., R. Mueller, and B. Ahrens. 2013. Evaluation of satellite-based and reanalysis precipitation data in the Tropical Pacific. J. Appl. Meteorol. Climatol. 52:634–644. doi:10.1175/JAMC-D-12-049.1

Riha, S.J., D.S. Wilks, and P. Simons. 1996. Impact of temperature and precipitation variability on crop model predictions. Clim. Change 32:293–311. doi:10.1007/BF00142466

Roy, B., C. Mruthyunjaya, and S. Selvarajan. 2002. Vulnerability of climate induced natural disasters with special emphasis on coping strategies of the rural poor in Coastal Orissa, India. Paper presented at the UNFCC COP 8 Conference organized by the Government of India, UNEP and FICCI, 23 October to 1 November 2002, New Delhi, India.

Ryu, J.H., M.D. Svoboda, J.D. Lenters, T. Tadesse, and C.L. Knutson. 2010. Potential extents for ENSO-driven hydrologic drought forecasts in the United States. Clim. Change 101:575–597. doi:10.1007/s10584-009-9705-0

Samra, J.S., G. Singh, and Y.S. Ramakrishna. 2003. Cold wave of 2002-03. Impact on agriculture. Indian Council of Agricultural Research, New Delhi, India.

Lecompte, E.L., A.W. Pong and J.W. Russell. 1998. Ice Storm'98. ICLR Research Paper Series No. 1, Institute for Catastrophic Loss Reduction, Toronto, Canada. p. 47.

Schmidt, N., E.K. Lipp, J.B. Rose, and M.E. Luther. 2001. ENSO influences on seasonal rainfall and river discharge in Florida. J. Clim. 14:615–628. doi:10.1175/1520-0442(2001)014<0615:EIOSRA>2.0.CO;2

Seneviratne, S.I., N. Nicholls, D. Easterling, C.M. Goodess, S. Kanae, J. Kossin, Y. Luo, J. Marengo, K. McInnes, M. Rahimi, M. Reichstein, A. Sorteberg, C. Vera, and X. Zhang. 2012. Changes in climate extremes and their impacts on the natural physical environment. In: C.B. Field, V. Barros, T.F. Stocker, D. Qin, D.J. Dokken, K.L. Ebi, M.D. Mastrandrea, K.J. Mach, G.-K. Plattner, S.K. Allen, M. Tignor, and P.M. Midgley, editors, Managing the risks of extreme events and disasters to advance climate change adaptation. A Special Report of Working Groups I and II of the Intergovernmental Panel on Climate Change (IPCC). Cambridge Univ. Press, Cambridge, U.K., and New York, NY. p. 109–230.

Sharda, V., P. Srivastava, L. Kalin, K. Ingram, and M. Chelliah. 2013. Development of Community Water Deficit Index: Drought forecasting tool for small- to mid-size communities of the southeastern United States. J. Hydrol. Eng. 18:846–858. doi:10.1061/(ASCE)HE.1943-5584.0000733

Sharda, V., and P. Srivastava. 2016. Value of enso-forecasted drought information for the management of water resources of small to mid-size communities. Trans. ASABE 59:1733–1744.

Sheffield, J., and E.F. Wood. 2011. Drought: Past problems and future scenarios. Earthscan, London, p. 210.

Sillmann, J., T. Thorarinsdottir, N. Keenlyside, N. Schaller, L.V. Alexander, G. Hegerl, S.I. Seneviratne, R. Vautard, X. Zhang, and F.W. Zwiers. 2017. Understanding, modeling and predicting weather and climate extremes: Challenges and opportunities. Weather and Climate Extremes 18:65–74. doi:10.1016/j.wace.2017.10.003

Sinha, S.K. 1994. Response of tropical agrosystems to climate change. In: Proceedings of the International Crop Science Congress 1:281-289.

Sivakumar, M.V.K. 1989. Agroclimatic aspects of rainfed agriculture in the Sudano-Sahelian zone. In: Soil, crop and water management systems for rainfed agriculture in the Sudano-Sahelian zone: Proceedings of an international workshop, ICRISAT Sahelian Center, Niamey, Niger, 7-11 Jan 1987. ICRISAT, Patancheru, India. p. 17-38.

Sivakumar, M.V.K., editor. 2000. Climate prediction and agriculture. Proceedings of the START/WMO international workshop held in Geneva, Switzerland, 27-29 September 1999. International START Secretariat, Washington D.C.

Sivakumar, M.V.K., R.P. Motha, D.A. Wilhite, and J.J. Qu. 2011. Towards a compendium on National Drought Policy: Proceedings of an expert meeting, July 14–15, 2011. Washington, D.C. AGM-12/WAOB-2011. World Meteorological Organization, Geneva.

Skees, J.R. 2003. Risk management challenges in rural financial markets: Blending risk management innovations with rural finance. Thematic papers presented at the USAID Conference: Paving the way forward for rural finance: An international conference on best practices, June 2–4, Washington, D.C. FAO, Rome, Italy.

Squires, V.R. 2001. Dust and sandstorms: An early warning of impending disaster. In: Y. Yang, V. Squires, and L. Qi, editors, Global alarm: Dust and sandstorms from the world's drylands. Asia RCU of the UNCCD, Bangkok.

Thorpe, A. 2016. The Weather enterprise: A global public-private partnership. WMO Bulletin 65:2.

UNDP. 2004. Reducing disaster risk: A challenge for development. Bureau for Crisis Prevention and Recovery, United Nations Development Programme, New York.

Waggoner, P.E. 1992. Preparing for climate change or projected global climate change and effects on crop production In: D.R. Buxton, R. Shibles, R.A. Forsberg, B.L. Blad, K.H. Asay, G.M. Paulsen and R.F. Wilson, editors, International crop science. Vol. I. CSSA, Madison, WI.

Wilhite, D.A. 2000. Drought as a natural hazard: Concepts and definitions. In: D.A. Wilhite, editor, Drought: A global assessment. Vol. 1. Routledge Publishers, London, UK. p. 3–18.

Wilhite, D.A. 2005. Drought and water crises: Science, technology, and management issues. Taylor and Francis, CRC Press, Boca Raton, FL. doi:10.1201/9781420028386

Wilhite, D.A. 2007. Preparedness and coping strategies for agricultural drought risk management: Recent progress and trends. In: M.V.K. Sivakumar and R. Motha, editors, Managing weather and climate risks in agriculture. Springer, Berlin, Heidelberg. p. 21–38. doi:10.1007/978-3-540-72746-0_2

WMO. 1994. Climate variability, agriculture and forestry. Technical Note 196, World Meteorological Organization, Geneva, Switzerland.

WMO. 1997. Extreme agrometeorological events. CAgM Report No. 73, TD No. 836, World Meteorological Organization, Geneva, Switzerland.

WMO. 2013. Global Climate 2001–2010: A decade of climate extremes-Summary report. WMO No.1119, World Meteorological Organization, Geneva, Switzerland.

WMO. 2016. Use of climate predictions to manage risks. WMO No. 1174. World Meteorological Organization, Geneva, Switzerland.

WMO. 2017. WMO Statement on the State of the Global Climate in 2016. WMO No. 1189, World Meteorological Organization, Geneva, Switzerland World Bank. 2013. World Development Indicators 2013. World Bank, Washington, D.C.

Xu, M., Q. Yang, and M. Ying. 2005. Impacts of tropical cyclones on Chinese lowland agriculture and coastal fisheries. In: M.V.K. Sivakumar, R.P. Motha, and H.P. Das, editors, Natural disasters and extreme events in agriculture. Springer, Berlin. p. 137–144. doi:10.1007/3-540-28307-2_8

Yaodong, D. 2005. Frost and high temperature injury in China. In: M.V.K. Sivakumar, R.P. Motha, and H.P. Das, editors, Natural disasters and extreme events in agriculture. Springer, Berlin. p. 145–157. doi:10.1007/3-540-28307-2_9

Yuxiang, Feng. 1990. Relationship between cucumber frost and ice nucleation active bacteria. Chinese Horticultural Journal 3:21–216.

Yangcai, Z., W. He, and S. Li. 1991. Outline of agricultural meteorology disasters in China. Chinese Meteorological Press, Beijing. p. 129–199.

Zhao, Y., S. Li, and Y. Zhang. 2005. Early detection and monitoring of drought and flood in China using remote sensing and GIS. In: M.V.K. Sivakumar, R.P. Motha, and H.P. Das, editors, Natural disasters and extreme events in agriculture. Springer, Berlin. p. 305–317. doi:10.1007/3-540-28307-2_17

Zheng, W., P.Z. Wang, and M.D. Zhu. 1989. Effects and grading of winter wheat freezing injury. (in Chinese). Bimonthly of Xinjiang Meteorology 7:29–34

Zheng, D., X. Yang, M.I. Minguez, C. Mu, Q. He, and X. Wu. 2018. Effect of freezing temperature and duration on winter survival and grain yield of winter wheat. Agric. For. Meteorol. 260-261:1–8. doi:10.1016/j.agrformet.2018.05.011

Zhu, X., and T.J. Troy. 2018. Agriculturally relevant climate extremes and their trends in the world's major growing regions. Earths Futur. 6:656–672. doi:10.1002/2017EF000687

Printed and bound by CPI Group (UK) Ltd, Croydon, CR0 4YY
17/04/2025
14658855-0001